		Date due	

Proceedings
of the Europ...
for Neurochemistry

WITHDRAWN

Proceedings of the European Society for Neurochemistry

Vol. 1

Second Meeting of the ESN
in Göttingen, August 1978

Edited by Volker Neuhoff
on behalf of the Organising Committee

H. S. Bachelard (U. K.), D. Biesold (GDR), V. Neuhoff (FRG), G. Porcellati (Italy), L. Svennerholm (Sweden), T. V. Waehneldt (FRG), V. P. Whittaker (FRG), H. Winkler (Austria)

Verlag Chemie
Weinheim · New York · 1978

Prof. Dr. Volker Neuhoff
Max-Planck-Institut für
Experimentelle Medizin
Forschungsstelle Neurochemie
Hermann-Rein-Straße 3
D-3400 Göttingen, Germany

This book contains 67 figures and 39 tables

CIP-Kurztitelaufnahme der Deutschen Bibliothek

European Society for Neurochemistry:
Proceedings of the European Society for Neurochemistry / ed. on behalf of the Organizing Committee for the Europ. Soc. for Neurochemistry. — Weinheim, New York: Verlag Chemie.

Vol. 1. Second meeting of the ESN in Göttingen, August 1978. — 1. Aufl. — 1978.
ISBN 3-527-25778-0 (Weinheim)
ISBN 0-89573-018-9 (New York)

© Verlag Chemie GmbH, D-6940 Weinheim, 1978
All rights reserved (including those of translation into foreign languages). No part of this book may be reproduced in any form — by photoprint, microfilm, or any other means — nor transmitted or translated into a machine language without written permission from the publishers.
Registered names, trademarks, etc. used in this book, even without specific indication thereof, are not to be considered unprotected by law.
Reproduction, Printer and Bookbinder: Zechnersche Buchdruckerei, D-6720 Speyer
Cover-Design: Prof. Dr. Volker Neuhoff and Verlag Chemie
Printed in West Germany

Preface

This volume is unusual inasmuch as all papers of the invited speakers and all poster abstracts submitted to the Second Meeting of the European Society for Neurochemistry are available as a printed book at the very beginning of the congress. This was achieved through the thoughtful cooperation of all the authors and through Verlag Chemie's readiness to print these proceedings at only a few weeks' notice. Thus, essentially new results are presented in print, such that they can be discussed during the congress. This is regarded as an attempt not only to prepare a solid foundation for the exchange of scientific facts, but also to promote and stimulate the development of ideas, however vague they may initially be. For this reason the organising committee — consisting of H. S. Bachelard (U. K.), D. Biesold (GDR), V. Neuhoff (FRG), G. Porcellati (Italy), L. Svennerholm (Sweden), T. V. Waehneldt (FRG), V. P. Whittaker (FRG), H. Winkler (Austria) — planned only a limited number of lectures, dispensing with parallel sessions commonly found in comparable meetings, in order that every participant will receive new information in his special field and will find time for exploratory discussions in other fields. In essence, further progress in the complex area of neurochemistry will either be inhibited by egocentric seclusion or be stimulated by the open exchange of results and ideas. It is my sincere wish that Volume 1 of the Proceedings of the European Society for Neurochemistry will serve to further international collaboration.

Quite naturally, the subjects of Volume 1 do not at all cover neurochemistry as a whole. Rather they were chosen by the organising committee of this particular congress. Further subjects will certainly be found in future issues of these Proceedings. This may also illustrate the fact that the European Society for Neurochemistry consists of scientists who are devoted to the elucidation of one of the most fascinating subjects of modern science and who regard Descartes' "cogito ergo sum" (1596—1650) as a provocative challenge.

It is a pleasure to thank the members of the organising committee and all the coworkers in my department for their cooperative efforts and sustained help, in particular Dr. T. V. Waehneldt and my secretary Mrs. Irene Fried.

Without generous financial support from the Deutsche Forschungsgemeinschaft and from the Niedersächsische Minister für Wissenschaft und Kunst for the scientific program and from the Fritz Thyssen Stiftung for fellowships, this congress would not have been possible. The congress fees could be kept rather low due to generous contributions from an Italian company and numerous German firms *via* the Stifterverband für die Deutsche Wissenschaft. On behalf of the European Society for Neurochemistry, the organising committee, and the participants I extend my sincere gratitude to these supporters. Without any doubt their means are productively invested, since they will not only serve basic research but also clinical neurochemistry, and thereby implement the aims of the European Society for Neurochemistry.

Göttingen, July 1978   Volker Neuhoff

Contents

Symposium
Myelin and myelin Disorders
Chairman: *L. Svennerholm*

Fine Structure of Myelin Sheaths 3
E. Mugnaini

Proteins of Central Nervous System Myelin 32
T. V. Waehneldt

Biosynthesis and Deposition of Myelin Lipids 48
N. Baumann

Myelin Catabolism 64
H. Woelk and *G. Porcellati*

Cellular and Humoral Immunity in Demyelinating Diseases. Multiple Sclerosis and Experimental Allergic Encephalomyelitis as a "Model" 78
E. A. Caspary

Immunological Studies in Multiple Sclerosis 91
A. N. Davison

Tissue Culture Studies of Experimental Allergic Encephalomyelitis . . . 100
F. J. Seil

Poster Abstracts

Studies on Cerebroside Organisation in Central Nerve Myelin 115
C. Linington and *M. G. Rumsby*

Probe Labelling Studies with the Myelin Basic Protein 116
A. G. Walker and *M. G. Rumsby*

Proteolipids of Rat Brain Myelin 117
H. Ch. Buniatian and *K. H. Manukian*

Relationship between Axonal Transport and the Wolfgram Proteins of Myelin . 118
P. P. Giorgi

Heterogeneous Organisation of Axons, Myelin and Glial Cells along the Optic Nerve and Tract 119
P. P. Giorgi and *H. Du Bois*

Biochemical Characterization of the Rabbit Optic Nerve during Ontogenetic Development 120
H. Tauber, V. Neuhoff, and *T. V. Waehneldt*

Comparison of CNP-ase Activity from White Matter, Myelin and Cultured Neuronal and Non-neuronal Cells 121
P. Clapshaw, J. Oey, and *W. Seifert*

Purification of 2′,3′-cyclic Nucleotide 3′-Phosphohydrolase 122
Y. Tsukada and *H. Suda*

The Activity of the Myelin-specific Cholesterol Ester Hydrolase in Human Cerebrospinal Fluid ... 123
H. Reiber and *W. Voss*

Some Chemical and Physical Studies on Myelin Isolated from Developing Rat Brain ... 124
F. Abulaban and *C. A. Lovelidge*

Studies on CNS Myelinogenesis ... 125
E. Zaprianova, M. Christova, M. Staykova, and *I. Goranov*

Metabolic Studies with Neurones, Astrocytes and Oligodendrocytes Isolated from Whole Rat Brain Tissue ... 126
S.-W. Chao and *M. G. Rumsby*

Biochemical Characterization of PNS Myelin Subfractions Obtained by Continuous Zonal Gradient Centrifugation ... 127
T. V. Waehneldt and *J.-M. Matthieu*

Myelin Consists of a Continuum of Particles. Comparison between Normal and Quaking Mice; Analysis in Human Pelizaeus-Merzbacher Disease and Infantile Neuronal Ceroid Lipofuscinosis ... 128
J. M. Bourre, S. Pollet, O. Daudu, and *N. Baumann*

Immunohistochemical Localization of Galactocerebroside during Development in the Cerebellum of Normal and Quaking Mouse ... 129
B. Zalc, M. Monge, P. Dupouey, and *N. Baumann*

The Protein Composition of PNS Myelin in Neurological Mutant Mice ... 130
J.-M. Matthieu

Shiverer Mouse; a Dysmyelinating Mutant with Absence of Major dense line and basic protein in myelin ... 131
C. Jacque, A. Privat, P. Dupouey, J. M. Bourre, T. D. Bird, and *N. Baumann*

Lysosomal Hydrolase in the Lymphocytes and Granulocytes of Patients with Multiple Sclerosis (MS) ... 132
P. Riekkinen, J. Palo, and *J. Wikström*

Prealbumin Content of Cerebrospinal Fluid and Serum in Persons with Multiple Sclerosis ... 133
Lj. Kržalić and *N. Kastrapeli*

Rat Myelin Proteins and Lipids in Early Period of Experimental Cyanide Encephalopathy ... 134
B. Zgorzalewicz and *J. Sedzik*

Specificity of Triethyllead toward Myelin Protein Synthesis ... 135
G. Konat, H. Offner, and *J. Clausen*

Myelin Composition and Lipogenic Enzyme Activity in Brain Tissue from Normal and B_{12}-deficient Bats ... 136
R. C. Cantrill, L. Kerr, and *J. van der Westhuyzen*

Alkane Biosynthesis by the Peripheral Nervous System ... 137
C. Cassagne, D. Darriet, and *J. M. Bourre*

The role of Soluble Cytoplasmic Phospholipid Exchange Proteins in Myelin Sheath Formation and Turnover of Myelin Phospholipids 138
E. M. Carey

The Myelinating Explant Culture as a Model for Myelin Maturation in vivo 139
G. E. Fagg, H. I. Schipper, and *V. Neuhoff*

Studies on Myelin Associated Glycolipids in Cultures of C_6 Glial and Dissociated Brain Cells 140
L. L. Sarlieve, G. Subba Rao, and *R. A. Pieringer*

Symposium
Factors Influencing Neuronal Development
Chairman: *D. Biesold*

The Development of Order in Neural Connections 143
G. Székely

Morphogenic Relations between Cell Migration and Synaptogenesis in the Neocortex of Rat 158
J. R. Wolff, B. M. Chronwall, and *M. Rickmann*

Membrane Glycoproteins in Synaptogenesis 174
G. Gombos, G. Vincendon, A. Reeber, M. S. Ghandour, and *J.-P. Zanetta*

Effect of Malnutrition on Ganglioside Development 189
I. Karlsson

Sensory Deprivation and Brain Development 199
V. Bigl and *D. Biesold*

Poster Abstracts

Glial Cells in Developing Rat Cerebellum: Biochemical and Immunohistochemical Study 216
M. S. Ghandour, G. Vincendon, E. Bock, D. Filippi, G. Laurent, J.-P. Zanetta, and *G. Gombos*

Glial Cells Markers in Adult Rat Cerebellum: Biochemical and Immunohistochemical Study 217
G. Vincendon, M. S. Ghandour, E. Bock, D. Filippi, G. Laurent, J.-P. Zanetta, and *G. Gombos*

Effect of Neonatal Hypothyroidism on Rat Cerebellum Ontogenesis . . . 218
F. Vitiello, M. S. Ghandour, J. Clos, J. Legrand, G. Vincendon, and *G. Gombos*

Regional Patterns of Glycolytic Enzymes in Developing Rat Brain 219
H. H. Gustke, D. Schiffer, and *S. L. Kowalewski*

Study of Glycosidases during the Development of Human Brain 220
P. Annunziata, G. C. Guazzi, and *A. Federico*

Acid Phosphatase during Development of Chicken Spinal Cord Anterior Horn Cells under Normal and Experimental Conditions 221
R. A. van Welsum and J. Drukker

Glutamate and Kainic Acid Binding to Synaptic Membranes of Cerebral Cortex from Developing Rats 222
C. Sanderson and S. Murphy

Availability of Amino Acids to the Developing Brain: Changes in the Kinetics and Specificity of Membrane Transport Systems in Separated Cells from the Rat Cerebral Cortex during Maturation 223
S. Murphy and A. Sinha

Brain and Plasma Amino Acid Pools in Early and Adult Proteincalorie Undernutrition 224
M. Rodés, J. Sabater, and F. Gonzales-Sastre

Some Aspects of the Regulation of GABA Level in Developing Rat Brain . 225
L. Ossola, M. Maitre, J. M. Blindermann, and P. Mandel

Glycoconjugates and other Biochemical Change in Aging Rat and Human Brain 226
A. Federico, R. M. Corona, and P. Annunziata

Effect of Cortisol on Choroidal Sodium-Potassium Adenosine Triphosphatase and Cisternal Cerebrospinal Fluid Pressure in Developing Chick Embryo 227
F. Stastný, Z. Rychter, and R. Jelinek

Prenatal Maternal Phenobarbital Alters Plasma Concentration of Corticosterone in Developing Offspring 228
J. W. Zemp, L. D. Middaugh, and W. O. Boggan

Use of Arrhenius Plots of Na-K-ATPase and Acetylcholinesterase as a Tool for Studying Changes in Lipid-Protein Interactions in Neuronal Membranes during Brain Development 229
M. Gorgani and E. Meisami

AchE and Ach in Rat Olfactory Bulb during Development and under Olfatory Deprivation 230
E. Meisami, R. Mousavi, and R. Safail

Localization and Activity of Cholinesterases in the Rat Developing Diaphragm 231
T. Kiauta and M. Brzin

Micro-analysis of Lipids in Peripheral Nerve Biopsies in Two Age Groups 232
S. Pollet, J. J. Hauw, J. C. Turpin, F. Le Saux, M. Monge, and N. Baumann

The Effect of Thyroid Hormone on Fatty Acid Activation during Myelination 233
R. C. Cantrill, L. Kerr, and E. M. Carey

Metabolism of 1-Alkyl- and 1-Alkenyl-sn-glycero-3-phosphoethanolamine in Subcellular Fractions of Myelinating Rat Brains 234
J. Gunawan, M. Vierbuchen, and H. Debuch

Muscarinic Receptor in Developing Rat Cerebellum 235
J. Mallol, C. Sarraga, M. Bartolomé, J.-P. Zanetta, G. Vincendon, and G. Gombos

Impairement of Brain Development in Man: Neurochemical Study of Two Diseases 236
A. Federico, R. M. Corona, I. D'Amore, P. Annunziata, and G. C. Guazzi

Symposium
Neurochemistry of Hypoxia
Chairman: *H. S. Bachelard*

Clinical Aspects of Anoxia in the Nervous System 239
L. Symon

Brain Energy Metabolism and Circulation in Hypoxia 253
L. Berntman and B. K. Siesjö

Effects of Low Oxygen on Brain Monoamine Metabolism 266
A. Carlsson

Arachidonic Acid Metabolism and Cyclic Nucleotides in the CNS during Hypoxia 271
C. Galli, C. Spagnuolo, L. Sautebin, and G. Galli

Phospholipids and its Metabolism in Ischemia 285
G. Porcellati, G. E. De Medio, C. Fini, A. Floridi, G. Goracci, L. A. Horrocks, J. W. Lazarewicz, C. A. Palmerini, J. Strosznajder, and G. Trovarelli

Poster Abstracts

Respiration of Synaptosomes 303
H. Wise, H. S. Bachelard, and G. G. Lunt

Effect of Hypoxia on the *in vivo* Incorporation of 2-^{3}H Glycerol and 1-^{14}C Palmitate into Lipids of Various Subcellular Fractions Purified from Different Regions of Guinea Pig Brain 304
M. Alberghina, I. Serra, E. Geremia, and A. M. Giuffrida

The Effect on Hypoxia on the Metabolism of Labeled Glucose and Acetate in the Rat Brain 305
K. Domańska-Janik and T. Zalewska

Energy Utilization and Changes in Some Intermediates of Glucose Metabolism in Normal and Hypoxic Rat Brain after Decapitation 306
T. Zalewska and K. Domańska-Janik

Regulation of the Level of Cyclic AMP in the Brain during Hypoxia and Ischemia 307
L. Khatchatrian and K. Domańska-Janik

Mechanisms of Damage of Mitochondrial Lipid Chemical Structure after Ischemic-Anoxia 308
M. D. Majewska, L. Khatchatrian, J. Strosznajder, and J. W. Lazarewicz

The Role of CMP on the Regulation of Brain Choline Phosphotransferase 309
E. Francescangeli, G. Goracci, L. A. Horrocks, and G. Porcellati

Effect on Ischemia of Energy Metabolism and Catecholamine Levels in the
Gerbil Brain *in vivo* 310
S. Mazzari and M. Finesso

Changes in the Gerbil Brain Phophoglycerides during Bilateral Ischemia . 311
G. E. De Medio, G. Trovarelli, G. Goracci, C. A. Palmerini, A. Floridi, C. Fini,
S. Mazzari, M. Finesso, and G. Porcellati

Phospholipidases in Ischemic Gerbil Brain 312
L. A. Horrocks, W. R. Snyder, A. D. Edgar, M. E. Nesham, and J. N. Allen

Role of Phospholipids in Calcium Accumulation in Brain Mitochondria from
Adult Rat after Ischemic Anoxia and Hypoxic Hypoxia 312
J. Strosznajder

Effect of Ischemic Anoxia and Hypoxic Hypoxia on Acylation of Lysoglycerophospholipids in Rat Brain Subcellular Fraction 314
J. Strosznajder

Symposium

Neurochemistry of Addictive Drugs

Chairman: A. Herz

Biochemical Aspects of Tolerance to, and Physical Dependence on, Central
Depressants 317
H. Kalant

Pathways of Alcohol Metabolism 332
K. F. Tipton, A. J. Rivett, and I. L. Smith

Membrane Effects of Alcohol in the Nervous System 346
H. Wallgren and P. Virtanen

Effects of Dependence-Producing Drugs on Neurotransmitters and Neuronal Excitability 359
S. Liljequist

Biochemical Theories of Opioid Dependence: An Analysis 374
H. O. J. Collier

Opiate Receptors and Endorphins in Opiate Addiction 386
V. Höllt, J. Bläsig, J. Dum, R. Przewzocki, and A. Herz

Membrane Constituents and the Mechanisms of Morphine Actions . . . 404
H. H. Loh and R. J. Hitzemann

Neurotransmitters and Opiate Addiction 425
G. Pepeu, F. Casamenti, and F. Pedata

Poster Abstracts

Calcium Release in Platelets: Action of Psychotropic Drugs 439
H. W. Reading

Drug Induced Displacement of 5-HT and LSD from Specific Binding Sites of
Synaptic Membranes 440
N. Weiner and W. Wesemann

Acute Effects of Hallucinogens on the Dopamine-Turnover in Rat Brain
Regions 441
L. Hetey and W. Oelssner

Regulation of Catecholamine Metabolism in Rat Brain during Chronic Treatment with, and Withdrawal from, Methamphetamine 442
M. E. Bardsley and H. S. Bachelard

Paradoxical Brain Malate Response to Morphine 443
L. J. King, K. H. Minnema, and E. E. Dowdy, Jr.

The Effects of Ethanol on Development in the Rat 444
N. K. Deterin, P. T. Ozand, and A. Karavasan

Potassium Ethylxanthoqenate as an Inhibitor of Ethanol Metabolism . . . 445
S. G. Yanev

Round Table
The Role of Peptides in Brain Function
Chairman: D. de Wied

Immunohistochemical Identification of Extrahypothalamic Connections of
Peptide Hormone Producing Neurons 449
A. Weindl and M. V. Sofroniew

Neuropeptides as Neurotransmitters 450
L. L. Iversen

Opioid Peptides 452
J. Hughes

Peptides in Control of Behaviour and Metabolism 453
K. L. Reichelt, O. E. Trygstad, P. D. Edminson, I. Foss, G. Saelid, J. H. Johansen, and J. Bøler

Neuropeptides and Behaviour 455
D. de Wied

Poster Abstracts

Purification of Brain Enkephalins by Chromatography on LH-20 Sephadex 457
J. Wideman and S. Stein

Purity of Commercially Available Radioactive Enkephalins as Studied by
HPLC 458
J. Wideman and S. Stein

The Effect of Metenkephalin and (D-met^2,Pro^5)-Enkephalinamide on the
Adenylate Cyclase Activity of Rat Brain 459
M. Wollemann, A. Szebeni, and L. Gráf

Distribution and Properties of Lipotropin Activating Proteases in Rat Brain 460
B. M. Austen and D. G. Smyth

The Localization and Characterization of Aminopeptidases in the CNS and the Hydrolysis of Neuropeptides 461
S. G. Shaw and W. F. Cook

Peptides and Memory: A Speculative Working Model 462
K. L. Reichelt, P. D. Edminson, and G. Saelid

Peptides and Rat Brain Membrane Phosphoproteins 463
H. Zwiers, V. M. Wiegant, A. B. Oestreicher, P. Schotman, and W. H. Gipsen

Substance P: Two Different Binding Sites in Rat Brain 464
A. Saria, N. Mayer, R. Gamse, and F. Lembeck

Endorphins Release from Pituitary Cells 465
R. Simantov

Humoral Control of Appetite: Isolation and Characterisation of a Peptide Inducing Obesity in Mice 466
P. D. Edminson, I. Foss, O. Trygstad, and K. L. Reichelt

Sea Anemone Toxins: Tools in Neurobiology and Pharmacology 467
L. Béress

Round Table
Dentritic, Axonal and Transneuronal Transport
Chairman: M. Cuénod

Dendritic, Axonal and Transneuronal Transport: An Introduction . . . 471
M. Cuénod

Three Aspects of the Axonal Transport: Molecular Selectivity, Intraaxonal Compartmentation and Intercellular Exchange 472
B. Droz and J.-Y. Couraud

Correlative Aspects of Intra- and Transneuronal Transport 474
G. W. Kreutzberg and P. Schubert

Rapid Axoplasmic Transport of Low Molecular Weight Substances . . . 475
D. G. Weiss

Retrograde Axonal and Trans-Synaptic Transport of Macromolecules . . 476
M. E. Schwab, M. Dumas, and H. Thoenen

Poster Abstracts

Fast Axonal Transport in Rat Sciatic Nerves in vitro 478
M. Hanson, A. Edström, and S. Gershagen

Axonal transport and Incorporation of Radioactivity after Injection of a Sialic Acid Precursor into the Red Nucleus, an Autoradiographic Study 479
L. D. Loopuijt

Axonal Transport. Interaction of Rapidly Transported Glycoproteins with Lectins and Glycoproteins 480
J.-O. Karlsson

Axonal Transport of Taurine in the Visual System of the Developing Rabbit 481
J. A. Sturman

Effect of Sulfhydryl Reagents on Microtubule-Associated ATP-ase Activity and Axonal Transport 482
M. Wallin, H. Larsson, and A. Edström

Migration of RNA and RNA Precursors along Neonatal and Young Adult Rat Optic Axons 483
M. J. Politis and N. A. Ingoglia

A Study on the Axonal Flow of Phospholipids in the Ciliary Ganglion of the Chicken 484
M. Brunetti, B. Droz, L. Di Giamberardino, and G. Porcellati

Substrate Specificity of γ-Aminobutyric Acid Transport System in the Rat Thyroid 485
H. Gebauer

Round Table

The Value of Nerve Cell Cultures for Neurochemical Research
Chairman: P. Mandel

The Value of Nerve Cell Cultures for Neurochemical Research 489
P. Mandel, M. Sensenbrenner, and J. Ciesielski-Treska

Primary Cultures of Astrocytes from Mammalian Brain Hemispheres as a Tool in Neurochemical Research 492
A. Schousboe

Poster Abstracts

Transmembrane Potentials of Cultured Glia and Glioma Cells 493
M. Kanje, P. Arlock, B. Westermark, and J. Pontén

Molecular Properties of the Action Potential Na^+ Ionophore. Interactions with Neurotoxins 494
Y. Jacques, M. Fosset, and M. Lazdunski

The Effects of Muscarinic Agonists and Antagonists on Cyclic GMP Levels in Mouse Neuroblastoma Cells 495
P. G. Strange

Cyclic Nucleotides and ATP Content in a Cultured Human Glioma Cell Line 138 MG 496
Y. Sommarin, A. Wieslander, and B. Cederholm

Transport and Metabolism of Glucose in C-1300 Neuroblastoma (N2A) and Glioma (C-6) Cells 497
K. Keller, K. Lange, and M. Zeitz

Transport of [^{3}H]L-Glutamate and [^{3}H]L-Glutamine by Dissociated Glial and Neuronal Cells in Primary Culture 498
V. J. Balcar and K. L. Hauser

Kinetics of Glutamate, Glutamine and Leucine Transport in Cultured Neuroblastoma and Glioma Cells 499
E. Walum and C. Weiler

Lithium Transport and Toxicity in Brain Cell Cultures 500
I. Szentistványi, Z. Janka, F. Joó, A. Juhász, and A. Rimanóczy

On the Histochemistry of Cultured Endothelial Cells from Dissociated Rat Brain 501
P. Panula, F. Joó, and L. Rechardt

Morphological and Biochemical Differentiation of Neuronal Cells from Chick Embryo Brains Cultivated on Polylysine-Coated Surfaces 502
B. Pettmann, A. Porte, and M. Sensenbrenner

Morphology and Biochemistry of Rat Cortical Neurons in Dissociated Cell Culture 503
K. L. Hauser and J. Heid

Rat Glial Cells in Primary Culture. Effects of Brain Extracts on Astroglia Differentiation and on Oligodendroglia Proliferation 504
J. P. Delaunoy, B. Pettmann, G. Roussel, A. Porte, and M. Sensenbrenner

Characterization of Synaptosomes Isolated from Brain Explant Cultures . 505
M. Giesing, K. Kriesten, R. Müller, and F. Zilliken

Distribution of Acridine Orange Accumulating Particles in Neuroblastoma Cells during Differentiation and their Characterization by Subcellular Fractionation 506
M. Zeitz, K. Lange, K. Keller, and H. Herken

Mutual Influence on Glycerolipids Induced by Neuronal Contacts in Organotypic Cultures of Rat Cerebrum 507
M. Giesing, K. Tischner, and F. Zilliken

Acquisition of Cell Surface Components in Developing Rat Cerebral Cells in Tissue Culture 508
E. Yavin, Z. Yavin, and Y. Dudai

Primary Cultures from Newborn Mouse Brain. An Astroglial Cell Model? . 509
E. Hansson and A. Sellström

Effects of GABA Analogues of Restricted Conformation on GABA Recognition Sites Involved in the Synaptic Transmission Process 510
A. Schousboe, P. Thorbeck, L. Hertz, G. Svenneby, and P. Krogsgaard-Larsen

Neuron Specific Uptake of ^{3}H-GABA in Cell Cultures of Cerebellum and Olfactory Bulb 511
D. N. Currie and G. R. Dutton

GABA Metabolism in Cultured Glia Cells 512
M. Tardy, J. Bardakjian, and P. Gonnard

Characterisation of Taurine Uptake by Neuronal and Glia Cells 513
J. Borg, V. J. Balcar, J. Mark, and P. Mandel

Effect of Various Neurotransmitters on the *de novo* Biosynthesis of Brain Glycerolipids during Development in Culture 514
M. Giesing, U. Gerken, and F. Zilliken

The phosphate activated glutaminase activity and glutamine uptake into astrocytes in primary cultures 515
G. Svenneby, A. Schousboe, L. Hertz, and E. Kvamme

Development of Enzymes Involved in Glutamate Metabolism in Primary Cultures of Mouse Astrocytes 516
F. Fosmark, L. Hertz, I. Damgaard, and A. Schousboe

Alkaline Phosphate Activity in Cultured Glia and Glioma Cells 517
M. Kanje, F. Joó, and A. Edström

Activity and Isoenzyme Pattern of Lactate Dehydrogenase in Astrocytes Cultured from Brains of Newborn Mice 518
C. Nissen and A. Schousboe

Influence of Reduced Cholesterol Synthesis on the Activity of Cerebroside-Sulfotransferase in Cultured Glioblastoma Cells (C_6) 519
H. P. Siegrist, T. Burkart, U. Wiesmann, and N. Herschkowitz

Ceramide Galactosyltransferase Activity of Rat C6 Glial Cells 520
N. M. Neskovic, J. P. Delaunoy, G. Rebel, and P. Mandel

Could Ganglioside Patterns during Neuroblastoma Differentiation be an Argument for a Non-Nervous Aspect of this Type of Cell? 521
H. Dreyfus, L. Freysz, J. Robert, S. Harth, P. Mandel, and G. Rebel

The Involvement of Gangliosides in Growth and Development of Cultured Neuronal and Glial Cells 522
J. Morgan and W. Seifert

Incorporation of Unsaturated Fatty Acids in Nerve Cell Membranes: Effects on Choline Transport and Metabolism 523
T. Y. Wong, C. Froissart, J. Robert, P. Mandel, and R. Massarelli

Round Table
Techniques for the Assay of Amines and Their Metabolism

Chairman: E. Änggård

Techniques for the Assay of Catecholamines (CA), Indolamines (IA) and their Metabolites 527
E. Änggård

Poster Abstracts

A method for Measuring Catecholamines and their Non-O-Methylated Metabolites in Isolated Brain Regions 531
M. E. Bardsley and *H. S. Bachelard*

On the Metabolism of Tryptamine and Serotonin: Concurrent Extraction and Assay of Tryptophan, Tryptamine, Indole-3-acetic Acid, Serotonin and 5-Hydroxyindoleacetic Acid by a New GC-MS (MID) Method 532
F. Artigas and *E. Gelpi*

The Biosynthesis of Dopamine and Octopamine by *schistocerca gregaria* Nervous Tissue 533
A. K. Mir and *P. F. T. Vaughan*

6-Hydroxy-dopamine and the Opiate Receptors 534
F. Andrasi and *A. Ujvari*

MAO and COMT Activities in Fibroblasts from Normal Persons and Patients with Genetically Controlled Metabolic Defects with Neurological Involvement 535
S. Singh, I. Willers, E.-M. Kluß, and *H. W. Goedde*

Effects of the Monoamine Oxidase Inhibitors Tranylcypromine, Phenelzine and Pheniprazine on the Uptake of Catecholamines in Slices from Rat Brain Regions 536
G. B. Baker, H. R. McKim, D. G. Calverley, and *W. G. Dewhurst*

Studies on the Binding of the Dopamine Precursor l-Dopa to tRNA . . . 537
H. Bernheimer, G. Högenauer, and *G. Kreil*

Association of Some Biogenic Amines and Related Psychoactive Drugs with Adenosine-5'-triphosphate in Aqueous Solution 538
H. Sapper, W. Gohl, M. Matthies, I. Haas-Ackermann, and *W. Lohmann*

The Sensitivity of Embryonic Central Motor Output to Monoaminergic Transmitters and Related Drugs 539
J. Sedláček

Stereochemical Aspects in the Metabolism of 4-Aminobutyrate in Mouse Brain 540
M. Galli Kienle, E. Bosisio, A. Manzocchi, and *E. Santaniello*

Stimulation of Rat Striatal Na,K-ATPase by Catecholamines: Nature and Localization 541
J. A. Van der Krogt, R. D. M. Belfroid, and *W. F. Maas*

Inhibition of Ethanolamine and Choline Phosphoglyceride Synthesis by CMP 542
A. Radomińska-Pyrek, Z. Dabrowiecki, and *L. A. Horrocks*

Higher 5-HT Levels in Different Brain Areas of Mutant Han-Wistar Rats . 543
N. N. Osborne, V. Neuhoff, H. Cremer, and *K.-H. Sontag*

Immunochemical Relationship of Monoamine Oxidase from Human Liver, Placenta, Platelets and Brain Cortex 544
S. M. Russel, J. Davey, and *R. J. Mayer*

Vectorial Orientation of Monoamine Oxidase in the Mitochondrial Outer Membrane. Immunochemical Studies on Mitochondrial Preparations from Human Liver and Brain Cortex 545
S. M. Russel, J. Davey, and *R. J. Mayer*

Comparison between the Effect of the Hypothalamic Releasing Hormones, TRH and MIH, and Amphetamine on Presynaptic Striatal Dopaminergic Mechanisms 546
L. M. Shapiro and *P. F. T. Vaughan*

Formation of Tetrahydronorharmane and 6-OH-Tetrahydronorharmane: A New Pathway for Indolealkylamines in Mammals 547
H. Rommelspacher, H. Honecker, and *B. Greiner*

Tetrahydronorharmane (Tetrahydro-β-carboline) Modulates the K^+-Evoked Release of S-Hydroxytryptamine and Dopamine 548
H. Rommelspacher

Metabolism of γ-Aminobutyric Acid (GABA) in Rat-Brain Mitochondria . . 549
M. Lopes-Cardozo and *R. W. Albers*

The Activity of Substituted 4-Aminocrotonic Acids as Analogues of the Neurotransmitter GABA 550
R. D. Allan, G. A. R. Johnston, and *B. Twitchin*

High-affinity, Bicuculline-Sensitive GABA Binding Processes in a Synaptosome-Enriched Fraction of Rat Cerebral Cortex 551
F. V. De-Feudis, M. Maitre, L. Ossola, A. Elkouby, and *P. Mandel*

Free Communications

A Simple, Sensitive and Volume-Independent Method for Quantitative Protein-Determination which is Independent of Other External Influences . . 555
V. Neuhoff, K. Philipp, and *H.-G. Zimmer*

Determination of Amino Acids Using Dansyl Chloride 556
C. Neubach, E. Schulze, and *V. Neuhoff*

Some Biochemical and Morphological Characteristic of Bulk Isolated Neurones from Adult Rat Brain 557
H. H. Althaus, P. J. Gebicke, R. Meyermann, W. B. Huttner, and *V. Neuhoff*

Isoelectrofocusing of Crude and Purified Choline Acetyl Transferase from Human Placenta 558
C. Froissart, P. Basset, T. Y. Wong, P. Mandel, and *R. Massarelli*

Possible Role of Membrane-bound Glucosidase in the Processing of Calf Brain Glycoproteins 559
M. G. Scher and *C. J. Waechter*

Properties of Choline Acetyltransferase and its Presence in Individual Identified Neurones in the CNS of the Snail *Helix pomatia* 560
N. N. Osborne

Effect of Benzene Inhalation on the Levels of GABA and its Metabolizing Enzymes in Various Regions of Rat CNS 561
G. K. Kadyrov

Tissue Differences in the Humen N-Acetyl-β-D-hexosaminidase Isoenzimatic Forms 562
T. Pàmpols, J. Codina, M. Girós, J. Sabater, and F. Gonzáles-Sastre

Possible Synaptic Function of Glutamate Decarboxylase 563
A. Fleissner

The Separation of Extra- and Intrecellular Acetylcholinesterase Activity of Rat Sympathetic Ganglion by Mild Proteolytic Treatment 564
B. Klinar and M. Brzin

Exposure of 7 Week Old Visually Naive Rats to Diffuse and Structured Light: Effects on Visual Cortex Acetylcholinesterase 565
N. Wood and S. P. R. Rose

Light Exposure and Non-Spontaneous Locomotor Activity in Visually Deprived and Normally Reared Rats: Cortical Effects in Three Components of the Cholinergic System 566
N. Wood and S. P. R. Rose

The Action of Anestetics on the Kinetics and Activity of Acetylcholinesterase from Synaptosomal Membranes 567
A. Pastuszko

Is Acetylcholine Localized in the Cortico-Striatal Path? 568
P. Haug, N. Schröder, J. Kim, K. Paik, and R. Hassler

Purification of Cysteine Sulfinate Transaminase 569
M. Recasens and P. Mandel

Differences in the Properties of Human Leucocyte and Fibroblast Cerebroside Sulphate Sulphatase 560
A. Poulos and K. Beckman

Differences in Protein Kinase Activity between Neuronal and Glial Nuclei from Human Cerebral Cortex 571
K. Reichlmeier, H. P. Schlecht, K. Citherlet, and M. Ermini

Role of Non-Covalent Bonds for the Holding of Mitochondrial Brain Hexokinase 572
B. Broniszewska-Ardelt

Membrane-bound A_4, Lactate Dehydrogenase of Rat Brain and its Possible Relationship to Anaerobic Glycolysis 573
H. H. Berlet, T. Lehnert, and B. Volk

Subfractionation of Mouse Brain Microsomes on a Continous Sucrose Gradient: Isolation and Identification of the Membranes Containing Ceramide Galactosyltransferase and Cerebroside-Sulfotransferase 574
H. P. Siegrist, T. Burkart, U. Wiesmann, and H. Herschkowitz

Purification and Some Properties of Rat Brain Guanylate Cyclase 575
J. Zwiller, P. Basset, and *P. Mandel*

Biosynthesis and Subcellular Distribution of Disaturated Phosphatidylcholine in Rat Brain 576
L. Freysz and *H. Van den Bosch*

Effects of L-Leucine on Glutamate Synthesis in Mitochondrial and Synaptosomal Fractions from Rat Brain 577
W. Lysiak, A. Szutowicz, and *S. Angielski*

Some Properties of Synaptic Membranes in Binding Amino Acids 578
P. Lähdesmäki and *E. Kumpulainen*

Interactions of Isolated Synaptic Vesicles and Presynaptic Cell Membranes *in vitro* and Modulation of Vesicular Release by Calcium Ions 579
R. Schmidt and *H. Zimmermann*

Comparison of Presynaptic Plasma Membrane and Synaptic Vesicle Proteins   580
H. Stadler and *T. Tashiro*

Penetration of Glutamine in Rat Brain Non-Synaptosomal Mitochondria . 581
A. Minn and *J. Gayet*

Effects of Barbiturates on Calcium Metabolism in Synaptosomes Visualized by Chlorotetracycline as a Fluorescent Chelate Probe 582
J. W. Lazarewicz, A. Pastuszko, E. Bertoli, and *K. Noremberg*

Influx of Citrate to Synaptosomes and Synaptosomal Mitochondria from Rat Brain 583
H. Ksiezak, U. Rafalowska, and *J. W. Lazarewicz*

Uptake and Incorporation of Adenosine into Adenine Nucleotides by Guinea Pig Neocortex Synaptosomes 584
C. Barberis, A. Minn, and *J. Gayet*

Do Membrane Glycoproteins Play a Role in Synaptogenesis? 585
A. Reeber, J.-P. Zanetta, G. Vincendon, and *G. Gombos*

Protein Components of Synaptic Membranes Binding Taurine 586
E. Kumpulainen, M. Olkinuora, and *P. Lähdesmäki*

Artificial Taurine Release after Electrical Stimulation of Retina *in vitro* .. 587
E. Schulze and *V. Neuhoff*

Brain Capillary Permeability and Amino Acid Levels in Plasma and Cerebral Regions of the Rat after Porto-Caval Shunt 588
G. Zanschin, P. Rigotti, P. Vassanelli, and *L. Battistin*

Characterization of Experimentel Phenylketonuria 589
J. D. Lane, B. Schöne, and *V. Neuhoff*

Brain Protein Metabolism and Amino Acid Transport across the Bloo-Brain Barrier in Hyperphenylalaninemic Rats 590
R. Berger, Th. Dias, and *F. A. Hommes*

Evidence for Synthesis of Acidic Proteins Following Learning in Day Old Chicks 591
A. Longstaff and *S. P. R. Rose*

Purification of Human S-100 Protein 592
D. A. Hullin and *R. J. Thompson*

The Influence of Early Protein — Calorie Malnutrition on Levels of a Glial Brain Specific Protein (S-100) in Discrete Brain Areas 593
K. G. Haglid, L. Rosengren, L. Rönnbäck, P. Sourander, and *A. Wronski*

Identification of a Group of Octopus Brain Proteins as Histones 594
P. Cimarra and *A. Guiditta*

Release of Calcium from Mitochondria in the Electric Organ of Torpedo Marmorata during Nerve Activity 595
R. Schmidt, H. Zimmermann, and *F. Joó*

Species and Regional Variations in Endogenous Prostaglandin Formation by Brain Homogenates 596
E. Änggård and *M. S. Abdel-Halim*

The Localization and of ATP-Citrate Lyase /CCE/ in Cholinergic Neurons of Rat Brain 597
A. Szutowicz and *W. Lysiak*

A Dynamic Approach to Study the Interaction of GABA with Pyridoxal Kinase 598
J. E. Churchich and *F. Kwok*

A Study of the Mechanism of Action of 2-Oxo-1-pyrrolidine Acetamine on RNA and Protein Metabolism in Rat Brain 599
Z. S. Tencheva, S. V. Tuneva, and *N. Tijtijlkova*

Does the First Visual Stimulation Evoke an Increase of Incorporation of Labelled Leucine into Cerebral Cortex of Kittens? 600
J. Skangiel-Kramska, M. Kossut, and *K. Mitros*

N.N-Dimethylaminoethanol Acetate (P_2-agent) — The Active Agent of the Vegetative Nervous System 601
E. F. P. Szabó

Structural Feature of Porcine Brain α-Tubulin 602
H. Ponstingl, E. Krauhs, M. Little, R. Hofer-Warbinek, and *T. Kempf*

Polypeptide Pattern of Mammalian Skeletal Muscle Membrane before and after Denervation 603
G. Savettieri and *O. Lo Verde*

Effects of Polyamines on RNA Synthesis in Cell Nuclei Isolated from Rat Brain 604
S. Štipek, J. Crkovská, S. Trojan, and *J. Prokeš*

Brain Norepinephrine Involvement in the Antihypertensive Effect of Pargyline 605
J. A. Fuentes and *A. Ordaz*

Antibodies to Mesodermal Cells — a Tool in Cell Differentiation Studies? 606
L. Rönnbäck and *L. Persson*

Immunochemical Relationship between Cytosol and Synaptosomal Rat Brain
Glycoproteins 607
G. Gennarini, D. Iannelli, P. Corsi, and C. Di Benedetta

Isolatiom of Brain Proteins Reacting *in vitro* with Anti-Neuronal Antibodies in
Patients with Huntington's Disease 608
E. Wedege and G. Husby

Exogeneous (1-^{14}C) Stearic Acid Uptake by Neurons and Astrocytes 609
O. Morand, N. Baumann, and J. M. Bourre

Content of Brain Sialoglycoconjugates in Mucolipidosis I and II 610
B. Berra, S. di Palma, and C. Lindi

Occurrence in Mouse Brain of Ganglioside Carrying Alkali Labile Linkage .. 611
S. Sonnino, R. Ghidoni, N. Baumann, M. L. Harpin, and G. Tettamanti

Gangliosides in Different Brain Areas of Inbred Mice Strains 612
H. Dreyfus, A. Guiliani-Debernardi, G. Mack, S. Harth, P. F. Urban, and P. Mandel

Uptake and Intracellular Distribution in Different Tissues of Ganglioside GM1,
Intravenously Injected in the Mouse 613
P. Orlando, A. Leon, R. Ghidoni, P. Massari, and G. Tettamanti

Effect of Exogenously Added Gangliosides on Neuronal Membrane
Properties 614
D. Benvegnu', A. C. Bonett, A. Leon, and G. Toffano

Cerebrospinal Fluid Total Fatty Acid Pattern 615
J. Tichý and I. Skorkovská

Comparative Aspects of Energy Metabolism in Brain Tissue from Insects 616
G. Wegener

α-Albumin (GFA): Dosage and Localization in Human Nervous Tissue and
Cerebrospinal Fluid 617
M. Noppe, D. Karcher, J. Gheuens, and A. Lowenthal

Presence, Metabolism and Uptake of Pipecolic Acid in the Mouse Brain 618
T. Schmidt-Glenewinkel, E. Giocobini, Y. Nomura, Y. Okuma, and T. Segawa

In vivo Ethanolamine Incorporation into Brain Lipids by Base-Exchange and
Net Synthesis 619
P. Orlando, G. Arienti, L. Corazzi, P. Massari, S. Roberti, and G. Porcellati

Model Systems for the Study of Antiparkinson Drugs — Synaptosomes and
Blood Platelets 620
I. von Pusch, G. Muschalek, H. Stöltzing, and W. Wesemann

Di-, Mono-, and Nonphytanyl Serum Triglycerides in Refsum's Disease and
their Distribution in the Lipoproteins 621
B. Molzer, H. Bernheimer, and E. Koller

The Oestrogen Receptor in the Rat Hypothalamus and Uterus: The Effect of
Oestradiol Administration 622
J. O. White, A. C. Neethling, and L. Lim

Effects of Histamine-Agonists and -Antagonists (H_1 and H_2) on Ganglionic Transmission and on Cyclic Nucleotides (cAMP and cGMP) Accumulation in Rat Superior Cervical Ganglion *in vitro* 623
T. Lindl

Isolation and Labeling of Chondroitin-4-,-6-sulfate, Heparan Sulfate and Hyaluronate in Rat Brain with Radiosulfate and ^{3}H-Glucosamine *in vitro* . 624
T. O. Kleine and *U. Schwadtke*

Effect of Di-N-propylacetate on Rat Myelin GABA-T and SSA-DH 625
J. W. van der Laan, Th. de Boer, and *J. Bruinvels*

Author Index 627
Subject Index 635

Symposium

Myelin and Myelin Disorders

Chairman:
L. Svennerholm

FINE STRUCTURE OF MYELIN SHEATHS

Enrico Mugnaini
Laboratory of Neuromorphology
Dept. of Biobehavioral Sciences, Box U-154
The University of Connecticut, Storrs, Conn. 06268 (USA)

Foreward
   While it is obvious that much knowledge on the cell biology of myelin has been accumulated over the last thirty years, it is true that the avenues of the microscopist, the biophysicist, the neurochemist and the neurophysiologist are still quite separate and have met only at obvious crossroads. Meetings of the kind sponsored here in Goettingen by the European Society of Neurochemistry represent, therefore, a welcome effort in promoting the bridging of interdisciplinary barriers.

   In our effort to understand the biology of myelin, and ultimately its pathobiology, we must come closer, not only to the level of molecular models of the membrane, but also to a regulatory biological understanding of the myelin forming cell. An explanation of neuroglial architecture and biomechanics, and neuronal-glial interactions will then be at a closer reach. The cytoplasmic biochemical machinery of the myelin forming cells now begins to be explored through the development of procedures for isolating pure glial cell fractions, and a resetting of common ways of thinking may be appropriate.

   As an anatomist, primarily, and a structural cell biologist, secondarily, my most obvious role in promoting this important area of basic research is to emphasize the role of the "knowing by seeing" approach, demonstrating--and explaining as far as presently possible--its more recent accomplishments.

   In dealing with myelin at a meeting of neurochemists, one feels compelled to begin by stressing that most of the widely accepted biochemical data are derived from "pure myelin" fractions. These, I understand, consist predominantly of compact myelin, although the presence of small amounts of cytoplasm can generally not be excluded. As a consequence, we know a good deal about the compact myelin membrane, i.e., the portion of the myelin sheath from which cytoplasm has been almost completely extruded, but little about the whole myelin sheath. Presumably, being less "myelin-like" than the compact myelin domain, other portions of the myelin sheath are eliminated during the separation procedure to reduce the occurrence of non-myelinic lipid and protein contaminants.

The electron microscopic anatomist, however, using thin sections, can scan the complete cellular structure at 10 Å resolution. Moreover, with the introduction of freeze-fracture, the morphological armamentarium has been enriched considerably. As known, the two major advantages of freeze-fracture electron microscopy are the production of large membrane vistas which improve the tridimentional representation of the structures under scrutiny, and the revelation of the membrane interior. The principles of the technique are shown diagrammatically in Fig. 1. The fracture plane passes in the middle of the lipid bilayer forming two artificial surfaces now called P and E faces. The P face adheres to the protoplasm and the E face to the exoplasm. The lipid domains of the membrane appear smooth while the proteins, or the proteo-lipid complexes, are seen as particulate components of varying morphology.

Thus, a complex morphological picture of the myelin sheath starts to emerge, although high resolution electron microscopy of the membrane's interior and exterior surfaces is still pending.

As in other cell membranes, the particulate intramembrane components of the myelinic plasmamembrane revealed by freeze-fracturing are distributed in a random, or at least a quasi-random, manner except at the sites of specialized cell junctions. Awareness that the myelin sheath includes extensive cell junctional areas is largely a contribution of freeze-fracture although they can be recognized in thin sections also.

Moreover, subtle topographical variations in the freeze-fracture features of the myelin membrane suggest that membrane domains exist, which, as in other tissues, are related to the geometry of the cell and its mode of association with other tissue elements.

The structure of the myelin sheath in peripheral and central nervous tissues has been repeatedly reviewed in the last ten years. Extensive and exhaustive accounts have been published quite recently in four independently edited volumes (Peters et al., 1976; Landon, 1976; Morell, 1977; Waxman, 1978). Consequently, a complete survey of the literature is beyond the scope of this communication.

I have chosen to present first a brief review of the similarities and the differences of the architectural principles of central and peripheral myelin with the purposes of ensuring a detailed and common language and of distinguishing, at least tentatively, the essential properties of myelin from other features related to more general aspects of the cell biology of the myelin forming cells. The emphasis on morphological complexity is not intended to claim relevance to minor points but to counteract possibly deleterious oversimplification. In the second part of my account I will deal with the concepts of "myelin membrane domains," "intra- and interlamellar adhesion," "the myelin module" and "myelination," as these develop from the above

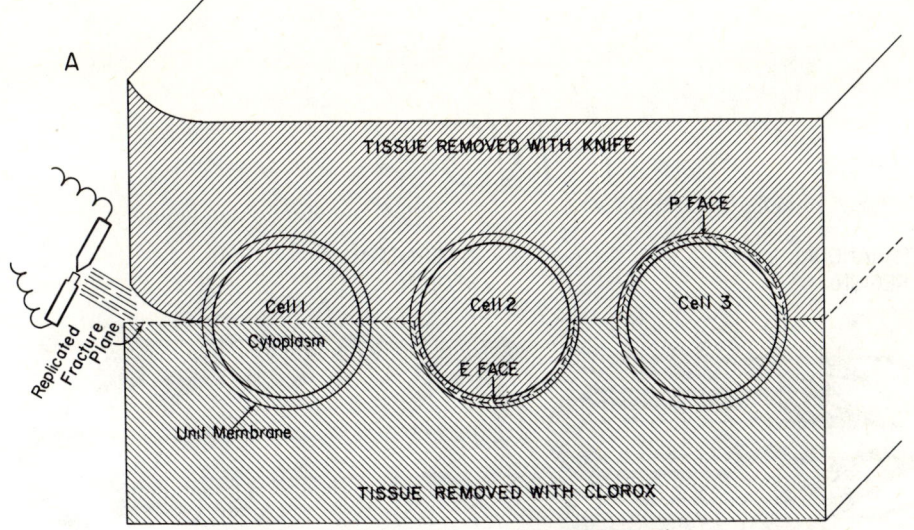

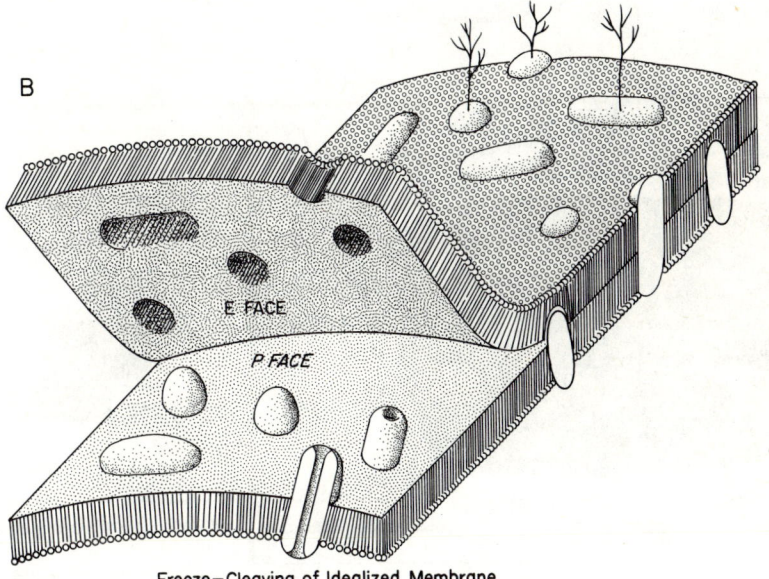

Fig. 1. Diagram of the freeze-fracture method. A: Fracturing of a tissue block is represented. Cell 1 is cut across in the fracture process, whereas in cells 2 and 3 the plasmalemma is cleaved in the middle, creating a fracture face E (cell 2) and a fracture face P (cell 3). When the replica is photographed, the membrane vista of cell 2 appears concave and that of cell 3 convex. B: The cleaving process of a cell membrane produces two complementary fracture surfaces. The E face is the internal membrane surface associated with the exoplasmic half of the membrane, and the P face the internal membrane surface associated with the protoplasmic half. Freeze-cleavage splits the lipid bilayer through the center of the hydrophobic region. The membrane proteins remain intact and cleave with one of the half membranes, leaving a pit on the complementary fracture surface. Two of the idealized protein particles are represented with a channel, or pore, through which ions and small solutes may pass. Glycoproteins are represented with their polysaccharidic tails extending into the extracellular space.

PNS

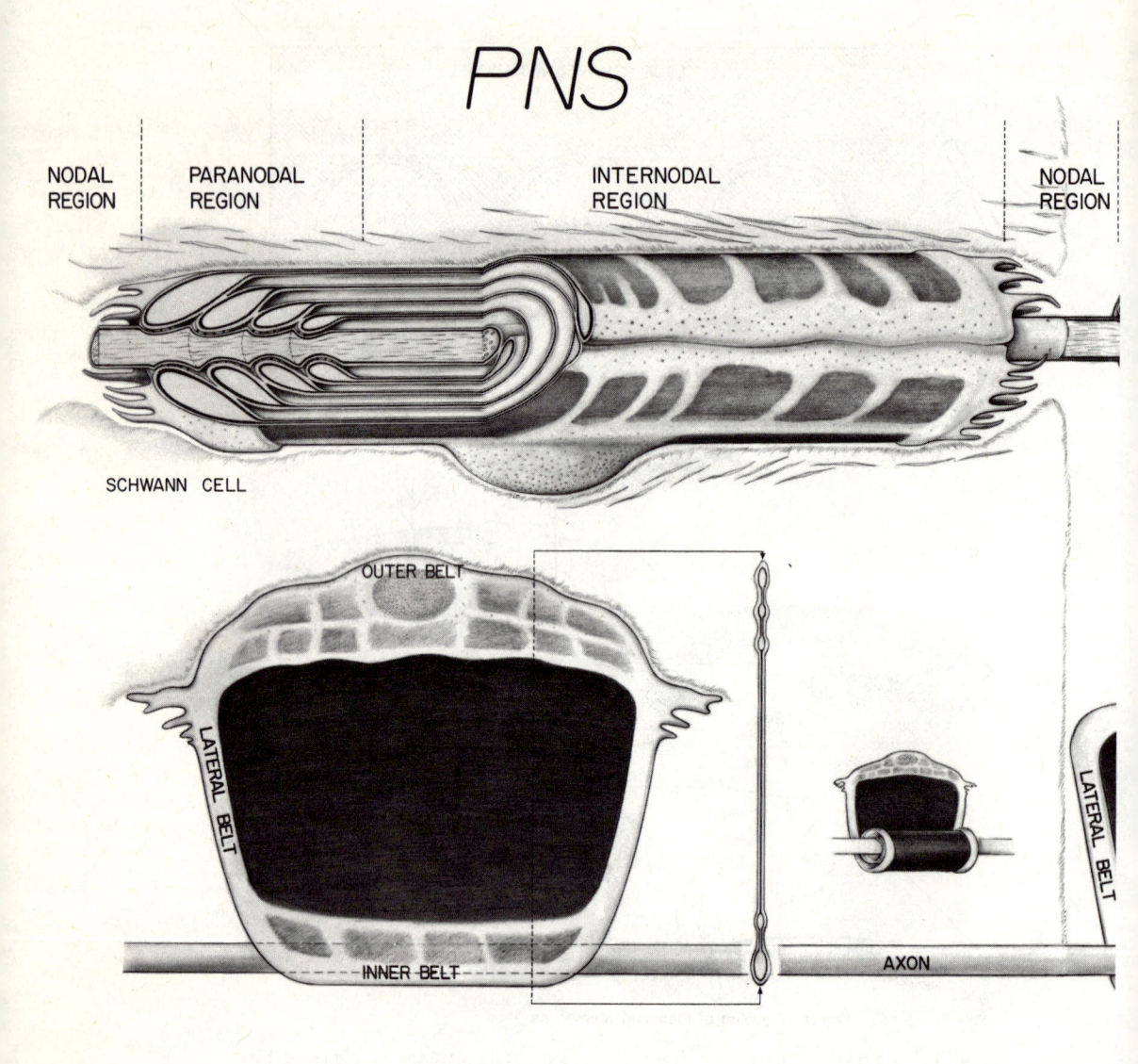

Fig. 2. Schematic representation of some of the principal features of central (CNS) and peripheral (PNS) myelin sheaths at an idealized nerve root. The ratio between sheath forming cells and myelinated axon segments is 1:1 at the periphery, but it is 1: 2 in the CNS (although a 1:1 relation may occur also centrally). The outer wall of the Schwann cell tube is covered by a basal lamina with adhering collagen fibers. The outer turn of the Schwann cell sheath forms nodal microvilli. In the lower half of the diagram the sheaths are represented unrolled to facilitate the visualization of their shape. Size is necessarily out of proportion, since, for an axon, say, 3 µm thick, the unrolled sheath would be approximately 300 µm long and 800 µm wide. The dark shaded area represents compact myelin, as also apparent in

CNS

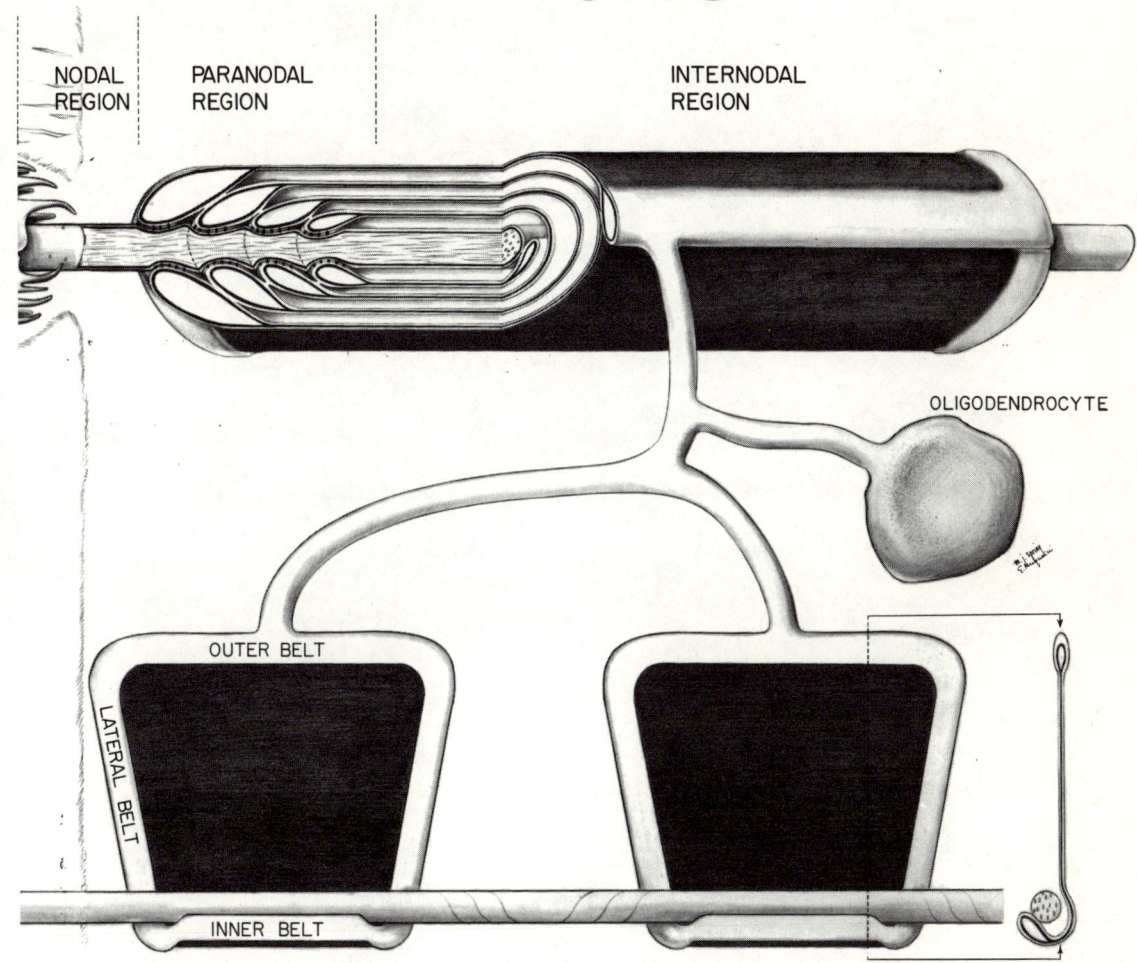

NODAL REGION | PARANODAL REGION | INTERNODAL REGION

OLIGODENDROCYTE

OUTER BELT

LATERAL BELT

INNER BELT

the idealized cross-section of the unrolled sheath, where the internal leaflets of the plasmamembrane form a major dense line. Cytoplasm (white) is shown here only where it is invariably found, i.e. at the marginal belt. The cytoplasmic incisures are shown in Fig. 14. Note also that the outer and inner turns of the sheath are entirely cytoplasmic in the PNS but not in the CNS. In the Schwann cell the cytoplasm is accumulated around the nucleus and in longitudinal and transversal channels (white) at the outer and inner turns, separated by regions where the cytoplasmic layer is extremely thin (softly shaded regions, referred to as semicompact by Mugnaini et al., 1978).

Figs. 3-6. Electron micrographs of myelinated axons from the cat cochlear nerve (PNS) (3) and the rat optic nerve (CNS) (4-6), illustrating cytoplasmic portions of the sheaths. The axon is labeled A, the outer belt (ob) and the inner belt (ib). Longitudinal cytoplasmic incisures are marked by small arrows and a Schmidt-Lanterman incisure by large arrows. Note that the smaller incisures are devoid of cytoplasmic organelles. The open arrowheads in Fig. 3 indicate the enlarged intramyelinic extracellular space between cytoplasmic incisures that occur in register in adjacent turns of the sheath. Where a cytoplasmic region faces a compact lamella, on the contrary, the intramyelinic space is narrow (dark arrowheads). The asterisks point to semi-compact regions of the outer and inner turns. At bl, basal lamina; cf, collagen fibers. In figs. 5 and 6 all axons have a compact inner turn, with cytoplasm restricted to the inner belt, except the axon labeled with a star, which has a cytoplasmic inner turn (x). The triangle in fig. 6 marks portions of two neighboring sheaths where the tissue extracellular space is as narrow as the intramyelinic extracellular space.

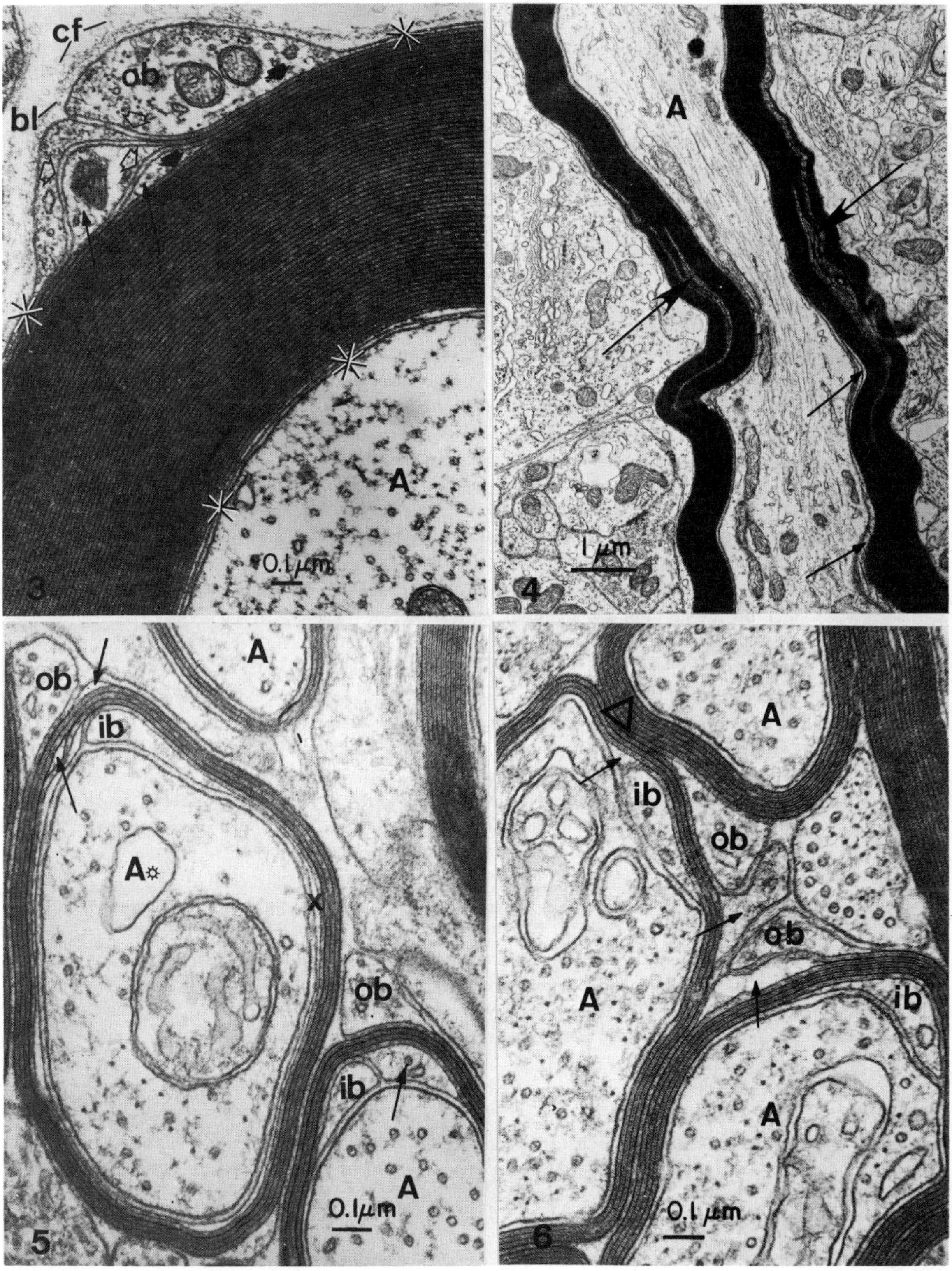
cf
ob
bl
A
0.1 µm
3
A
1 µm
4
ob
A
ib
A☼
ob
ib
A
0.1 µm
5
A
ib
ob
ob
A
ib
A
0.1 µm
6

analysis.

Common morphological features of central and peripheral sheaths

A combination of light microscopic techniques with thin section and freeze-fracture electron microscopy has helped in establishing several basic principles of architecture common to both central and peripheral sheaths (for details see Schnapp and Mugnaini, 1977) a) each myelin segment is formed by a single myelin forming cell and consists of a trapezoidal cellular flap of varying dimensions which is wrapped around the axon (Fig. 2).

b) The cellular flap contains little cytoplasm, confined to the periphery (marginal cytoplasmic belt) and to longitudinally and circumferentially oriented tubes (cytoplasmic incisures) (Figs. 3-6). The incisures communicate with the marginal belt and with each other thus forming a cytoplasmic reticulum bordering and intruding the compact myelin domain (Fig. 4).

c) A typical myelin membrane is formed where the cytoplasm has been excluded from the flap and the cytoplasmic leaflets become united to form the so-called major dense line (Fig. 2). Such a degree of membrane adherence is a unique phenomenon.

d) The regular apposition of the turns of the spiralled myelin membrane gives rise to a periodic structure where the thickness of the intercellular cleft is kept constant, and remains thinner than in all other membrane-to-membrane appositions with the exception of the acrosomal and the nuclear membranes in spermatids, the chloroplast thylaxoids, the retinal receptor discs and certain specializations of the endoplasmic reticulum in neurons. Previously postulated obliteration of the intramyelinic extracellular space is now known to be an artifact of dehydration. Well preserved myelin shows two intraperiod lines separated by a cleft (Fig. 7). Failure to see a double intraperiod line always means poor preservation of ultrastructure.

e) In spite of the close appostion of its exoplasmic and cytoplasmic leaflets, the freeze-fracture properties of the myelin lamellae resemble those of other membranes. Freeze-cleaving generates a protoplasmic (P face) and an exoplasmic (E face) surface (Fig. 8).

f) The intramembrane particles in the nonjunctional compact myelin membrane have a random distribution.

g) At the borders of each myelin segment (inner, outer and lateral mesaxons) and along the cytoplasmic incisures there is a zonula occludens characterized in freeze-fracture replicas, by linear, interrupted and non-anastomosed strands or rows of particles, cleaving mostly with the P face, and complementary to an easily demonstrable groove on the E face.

h) Adherent junctions and occasional gap junctions may be present between adjoining cytoplasmic regions of the sheath.

i) Each myelin segment adheres to the axon at both ends, by virtue of an extensive cell junction, the septate-like paranodal axo-glial junction (reviewed by Schnapp

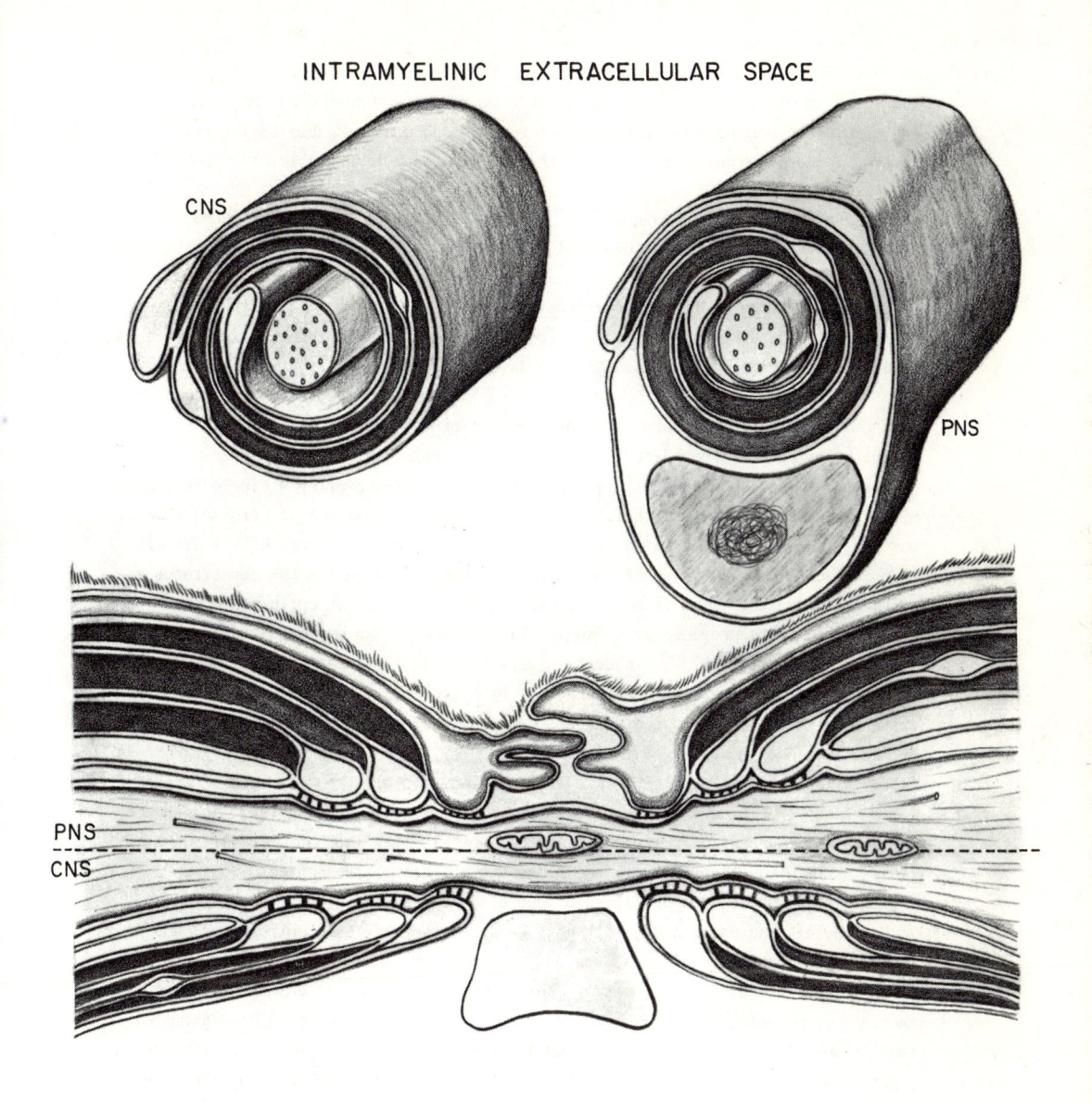

Fig. 7. Diagrammatic representation of internodal portions (top) of central and peripheral myelinated fibers in cross section and of their respective paranodal and nodal regions in longitudinal sections (bottom). The drawing emphasizes the intramyelinic extracellular space (dark shaded), whose thickness has been exaggerated. Note that at all borders this space is delimited by tight junctions.

et al., 1976). This is characterized by an oriented extracellular matrix (septa), indulated junctional membranes, and aligned intramembrane particles that, in the glial membrane, cleave with either fracture face and are incapable of lateral movement. Corresponding particles in the axolemma seem less stable.

The axo-glial junction is essential in keeping the myelin sheath stretched and drawn apposed to the axon (Blank et al., 1974; Yu and Bunge, 1975).

Differences between central and peripheral sheaths

A conspicuous number of differences between central and peripheral axons exist in spite of the common architectural principles listed above.

These differential features are analyzed one by one in the following:
1) While a single oligodendrocyte may provide a myelin segment to a number of different axons, a Schwann cell regularly forms a sheath covering a single internode. Since peripheral nerves and their roots are subjected to mechanical stresses that must be of an order of magnitude greater than that in most portions of the CNS, it seems reasonable to assume that the one-to-one ratio seen in the PNS constitutes a mechanical advantage with respect to lower ratios. In the rat optic nerve the number of segments provided by a single oligodendrocyte was estimated to be about 40 (Davison and Peters, 1970). In certain regions of the rat brain stem, 10 myelin segments per oligodendrocyte were observed by Sternberger et al. (1978). Near the entrance of the VIII nerve root, in the cat, this number is only 2-3 (McFarland and Friede, 1971) on the average, thus approaching the situation in the nerves. Hortega (1928) illustrated Schwann cell-like oligodendrocytes (his type IV) some of which may myelinate a single axon, as also shown in Fig. 9. The varying ratio in the CNS may have a biomechanical correlate. A second possibility is that there are limitations in the production of myelin membrane and a single large central axon may saturate the synthetic capabilities of one oligodendrocyte.
2) Peripheral sheaths include, in general, more cytoplasm than central ones, and commonly show longitudinal incisures and circumferential incisures (Schmidt-Lanterman clefts). Both kinds of incisures are not as rare in the CNS as commonly thought, and they are more common in large fibers than in small ones (Hortega, 1928; Blakemore, 1969) (Figs. 3-6). Also, the outer and inner turns of the peripheral sheath contain a cytoplasmic reticulum interrupted by semicompact domains (i.e. regions where the facing inner leaflets of the myelin membrane come very close but do not form a major dense line (Mugnaini et al., 1977). The central sheath, with the exception of the marginal belt and a longitudinal incisure, aligned with the outer and the inner mesaxon of the same fiber or with the outer mesaxon of a neighboring sheath, usually shows a compacted outer turn and, in a reasonably high number of cases, a compacted inner turn.

Regional and species variations in the number and extent of cytoplasmic incisures in central and peripheral sheaths have not been explored adequately. Possible differences may be related to scarcely understood metabolic, developmental, or biomechanical factors. If the amount and number of cytoplasmic proteins in pure myelin fractions is related to the incisures, their variation may explain certain divergences among biochemical studies.

3) At areas rich in subjacent cytoplasm, the Schwann cell plasmalemma of the outer turn of the sheath displays plasmalemmal vesicles (caveolae) (Fig. 11) which may have a concentration of up to $30/\mu m^2$, thus almost doubling the real surface membrane area in these regions (Mugnaini et al., 1977). Plasmalemmal vesicles are absent in central sheaths. In smooth and striated muscle fibers plasmalemmal vesicles have been shown to disappear reversibly with excessive stretch. Their functions in Schwann cells remains to be ascertained.

4) While central sheaths do not show surface specializations visible in routinely stained thin sections, the outer surface of the Schwann cell is covered with a basal lamina whose outer aspect faces bundles of collagen fibers (Fig. 3).

5) At the Ranvier nodes peripheral fibers usually show an array of microvilli (Fig. 2), formed by the outermost paranodal loop, and are surrounded by an electron dense, floccular extracellular substance. The microvilli are regularly spaced in most fibers larger than $2\,\mu m$. The oligodendrocyte do not form typical microvilli at central nodes.

Facing the naked nodal axolemma, one finds rather commonly an enlarged astrocytic process, in the CNS, and a flocculent extracellular matrix, in the PNS.

6) The presence of tight junctions in register from one lamella to the next between compact myelin lamellae of central sheaths is now widely recognized. It is probable, though not yet proven, that these tight junctions correspond to the "radial component" of Peters (1961) seen in thin sections (Fig. 9). The term "radial" refers to the repeat of the junction from one lamella to the next. Since, however, the tight junction in each myelin membrane is oriented parallel to the long axis of the sheath, the structural component of the junction is, in fact, tangential. Since the tight junctional proteins in one membrane are interlocked with those on the facing membrane, the lamellae must be rather firmly bound together at these sites. The formation of these repeating tight junctions indicates complex influences in protein assembly between the lamellae. Schnapp and Mugnaini (1977) have presented pictures suggesting turnover or de novo assembly of tight junctions in sheaths from adult animals.

In a preliminary survey of various fiber tracts in the rat CNS, we have found that the radial component occurs with different frequency and may be more common in sheaths of spinal cord, medulla and optic nerve than in those of the forebrain and the cerebellum. This differential distribution remains unexplained. It may account, in part, for regional differences in the protein composition of myelin fractions. Perhaps the interlamellar tight junctions counteract the effects of shearing forces on interlamellar adhesion?

A typical "radial component" has not been demonstrated in the PNS, although spirally or longitudinally directed tight junctions are very frequent in all peripheral sheaths. These accompany the longitudinal and the Schmidt-Lanterman incisures, and may have a role analogous to that of the radial component in central myelin.

7) The existence of dissimilarity in the chemical makeup of central and peripheral myelin is well known. Regional differences have also been reported. The spiralled membrane itself has a larger repeat period in the peripheral than in central sheaths. The largest disparity is in the width of the intramyelinic extracellular space that is larger in peripheral than in central myelin. Also, peripheral myelin contains more glycoproteins than central myelin. With freeze-fracture (Fig. 8), it has been shown that while the E face of the central compact myelin membrane is always relatively smooth, the peripheral E face is covered with a large number of randomly distributed particles (Schnapp and Mugnaini, 1975;1978). Perhaps some of these features are correlated. The small particles may cleave with the E face by virtue of an asymmetric location in the membrane and may represent glycoproteins. Polysaccaridic tails in the extracellular space may maintain a wider interlamellar distance.

Membrane domains of the myelin forming cell

As mentioned previously, freeze-fracturing has revealed in the myelin membrane intramembrane particles with different cleaving properties and distribution. The glial intramembrane particles fall into five different main categories, presumably corresponding to an equal number of protein classes:

  i) elongated P face particles seen in both central and peripheral compact myelin
  ii) round P face particles seen in both central and peripheral compact myelin
  iii) round E face particles seen mainly in peripheral compact myelin
  iv) tight junctional P face particles forming strands, seen in both central and peripheral sheaths at the mesaxons and between successive turns of the myelin sheath.
  v) junctional particles seen only in the paranodal glial membrane of both central and peripheral sheaths. These are distributed in parallel rows and cleave with either face.

The particulate components of central sheaths (i,ii,iv, and v) are shown in Fig.10.

Subclasses of particles may exist, but firm criteria for their distinction have not been established so far. While it is probable that proteins corresponding to classes i-iv are present in the purified myelin fraction, the paranodal glial membrane that contains the fifth kind of particulate component may be confined to another portion of the tissue homogenate, since it is junctionally attached to the axolemma.

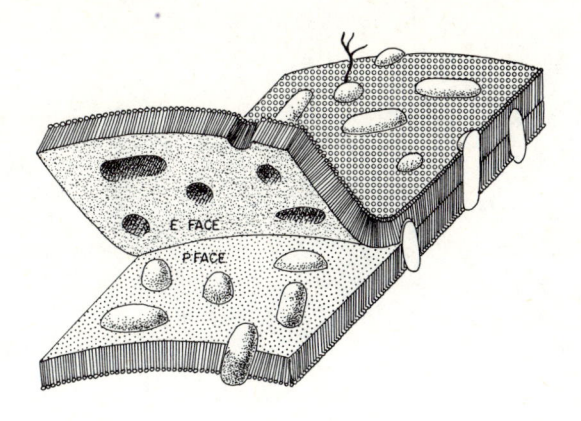

A. FREEZE-CLEAVED CENTRAL MYELIN

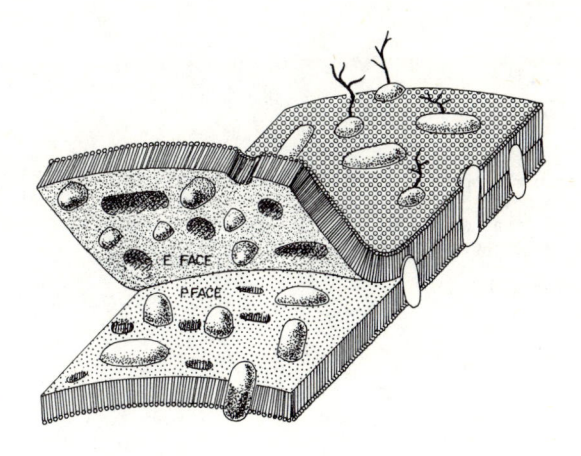

B. FREEZE-CLEAVED PERIPHERAL MYELIN

Fig. 8. Diagrammatic representation of partially freeze-cleaved central (A) and peripheral (B) compact myelin membranes. While in central myelin round and elongated particles cleave mainly with the P face, in peripheral myelin many rounded particles are seen, in addition, on the E face. Based on biochemical data, more glycoproteins have been represented in peripheral than in central myelin.

Fig. 9. Electron micrograph from the rat spinal cord showing a Schwann-like oligodendrocyte ensheathing a single axon (A). The arrowheads indicate the radial component of Peters, that coincides with (or at least accompanies) interlamellar tight junctions. At ob, outer belt; om, outer mesaxon.

Fig. 10. Electron micrograph of a freeze-fractured central myelinated fiber showing at mp, round and elongated P face particles of the compact myelin membrane; t, tight junctional E face grooves of the lateral, paranodal belt; jp, aligned intramembrane particles of the lateral belt membrane involved in the axo-glial junction; my, compact myelin turns; and A, cross-fractured axon (frog spinal cord, Schnapp and Mugnaini, 1977).

0.1 μm

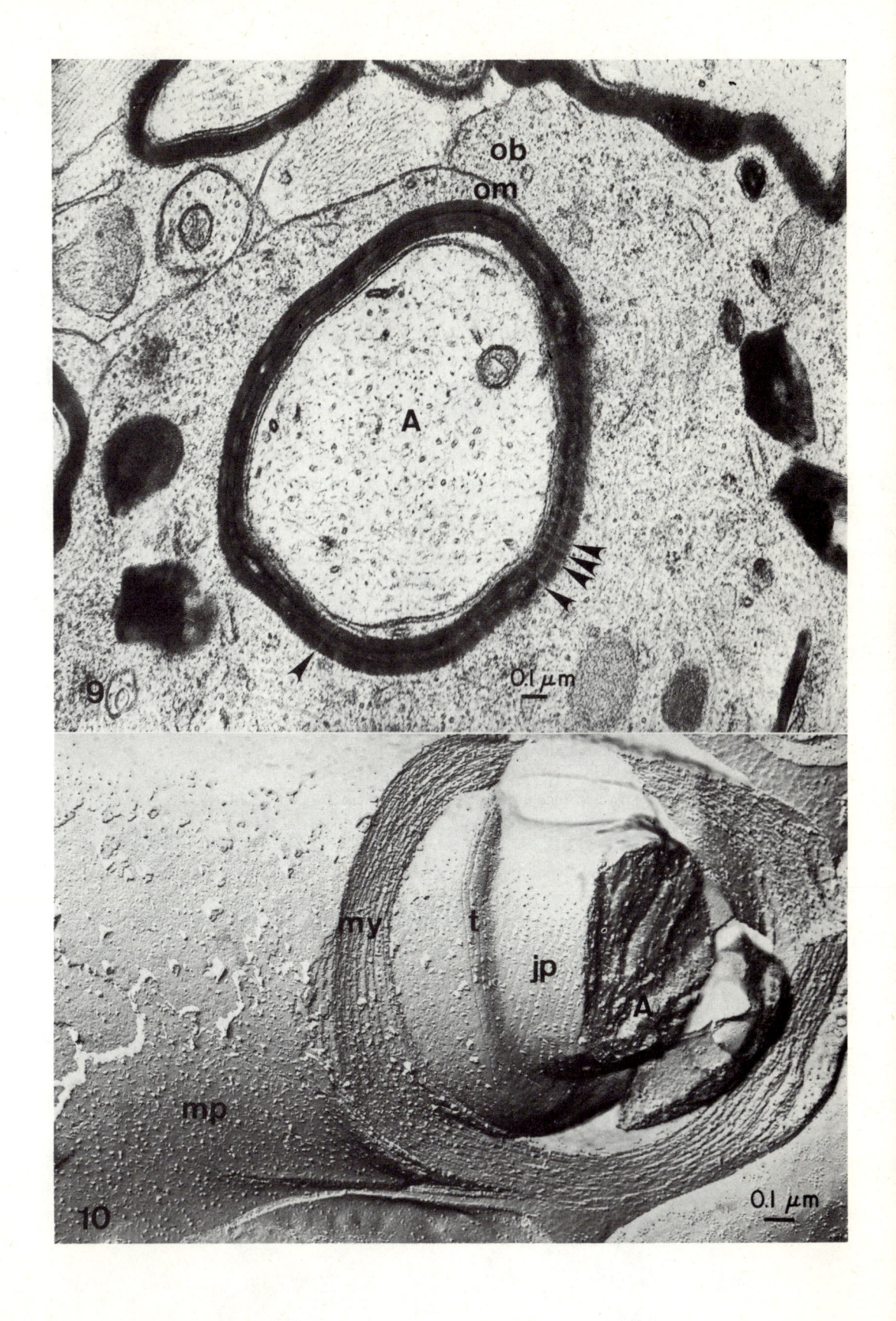

ob
om
A

0.1 μm

my
t
jp
A
mp

Some conclusions arising from the observations presented so far are that the myelin membrane is far from homogeneous, the extent of the specialized domains differs somewhat in central and peripheral sheaths, and that regional variations occur in both tissues.

A closer analysis of inhomogeneity in the myelin membrane, therefore, may offer some useful biological insight.

We know that a basal lamina is produced at the Schwann cell outer surface only, a paranodal axo-glial junction is formed at both ends of the myelin segment only, and villi are formed at the node of Ranvier in peripheral nerves where the nodal gap is occupied by a stainable material. Moreover, we have observed that tight junctions accompany cytoplasmic containing incisures, and the intramyelinic extracellular space is larger at sites where cytoplasmic regions on successive turns of the sheath face one another than where one cytoplasmic incisure faces compact myelin (Fig. 3). In the latter case the extracellular space is as thin as in between compact myelin turns. Longitudinal incisures usually occur in register from one lamella to the next, as do the tight junctions of the radial component. In the CNS, adjoining myelin sheaths at points are separated by a gap similar in thickness to the intramyelinic extracellular space (Fig. 6).

It seems plausible, therefore, that inhomogeneity develops in the myelin membrane proper, and in the plasmamembrane of the myelin forming cell in general, in relation to its varying associations on either the exoplasmic and protoplasmic aspects, even encompassing the next neighboring cellular element. These variants are quite numerous and are listed in table I.

The term myelin membrane is used here, and in the following, in its widest sense (plasmamembrane of the myelin forming cell) and includes the compact myelin proper.

The differences in the microenvironment of the myelin membrane may be related to variation in its lipid and protein composition thus generating different myelin membrane domains. A hypothetic list of myelin membrane domains is given in table II.

It is conceivable that some differences in the protein makeup of the various domains of the myelin membrane can be revealed by freeze-fracture. Patterns of distribution of intramembrane particles should, therefore, be studied in detail in the various portions of the sheath. This has not been accomplished as yet. Subtle differences are difficult to circumstantiate, because of sampling problems and preparative artifacts. The latter are quite frequent, especially when dealing with frozen, unfixed myelin; fixation and glycerination on the other hand may alter the cleaving properties of the protein particles. As shown in Fig. 11, one occasionally observes in mature sheaths an immediately evident difference in the distribution of

Microenvironment of myelin membrane

CNS

1) Exoplasmic aspect
- general extracellular space
- intramyelinic extracellular space
- internodal periaxonal extracellular space
- paranodal periaxonal extracellular space
- central nodal substance (scarce)

2) Protoplasmic aspect
- thick cytoplasmic mass
- thin cytoplasmic layer
- moiety of major dense line

3) Next neighbor

outside the sheath:
- myelinated fiber — compact portion of outer myelin turn*
- cytoplasmic portion of outer myelin turn**
- oligodendrocytic cell body or process**
- astrocytic cell body or process
- microglial cell body or process
- nerve cell body or dendrite or myelinated axon

within the sheath:
- compact portion of myelin lamella*
- cytoplasmic portion of myelin lamella**
- internodal axolemma
- paranodal axolemma
- nodal axolemma

PNS

1) Exoplasmic aspect
- general extracellular space
- intramyelinic extracellular space
- internodal periaxonal extracellular space
- paranodal periaxonal extracellular space
- peripheral nodal substance

2) Protoplasmic aspect
- thick cytoplasmic mass
- thin cytoplasmic layer
- moiety of major dense line

3) Next neighbor

outside the sheath:
- basal lamina and collagen fibers

within the sheath:
- compact myelin lamella
- cytoplasmic myelin lamella
- internodal axolemma
- paranodal axolemma
- nodal axolemma

*,** Single and double asterisks mark membrane domains of presumably similar composition.

TABLE I. Tentative list of the parameters which may correspond to differences in the microenvironment of the myelin membrane.

THE UNIT MEMBRANE OF MYELIN SHEATH CAN FACE:

CNS

EXOPLASMIC ASPECT		CYTOPLASMIC ASPECT	
EXTRACELLULAR SPACE	NEXT NEIGHBOR		SITE OF SHEATH
T.E.S.	Other cell	Cytoplasm	Outer belt
I.E.S.	Compact lamella	Cytoplasm	Outer belt or cytoplasmic incisure
I.E.S.	Cytoplasmic lamella	Cytoplasm	Cytoplasmic incisure
T.E.S.	Other cell	Major dense line	Compact outer turn
I.E.S.	Minor dense line	Major dense line	Compact myelin proper
I.P.S.	Intermodal axon	Major dense line	Compact inner turn
I.P.S.	Intermodal axon	Cytoplasm	Inner belt
P.P.S.	Paranodal axon	Cytoplasm	Paranodal junction
N.E.S.	Astrocytic process (or other cell process)	Cytoplasm	Last paranodal loop

PNS

T.E.S.	Basal lamina	Thick cytoplasm	Outer turn
T.E.S.	Basal lamina	Thin cytoplasm	Semicompact outer turn
I.E.S.	Compact lamella	Cytoplasm	Cytoplasmic incisure
I.E.S.	Cytoplasmic lamella	Cytoplasm	Cytoplasmic incisure
I.E.S.	Minor dense line	Major dense line	Compact myelin proper
I.P.S.	Intermodal axon	Thick cytoplasm	Inner turn
I.P.S.	Intermodal axon	Thin cytoplasm	Semicompact inner turn
P.P.S.	Paranodal axon	Cytoplasm	Paranodal junction
N.E.S.	Other microvilli or nodal axon	Cytoplasm of microvillus	Microvilli

TABLE II. List of possible myelin membrane domains (Cp. Figs. 3-6, and 14) in central and peripheral sheaths. T.E.S.: tissue extracellular space; I.E.S.: intramyelinic extracellular space; I.P.S.: intermodal periaxonal space; P.P.S.: paranodal periaxonal space; N.E.S.: nodal extracellular space.

intramembrane particles at the two sides of the tight junctional strands. Also, the membrane adjacent cytoplasmic masses seem to contain a higher concentration of large, round particles (> 80 Å in diameter) than the compact myelin membrane (Fig. 12). The tight junctions, in addition to the subjacent cytoplasm, may contribute to a parcellation of the myelin membrane. This could be accomplished in different ways: 1) by reducing paracellular diffusion and thus introducing a restraint in the equilibration of the solutes in the extramyelinic and the intramyelinic extracellular spaces, and 2) by limiting diffusion within the membrane interior itself, i.e. in the plane of the lipid bilayer.

The possibility that by virtue of the tight junction, materials may remain sequestered within the intramyelinic extracellular space has been suggested by Mugnaini and Schnapp (1974). This hypothesis has not yet been circumstantiated by experimental proofs.

Intra-and Interlamellar adhesion
Since the compact myelin membrane constitutes the main portion of the mature myelin sheath, its features remain of paramount importance. Two of the main characteristics of this domain of the sheath are: i) the protoplasmic membrane leaflets of the myelin membrane appear fused, and ii) the extracellular space between adjacent turns of the sheath remains extremely narrow. The fluid spaces within and in between the myelin lamella (e) are therefore exceedingly thin (Worthington, 1971). The extracellular cleft in myelin is even smaller than that observed at the spot-like gap junctions (20-30 Å), where a special protein is present at high density, and particles occur in a quasi-crystalline order forming pairs across the extracellular space (Gilula 1977).

In the myelin membrane the protein-to-lipid ratio is quite low. Which are the substrates for the unique membrane relations in myelin? Why is compact myelin formed at the inner and outer turns of central, but not peripheral sheaths?

It has been suggested that the apparent fusion of the protoplasmic leaflets to form a major dense line (intralamellar adhesion) is a consequence of interactions between adjacent molecules of basic protein, whose coiled chain is presumed to be largely peripheral to the lipid bilayer (Braun, 1977). If this is the case, is there less basic protein at the inner and outer turns of peripheral sheaths, where a major dense line is not formed ("semicompact" membrane domains of Mugnaini et al., 1977)? Involvement of the proteolipids in interlamellar adhesion and spacing has also been suggested (Mateu et al., 1973; Vail, et al.,1974; Pinto da Silva and Miller, 1975; Schnapp and Mugnaini, 1977; Braun, 1977).

Since in the compact myelin membrane the intramembrane particles, with the exception of those constituting the interlamellar tight junctions, appear distributed at random, pairing of individual protein particles on adjacent protoplasmic and exoplasmic leaflets of the myelin membrane is possible but unproven.

Speculatively, one may also conceive that the myelinating cell synthetizes or accumulates a bivalent ligand which diffuses and concentrates into the secluded intramyelinic extracellular space and into the exceedingly narrow cytoplasmic fluid layer of the compact lamella. Specialized glial cells that form the lining of the choroid plexus do control cerebrospinal fluid production and share with Schwann cells the properties of forming tight junctions and microvilli (Mugnaini, 1977).

Schmidt-Glenewinkel (personal communication) has brought to my attention that the polyamine spermidine has a high concentration in both central white matter and peripheral myelin nerve fractions (Seiler and Schmidt-Glenewinkel, 1975; Seiler and Deckhart, 1976). The concentration of spermidine increases during the period of myelination (Seiler and Lamberty, 1973; 1975), is lower in the quaking mouse (Russel and Meier, 1975), which has an extensive myelin deficit (Friedrich, 1974), and decreases during Wallerian degeneration (Seiler and Schroeder, 1970) relative to the number of cells.

Spermidine has been localized autoradiographically in the myelin sheath and in myelin forming cells (Fisher et al., 1972). As known, spermidine is synthesized from ornithine via putrescine. Ornithine decarboxylase is present in nervous tissue. The enzyme in 3-week-old rats is low in isolated neuronal fractions but relatively high in glial fractions (Schmidt-Glenewinkel, personal communication). Also, the putrescine converting enzymes have relatively high activity in nervous tissue (Raina et al., 1976).

The spermidine molecule is approximately 11 $\overset{\circ}{A}$ long (estimated from data of Liquori et al., 1967), but can assume a conformation which allows it to adapt to a space shorter than its length. The molecule is positively charged at its ends and at an intermediate nitrogen. One of the oxygens of phosphate groups is negatively charged. Carboxyl groups of glutamate and aspartate in proteins are also negatively charged. A high density of negative charges in myelin has been shown electron microscopically by Bittinger and Heid (1977) using cationic ferritin.

Because of different strengths of the corresponding acids, the electrostatic interaction with the nitrogen of spermidine might be stronger in the case of the phosphate groups than it is for the carboxyl groups. Phosphorylation of the membrane proteins might, thus, influence binding of a ligand.

In Fig. 13 I have illustrated, in a cartoon like fashion, a hypothetical situation for the myelin membrane that would allow a strong, short range, attractive force between adjacent membranes with a random protein distribution of proteins (i.e., without transmembrane pairing), and with some degree of freedom of

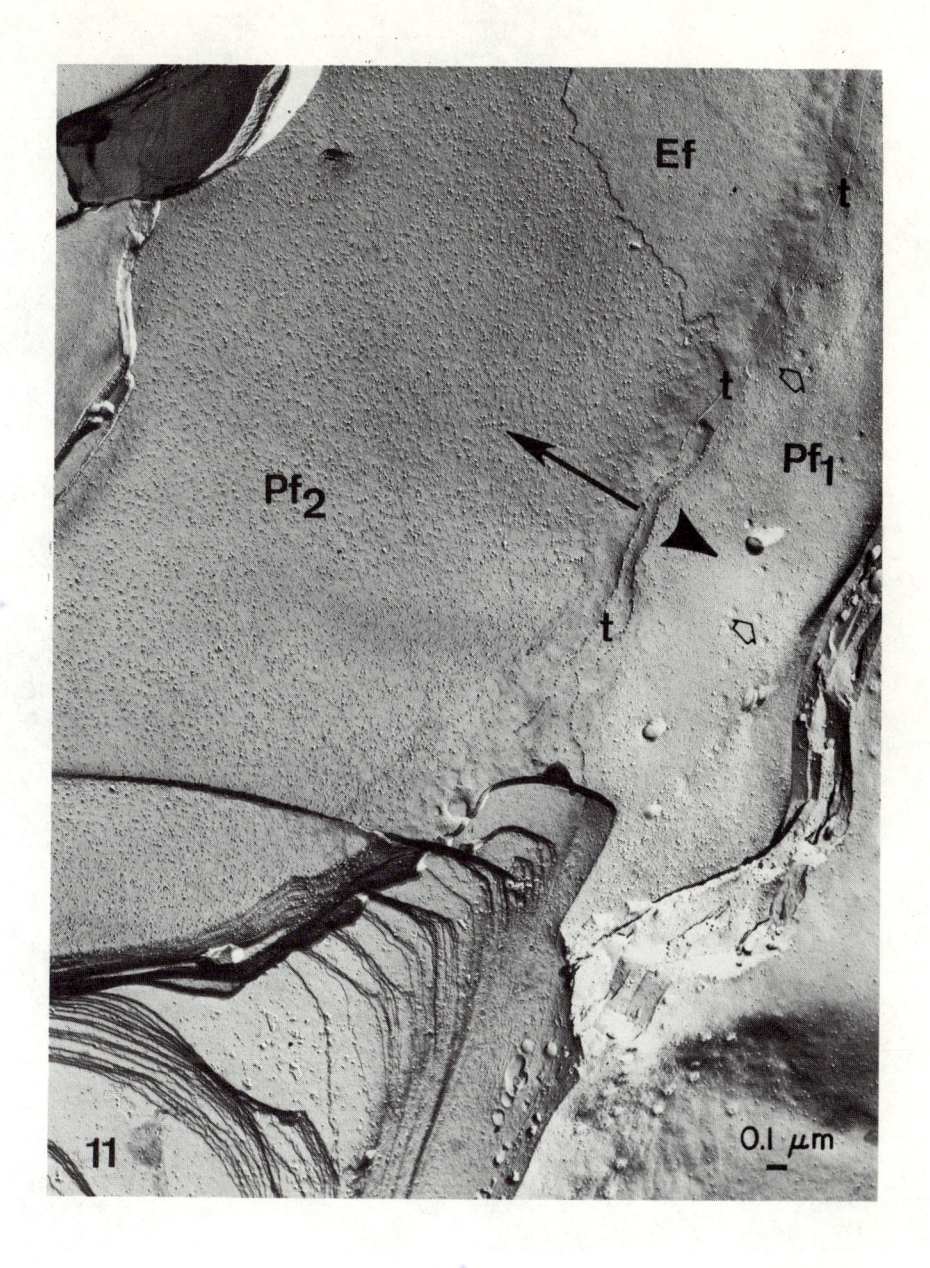

Fig. 11. Electron micrograph of a freeze-fractured fiber from the rat sciatic nerve. The tight junctions at the outer mesaxon are labeled $\underline{t}$. The P face of the cytoplasmic outer belt is labeled $\underline{Pf}_1$ and the P face of the second myelin turn $\underline{Pf}_2$. Caveolar stomata are indicated by open arrowheads. Notice the asymmetry of the pattern of intramembrane particles on opposite sides of the tight junction (large arrow/dark arrowhead). At $\underline{cf}$, collagen fibers (Schnapp and Mugnaini, unpublished).

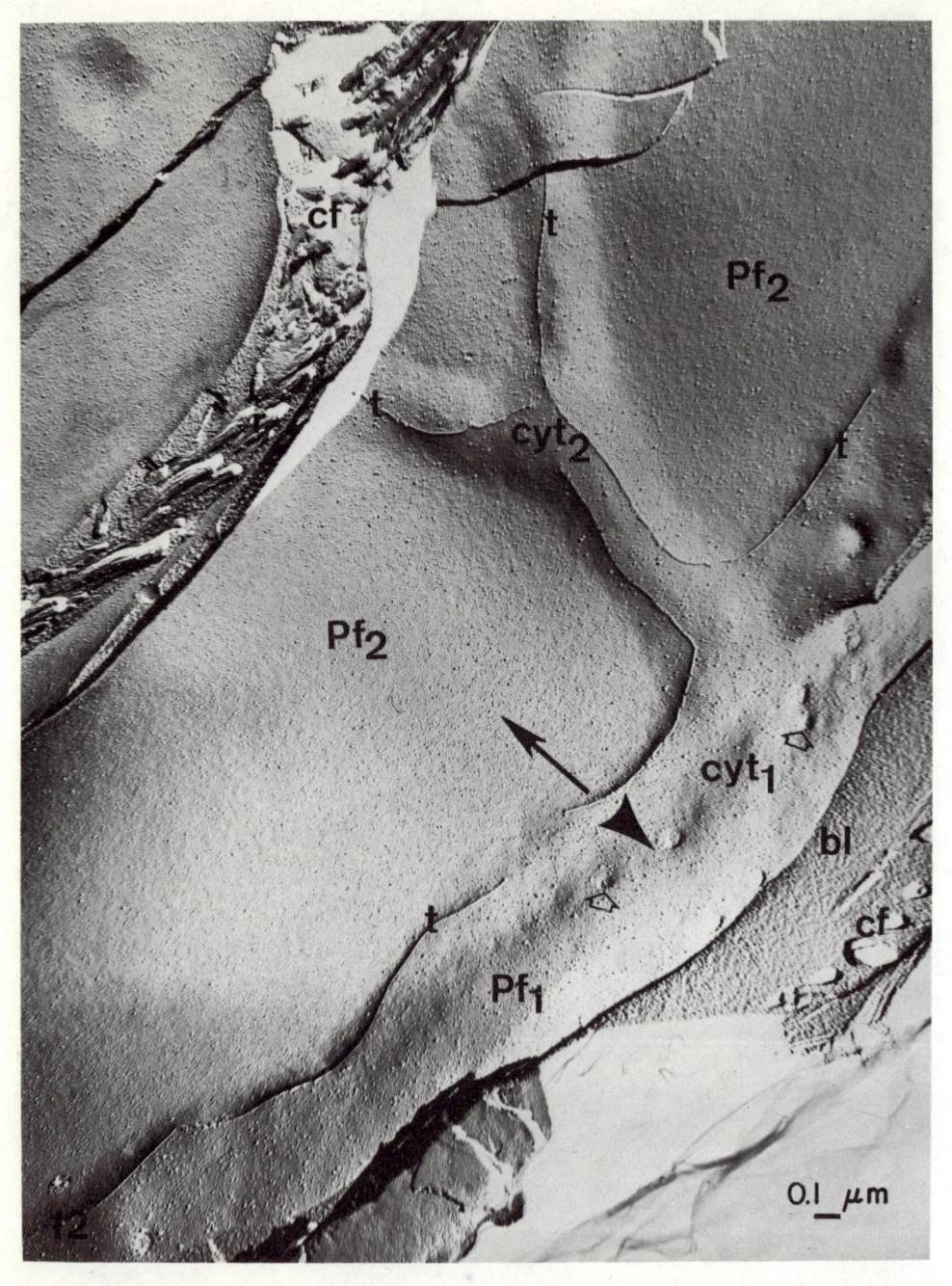

Fig. 12. Electron micrograph of a freeze-fractured fiber from the chicken ciliary ganglion. The peripheral portion of the myelin sheath has been exposed by the fracturing nearly the same as in fig. 11, and the labeling is, therefore, similar. A cytoplasmic incisure (cyt$_2$) departs from the outer belt (cyt$_1$), with accompanying interrupted tight junctional strands. Note that large intramembrane particles are more numerous on the outbulging membrane face (corresponding to a thick cytoplasmic tube) than in the flat portions (Pf) of the myelin lamella. At bl, basal lamina.

slippage. The involvement of a ligand is not incompatible with the idea that proteins in the membrane are essential for the lamellar adhesion, although phospholipids, theoretically, may be at least as important.

Studies of artifical membranes with inserted myelin components, with and without ligands, seems warranted.

A fascinating prospect is that, once the precise nature of the lamellar adhesion is solved, one may be able to unwrap, swell and contract the myelin spiral, much as it is done now by the artist at the drawing table, and thus make the membrane surfaces accessible to differential analysis.

The Myelin Module

As a conclusion of the above considerations, I wish to introduce a reductionistic and speculative concept of the myelin membrane, applicable, with some distinctions based on the differential features already described, to both central and peripheral sheaths. It is hoped that this simplified concept will help the biochemist in relating cellular events with the structure of the myelin both in adult and developing animals.

Due to the presence of a marginal cytoplasmic belt at the periphery of the sheath and cytoplasmic incisures interrupting the compact myelin domain, the spiralled cellular flap that forms the myelin sheath can be thought of as consisting of structural and functional modules of varying shape and dimension and comprising a cytoplasmic channel continuous with a compacted membrane area. In two dimensions (Fig. 14) this functional unit would appear as a membrane bound loop containing cytoplasm and attached on one or both sides to a membranous stalk made up of two adjoined membranes, whose internal leaflets come together to form the myelin major dense line. The border between the loop and the stalk is often marked by a number of tight junctional particles which span the lipid bilayer slowing down both the lateral movement of proteins in the membrane interior and the diffusion of large molecules from one side to the other of the extracellular compartment. The latter is thinner than 20 Å (labeled +) at the stalk and in the order of 100 Å (labeled X) at the loop. Due to the presence of streaming cytoplasm in the loop, but not in the stalk, components of the membrane covering the loop would turn over more quickly than those in the stalk. Lateral diffusion from the loop to the stalk, slowed down at points by the tight junction, would ensure some turnover in the compact membrane. In the diagram the myelin module labeled A is located at the inner, outer and lateral borders of the myelin sheath, while the module labeled B is located in the central and intermediate turns (Fig. 14). As noted by Schnapp and Mugnaini (1977), while the cytoplasmic belt is always present, the cytoplasmic incisures may open and close from time to time, although the latter point remains to be proven. It is possible

MEMBRANE ADHESION IN MYELIN

Fig. 13. A cartoon of the compact myelin lamellae where unknown mechanisms (monkeys) provide for adhesion between the adjoining membranes of one lamella (intralamellar adhesion) at the major dense line (M.D.L.) and between successive lamellae (interlamellar adhesion) at the paired minor dense lines bordering the intramyelinic extracellular space (I.E.S.).

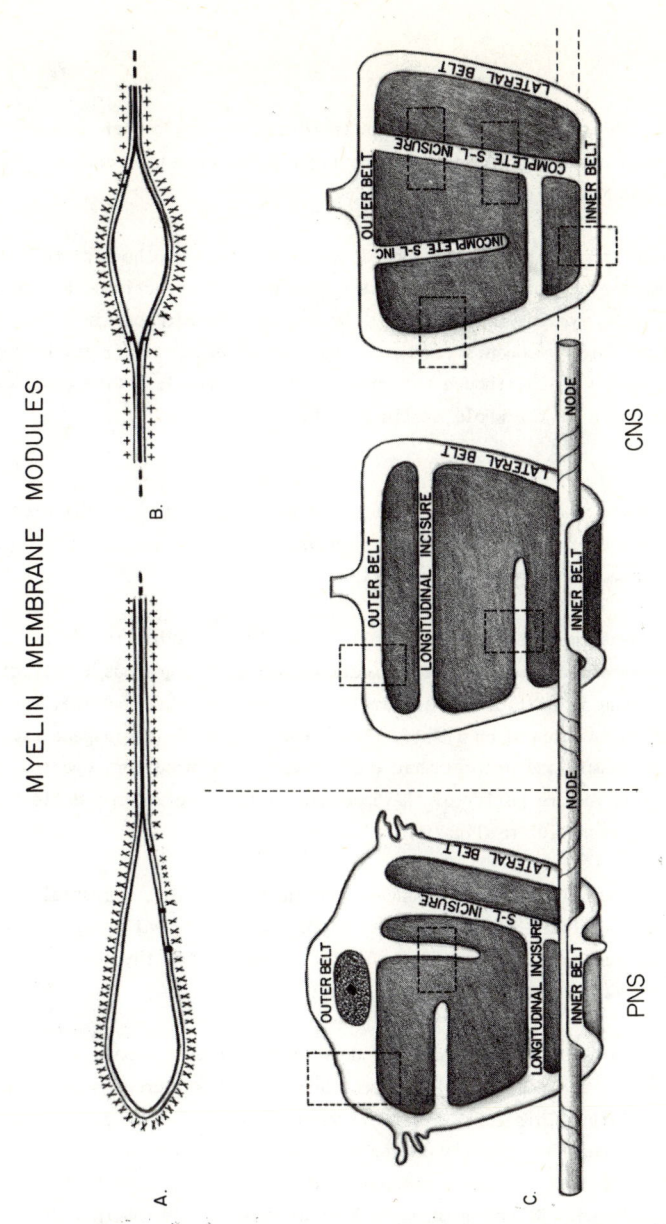

Fig. 14. Diagrammatic representation of myelin modules at marginal (A) and central (B) portions of the sheaths. The glial plasmamembrane is shown by a double line where cytoplasm is present in the lamella and by a triple line where compact myelin is formed. The dots represent tight junctional membrane particles. The Xs mark the extracellular aspect of the cytoplasmic region of the module, while the +s mark the extracellular aspect of the compact myelin membrane. C: the location of the modules is outlined in figuratively unrolled sheaths. The marginal cytoplasmic belt and the complete and incomplete longitudinal and circumpherential incisures that form the cytoplasmic reticulum of the sheaths are shown in white. The compact myelin membrane is shaded.

The model of the myelin module ignores diffusion across the myelin lamellae, about which little is known at present.

27

that the half life of the incisures is in the order of days rather than hours, and may be accompanied by a low turnover of the tight junctional proteins (see Figs. 14-17 of Schnapp and Mugnaini, 1977).

The concept of the myelin module should encourage the neurochemist to consider the myelin membrane as a more dynamic entity than commonly practiced, and to verify the idea that "pure myelin" fractions may, or should indeed, contain a few, yet critical, cytoplasmic components. Some "contaminants" of myelin fractions may arise from myelin domains which, though not compact, are nevertheless true and functionally essential parts of the whole myelin sheath.

Myelination

Evidently, the myelin sheath, beginning with the early stages of differentiation, represents a degree of complexity at least an order of magnitude higher than many others of the common cell biological mdoels.

While the plasmamembrane of the red cell is easily accessible to various membrane probes, can be obtained in high purity fractions and is homogeneous in structure, the myelin membrane is part of a thin lamella and has the disadvantage, for the researcher, of being wrapped on itself, thus presenting a limited aspect exposed to the general exoplasmic and protoplasmic compartment. The need for repeated centrifugations to insure purity of the immature myelin fraction may deplete it of constituents important for myelination.

It seems obvious, therefore, that understanding myelination, a crucial step in the evolution and ontogeny of vertebrates, represents an unusual challenge and that renewed, special and imaginative efforts will be required for the solution of the problems considered in this meeting.

Preliminary freeze-fracture studies of developing nerves (Schnapp and Mugnaini, 1978) indicate that the number of intramembrane particles increases during maturation of the myelin sheath. This correlates well with biochemical data that show an increase in the amount of proteins during myelination. Changes in the distribution of individual classes of intramembrane particles during maturation of the sheath remain to be analyzed. Recent immunocytochemical studies (Sternberger et al., 1978) have shown that the basic protein is first concentrated in the perikaryon of the myelin forming cell.

Studies on the freeze-fracture feature of myelin and myelin forming cells in experimental conditions, accompanied by demyelination or hypomyelination, and in mutant strains of animals with neurologic disorders are still in their infancy.

Special problems, such as the interaction between the myelin forming cell, the axon and the surrounding tissue; the control of glial cell proliferation; the determination of the ratio between number of myelin segments and glial cells; the length of the internodes and the number of myelin lamellae formed during the wrapping process; and the biomechanical properties of myelin, can only be alluded to in this account and require far more complex approaches than membrane isolation and chemical analysis.

In this context, I wish to limit myself to consideration of the wrapping process by which the myelin sheath attains its mature features during development.

Our own preliminary studies indicate that the tight junctions at the outer mesaxon appear before the wrapping process begins, while those at the inner mesaxon develop after compaction of the lamellae has initiated (Schnapp and Mugnaini 1978). It seems, therefore, that while the outer mesaxonal tight junction may be involved in the initiation of myelination, the non-anchored, microfilament containing inner glial loop may act as a growth cone. Membrane growth, however, may obviously take place also by incorporation of new elements at all cytoplasmic borders, as considered in the section on the myelin module (see also Hendelman and Bunge, 1969; Hedley-Whyte et al., 1969; Webster, 1971).

I consider it improbable that the outer mesaxonal region is actively engaged in the wrapping process. In fact, I have observed examples where two central axons share the outer tongue process (Mugnaini, in preparation). Interestingly, enough, in one such case, the two internodes differ in the number of myelin turns. This suggests that the axon may regulate locally the parameters of myelination, with a degree of independence from the biology of the glial cell body.

Acknowledgements

I wish to thank Drs. Victor L. Friderich, Jr. and Thomas Schmidt-Glenewinkel for suggestions; Dr. Bruce Schnapp for invaluable help with the freeze-fracturing; Ms. Mary-Jane Spring and Ms. Anne-Lise Dahl for assistance in preparing the illustrations; and Ms. Edith Murphy for typing. This work was supported by NIH grant NS 09904-08.

References

Bittinger, H. and Heid, J. (1977) The subcellular distribution of particle-bound negative charges in rat brain, J. Neurochem. 28: 917-922.

Blakemore, W. F. (1969) Schmidt-Lanterman incisures in the central nervous system, J. Ultrastruct. Res. 29: 496-498.

Blank, W. F., Bunge, M. B. and Bunge, R.P. (1974) Sensitivity of the myelin sheath, particularly the Schwann cell-axolemmal junction, to lowered calcium levels in cultured sensory ganglia, Brain Res. 67: 503-518.

Braun, P. (1977) Molecular architecture of myelin In: Myelin, edited by P. Morell, pp. 91-115, Plenum Press, New York.

Davison, A.N. and Peters, A. (1970) Myelination, C. C. Thomas, Springfield, Ill.

Fischer, H.A., Schröder, J. M. and Seiler, N. (1972) Interrelationships between polyamines and nucleic acids, Z. Zellforsch. 128: 393-405.

Friedrich, V. L., Jr. (1974) The myelin deficit in quaking mice, Brain Res. 82: 168-172.

Gilula, N. (1977) Gap junctions and cell communication. In: International Cell Biology 1976-1977, edited by B. R. Brinkley and K. R. Porter, Rockefeller University Press, New York, pp. 61-69.

Hedley-Whyte, E. T., Rawlins, T. A., Salpeter, M. M. and Uzman, B. G. (1969) Distribution of cholesterol -1,2-H^3 during maturation of mouse peripheral nerve, Lab. Invest. 21: 536-547.

Hendelman, W. J. and Bunge, R. P. (1969) Radioautographic studies of choline incorporation into peripheral nerve myelin, J. Cell Biol. 40: 190-208.

Hortega, P. Del Rio (1928) Tercera aportacion al conocimiento morfologico e interpretacion functional de la oligodendroglia, Mem. Real Soc. Españ. Hist. Nat. 14: 1-118.

Landon, D. N. (Ed.) (1976) The Peripheral Nerve, Chapman and Hall, London.

Liquori, A. M., Constantino, L., Crescenzi, V., Elia, V., Giglio, E., Puliti, R., Desantis-Savino, M. and Vitagliano, V. (1967) Complexes between DNA and polyamines: a molecular model, J. Mol. Biol. 24: 113-122.

Mateu, L., Luzzati, V., London, Y., Gould, R. M., V$_o$ssenberg, F. G. A. and Olive, J. Jr. (1973) X-ray diffraction and electron microscope study of the interactions of myelin components: The structure of a lamellar phase with a 150 to 180 Å repeat distance containing basic proteins and acidic lipids, J. Mol. Biol. 75: 697-709.

McFarland, D. E. and Friede, R. L. (1971) Number of fibers per sheath cell and internodal length in cat cranial nerves, J. Anat. (Lond.) 100: 169-176.

Morell, P. (Ed.) (1977) Myelin, Plenum Press, New York.

Mugnaini, E., Osen, K. K., Schnapp, B. and Friedrich, V. L., Jr. (1977) Distribution of Schwann cell cytoplasm and plasmalemmal vesicles (caveolae) in peripheral myelin sheaths: An electron microscopic study with thin sections and freeze-fracturing, J. Neurocytol. 6: 647-688.

Mugnaini, E. (1977) Astrogliogenesis. In: Brain Information Service Report 46: 91-97. Tenth Annual Winter Conference on Brain Research, Keystone, Colo.

Mugnaini, E. and Schnapp, B. (1974) The zonula occludens of the myelin sheath: Its possible role in demyelinating conditions, Nature (Lond.) 251: 725-726.

Peters, A. (1961) A radial component of central myelin sheaths, J. Biophys. Biochem. Cytol. 11: 733-735.

Peters, A., Palay, S. L. and Webster, H. deF. (1976) The Fine Structure of the Nervous System, W. B. Saunders, Philadelphia.

Pinto Da Silva, P. and Miller, R. G. (1975) Membrane particles on fracture faces of frozen myelin, Proc. Nat. Acad. Sci. USA 72: 4046-4050.

Raina, A., Pajula, R. L. and Eloranta, T. (1976) A rapid assay method for spermidine and spermine synthases. Distribution of polyamine-synthesizing enzymes and methionine adenosyltransferase in rat tissues, FEBS Letters 67: 252-255.

Russel, D. H. and Meier, H. (1975) Alterations in the accumulation patterns of polyamines in brains of myelin-deficient mice, J. Neurobiol. 6: 267-275.

Schnapp, B. and Mugnaini, E. (1978) Membrane architecture of myelinated fibers as seen by freeze-fracture. In: Physiology and Pathobiology of Axons, Stephen G. Waxman, Ed., Raven Press, N. Y. pp. 83-123.

Schnapp, B. and Mugnaini, E. (1977) Freeze-fracture properties of central myelin in the bullfrog, Neuroscience 2: 459-467.

Schnapp B., Peracchia, C. and Mugnaini, E. (1976) The paranodal axo-glial junction in the central nervous system studied with thin sections and freeze-fracture, Neuroscience 1: 181-190.

Schnapp, B. and Mugnaini, E. (1975) The myelin sheath: electron microscopic studies with thin sections and freeze-fracture. In: Golgi Centennial Symposium: Proceedings, M. Santini, Ed., Raven Press, New York, pp. 209-233.

Seiler, N. and Deckart, K. (1976) Association of putrescine, spermidine, spermine and 4-aminobutyric acid with structural elements of brain cells, Neurochem. Res. 1: 451-467.

Seiler, N. and Schmidt-Glenewinkel, T. (1975) Regional distribution of putrescine, spermidine and spermine in relation to the distribution of RNA and DNA in the rat nervous system, J. Neurochem. 24: 791-795.

Seiler, N. and Lamberty, W. (1975) Interrelations between polyamines and nucleic acids: changes of polyamine and nucleic acid concent-ations in the developing rat brain, J. Neurochem. 24: 5-13.

Seiler, N. and Lamberty, W. (1972) Interrelationships between polyamines and nucleic acids. Changes of polyamine and nucleic acid concentrations in the growing fish brain (Salmoirideus gibb.) J. Neurochem. 20: 709-717.

Seiler, N. and Schroeder, J.M. (1970) Beziehungen zwischen Polyaminen und Nuclein-säuren, II. Biochemische und feinstrukturelle Untersuchungen am peripheren Nerven während der Wallerschen Degeneration, Brain Res. 22: 81-103.

Sternberger, N. H., Itoyama, Y., Kies, N. W. and Webster, H. deF. (1978) Immuno-cytochemical method to identify basic protein in myelin-forming oligodendrocytes of newborn rat CNS, J. Neurocytol. 7: 251-263.

Vail, W. J., Papahadjopoulos, D. and Moscareleo, M. A. (1974) Interaction of a hydrophobic protein with liposomes. Evidence for particles seen in freeze-fracture as being proteins, Biochem. Biophys. Acta. 345: 463-467.

Waxman, S. G. (Ed.) (1978) Physiology and Pathobiology of Axons, Raven Press, New York.

Webster, H. deF. (1971) The geometry of peripheral myelin sheaths during their formation and growth in rat sciatic nerves, J. Cell Biol. 48: 348-367.

Worthington, C.R. (1971) X-ray analysis of nerve myelin. In: Biophysics and Physiology of Excitable Membranes, W. J. Aldeman, Ed., Van Nostrand Reinhold, New York, pp. 1-46.

Yu, R. and Bunge, R.P. (1975) Damage and repair of the peripheral myelin sheath and node of Ranvier after treatment with trypsin, J. Cell Biol. 64: 1-14.

PROTEINS OF CENTRAL NERVOUS SYSTEM MYELIN

T.V. Waehneldt
Max-Planck-Institut für experimentelle Medizin
Forschungsstelle Neurochemie
3400 Göttingen, West Germany

This short review on the proteins of CNS myelin is written as an introduction for the reader who is little familiar with the field. By necessity, it cannot be comprehensive or exhaustive. Therefore, this article will often refer to other reviews, in particular to those in the excellent book "Myelin", edited by Pierre Morell (Plenum Press, New York and London, 1977), which cover many biochemical aspects up to 1975, let aside morphological, physical and physiological data, and the pathology of myelin. Attempts will be made, however, to include more recent publications which have appeared after "Myelin" went to press.

(1) Isolation of CNS myelin and of its proteins

Since brain tissue contains additional proteins with solubility properties similar to those of myelin, it is advisable to isolate and purify myelin prior to extraction of its membrane-bound proteins. Due to its abundance in CNS tissue - about 20-25% of the dry weight of total rat brain is myelin (Norton and Poduslo, 1973) -, due to its low density - myelin has a buoyant densitiy lower than that of all other subcellular particles with the exception of synaptic vesicles (Whittaker, 1964) and corresponding to 0.5-0.8 M sucrose (Waehneldt, 1978a) - and due to its compaction as revealed by electron microscopy (see the preceding article by Mugnaini) which represents a fortuitous in situ enrichment, myelin can be easily isolated from CNS tissue of adult animals. Commonly, an iso-osmotic homogenate is placed on 0.8-0.9 M sucrose for density gradient centrifugation, the resulting floating layer of crude myelin is hypo-osmotically shocked and is finally subjected to differential centrifugation at moderately high speed to sediment only large compact myelin particles. Repetition of any of these steps eventually results in myelin of high purity. Basically this procedure represents the often applied method of Norton and Poduslo (1973). For further reading see Norton (1976, 1977).

Attempts have been made to extract proteins from myelin and to separate them by employing solubility differences in several media such as chloroform-methanol (2 : 1, v/v) (Gonzalez-Sastre, 1970), rendering the high molecular weight Wolfgram-type proteins insoluble, or addition of

Table 1. Classification of CNS myelin proteins

Protein Species	Symbol	Mol. W.[x]	B/A[y]	Spec.	Prop.[*]	Ref.
Myelin-Associated Glycoprot.	mGP	110,000		rat	0.3-2.0	(1)
Wolfgram Prot.	WP	51,340		rat	7-18	(2)
	W1	54,000	0.76	rat		(3)
	W2	62,000	0.67	rat		(3)
Proteolipid Prot.	PLP	24,760		rat	30-40	(2)
	P7	23,500	0.89[+]	rat		(4)
	N-2		0.95[+]	human		(5)
		25,100	0.89[θ]	bovine		(6)
DM-20 or Intermediate Prot.	DM-20	20,540		rat	7-9	(2)
	I			mouse		(7)
	P8M			mouse		(8)
		20,700	0.89[θ]	bovine		(6)
	LI, SI			rat		(9)
Large Basic Prot.	L	18,400	1.79	rat	15-23	(10)
	(L)BP	16,730		rat		(2)
Small Basic Prot.	S	14,400	1.98	rat	18-34	(10)
	(S)BP	13,420		rat		(2)
Prelarge Basic Prot.	preL	21,500		mouse		(11)
Presmall Basic Prot.	preS	17,000		mouse		(11)

x) Apparent molecular weights on SDS-PAGE, except ref. (10); y) Basic amino acid residues/acidic amino acid residues; *) Proportions of protein components, expressed as per cent of total dye binding capacities (Zgorzalewicz et al., 1974); +) 48 h hydrolysis; θ) 22 h hydrolysis, (1) Quarles et al., 1973a; (2) Agrawal et al., 1972; (3) Nussbaum et al., 1977; (4) Nussbaum et al., 1974; (5) Gagnon et al., 1971; (6) Nicot et al., 1973; (7) Morell et al., 1972; (8) Nussbaum and Mandel, 1973; (9) Cammer and Norton, 1976; (10) Martenson et al., 1972; (11) Barbarese et al., 1977.

salt or dilute acid to the chloroform-methanol extract, which precipitates the basic protein while leaving the proteolipid protein in solution (for nomenclature and some characteristic features, see Table 1). Eng et al. (1968) used a mixture of the non-ionic detergent Triton X-100 and ammonium acetate to extract all proteins except the Wolfgram-type proteins. By removing Triton X-100 the proteolipid protein became insoluble whereas the basic protein remained in solution. Although these procedures allow for the processing of large quantities of myelin they have, however, not been carried beyond the stage of a certain enrichment of individual components as revealed by high resolution SDS-PAGE (sodium dodecylsulfate-polyacrylamide gel electrophoresis). Only the basic protein can be selectively extracted free from other protein components with diluted acid (Carnegie and Dunkley, 1975) or even salt (Eng et al., 1968) provided the isolated myelin is of high purity, i.e., is histone-free, and has not been subjected to degradative conditions (Matthieu et al., 1977).

The chloroform-methanol-insoluble residue of myelin has been found to be enriched in high molecular weight glycoproteins besides other high molecular weight proteins. This feature has been exploited by Quarles and Pasnak (1977) who extracted the major myelin glycoprotein with lithium 3,5-di-iodosalicylate, and partitioned with phenol-water. Although this method appears comparable in its simplicity to that of the acid extraction of basic protein the major myelin glycoprotein is not entirely pure on subsequent SDS-PAGE (Quarles and Pasnak, 1977). From this result, as well as from numerous others, it is apparent that the final purifying "touch" to the proteins and glycoproteins of myelin can to date be given only by preparative SDS polyacrylamide gel electrophoresis of appropriate gel and buffer compositions (Allison et al., 1974; Magno-Sumbillo and Campagnoni, 1977).

(2) <u>Classification and biochemical characterization of CNS myelin proteins</u>

It is apparent from the protein profile on SDS-PAGE (Fig. 1) that CNS myelin proteins can be arbitrarily divided into two groups according to their molecular weights and their relative proportions. One is the group of "typical", "major", or "low molecular weight" myelin proteins, comprising basic protein(s), DM-20 or intermediate protein, and proteolipid protein, with molecular weights below about 30,000 D. The other group with molecular weights above approximately 30,000 D consists of a large number of minor proteins, labelled as "Wolfgram-type" (Eng et al., 1968), "minor" (Zanetta et al., 1977b) or "high molecular weight" (Waehneldt, 1978a) proteins, among which the Wolfgram proteins are predominant, besides the PAS-staining major myelin glycoprotein. The protein pattern is relatively simple; nevertheless, the analytical data reveal compositional extremes between different protein classes as well as heterogeneity within one class. In the following, brief descriptions will be given for the proteins so far isolated and characterized. The reader is also referred to the reviewing chapters of Braun and Brostoff (1977) and of Braun (1977) which give further details for PLP and BP.

<u>Proteolipid protein</u> (PLP), also named "proteolipid P7 apoprotein" (Nussbaum et al., 1974), "Folch-Lees proteolipid protein" (Braun, 1977) or "Lipophilin" (Boggs et al., 1976). This protein is very peculiar

Fig. 1. Schematic drawing of SDS-PAGE of myelin proteins from adult rat brain. HMW, high molecular weight proteins; LMW, low molecular weight proteins. For nomenclature and molecular weights see Table 1.

because of its high content of hydrophobic amino acid residues and because of covalently bound fatty acids (Stoffyn and Folch, 1971) which render it soluble in chloroform-methanol (2 : 1, v/v) (see Fig. 2). Upon SDS-PAGE the molecular weight has been found to be about 24000, a value recently confirmed by Jollès et al. (1977) on the basis that all arginine residues in the native molecule (6 in total) were found in 6 different arginine-containing tryptic peptides. Thus, the notion that PLP is an oligomer of a small 5000 D subunit (Chan and Lees, 1974; see also Vacher-Leprêtre et al., 1976) appears rather unlikely. However, the delipidated protein has a tendency to form higher molecular weight aggregates which do not enter SDS gels (Morell et al., 1975; for further discussion, see also Braun and Brostoff, 1977) or possibly dimers after reduction plus carboxymethylation (Vacher-Leprêtre et al., 1976). The latter has been ascribed to hydrophobic bonding alone. The proteolipid protein is conveniently purified to homogeneity by preparative SDS-PAGE (Waehneldt, 1971) which permits amino acid (see Fig. 2) and partial sequence analyses (Nussbaum et al., 1974). The N-terminal sequence of 20 amino acid residues is characterized by an extremely high proportion of hydrophobic amino acid residues which has recently been reconfirmed (Jollès et al., 1977). Moreover, one Cys-His-Cys and two Cys-Cys clusters have been found, possibly explaining the initial difficulties of enzymatic cleavage of the native molecule which has been avoided by performic acid oxidation prior to tryptic digestion (Jollès et al., 1977).

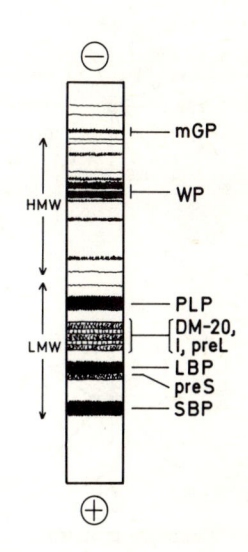

Intermediate or DM-20-protein (proteolipid P8M apoprotein). Due to its highly apolar character this protein belongs also to the class of the proteolipid proteins (Fig. 2) (Agrawal et al., 1972; Nussbaum and Mandel, 1973). Recent analyses (Vacher-Leprêtre et al., 1976) on bovine proteolipid apoproteins (amino acid composition, N-terminal sequence and C-termini) have shown substantial similarities between the 25,000 and 20,000 components of these authors (the nomenclature of their components corresponds to respectively PLP and DM-20, as used here). Moreover, both

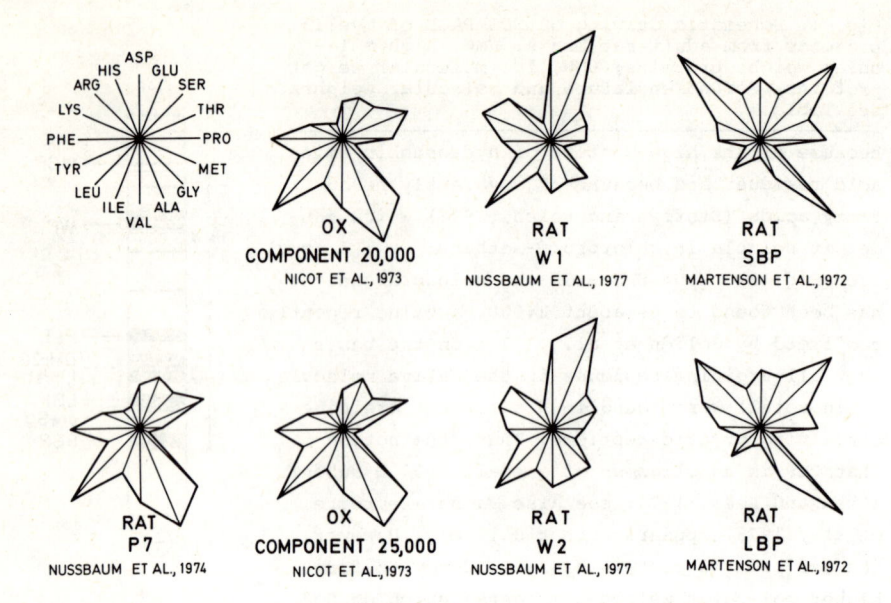

Fig. 2. Amino acid composition of purified myelin proteins (Table 1). Following the code at the upper left hand corner, molar percentages of individual amino acids are shown as distances from the center.

---

proteins tend to form polymers inaccessible to the disruptive forces of SDS. This was taken as an indication that a smaller monomeric protein would give rise to two oligomeric forms (20,000 and 25,000 D) upon transfer from the initial organic solution to an aqueous SDS solution. As with PLP such a suggestion appears not likely (Jollès et al., 1977); rather, these observations are congruent with two independent proteins of 20,000 and 25,000 D having some sequences in common. Possibly a deletion of an internal sequence from PLP could lead to a shorter protein (DM-20), a situation similar to that of the basic proteins in rodents (Martenson et al., 1971; 1972). This question will however be answered only by the complete sequential analysis of both proteins.

Basic protein (BP). Owing to the ease of extraction with dilute acid or salt media, myelin basic protein is one of the best characterized nervous tissue proteins. It is histone-like in its charge properties and it has a molecular weight of around 18,000 (∼170 amino acid residues per mol), with slight variations depending on species (for references, see Carnegie and Dunkley, 1975). Certain rodents possess an additional smaller basic protein (14,000, ∼130 amino acid residues per mol) which is described as an internal deletion of around 40 amino

acid residues from the large basic protein (residues 118-158, Carnegie
and Dunkley, 1975). The complete sequence has been established for
bovine (Eylar et al., 1971) and human (Carnegie, 1971) basic protein
as well as for rat small basic protein (Dunkley and Carnegie, 1974).
Certain structural features are characteristic of the basic protein
(Braun and Brostoff, 1977): aside from an N-terminal acetylated alanine
and a C-terminal arginine there is a single tryptophan residue, two
Phe-Phe sequences, one tri-Pro sequence around amino acid residue 100
and a methylated arginyl residue in the neighborhood. With the exception of frog myelin basic protein (Martenson et al., 1975) no
myelin basic protein has yet been described to carry a cysteinyl
residue. Guinea pig basic protein (Deibler et al., 1975) as well as
bovine basic protein (Chou et al., 1976) showed multiple bands upon
high pH disc gel electrophoresis and chromatography. This microheterogeneity was found to be due to phosphorylation each of a single serine
and threonine residue (for further references regarding phosphorylation,
see Williams and Rodnight, 1977), to deamination of glutamine residues
which are mostly amidated in contrast to aspartyl residues, and due to
loss of the C-terminal arginine in the case of guinea pig basic protein. Since only specific residues appear to be involved in the modifications the action of highly specific enzymes is suggested. It has to
be kept in mind, however, that these relatively small changes may also
occur during the isolation of the protein (Martenson et al., 1976), in
particular that of C-terminal loss of arginine, less so that of deamination (Chou et al., 1977).

Substantially more pronounced alterations involving the proteinaceous backbone are found in studies on autolytic degradation of
the basic protein. Incubation of previously frozen bovine white matter
at room temperature for several hours resulted in almost complete loss
of basic protein (Ansari et al., 1975) while no change was found when
unfrozen tissue was used; however, higher losses of the basic proteins
than of DM-20 or PLP were observed when rat brains were incubated and
myelin prepared thereof (Fishman et al., 1977). These losses were
higher in 15-day-old rats than in adult animals. Smaller molecular
weight degradation products could not be detected by this group which
was ascribed to either loss *in situ* or loss during isolation; by contrast, Matthieu et al. (1977a) showed the appearance of a fast-migrating band concomitant with a uniform decrease of the two basic proteins
from mouse brain myelin isolated after *in situ* autolysis. It is worth
noting that bovine brain stem myelin basic protein was not affected
in situ after 24 h at room temperature (Matthieu et al., 1977).

The only change found was a 21% decrease in the major myelin glycoprotein. Taken together, these results suggest that unless the tissue is grossly disrupted by freezing and thawing, myelin is only gradually invaded by lysosomal enzymes first attacking the externally exposed glycoprotein and thereafter the cytoplasmically orientated basic protein before acting on the hydrophobic proteins DM-20 and PLP (Braun, 1977).

In recent elegant work two additional basic proteins were identified in mouse myelin (Barbarese et al., 1977). They were designated "prelarge" and "presmall" basic proteins (see Table 1). A radioimmunoassay after SDS-PAGE clearly indicated the presence in all four proteins of antigenic sites existing in the small basic protein. Tryptic fingerprints of the isolated proteins as well as tryptophan-specific cleavage demonstrated a single peptide of a molecular weight of approximately 3000, extending from the N-terminus common to both large and small basic proteins. Since this 3000 peptide contained a single tryptophan residue and since tryptophan-specific splitting of the intact prelarge and presmall basic protein produced only two fragments of which the sum equals that of the intact molecule this tryptophan residue is either not involved in cleavage or must be located at, or very close to, the N-terminus of the protein.

It is interesting to note that Allison et al. (1974) observed an extra band beside the large basic protein which was also observed in the recent autolytic study by Fishman et al. (1977). The question arose if this protein was either a degradation product of the large basic protein or another myelin protein. With the results of Barbarese and collegues at hand (1977) we can say that this extra band was the presmall basic protein (Fig. 1). In this context, attention should also be paid to a putative basic protein (SI) migrating between the large basic protein and PLP (see discussion of Cammer and Norton, 1976) which may turn out to be the prelarge protein in rat and which may decrease during development in parallel with LBP (Zgorzalewicz et al., 1974). The hydrophobic nature of LI is underlined by the parallel disappearance of LI and PLP upon boiling of delipidated myelin samples in mercaptoethanol (Morell et al., 1975). No mention was made by Barbarese et al. (1977) of the DM-20 protein.

<u>Wolfgram protein</u> (WP). Wolfgram and Kotorii (1968) concluded on the basis of differential amino acid analyses that a third protein fraction rich in dicarboxylic amino acids must exist in myelin besides Folch-

Lees proteolipid protein and basic protein. In the following this
fraction has been labelled "Wolfgram-type" fraction (Eng et al., 1968)
or "acidic proteolipid protein of Wolfgram" (Gonzalez-Sastre, 1970).
High resolution SDS-PAGE revealed the existence of a multitude of high
molecular weight bands in the chloroform-methanol-insoluble residue
(Zanetta et al., 1977a, 1977c), with however a prominent protein-
staining region around 50,000-65,000 D in SDS-PAGE. Quite tacitly, the
protein bands of this region, mostly a doublet or a triplet, have come
to be labelled "Wolfgram protein(s)". Wolfgram proteins have been isolated from bovine white matter (Wiggins et al., 1974) and from rat
brain (Nussbaum et al., 1977) by preparative SDS-PAGE and their amino
acid compositions determined. They were characterized by high contents
of acidic amino acids and by absence (Wiggins et al., 1974) or near
absence (Nussbaum et al., 1977) of cysteine. Thus, the Wolfgram proteins
differ entirely from the hydrophobic PLP and DM-20 proteins and from
the basic proteins (Fig. 2).

Myelin-associated glycoprotein (mGP), also "major myelin glycoprotein".
Intracerebral injection of radioactive fucose resulted in specific
labeling of a glycoprotein of 110,000 apparent molecular weight (SDS-
PAGE) in purified myelin from rat (Quarles et al., 1973a) or other
species (Matthieu et al., 1974a; Everly et al., 1977). This glycoprotein was not prominent in other subcellular particles; in addition, it
was reduced in the myelin-deficient neurological mouse mutant Quaking
and virtually absent in Jimpy (Matthieu et al., 1974b, 1974c; Matthieu
and Waehneldt, 1978), implying that it was closely associated with
myelin and increased in parallel with myelin. The glycoprotein was also
labelled with radioactive glucosamine and N-acetylmannosamine (Quarles
et al., 1973a) and rather selectively with ^{35}sulfate (Matthieu et al.,
1975a). There was a shift toward a higher apparent molecular weight of
mGP when myelin from young (12-15 days) was compared to more mature
(older than 25 days) rats using the double label technique (Quarles et
al., 1973b; Matthieu et al., 1975a). Differences in apparent molecular
weights could also be demonstrated using the glycoprotein PAS stain
(Matthieu et al., 1975a). For a general review, see Quarles (1975).

A heterogenous population of glycopeptides, derived by extensive
pronase digestion from partially purified mGP from immature rat myelin,
were found to be enriched in the highest molecular weight class when
compared with that of adult rat myelin (Quarles, 1976). Neuraminidase
treatment of mGP prior to pronase digestion decreased the high molecular weight class glycopeptides, largely eliminating the developmental
differences, but changes more complex than sialic acid residue elimi-

nation could be involved (Quarles, 1976; Quarles and Everly, 1977) as well as shortening of the polypeptide chain. The myelin-associated glycoprotein (component A, see Zanetta et al., 1977c), enriched by sequential lectin affinity chromatography, was recently shown to consist of two closely spaced glycopeptides as revealed by Coomassie blue staining. This result and particularly the promising approach of Quarles and Pasnak (1977) should allow for the complete purification of mGP from young and adult animals so that unequivocal results will eventually be obtained for the carbohydrate moieties and for the protein backbone.

(3) Protein heterogeneity of CNS myelin subfractions

Large-sized rat brain myelin particles, free from microsomal material and myelin-related membranes, distributed in a bell-shaped mode upon sucrose density centrifugation in a zonal gradient (Waehneldt, 1978a). The maximum was at a lower density in myelin from young rats when compared to that from adult animals. Similar findings were made by Adams and Fox (1969) for rat brain myelin and by Sheads et al. (1977) for mouse brain myelin employing however discontinuous sucrose gradients. Furthermore, myelin from rat spinal cord had a sharp maximum at a density lower than that of the more broadly distributing myelin from forebrain (Waehneldt, 1978a; see also Sheads et al., 1977). Isolated subfractions floated at their original density on recentrifugation, implying that intermittent changes in composition were very unlikely. The density of the particles was largely governed by the protein/lipid ratio which increased from the light to the heavy side (Bourre et al., 1977; Reiber and Waehneldt, 1978).

Substantial differences in protein composition were noted when these myelin subfractions were compared (Matthieu et al., 1973; Zimmerman et al., 1975; Fujimoto et al., 1976; Benjamins et al., 1976; Quarles, 1977; Waehneldt, 1978a, 1978b; Reiber and Waehneldt, 1978). In general, high molecular weight proteins including WP increased towards the heavy side; basic protein(s) decreased, while PLP assumed an intermediate position, rising to a plateau in the heavy fractions of spinal cord or decreasing again towards the heavy side of the gradient (Waehneldt, 1978a; Reiber and Waehneldt, 1978). Disregarding the high molecular weight proteins, substantial changes in the ratio of the hydrophobic proteins vs. the hydrophilic proteins (PLP + DM-20/SBP + LBP, Waehneldt, 1978a; PLP/BP, Reiber and Waehneldt, 1978) were observed as well as increases of the SBP/LBP ratio, not only between adult and young animals and different anatomical regions (Zgorzalewicz et al., 1974; Magno-Sumbilla and Campagnoni, 1977; Waehneldt, 1978a) but also from the heavy towards the light end of the zonal gradient (Waehneldt, 1978a). Moreover, the levels of 2',3'-cyclic nucleotide 3'-phosphohydrolase (CNP), a marker regarded as specific for myelin (Kurihara and Tsukada, 1968) increased drastically from the light towards the heavy fractions (Matthieu et al., 1973), with values in the light fractions of spinal cord myelin below those of the whole homogenate (Waehneldt, 1978a). Taken together, these results demonstrate substantial heterogeneity of compact myelin particles which resides in differences of species, age and regions examined (Zimmermann et al., 1975; Waehneldt, 1978b).

By careful removal of microsomal fractions prior to any hypo-osmotic condition during the preparation of myelin (Waehneldt and Mandel, 1972) an additional fraction of small-sized material (SN 4) could be obtained which is myelin-related in having the same protein components as myelin but of substantially differing proportions (Waehneldt, 1975, 1978; Waehneldt et al., 1977). This fraction clearly falls into the category of Davison's concept of a "myelin-like" fraction (Banik and Davison, 1969; Agrawal et al., 1970, 1973, 1974). In SN 4 the high molecular weight proteins were enhanced (WP and mGP) with CNP activities about twice as high as those of the heavy myelin subfractions (Waehneldt et al., 1977; Waehneldt, 1978b). A microsome-depleted fraction ($W_1 3$), very similar, if not identical, to SN 4 has recently been isolated by McIntyre et al. (1978) from rat brain. Moreover, a fraction F was isolated by these authors, which was very light and which floated on 0.32 M sucrose upon initial density gradient centrifugation steps. It was characterized by CNP activities similar to myelin and by low levels of PLP and by deficiency in mGP, while the basic proteins (and the ratio SBP/LBP) were elevated, together with high molecular weight components. Fractions such as SN 4, $W_1 3$ and I2 (Toews et al., 1976) on the one hand and F on the other hand are ordinarily discarded during the isolation of compact myelin; they are, however, of particular interest in that they represent extremes of membranes which are linked to myelination and as such may be examples of that which Quarles (1977) has labelled "oligo-dendroglia-derived" membranes.

In membranes such as SN 4 the possibility of contamination must be seriously considered, more than in the case of multilayered myelin (see Norton (1977), Zanetta et al. (1977b) and Quarles (1977) for discussion). Contaminating membranes may be entirely unrelated to myelin (Matthieu and Waehneldt, 1978) or may be of axolemmal origin (Matthieu et al., 1977b, 1978b); however, owing to the gradual remodelling into myelin, oligodendroglial membranes should not be regarded as contaminants. The unequivocal assigning of proteins to the oligodendroglia-myelin-continuum is a difficult task and has only recently been tackled for PLP (Agrawal et al., 1977), WP (Nussbaum et al., 1977) and mGP (Matthieu et al., 1978b).

Virtually no information is as yet available concerning the precise functions of myelin proteins. However, more recent investigations have attempted to elucidate the disposition of proteins in the periodical myelin structure, thus finding initial clues to their potential roles. General concensus appears now to exist about the localization of the myelin basic protein at the cytoplasmic side (Poduslo and Braun, 1975), possibly forming the major dense line by dimerization (Moore et al., 1977) or another mechanism (Crang and Rumsby, 1977), while the carbohydrate moiety of the myelin-associated glycoprotein is directed toward the external side of the membrane and may serve to line up these sides for subsequent fusion (Poduslo et al., 1976; see also Braun, 1977, for a broader discussion of the organization of other myelin proteins). These promising approaches as well as protein cross-linking studies and immunological investigations will ultimately lead to an understanding of the complex protein-protein and protein-lipid interactions operating in the formation of myelin and in the mature sheaths.

Acknowledgements

The author is indebted to Dr. G.E. Fagg, Dr. J.-M. Matthieu and Prof. V. Neuhoff for discussion and critical comments. Supported in part by the Deutsche Forschungsgemeinschaft (SFB 33).

References

Adams, D.H. and Fox, M.E. (1969) The homogeneity and protein composition of rat brain myelin, Brain Res. 14: 647-661.

Agrawal, H.C., Banik, N.L., Bone, A., Davison, A.N., Mitchell, R.F. and Spohn, M. (1970) The identity of myelin-like fraction isolated from developing brain, Biochem. J. 120: 635-642.

Agrawal, H.C., Burton, R.M., Fishman, M.A., Mitchell, R.F. and Prensky, A.L. (1972) Partial characterization of a new myelin protein component, J. Neurochem. 19: 2083-2089.

Agrawal, H.C., Trotter, J.L., Mitchell, R.F. and Burton, R.M. (1973) Criteria for identifying a myelin-like fraction from developing brain, Biochem. J. 136: 1117-1119.

Agrawal, H.C., Trotter, J.L., Burton, R.M. and Mitchell, R. (1974) Metabolic studies on myelin: evidence for a precursor role of a myelin subfraction, Biochem. J. 140: 99-109.

Agrawal, H.C., Hartman, B.K., Shearer, W.T., Kalmbach, S. and Margolis, F.L. (1977) Purification and immunohistochemical localization of rat brain myelin proteolipid protein, J. Neurochem. 28: 495-508.

Allison, J.H., Agrawal, H.C. and Moore, B.W. (1974) Effect of N,N,N',N'-Tetramethylethylenediamine on the migration of proteins in SDS polyacrylamide gels, Analyt. Biochem. 58: 592-601.

Ansari, K.A., Hendrickson, H., Sinha, A.A. and Rand, A. (1975) Myelin basic protein in frozen and unfrozen bovine brain: a study of autolytic changes in situ, J. Neurochem. 25: 193-195.

Banik, N.L. and Davison, A.N. (1969) Enzyme activity and composition of myelin and subcellular fractions in the developing rat brain, Biochem. J. 115: 1051-1062.

Barbarese, E., Braun, P.E. and Carson, J.H. (1977) Identification of prelarge and presmall basic proteins in mouse myelin and their structural relationship to large and small basic proteins, Proc. Natl. Acad. Sci., USA, 74: 3360-3364.

Benjamins, J.A., Gray, M. and Morell, P. (1976) Metabolic relationships between myelin subfractions: entry of proteins, J. Neurochem. 27: 571-575.

Boggs, J.M., Vail, W.J. and Moscarello, M.A. (1976) Preparation and properties of vesicles of a purified myelin hydrophobic protein and phospholipid, Biochim. Biophys. Acta 448: 517-530.

Bourre, J.M., Pollet, S., Daudu, O., Le Saux, F. and Baumann, N. (1977) Myelin consists of a continuum of particles of different density with varying lipid composition: major differences are found between normal mice and Quaking mutants, Biochimie 59: 819-824.

Braun, P.E. (1977) Molecular architecture of myelin. In: Myelin (Morell, P., ed.) Plenum Press, New York, pp. 91-115.

Braun, P.E. and Brostoff, S.W. (1977) Proteins of myelin. In: Myelin (Morell, P., ed.) Plenum Press, New York and London, p. 201-231.

Cammer, W. and Norton, W.T. (1976) Disc gel electrophoresis of myelin proteins: new observations on development of the intermediate proteins (DM-20), Brain Res. 109: 643-648.

Carnegie, P.R. (1971) Amino acid sequence of the encephalitogenic basic protein of human myelin, Biochem. J. 123: 57-67.

Carnegie, P.R. and Dunkley, P.R. (1975) Basic proteins of central and peripheral nervous system, Adv. Neurochem. 1: 95-135.

Chan, D.S. and Lees, M.B. (1974) Gel electrophoresis studies of bovine brain white matter proteolipid and myelin proteins, Biochemistry 13: 2704-2709.

Chou, F.C-H., Chou, C-H. J., Shapira, R. and Kibler, R.F. (1976) Basis of microheterogeneity of myelin basic protein, J. Biol. Chem. 251: 2671-2679.

Chou, F.C-H., Chou, C-H. J., Shapira, R. and Kibler, R.F. (1977) Modifications of myelin basic protein which occur during its isolation, J. Neurochem. 28: 1051-1059.

Crang, A.J. and Rumsby, M.G. (1977) Molecular organisation of lipid and protein in the myelin sheath, Biochem. Soc. Trans. 5: 1431-1434

Deibler, G.E., Martenson, R.E., Kramer, A.J., Kies, M.W. and Miyamoto, E. (1975) The contribution of phosphorylation and loss of COOH-terminal arginine to the microheterogeneity of myelin basic protein, J. Biol. Chem. 250: 7931-7938.

Dunkley, P.R. and Carnegie, P.R. (1974) Amino acid sequence of the smaller basic protein from rat brain myelin, Biochem. J. 141: 243-255.

Eng, L.F., Chao, F.-C., Gerstl, B., Pratt, D. and Tavaststjerna, M.G. (1968) The maturation of human white matter myelin. Fractionation of the myelin membrane proteins, Biochemistry 7: 4455-4465.

Everly, J.L., Quarles, R.H. and Brady, R.O. (1977) Proteins and glycoproteins in myelin purified from the developing bovine and human central nervous system, J. Neurochem. 28: 95-101.

Eylar, E.H., Brostoff, S., Hashim, G., Caccam, J. and Burnett, P. (1971) Basic A$_1$ protein of the myelin membrane: the complete amino acid sequence, J. Biol. Chem. 246: 5770-5784.

Fishman, M.A., Trotter, J.L. and Agrawal, H.C. (1977) Selective loss of myelin proteins during autolysis, Neurochem. Res. 2: 247-257.

Fujimoto, K., Roots, B.I., Burton, R.M. and Agrawal, H.C. (1976) Morphological and biochemical characterization of light and heavy myelin isolated from developing rat brain, Biochim. Biophys. Acta 426: 659-668.

Gagnon, J., Finch, P.R., Wood, D.D. and Moscarello, M.A. (1971) Isolation of a highly purified myelin protein, Biochemistry 10: 4756-4763.

Gonzalez-Sastre, F. (1970) The protein composition of isolated myelin, J. Neurochem. 17: 1049-1056.

Kurihara, T. and Tsukada, Y. (1968) 2',3'-cyclic nucleotide 3'-phosphohydrolase in the developing chick brain and spinal cord, J. Neurochem. 15: 827-832.

Jollès, J., Nussbaum, J.-L., Schoentgen, F., Mandel, P. and Jollès, P. (1977) Structural data concerning the major rat brain myelin proteolipid P7 apoprotein, FEBS Lett. 74: 190-194.

Magno-Sumbilla, C. and Campagnoni, A.T. (1977) Factors influencing the electrophoretic analysis of myelin proteins: application to changes occurring during brain development, Brain Res. 126: 131-148.

Martenson, R.E., Deibler, G.E. and Kies, M.W. (1971) The occurrence of two myelin basic proteins in the central nervous system of rodents in the suborders Myomorpha and Sciuromorpha, J. Neurochem. 18: 2427-2433.

Martenson, R.E., Deibler, G.E., Kies, M.W., McKneally, S.S., Shapira, R. and Kibler, R.F. (1972) Differences between the two myelin basic proteins of the rat central nervous system, Biochim. Biophys. Acta 263: 193-203.

Martenson, R.E., Deibler, G.E. and Kramer, A.J. (1975) The presence of cysteine in frog myelin basic protein, J. Neurochem. 24: 959-962.

Martenson, R.E., Kramer, A.J. and Deibler, G.E. (1976) Microheterogeneity and phosphate content of myelin basic protein from "freeze-blown" guinea pig brains, J. Neurochem. 27: 1529-1531.

Matthieu, J.-M., Quarles, R.H., Brady, R.O. and Webster, H. deF. (1973) Variations of proteins, enzyme markers and gangliosides in myelin subfractions, Biochim. Biophys. Acta 329: 305-317.

Matthieu, J.-M., Brady, R.O. and Quarles, R.H. (1974a) Develomental change in a myelin-associated glycoprotein: a comparative study in rodents, Develop. Biol. 37: 146-152.

Matthieu, J.-M., Brady, R.O. and Quarles, R.H. (1974b) Anomalies of myelin-associated glycoproteins in Quaking mice, J. Neurochem. 22: 291-296.

Matthieu, J.-M., Quarles, R.H., Webster, H. deF., Hogan, E.L. and Brady, R.O. (1974c) Characterization of the fraction obtained from the CNS of Jimpy mice by a procedure for myelin isolation, J. Neurochem. 23: 517-523.

Matthieu, J.-M., Brady, R.O. and Quarles, R.H. (1975a) Change in a myelin-associated glycoprotein in rat brain during development: metabolic aspects, Brain Res. 86: 55-65.

Matthieu, J.-M., Koellreutter, B. and Joyet, M.-L. (1977a) Changes in CNS myelin proteins and glycoproteins after in situ autolysis. Brain Res. Bull. 2: 15-21.

Matthieu, J.-M., Webster, H. deF., Beny, M. and Dolivo, M. (1977b) Characterization of two subcellular fractions isolated from myelinated axons, Brain Res. Bull. 2: 289-298.

Matthieu, J.-M., Webster, H. deF. and De Vries, G.H. (1978b) Glial versus neuronal origin of myelin proteins and glycoproteins studied by intraocular and intracranial labelling, J. Neurochem., in press.

Matthieu, J.-M. and Waehneldt, T.V. (1978) Protein and enzyme distribution in microsomal and myelin fractions from rat and Jimpy mouse brain, Brain Res. 156: 000-000.

McIntyre, R.J., Quarles, R.H., Webster, H. deF. and Brady, R.P. (1978) Isolation and characterization of myelin-related membranes, J. Neurochem., in press.

Moore, W.J., Chapman, B.E. and Littlemore, L. (1977) Conformation of myelin basic protein and its role in myelin formation, Sat. Symp. Intl. Soc. Neurochem., Helsinki, p. 14.

Morell, P., Greenfield, S., Costantino-Ceccarini, E. and Wisnieswki, H. (1972) Changes in the protein composition of mouse brain myelin during development, J. Neurochem. 19: 2545-2554.

Morell, P., Wiggins, R.C. and Jones Gray, M. (1975) Polyacrylamide gel electrophoresis of myelin proteins: a caution, Analyt. Biochem. 68: 148-154.

Nicot, C., Nguyen Le. T., Lepêtre, M. and Alfsen, A. (1973) Study of Folch-Pi apoprotein. I. Isolation of two components, aggregation during delipidation, Biochim. Biophys. Acta 322: 109-123.

Norton, W.T. and Poduslo. S.E. (1973) Myelination in rat brain: Method of myelin isolation, J. Neurochem. 21: 749-758.

Norton, W.T. (1976) Formation, structure and biochemistry of myelin. In: Basic Neurochemistry (Siegel, G.J., Albers, R.W., Katzman, R. and Agranoff, B.W., eds.) Little, Brown and Company, Boston, 2nd ed., pp. 74-99.

Norton. W.T. (1977) Isolation and characterization of myelin. In: Myelin (Morell, P. ed.) Plenum Press, New York, pp. 161-199.

Nussbaum, J.L. and Mandel, P. (1973) Brain proteolipids in neurological mutant mice, Brain Res. 61: 295-310.

Nussbaum, J.L., Rouayrenc, J.F., Mandel, P., Jollès, J. and Jollès, P. (1974) Isolation and terminal sequence determination of the major rat brain myelin proteolipid P7 apoprotein, Biochem. Biophys. Res. Comm. 57: 1240-1247.

Nussbaum, J.L., Delaunoy, J.P. and Mandel, P. (1977) Some immunochemical characteristics of W1 and W2 Wolfgram proteins isolated from rat brain myelin, J. Neurochem. 28: 183-191.

Poduslo, J.F. and Braun, P.E. (1975) Topographical arrangement of membrane proteins in the intact myelin sheath, J. Biol. Chem. 250: 1099-1105.

Poduslo, J.F., Quarles, R.H. and Brady, R.O. (1976) External labeling of galactose in surface membrane glycoproteins of the intact myelin sheath, J. Biol. Chem. 251: 153-158.

Quarles, R.H., Everly, J.L. and Brady, R.O. (1973a) Evidence for the close association of a glycoprotein with myelin in rat brain, J. Neurochem. 21: 1177-1191.

Quarles, R.J., Everly, J.L. and Brady, R.O. (1973b) Myelin-associated glycoprotein: a developmental change, Brain Res. 58: 506-509.

Quarles, R.H. (1975) Glycoproteins in the nervous system. In: The Nervous System (Tower, D.B. and Brady, R.O., eds.) Vol 1 of The Basic Neurosciences, Raven Press, New York, pp. 493-501.

Quarles, R.H. (1976) Effects of pronase and neuraminidase treatment on a myelin-associated glycoprotein in developing brain, Biochem. J. 156: 143-150.

Quarles, R.H. and Pasnak, C.F. (1977) A rapid procedure for selectively isolating the major glycoprotein from purified rat brain myelin, Biochem. J. 163: 635-637.

Quarles, R.H. and Everly, J.L. (1977) Glycopeptide fractions prepared from purified central and peripheral rat myelin, Biochim. Biophys. Acta 466: 176-186.

Quarles, R.H. (1977) The biochemical and morphological heterogeneity of myelin and myelin-related membranes. In: Biochemistry of Brain (Kumar, S., ed.) Oxford: Pergamon Press Ltd., pp. 000-000.

Reiber, H. and Waehneldt, T.V. (1978) Biochemical characterization of myelin subfractions obtained from bovine brain stem by zonal centrifugation, Neuroscience Lett., in press.

Sheads, L.D., Eby, M.J., Sampugna, J. and Douglass, L.W. (1977) Myelin subfractions isolated from mouse brain. Studies of normal mice during development, Quaking mutants, and three brain regions, J. Neurobiol. 8: 67-89.

Stoffyn, P. and Folch-Pi, J. (1971) On the type of linkage binding fatty acids present in brain white matter proteolipid apoprotein, Biochem. Biophys. Res. Comm. 44: 157-161.

Toews, A.D., Horrocks, L.A. and King, J.S. (1976) Simultaneous isolation of purified microsomal and myelin fractions from rat spinal cord, J. Neurochem. 27: 25-31.

Vacher-Leprêtre, M., Nicot, C., Alfsen, A., Jollès, J. and Jollès, P. (1976) Study of the apoprotein of Folch-Pi bovine proteolipid. II. Characterization of the components isolated from sodium dodecyl sulfate solutions, Biochim. Biophys. Acta 420: 323-331.

Waehneldt, T.V. (1971) Preparative isolation of membrane proteins by polyacrylamide gel electrophoresis in the presence of ionic detergent (SDS): proteins of rat brain myelin, Analyt. Biochem. 43: 3o6-312.

Waehneldt, T.V. and Mandel, P. (1972) Isolation of rat brain myelin, monitored by polyacrylamide gel electrophoresis of dodecylsulfate-extracted proteins, Brain Res. 40: 419-436.

Waehneldt, T.V. (1975) Ontogenetic study of a myelin-derived fraction with 2',3'-cyclic nucleotide 3'-phosphohydrolase activity higher than that of myelin, Biochem. J. 151: 435-437.

Waehneldt, T.V., Matthieu, J.-M. and Neuhoff, V. (1977) Characterization of a myelin-related fraction (SN 4) isolated from rat forebrain at two developmental stages, Brain Res. 138: 29-43.

Waehneldt, T.V. (1978a) Density and protein profiles of myelin from two regions of young and adult rat CNS, Brain Res. Bull. 3: 37-44.

Waehneldt, T.V. (1978b) Protein heterogeneity in rat CNS myelin subfractions, Adv. Exp. Med. Biol. 100: 000-000.

Whittaker, V.P., Michaelson, I.A. and Kirkland, R.J.A. (1964) The separation of synaptic vesicles from nerve-ending particles (synaptosomes), Biochem. J. 90: 293-303.

Wiggins, R.C., Joffe, S., Davidson, D. and Del Valle, U. (1974) Characterization of Wolfgram proteolipid protein of bovine white matter and fractionation of molecular weight heterogeneity, J. Neurochem. 22: 171-175.

Williams, M. and Rodnight, R. (1977) Protein phosphorylation in nervous tissue: possible involvement in nervous tissue function and relationship to cyclic nucleotide metabolism, Progr. Neurobiol. 8: 183-250.

Wolfgram, F. and Kotorii, K. (1968) The composition of the myelin proteins of the central nervous system, J. Neurochem. 15: 1281-1290.

Zanetta, J.-P., Sarliève, L.L., Reeber, A., Vincendon, G. and Gombos, G. (1977a) A protein fraction enriched in all myelin associated glycoproteins from adult rat central nervous system, J. Neurochem. 29: 355-357.

Zanetta, J.-P., Ghandour, M.S., Vincendon, G., Eberhart, R., Sarliève, L.L. and Gombos, G. (1977b) Minor proteins of CNS myelin fractions: are they intrinsic to the myelin sheath? J. Neurochem. 29: 359-363.

Zanetta, J.-P., Sarliève, L.L., Mandel, P., Vincendon, G. and Gombos, G. (1977c) Fractionation of glycoproteins associated to adult rat brain myelin fractions, J. Neurochem. 29: 827-838.

Zgorzalewicz, B., Neuhoff, V. and Waehneldt, T.V. (1974) Rat myelin proteins. Compositional changes in various regions of the nervous system during ontogenetic development, Neurobiology 4: 265-276.

Zimmerman, A.W., Quarles, R.H., Webster, H. deF., Matthieu, J.-M. and Brady, R.O. (1975) Characterization and protein analysis of myelin subfractions in rat brain: developmental and regional comparison, J. Neurochem. 25: 749-757.

BIOSYNTHESIS AND DEPOSITION OF MYELIN LIPIDS

N. Baumann

Laboratoire de Neurochimie INSERM U. 134, CNRS ERA 421, Hôpital de la Salpêtrière, 75634 Paris Cédex 13, France

Peters (1960) showed for the first time by electron microscopy that myelin membranes originate from oligodendrocytes in the central nervous system. Nevertheless intercellular interactions between neurons and glial cells are necessary although unclear for this developmental process to occur.

It has been known for many years that myelin is a lipid rich structure and recent reviews have been published on its composition (Norton, 1975 ; Morell, 1977 ; Rumsby, 1978). Analysis of isolated myelin show that it has a much higher lipid:protein ratio than any other subcellular fractions. Its dry weight consists of 70 to 85 % lipid. Myelin like all membranes contains phospholipids ; among them, there is a particular enrichment in ethanolamine plasmalogens and di and tri phosphoinositides. Cholesterol constitutes the largest proportion of lipid molecules with transiently cholesterol esters at the onset of myelination. Myelin is especially notable for its high content of glycosphingolipids. Galactocerebrosides and sulphatides comprise some 20 % of the total myelin lipids and are characterized by very long chain saturated and monoenoic fatty acids some of which have an additional hydroxy group on C-2. No other animal membrane system has such high concentrations of these glycolipids and it must be presumed that they are concentrated in myelin for some structural purpose that may be related in part to the long acyl chains as their presence in the myelin sheath confers to this membrane increasing stability (Vandenheuvel, 1963). G_{M1}[1,2] is the major ganglioside in brain mature myelin (Suzuki et al., 1967) ; another ganglioside G_{M4}[1] (called also G_7[3]) has been identified in human myelin (Ledeen et al., 1973) ; it is less abundant in other species and increases throughout the evolutionary scale of vertebrates (Cochran et al., 1977). Other glycolipids have been found associated with myelin: monogalactosyldiglyceride (Desmukh et al., 1971) and sulphogalactosylglycerolipid (Pieringer et al., 1977). Alkanes are also present in myelin (Bourre et al., 1977a).

[1] According to Svennerholm's nomenclature (1964)
[2] $3II^3$ NeuAc - Gg Ose$_4$ - Cer according to the IUPAC-IUB recommendation
[3] According to the nomenclature of Korey and Gonatas (1963)

I - <u>GENETICS</u>

Myelination requires active biosynthesis of lipids. Occurring at a particular time in a given species, its timing is evidently specified in the genetic code. At this critical period numerous factors interact so that the process which has this once-and-for-all opportunity develops normally. If the right conditions are not present, myelin formation is impaired (Davison and Dobbing, 1966). Thus undernutrition can lead to a reduction in the extent of myelination, which nevertheless still occurs at its normal time. To understand myelination, information is required on the histological events, their biochemical correlated and also the metabolic and hormonal influences on this process and its genetic control. Experiments on mice and rats in which myelination occurs after birth in brain are particularly useful to follow its development ; furthermore, investigations in these species can be made under strictly controlled conditions of genetic background, rearing, housing and food. In addition, there are myelin deficient mutants in mice (Sidman et al., 1964 ; Bird et al., 1977) blocked at specific and different steps of myelination.

In the past ten years, the study of dysmyelinating mutants in numerous laboratories has given arguments as to the cellular origin of certain myelin components, confirmed the sequential nature of myelination in relation to biosynthesis and deposition of myelin lipids and proteins. Although dysmyelinating mutant mice have the same apparent phenotype manifesting abnormal gait, tremor and seizures at the time of myelination around the 10th or 12th day after birth, they have different genotypes. The Quaking mutation is autosomal recessive (chromosome 17) ; the Jimpy mutation is X-linked recessive. The Shiverer mutation has not yet been mapped but differs genetically from both. This implies that genes located on several chromosomes are involved in the formation of myelin and interfere directly or indirectly with the synthesis and (or) the deposition of its lipid components.

Some specific steps of myelination, blocked in the mutants, seem to differ in their impact on the central nervous system (CNS) and the peripheral nervous system (PNS). For instance, the Jimpy mutant is nearly devoid of myelin in the CNS (Sidman et al., 1964 ; Nussbaum et al., 1969 ; Privat et al., 1972) and not in the PNS. Also, the effects of the Quaking mutation on myelin formation (Berger, 1971 ; Wisniewski and Morell, 1971 ; Suzuki and Zagoren, 1977 ; Aguayo et al., 1978) and lipid content (Baumann et al., 1967 ; Jacque et al., 1969 ; Kishimoto, 1971) are more intense in the CNS than in the PNS. In both mutants, there is a proliferation of astrocytes (Berger, 1971 ; Skoff, 1976 ;

Jacque et al., 1974, 1976a) and no defect in the maturation of the neurons at least in relation to the development of some specific cell markers (Jacque et al., 1976b). The phenotypes involve essentially the oligodendrocyte cell line (Berger, 1971 ; Wisniewski and Morell, 1971 ; Farkas-Bargeton et al., 1972 ; Meier and Bischoff, 1975). In Jimpy, there are no mature oligodendrocytes ; precursor cells are present at an abnormally late stage and myelination does not occur ; it is therefore possible to relate any profound deficiency in lipids to the lack of oligodendrocytes. Whatever is the cause of the mutation this objective defect is sufficient to use this mutant as a tool to localize in the oligodendrocyte, lipid components and enzymatic systems related to their synthesis. In the Quaking homozygotes, the oligodendrocytes are present but not well differenciated ; myelin is formed although immature in its protein and lipid composition (Gregson and Oxberry, 1972 ; Baumann et al., 1973) ; there is also a defect in compaction of the myelin layers. Here again, whatever is the cause of the mutation, this mutant is a useful tool to study myelin maturation, metabolism, deposition and assembly of myelin protein and lipid components, and some aspects related to their regulation. Another mutant, Shiverer, deprived of the major dense line of myelin (Jacque et al., 1978) may help to understand what are the specific lipid and protein constituents involved in this particular process and how compaction occurs normally.

II - <u>BIOSYNTHESIS OF MYELIN LIPIDS</u>

If enzymatic systems involved in the synthesis of myelin lipids can be located by procedures involving isolations of the different cell types of brain, mutants are also a useful tool. By the fact that intercellular interactions are not destroyed, they have been also a help in identifying among lipids common to all cellular membranes, specific pools involved in myelin lipid formation.

Synthesis of myelin lipids in the mouse occurs mainly between the 10th and the 50th day after birth with a peak at 18 days. This is also the case for the rat. In both mice and rats, there seem to be small variations according to the strains.

1. <u>Fatty acid biosynthesis</u>

Myelin sphingoglycolipids are very characteristics and consist essentially of galactocerebrosides, sulphatides and G_{M1}. Galactocerebrosides and sulphatides contain a great amount of long (C_{18}) and very long (C_{22}, C_{24}) chain fatty acids, saturated, mono-unsaturated and alpha-hydroxylated.

Myelin fatty acids can be derived from endogenous fatty acids but, at least for stearic acid, there is an uptake from the circulation into myelin lipids (Gozlan-Devillierre et al., 1978) ; this fatty acid can be either directly incorporated into myelin, elongated inside brain or metabolized into acetate units used for synthesis of medium chain fatty acids as palmitic acid and elaboration of cholesterol ; these products of metabolism are partly incorporated into myelin lipids. Nevertheless, endogenous biosynthesis remains the most important pathway.

Whereas most enzymes involved in lipid synthesis show increased enzyme activity during the period of most active myelination and lipid deposition, the activity of rat brain soluble fatty acid synthetase has been found to decline a few days after birth (Goldberg et al., 1973 ; Volpe et al., 1973 ; Cantrill and Carey, 1975) ; the reaction products are medium chain fatty acids, mainly palmitic acid (Brady, 1960 ; Pollet et al., 1969). However total fatty acid biosynthesis seems to be maximum at time of rapid myelination (Aeberhard et al., 1969 ; Carey and Parkin, 1975).

Formation of long chain and very long chain fatty acids necessary to myelin occurs in microsomes (Bourre et al., 1970, 1973a ; Pollet et al. 1973 ; Goldberg et al., 1973) by at least two elongating systems, one leading to the synthesis of stearic acid from palmityl-CoA (C_{16} elongase), the second building up fatty acids with longer chain lengths (especially with 24 carbon atoms) from stearyl-CoA (C_{18} elongase). Study of the Quaking mutant, only deficient in the second system, lead to their individualization. Both require acyl-CoA, malonyl-CoA and NADPH. Saturated and mono-unsaturated very long chain fatty acids share a common elongating pathway (Bourre et al., 1976). This finding explains why myelin mono-unsaturated fatty acids are n-9 and elucidates the reduction of both mono-unsaturated and saturated very long chain fatty acids in the Quaking mutant. In microsomes, elongation of these mono-unsaturated fatty acids may occur (Bourre et al., 1977b) after desaturation of stearic acid as there is a stearyl-CoA desaturase in brain microsomes (Pullarkat and Reha, 1975).

In mitochondria, very long chain fatty acids can be synthesized (Boone and Wakil, 1970 ; Paturneau-Jouas et al., 1976), involving both NADPH and NADH and acetyl-CoA as the donor of the two-carbon unit for elongation. These systems increase regularly in activity during development and are not related to myelination (Bourre et al., 1977c). Contrary to the microsomal systems, they are not affected in the Quaking mutant.

The microsomal fatty acid elongating systems related to myelination are presumably localized in the oligodendrocytes as they are totally deficient in the Jimpy mice and decreased in their specific and total activities in the Quaking mutant (Bourre et al., 1977c).

Alpha-hydroxy fatty acids are necessary components of galactolipids and are synthesized from the corresponding non hydroxy fatty acids (Hajra and Radin, 1963). The synthesis of cerebronic acid from lignoceric acid by an alpha-hydroxylase is particularly important at the time of myelination (Murad and Kishimoto, 1975). Its absence in the Jimpy mutant indicates that the enzyme is located in glial cells, presumably oligodendrocytes. Synthesis of stearyl-alcohol by reduction of stearyl-CoA in the presence of NADPH increases during development and parallels the plasmalogen deposition ; it is reduced in the dysmyelinating mutants (Bourre and Daudu, 1978).

2. Ceramide

Ceramides are made by a group of different microsomal acyltransferases, one for the stearate cluster, one for the lignocerate cluster and two for the α-hydroxy acid clusters (Ullman and Radin, 1972 ; Morell and Radin, 1970). The enzyme that synthesizes lignoceroylceramide is primarily located in glia while the one making stearoylceramide is primarily located in neurons (Morell and Radin, 1970). In the Quaking mouse, only the enzyme which gives rise to very long chain lignoceroylceramide is diminished while the synthesis of long chain stearoylceramide is normal (Zalc et al., 1974), indicating that the normal enzymatic activity for lignoceroylceramide synthesis may be located in oligodendrocytes, just as the synthesis of the substrate, lignoceric acid.

3. Galactocerebrosides

Galactocerebrosides are essentially constituents of myelin, contrary to glucocerebrosides which seem to be mostly located in neurons. The UDP-galactose:ceramide galactosyl transferase is located in microsomes from white matter (Shah, 1971). The enzyme is mostly active at the time of myelination, whether it leads to the synthesis of non hydroxy or hydroxy fatty acids (Costantino-Ceccarini and Morell, 1972). The enzyme activity is very low in Jimpy and reduced in Quaking (Neskovic et al., 1972), suggesting an oligodendrocyte location which has been confirmed on intact oligodendroglia maintained in culture (Poduslo and McKhann, 1977). Nevertheless, several authors indicate that this transferase is also present in the myelin membrane (Neskovic et al., 1973 ; Costantino-Ceccarini and Suzuki, 1975) and in neuronal and astroglial fractions (Neskovic et al., 1972).

4. Sulfatides

Sulfatides are highly enriched in myelin and accumulate rapidly as myelination begins. The PAPS cerebroside sulfotransferase increases rapidly in activity while myelin is formed. Compared to neurons, this enzyme is highly enriched in isolated oligodendroglia (Benjamins et al., 1974). Further evidence for close association of this enzyme with myelination is the finding of low activity in the myelin mutants Quaking and Jimpy (Neskovic et al., 1972 ; Sarlieve et al., 1972).

5. Ganglioside G_{M4} (G_7)

This specific myelin compound is synthesized using galactocerebroside as substrate and CMP-NeuAc (Yu and Lee, 1976).

6. Ganglioside G_{M1}

There is a special fraction of G_{M1} which is located in myelin (Suzuki et al., 1967). In mouse, there is a doubling of the total concentration between 30 and 60 days (Baumann, unpublished) and an absence of this late rise in the Quaking mutant (Baumann et al., 1978). This fact and other data related to the evolution of its sphingosine composition during development (Mansson et al., 1978) may indicate the existence of two pools, one of them being synthesized in the oligodendrocyte.

7. Galactosyl diglyceride (galactosyl-diacylglycerol)

Monogalactosyl diglyceride increases normally with myelination (Desmukh et al., 1971). Of the three cell types, the oligoglia has the highest and the neurons the lowest capacity to synthesize and to accumulate this lipid (Desmukh et al., 1974) ; although the biosynthesis of digalactosyl diglyceride parallels exactly the monogalactosyl diglyceride biosynthetic enzyme, this lipid has never been detected up to now in vivo.

8. Sulfogalactosyl glycerolipid

The sulfogalactosyl glycerolipid of rat brain (Pieringer et al., 1977) is closely associated with myelination. Galactosyl-diacylglycerol is its direct precursor. The enzyme is the adenosine 3'-phosphate 5'-sulfatophosphate-galactosyl-diacylglycerol sulphotransferase.

9. Phospholipids

While most phospholipids, such as phosphatidyl-choline and phosphatidyl-ethanolamine, are ubiquitous components of most membranes, two phospholipids, triphospho-inositide and ethanolamine plasmalogen, are enriched in myelin. Except for the enzymes which synthesize phosphatidyl-choline (McCaman and Cook, 1966) and phosphatidyl-inositol (Salway et al., 1968), developmental changes in many of the enzymes which metabolize individual phospholipids and their cellular and subcellular location in brain have not been extensively documented. The biosynthesis of polyphospho-inositides is reduced in the Quaking mouse (Hauser et al. 1971). Several enzymes rise and fall in parallel with rate of myelination among tham glycerolphosphate dehydrogenase (Laatsch, 1962), acyltransferases (Benes et al., 1973).

For sphingomyelin, there seem to be different pools (Freysz et al., 1976) according to molecular species which might be related to their cellular and subcellular location, oligodendroglial cells and myelin being richer in long chain fatty acid sphingomyelins.

Although ethanolamine plasmalogen content is depressed in the Jimpy brain, phospholipid biosynthesis in the mutant brain is potentially normal (Dorman et al., 1977).

Although there are at least two different pools (Gaiti et al., 1976) of phosphatidyl-serine, phosphatidyl-ethanolamine and phosphatidyl-choline in rat brain microsomes, their relation to myelin has not been established. Moreover, lipids are not readily transferred to myelin by phospholipid exchange proteins (Carey and Foster, 1977).

10. Cholesterol

Enzymes in the synthetic pathway for cholesterol appear to increase in activity during myelination, then decrease (Jones et al., 1975). Although cholesterol esters are mainly present at early time of myelination, the cholesterol-esterifying enzyme increases and remains at maximal level (Eto and Suzuki, 1972) and a cholesterol-esterifying enzyme has been identified in rat CNS myelin (Choi and Suzuki, 1978).

III - REGULATION OF LIPID BIOSYNTHESIS

Several arguments suggest strongly that an enzyme activity is related to myelination and located in mature oligodendrocytes :

1) if activity is absent before 10 days of age in the rat or the mouse, attains a maximum around 18-20 days and slowly decreases thereafter with increasing age with a low level around 50-60 days,

2) if activity is extremely low in the Jimpy mouse (which is nearly devoid of myelin) and decreased in Quaking,

3) if activity is maximum in oligodendrocytes separated by cell isolation techniques.

This is indeed the case for synthesis of very long chain fatty acids (saturated, mono-unsaturated and alpha-hydroxylated), very long chain lignoceroylceramide, galactocerebrosides, galactosyl-diacylglycerol ; their activity is generally reduced by 90 % in Jimpy.

Thus oligodendrocytes contain the enzyme systems necessary for the synthesis of galactocerebrosides with very long chain fatty acids. Very long chain fatty acids (when activated as acyl-CoA derivatives), saturated, mono-unsaturated and alpha-hydroxylated fatty acids, are substrates for very long chain ceramides. These ceramides are substrates for the galactosyl transferase. Although there are precursor-product relationships, these enzymes are less active in Quaking by the same order of magnitude (30-50 %) ; one would expect them to be more and more reduced at the end of a metabolic pathway. There may be, as already suggested for alpha-hydroxylation (Murad and Kishimoto, 1975), a coordinated inducibility for all these enzymes which would be correlated by a single operational genetic unit, this activation being maximum at 18 days and proper to brain. Also for instance, in this organ the activity of the PAPS cerebroside sulfotransferase as well as the activity of the C_{18} elongase peak normally at 18 days are reduced to 50 % of controls in Quaking and extremely low (10 % control) in Jimpy, at the same age. On the other hand, in the kidney, when sulfatides containing very long chain fatty acids are present, these enzymes are not activated and identical to the controls in the mutants (Sarlieve et al., 1971 ; Bourre et al., 1975). This brain inducibility may be dependent on hormones. Several enzymes of glycolipid synthesis induced at myelination time seem to respond to thyroid hormones or have lower levels if altered thyroid functions (Walravens and Chase, 1969 ; Wysocki and Segal, 1972 ; Murad et al., 1976 ; Flynn et al., 1977).

Most enzymes are located in microsomes. For the synthesis of very long chain fatty acids, mitochondria possess systems for elongating long chain fatty acids to very long chain fatty acids but in no way can they compensate for a deficiency in microsome synthesizing systems (Bourre et al. 1977c).

Several hydrolyzing enzymes seem to be present in myelin (Eto and Suzuki, 1973 ; Yamaguchi et al., 1978 ; Suzuki, 1978) or activated during myelinogenesis (Mickel and Gilles, 1970 ; Ansell, 1973 ; Sarlieve et al., 1976 ; Dorman et al., 1978). Although their function remains unclear, they probably have a role in maintaining the exact requirement for myelin lipids as it is known from human pathology that an excess of lipids (for instance in sulfatidoses) can give rise to defect in myelination.

In general, lipids common to all membranes as phospholipids are not deficient in dysmyelinating mutants. Surprisingly, cholesterol biosynthesis seems to be affected as its level is extremely low in Jimpy (Kandutsch and Saucier, 1972). This could indicate that cholesterol from nutrition may be necessary for cholesterol deposition in brain membranes other than myelin, as other brain structures containing cholesterol are considered as normal in Jimpy brain.

IV - DEPOSITION OF MYELIN LIPIDS

Most of myelin specific lipids are synthesized in microsomes. There exist a precursor-product relationship between microsomes and myelin for galactocerebrosides and very long chain fatty acids (Bourre et al., 1973b ; Hayes and Jungalwala, 1976 ; Desmukh and Bear, 1977), for galactosyl-diacylglycerol (Desmukh and Bear, 1977), for sulfatides (Herschkowitz et al., 1969).

The mode of transport from endoplasmic reticulum to myelin remains unclear. Mechanisms for the synthesis and movement of lipids and proteins to the plasma membrane of the oligodendrocyte almost certainly follow the general principles established for the flow of membrane material from the endoplasmic reticulum and Golgi apparatus to the surface membrane (Palade, 1975). It can be envisaged (Rumsby, 1978) that lipids and proteins are incorporated in the plasma membrane (of an oligodendrocyte) as vesicles that fuse within the membrane. The oligodendroglia cell process which recognizes the axon acts as a template for the addition of myelin specific proteins and lipids. Arguments developed from subfractionation procedures (Agrawal et al., 1970 ; Agrawal et al., 1974 ; Zimmerman et al., 1975 ; Benjamins et al., 1976 ; Hofteig and Druse, 1976 ; Bourre et al., 1977d, Waehneldt et al., 1977) indicate variations in myelin composition during development ; myelin from younger animals shows more protein in the denser fraction than that from older animals. When using labelled precursors (Benjamins et al., 1976), there is also a precursor-product relationship for individual lipids from the densest to the lightest subfraction ; phosphatidyl-ethanolamine and ethanolamine plasmalogen are first added to the densest fraction than to the lightest. In contrast, phosphatidyl-choline and its plasmalogen analogue are added simultaneously just as cerebroside, sulfatide and galactosyl diglyceride. The assembly of the myelin sheath involves an obligate addition of certain lipids while other lipids are probably added in a random order.

The study of the Quaking mutant using a continuous sucrose density gradient after myelin isolation (Bourre et al., 1977d) indicates an arrest in myelin maturation with persistance of a very dense fraction and no formation of the major subfraction (concentrated between 0.6 and 0.7 M sucrose) which contains more cerebrosides, ethanolamine phosphatides and phosphatidyl-serine and less proteins. The abnormally high amount of heavy material found in mutant myelin is not directly related to microsomes but presumably to a pre-myelin membrane which does not mature normally. This view is supported by myelin subfractionations in young normal animals (13 days old) which present mature myelin as well as a large amount of the densest fraction.

Investigations on the possible assymetric localization of lipids in the myelin lamellae and on protein-lipid interactions are in progress in several laboratories ; they are complicated by the fact that isolated myelin may not be truly representative of the intact form (Rumsby, 1978). The myelin deficient mutant Shiverer, devoid of the major dense line and basic protein (Jacque, 1978), may be a help in the understanding of some of the lipid-protein associations.

REFERENCES

Aeberhard, E., Grippo, J. and Menkes, J.M. (1969) Fatty acid synthesis in the developing brain, Pediat. Res. 3 : 590-596.

Agrawal, H.C., Banik, N.L., Bone, A.H., Davison, A.N., Mitchell, R.F. and Spohn, M. (1970) The identity of a myelin-like fraction isolated from developing brain, Biochem. J. 120 : 635-642.

Agrawal, H.C., Trotter, J.L., Burton, R.M. and Mitchell, R.F. (1974) Metabolic studies on myelin. Evidence for a precursor role of a myelin subfraction, Biochem. J. 140 : 99-109.

Aguayo, A.J., Bray, G.M. and Perkins, S.C. (1978) Axon-Schwann cell relationships in neuropathies of mutant mice, Ann. N.Y. Acad. Sci. in press.

Ansell, G.B. (1973) Phospholipids in the nervous system in "Form and function of phospholipids" (G.B. Ansell, R.M.C. Dawson, J.N. Hawthorne Eds) BBA Library, Elsevier N.Y. 3 : 377-422.

Baumann, N.A., Jacque, C., Pollet, S. and Harpin, M.L. (1967) Etude biochimique de la mutation Quaking chez la souris. Analyse des lipides et des acides gras du cerveau, C.R. Acad. Sci. Paris 264 : 2953-2956.

Baumann, N., Bourre, J.M., Jacque, C. and Harpin, M.L. (1973) Lipid composition of Quaking mouse myelin : comparison with normal mouse myelin in the adult and during development, J. Neurochem. 20 : 753-759.

Baumann, N., Sonnino, S., Ghidoni, R., Harpin, M.L. and Tettamanti, G. (1978) Etude des gangliosides chez les souris normales (C57 Bl/6 J) et Quaking. Presence et caractérisation d'un ganglioside labile en milieu alcalin, C.R. Acad. Sci. Paris in press

Benes, F.R., Higgins, J.C. and Barnett R.J. (1973) Ultrastructural localization of phospholipid synthesis in the rat trigeminal nerve during myelination, J. Cell Biol. 57 : 613.

Benjamins, J.A., Guarnieri, M., Sonneborn, M. and McKhann, G.M. (1974) Sulfatide synthesis in isolated oligodendroglial and neuronal cells, J. Neurochem. 23 : 751-758.

Benjamins, J.A., Miller, S.L. and Morell, P. (1976) Metabolic relationships between myelin subfractions : entry of galactolipids and phospholipids, J. Neurochem. 27 : 565-570.

Berger, B. (1971) Quelques aspects ultrastructuraux de la substance blanche chez la souris Quaking, Brain Res. 25 : 35-53.

Bird, T.D., Farrel, D.F. and Sumi, S.H. (1977) Genetic developmental myelin defect in Shiverer mouse, Transact. Amer. Soc. Neurochem. 8 : 153.

Boone, S.C. and Wakil, S.J. (1970) In vivo synthesis of lignoceric acid and nervonic acid in mammalian liver and brain, Biochemistry 17 : 1470-1479.

Bourre, J.M., Pollet, S.A., Dubois, G. and Baumann, N.A. (1970) Biosynthèse des acides gras à longue chaîne dans les microsomes de cerveau de souris, C.R. Acad. Sci. Paris 271 : 1221-1223.

Bourre, J.M., Pollet, S., Chaix, G., Daudu, O. and Baumann, N. (1973a) Etude in vitro des acides gras synthétisés dans les microsomes de cerveau de souris normales et Quaking, Biochimie 55 : 1473-1479.

Bourre, J.M., Pollet, S., Daudu, O. and Baumann, N. (1973b) Evolution in mice brain microsomes of lipids and their constituents during myelination, Brain Res. 51 : 225-239.

Bourre, J.M., Daudu, O. and Baumann, N. (1975) Fatty acid biosynthesis in mice brain and kidney microsomes : comparison between Quaking mutant and control, J. Neurochem. 24 : 1095-1097.

Bourre, J.M., Daudu, O. and Baumann, N. (1976) Nervonic acid biosynthesis by erucyl-CoA elongation in normal and Quaking mouse brain microsomes. Elongation of other unsaturated fatty acyl-CoAs (mono and poly-unsaturated), Biochim. Biophys. Acta 424 : 1-7.

Bourre, J.M., Cassagne, C., Larrouquere-Regnier, S. and Darriet, D. (1977a) Occurence of alkanes in brain myelin. Comparison between normal and Quaking mouse, J. Neurochem. 29 : 645-648.

Bourre, J.M., Pollet, S., Paturneau-Jouas, M. and Baumann, N. (1977b) Saturated and mono-unsaturated fatty acid biosynthesis in brain : relation to development in normal and dysmyelinating mutant mice in "Function and biosynthesis of lipids" (N.G. Bazan, R.R. Brenner, N.M. Giusto Eds) Plenum Press N.Y. p. 103-109.

Bourre, J.M., Paturneau-Jouas, M.Y., Daudu, O.L. and Baumann, N.A. (1977c) Lignoceric acid biosynthesis in the developing brain. Activities of mitochondrial acetyl-CoA-dependent synthesis and microsomal malonyl-CoA chain elongating system in relation to myelination, Eur. J. Biochem. 72 : 41-47.

Bourre, J.M., Pollet, S., Daudu, O., Le Saux, F. and Baumann, N. (1977d) Myelin consists of a continuum of particles of different density with varying lipid composition : major differences are found between normal mice and Quaking mutants, Biochimie 59 : 819-824.

Bourre, J.M. and Daudu, O. (1978) Stearyl-alcohol biosynthesis from stearyl-CoA in mouse brain microsomes in normal and dysmyelinating mutants Quaking and Jimpy, Neuroscience Letters 7 : 225-230.

Brady, R.O. (1960) Biosynthesis of fatty acids : II-studies with enzymes obtained from brain, J. Biol. Chem. 23 : 3099-3103.

Cantrill, R.C. and Carey, E.M. (1975) Changes in the activities of de novo fatty acid synthesis and palmityl-CoA synthetase in relation to myelination in rabbit brain, Biochim. Biophys. Acta 380 : 165-175.

Carey, E.M. and Parkin, L. (1975) Fatty acid metabolism in the microsomal fraction of developing rabbit brain, Biochim. Biophys. Acta 380: 176-189.

Carey, E.C. and Foster, P.C. (1977) Protein mediated transfer of phosphatidyl-choline to myelin, Biochem. Soc. Transac. 6 : 1412-1414.

Choi, M. and Suzuki, K. (1978) Presence of a cholesterol-esterifying enzyme in rat CNS myelin, Trans. Amer. Soc. Neurochem. 9 : 177.

Cochran, F.B., Yu, R.K. and Ledeen, R.W. (1977) Comparison of CNS myelin gangliosides in several vertebrate species, Proc. Internat. Soc. Neurochem. 6 : 496.

Costantino-Ceccarini, E. and Morell, P. (1972) Biosynthesis of brain sphingolipids and myelin accumulation in the mouse, Lipids 7 : 656-659.

Costantino-Ceccarini, E. and Suzuki, K. (1975) Evidence for presence of UDP-galactose : ceramide galactosyl transferase in rat myelin, Brain Res. 93 : 358-362.

Davison, A.N. and Dobbing, J. (1966) Myelination as a vulnerable period in brain development, Brit. Med. Bull. 22 : 40-44.

Desmukh, D.S., Inoue, T. and Pieringer, R.A. (1971) The association of the galactosyldiglycerides of brain with myelination, J. Biol. Chem. 246 : 5688-5694.

Desmukh, D.S., Flynn, T.J. and Pieringer, R.A. (1974) The biosynthesis and concentration of galactosyldiglyceride in glial and neuronal enriched fractions of actively myelinating rat brain, J. Neurochem. 22 : 479-485.

Desmukh, D.S. and Bear, W.D. (1977) The distribution and biosynthesis of the myelin galactolipids in the subcellular fractions of brains of Quaking and normal mice during development, J. Neurochem. 28 : 987-994.

Dorman, R.V., Freysz, L. and Horrocks, L.A. (1977) Synthesis of ethanolamine phosphoglycerides by microsomes from the brains of Jimpy and Quaking mice, J. Neurochem. 29 : 231-233.

Dorman, R.V., Freysz, L., Mandel, P. and Horrocks, L.A. (1978) Plasmalogenase activities in the brains of Jimpy and Quaking mice, J. Neurochem. 30 : 157-159.

Eto, Y. and Suzuki, K. (1972) Cholesterol esters in developing rat brain : enzymes of cholesterol ester metabolism, J. Neurochem. 19 : 117-122.

Eto, Y. and Suzuki, K. (1973) Developmental change of cholesterol hydrolases localized in myelin and microsomes of rat brain, J. Neurochem. 20 : 1475-1478.

Farkas-Bargeton, E., Robain, O. and Mandel, P. (1972) Abnormal glial maturation in the white matter in Jimpy mice, Acta Neuropath. (Berl.) 21 : 272-281.

Flynn, T.J., Desmukh, D.S. and Pieringer, R.A. (1977) Effects of altered thyroid function on galactosyl-diacylglycerol metabolism in myelinating rat brain, J. Biol. Chem. 252 : 5864-5870.

Freysz, L., Lastennet, A. and Mandel, P. (1976) Metabolism of brain sphingomyelins : half lives of sphingosine fatty acids and phosphate from two types of rat brain sphingomyelin, J. Neurochem. 27 : 355-359.

Gaiti, A., Brunetti, M., Woelk, H. and Porcellati, G. (1976) Relationships between base-exchange reaction and the microsomal phospholipid pool in the rat brain in vitro, Lipids 11 : 823-829.

Goldberg, I., Schechter, I. and Bloch, K. (1973) Fatty acyl-coenzyme A elongation in brain of normal and Quaking mice, Science 182 : 497-499.

Gozlan-Devillierre, N., Baumann, N. and Bourre, J.M. (1978) Distribution of radioactivity in myelin lipids following subcutaneous injection of (^{14}C) stearate, Biochim. Biophys. Acta 528 : 490-496.

Gregson, N.A. and Oxberry, J.M. (1972) The composition of myelin from the mutant mouse Quaking, J. Neurochem. 18 : 2119-2128.

Hajra, A.L. and Radin, N.S. (1963) In vivo conversion of labelled fatty acid to the sphingolipid fatty acid in rat brain, J. Lipid Res. 4 : 448-453.

Hauser, G., Eichberg, J. and Jacobs, S. (1971) Poly-phosphoinositide levels and biosynthesis in Quaking mouse brain, Biochem. Biophys. Res. Commun. 43 : 1072-1080.

Hayes, L.W. and Jungalwala, F.B. (1976) Synthesis and turnover of cerebrosides and phosphatidyl-serine of myelin and microsomal fractions of adult and developing rat brain, Biochem. J. 160 : 195-204.

Herschkowitz, N., McKhann, G.M., Saxena, S., Shooter, E.M. and Herndon, R.M. (1969) Synthesis of sulphatide-containing lipoproteins in rat brain, J. Neurochem. 16 : 1044.

Hofteig, J.H. and Druse, M.J. (1976) Metabolism of three subfractions of myelin in developing rats, Life Sci. 18 : 543-552.

Jacque, C., Harpin, M.L. and Baumann, N.A. (1969) Brain lipid analysis of a myelin deficient mutant, the Quaking mouse, Eur. J. Biochem. 11 : 218-224.

Jacque, C., Jorgensen, O. and Bock, E. (1974) Quantitative studies of the brain specific antigens S 100, GFA, 14-3-2, D1, D2, D3 and C1 in Quaking mouse, FEBS Lett. 49 : 264-266.

Jacque, C., Baumann, N. and Bock, E. (1976a) Quantitative studies of the brain specific antigens GFA, 14-3-2, synaptin C1, D1, D2, D3 and D5 in Jimpy mouse, Neuroscience Letters 3 : 41-44.

Jacque, C., Jorgensen, O., Baumann, N. and Bock, E. (1976b) Brain specific antigens in the Quaking mouse during ontogeny, J. Neurochem. 27 : 905-909.

Jacque, C., Privat, A., Dupouey, P., Bourre, J.M. and Baumann, N. (1978) Shiverer mouse : a dysmyelinating mutant with absence of major dense line and basic protein in myelin, ESN 2nd meeting at Göttingen.

Jones, J., Rios, A., Nicholas, H. and Ramsey, R. (1975) The biosynthesis of cholesterol and other sterols in brain tissue : distribution in subcellular fractions as a function of time after injection of (2-^{14}C) mevalonic acid sodium (2-^{14}C) acetate and (U-^{14}C) glucose into 15 day-old rats, J. Neurochem. 24 : 117-121.

Kandutsch, A.A. and Saucier, S.E. (1972) Sterol and fatty acid synthesis in developing brains of three myelin-deficient mutants, Biochim. Biophys. Acta 260 : 26-34.

Kishimoto, Y. (1971) Abnormality in sphingolipid fatty acids from sciatic nerve and brain of Quaking mice, J. Neurochem. 18 : 1365-1368.

Korey, S.R. and Gonatas, J. (1963) Separation of human brain gangliosides, Life Sciences 5 : 296-302.

Laatsch, R.H. (1962) Glycerol phosphate dehydrogenase activity of developing rat central nervous system, J. Neurochem. 14 : 1167.

Ledeen, R.W., Yu, R.K. and Eng, L.F. (1973) Gangliosides of human myelin : sialosylgalactosyl ceramide (G7) as a major component, J. Neurochem. 21 : 829-839.

Mansson, J.E., Vanier, M.T. and Svennerholm, L. (1978) Changes in the fatty acid and sphingosine composition of the major gangliosides of human brain with age, J. Neurochem. 30 : 273-275.

McCaman, R.E. and Cook, K. (1966) Intermediary metabolism of phospholipids in brain tissue III phosphocholine-glyceride transferase, J. Biol. Chem. 241 : 3390.

Meier, C. and Bischoff, A. (1975) Oligodendroglial cell development in Jimpy mice and controls, J. Neurol. Sci. 26 : 517-528.

Mickel, H.S. and Gilles, F.H. (1970) Changes in glial cells during human telencephalic myelinogenesis, Brain Res. 93 : 337-346.

Morell, P. and Radin, N.S. (1970) Specificity in ceramide biosynthesis from long chain bases and various fatty acyl-CoA's in brain microsomes, J. Biol. Chem. 245 : 342-350.

Morell, P. (1977) Myelin, Plenum Press, N.Y.

Murad, S. and Kishimoto, Y. (1975) α-hydroxylation of lignoceric acid to cerebronic acid during brain development, J. Biol. Chem. 250 : 5841-5846.

Murad, S., Strycharz, G.D. and Kishimoto, Y. (1976) α-hydroxylation of lignoceric and nervonic acids in the brain. Effects of altered thyroid function on postnatal development of the hydroxylase activity, J. Biol. Chem. 251 : 5237-5241.

Neskovic, N.M., Sarlieve, L.L. and Mandel, P. (1972) Biosynthesis of glycolipids in myelin deficient mutants : brain glycosyl transferases in Jimpy and Quaking mice, Brain Res. 42 : 147-157.

Neskovic, N.M., Sarlieve, L.L. and Mandel, P. (1973) Subcellular and submicrosomal distribution of glycolipid synthesizing transferases in Jimpy and Quaking mice, J. Neurochem. 20 : 1419-1430.

Norton, W.T. (1975) Myelin : structure and biochemistry in "The nervous system. I. The basic neurosciences (D.B. Tower Ed.) Raven Press, N.Y., p. 467-481.

Nussbaum, J.L., Neskovic, N. and Mandel P. (1969) A study of lipid components in brain of the Jimpy mouse, a mutant with myelin deficiency, J. Neurochem. 16 : 927.

Palade, G. (1975) Intracellular aspects of the process of protein synthesis, Science 189 : 347-358.

Paturneau-Jouas, M., Baumann, N. and Bourre, J.M. (1976) Elongation of palmityl-CoA in mouse brain mitochondria : comparison with stearyl-CoA, Biochem. Biophys. Res. Commun. 71 : 1326-1334.

Peters, A. (1960) The formation and structure of myelin sheaths in the central nervous system, J. Biophys. Biochem. Cytol. 8 : 431-446.

Pieringer, J., Subbarao, G., Mandel, P. and Pieringer, R.A. (1977) The association of the sulphogalactosylglycerolipid of rat brain with myelination, Biochem. J. 166 : 421-428.

Poduslo, S.E. and McKhann G.M. (1977) Synthesis of cerebrosides by intact oligodendroglia maintained in culture, Neuroscience Letters 5 : 159-163.

Pollet, S.A., Bourre, J.M. and Baumann, N.A. (1969) Biosynthèse des acides gras dans les microsomes de cerveau de souris, C.R. Acad. Sci. Paris 268 : 1426-1429.

Pollet, S., Bourre, J.M., Chaix, G., Daudu, O. and Baumann, N. (1973) Biosynthèse des acides gras dans les microsomes de cerveau de souris, Biochimie 55 : 333-341.

Privat, A., Robain, O. and Mandel, P. (1972) Aspects ultrastructuraux du corps calleux chez la souris Jimpy, Acta Neuropath. 21 : 282-295.

Pullarkat, R.K. and Reha, H. (1975) Stearyl-CoA desaturase activities in rat brain microsomes, J. Neurochem. 25 : 607-610.

Rumsby, M.G. (1978) Organization and structure in central nerve myelin, Biochem. Soc. Transactions 6 : 448-462.

Salway, J.G., Harwood, J.L., Kai, M., White, G.L. and Hawthorne, J.N. (1968) Enzymes of phosphoinositide metabolism during rat brain development, J. Neurochem. 15 : 221.

Sarlieve, L.L., Neskovic, N.M. and Mandel, P. (1971) PAPS-cerebroside sulfotransferase activity in brain and kidney of neurological mutants, FEBS Letters 19 : 91.

Sarlieve, L.L., Neskovic, N.M., Rebel, G. and Mandel, P. (1972) Some properties of brain PAPS-cerebroside sulfotransferase in Jimpy and Quaking mice, Neurobiology 2 : 70-82.

Sarlieve L.L., Farooqui, A.A., Rebel, G. and Mandel, P. (1976) Arylsulphatase A and 2',3'-cyclic nucleotide 3'-phosphohydrolase activities in the brains of myelin deficient mutant mice, Neuroscience 1 : 519-522.

Shah, S.N. (1971) Glycosyl transferases of microsomal fractions from brain : synthesis of glucosyl ceramide and galactosyl ceramide during development and the distribution of glucose and galactose transferase in white and grey matter, J. Neurochem. 18 : 395-402.

Sidman, R.L., Dickie, M.M. and Appel, S.H. (1964) Mutant mice (Quaking and Jimpy) with deficient myelination in the central nervous system, Science 144 : 309-311.

Skoff, R.P. (1976) Myelin deficit in the Jimpy mouse may be due to cellular abnormalities in astroglia, Nature 264 : 560-562.

Smith, M.E., Hasinoff, C.M. and Fumagalli, R. (1970) Inhibitors of cholesterol synthesis and myelin formation, Lipids 5 : 665-671.

Suzuki, K., Poduslo, S.E. and Norton, W.T. (1967) Gangliosides in the myelin fraction of developing rats, Biochim. Biophys. Acta 144 : 375-381.

Suzuki, K. and Zagoren, J.C. (1977) Quaking mouse : an ultrastructural study of the peripheral nerves, J. Neurocytol. 6 : 71-84.

Suzuki, K. (1978) A novel magnesium independent neutral sphingomyelinase associated with rat central nervous system myelin, J. Biol. Chem. in press.

Svennerholm, L. (1964) The gangliosides, J. Lipid Res. 5 : 145.

Ullman, M.D. and Radin, N.S. (1972) Enzymatic formation of OH-ceramides and comparison with enzymes forming non OH-ceramides, Arch. Biochem. 152 : 767-777.

Vandenheuvel, F.A. (1963) Study of biological structure at the molecular level with stereomodel projections. I - The lipids in the myelin sheath of nerve, J. Am. Oil Chem. Soc. 40 : 455-461.

Volpe, J.J., Lyles, T.O., Roncari, D.A. and Vagelos, P.R. (1973) Fatty acid synthesis of developing brain and liver content synthesis and degradation during development, J. Biol. Chem. 7 : 2502-2513.

Waehneldt, T. (1977) Characterization of a myelin-related fraction (SN 4) isolated from rat forebrain at two developmental stages, Brain Res. 138 : 29-45.

Walravens, P. and Chase, H.P. (1969) Influence of thyroid on formation of myelin lipids, J. Neurochem. 16 : 1477-1484.

Wisniewski, H. and Morell, P. (1971) Quaking mouse : ultrastructural evidence for arrest of myelinogenesis, Brain Res. 29 : 63-73.

Wysocki, S.J. and Segal, W. (1972) Influence of thyroid hormones on enzyme activities of myelinating rat central nervous tissues, Eur. J. Biochem. 28 : 183-189.

Yamaguchi, S., Hanada, E. and Suzuki, K. (1978) Close association of selective acid hydrolases with rat CNS myelin, Neuroscience Meeting at Florence.

Yu, R.K. and Lee, S.H. (1976) In vitro biosynthesis of sialosylgalactosyl ceramide (G7) by mouse brain microsomes, J. Biol. Chem. 251 : 198-203.

Zalc, B., Pollet, S.A., Harpin, M.L. and Baumann, N.A. (1974) Ceramide biosynthesis in mouse brain microsomes : comparison between C57/Bl controls and Quaking mutants, Brain Res. 81 : 511-518.

Zimmerman, A.W., Quarles, R.H., Webster, H. de F., Matthieu, J.M. and Brady, R.O. (1975) Characterization and protein analysis of myelin subfractions in rat brain : developmental and regional comparisons, J. Neurochem. 25 : 749-757.

MYELIN CATABOLISM

H. Woelk and G. Porcellati
Department of Neurology, Unit of Neurochemistry, University of Erlangen-Nürnberg (GFR) and Department of Biochemistry, The Medical School, University of Perugia, Perugia (Italy)

Introduction

   Myelin differs from other membranes in its high lipid and low protein content, its relative stability and low enzyme activity. Purified myelin membranes contain about 75 per cent of the dry weight as lipid and the myelin sheath has a common pattern of cholesterol:phospholipid:galactolipid with a molar ratio of about 2:2:1. Particularly characteristic for the myelin membrane is the high concentration of cerebrosides, sulfatides and ethanolamine-plasmalogen, the latter accounting for approximately 80 per cent of the ethanolamine containing phosphatides in myelin. The galactolipids contain non-hydroxy as well as hydroxy fatty acids with hydrocarbon chains four to six carbon atoms longer than the average chain length found in other membranes.
There are two major protein constituents occurring in almost equal amounts in the myelin membrane: chloroform/methanol soluble proteolipid protein and basic protein. Variable amounts of high molecular weight proteins and small amounts of glycoprotein and ganglioside are also present in the myelin membrane. Basic protein has a molecular weight of about 1 8000 and since the discovery of the encephalitogenic property of the protein an important role of the basic protein in the pathogenesis of demyelinating diseases has been assumed. The basis for the encephalitogenicity of basic protein has been linked to its absence from the brain during the neonatal period of immunological tolerance. The proteolipid protein comprises about 50 per cent of the myelin proteins of the central nervous system (Mehl and Wolfgram, 1969). Recent advances in the technique of solubilizing membrane lipid-protein complexes in sodium dodecyl sulphate (SDS) revealed that the proteolipid protein is a group of proteins, the molecular weights of which ranging between 10 000 and 50 000. Whaehneldt and Mandel (1972) monitored the purification of myelin by polyacrylamide gel electrophoresis of SDS-extracted membrane proteins. Myelin showed fast-migrating proteins, proteolipid protein, a slow-migrating protein fraction and other unidentified minor proteins.
Electron microscopy and x-ray diffraction studies have forwarded experimental evidence that the myelin sheath is built up of regular arrays of bimolecular lipid layers with a hydrophobic interior sand-

wiched between proteins. Basic protein seems to be located exclusively on the interior or cytoplasmic site, with other myelin proteins localized on the outer surface of the membrane. Acidic lipids such as phosphatidylinositol, phosphatidylserine and sulfatides have been found as components of isolated encephalitogenic basic protein and it has been demonstrated that these lipids protect the myelin basic protein from hydrolysis by trypsin. The formation of complexes between myelin basic protein and triphosphoinositide, sulphatide or other anionic phospholipids has also been shown in a biphasic system and the electrostatic nature of the bonds is indicated by dissociation at high salt concentrations.

Experimental evidence has been presented suggesting that the centre of the hydrophobic region can exist in a fluid state. Infrared absorption spectra, proton magnetic resonance spectra and electron spin resonance data all point to fluidity of apolar groups of lipids in myelin, resulting from apolar associations between membrane lipids and proteins. Apolar proteolipid protein may serve to penetrate the lipid layer and interact across both the cytoplasmic and extracellular surfaces of the myelin lamellae. The metabolically stable proteolipid protein may serve as the permanent membrane framework linking the more fluid lipid bilayer with the metabolically more active high molecular weight and basic protein layer.

In the early stages of myelin biosynthesis in the central nervous system, connexions between the tongue of cytoplasm on the outside of the myelin sheath and the processes coming from oligodendroglial cells have been observed. Ultrastructural studies by Poduslo and Norton (1972) of isolated bovine oligodendroglia show that whorls of loosely packed myelin lamellae are frequently associated with and extended from the plasma membrane. It is postulated that the myelin-like material is largely derived from glial cell plasma membrane constituting the loosely wrapped membrane surrounding myelinating fibres before compact myelin lamellae are established (Davison and Cuzner, 1977).

Catabolism of Myelin Proteins

Although early investigations concluded that white matter and myelin membrane proteins were metabolically stable, this view has been widely challenged. In an extensive investigation Sabri et al. (1974) examined the protein metabolism of myelin from developing rat brain for periods ranging from 5 h to 210 days after i. p. injection of labelled lysine and glucose. The catabolism of purified myelin proteins was studied and the half-lives of individual myelin proteins were calculated. Sabri et al. (1974) could demonstrate that myelin basic proteins turned over at two different rates. The half-life of the fast component of myelin

basic proteins was about 20 days and the slow component exhibited a
high degree of metabolic stability. The proteolipid protein underwent
a slow turnover, whereas the high molecular weight Wolfgram proteins
turned over at a faster rate (Sabri et al., 1974). From these data it
was concluded that the Wolfgram proteins are localized on the membrane
surface, whereas the proteins with a high degree of metabolic stability
could provide an inert structural framework for the membrane lipids.
Agrawal et al. (1976) studied the accumulation and turnover of the proteolipid proteins in developing and adult rat brain after administration of labelled tryptophan. Evidence was obtained that the proteolipid protein fraction is not metabolically inert, as had been judged
earlier, but that it has an apparent half-life of 188 days when animals
are injected at 40 days of age. In contrast, in developing brain the
half-life of proteolipid protein varies from 7 to 188 days.

To account for the active metabolic processes of myelin, it could be
postulated either that the digestive enzymic activities should be present as an essential part of the membrane structure, or, as has been
observed by electron microscopy, that in some areas cytoplasmic spaces
with lysosomal properties are present within the myelin membrane. With
the exception of 2', 3'-cyclic nucleotide 3'-phosphohydrolase being
localized mainly in the myelin sheath of the central nervous system
and which is regarded to be a marker enzyme for the myelin membrane,
the enzymic activity of purified myelin seems to be very low. Myelin
basic proteins undergo rapid proteolysis within a few hours after death
when neural tissues are allowed to autolyze at room temperature. The
proteolytic degradation depends on the pH-value and proceeds at different rates in the various regions of the brain. The degradation of
myelin basic proteins was faster in the spinal cord than in the forebrain (Sammeck and Brady, 1972). A considerable rate of breakdown of
basic proteins could be observed in crude myelin preparations from various regions of the brain and spinal cord. When purified spinal cord
myelin was kept under similar conditions for as long as 16 days, no
breakdown of the myelin basic proteins could be detected, indicating
that acidic proteinases are not an integral part of the myelin sheath.
Riekkinen et al. (1970) prepared myelin by different gradient systems
and assessed the purity of the preparations by electron microscopy and
cytoplasmic, mitochondrial and lysosomal marker enzymes. 7 to 14 per
cent of the neutral proteinase activity of the crude brain homogenate
was found in purified myelin, favouring the possibility that a part of
the neutral proteinase may be closely linked to the myelin sheath. The
data presented by Riekkinen et al. (1970) showed that binding of the
neutral proteinase to myelin was insensitive to cation or anion treatment alone, but that the liberation of the enzymic activity required

the addition of Triton-X-100, a detergent known to be effective in splitting proteolipid bonds.

The rapid enzymatic degradation of the myelin proteins seems to be dependent on the presence of enzyme containing subcellular particles. The catabolism of myelin proteins might be mediated by an opening along the tightened myelin sheaths and infusion of oligodendroglial cytoplasm and lysosomes. Another possible transport mechanism of the enzymes across the membrane would be the formation of transient holes in the membrane structure through which the enzyme can enter the interior of the myelin sheath. Subsequently, the enzyme would diffuse along the interstitial channels to reach the membrane proteins. There could also be the formation at the water-membrane interface of an enzyme-phospholipid complex, which diffuses or rotates through the membrane as has been suggested by Wood et al. (1974). Sammeck and Brady (1972) presumed that the breakdown of myelin basic proteins, which occurred in the crude myelin fractions was due to the presence of lysosomal enzymes. In order to further elucidate this question, Sammeck and Brady (1972) examined the catabolism of the large and small myelin basic proteins in the presence of lysosomal preparations from spinal cord, forebrain and brain stem. Evidence for the degradation of the basic proteins was obtained by the appearance of the breakdown products similar to those observed in the electrophoresis pattern after incubation of the crude myelin preparation. Heating of the lysosomal fraction destroyed the enzymic activity. Röyttä et al. (1974) showed a profuse breakdown of basic protein and changes of the proteolipid protein when myelin was incubated in the presence of a partially purified lysosomal, acid proteinase - containing fraction. On the contrary, when purified myelin was incubated in the presence of neutral proteinase-containing fraction, isolated from brain tissue, no breakdown of myelin proteins could be observed. Experimental evidence has been presented that acidic conditions activate proteolytic enzyme of lysosomal origin, suggesting that the maintenance of pH-values near neutrality prevents or limits the degradation of proteins. The degradation products of myelin basic proteins described for the crude myelin preparations by Sammeck and Brady (1972) were similar to those, when the proteins had been incubated with a partially purified acid proteinase obtained from brain tissue. These findings suggest that the proteinases and peptidases involved in the catabolism of myelin basic proteins might be closely linked to the myelin membrane.

In an extensive investigation on the effect of proteolytic attack on the structure of central nervous system myelin Wood et al. (1974) considered the changes occurring in the interaction of the remaining proteins and lipids after selectively digesting away the basic proteins.

It could be demonstrated that by using very low concentrations of trypsin it is possible to incubate isolated myelin particles and substantially remove the basic protein, while leaving the Wolfgram and proteolipid protein virtually intact (Wood et al., 1974). With higher concentrations of trypsin and longer periods of incubation, the Wolfgram protein is also removed and the proteolipid partially disappears from the membrane. The myelin basic protein fraction is substantially broken down with the appearance of peptides in the supernatant fluid which are similar to those obtained by direct tryptic action on isolated basic protein. Wood et al. (1974) found no evidence that any substantial part of the basic protein molecules had been protected from hydrolysis. It seems that trypsin becomes attached to the myelin membrane and cannot be detached by repeated washing. Although the basic proteins are removed from the myelin membrane almost as effectively by acetylated trypsin, the enzyme does not become associated with the sedimented membrane. The rapidity of the digestion of myelin basic protein by trypsin leads to the conclusion that the basic protein is not evenly distributed over all the myelin lamellae, but accumulated in the outer layers. If basic proteins were not uniformly distributed over the myelin membrane, it could help to explain the rapid metabolism of polyphosphoinositides. It has been suggested that these very acidic lipids are linked with basic proteins and their location in a more accessible outer membrane would allow the substrate and enzymes involved to operate more effectively. The observations presented by Wood et al. (1974) do not preclude the possibility that the early loss of basic protein during demyelination results in an eventual disruption of the myelin structure. However, one would have to assume that this would be more in the nature of a triggering action, since the results obtained by Wood et al. (1974) indicate that the membrane components other than the basic proteins are responsable for the integrity of the bulk of the myelin membrane structure.

Destruction of myelin in demyelinating conditions is probably indicated through increased acid hydrolase activity. Changes in lysosomal enzyme activities in the course of demyelination in multiple sclerosis have been published by Davison and Cuzner (1977). In acute cases of multiple sclerosis there is a generalized increase in the lesions of acid proteinase and ß-glucuronidase (Table 1), whereas in chronic cases acid proteinase activity alone is raised, particularly at the rim of plaques. It has been postulated that lysosomal enzymes are released in the early stages of demyelination and that the myelin basic protein is preferentially digested by acid proteinases (Riekkinen et al., 1972). They were able to demonstrate increased activity of acid hydrolases in areas with normal or only slightly decreased values of

TABLE 1

MEAN LYSOSOMAL ENZYME ACTIVITY IN ACUTE MULTIPLE SCLEROSIS

Tissue	Cathepsin A* (pH 5.5)	Cathepsin D* (pH 3.6)	β-Glucuronidase* (pH 5.5)	Protein+
Plaque	2.91	49	0.55	73
Normal appearing white matter in acute cases	1.56	23.8	0.45	91.6
Normal white matter	1.14	20.34	0.22	96

* Results are expressed as umoles/g wet wt. per hour on 2-11 samples

+ Protein results are expressed as mg/g wet wt. using bovine serum albumin standard

(According to Davison and Cuzner, 1977)

cerebroside and the myelin marker enzyme 2',3'-cyclic nucleotide 3'-phosphohydrolase, suggesting that early lesions are outside the myelin sheath and that demyelination might be only a result of disease and not a primary disturbance in the disease process. In an electron microscopic and neurochemical study on the demyelination process during subacute sclerosing panencephalitis Nevalainen et al. (1972) showed that macrophages containing different types of inclusions, especially lysosomes, were the most active cells in the demyelinating process. Macrophages proved to be the main source of increased lysosomal hydrolase activity playing an important role in the demyelinating process. Also in multiple sclerosis plaques, macrophages contribute to the increase of acid hydrolases and myelin is digested mainly inside the cells. Various glial cells showing reactive changes in the form of increased lysosomes, seemed also to participate in the phagocytosis of myelin. On the contrary, plasma cells, which are constantly found in areas of active demyelination, did not show any phagocytic activity or conspicuous lysosomal changes (Nevalainen et al., 1972).

Catabolism of Myelin Lipids

During the last years experimental evidence has been presented for a certain turnover of the phospholipid components of the myelin membrane. Jungalwala and Dawson (1971) injected different labelled compounds into the lateral ventricles in the brains of adult rats and followed the labelling of individual phospholipid classes in purified myelin. Jungalwala and Dawson (1971) concluded from their experiments that a substantial part of the myelin phospholipids turnover rather readily, although a small pool of slowly exchangeable material also exists. There is good evidence that serine, inositol and choline containing phosphatides in the myelin sheath turned over faster than ethanolamine phosphoglycerides and sphingomyelin. It may be concluded that most of the myelin phospholipid molecules are in fairly rapid equilibrium with other components of the brain tissue. That this renewal does not represent a rapid exchange of parts of the phospholipid molecules, e.g. base or fatty acid exchange rather than whole molecule turnover, is suggested by the phospholipid precursors showing a similar pattern of incorporation and elimination from the myelin phospholipids.
It seems that two metabolic pools exist in myelin, one rapidly exchanging pool and the other a more persistent and slowly turning over part, which can only be significantly labelled in the developing animal. It has been suggested that the rapid exchange occurs at the surface of the myelin sheath and that during myelination the myelin layers become gradually enveloped in fresh myelin and so become unavailable to dynamic metabolic processes.

TABLE 2

PHOSPHOLIPASE A$_2$ ACTIVITY IN RAT BRAIN PARTICULATE FRACTIONS, MEASURED AT pH 8.4

Fraction	Protein (mg/g)	Activity (a)	(b)	(c)
Homogenate	112	8.03	73.7	(1.00)
Nuclei	10.0	0.12	12.0	0.16
Myelin	10.9	0.29	27.3	0.36
Synaptosomes	11.2	0.90	80.6	1.08
Mitochondria	15.6	5.10	326.6	4.40
Microsomes	19.1	0.15	8.0	0.11

(a) umol product x (g of fresh homogenate)$^{-1}$ or umol x (equivalent amount of each fraction)$^{-1}$
(b) nmol product/mg protein
(c) specific activity of each fraction relative to the specific activity of the original homogenate (1.00)

(According to Woelk and Porcellati, 1972)

Although myelin contains approximately 50 per cent of the total phospholipids of brain, there is no evidence for the existence of greater amounts of enzymes within the myelin membrane itself catalyzing the degradation of lipids. As to our knowledge, there is only a small but consistent activity of phospholipase A_2 in the purified myelin membrane (Table 2), the bulk of activity being located in mitochondria and synaptosomes (Woelk and Porcellati, 1973). Phospholipase A_1 and A_2 hydrolyze the fatty acid ester linkages at the 1 or 2 position of phosphoglycerides, respectively, giving rise to the production of lysophosphatides. Together with the action of acyl-transferases the phospholipases A are responsable for the fatty acid pattern of glycerophosphatides.

Treatment of pure bovine central nervous system myelin by highly purified phospholipase A_2 from the snake venom of Crotalus Atrox showed that phosphatidylethanolamine was hydrolysed at a faster rate than the corresponding ethanolamine plasmalogen (Coles et al., 1974). Within the myelin membrane the enzyme had the highest specific activity for phosphatidylserine and the lowest one for ethanolamine containing phosphoglycerides, whereas phosphatidylcholine was cleaved at an intermediate rate. The rates of hydrolysis of the phosphoglycerides associated with the myelin sheath were considerably slower than for the pure phosphoglycerides, and a high enzyme concentration was required to degrade all the phosphatide substrates in myelin. A partial explanation for the lower susceptibility of the myelin-bound substrates towards the enzyme may lie in the architecture of the myelin membrane, where different phosphatides are closely linked to myelin proteins. The products of the enzymic reaction, saturated and unsaturated fatty acids and lysophosphatides, which remain associated with the membrane, may alter the properties of the myelin sheath or may have a direct action on the enzyme itself (Coles et al., 1974).

In a biochemical study on myelin in Wallerian degeneration of the rat optic nerve Bignami and Eng (1973) demonstrated that the yield of myelin from the degenerated nerves was decreased, but the isolated myelin appeared to be morphologically normal. The proportion of cholesterol in the myelin lipids was slightly increased, whereas that of the ethan - olamine phosphoglycerides was decreased and galactolipids were normal. Measuring lipid and protein changes in sciatic nerve during Wallerian degeneration, Wood and Dawson (1974) observed cholesterol esters 3 days after sectioning, while cholesterol and probably cerebrosides were reduced. The loss of phospholipids from the nerve membrane was a later event, whereas the conversion of membrane cholesterol to cholesterol esters appeared to be a very early event. There is some indication that lipid changes during Wallerian degeneration may be preceeded by some breakdown of the myelin proteins, but it is not known, if protein breakdown simply precedes or actually allows the initiation of lipid breakdown. In an investigation on the activities of glycerophosphatide hydrolyzing enzymes, Webster (1973) measured the phospholipase A_1 and A_2 activities in normal and sectioned rat sciatic nerve; the presence of significant phospholipase A activities in the normal nerve and the very early increases in the distal degenerating portion, within two days of transection, suggested that both types of enzyme are associated, at least partially, with Schwann cells (Webster, 1973). It is possible, however, that some of the increase is due to macrophages invading the nerve at a later stage after sectioning, and which are known to exhibit phospholipase A activity.

The lysophosphatides, produced by the action of phospholipase A during demyelination, may play an important role in the break-up of the myelin sheath by its cytolytic and solubilizing properties. Normal peripheral nerve can reacylate lysophosphatidylcholine in vitro and it may be that lysophosphatides produced endogeneouly in degenerating nerve are acylated more actively than under normal conditions, thus preventing their accumulation. In this connection it is of considerable interest that studies on the effect of lysophosphatides after intraneural injection of very small amounts of lysophosphatidylcholine have shown a complete demyelination.

Both glia and neuronal-cell enriched fractions have been shown by Woelk et al. (1973) to contain phospholipase A_1 and A_2 activities. Neuronal and glial phospholipase A_1 had optimal activities at pH 7.2 and 5.4, respectively, whereas phospholipase A_2 activities in neurons and glial cells were optimal at pH 5.4 and 8.0, respectively. Pronounced differences in the enzyme activities of phospholipase A_1 and A_2 were found in neurons and glia, phospholipase A_1 activity being 8-fold and A_2 activity 5-fold higher in neurons than in glia. Determination of kinetic constants (K_m and V) of the neuronal phospholipase A_1 acting on different phosphatide substrates showed that the enzyme was most active with phosphatidylcholine, whereas phosphatidylserine was less extensively hydrolysed. Taking into consideration the different pH optima described for the brain phospholipases A (Woelk and Porcellati, 1973; Woelk et al., 1973) it is quite possible that each enzyme activity has a distinct cellular and subcellular localization. The acidic phospholipase A_1 may be localized in glial lysosomes and may play an important role in the pathogenesis of demyelination, since invasion and hypertrophy of glial cells has been observed in demyelinating conditions (Nevalainen, 1972). Several studies on the activity of lipid-hydrolyzing enzymes in the course of demyelinating disorders have been published in the last years (Woelk and Kanig, 1974; Woelk and Peiler-Ichikawa, 1974). In order to

TABLE 3

PHOSPHOLIPASE A_2 ACTIVITY IN NORMAL AND MULTIPLE SCLEROSIS BRAINS TOWARDS DIFFERENT CHOLINE GLYCEROPHOSPHATIDES

Substrate	Control brains	MS-brains	Difference (%)
1-Alk-1'-enyl-2-(^{14}C)linoleoyl-sn-glycero-3-phosphorylcholine	82.6 ± 1.4	122.8 ± 3.0	+ 48.7
1-Alkyl-2-(^{14}C)linoleoyl-sn-glycero-3-phosphorylcholine	84.2 ± 1.7	128.4 ± 1.8	+ 52.5
1-Alk-1'-enyl-2-(^{14}C)linolenoyl-sn-glycero-3-phosphorylcholine	78.5 ± 2.0	119.1 ± 2.5	+ 51.7
1-Alk-1'-enyl-2-(^{14}C)arachidonoyl-sn-glycero-3-phosphorylcholine	53.4 ± 1.1	81.4 ± 1.5	+ 52.4

Values are expressed as nmol fatty acid hydrolysed × mg^{-1} prot. × h^{-1}. Each value is the mean of 9 brains ± standard error of the mean.

investigate the phospholipid metabolism in experimental allergic encephalomyelitis (EAE), Woelk and Kanig (1974) measured the activity of brain phospholipase A_1 against various labelled glycerophosphatide substrates. In the acute stage of the demyelinating disorder a significant increase in the enzyme activity could be observed. The enhanced phospholipase A_1 activity was of the same order of magnitude for all glycerophosphatides investigated (Woelk and Kanig, 1974). Somewhat conflicting results have been obtained regarding arylsulphatase A activity in the rat central nervous system during EAE. A two- to threefold increase of arylsulphatase A activity in the preacute stage of this demyelinating disease with a 60 per cent decrease of the sulphatide content in the acute stage of EAE has been observed. On the other hand, Maggio et al. (1973) did not find any significant increase in the enzyme activity in the preacute stage of the disease, whereas an increase of about 20 per cent in the acute stage of the demyelinating disease was found. Choline plasmalogen and the corresponding alkyl-ether, specifically labelled at the two position with different fatty acids, were subjected to hydrolysis by phospholipase A_2 prepared from multiple sclerosis and control brains (Woelk and Peiler-Ichikawa, 1974). Alkyl- and alkenyl-derivatives were hydrolysed at similar rates by the enzyme and in brain tissue from multiple sclerosis patients an increase of approximately 50 per cent in the phospholipase A_2 activity was observed for all substrates investigated (Table 3). The significant decrease of the myelin typical lipid ethanolamine plasmalogen in white matter and myelin from multiple sclerosis patients may be explained, at least in part, by the enhanced phospholipase A_2 activity observed during this demyelinating disorder (Woelk and Peiler-Ichikawa, 1974).

Conclusions and Summary

Though a part of the brain neutral proteinase and phospholipase A_2 activity seems to be closely associated with the myelin membrane, the degradation of the bulk of myelin proteins and lipids is probably mediated by proteinases and lipid-hydrolyzing enzymes of lysosomal origin. Experimental evidence has been forwarded by electron microscopy that in some areas cytoplasmic spaces with lysosomal properties are present within the myelin membrane. The greater part of the hydrolytic activity being responsable for the degradation of myelin proteins and lipids, might be mediated, however, by an opening along the tightened myelin sheaths and infusion of cytoplasm and lysosomes from oligodendroglial cells or macrophages.

Myelin basic protein seems to be located exclusively on the interior or cytoplasmic site, with other myelin proteins localized on the outer surface of the membrane. Acidic lipids such as phosphatidylinositol, phosphatidylserine and sulfatides have been found as components of isolated myelin basic protein and it has been demonstrated that these lipids protect the protein from hydrolysis by proteinases. On the other hand, the rates of the phospholipase A_2 catalyzed hydrolysis of phosphoglycerides associated with the myelin sheath were considerably slower than for the pure, isolated substrates. A partial explanation for the lower susceptibility of the myelin-bound substrates towards the enzyme may lie in the architecture of the myelin membrane, where different phosphatides are closely linked to myelin proteins.

In demyelinating conditions it has been observed that lipid changes of the myelin membrane may be preceeded by some breakdown of the myelin proteins, but as yet it is not known, if protein breakdown simply precedes or actually allows the initiation of lipid breakdown. There is now good evidence for the cooperative action of proteinases and lipid-hydrolyzing enzymes in the breakdown of the myelin sheath.

The decrease of phosphatides including ethanolamine plasmalogen and the production of lysophosphatides as well as the loss of fatty acids in the white matter and myelin in the course of demyelinating diseases may be due to increased phospholipase A_1, A_2 and plasmalogenase activity. The competitive inhibition of the phospholipase A_1 catalyzed hydrolysis of phosphatidylcholine by the vinyl-ether groups of the corresponding plasmalogen (Woelk and Porcellati, 1973) may indicate a possible role of this myelin typical lipid compound. This is particularly important for the high content of ethanolamine plasmalogen in myelin and white matter and for its possible protective role in maintaining a certain degree of metabolic stability of myelin itself by competitively inhibiting the hydrolysis of the various diacyllipids by phospholipases A.

References

Agrawal, H.C., Fujimoto, K. and Burton, R.M. (1976) Accumulation and turnover of the classical Folch-Lees proteolipid proteins in developing and adult rat brain, Biochem. J. 154: 265-269.

Bignami, A and Eng, L.F. (1973) Biochemical studies of myelin in Wallerian degeneration of rat optic nerve, J. Neurochem. 20: 165-173.

Coles, E., MCIlwain, D.L. and Rapport, M.M. (1974) The activity of pure phospholipase A_2 from CROTALUS ATROX venom on myelin and on pure phospholipids, Biochim. Biophys. Acta 337: 68-78.

Davison, A.N. and Cuzner, M.L. (1977) Immunochemistry and biochemistry of myelin, Brit. Med. Bull. 33: 60-66.

Jungalwala, F.B. and Dawson, R.M.C. (1971) The turnover of myelin Phospholipids in the adult and developing rat brain, Biochem. J. 123: 683-693.

Maggio, B., Maccioni, H.J. and Cumar, F.A. (1973) Arylsulphatase A activity in rat central nervous system during experimental allergic encephalomyelitis, J. Neurochem. 20: 503-510.

Mehl, E. and Wolfgram, F. (1969) Myelin types with different protein components in the same species, J. Neurochem. 16: 1091-1097.

Nevalainen, T.J., Riekkinen, P.J., Rinne, U.K., Frey, H.J. and Arstila, A.U. (1972) Electron microscopic and neurochemical studies on demyelination in subacute sclerosing panencephalitis, Europ. Neurol. 7: 297-312.

Poduslo, S.E. and Norton, W.T. (1972) Isolation and some chemical properties of oligodendroglia from calf brain, J. Neurochem. 19: 727-736.

Riekkinen, P.J., Clausen, J. and Arstila, A.U. (1970) Further studies on neutral proteinase activity of CNS myelin, Brain Res. 19: 213-227.

Riekkinen, P.J., Rinne, U.K., Arstila, A.U., Kurihara, T. and Pelliniemi, T.T. (1972) STUDIES ON THE PATHOGENESIS OF MULTIPLE SCLEROSIS 2',3'-cyclic nucleotide 3-phosphohydrolase as a marker of demyelination and correlation of findings with lysosomal changes, J. neurol. Sci. 15: 113-120.

Röyttä, M., Frey, H., Riekkinen, P.J., Laaksonen, H. and Rinne, U.K. (1974) Myelin Breakdown and basic protein, Exp. Neurol. 45: 174-185.

Sabri, M.I., Bone, A.H. and Davison, A.N. (1974) Turnover of myelin and other structural proteins in the developing rat brain, Biochem. J. 142: 499-507.

Sammeck, R. and Brady, R.O. (1972) Studies of the catabolism of myelin basic proteins of the rat IN SITU and IN VITRO, Brain Res. 42: 441-453.

Waehneldt, T.V. and Mandel, P. (1972) Isolation of rat brain myelin, monitored by polyacrylamide gel electrophoresis of dodecyl sulfate-extracted proteins, Brain Res. 40: 419-436.

Webster, G.R. (1973) Phospholipase A activities in normal and sectioned rat sciatic nerve, J. Neurochem. 21: 873-876.

Woelk, H. and Porcellati, G. (1973) Subcellular distribution and kinetic properties of rat brain phospholipases A_1 and A_2, Hoppe-Seyler's Z. Physiol. Chem. 354: 90-100.

Woelk, H., Goracci, G., Gaiti, A. and Porcellati, G. (1973) Phospholipase A_1 and A_2 activities of neuronal and glial cells of the rabbit brain, Hoppe-Seyler's Z. Physiol. Chem. 354: 729-736.

Woelk, H. and Kanig, K. (1974) Phospholipid metabolism in experimental allergic encephalomyelitis: activity of brain phospholipase A_1 towards specifically labelled glycerophospholipids, J. Neurochem. 23: 739-743.

Woelk, H. and Peiler-Ichikawa, K. (1974) Zur Aktivität der Phospholipase A_2 gegenüber verschiedenen 1-Alk-1'-enyl-2-acyl und 1-Alkyl-2-acyl-Verbindungen während der Multiplen Sklerose, J. Neurol. 207: 319-326.

Wood, J.G. and Dawson, R.M.C. (1974) Lipid and protein changes in sciatic nerve during Wallerian degeneration, J. Neurochem. 22: 631-635.

Wood, J.G., Dawson, R.M.C. and Hauser, H. (1974) Effect of proteolytic attack on the structure of CNS myelin membrane, J. Neurochem. 22: 637-643.

CELLULAR AND HUMORAL IMMUNITY IN DEMYELINATING DISEASES. MULTIPLE SCLEROSIS AND EXPERIMENTAL ALLERGIC ENCEPHALOMYELITIS AS A 'MODEL'

E. A. Caspary
M.R.C. Demyelinating Diseases Unit, Newcastle General Hospital,
Westgate Road, Newcastle upon Tyne NE4 6BE, U.K.

Introduction

Demyelination in the human central nervous system (CNS) can be considered either as primary or secondary to the disease process. In some conditions though the initial clinical symptoms may arise as a consequence of one pathological mechanism, e.g. inflammation which could arise from a multiplicity of causations. This may be followed by demyelination consequent on the primary damage and leading ultimately to permanent loss of neurotransmission in the affected areas. In man few of these conditions can be attributed to specific causes with the exception of demyelination in disorders of metabolism such as subacute combined degeneration of the cord or arising from a variety of chemical toxins or infection with neurotropic viruses.

Multiple sclerosis (MS) described over 100 years ago by Cruveilhier (1829) and comprehensively reviewed by others since (Vinken and Bruyn, 1970; McAlpine et al., 1972) is by far the commonest of the demyelinating diseases in the Western world though its cause remains unknown. At present it is suggested that the process is auto-immune to specific CNS antigens as in post-rabies vaccine encephalomyelitis (Hurst, 1932; Finley, 1938; Rivers et al., 1933) or immune mediated with or without CNS antigen specificity (Wisniewski, 1977) both possibly initiated by viral infection (reviewed Johnson, 1975) followed by an immune response to CNS destruction by those agents or their expression as a "foreign" antigen on the cell surface. There is little evidence of direct viral attack in MS even though viruses or virus-like particles have been reported (Prineas, 1972; ter Meulen et al., 1972) and therefore only genomic incorporation of a virus, or any virus, appears as the most rational mechanism. In this context the association with the major histocompatibility antigens (Alter et al., 1976; Editorial, 1976) especially the finding that the same clinical disease may be associated with different genetic markers on a geographical basis (Kurdi et al., 1977).

The relation of the rarer demyelinating diseases to MS presents a major problem. It would not be unreasonable to consider neuromyelitis optica (Devic's disease) (reviewed Clays and Netsky, 1970) and concentric sclerosis (reviewed Courville, 1970), as special variants of MS. Similarly myelinoclastic diffuse sclerosis (Schnilder's disease) (Schilder, 1912; Poser, 1957) with other 40% of patients under the age of 10 may fit in as a juvenile form of MS. On the other hand progressive multifocal leukoencephalopathy (reviewed Richardson, 1970) occurs primarily with lymphoproliferative diseases (Hodgkins; leukaemia) and appears to be associated with the presence of viruses such as SV40. Some demyelination may also occur in subacute sclerosing panencephalitis (see Johnson, 1975; Katz and Koprowski, 1973) in association with high measles antibody in serum and cerebrospinal fluid and is presumably caused by measles variants following an earlier measles infection. Similar viral diseases are also seen in rare patients on cytotoxic therapy for leukaemia, etc., or on severe immunosuppression.

Allergic encephalomyelitis and its animal experimental model have been comprehensively reviewed by Alvord (1970) and has formed the basis for much of the immune approach to demyelination.

A number of similar disorders occur in animals (reviewed Fauchiger and Fankhauser, 1970). One of these, old dog distemper has recently come into some prominence in view of the suggested association between distemper and MS (Cook et al., 1978; Krakowka and Koestner, 1978) though this is not universally accepted.

Immune mediated demyelination in peripheral nerves as exemplified in the Guillain-Barre syndrome and the model disease of experimental allergic neuritis will be considered only briefly.

The present review proposes to examine the cellular and humoral immunology of CNS demyelinating disease primarily using MS and the model disease experimental allergic encephalomyelitis (EAE) as examples.

Possible Mechanism of Demyelination

In allergic demyelination following a response to CNS antigens usually in Freund's complete adjuvant the best information is obtained by observing the morphological changes in sequence (Wisniewski et al., 1969). Demyelination follows the appearance of inflammatory cells and consists in essence of the direct attack by mononuclear cells (macrophages) on apparently normal nerve fibres. The pseudopodia of these cells then split off the myelin lamellae leading to a picture of naked axons surrounded by myelin laden macrophages. The process initiating this attack on apparently normal fibres is not understood though it relates to the nature and composition of the experimental antigenic challenge.

While it is evident that little or no demyelination occurs with myelin basic protein (MBP) gross myelin damage results from immunisation with homologous spinal cord (Wisniewski and Keith, 1977). Direct attack by macrophages is also seen in Wallerian peripheral demyelination though again the initiating process remains obscure (Madrid and Wisniewski, 1976).

Viruses may induce demyelination in several ways. Firstly by direct attack as in PML on the myelin supporting cells leading to a form of 'metabolic' demyelination. All the remaining mechanisms appear to contain an immune vector.

Myelin may be damaged by immune response to virus in or on oligodendroglial cells, a 'bystander' reaction to free virus, immune attack on modified surface antigens on the nerve cell arising from genomic viral nucleic acid or less likely self-sensitisation to CNS arising from tissue destruction. Several of these factors could operate simultaneously.

Immune demyelination by pure 'bystander' disease in response to local reactions with any antigen remains a final possibility though this is again most likely to fit into the virus category. This response has been demonstrated in the rabbit eye model (Wisniewski and Bloom, 1975) though attempts to produce the effect within the CNS have been unsuccessful to date (Caspary, unpublished).

In summary the pathological mechanism of demyelination with some obvious exceptions, e.g. PML will contain an immune element.

Multiple Sclerosis

An immune mediated pathogenesis for this disease has been the underlying hypothesis for much research on MS over many years. The decision as to whether this is autoimmune or a response to induced 'non-self' remains to be made, however the end stage mechanism appears the same and further studies offer some hope of possible therapeutic intervention in at least part of the progressive pathological course of MS.

Cellular Immunity in MS

Variations in the proportion of T and B lymphocytes in MS have been described though again there is no universal agreement and differences between controls and patients are small even if statistically significant (Lisak et al., 1975; Oger et al., 1975; Symington et al., 1978). It would appear that the number of cells with B-cell markers is slightly increased particularly during clinical relapse (Lisak et al., 1975) while the T cell population forming E rosettes remains unchanged. The only abnormality in the latter may lie in the number of receptor sites since they bind fewer red cells in rosettes (Oger et al., 1975). Santoli et al. (1978) have shown that about half their patients had an excess of a T sub-population (T-G cells) able to bind Ig-G immune complex and point out that this also occurs in viral infection or as a result of a strong immunological stimulus. They also suggest that the T-G cells may split into two groups respectively with cytotoxic or suppressor function. Final clarification of these points must await identification of specific cell markers and related immune function.

This must be considered from two aspects, immunity to brain or CNS derived antigens and the ability of lymphoid cells to response to non-specific mitogens or non-self antigens, e.g. purified protein derivative of tubercle, viral antigens, etc.

Earlier work on blastogenic lymphocyte response to CNS antigen is reviewed by Lumsden (1972) and even with more refined methodology the situation has not changed appreciably. Both negative and positive findings have been reported in blastogenic transformation (Hughes et al., 1968; Dan and Peterson, 1970; Behan et al., 1972; Symington et al., 1978) using MBP as stimulating antigen and even here differences are small and inconsistent. When patients are followed sequentially there is no significant variation in the response to MBP with the clinical conditions of the patients (Wilcox and Caspary, unpublished). Webb et al. (1974) claim that a group of MS patients not responsive to steroid treatment had positive transformation with MBP while those responding to treatment had no lymphocyte sensitisation. It would therefore appear that these findings are not significant in the pathogenesis of MS.

Lymphocyte responses to mitogen, in particular, phytohaemagglutinin (PHA) also give discordant results in MS. Reports of depressed T lymphocyte function with PHA particularly in active phases of MS (Jenson, 1968), have not been confirmed by other workers (Knight et al., 1975; Lamoureux et al., 1976; Symington et al., 1978). The ratio of stimulation by pokeweed mitogen (T-plus B-cell responses) to PHA response has been claimed to be elevated (Arnason, 1975) though this was not confirmed in the studies of Knight (1977). Thus the functional aspects of circulating lymphocytes can be accepted as essentially within normal limits and again do not vary with the clinical condition (Wilcox and Caspary, unpublished).

The mixed lymphocyte reaction (MLR) appears to be reduced in MS (Hadberg et al., 1971, 1973; Cazzulo and Severaldi, 1974; Knight et al., 1975; Knight, 1977) similar to that found in rheumatoid arthritis. One may then speculate that a disease process may relate to some alteration at recognition sites leading to defective function such as inability to recognise 'self'.

A number of studies on cellular immunity to viral antigens suggest some deficit in response to these and focussed attention on suppression systems. Utermohlen and Zabriskie (1973) who showed while reactivity to measles was suppressed MS leukocytes these reacted normally with other viral antigens. They suggest that this may result from loss of measles responding lymphocytes or, more likely, relate to the specific genetic constitution of MS patients in terms of HL-A markers. This approach was taken one stage further by Arnason and Antel (1978) who demonstrated true S-cell variations at the beginning and end of a clinical attack. Their test system measures a more general effect not related to any specific antigen and may tempt one to the speculation that even a temporary variation in immuno-regulation (S-cells?) may initiate a destructive attack in a primed system.

Cellular immune function in MS has been examined by a number of methods (see Lisak, 1975; Knight, 1977) but only two have suggested either a direct relationship to the clinical course of the disease or diagnostic significance.

Mertin et al. (1973) in a preliminary communication showed that polyunsaturated fatty acids exerted a greater inhibitory effect on MS lymphocytes than those from other neurological disorders not related to the nature of the stimulating antigen and measured by the macrophage electrophoretic mobility test. This sensitive method had previously failed to detect any difference in response to CNS antigen (MBP) in patients and controls (Caspary and Field, 1970). The underlying hypothesis of the preliminary work was that unsaturated fatty acid as a prostaglandin precursor had a weak immunoregulatory action. Experiments using graft rejection have yielded discordant results in the hands of different workers and in EAE the inhibitory effect is small (Meade et al., 1978; Hughes, Caspary and Mertin, unpublished). However blast transformation experiments using mitogen (PHA) or antigen (PPD) clearly showed an inhibition of lymphocyte response in vitro (Mertin and Hughes, 1975). The fatty acid studies were extended by two other groups of workers (see Field and Shenton, 1975) showing high disease specificity though a double-blind trial (Mertin and Caspary, 1975) failed to confirm this. It is of interest that the later workers based their investigation on R. H. S. Thompson's hypothesis of an inborn error in handling unsaturated fatty acids leading to systemic abnormal cell membrane composition. However this latter hypothesis has not gained universal acceptance.

Leucocyte sensitivity as measured by migration inhibition to CNS antigen has also been comprehensively investigated in recent years. The measured response is increased in MS, especially around the period of exacerbation and is distinguished from a similar increase in cases of stroke by its persistence (Rocklin et al., 1971; Sheremata et al., 1977). The essential stimulating antigen appears to be the encephalitogen (MBP) active in primates (Sheremata et al., 1977). It is surprising that this response was not detectable by other methods for detecting direct cellular sensitisation though it could reflect an activated immune status in the lymphocyte population. To date these experiments have not been confirmed.

Humoral immunity in MS

The vast bulk of early studies seeking circulating antibody to CNS antigens is reviewed by Lumsden (1972). The results are variable, not only by using different methodology but indeed between workers using the same techniques. A more recent review (Caspary, 1977) brings this further up to date. To add to the difficulties antibody to CNS can be found in normal subjects (Field et al., 1963; Uyeda and Murphy, 1976) and it therefore is unlikely to be of sole pathogenic significance. The much more precise and sensitive radioimmunoassay methods have now added to our knowledge of specific humoral response to MBP in MS, several workers failed to detect any difference from controls and a recent study by Biggins and Caspary (unpublished) showed no relation between titre and clinical condition in sequential samples. Conventional methods of antibody detection appear of little value in MS and some form of functional testing is required. The use of the rabbit eye demyelinating model with patients serum and/or cells (Wray et al., 1975) may prove to be a valuable approach when related to the demonstration of antibody mediated cytotoxicity (Stoner et al., 1977) in the same model. There is a need for a true in vitro cytotoxic assay to simplify and quantitate this system. The vast bulk of recent serology in MS has been to viral antibody and though of import for the pathogenesis of the disease probably does not relate to demyelination (Fraser, 1977). Bornstein (1973) has summarised some of these functional aspects of 'antibody' in relation to organotypic cultures of CNS. These show demyelination and remyelinate when serum is removed similarly serum produces a transmission defect in culture. Other humoral factors acting non-specifically on cellular immune responses remain ill defined at present.

Experimental allergic encephalomyelitis (EAE) - a 'model' for MS

The pathology of rabies vaccine encephalomyelitis and the similarity of the lesions to those of MS has given a marked stimulus to this investigative model (Kies and Alvord, 1959). In brief EAE now falls into 3 categories. Firstly the acute form which occurs as a single clinical episode with recovery depending on severity. The pathology is essentially perivascular inflammatory cuffing with minimal myelin damage in rodents though monkeys for example show demyelinated plaques. A chronic form was produced by Stone and Lerner (1965) and this was later modified to give a relapsing form (Wisniewski and Keith, 1977) as a closer clinical parallel to MS. In general the slow forms of EAE show greater demyelination to the extent of naked eye lesions in the relapsing form as well as extensive inflammatory perivascular change. The clinical presentation remains that of EAE, thus the individual relapses of the relapsing form appear like individual attacks of acute disease and recover at the same rate (Keith - personal communication). The disease progresses through several relapses and remissions and then appears to burn itself out leaving the subject in a static neurological condition. The relation between myelin damage and symptoms could be taken ultimately as permanent blockage permitting by-pass and showing some remyelination (Keith and Eastman - personal communication). It thus follows that as in acute EAE relapsing disease is inflammatory with fast onset and recovery, indeed it parallels optic (retrobulbar) neuritis in this respect. To date no manipulation or therapy has been successful in either causing or preventing relapses, on the other hand treatment with MBP on a synthetic polymer (Arnon, 1975) reduces the severity of the disease (Keith - personal communication). Although it is possible to induce relapsing disease in random-bred animals this has high mortality and implies that a suitable genetic background is required, perhaps on a parallel with human disease. Most of the experimental immunology has been studied in acute disease and much remains to be learnt from the new model.

Immunology of EAE

Much of the background of this subject is covered by the excellent and comprehensive reviews of Paterson (1965) and Alvord (1970). In brief EAE can be induced by a small peptide (Eylar et al., 1972) derived from the basic protein of myelin (see Kies et al., 1965) as well as MBP, myelin or myelin containing CNS homogenates. Disease severity follows the same order (McDermott and Caspary, 1976). Animals treated or pretreated with antigen show a protective effect and if treated at birth may acquire tolerance (Paterson, 1965). Purified myelin basic protein has been the antigen of choice for many immunological studies.

Cellular sensitisation as measured by antigen inhibition of migration of peritoneal macrophages rises after sensitisation with or without adjuvant and persists; it does not however relate to clinical disease (see Hughes and Newman, 1968). Studies on blast transformation tend to give small responses with MBP and although sensitisation can be demonstrated one hesitates to attach much significance to these findings. Webb et al. (1973) studied transformation in guinea pigs and were able to show differences between strains - similar differences between the immune responses in resistant and susceptible strains have also been found in guinea pigs (Kies et al., 1975) and rats (McFarlin et al., 1975). Allbritton and Loan (1975) claimed that a cytotoxic assay using spleen cells correlated with disease though perhaps this may reflect the rise in cellular and humoral immunity to a plateau level following challenge (Caspary, unpublished). Other immune responses consequent on variation in species, adjuvants or techniques have offered more data, but not directly related to disease. Perhaps the most important information in EAE is that the disease may be passively transferred with lymph node cells but not with serum, further that the required cell is the T cell (see Knight, 1977; Caspary, 1977). The very interesting observations of Swierkosz and Swanborg (1975) of immunologically specific control of EAE by S-cells are probably our first break through to the nature of the regulating mechanism in EAE. Their data may also provide the explanation for antigen protection against passive transfer (Driscoll et al., 1975).

EAE and MS

The immunology of these two disorders have so far provided little concrete information either to justify the model or to relate directly to the course of clinical disease. In the model disease is related to T-cell function and controlled by S-cells, though B cell depletion enhances disease and 'antibody' can also influence (inhibit ?) the process. EAE is susceptible to treatment with immunosuppressive drugs and anti-lymphocyte serum and this distinguishes it from MS which in general fails to respond significantly to this form of therapy. The appearance of the early lesion in MS closely resembles that of EAE (Prineas, 1975) and thus provides some morphological justification for the EAE model. The known structure of MBP or peptide induces the clinical signs of inflammatory EAE, demyelination appears to require other constituents of the CNS. It is suggested that in MS a non-self modification fulfills the same function and may perhaps be attributed to a viral infection. The precise stimulus for immune demyelination remains in doubt requiring detailed chemistry of the relevant CNS constituents and their immunological reactions.

References

Allbritton, A. R. and Loan, R. W. (1975) Correlation of cytotoxicity with experimental allergic encephalomyelitis, Cell. Immunol. 19: 91-98.

Alter, M., Harshe, M., Anderson, V. E., Emme, L. and Young, E. J. (1976) Genetic association of multiple sclerosis and HL-A determinants, Neurology 26: 31-36.

Alvord, E. C. (1970) Acute disseminated encephalomyelitis and 'allergic' neuroencephalopathies. Handbook of Clinical Neurology, Vol. 9, Multiple Sclerosis and other demyelinating diseases. North-Holland Publishing Co., Amsterdam, p.500-511.

Arnason, B. G. W. (1975) Histocompatibility testing in multiple sclerosis. Multiple Sclerosis Research, H.M.S.O., London, p.80-84.

Arnason, B. G. W. and Antel, J. (1978) Suppressor cell function in multiple sclerosis. Ann. Immunol. (Inst. Pasteur) 129C: 159-170.

Arnon, R. (1975) Immunological approaches to control of multiple sclerosis desensitisation studies. Multiple Sclerosis Research, H.M.S.O., London, p.271-283.

Behan, P. O., Behan, W. M. H., Feldman, R. E. and Kies, M. W. (1972) Cell mediated hypersensitivity to normal antigens in humans and non-human primates with neurological disease. Arch. Neurol. (Chicago) 27: 145-152.

Bornstein, M. B. (1973) The immunopathology of demyelinative disorders examined in organotypic cultures of mammalian central nerve tissue. Progr. Neuropath. Vol. II (Ed. H. M. Zimmerman), Grune & Stratton, Inc., New York, p.69-89.

Burnett, P. R. and Eylar, E. H. (1971) Allergic encephalomyelitis: oxidation and cleavage of the single tryptophan residue of the A1 protein from bovine and human myelin. J. Biol. Chem. 246: 3425.

Caspary, E. A. (1977) Humoral factors involved in immune processes in multiple sclerosis and allergic encephalomyelitis. Brit. Med. Bull. 33: 50-53.

Caspary, E. A. and Field, E. J. (1970) Sensitisation of blood lymphocytes to possible antigens in neurological disease. Europ. Neurol. 4: 257-266.

Cazzulo, C. L. and Severaldi, E. (1974) Multiple sclerosis immunogenetics a) Association between HL-A antigens and measles antibody levels, b) Existence of common MLR/immune response determinants in MS patients. Bull. Inst. Sterater Milanese 53: 615-624.

Cloys, D. E., and Netsky, M. G. (1970) Neuromyelitis optica. Handbook of Clinical Neurology, Vol. 9, Multiple Sclerosis and Other Demyelinating Diseases. North-Holland Publishing Co., Amsterdam, p.426-436.

Cook, S. D., Dowling, P. C. and Russell, W. C. (1978) Multiple Sclerosis and Canine Distemper, Lancet, March 18, p.605-6.

Courville, C. B. (1970) Concentric sclerosis, Handbook of Clinical Neurology, Vol. 9, Multiple Sclerosis and Other Demyelinating Diseases, North-Holland Publishing Co., Amsterdam, p.437-451.

Cruveilhier, J. (1835-1842) Atlas d'anatomie pathologique, Livre 32, Paris, Balliere, p.38.

Dan, P. C. and Peterson, R. D. A. (1970) Transformation of lymphocytes from patients with multiple sclerosis, Arch. Neurol. (Chicago), 23: 32-40.

Driscoll, B. F., Kies, M. W., Alvord, E. C. (1975) Adoptive transfer of experimental allergic encephalomyelitis (EAE): Prevention of successful transfer by treatment of donors with myelin basic protein, J. Immunol. 114: 1, 2, 291-292.

Editorial (1976) Histocompatibility antigens and Multiple Sclerosis, Lancet ii: p.11.

Field, E. J., Caspary, E. A. and Ball, E. J. (1963) Some biological properties of a highly active encephalitogenic factor isolated from human brain, Lancet 2: 11.

Field, E. J. and Shenton, B. K. (1975) Inhibitory effect of unsaturated fatty acid on lymphocyte antigen interaction with special reference to multiple sclerosis, Acta Neurol. Scand. 52: 121-136.

Finley, K. H. (1938) Pathogenesis of encephalitis occurring with vaccination, variola, and measles, Arch. Neurol. Psychiat. (Chic.) 39: 1047-1054.

Fraser, K. B. (1977) Multiple Sclerosis: a virus disease? Brit. Med. Bull. 33: 1, 34-39.

Frauchiger, E. and Fankhauser, R. (1970) Demyelinating Diseases in Animals. Their relevance to the pathogenesis of multiple sclerosis. Handbook of Clinical Neurology, Vol. 9, Multiple Sclerosis and Other Demyelinating Diseases. North-Holland Publishing Co., Amsterdam, p.664-689.

Hedberg, H., Kallen, B., Low, B. and Nilsson, O. (1971) Impaired mixed leukocyte reaction in some different diseases notably multiple sclerosis and various arthritides, Clin. exp. Immunol. 9: 201-207.

-ibid.- (1973) Further studies on the mixed leucocyte reaction in multiple sclerosis and various arthritides. Scand. J. Immunol. 2: 291-297.

Hughes, D., Caspary, E. A. and Field, E. J. (1968) Lymphocyte transformation induced by encephalitogenic factor in multiple sclerosis and other neurological diseases, Lancet ii: 1205-1207.

Hughes, D. and Newman, S. (1968) Lymphocyte sensitivity to encephalitogenic factor in guinea pigs with experimental allergic encephalomyelitis as shown by *in vitro* inhibition of macrophage migration, Int. Arch. Allergy 34: 237-256.

Hurst, E. W. (1932) The effects of the injection of normal brain emulsion into rabbits, with special reference to the aetiology of the paralytic accidents of anti-rabic treatment, J. Hyg. (Lond.) 32: 33-34.

Jensen, M. K. (1968) Lymphocyte transformation in multiple sclerosis.

Acta Neurol. Scand. 44: 200-206.

Johnson, R. T. (1975) Virological data supporting the viral hypothesis in multiple sclerosis, Multiple Sclerosis Research, H.M.S.O., London, p.155-177.

Katz, M. and Koprowski, H. (1973) The significance of failure to isolate infectious viruses in cases of subacute sclerosing panencephalitis, Archiv. für die gesamte Virusforsch. 41: 390-393.

Kies, M. W. and Alvord, E. C. (1959) 'Allergic' encephalomyelitis, Springfield, Ill., Thomas.

Kies, M. W., Driscoll, B. F., Lisak, R. P. and Alvord, E. C. (1975) Immunologic activity of myelin basic protein in strain 2 and strain 13 guinea pigs, J. Immunol. 115: 1, 75-79.

Kies, M. W., Thompson, E. B. and Alvord, E. C. (1965) The relationship of myelin proteins to experimental allergic encephalomyelitis, Ann. N.Y. Acad. Sci. 124: 148-160.

Knight, S. C. (1977) Cellular immunity in multiple sclerosis, Brit. Med. Bull. 33: 45-49.

Knight, S. C., Lance, E. M., Abbosh, J., Munro, A. and O'Brien, J. (1975) Intensive immunosuppression in patients with disseminated sclerosis, Clin. exp. Immunol. 21: 23-31.

Krakowka, S. and Koestner, A. (1978) Canine distemper virus and multiple sclerosis, Lancet, May 27th, p.1127-8.

Kurdi, A., Ayesh, I., Aboaliat, A., McDonald, W. I., Compston, D. A. S. and Bachelor, J. R. (1977) Different B lymphocyte alloantigens associated with multiple sclerosis in Arabs and North Europeans, Lancet i: 1123-1124.

Lamoureux, G., Giard, N., Jolicoeur, N., Toughlian, V. and Desrosiers, M. (1976) Immunological features in multiple sclerosis, Brit. Med. J. i: 183-186.

Lisak, P. P. (1975) Multiple sclerosis: immunologic aspects, Ann. clin. lab. Science 5: 324-329.

Lisak, R. P., Levinson, A. L., Zweiman, B. and Abdou, N. I. (1975) T and B lymphocytes in multiple sclerosis, Clin. exp. Immunol. 22: 30.

Lumsden, C. E. (1972) The Clinical Pathology of Multiple Sclerosis, Multiple Sclerosis: A Reappraisal, Churchill Livingstone, p.311-621.

Madrid, R. E. and Wisniewski, H. M. (1977) Axonal degeneration in demyelinating disorders, J. Neurocytol. 6: 103-117.

McAlpine, D., Lumsden, C. E. and Acheson, E. D. (1972) Multiple Sclerosis: A Reappraisal, Churchill Livingstone, Edinburgh & London.

McDermott, J. R. and Caspary, E. A. (1975) A comparison of the encephalitogenic activities of human myelin basic protein and the synthetic peptide determinant comprising residues, J. Neurochem. 25: 711-713.

Meade, C. J., Mertin, J., Sheena, J. and Hunt, R. (1978) Reduction by linoleic acid of the severity of experimental allergic encephalomyelitis in the guinea pig, J. Neurol. Sci. 35: 291-308.

Mertin, J. and Hughes, D. (1975) Specific inhibitory action of polyunsaturated fatty acids on lymphocyte transformation induced by PHA and PPD, Int. Arch. All. 48: 203-210.

Mertin, J., Shenton, B. K. and Field, E. J. (1973) Unsaturated fatty acids in multiple sclerosis, Brit. Med. J. 2: 777-778.

McFarlin, D. E., Hsu, S. C-L., Slemenda, S. B., Chou, F. C-H., and Kibler, R. F. (1975) The immune response against myelin basic protein in two strains of rat with different genetic capacity to develop experimental allergic encephalomyelitis, J. Exp. Med. 141: 1, 72-81.

Oger, J. F., Arnason, B. G. W., Wray, S. H. and Kistler, P. (1975) A study of B and T cells in multiple sclerosis, Neurology 25: 444.

Paterson, P. Y. (1966) Experimental allergic encephalomyelitis and autoimmune disease, Advances in Immunology, Academic Press, New York, Vol. 5, p.131-208.

Poser, C. M. (1957) Diffuse disseminated sclerosis in the adult, J. Neuropath. exp. Neurol. 16: 61-78.

Prineas, J. (1972) Paramyxovirus-like particles associated with acute demyelination in chronic relapsing multiple sclerosis, Science 178: 760-763.

Prineas, J. (1975) Pathology of the early lesion in multiple sclerosis, Human Pathology 6: 5, 531-554.

Richardson, E. P. (1970) Progressive multifocal leukoencephalopathy, Handbook of Clinical Neurology, Vol. 9, Multiple Sclerosis and Other Demyelinating Diseases, North-Holland Publishing Co., Amsterdam, p.485-499.

Rivers, T. M., Sprunt, D. H. and Berry, G. P. (1933) Observations on attempts to produce acute disseminated encephalomyelitis in monkeys. J. exp. Med. 58: 39-53.

Rocklin, R. E., Sheremata, W. A., Feldman, R. G., Kies, M. W. and David, J. R. (1971) Guillain-Barre and multiple sclerosis - *in vitro* cellular responses to nervous tissue antigens. New Eng. J. Med. 283: 804-808.

Santoli, D., Moretta, L., Lisak, R., Gilden, D. and Koprowski, H. (1978) Imbalances in T cell subpopulation in multiple sclerosis patients, J. Immunol. 120: 1369-1371.

Schilder, P. (1912) Zur Kenntris de sogenannten diffusen Sklerose, Z. ges. Neurol. Psychiat. 10: 1-60.

Sheremata, W., Eylar, E. H. and Cosgrove, J. B. R. (1977) Multiple Sclerosis: sensitisation to a myelin basic protein fragment (peptide 7) encephalitogenic to primates, J. neurol. Sci. 32: 255-263.

Sheremata, W., Triller, H., Cosgrove, J. B. R. and Eylar, E. H. (1977) Direct leukocyte migration inhibition by myelin basic protein in exacerbations of multiple sclerosis. Can. med. ass. J. in press.

Stone, S. H. and Lerner, E. M. (1965) Chronic disseminated allergic encephalomyelitis in the guinea pig, Ann. N.Y. Acad. Sci. 124: 227-241.

Stoner, G. L., Brosnan, C. F., Wisniewski, H. M. and Bloom, B. R. (1977)

Studies on demyelination by activated lymphocytes in the rabbit: 1. Effects of a mononuclear cell infiltrate induced by products of activated lymphocytes. 2. Antibody-dependent cell-mediated demyelination, J. Immunol. 118: 2094-2110.

Swierkosz, J. E. and Swanborg, R. H. (1975) Suppressor cell control of unresponsiveness to experimental allergic encephalomyelitis, J. Immunol. 115: 3, 631-633.

Symington, G. R., Mackay, I. R., Whittingham, S., White, J. and Buckley, J. D. (1978) A profile of immune responsiveness in multiple sclerosis. Clin. exp. Immunol. 31: 141-149.

ter Meulen, V., Koprowski, H., Iwasaki, F., Kackell, Y. M. and Mullar, D. (1972) Fusion of cultured multiple sclerosis brain. Celsl with indicator cells; presence of nucleocapsids and virions and isolation of a parainfluenza type virus, Lancet ii: 1-5.

Utermohlen, V. and Zabriskie, J. B. (1973) A suppression of cellular immunity in patients with multiple sclerosis, J. exp. Med. 138: 1591-1596.

Uyeda, C. T. and Murphy, P. O. (1976) Hypersensitivity to purified brain proteins in healthy individuals, Nature 264: 650-652.

Vinken, P. J. and Bruyn, G. W. (1970) Multiple Sclerosis and Other Demyelinating Diseases, Handbook of Clinical Neurology, North-Holland Publishing Co., Amsterdam, Vol. 9.

Webb, C., Teitelbaum, D., Abramsky, O., Arnon, R. and Sela, M. (1974) Lymphocytes sensitised to basic encephalitogen in patients with multiple sclerosis unresponsive to steroid treatment, Lancet ii: 66-68.

Webb, C., Teitelbaum, D., Arnon, R. and Sela, M. (1973) Correlation between strain differences in susceptibility to experimental allergic encephalomyelitis and the immune response to encephalitogenic protein in inbred guinea pigs, Immunological Communications 2(2), 185-192.

Wisniewski, H. M. (1977) Immunopathology of demyelination in autoimmune diseases and virus infections, Brit. Med. Bull. 33: 1, 54-59.

Wisniewski, H. M. and Bloom, B. R. (1975) Experimental allergic optic neuritis (EAON) in the rabbit, J. Neurol. Sci. 24: 257-263.

Wisniewski, H. M. and Keith, A. B. (1977) Chronic relapsing encephalomyelitis: an experimental model of multiple sclerosis, Ann. Neurol. 1: 144-148.

Wisniewski, H. M., Prineas, J. W. and Raine, C. S. (1969) An ultrastructural study of experimental allergic encephalomyelitis in the peripheral nervous system, Lab. Invest. 21: 105-118.

Wray, S. H., Cogan, D. G. and Arnason, B. G. W. (1974) Paper read at the Association for Research in Vision and Ophthalmology, Sarasota, Fla, 26th April 1976.

IMMUNOLOGICAL STUDIES IN MULTIPLE SCLEROSIS

Alan N. Davison
Department of Neurochemistry, Institute of Neurology, The National
Hospital, Queen Square, London, WC1N 3BG, UK.

I. Introduction

Multiple sclerosis (MS) or disseminated sclerosis is a relatively common neurological disease in which there are various signs of damage to the insulating myelin sheath of the central nervous system. The lesions are characterised by their disseminated nature in both time and space. In Northern Europe the incidence of MS is about 1 per 2,000 of the population; lower rates have been published for southern France, Italy and the Balkans. Data from epidemiological studies on migrants from high-risk areas (e.g. Europe) to low-risk areas (e.g. Israel or South Africa), have been interpreted as showing that the migrant carries with him, after the age of 15, the high risk of his country of origin. This suggests that an infective agent is acquired by MS sufferers during childhood; but alternative explanations should be considered. For example, Orientals and many south Europeans and black Africans may be relatively unsusceptible to the disease (Acheson, 1977).

II. Immunogenetics

There is some evidence, from familial studies, that an inherited factor is implicated in MS, for the proportion of first degree relatives of MS patients is about 20 times that expected from the general population. Progress has been made as a result of the examination of the tissue-antigen types (first identified as transplantation antigens, associated with cell rejection). The specific antigens carried on the human leucocyte (HLA) have been found to be associated with increased susceptibility to different diseases including MS. The gene locus on chromosome six may itself directly control factors affecting disease susceptibility, or the HLA gene may be in linkage disequilibrium with the immune regulatory gene (Ir), and hence reflect a difference in the immune system of the patient. Research on B-lymphocytes has shown that about 83% of MS patients have one of the antigens DRW 2 (BT 101) compared with 33% in the normal population (Compston et al, 1976), so increasing their risk by 9.8 times. Susceptibility to MS (associated with HLA antigen) may, therefore, be due to an inherited defect in immunity.

It is noteworthy that MS patients in Jordan have a different antigen type DRW 4 (BT 102), (Kurdi et al, 1977); this altered distribution probably relates to a different linkage disequilibrium with a specific Ir gene. Such a congenital defect may relate to altered function of T-cells. There is, for example, evidence of a failure of lymphocytes to react against measles (Zabriskie, 1975), and there is an apparent abnormality in the production of antibody in the CSF. However, the low concordance rate in monozygotic twins shows that a simple genetic explanation of MS is not applicable; there may in addition to a genetic predisposition, be common exposure to an infective agent.

III. Antibodies in the cerebrospinal fluid (CSF)

Although the total protein content of cerebrospinal fluid is 200 times less than that of serum, the level of immunoglobulin G (IgG) is disproportionately lower (15% in serum, 3% in cerebrospinal fluid).

Since serum proteins are able to leak into spinal fluid as a result of damage to capillaries (e.g. plaques near the ventricles), correction must be made for any such passive transfer by relating IgG values to another high-molecular-weight protein, e.g. macroglobulin or albumin (Tourtellotte, 1975). Having allowed for this, MS is clearly still a disease in which the amount of IgG in the CSF is much higher than amounts that can be explained by simple 'transfer' of the substance from serum. About 75% of MS patients have an increased concentration of IgG in the CSF and this increase persists throughout their illness. It seems probable that the IgG is synthesized within the brain, for many plasma cells, isolated from the CSF, when kept in tissue culture produce antibody similar to that found in the CSF. In the normal subject, the IgG region of separated CSF has a diffuse homogenous appearance, but in MS and in diseases of the nervous system where there is a known 'infectious aetiology' (e.g. measles encephalitis or sarcoid), a restricted heterogeneity of the IgG is seen. These oligoclonal bands of antibody are relatively diffuse in themselves, and probably represent the immunodominant antigens, each of which has stimulated several clones of lymphocytes to produce antibody. In 90% of MS patients, oligoclonal bands can be detected on electrophoresis (Thompson et al, 1978). In addition, in about 53% of patients (Link, 1973) there is an alteration in the ratio of kappa:lambda light chains, suggesting some abnormality in the nature of the IgG. Thus the detection of oligoclonal bands in the CSF is of considerable value, not only in confirming the diagnosis of MS, but also in demonstrating an important immunological component in the pathogenesis.

IV. Viruses and multiple sclerosis

Epidemiological, clinical and serological research suggest that a paramyxovirus might be the causative agent in MS. If there is an infective agent, no one has yet been able to grow it reproduceably in tissue culture, or transfer the disease experimentally to primates. Measles-like inclusions sometimes seen in the glial cells of MS sufferers and rarely in white cells, may well turn out to be artefact. Nevertheless, ultrastructural and biochemical changes are detectable in otherwise apparently normal white matter before overt damage occurs, suggesting early damage to glial cells by a latent infective process. Thus away from the lesion in areas of apparently normal white matter, it has been reported that there is astrocytic hypertrophy (Andrews, 1972; Arstila et al, 1973) in the absence of a cellular reaction. In the areas of white matter Cuzner et al (1976) found increased β-glucuronidase, arylsulphatase and acetylcholinesterase activity, reflecting possible lysosomal enzyme changes in affected glial cells. Comparable alterations are seen in the early stages of scrapie, the slow virus disease of sheep (Kimberlin, 1973; Millson and Bountiff, 1973).

V. Pathology

In areas of established demyelination there is, as expected, loss of myelin proteins and lipids, and increased deposition of glial fibrillary acidic protein derived from activated astrocytes. In active lesions plasma cells (antibody producing lymphocytes) and lipid containing phagocytes are found. They are presumably derived from a zone of lymphocytic and macrophage infiltration around the central vein from which the lesions expand. At the edge of active plaques, one of the earliest biochemical changes occurring is a selective loss of myelin basic protein, due primarily to the action of localised proteinases with the co-operative action of other hydrolases. This loss increases towards the centre of the damage. Since this basic protein (M.Wt. 18,000) and its derived peptides are powerful encephalitogens, its release into the circulation has been thought to sensitize lymphocytes and to be associated with an autoimmune response.

A. Experimental allergic encephalomyelitis (EAE)

About two weeks after injection of the basic protein (or encephalitogenic peptides) with appropriate adjuvants, sharply circumscribed areas of perivascular inflammation are seen within the CNS. The infiltrating cells are predominantly mononuclear with macrophages and lymphocytes and few plasma cells, but in monkeys and dogs polymorphonuclear leucocytes are numerous. Unlike MS, there is little gliosis and relatively minor demyelination. Where inflammatory cells are present, the outstanding pattern of demyelination involves direct attack by mononuclear cells (Wisniewski, 1977), although such attack is rarely seen in MS. Much more comparable lesions to MS are seen in primates and in chronic EAE where there are intermittent relapses and remission of the paralysis (Stone and Lerner, 1965; Wisniewski and Keith, 1977; Raine and Stone, 1977). Antibody to basic protein does not appear until after the clinical response in experimental animals, and indeed recovery appears to follow re-establishment of immunological tolerance to myelin basic protein. Thus, high levels of antibody appear to protect animals (Lennon and Dunkley, 1974). In the CSF of sheep with EAE, antibody to the basic protein does not appear until within two days of the onset of clinical symptoms. Basic protein bound to antibody is found at about the same time (Gutstein and Cohen, 1978). There is evidence that the CSF antibody originates from the blood.

Since EAE can be transferred to donors from lymph node suspension, it is an example of a T-cell mediated condition, so that lymphocytes sensitized to myelin antigens carry the information for initiating the perivascular mononuclear cell infiltration of the CNS. Study of the lymphocytes from the blood of guinea pigs with acute EAE shows that the percentage of early (active or high affinity rosetting) T-cells decreases dramatically during the period of paralysis (Traugott et al, 1978). Early T-cells could be recovered from the CNS of the paralysed guinea pigs, suggesting that their decreased concentration in the blood was due to migration into the target organ, the CNS. However, there does seem to be a humoral factor present in the blood of animals with EAE (and some MS patients), which is able to induce phagocytic activity when added to myelinated explants. Recent work by Stoner et al, (1977) demonstrates that products of activated lymphocytes induce mononuclear cell infiltration in the eye only of rabbits pre-sensitized to spinal cord. Thus, antigenic stimulus of the lymphocyte can be non-specific and need not be restricted to myelin basic protein. In non-sensitized rabbits, demyelination can be induced by injecting lymphocytic products, together with immune serum from an EAE rabbit. These results, therefore, suggest that an antibody-dependent cell mediated mechanism operates in this experimental primary demyelination reaction (Brosnan et al, 1977). Although antibodies to basic protein have often been implicated as an important factor in MS so far, in contradistinction to EAE, such antibody to basic protein has not been detected. For example, Lennon and Mackay (1972) using labelled basic protein as antigen, failed to find antibody in serum or CSF of MS patients. However, antibody directed against oligodendroglia have been found in the sera of 19 out of 21 patients with MS (Abramsky et al,

1977); as yet the relation of this to disease state is not established. This, too, contrasts with EAE in which there is evidence that the antigen is the myelin basic protein rather than a specific oligodendroglial antigen (Raine et al, 1978).

TABLE 1. Lysosomal enzyme activity of mononuclear and polymorphonuclear cells

Cell	Cathepsin D* (pH3.6)	Neutral proteinase* (pH7.6)	β-Glucuronidase* (pH5.5)	Phospholipase A_2* (pH5.5)
Polymorphonuclear leucocyte	0.14	2.09	0.04	+
Macrophage	2.50	0.10	0.20	+++
Lymphocyte	0.16	0.10	0.04	+

*Results are expressed as μmols/mg protein per hour

B. Changes in blood and CSF of MS patients

During a relapse in multiple sclerosis, there are changes in lymphocyte reactivity. However, the proteolytic enzyme activity of lymphocytes is low, and neutral proteinase is primarily found within polymorphonuclear leucocytes (PML) and acid proteinases in macrophages (Table 1). In an exacerbation (Table 2), there is an increase in the specific activity of PML neutral proteinase activity (Cuzner et al, 1975), which is largely soluble enzyme for bound proteinase decreases (Tchorzewski et al, 1976) during an attack. In the CSF we have found a small but significant increase in cell bound neutral proteinase activity probably due to the appearance of a small number of macrophages (or PML) during an exacerbation of the disease, when there is also release of basic protein or its peptides from the degenerating myelin sheath (Cohen et al, 1976). Since one main function of the PML is to digest antigen-antibody complexes, the concentration of circulating immune complex was measured at different stages in the progress of MS. No significant differences were seen in MS compared to control patients (Cuzner et al, 1978). There is, however, evidence of localised immune complex deposition with the central nervous system and continuous endogenous synthesis of IgG. This suggests that an inflammatory reaction is mediated through the presence of deposited immune complex at restricted sites within the CNS. A slightly increased concentration of lysozyme in the CSF of MS patients (Hansen et al, 1977), also indicates continuous inflammatory reaction for this bacteriolytic protein is typically secreted by the monocyte/

TABLE 2. NEUTRAL AND ACID PROTEINASE ACTIVITY IN LEUCOCYTES FROM MULTIPLE SCLEROSIS PATIENTS AND CONTROLS

Subjects	Neutral proteinase (μmol/mg protein /hr/PMN)[a]	Neutral proteinase	Acid proteinase	Protein mg/10⁶ leucocytes	White cells total x 10⁻³.mm⁻³	polymorphs %
Multiple Sclerosis (age 18-54)						
In acute exacerbation (15)	2.01 ± 0.65	17.05 ± 4.39		0.065 ± 0.03	7.68 ± 2.68	63.7
In remission (25)*	1.67 ± 0.50	12.20 ± 3.65		0.063 ± 0.03	7.95 ± 1.72	64.5
Progressive (5)*	2.25 ± 0.31	16.14 ± 4.31		0.049 ± 0.02	6.37 ± 1.69	60.5
Other Neurological Diseases (age 11-66) (25)*	1.50 ± 0.42	11.18 ± 2.99		0.067 ± 0.03	8.30 ± 3.77	64.2
Rheumatoid Arthritis (age 40-60) (19)*	1.26 ± 0.48	9.11 ± 3.49		0.037 ± 0.02	7.90 ± 2.39	71.0
Normal Controls (age 20-50) (10)*	1.54 ± 0.24	10.27 ± 2.22		0.041 ± 0.01	6.88 ± 2.83	54.0

a = Neutral proteinase specific activity is corrected for the proportion of PMN cells, as the enzyme is concentrated in these cells. Acid proteinase activity is uniformly distributed between lymphocytes and PMN cells.
* = Number of patients.
(Data after Cuzner, et al., 1978)

macrophage system, and by polymorphonuclear leucocytes. In addition, there is recent evidence of increased T-cell activity in the CSF of MS patients only during or close to an exacerbation (Allen et al, 1976; Naess and Nyland, 1978).

VI. Therapy

It may be possible to use drugs to modify the immunological reaction in those instances where perivascular cuffing and cellular infiltration is associated with demyelination. Thus, immunosuppression has been attempted as a rationale therapy, and more recently there has been interest in the possibility of using anti-inflammatory drugs as in rheumatoid arthritis. There are several points of action for these drugs (Morley, 1976). Our own studies and those of McIlhenny et al (1978) on attempted suppression of EAE suggest that prostaglandin inhibitors (flumizole), may even make guinea pigs worse. However, myelin basic protein (1.4mg/kg) given intravenously eleven days after immunization, gives some protection to animals. Similarly, some of the effects of EAE can be suppressed by the anti-inflammatory drug EN 3638, as reported by Levine and Sowinski (1978). Niridazole, which selectively blocks the T-cell system, also suppresses EAE in mice (Bernard et al, 1977). Since active T-cells have been identified in the CSF of MS patients during an exacerbation, there is hope that it may be possible to modify the inflammatory reaction in patients during the course of a relapse.

References

Abramsky, O., Lisak, R.P., Silberberg, D.H. and Pleasure, D. (1977) Antibodies to oligodendroglia in patients with multiple sclerosis, New Engl. J. Med. 298: 1207-1211.

Acheson, E.D. (1977) Epidemiology of multiple sclerosis, Brit. Med. Bull. 33: 9-14.

Allen, J.C., Sheremata, W., Cosgrove, J.B.R., Osterland, K. and Shea, M. (1976) Cerebrospinal fluid T and B lymphocyte kinetics related to exacerbations of multiple sclerosis, Neurology, 26: 579-583.

Andrews, J.M. (1972) The ultrastructural neuropathology of multiple sclerosis, in Multiple sclerosis (ed. Wolfgram, F., Ellison, G.W., Stevens, J.G. and Andrews, J.M.), pp. 23-52, Academic Press, N.Y..

Arstila, A.U., Riekkinen, P., Rinne, U.K. and Laitinen, L. (1973) Studies on the pathogenesis of multiple sclerosis, European Neurol. 9: 1-20.

Bernard, C.C.A., Leydon, J. and MacKay, I.R. (1977) Anti-T-cell activity of niridazole in experimental autoimmune encephalomyelitis, Int. Archs. Allergy Appl. Immun. 53: 555-559.

Brosnan, C.F., Stoner, G.L., Bloom, B.R. and Wisniewski, H.M. (1977) Studies on demyelination by activated lymphocytes in the rabbit eye. II. Antibody-dependent cell-mediated demyelination, J.Immunol. 118: 2103-2110.

Cohen, S.R., Herndon, R.M. and McKhann, G.M. (1976) Radioimmunoassay of myelin basic protein in spinal fluid: an index of active demyelination, New Engl.J.Med. 295: 1455-1457.

Compston, D.A.S., Batchelor, J.R. and McDonald, W.I. (1976) B lymphocyte alloantigens associated with multiple sclerosis, Lancet ii: 1261-1265.

Cuzner, M.L., McDonald, W.I., Rudge, P., Smith, M., Borshell, N.J., and Davison, A.N. (1975) Leucocyte proteinase activity in acute multiple sclerosis, J.Neurol.Sci. 26: 107-111.

Cuzner, M.L., Barnard, R.D., MacGregor, B.J.L., Borshell, N.J., and Davison, A.N. (1976) Myelin composition in acute and chronic multiple sclerosis in relation to cerebral lysosomal activity, J.Neurol.Sci. 29: 323-334.

Cuzner, M.L., Davison, A.N. and Rudge, P. (1978) Proteolytic enzyme activity of blood leucocytes and cerebrospinal fluid in multiple sclerosis, Ann.Neurol. (in press).

Gutstein, H.S. and Cohen, S.R. (1978) Spinal fluid differences in experimental allergic encephalomyelitis and multiple sclerosis, Science. 199: 301-303.

Hansen, N.E., Karle, H., Jensen, A. and Bock, E. (1977) Lysozyme activity in cerebrospinal fluid, Acta.Neurol.Scand. 55: 418-424.

Kimberlin, R.H. (1973) Subacute spongiform encephalopathies in domestic and laboratory animals, Biochem.Soc.Trans. 1: 1058-1061.

Kurdi, A., Ayesh, I., Abdallat, A., Maayta, U., McDonald, W.I., Compston, D.A.S., and Batchelor, J.R. (1975) Different 'B' lymphocyte alloantigens associated with multiple sclerosis in Arabs and North Europeans, Lancet, i: 1123-1125.

Lennon, V.A. and Dunkley, P.R. (1974) Humoral and cell-mediated immune responses of Lewis rats to syngeneic basic protein of myelin, Int.Arch.Allergy Appl.Immunol. 47: 598-608.

Lennon, V. and MacKay, I.R. (1972) Binding of ^{125}I myelin basic protein by serum and cerebrospinal fluid, Clin.exp.Immunol. 11: 595-603.

Levine, S. and Sowinski, R. (1978) Suppression of experimental allergic encephalomyelitis by 6-hydroxyphthalaldehydic acid, O-(p-chlorobenzyl) oxime (EN 3638), J.Immunol. 120: 602-606.

Link, H. (1973) Immunoglobulin abnormalities in multiple sclerosis, Ann.Clin.Res. 5: 330-336.

McIlhenny, H.M., Levine, S., Wiseman, E.H. and Sowinski, R. (1978) Disposition and activity in experimental allergic encephalomyelitis of flumizole, a non-acidic, non-steroidal, anti-inflammatory agent, Exper.Neurol. 58: 126-137.

Millson, G.C. and Bountiff, L. (1973) Glycosidases in normal and scrapie mouse brain, J.Neurochem. 20: 541-546.

Morley, J. (1976) Prostaglandins as regulators of lymphoid cell function in allergic inflammation: a basis for chronicity in rheumatoid arthritis. In: WHO/ARC Symposium on Infection and Immunology in the Rheumatic Diseases. Ed.Dumonde, D.C., Blackwells, London.

Naess, A. and Nyland, H. (1978) Multiple Sclerosis. T Lymphocytes in Cerebrospinal Fluid and Blood, Europ.Neuro. 17: (2), 61-67.

Raine, C.S. and Stone, S.H. (1977) Animal model for multiple sclerosis, New York State J.Med. 77: 1693-1696.

Raine, C.S., Traugott, V., Iqbal, K., Snyder, D.S., Cohen, S.R., Farooq, M. and Norton, W.J. (1978) Encephalitogenic properties of purified preparations of bovine oligodendrocytes tested in guinea pigs, Brain Res. 142: 85-96.

Stone, S.H. and Lerner, E.M. (1965) Chronic disseminated allergic encephalomyelitis in guinea pigs, Ann.New York Acad.Sci. 122: 227-241.

Stoner, G.L., Brosnan, C.F., Wisniewski, H.M. and Bloom, B.R. (1977) Studies on demyelination by activated lymphocytes in the rabbit eye. I.Effects of a mononuclear cell infiltrate induced by products of activated lymphocytes, J.Immunol. 118: 2094-2102.

Tchorzewski, H., Czernicki, J. and Maciejek, Z. (1976) Polymorphonuclear leukocyte lysosome activities and lymphocyte transformation in multiple sclerosis and some other central nervous system chronic diseases, Eur.Neurol. 14: 386-396.

Thompson, E.J., Kaufmann, P., Shortman, R.C., Rudge,P., and McDonald, W.I. (1978) Oligoclonal immunoglobulins and plasma cells in spinal fluid of patients with multiple sclerosis,(in preparation).

Tourtellotte, W.W. (1975) What is multiple sclerosis? Laboratory criteria for drugs. in Multiple Sclerosis Research (Davison,A.N., Humphrey, J.H., Liversedge, A.L., McDonald, W.I., and Porterfield, J.S., ed.) pp. 9-25. H.M.S.O. London.

Traugott, U. and Raine, C.S. (1977) Experimental allergic encephalomyelitis in inbred guinea pigs. Correlation of decrease in early T-cells with clinical signs in suppressed and unsuppressed animals, Cell.Immunol. 34: 146-155.

Traugott, U., Stone, S.H. and Raine, C.S. (1978) Experimental allergic encephalomyelitis - Migration of early T-cells from the circulation into the central nervous system, J.Neurol.Sci., 36: 55-61.

Wisniewski, H.M. (1977) Immunopathology of demyelination, Brit.Med. Bull. 33: 54-59.

Wisniewski, H.M. and Keith, A.B. (1977) Chronic relapsing experimental allergic encephalomyelitis: an experimental model of multiple sclerosis, Ann.Neurol. 1: 144-148.

Zabriskie, J.B. (1975) Cell-mediated immunity to viral antigens in multiple sclerosis. in Multiple Sclerosis (Ed. Davison, A.N., Humphrey, J.H., Liversedge, A.L., McDonald, W.I., and Porterfield, J.S.), pp. 142-150. H.M.S.O., London.

TISSUE CULTURE STUDIES OF EXPERIMENTAL ALLERGIC ENCEPHALOMYELITIS

Fredrick J. Seil
Research Service, Veterans Administration, Hospital and Department of
Neurology, University of Oregon Health Sciences Center, Portland, Oregon
(U.S.A.)

Introduction

A survey will be presented of some of the studies employing a tissue culture approach to the investigation of the experimental demyelinating disease, experimental allergic encephalomyelitis (EAE), and the relationship of these studies to the human demyelinating disease, multiple sclerosis (MS), will be examined. The focus of this review will be on serum factors which either demyelinate already myelinated central nervous system (CNS) explants (Bornstein and Appel, 1961) or which inhibit myelin formation in unmyelinated cultures (Bornstein and Raine, 1970). These serum factors are collectively referred to as "antimyelin factors" or "antimyelin antibodies." The associated subjects of cell-induced demyelination of tissue cultures, peripheral nervous system demyelinating factors, serum factors which block neuroelectric activity in vitro and studies of cell and serum-induced gliotoxic effects in simple culture systems have been discussed recently in a more detailed review (Seil, 1977).

Antimyelin Factors in EAE

In a pioneering study, Bornstein and Appel (1961) demonstrated that sera from rabbits with EAE induced by sensitization with whole CNS tissue in combination with Freund's complete adjuvant (FCA) reversibly demyelinated rat cerebellar cultures. The demyelinating effect was lost upon prior heating of the sera to 56°C and restored upon addition of fresh normal guinea pig serum as a source of complement. Axons were spared and remyelination occured after removal of the test serum and replacement with normal nutrient medium. In a sequel to this study, Appel and Bornstein (1964) were able to remove demyelinating activity of sera from rabbits sensitized with whole CNS by absorbing the sera with brain tissue, but not by absorption with lung, liver, kidney or red blood cells. Demyelinating activity was found in the 7S gamma 2-globulin (IgG2) fraction of the sera. Globulins in demyelinating sera were localized on myelin sheaths and cell membranes of CNS cultures, as determined by fluorescent antibody techniques. Recently Johnson et al. (1977) exposed mouse spinal cord explants to demyelinating rabbit sera, subsequently added

peroxidase-labeled anti-rabbit immunoglobulin and reacted with diamino-
benzidine. Myelin and oligodendrocyte plasma membranes in the cultures were
stained, while cultures exposed to control sera were unstained. These results
were consistent with the findings of Appel and Bornstein's earlier study.

The concept that demyelinating activity in sera from whole CNS-sensitized
animals is associated with the immunoglobulin fraction of serum gained further
support from a study by Grundke-Iqbal and Bornstein (1977) in which
demyelinating activity was reduced by removing most of the IgG from the sera.
The isolated IgG fraction was found to have demyelinating activity. Lebar et
al. (1976) demonstrated demyelinating antibodies of the IgG2 class in sera
from guinea pigs sensitized with whole guinea pig CNS applied to myelinated
guinea pig cerebellar cultures, ie. an entirely homologous system.
Demyelinating activity was not found in IgG1 or IgM fractions of serum.

In addition to demyelinating CNS cultures, Bornstein and Raine (1970)
observed that sera from animals sensitized with whole CNS prevented
remyelination, even if reduced to very low concentrations which would be
insufficient to produce the initial demyelination. They also observed that if
CNS explants were exposed to low concentrations of such sera prior to
myelination, glial differentiation and myelin formation were inhibited.
Sulfatide synthesis was also inhibited in cultures exposed to anti-CNS
antisera prior to myelination (Fry et al., 1972). Bornstein et al. (1977)
noted that heated (to destroy complement) anti-CNS antisera did not prevent
glial differentiation in vitro, but that few of the processes of
oligodendrocytes exposed to such sera found and ensheathed axons to form
aberrant swollen lamellae. Removal of the antisera after three weeks in vitro
resulted in differentiation of other oligodendrocytes and normal myelin
formation. On the other hand, prolonged exposure of CNS cultures to high
concentrations of unheated anti-CNS antisera resulted in failure of
remyelination upon removal of the antisera, and in excess production of
astrocytic fibers, simulating a state of "sclerosis" (Raine and Bornstein,
1970b).

In aggregate, these findings could be intepreted as indicating a
pathogenetic role for circulating antibodies in the production of
demyelination and eventually of sclerosis in EAE. In all of these studies EAE
had been induced by sensitization with whole CNS tissue or white matter. EAE,
however, could also be induced by myelin (Laatsch et al., 1962), by a basic
protein (BP) fraction of myelin (Kies and Alvord, 1959; Laatsch et al., 1962),
or by some peptide fragments of the whole BP molecule (Carnegie et al., 1967;
Chao and Einstein, 1968; Eylar and Hashim, 1968; Eylar et al., 1970). The
encephalitogenic protein (BP) produced EAE identical by clinical, histological

and ultrastructural criteria to that produced by sensitization with whole CNS (Lampert and Kies, 1967).

Lumsden (1966) induced EAE in guinea pigs by sensitization with a diffusible "protein free" peptide. Sera from these animals failed to demyelinate myelinated CNS explants. Lumsden attributed this failure to a lack of antigenicity of his peptide. We (Seil et al., 1968) therefore exposed myelinated mouse cerebellar cutures to sera from guinea pigs sensitized or hyperimmunized with the whole BP molecule, as well as to sera from whole CNS-sensitized guinea pigs. Animals were sensitized with a single intradermal injection of either homologous BP or whole CNS combined with FCA. Guinea pigs were hyperimmunized or "protected" by interval injections of BP plus incomplete Freund's adjuvant (without Mycobacterium tuberculosis), followed by a challenge dose of BP and FCA. Such animals developed high titers of anti-BP antibody without develping EAE. Sera from whole CNS-sensitized guinea pigs (anti-CNS antisera) demyelinated cerebellar cultures, consistent with Bornstein and Appel's (1961) results with rabbits. Such sera did not contain detectable levels of anti-BP antibody. Sera from guinea pigs sensitized or hyperimmunized with BP failed to demyelinate CNS explants, although such sera contained antibody to BP in three gamma-globulin classes. These results were interpreted as demonstrating that anti-BP antibody was not the demyelinating factor, and doubt was raised about the role of demyelinating antibody in the pathogenesis of EAE, as EAE could be induced in the absence of demyelinating antibody.

Our results conflicted with those of a study reported by Yonezawa et al. (1969), who obtained in vitro demyelinating activity by sera from rabbits and guinea pigs sensitized with BP. However, the BP employed in this study was found to be contaminated with other myelin fractions, and when the study was repeated with BP prepared by M.W. Kies, results consistent with ours were obtained (Yonezawa, T., personal communication, 1974). Lebar et al. (1976) also confirmed a lack of demyelinating activity in guinea pig anti-BP antisera. There have been some inconsistencies in the reports from Bornstein's laboratory. Kristensson et al. (1976) stated that sera from five rabbits inoculatd with BP demyelinated CNS cultures, and promised a future publication in which details would be provided to support this claim. In a paper published in the same year, Bornstein and Raine (1976) stated that, "demyelination is not usually induced in cultures exposed to serum from EAE-affected animals challenged with myelin basic protein in complete Freund's adjuvant." Eylar and Bornstein (Eylar, E.H., presented at the Tenth Annual Winter Conference on Brain Research, Keystone, Colorado, January, 1977) obtained results which convincingly demonstrated a lack of demyelinating activity in sera from animals sensitized with BP.

In that myelination inhibition appeared to be a more sensitive and objectively assayable in vitro index of serum antimyelin activity (Bornstein and Raine, 1970), we also compared the myelination inhibiting properties of anti-CNS and anti-BP antisera in a series of experiments (Seil et al., 1976a). In the first (Kies et al., 1973), sera from guinea pigs sensitized with equivalent amounts of homologous CNS and heterologous (bovine) BP were compared for myelination inhibiting capability and quantitative levels of antibody to BP. Most of the anti-CNS antisera, which contained lower levels of anti-BP antibody, inhibited myelination of cerebellar cultures, while most anti-BP antisera, which contained higher levels of anti-BP antibody, failed to inhibit myelin formation. We then extended these studies to another species, subhuman primates, selected because of their exquisite sensitivity to EAE induction (Seil et al., 1973). As with guinea pig sera, anti-CNS antisera from subhuman primates inhibited myelination, while anti-BP antisera did not. Sera from Lewis rats sensitized with guinea pig CNS and BP were similarly evaluated with similar results (Seil et al., 1975a). Finally, rabbits were sensitized with bovine CNS and hyperimmunized with BP of different species, including bovine, human, monkey, guinea pig and rabbit BP (Seil et al., 1975b). Five rabbits sensitized with whole CNS all developed EAE, and their sera all inhibited myelination, but none had measurable levels of precipitating antibody to BP. Nine rabbits hyperimmunized with five species of BP demonstrated the entire spectrum of possible combinations of EAE and precipitating anti-BP antibody. The serum from only one of these animals was positive for myelination inhibition, and this animal developed neither EAE nor precipitating anti-BP antibody. Thus a complete dissociation of EAE induction, anti-BP antibody and myelination inhibiting factor was demonstrated. Yonezawa et al. (1976) also reported failure of myelination inhibition by anti-BP antisera. In a preliminary study, we (Kies, M.W. and Seil, F.J., unpublished observations) found that if guinea pigs were given interval inoculations of BP plus incomplete Freund's adjuvant followed by injection of whole CNS combined with FCA, EAE was prevented. However, sera from the majority of these animals inhibited myelination, thus supporting the notion that myelination inhibiting factor is provoked by some agent other than BP. These combined studies provided further evidence that BP, the protein that induces EAE, is not the antigen that induces serum antimyelin factors.

Other possible antigens (or haptens) have been investigated. Purified myelin itself induced both myelination inhibiting factor (Seil et al., 1975a) and demyelinating antibody (Lebar et al., 1976). In the latter study, demyelinating activity was abolished by absorption of anti-myelin antisera with either purified myelin or whole CNS. Absorption with both of these substances also removed demyelinating activity from anti-CNS antisera.

Dubois-Dalcq et al. (1970) observed that rabbits inoculated with cerebroside did not develop EAE, but their sera demyelinated cerebellar cutures. Fry et al. (1974) confirmed this observation and also demonstrated parallel myelination inhibition and inhibition of sulfatide synthesis in CNS explants by anti-cerebroside antisera from rabbits. Absorption with cerebroside abolished the antimyelin activity. Antibody to cerebroside was also found in anti-whole CNS rabbit antisera, and the antimyelin activity of these sera was also removed by absorption with cerebroside. Dorfman et al. (1977) demonstrated a correlation between the minimum effective concentration of anti-CNS rabbit antisera that inhibited myelination in cerebellar cultures and titers of antibody to galactocerebroside. These investigators also found control levels of antibody to ganglioside, sphingomyelin and ceramide in the same rabbit sera.

Lebar et al. (1976) did not find demyelinating activity in serum from guinea pigs inoculated with cerebroside and we (Seil et al., 1975a) observed no inhibition of myelination in vitro by sera from Lewis rats injected with cerebroside mixed with either bovine serum albumin or with BP. We were, however, able to demonstrate inhibition of myelination in cerebellar cultures by sera from rabbits inoculated with synthetic galactocerebrosides (Hruby et al., 1977). This study provided evidence that synthetic galactocerebrosides not possibly contaminated with other CNS components could evoke myelination inhibiting antibodies in rabbits. To date, serum antimyelin activity induced by inoculation with natural or synthetic cerebrosides has been documented only in this species.

Yonezawa et al. (1976) described inhibition of myelination in cerebellar explants by sera from four rabbits inoculated with ganglioside. From the published data it would appear that their interpretation is acceptable in the case of two of the animals. With regard to the other two sera, however, some differences in interpretation are possible, as 53% of the cultures exposed to one of these sera myelinated, while 64% of the cultures exposed to the other serum formed myelin. According to some investigators' criteria (Kies et al., 1973), these sera might be considered negative for myelination inhibition. In any case, a larger series of animals might provide more definitive results with regard to the antimyelin activity of anti-ganglioside antisera.

In a recent development, Saida et al. (1977) found that sera from rabbits serially injected with purified oligodendrocytes and FCA both demyelinated and inhibited myelination in cerebellar cultures. These animals did not develop EAE. Serum antimyelin activity was abolished by absorption with oligodendrocytes, but only partially decreased by absorption with myelin. Antibodies to galactocerebroside and to ganglioside were not elevated in these sera. The results of this study indicated that anti-oligodendrocyte antibodies, which appeared to be distinct from anti-galactocerebroside antibodies, were also active against myelin or myelin formation in vitro.

Other antigenic candidates about which there is incomplete data include myelin proteolipid protein and fragments of myelin basic protein. In a preliminary study (Seil, F.J. and Agrawal, H.C., unpublished observations), rabbit sera containing low titers of anti-proteolipid protein antibody did not inhibit myelin formation in CNS explants. Sera containing high titers of anti-proteolipid protein antibody from a rabbit and a goat were also negative, while another rabbit serum with a high antibody titer inhibited myelination in vitro. The single positive serum could be fortuitous, as occasional control sera also demonstrate antimyelin effects, but a larger series of animals needs to be tested before conclusions can be drawn.

An unexpected finding in one of our studies (Seil et al., 1975b) was that of myelination inhibition in vitro by serum from one of two rabbits sensitized with a peptide containing residues "43-88" of guinea pig BP, and by serum from one of two rabbits sensitized with rat S basic protein. Rat S is the smaller of two rat myelin basic proteins, and it resembles guinea pig BP minus residues "117-156." It is conceivable that an antigenic site which can induce antimyelin factors is exposed in a fragment of BP, whereas the site may be covered in the intact BP molecule. This would, of course, need to be a different antigenic site than that which induces EAE. These findings also need verification with a larger series of animals before they can be considered acceptable. Other antigenic candidates await evaluation.

The results of these various studies, minus the preliminary data which need further investigation, are summarized in Table 1. It is evident from this table that EAE can be induced without induction of demyelinating and myelination inhibiting antibodies, and that demyelinating and myelination inhibiting antibodies can be evoked without induction of disease. The question must therefore be raised as to whether or not these antibodies have a significant role in EAE induction. It is also apparent that more than a single CNS component can evoke antibodies that affect myelin in vitro. The full range of antigenic possibilities remains to be elucidated.

Table 1. EAE and antibody induction by CNS antigens or haptens. NT, not tested. BP, myelin basic protein. GC, galactocerebroside. GS, ganglioside. FCA, Freund's complete adjuvant.

Antigen (or hapten)	EAE	Antibodies Demyel- inating	Myelination Inhibiting	Other
Whole CNS (or white matter)	+	+[a]	+[a]	Anti-BP low or − Anti-GC + Anti-GS −[b]
Purified myelin	+	+[c]	+	NT
Myelin basic protein	+	−	−	Anti-BP +
BP protected (challenge with BP + FCA)	−	−	−	Anti-BP ++
Galactocerebroside (rabbits)	−	+[d]	+[d]	Anti-GC +
Purified oligodendrocytes (rabbits)	−	+[e]	+[e]	Anti-GC − Anti-GS −

a) Antimyelin activity removed by absorption of sera with whole CNS, purified myelin and cerebroside, but not with BP.
b) Anti-sphingomyelin and anti-ceramide antibodies also at control levels.
c) Demyelinating activity abolished by absorption of sera with purified myelin or whole CNS.
d) Antimyelin activity removed by absorption of sera with cerebroside.
e) Antimyelin activity removed by absorption of sera with purified oligodendrocytes and partially reduced by absorption with purified myelin.

Antimyelin Factors in MS

   Bornstein (1963) observed that sera from approximately 60% of patients with multiple sclerosis in exacerbation demyelinated CNS cultures, an observation confirmed in other laboratories (Hughes and Field, 1967; Lumsden, 1971). Bornstein's most recently published figures (Bornstein and Hummelgard, 1976) indicate that sera from 64% of patients with active disease were positive for demyelinating activity. Sera from 41% of patients with questionable disease activity, from 11% of patients with no active disease and from 7% of normal controls also demyelinated CNS explants. When MS patients were divided into exacerbating-remitting and chronic-progressive groups, the resulting data indicated that in the exacerbating-remitting group, sera from 71% of patients with active disease, 65% of patients with questionably active disease and 19% of patients whose disease was inactive contained demyelinating factors. Sera from 48% of the chronic-progressive group were positive for demyelinating activity. In three patients with an exacerbating-remitting course, the occurrence of exacerbations was correlated with the appearance of serum demyelinating activity, whereas serum demyelinating factor was not present during remissions. Serum demyelinating activity was constantly present in two patients with chronic-progressive MS who were monitored over a period of time. Demyelinating activity in MS sera was complement dependent and was abolished by absorption of the sera with brain tissue, but not by absorption with red blood cells.

   Although a very similar pattern of demyelination of myelinated CNS cultures was produced by MS sera and anti-CNS antisera, even at the ultrastructural level (Raine and Bornstein, 1970a; Raine et al., 1973), some interesting differences in immunological properties of these sera have begun to be revealed by recent studies. Wolfgram and Duquette (1976) noted that demyelinating activity of MS sera was poorly absorbed by purified myelin, but was abolished by absorption with a nonmyelin CNS tissue pellet which, along with many other components, contained oligodendrocytes. By contrast, demyelinating activity was removed from anti-CNS antisera by absorption with purified myelin (Lebar et al., 1976). Moreoever, demyelinating antibody could be induced by sensitization of animals with purified myelin. In a study not involving a tissue culture system, Abramsky et al. (1977) demonstrated antibodies to oligodendrocytes in 19 of 21 MS patients. The antibodies were absorbed by exposure of sera to isolated oligodendrocytes or whole white matter, but not by absorption with purified myelin. These data, while obtained from a different system, are in keeping with Wolfgram's and Duquette's observations. Both sets of observations suggest that antibodies in MS are not directed at myelin, and the study of Abramsky et al. suggests that MS sera contain anti-oligodendrocyte antibodies.

A further difference in demyelinating sera from MS patients and anti-CNS antisera from animals with EAE was indicated by two studies of the association of serum demyelinating activity with immunoglobulins. Grundke-Iqbal and Bornstein (1977) demonstrated that while removal of 90% of the IgG reduced demyelinating activity in anti-CNS antisera, a similar procedure produced no significant change in the demyelinative capacity of MS sera. Demyelinating activity was found in the isolated IgG fraction of anti-CNS antisera, but only minimal demyelinating activity was present in the isolated IgG fraction of sera from MS patients. Johnson et al. (1977) found evidence for an association of demyelinating activity in anti-CNS antisera with an immunoglobulin by an immunoperoxidase method, but failed to find such an association in demyelinating MS sera. These recent studies are at variance with an earlier report by Dowling et al. (1968) that demyelinating activity was associated with both IgG and IgM complement dependent antibodies in sera from two patients with acute MS.

While anti-CNS antisera from animals with EAE have generally been reported to both demyelinate already myelinated CNS explants and to inhibit myelination if applied to cultures prior to formation of myelin (see Table 1), there may be some dissociation of these properties in MS sera. Yonezawa et al. (1976) reported myelination inhibition by sera from two MS patients in exacerbation. Sera from these same patients in remission did not inhibit myelination. Questions can again be raised about these investigators' criteria for considering a serum positive for myelination inhibition, particularly with regard to one of the MS sera that was interpreted as positive when 60% of the cultures exposed to this serum myelinated. Ulrich has not found myelination inhibiting activity in 40 sera from MS patients (Ulrich, J., personal communication, 1976). Our experience (Seil, F.J., unpublished observations) is in agreement with Ulrich's, as we have not observed myelination inhibition by sera from four acute and two chronic-progressive cases of MS. If these observations are confirmed by further studies, and MS sera indeed demyelinate CNS cultures without inhibiting myelin formation, then this would constitute further evidence that demyelinating factors in MS sera are directed at different CNS components than antibodies in anti-CNS antisera, which appear to be directed at myelin or some fraction thereof. A failure of MS sera to inhibit myelination might also indicate some difference between MS sera and anti-oligodendrocyte antisera from animals inoculated with purified oligodendroglial cells, as the latter have been reported to cause both demyelination and myelination inhibition in CNS cultures (Saida et al., 1977).

Although there has been some observed correlation between demyelinating activity in MS sera and disease activity, the significance of this correlation with regard to pathogenesis of the disease is not clear. An observation which has complicated interpretation of the relevance of serum demyelinating factor to the pathogenesis of MS is the finding that approximately 60% of sera from patients with amyotrophic lateral sclerosis also demyelinated CNS tissue cultures (Bornstein, 1963; Field and Hughes, 1965). In this disease, demyelination occurs secondarily to axonal degeneration, and is not a primary event as in MS. Yonezawa et al. (1976) demonstrated that demyelinating sera from patients with amyotrophic lateral sclerosis did not inhibit myelin formation in vitro.

Table 2. Comparison of several characteristics of demyelinating MS sera and experimental anti-CNS and anti-oligodendrocyte antisera. NT, not tested.

Effects on CNS Cultures	MS	Anti-CNS	Anti-Oligodendrocyte
In vitro demyelination	+	+	+
Demyelination after absorption with CNS	-	-	NT
Demyelination after absorption with purified myelin	+	-	Partially reduced
Demyelination after absorption with nonmyelin CNS components	-	NT	-
Demyelination after removal of 90% of IgG	+	Reduced	NT
Demyelination by isolated IgG fraction	Minimal	+	NT
In vitro myelination inhibition	-	+	+

A comparison of various properties of demyelinating MS sera and experimental anti-CNS and anti-oligodendrocyte antisera is summarized in Table 2. Although all of these sera demyelinated CNS tissue cultures, the

demyelinating capacity of anti-CNS antisera was abolished by absorption with
purified myelin, while MS sera and anti-oligodendrocyte antisera continued to
demonstrate demyelinating activity after such treatment. In these latter two
sera, demyelinating activity was removed by absorption with nonmyelin CNS
components, namely a nonmyelin CNS pellet in the case of MS sera, and purified
oligodendrocytes in the case of anti-oligodendrocyte antisera. Anti-
oligodendrocyte antibodies were shown to be present in MS sera (Abramsky et
al., 1977). While demyelinating activity in MS sera correlates to some degree
with disease activity, demyelinating anti-oligodendrocyte antisera were
induced without induction of disease. Both of the experimental demyelinating
antisera also inhibited myelination in vitro, a property which is apparently
not shared by MS sera.

Synthesis and Conclusions

The key question to be considered relative to these studies of serum
antimyelin factors in EAE and MS is whether they are significant factors in
the pathogenesis of experimental and human demyelinating diseases. It has
been shown that EAE can be induced without induction of demyelinating and
myelination inhibiting antibodies by sensitizing animals with the
encephalitogenic myelin basic protein. At least two agents, namely
galactocerebroside and purified oligodendrocytes, have been shown to evoke
demyelinating and myelination inhibiting antibodies in rabbits. Neither the
antibody directed against galactocerebroside nor the antibody directed against
oligodendrocytes is associated with induction of EAE. It therefore seems
doubtful that these antibodies have a major role in the pathogenesis of EAE.

Lebar et al. (1976) suggested a possible augmenting role for antimyelin
antibodies in EAE induced by sensitization with whole CNS. This suggestion
was based on their experience that EAE induced by whole CNS is of earlier
onset and can be passively transferred earlier than EAE induced by BP, as well
as their experience that BP combined with incomplete Freund's adjuvant is able
to prevent BP-induced EAE more completely than whole CNS-induced EAE. While
such an augmenting role may be possible, there is no experimental evidence to
support this concept. In any event, what seems of greater significance is the
fact that antibodies which affect myelin in vitro can be totally dissociated
from disease induction, and therefore a cause-effect relationship becomes
improbable. These antibodies remain valuable, however, as tools for
investigating mechanisms by which myelin is formed (Ulrich and Bornstein,
1973; Bornstein et al., 1977).

Upon comparison of serum antimyelin factors in EAE and MS, it is becoming
clear from recent studies that MS sera and anti-CNS antisera are dissimilar.

An important difference is that anti-CNS antisera appear to be directed against myelin or a component(s) of myelin, while demyelinating MS sera appear to be directed against a nonmyelin CNS fraction, possibly oligodendrocytes. Another difference is that demyelinating activity in anti-CNS antisera appears to be associated with an immunoglobulin, while doubt has been raised that demyelinating factors in MS sera are immunoglobulins. When comparing MS sera with anti-oligodendrocyte antisera, it is worthy of note that anti-oligodendrocyte sera are not associated with induction of disease.

In speculating about the possibiity that a serum demyelinating factor has a role in the etiology of MS, one must, in addition to what has been learned from the animal studies, consider the facts that sera from 60% of patients with amyotrophic lateral sclerosis, a disease in which demyelination is secondary, also demyelinate tissue cultures, and that sera from 30-40% of patients with acute MS are not demyelinative. With regard to the former, the presence of demyelinating factor in a disease in which myelin is secondarily destroyed could indicate that the demyelinating factor is induced in response to myelin degeneration, and does not itself cause demyelination. It is not clear that serum demyelinating factors in MS and amyotrophic lateral sclerosis are the same. In MS, the demyelinating factor appears to be directed against oligodendrocytes. The possibility remains, however, that demyelinating factor in MS could be secondarily induced, eg. in response to oligodendrocyte destruction.

With regard to non-demyelinating sera from acute MS patients, an explanation might simply be that demyelinating factor was not produced in sufficient titer to be detectable in vitro after having been bound in vivo. Alternatively, there may not have been sufficient oligodendrocyte destruction to secondarily induce demyelinating factor, if that is the mechanism of its induction. Still another alternative is that demyelinating factor may be an epiphenomenon not at all related to the pathogenesis of MS, and therefore inconstantly present in association with the disease. Bornstein and Hummelgard (1976) theorized that non-demyelinating sera from acute MS patients may contain a serum factor which blocks electrical activity in CNS cultures (Bornstein and Crain, 1965), thus accounting for clinical signs and symptoms in the absence of demyelinating factor. In that similar blocking factors have been found in sera from normal subjects (Crain et al., 1975; Seil et al., 1976b), this seems an unlikely postulate.

It cannot be stated at this time that a serum demyelinating factor which is associated with disease activity in some cases of MS does or does not have a role in the pathogenesis or perhaps in the perpetuation of the disease. On the basis of what is presently known, the concept of such a role must be regarded with reservations.

References

Abramsky, O., Lisak, R.P., Silberberg, D.H. and Pleasure, D.E.(1977) Antibodies to oligodendroglia in patients with multiple sclerosis, N. Engl. J. Med. 297:1207-1211.

Appel, S.H. and Bornstein, M.B. (1964) The application of tissue culture to the study of experimental allergic encephalomyelitis: II. Serum factors responsible for demyelination, J. Exp. Med. 119:303-312.

Bornstein, M.B. (1963) A tissue culture approach to demyelinative disorders, Natl. Cancer Inst. Monogr. 11:197-214.

Bornstein, M.B. and Appel, S.H. (1961) The application of tissue culture to the study of experimental "allergic" encephalomyelitis: I. Patterns of demyelination, J. Neuropathol. Exp. Neurol. 20:141-157.

Bornstein, M.B and Crain, S.M., (1965) Functional studies of cultured brain tissues as related to "demyelinative" disorders, Science 148:1242-1244.

Bornstein, M.B. and Hummelgard, A. (1976) Multiple sclerosis: serum induced demyelination in tissue culture, in Shiraki, H., Yonezawa, T. and Kuroiwa, Y. (eds.), The Aetiology and Pathogenesis of the Demyelinating Diseases, Japan Science Press, Tokyo, pp. 341-350.

Bornstein, M.B. and Raine, C.S. (1970) Experimental allergic encephalomyelitis: antiserum inhibition of myelination in vitro, Lab. Invest. 23:536-542.

Bornstein, M.B. and Raine, C.S. (1976) The initial lesion in serum-induced demyelination in vitro, Lab. Invest. 35:391-401.

Bornstein, M., Diaz, M. and Raine, C. (1977) Dissociation of myelinogenesis in tissue culture by anti-CNS antiserum, J. Neuropathol. Exp. Neurol. 36:594.

Carnegie, P.R., Bencina, B. and Lamoureux, G. (1967) Experimental allergic encephalomyelitis. Isolation of basic proteins and polypeptides from central nervous tissues, Biochem. J. 105:559-568.

Chao, L.P. and Einstein, E.R. (1968) Isolation and characterization of an active fragment from enzymatic degradation of encephalitogenic protein, J. Biol. Chem. 243:6050-6055.

Crain, S.M., Bornstein, M.B. and Lennon, V.A. (1975) Depression of complex bioelectric discharges in cerebral tissue cultures by thermolabile complement-dependent serum factors, Exp. Neurol. 49:330-335.

Dorfman, S.H., Fry, J.M., Silberberg, D.H., Grose, C. and Manning, M.C. (1977) Antigalactocerebroside antibody titer correlates with in vitro activity of myelination inhibiting antisera, J. Neuropathol. Exp. Neurol. 36:600.

Dowling, P.C., Kim, S.U., Murray, M.R. and Cook, S.D. (1968) Serum 19S and 7S demyelinating antibodies in multiple sclerosis, J. Immunol. 101:1101-1104.

Dubois-Dalcq, M., Niedieck, B., and Buyse, M. (1970) Action of anti-cerebroside sera on myelinated tissue cultures, Pathol. Eur. 5:331-347.

Eylar, E.H. and Hashim, G.A. (1968) Allergic encephalomyelitis: the structure of the encephalitogenic determinant, Proc. Natl. Acad. Sci. USA 61:644-650.

Eylar, E.H., Caccam, J., Jackson, J.J., Westall, J.C. and Robinson, A.B. (1970) Experimental allergic encephalomyelitis: synthesis of disease-inducing site of the basic protein, Science 168:1220-1223.

Field, E.J. and Hughes, D. (1965) Toxicity of motor neurone disease serum for myelin in tissue culture, Brit. Med. J. 2:1399-1401.

Fry, J.M., Lehrer, G.M. and Bornstein, M.B. (1972) Sulfatide synthesis: Inhibition by experimental allergic encephalomyelitis serum, Science 175:192-194.

Fry, J.M., Weissbarth, S., Lehrer, G.M. and Bornstein, M.B. (1974) Cerebroside antibody inhibits sulfatide synthesis and myelination and demyelinates in cord tissue cultures, Science 183:540-542.

Grundke-Iqbal, I. and Bornstein, M.B. (1977) Are demyelinating serum factors in MS and EAE gamma globulins? J. Neuropathol. Exp. Neurol. 36:625.

Hruby, S., Alvord, E.C., Jr. and Seil, F.J. (1977) Synthetic galacto-cerebrosides evoke myelination inhibiting antibodies, Science 195:173-175.

Hughes, D. and Field, E.J. (1967) Myelinotoxicity of serum and spinal fluid in multiple sclerosis: a critical assessment, Clin. Exp. Immunol. 2:295-309.

Johnson, A.B., Blum, N.R. and Bornstein, M.B. (1977) Immunoperoxidase studies on allergic encephalomyelitis and multiple sclerosis serum factors in tissue culture, J. Neuropathol. Exp. Neurol 36:607.

Kies, M.W. and Alvord, E.C., Jr. (1959) Encephalitogenic activity in guinea pigs of water-soluble protein fractions of nervous tissue, in Kies, M.W. and Alvord, E.C., Jr. (eds.), "Allergic" Encephalomyelitis, Charles C. Thomas, Springfield, Ill., pp. 293-299.

Kies, M.W., Driscoll, B.F., Seil, F.J. and Alvord, E.C., Jr. (1973) Myelination inhibition factor: dissociation from induction of experimental allergic encephalomyelitis, Science 179:689-690.

Kristensson, K., Wisniewski, H.M. and Bornstein, M.B. (1976) About demyelinating properties of humoral antibodies in experimental allergic encephalomyelitis, Acta Neuropathol. 36:307-314.

Laatsch, R.H., Kies, M.W., Gordon, S. and Alvord, E.C., Jr. (1962) The encephalomyelitic activity of myelin isolated by ultracentrifugation, J. Exp. Med. 115:777-778.

Lampert, P.W. and Kies, M.W. (1967) Mechanism of demyelination in allergic encephalomyelitis of guinea pigs - an electron microscopic study, Exp. Neurol. 18:210-223.

Lebar, R., Boutry, J.M., Vincent, C., Robineaux, R. and Voisin, G.A. (1976) Studies on autoimmune encephalomyelitis in the guinea pig. II. An in vitro investigation on the nature, properties and specificity of the serum-demyelinating factor, J. Immunol. 116:1439-1446.

Lumsden, C.E. (1966) Immunopathological events in multiple sclerosis, in Proceedings of the Fifth International Congress of Neuropathology, Zurich, 1965 (Excerpta Medica International Congress Series No. 100), Excerpta Medica, Amsterdam, pp. 231-239.

Lumsden, C.E. (1971) The immunogenesis of the multiple sclerosis placque, Brain Res. 28:365-390.

Raine, C.S. and Bornstein, M.B. (1970a) Experimental allergic encephalomyelitis: an ultrastructural study of experimental demyelination in vitro, J. Neuropathol. Exp. Neurol. 29:177-191.

Raine, C.S. and Bornstein, M.B. (1970b) Experimental allergic encephalomyelitis: a light and electron microscopic study of remyelination and "sclerosis" in vitro, J. Neuropathol. Exp. Neurol. 29:552-574.

Raine, C.S., Hummelgard, A., Swanson, E. and Bornstein, M.B. (1973) Multiple sclerosis: serum-induced demyelination in vitro: a light and electron microscopic study, J. Neurol. Sci. 20:127-148.

Saida, T., Abramsky, O., Silberberg, D.H., Pleasure, D., Lisak, R.P. and Manning, M. (1977) Anti-oligodendrocyte serum demyelinates cultured CNS tissue, in Society for Neuroscience, 7th Annual Meeting, p. 527.

Seil, F.J. (1977) Tissue culture studies of demyelinating disease: a critical review, Ann. Neurol. 2:345-355.

Seil, F.J., Falk, G.A., Kies, M.W. and Alvord, E.C., Jr. (1968) The in vitro demyelinating activity of sera from guinea pigs sensitized with whole CNS and with purified encephalitogen, Exp. Neurol. 22:545-555.

Seil, F.J., Rauch, H.C., Einstein, E.R. and Hamilton, A.E. (1973) Myelination inhibition factor: its absence in sera from subhuman primates sensitized with myelin basic protein, J. Immunol. 111:96-100.

Seil, F.J., Smith, M.E., Leiman, A.L. and Kelly, J.M. (1975a) Myelination inhibiting and neuroelectric blocking factors in experimental allergic encephalomyelitis, Science 187:951-953.

Seil, F.J., Kies, M.W. and Bacon, M. (1975b) Neural antigens and induction of myelination inhibition factor, J. Immunol. 114:630-634.

Seil, F.J., Kies, M.W., Rauch, H.C. and Smith, M.E. (1976a) Myelination inhibition factor: its absence in sera from basic protein-sensitized animals, in Shiraki, H., Yonezawa, T. and Kuroiwa, Y. (eds.), The Aetiology and Pathogenesis of the Demyelinating Diseases, Japan Science Press, Tokyo, pp. 243-253.

Seil, F.J., Leiman, A.L. and Kelly, J.M. (1976b) Neuroelectric blocking factors in multiple sclerosis and normal human sera, Arch. Neurol. (Chicago) 33:418-422.

Ulrich, J. and Bornstein, M.B. (1973) Experimental allergic encephalomyelitis (EAE): delayed myelination-inhibition in vitro with EAE-serum: changes in vulnerability of oligodendroglia and newly formed myelin sheaths, Acta Neuropathol. 25:138-148.

Wolfgram, F. and Duquette, P. (1976) Demyelinating antibodies in multiple sclerosis, Neurology (Minneap.) 26 (no. 6, part 2):68-69.

Yonezawa, T., Ishihara, Y. and Sato, Y. (1969) Demyelinating antibodies of experimental allergic encephalomyelitis and peripheral neuritis, represented by demyelinating pattern in vitro, J. Neuropathol. Exp. Neurol. 28:180-181.

Yonezawa, T., Saida, T. and Hasegawa, M. (1976) Myelination inhibiting factor in experimental allergic encephalomyelitis and demyelinating diseases, in Shiraki, H., Yonezawa, T. and Kuroiwa, Y. (eds.), The Aetiology and Pathogenesis of the Demyelinating Diseases, Japan Science Press, Tokyo, pp. 255-263.

STUDIES ON CEREBROSIDE ORGANISATION IN CENTRAL NERVE MYELIN.

C. Linington and M.G.Rumsby,
Department of Biology, University of York,
Heslington, York, YO1 5DD, U.K.

The high proportions of cerebroside, accounting for some 17% of the total lipid of the membrane, are a distinctive feature of the myelin sheath in central nerve tissue. However, the function and molecular arrangement of this glycosphingolipid in myelin is not clearly defined (1,2). We are interested in understanding how cerebroside is distributed between the external and cytoplasmic apposition surfaces of compact myelin and in determining the function of the lipid in the system. Our approach is to use a combination of enzymatic and chemical labelling techniques with histochemical observations. Differential scanning calorimetry has been employed to provide data on the way in which cerebroside interacts with other myelin lipids and with myelin proteins.

Myelin is isolated by standard procedures to give preparations in which the multilamellar structure of the membrane is preserved and in which the external apposition surface in individual membrane fragments is exposed to the environment. Studies with galactose oxidase and sodium metaperiodate suggest that there may be two pools of cerebroside in the myelin membrane which differ in their accessibility. One pool, containing about 40% of the total cerebroside and including hydroxy and non - hydroxy fatty acid cerebrosides equally, is easily accessible to galactose oxidase (MW 60,000 daltons) at 20° and to periodate at 0°. The remaining cerebroside can be oxidised very rapidly by periodate at 20°. Comparative studies with cerebroside - containing liposomes suggest that periodate is impermeable to membranes under the conditions used. Our conclusions from these experiments will be reported. Ferritin - conjugated galactose - specific lectins and anticerebroside antisera used as electron dense histochemical markers indicate that the external surface of myelin is cerebroside - rich. Such markers are now being used with isolated oligodendrocytes to look at the disposition of cerebroside in the external and cytoplasmic faces of the plasma membrane which is in continuity with compact myelin *in situ*.

The normally very high endothermic phase transition of hydrated cerebrosides can be lowered by interaction with cholesterol (3) or other glycerophospholipids (4). Results showing how different myelin lipids and proteins influence the phase transition of hydrated cerebroside will be presented.

1. Rumsby, M.G. & Crang, A.J.(1977) Cell Surf. Rev. 4: 247 - 362
2. Rumsby, M.G. (1978) Biochem. Soc. Trans. 6: 448 - 462
3. Oldfield, E. and Chapman, D. (1972) FEBS Lett. 23: 285 - 297
4. Clowes, A.W., Cherry, R.J. and Chapman, D.(1971) Biochim.Biophys.Acta 249: 301 - 307

PROBE LABELLING STUDIES WITH THE MYELIN BASIC PROTEIN.

A. G. Walker and M. G. Rumsby,
Department of Biology, University of York,
Heslington, York, YO1 5DD, U.K.

A summary of recent results from a number of different sources has concluded (1) that the myelin basic protein (MBP) is located on the cytoplasmic apposition surface in myelin where it may have an important role in maintaining the compaction of this apposition region by a combination of ionic and hydrophobic interactions. The structural form of the basic protein at this cytoplasmic apposition site in compact myelin is of considerable interest and there are suggestions that it could link the two adjacent surfaces of the apposition in either a monomeric (1, 2) or a dimeric (3) form. Our present probe labelling studies with the MBP are designed to resolve this question in part.

Several small covalently - reacting probe molecules are being used to look at structural features of the MBP. Initially we have concentrated on the use of trinitrobenzene sulphonic acid (TNBS) which reacts mainly with the ε- amino groups on lysyl residues in the protein. Conditions for TNBS reaction with the MBP have been worked out as being low ionic strength, $25°$, pH 7.8, stirring under nitrogen and in the dark. A 40 - fold excess of probe to available reactive groups on the protein is used. Because the MBP is known to aggregate at alkaline pH and reaction conditions are pH 7.8 for TNBS labelling we have examined in detail the increase in light scattering which occurs with the MBP in solution in water with increasing pH. Reversible aggregation is most apparent above pH 8.5 but we notice an earlier intermediate phase of aggregation between pH 6 and 8. Titration of the basic protein in solution with base reveals that there is no buffering capacity of the protein at the pH used for probe labelling. With labelling under nitrogen the amount of base used up in such reactions can be used to monitor TNBS binding to the protein.

Initial results from spectroscopic methods of quantitating TNBS reaction with the MBP indicate that some 5 lysyl residues out of the total possible of 13 in the whole molecule are labelled. This contrasts with the 8 residues labelled by acetylation (4). Our preliminary figures have to be re - examined by other methods because of possible photodegradative effects. Current results looking at which lysyl residues along the polypeptide chain are labelled will be presented and conclusions on the structural role of the MBP at the cytoplasmic apposition will be considered.

1. Rumsby, M.G. and Crang, A.J. (1977) Cell Surf. Rev. 4: 247 - 362
2. Jones, A.J.S. and Rumsby. M.G. (1977) Biochem. J. 167: 583 - 591
3. Smith, R. (1977) Biochim.Biophys.Acta 470: 170 - 184
4. Steck, A.J. et al.(1976) Biochim.Biophys.Acta 455: 343 - 352

PROTEOLIPIDS OF RAT BRAIN MYELIN

H. Ch. Buniatian and K.H. Manukian
Institute of Biochemistry, Academy of Sciences of the Armenian SSR,
Erevan, USSR.

The proteolipids (PL) of myelin accounted for 71 % of PL of rat brain homogenates and 30 % of total proteins of myelin. In sodium dodecylsulfate disc electrophoresis the purified PL of myelin showed one major and one minor band with molecular weights of 34 - 36000 and 28 - 30000 respectively. The first major band corresponded to the main band of PL isolated from white matter of bovine brain. The second band resembled protein DM-20 discovered in myelin by Agrawal (1972). These proteins were also obtained from myelin.

The phospholipid content of the crude PL fraction comprised 30 %. Some phospholipids were lost in the course of purification of PL, their content decreasing from 25 to 7 %. The main phospholipids of the crude PL fraction were phosphatidyl-ethanolamine, phosphatidyl choline, the content of phosphatidyl-serine was somewhat lower. The neutral phospholipids of PL which were loosely bound, gradually decreased during purification while the percentage of acid phospholipids, especially that of phosphatidyl-serine increased from 25 to 63 %. PL of myelin that have been washed with ethanol-ether at 20° C contained only acid phospholipids of which 62 % were phosphatidyl-serine, 12 - 14 % - monophosphoinositol and 11 % cardiolipin.

RELATIONSHIP BETWEEN AXONAL TRANSPORT AND THE WOLFGRAM PROTEINS OF MYELIN

P.P. Giorgi
Institute of Anatomy, University of Lausanne, 1011 Lausanne CHUV, Switzerland.

We are investigating the possible metabolic relationship between the axon and its myelin sheath (1). Radioactive amino acids injected into the eyes of adult rabbits are incorporated into proteins by retinal ganglion cells and transported along the optic nerve and tract (2). Proteins associated with myelin purified from the optic tract become considerably radioactive (14.2% $\pm$ 1.8 of total tissue protein radioactivity) with a peak at 11-15 days after injection of ^{3}H-leucine or ^{3}H-glycine. Labelling of myelin due to the supply of radioactive precursors through the blood stream is too low to explain this phenomenon (1). Contamination of purified myelin by radioactive intra-axonal proteins is under investigation. The characterisation of myelin proteins (by myelin subfractionation and protein electrophoretic separation) showed that the basic protein and the proteolipid protein had very low levels of radioactivity (probably due to systemic labelling). About 60% of the myelin radioactivity was associated with three polyacrylamide gel bands behaving as Wolfgram proteins. Confirmation that one (or more) of these bands corresponds to two antigenically similar Wolfgram proteins recently found to be localized in the myelin sheath (3) would rule out the contamination hypothesis.

Future work will be based on the following working hypothesis. a) A certain amount of myelin-associated radioactivity may be due to axonal contamination (probably neurofilament protein). b) A certain amount of myelin-associated radioactivity would be present in the intact myelin sheath. Degradation of slowly transported axonal proteins could supply radioactive amino acids to a local glial compartment specialised in synthesizing Wolfgram proteins and dependent on the axon for supply of precursors. (Supported by grant 3.064-0.76 from the Swiss National Science Foundation).

(1) Giorgi, P.P. et al (1973) Nature New Biol. 244: 121
(2) Karlsson, J.O. and Sjöstrand, J. (1971) J. Neurochem. 18: 749
(3) Roussel, G. et al (1978) J. Neurocytol. 7: 155

HETEROGENEOUS ORGANISATION OF AXONS, MYELIN AND GLIAL CELLS ALONG THE OPTIC NERVE AND TRACT

P.P. Giorgi and H. Du Bois
Institute of Anatomy, University of Lausanne, 1011 Lausanne CHUV, Switzerland.

The optic nerve is commonly considered to be uniform along its length. This is in spite of biochemical (1) and morphological (2) evidence for a distal-proximal (eye to brain) heterogeneity in the degree of myelination. Because of conflicting reports on this evidence, we sought more information in adult rabbits by combining several approaches.

After subcellular fractionation we obtained an increasing distal-proximal yield of purified myelin (freeze-dried myelin weight / tissue wet weight). Because of possible differences in water content and volume of the connective tissue compartment, these results are not sufficient evidence for heterogeneity in the degree of myelination. From electron micrographs of transverse sections taken at different levels of the optic pathway, we measured axon area (AA), myelin area (MA) and myelin thickness (MT) using a graphics tablet linked to a Nova 2 computer. An increase in the MA/AA ratio and the gradual establishment of a bimodal MT spectrum along the distal-proximal direction of the optic nerve were observed, without substantial changes in AA.

To integrate these results with information on the myelin synthesizing cells, we incubated different segments of the optic pathway in vitro (3) in the presence of ^{3}H-leucine. After three hours incubation, the radioactivity associated with the TCA-soluble, TCA-insoluble and the purified myelin fraction was determined for each segment. Data showed a distal-proximal decrease in the net amino acid incorporation into total protein. However, the lower incorporation of the proximal segments was directed more specifically towards myelin proteins. This is despite the fact that the specific activity of the myelin fraction decreases along the distal-proximal direction of the pathway.

Our results suggest that the axon does not change in calibre along its course from the eye to the chiasma (at least) but is surrounded by a sheath of myelin which gradually becomes thicker and metabolically less active. This would suggest a heterogeneity of oligodendrocytes along the optic pathway (in terms of relative number and/or metabolic activity) which deserves further investigation.
(Supported by grant 3.064-0.76 from the Swiss National Science Foundation to P.P.G.)

(1) Friede, R.L. et al (1971) J. Anat., 108: 365
(2) Treff, W.M. et al (1972) J. Microscopy, 95: 337
(3) Rawlins, F.A. and Smith, M.E. (1971) J. Neurochem., 18:1861

BIOCHEMICAL CHARACTERIZATION OF THE RABBIT OPTIC NERVE DURING ONTO-
GENETIC DEVELOPMENT

H. Tauber, V. Neuhoff and T.V. Waehneldt
Max-Planck-Institut für experimentelle Medizin, D-3400 Göttingen

Portions of total homogenates (TH) of optic nerves from rabbits (1-70 days, postnatal) were subjected to high speed centrifugation to yield a supernatant fraction (SOL) and a particulate fraction (PART). The three fractions were analyzed for proteins by SDS-PAGE and the activities of 2',3'-cyclic nucleotide 3'-phophohydrolase (CNP) and of acetylcholinesterase (AChE) were determined.

In the SOL fraction no myelin proteins could be identified during the entire 70 days. Substantial similarities were found comparing the TH with the PART fractions. A doublet migrating in the region of the Wolfgram protein was detected as early as 3-4 days of which the upper band showed a decrease with age and was identical in rate of migration with the axolemmal V band of Matthieu et al. (1977). Traces of proteolipid protein (PLP) were first seen on the 7th day with a gradual increase thereafter. The DM-20 protein slowly appeared from the 8th to the 9th day while the basic protein (BP) was seen abruptly on the 8th day. These results are indicative of an overall synchronization of protein deposition with however a slightly staggered appearance of PLP followed by BP.

The levels of AChE did not change with age in all three fractions examined. By contrast, CNP was low up to the 7th day (around 50 µmol/mg protein . h in TH) to increase significantly after day 8. At day 25 specific activities of about 1100 were found which decreased to 700 at adulthood. The CNP levels in the SOL fractions were negligible.

In particulate fractions banding between 0.32 and 1.0 M sucrose from the optic pathway of adult rabbits, following CNP values were found for optic nerve, chiasm, and optic tract, respectively: 750 $\pm$ 7, 680 $\pm$ 10, 600 $\pm$ 18, while the TH of these regions were virtually identical, i.e., 380 $\pm$ 10.

Matthieu, J.-M., Webster, H. deF., Bény, M. and Dolivo, M. (1977)
    Brain Res. Bull. 2: 289-298

COMPARISON OF CNP-ASE ACTIVITY FROM WHITE MATTER, MYELIN AND
CULTURED NEURONAL AND NON-NEURONAL CELLS
P.Clapshaw, J.Oey, W.Seifert
Max-Planck-Institute, Friedrich-Miescher-Laboratorium
Tübingen, W-Germany

Our studies are aimed at elucidating the biological function of the "myelin-specific" enzyme 2´3´- Cyclic Nucleotide -3´- Phosphohydrolase (CNPase). CNPase has been purified in our laboratory by a modification of the detergent free method of Guha and Moore (1975) from white matter of bovine brain as well as from myelin (purified by the method of Waehneldt, Matthieu and Neuhoff, 1977). This isolated CNPase has been further analyzed by sucrose gradient centrifugation and polyacrylamide gel electrophoresis. The molecular weight is between 40 and 50.000 (Clapshaw and Seifert, 1977).

Both stimulatory and inhibitory conditions for CNPase have been investigated in a series of experiments. A surprising 10-fold non-specific stimulation by serum from animals injected or not injected with isolated CNPase may account for the fact that this enzyme has been isolated by several groups but no antiserum has been reported to date. We are currently attempting to raise antibodies against CNPase by several methods designed to circumvent this difficulty.

Kinetic analysis of our isolated enzyme reveals a Km of approximately 8 mM which is consistent with that reported by Drummond et al. (1962). This high Km has led us to investigate several possible sources for substrate activity, notably whole cell soluble fraction as well as isolated myelin.

We have examined the activity of CNPase in several neuronal, glial and fibroblastic cell lines. For each cell type the specific activity of intact cells (ecto-enzyme activity) was compared with that of cell homogenates (total activity) as well as that of membrane fractions (membrane-bound activity). These studies reveal that CNPase is not restricted to myelin and oligodendroglial cells, but a rather ubiquitous enzyme which must have an important general function in the cellular metabolism.

Clapshaw, P. and Seifert, W. (1977) Hoppe-Seylers Z.Physiol. Chemie
   358: 1189 - 1189
Drummond, G., Iyer, N. and Keith, J.(1962) J. Biol. Chem.
   237: 3535 - 3539
Guha, A. and Moore, S. (1975) Brain Research
   89: 279 - 286
Waehneldt, T.V., Matthieu, J.-M. and Neuhoff, V. (1977) Brain Research
   138: 29 - 43

PURIFICATION OF 2',3'-CYCLIC NUCLEOTIDE 3'-PHOSPHOHYDROLASE (CNPase) FROM BOVINE
CEREBRAL WHITE MATTER

Y. TSUKADA and H. SUDA
Department of Physiology, KEIO University, School of Medicine,
Shinanomachi 35, Shinjuku-ku, Tokyo, JAPAN

2',3'-Cyclic nucleotide 3'-phosphohydrolase (CNPase) is found with its high concentration in the CNS and well known as a good marker enzyme of myelin sheath(4). As to the extraction of CNPase from cerebral white matter, it had been reported that the treatment with some organic solvents(1) or guanidium chloride(3) was effective. But we showed that guanidium chloride of 1.0 M led to lose CNPase activity(5).

In this paper, we reported on the solubilization and purification of CNPase from the white matter of the bovine cerebrum. To remove easily extractable protein, the tissue was homogenized in 1.0% Triton X-100·10 mM tris HCl pH 6.9 , stirred for 30 min and centrifuged at 105,000g for 60 min. 70% protein in the homogenate was released into the supernatant, but CNPase was remained in the precipitate. Then the precipitate was treated twice with 1.0% Triton X-100·1.0 M ammonium acetate·10 mM tris HCl pH 8.2 to solubilize CNPase. By this treatment, CNPase was obtained over 90% in the 105,000g X 60 min supernatant. In this supernatant, PLP and BP from myelin were also present(2). On the next step, CNPase fraction was chromatographed on Sepharose 6B column. CNPase was recovered with wide distribution. The fractions containing CNPase were collected and rechromatographed on Sephadex G-100 column. On this step, CNPase was sharply fractionated with a single peak. CNPase fraction after Sephadex-chromatography was subjected again on Phenyl-Sepharose CL-4B column. CNPase was eluted with a single peak on 1.75 M ammonium acetate·3.5% Triton X-100. The specific activity of CNPase in this fraction was 800 times higher than the original homogenate of the white matter. The yield of CNPase was over 85% of original homogenate. The CNPase fraction finally obtained was applied on Reisfeld disc gel electrophoresis containing 0.5% Triton X-100. CNPase activity was found in a single band of Rf 0.25, but main and minor protein bands were at Rf 0.25 and 0.40 respectively. We assumed that Rf 0.40 band was the contamination of PLP.

In conclusion, CNPase was able to solubilize by the use of detergent and neutral salt, and it was highly purified further by the column chromatography. Then, it was assumed that CNPase was a proteolipid and basic protein having molecular weight of 20,000 - 30,000 and it could be different from PLP and BP by the data resulting from the SDS gel electrophoresis. Km value of CNPase was found to be 3.5 mM.

1)DRUMOND,G.I.,IYER,N.T. and KEITH,J., J.Biol.Chem. 237 3535 (1962) 2)ENG,L.F.,CHAO, F.C.,GERSTL,B.,PRATT,D.and TAVASTSTJERNA,M.G., Biochemistry 12 4455 (1968) 3)GUHA, A.and MOORE,S., Brain Res. 89 279 (1975) 4)KURIHARA,T.and TSUKADA Y., J.Neurochem. 14 1167 (1967) 5)TSUKADA,Y.,NAGAI,K.and SUDA,H., Bull.Jap.Neurochem.Soc.16 85 (1977)

THE ACTIVITY OF THE MYELIN-SPECIFIC CHOLESTEROL ESTER HYDROLASE IN
HUMAN CEREBROSPINAL FLUID

H. Reiber and W. Voss
Neurochem. Labor d. Neurologischen Klinik and Max-Planck-Institut für
experimentelle Medizin, Forschungsstelle Neurochemie, Göttingen (FRG)

The cerebrospinal fluid (CSF) of patients afflicted with various
neurological diseases was analysed for its myelin-specific cholesterol
ester hydrolase (EC 3.1.1.13, Eto and Suzuki). A significant difference
between "normal" CSF and that taken from multiple sclerosis patients
was observed.

In a recently published study (Shah and Johnson) the activity of the
microsomal species of the cholesterol ester hydrolase was measured in
CSF. In this investigation the myelin-specific species could not be determined since the detergent employed was unsuitable. We therefore used
the method of Igarashi and Suzuki where sodium-taurocholate as detergent
and phosphatidylserine (bovine brain) was used. Furthermore the CSF was
concentrated 5-10 fold.

Control CSF (normal protein levels, normal cell numbers and normal
immunglobulin fractions) had a cholesterol ester hydrolase activity
range between $\underline{80 - 170} \times 10^{-11}$ mol/h/ml CSF (n= 10 cases). In contrast
a cholesterol ester hydrolase activity of $\underline{3 - 36} \times 10^{-11}$ mol/h/ml CSF
was found in CSF of patients suffering from multiple sclerosis (n= 16
confirmed cases). The difference between enzyme activity in CSF of control and MS patients is significant and no overlap in the two value
ranges is observed. The enzyme activity has also been determined in CSF
from patients (n=12) suffering from other neurological diseases (encephalitis, meningitis, Lues cerebrospinalis etc.) where it was found to
be in the control range.

Although the present study involved only a limited number of cases
the results strongly suggest that the activity of the myelin-specific
cholesterol ester hydrolase in the CSF may be used as a diagnostic procedure in detecting MS disease. Furthermore these findings may contribute
to the understanding of the events in demyelinating diseases.

Eto, Y. and Suzuki, K. (1973) J. Biol. Chem. 248: 1986-1991
Shah, S.N. and Johnson, R.C. (1978) Exp. Neurology 58: 68-73
Igarashi, M. and Suzuki, K. (1977) J. Neurochem. 28: 729-738

SOME CHEMICAL AND PHYSICAL STUDIES ON MYELIN ISOLATED FROM DEVELOPING RAT BRAIN

F. Abulaban and C.A. Lovelidge
Biochemistry Department, Guy's Hospital Medical School, London SE1 9RT, England.

It has been shown that, of the myelin-associated proteins, the Wolfgram protein (WP) decreases and the Proteolipid protein (PLP) and Basic proteins (BP) increase in relative proportion during development (Waehneldt and Neuhoff (1974), Banik and Smith (1977) etc.).

Myelin has been prepared from whole brains of developing rats aged 13 days to 4 months and it has been found that the protein content of the myelins, as a proportion of the dry weight, increases from 3.8% at 13 days to 20% at 60 days, from which time it remains constant. Similar results were reported by Eng and Noble (1968). Such observations suggest that the PLP and BP are added to myelin during development, diluting out, rather than replacing the proteins present in the immature membrane. The results may also explain, at least in part, the observed increase in brain protein as a proportion of brain weight during development (Waenheldt and Neuhoff (1974)).

Further support for the validity of the results has come from density measurements, from solubilization studies and from observations of the effect of sonication on the developing myelins.

Flotation density in sucrose was found to vary from 1.0608 for myelin containing 12.2% protein from 15 days old rats, to 1.0800 for myelin containing 20% protein isolated from adult rats.

The minimum amount of lysophosphatidylcholine (LPC) required to fully solubilize myelin (W_{min}) (Gent et al. (1961)) was also found to depend on the protein content of the myelins (2.08$\pm$0.30 mg LPC/mg myelin protein). Further, the lipid-protein complexes, products of LPC-solubilised myelin (Gent et al. (1971)), varied in proportion during development.

Sonication produced major changes in the myelins isolated from animals up to 25 days, but from 30 days onwards the myelins were extraordinarily resistant to this treatment.

The results suggest that myelin may be synthesized initially as an extensive lipid bilayer containing very small quantities of WP and that to this "base" are added protein-rich lipid-protein complexes containing PLP and BP, which bring about compaction and physical stability of the myelin.

Banik, N.L. and Smith, M.E. (1977) Biochem. J. 162: 247-255
Eng, F.E. and Noble, E.P. (1968) Lipids 3: 157-162
Gent,W.L.G.,Gregson, N.A.,Gammack,D.B. and Raper,J.H.(1964)Nature 204: 553-555
Gent,W.L.G.,Gregson,N.A.,Lovelidge,C.A. and Winder,A.F. (1971) Biochem.J. 122:63p
Waehneldt, T.V. and Neuhoff, V. (1974) J. Neurochem. 23: 71-77

STUDIES ON CNS MYELINOGENESIS

E. Zaprianova, M. Christova, M. Staykova, I. Goranov
Institute of Morphology, Bulgarian Academy of Sciences,
Sofia, Bulgaria

The initiation and the mechanism of myelination remain an enigma (Raine, 1977).

We have studied the localization and the sites of synthesis of myelin-typical lipids and proteins during myelination of CNS using biochemical, histochemical, electron microscopic and autoradiographic methods (Zaprianova, 1970; Zaprianova, 1976). Our results showed that prior and at the very begining of myelination myelin lipids occured in the oligodendrocytes. During active myelination no myelin lipids could be detected histochemically in these cells but such lipids appeared in the neurones. The nerve cells displayed also an intensive activity of the enzymes involved in the lipid synthesis. They showed ultrastructural signs characteristic of an intense lipoprotein synthesis, observed in the myelinating oligodendrocytes too. This is in agreement with the results of Binaglia et al. (1973) who founded that neuronal cell-enriched fraction posseses a much higher rate of synthesis of phosphatidylcholine and phosphatidylethanolamine than glial cells. Our recent immunofluorescent studies visualized myelin basic protein in the neuronal pericarya of the guinea pig brain during active myelination. It has been reported that myelin basic protein is not present in the neurons in the brain of newborn rats (Sternberger et al., 1978). A possible species differences should be taken into consideration since we could not observe myelin lipids in the neurones of the rat brain as it was demonstrated in the rabbit and guinea pig brains.

The experimental data reported point to a cooperation of neurone and oligodendrocyte in CNS myelinogenesis. Some of myelin-typical lipids and proteins might be synthesized in the neurones and transported to the oligodendrocyte membrane in the period of myelination.

Binaglia, L., Goracci, G., Porcellati, G., Roberti, R and Woelk, H.
    (1973) J. Neurochem. 21: 1068-1076
Raine, C.S. In Myelin, Plenum Press (1977) 1-41
Sternberger, N.H., Itoyama, J., Kies, M.W. and Webster, H. de F.
    (1978) J. Neurocytol. 7: 251-263
Zaprianova, E. (1970) Acta Anat. 75: 276-300
Zaprianova, E. (1976) Ann. Histochim. 21: 223-227

METABOLIC STUDIES WITH NEURONES, ASTROCYTES AND OLIGODENDROCYTES ISOLATED FROM WHOLE RAT BRAIN TISSUE.

S.-W. Chao and M. G. Rumsby,
Department of Biology, University of York,
Heslington, York, YO1 5DD, U.K.

We have developed a method which allows us to isolate neurone, astrocyte and oligodendrocyte cell fractions from whole rat brain tissue (1). This method is now being used in studies looking at the chemical composition and metabolic properties of these different cell lines in brain tissue especially in relation to the synthesis of myelin - specific lipids and proteins. That the myelin sheath is in direct continuity with the plasma membrane of the oligodendrocyte is well established (2) but there is still some doubt as to whether cell types other than oligodendrocytes contribute to the synthesis of some of the lipids and proteins of mature myelin (3). Our present work is partly designed to obtain answers to this problem.

The uptake and utilisation of radiolabelled substrates by neurones, astrocytes and oligodendrocytes isolated from whole rat brain tissue is being examined in depth with experiments in which the _in vivo_ and _in vitro_ behaviour of the different cell types is compared. The peak of glucose uptake for all three cell types occurs at 17 days of age. This corresponds well with the period when myelination is most active. At this stage, however, glucose uptake by oligodendrocytes is some 6 times greater on a per cell basis than is found for neurones and astrocytes. In 17 day old brain tissue neurones and astrocytes show a 2 - fold increase in glucose uptake compared with 3 months of age; for oligodendrocytes the same ratio is 8 - fold. Incorporation of radioactivity from glucose into myelin is some 7 times greater at 17 days of age compared with 3 months. With 30 day old animals glucose uptake into neurones, astrocytes and oligodendrocytes is maximal two hours after intraperitoneal injection while for myelin the peak uptake is at 3 days. Incorporation of radioactivity from glucose into lipid by oligodendrocytes and neurones decreases as animal age increases from day 12 to day 30; astrocytes show a 2 - fold increase in incorporation into lipid over this same period. For oligodendrocytes and myelin some 80 - 90% of the total radioactivity incorporated into lipid from glucose is accounted for in cholesterol + cerebroside + phospholipid; for astrocytes and neurones the corresponding figures are 62% and 64% respectively. The significance of the results and data from experiments with palmitic acid and leucine will be described in full.

1. Chao, S.-W. and Rumsby, M. G. (1977) Brain Res. _124_: 347 - 351
2. Bunge, R.B. (1968) Physiol. Rev. _48_: 197 - 251
3. Giorgi, P.P. and Field, E.J. (1973) IRCS Med. Sci. _2_: 2 - 4

BIOCHEMICAL CHARACTERIZATION OF PNS MYELIN SUBFRACTIONS OBTAINED BY CONTINUOUS ZONAL GRADIENT CENTRIFUGATION

T.V. Waehneldt and J.-M. Matthieu
Max-Planck-Institut für experimentelle Medizin, D-3400 Göttingen, and Service de Pédiatrie, CHUV, CH-1011 Lausanne

Aqueous homogenates of frozen sciatic nerves from adult rabbits were centrifuged on continuous zonal gradients (0.1-1.2 M sucrose, 90,000 g_{av}, 3.5 h). Based on optical density at 300 nm, three peaks were obtained, with the following protein proportions and sucrose molarities: A, 1.5%, 0.102; B, 6.7%, 0.308; C, 91.9%, 0.571. While the central portion of peak C showed low values both for AChE and CNP (0.57 ± 0.23 and 82 ± 17 µmol/mg protein . h, respectively), peak B had comparable values for AChE with however elevated CNP levels. In contrast, the low amount of material in peak A showed an opposite trend. Neglecting proteins of high molecular weights, analyses of the 6 fractions by SDS-PAGE (A = fraction I; B = fraction II; C = fraction III-VI) demonstrated an increase of the P_O glycoprotein from fraction III to VI of peak C whereas the sum of the two basic proteins P_1 + P_2 decreased from the lightest fraction III to the heaviest fraction VI. The intermediate proteins P_3 and P_4 showed slight increases toward the heavy side of peak C. While the proteins of A displayed no significant differences when compared to C the proteins of B showed low values of P_1, an increase of P_3 and some enhancement of P_O.

Besides the occurrence of two smaller and lighter peaks this study demonstrated that the bulk of rabbit PNS myelin banded at a density slightly lower than that of rabbit CNS myelin and that the CNP distribution was very similar to that of rat spinal cord with the specific acitivities about 10 times lower. Although these peaks contained all the typical myelin proteins their proportions showed significant variations in the different peaks indicating heterogeneity in the overall membrane composition of PNS myelin. Whether this particle heterogeneity reflects the in situ situation or is the product of the isolation procedure requires further investigation.

Supported by a twinning grant from the European Training Programme in Brain and Behaviour Research.

MYELIN CONSISTS OF A CONTINUUM OF PARTICLES. COMPARISON BETWEEN NORMAL AND QUAKING MICE ; ANALYSIS IN HUMAN PELIZAEUS-MERZBACHER DISEASE AND INFANTILE NEURONAL CEROID LIPOFUSCINOSIS

J. M. Bourre, S. Pollet, O. Daudu and N. Baumann
Laboratoire de Neurochimie INSERM U. 134, CNRS ERA 421, Hôpital de la Salpêtrière, 75634 Paris Cédex 13, France.

Density gradient centrifugation was used by a number of investigations to separate myelin into subfractions, but its density profile was not determined. Myelin isolated according to Norton and Poduslo consists of a continuum of particles of different densities, as shown by discontinuous sucrose density gradient centrifugation. In normal animals most of the material (65 per cent) is concentrated between 0.6 and 0.7 M sucrose (the maximum being found at 0.66 M sucrose, corresponding to 23 per cent). The density differences among various myelin fractions are related to their protein/lipid ratios, as lighter fractions contain less protein and more lipid. Lipid analysis shows a decrease in the amount of every lipid from the lightest to the heaviest fraction : the light fraction is richer in phosphatidyl-ethanolamine, phosphatidyl-serine and cerebrosides. The distribution is highly abnormal in purified myelin from Quaking mutant ; very low quantities of myelin with normal density are found, but unexpected large amount of high density particles are present, possibly related to a "pre-myelin" material (oligodendroglial) processes which are not maturing into normal myelin.

Human myelin isolated from frozen brain consists also of a continuum of particles as shown by sucrose continuous gradient (between 0.4 and 0.85 M) on zonal rotor. This was determined by measuring the O.D. at 260 and 280 mμ, the protein concentration by Lowry method and sucrose concentration by polarimetry in 60 fractions. Most of the material is concentrated between 21-24 % sucrose, the maximum being found at 22,5 % (0.66 M). The fractions are dialyzed for 62 hrs to remove the sucrose and the pellet obtained after centrifugation shows that the protein concentration increases from the lighter fraction to the heavier.

Myelin from Pelizaeus-Merzbacher diseased child (from Pr. Araoz, U.S.A.) presents a density pattern similar to the Quaking. However, in infantile neuronal ceroïd lipofuscinosis (from Pr. Haltia, Finland) myelin has a nearly normal density profile, although its yield is extremely low ; it appears to reflect a wallerian degeneration in the CNS. In contrast, the defect in Pelizaeus-Merzbacher disease is in the myelin biosynthesis, possibly at the level of very long chain fatty acids. Thus density profile of myelin in an useful tool to determine if dysmyelination is due to defective biosynthesis of myelin or secundary to extra-myelin abnormal events.

IMMUNOHISTOCHEMICAL LOCALIZATION OF GALACTOCEREBROSIDE DURING DEVELOPMENT IN THE CEREBELLUM OF NORMAL AND QUAKING MOUSE.

B. Zalc, M. Monge, P. Dupouey[+] and N. Baumann
Laboratoire de Neurochimie INSERM U. 134, Hôpital de la Salpêtrière, 75634 Paris Cédex 13, France and [+] Laboratoire de Biochimie des Antigènes, Institut Pasteur, 75015 Paris, France.

An immunohistochemical technique has been used to carry out a study on the cellular localization of galactosyl ceramide, biochemically shown to be primarily or exclusively glial and myelinic. The present study was concerned with the indirect immunofluorescent localization of this lipid, during different stages of development in the cerebellum of normal mouse and of the myelin deficient mutant Quaking. Frozen sections of cerebellum, cut in a cryostat, were fixed in cold 4 % buffered formaldehyde. Incubation with antigalactosyl ceramide purified on an immunoadsorbant was required in order to circumvent non specific staining. In the early stages of development, before myelination has started, in both the normal and Quaking mutant, only the fiber tracts were stained. After 18 days of age, the general aspect was unchanged in the Quaking mouse. Only in the normal control appeared a deposition of galactosyl ceramide in the neuronal membranes of the deep granular layer which raises the problem of intercellular interaction in relation to the synthesis of these lipids. There was also a partial negativation of the myelinated fiber tracts. This paradoxical aspect observed in the normal mouse was attributed, either to a non penetration of the antibodies caused by the high degree of compaction of the myelin, or to a masking phenomenon of the cerebroside at the surface of the myelin sheath. This hypothesis has been further studied by submitting the cerebellar sections to various treatments as trypsinisation or alkaline methanolysis or alkaline hydrolysis, prior to the incubation with the antibodies.

THE PROTEIN COMPOSITION OF PNS MYELIN IN NEUROLOGICAL MUTANT MICE

J.-M. Matthieu

Laboratoire de Neurochimie, Service de Pédiatrie, Centre Hospitalier Universitaire Vaudois, CH-1011 Lausanne, Switzerland

The Quaking mutation in mice is characterized by a severe myelin deficit secondary to an arrest of myelinogenesis. Recently, we suggested that the mutation results in dysmyelination rather than hypomyelination (Matthieu et al.,1978). Although less affected than CNS, PNS from Quaking mice is also hypomyelinated (Suzuki and Zagoren, 1977) and this prompted us to investigate the protein composition of PNS myelin from Quaking mice and compare it with that of littermate controls and other myelin-deficient mice, the Jimpy mutants.

Myelin was isolated according to Norton and Poduslo (1973) from pooled sciatic nerves grinded in liquid nitrogen. The Quaking and Jimpy mice were 25 and 16 days of age, respectively. After partial delipidation with ether-ethanol, proteins were solubilized and separated in the presence of sodium dodecyl sulfate on 15% polyacrylamide gels by electrophoresis. The gels were stained with 1% Fast Green and scanned on a spectrophotometer at 580 nm. The relative amounts of protein in the different bands are expressed as a percentage of the dye binding capacity. Both disc and slab gels were used.

The amount of myelin isolated from sciatic nerves of Quaking mutants was only 50% of that from littermate controls. Among the high molecular weight proteins, a band labelled H_3 was decreased in Quaking PNS myelin. P_1 and P_2 were drastically decreased in Quaking mice, while P_0 the major PNS myelin protein was present in normal amounts. The other minor components were relatively increased. The ratio P_0/P_1 was 3.7 in control and 10.0 in Quaking myelin. Similar results were found for the P_0/P_2 ratio. This contrasted with the protein composition of PNS myelin in Jimpy mice which showed no anomaly.

Supported by the Swiss National Science Foundation, Grant 3.684.76.

Matthieu, J.-M., Koellreutter, B. and Joyet, M.-L. (1978) J. Neurochem. 30: 783-790
Norton, W. T. and Poduslo, S. E. (1973) J. Neurochem. 21: 749-757
Suzuki, K. and Zagoren, J. C. (1977) J. Neurocyt. 6: 71-84

SHIVERER MOUSE ; A DYSMYELINATING MUTANT WITH ABSENCE OF MAJOR DENSE LINE AND BASIC PROTEIN IN MYELIN

C. Jacque, A. Privat[+], P. Dupouey[++], J. M. Bourre, T. D. Bird[+++] and N. Baumann
Laboratoire de Neurochimie INSERM U.134, Hôpital de la Salpêtrière, 75634 Paris Cédex 13, France and [+] Laboratoire de Culture du Tissu Nerveux INSERM U.106, Hôpital de Port-Royal, 75014 Paris, France and [++] Laboratoire de Biochimie des Antigènes, Institut Pasteur, 75015 Paris, France and [+++] Veterans Administration Hospital, Division of Neurology and Medical Genetics, Seattle, U.S.A.

Shiverer is a recessive autosomal mutant with defective myelination of the central nervous system (1). Three months old animals were examined. With the electron microscope, the myelin appeared sparse and abnormal. The major dense line was absent ; the oligodendrocytes were packed with vacuoles, lined by an unit membrane whose inner leaflet was thickened. Myelin lipids were drastically reduced, especially glycolipids : cerebrosides and sulfatides were 20 % and G_{M1} 50 % of normal. Very long chain fatty acids were reduced by 80 %. Radioimmunoassay of basic protein showed a 20 fold decrease in brain compared to controls. Immunohistochemistry of basic protein demonstrated a total absence of fluorescence in myelin in contrast with controls. An increased label of astrocytes was shown by immunohistofluorescence using GFA antibodies. A three fold rise in GFA level was evidenced by quantitative immunoelectrophoresis in all structures of CNS (cerebellum, forebrain, brain stem and spinal cord). S 100 was not significantly modified.

These results give evidence that basic protein is located in the major dense line of myelin.

(1) Bird, T.D., Farrel, D.F. and Sumi, S.H. (1977) Transact. Amer. Soc. Neurochem. 8 : 153

LYSOSOMAL HYDROLASES IN THE LYMPHOCYTES AND GRANULOCYTES OF PATIENTS WITH MULTIPLE SCLEROSIS (MS)

P. Riekkinen, J. Palo and J. Wikström
Departments of Neurology, University of Kuopio and University of Helsinki, Finland

Since proteolytic enzymes are activated in the demyelinative process in MS the activities of acid proteinase, beta-glucuronidase and acid phosphatase were measured from extensively purified (up to 99% pure) preparations of peripheral lymphocytes (including monocytes) and granulocytes of 20 MS patients and 10 healthy controls. The results were compared to those reported earlier for lysosomal hydrolases in the total leukocyte extract (Cuzner et al., 1975), in lymphocytes and granulocytes (Riekkinen et al., 1977), and in the serum and CSF of MS patients (Hultberg and Olsson, 1978).

The activities of all three enzymes were higher in the control lymphocytes than in MS lymphocytes. The activity of acid proteinase was higher in MS granulocytes (54.9 $\pm$ 6.29 nM/ mg protein/ h, mean $\pm$ S.E.M.) than in the controls (43.8 $\pm$ 2.19) while no differences were found for the other enzymes. The results were similar when they were expressed per 10^6 cells. However, the activity of acid proteinase decreased and became even lower (23.6 $\pm$ 4.22) than that of the controls (33.2 $\pm$ 6.74). Except for acid proteinase, all activities were on the same level or slightly higher than those reported earlier from our laboratory. Five MS patients were in relapse and had the highest activities in the MS series.

The activities of lysosomal hydrolases are low in MS lymphocytes but, in spite of great individual variations, both acid and neutral proteinase may be activated in MS granulocytes. New specimens will be taken from all patients and controls within a few months to follow up longitudinal changes in the enzyme activities and to correlate the findings to the clinical condition of the patients.

Cuzner, M.L., McDonald, W.I., Rudge, P., Smith, M. and Borshell, N. (1975) J. neurol. Sci. 26: 107-111

Hultberg, B. and Olsson, J.-E. (1978) Acta neur. scand. 57: 201-215

Riekkinen, P., Palo, J. and Asikainen, I. (1977) Acta neur. scand. 56: 83-86

PREALBUMIN CONTENT OF CEREBROSPINAL FLUID AND SERUM IN PERSONS
WITH MULTIPLE SCLEROSIS

Lj.Kržalić and N.Kastrapeli
Institute for Pathological Physiology and Depatement of Neuropsychiatry,Faculty of Medicine,Belgrade,Yugoslavia

The high prealbumin content of the cerebrospinal fluid(CSF), its changes in some neurological diseases as well as the selective diffusion of this protein through cell membranes prompted us to examine the prealbumin level in the CSF in persons affected with multiple sclerosis(MS). As changes in the CSF protein can be the result of its migration from the serum, we also determinated the prealbumin content in the serum of subjects with multiple sclerosis.

CSF and sera were obtained from patients of either sex from the Neuropsychiatric Clinic in Belgrade. Prealbumin was determinated quantitatively in nonconcentrated CSF and serum by electroimmunodiffusion using the Laurell method. Total proteins were determinated by the method of Lowry et al.

The results of our determinations showed a statistically significant increase of CSF prealbumin in persons affected with MS. The serum prealbumin content was the same as in healty persons.

On the basis of our results we assume that: (1) increased CSF prealbumin is not the result of migration of serum prealbumin,(2)increased CSF prealbumin in subjects affected with MS may serve as a contribution to the diagnosis of this disease, and finally (3)changes in CSF prealbumin can be of importance for studying protein changes in the first phase of chemical degradation in the process of demyelination. Increased prealbumin may due to increased intrathecal synthesis or degradation of degenerated tissue of the central nervous system.

RAT MYELIN PROTEINS AND LIPIDS IN EARLY PERIOD OF EXPERIMENTAL
CYANIDE ENCEPHALOPATHY

B. Zgorzalewicz, J. Sędzik
Institute of Neurology and Diseases of Sensory Organs, Academy of
Medicine, Poznań, Poland

The composition of myelin proteins and lipids was studied in the
predemyelinating period of experimental cyanide encephalopathy. The
experiments were performed on Wistar rats in which cyanide intoxication was produced by inhaling of hydrogen cyanide as described by Levine and Stypulkowski /1959/. The animals were sacrificed at different
intervals following HCN poisoning /after 4 hours, 2, 4, and 7 days/.
Myelin was isolated according to the procedure of Norton and Poduslo
/1973/ and myelin protein fractions were obtained by means of SDS
polyacrylamide gel electrophoresis as previously described by Zgorzalewicz et al. /1974/. Myelin lipids were estimated after chromatographic separation - according to the method of Svennerholm /1964/.

An increase of total myelin protein and lipid content, with reciprocal dehydratation of myelin sheath was observed. The most intensive
changes were found at the second day after HCN poisoning. The relative
distribution of particulate myelin proteins and lipids remain unchanged during the time span studied.

The early stage of cyanide induced encephalopathy seems to be the
result of changes in space conformation of myelin proteins and lipids,
which may be connected with the rearrangement of the water molecules
in the myelin membranes, preeceeding further decompositional events,
characteristic for late period of this experimental disease. These
results may speak strongly in favour of the findings of Wender and
Sędzik /1977/ in the X-ray diffraction studies of myelin pattern in
predemyelinating period of experimental cyanide encephalopathy. The
alterations in the profiles of electron density indicate that in the
early period of cyanide induced encephalopathy occurred deep changes
in the physical structure of myelin membrane, as well in its protein
as in the lipid layers.

Levine, S., Stypulkowski, W. /1959/ Arch. Pathol.
  67: 459-469
Norton, W.T., Poduslo, S.K. /1973/ J. Neurochem.
  21: 749-759
Svennerholm, L. /1964/ J. Neurochem.
  11: 839-853
Wender, M., Sędzik, J. /1977/ Zbl. alg. Path.
  121: 281
Zgorzalewicz, B., Neuhoff, V. and Waehneldt, T.V. /1974/ Neurobiology
  4: 264-276

SPECIFICITY OF TRIETHYLLEAD TOWARD MYELIN PROTEIN SYNTHESIS

G. Konat, H. Offner and J. Clausen
The Neurochemical Institute, 58, Rådmandsgade,
2200 Copenhagen N, Denmark.

Intoxication with triethyllead ($PbEt_3$) effectively suppresses the myelin deposition in the developing rat brain (Konat & Clausen 1974). Among different myelin constituents, availability of proteins seems to be the rate limiting factor in the membrane assembly (Konat et al. 1976). The present investigation has been extended to evaluate the effect of $PbEt_3$ on the cerebral protein synthesis.

Young rats were injected intraperitoneally with $PbEt_3$ according to the schedule: Day 20 - 8mg/kg and Day 24 - 5mg/kg body weight. All the lead in the forebrain was found in the form of $PbEt_3$. The $PbEt_3$ tissue concentrations fluctuated between 36 and 27 nmol/g w.w. during the first seven days of intoxication. Thereafter, the $PbEt_3$ level decreased steadily and reached 3nmol/g on Day 34.

The incorporation of ^{14}C leucine into the acid-insoluble protein in the forebrain slices was studied both in the total homogenate and in the purified myelin fraction. The protein synthesis in the slices prepared from the intoxicated forebrains was significantly suppressed as compared to the control tissue. The extent of inhibition was positively correlated with the $PbEt_3$ concentration found in the forebrains. The synthesis of total protein was maximally restrained by 17%, whereas the maximal inhibition of myelin protein synthesis was 49%. The myelin protein synthesis was also inhibited more than the total protein synthesis when $PbEt_3$ was added in vitro to the slices prepared from 27-day-old control forebrain. Thus, 3µM $PbEt_3$ caused 19 and 39% inhibition of the total and myelin protein synthesis respectively.

The results indicate that $PbEt_3$ present in the forebrain is per se toxic to the CNS cells and is responsible for the depressed metabolic activity as exemplified by the protein synthesis. Furthermore, $PbEt_3$ reveals a specificity either toward the myelin-forming cells (oligodendrocytes) or toward processes involved in the furnishing of the membrane proteins regardless of the cell type.

Konat, G. & Clausen, J. (1974) Environ. Physiol. Biochem.
  4: 236-242
Konat, G., Offner, H. & Clausen, J. (1976) Exp. Neurol.
  52: 58-65

MYELIN COMPOSITION AND LIPOGENIC ENZYME ACTIVITY IN BRAIN
TISSUE FROM NORMAL AND B_{12}-DEFICIENT BATS.

R.C. CANTRILL*, L. KERR* AND J. VAN DER WESTHUYZEN**

* Department of Medical Biochemistry, and
** Department of Haematology,
   School of Pathology of the South African
   Institute for Medical Research and
   The University of the Witwatersrand,
   Johannesburg, South Africa.

The Egyptian fruit bat <u>Rousettus aegyptiacus</u> has been proposed as an animal model for the investigation of the neurological complications of vitamin B_{12} deficiency (Green et al, 1975). These animals develop ataxia, dyskenesis and hind limb paralysis; they exhibit climbing difficulties and changes in the flight cycle. Patchy spongiose changes in the white matter of the lower cervical and upper thoracic regions suggestive of early demyelination have also been described (Green et al, 1975). The relationship between vitamin B_{12} deficiency and neurological damage is still unclear.

Vitamin B_{12} is required for the enzymatic conversion of L-methylmalonyl coenzyme A to succinyl coenzyme A. Vitamin B_{12}-deficient patients show increased biosynthesis of odd-chain fatty acids (Barley et al, 1972 Frenkel, 1973) as well as the formation of methyl branched fatty acids (Cardinale et al, 1970).

We have investigated the composition of whole brain and myelin lipids from normal and B_{12}-deficient bats. The substrate specificity of bat brain fatty acid synthetase and acyl-CoA synthetase has also been determined in sub-cellular fractions from normal and B_{12}-deficient animals. The possible role of changes in lipid metabolism in demyelination are discussed.

Barley, F.W., M.G. Sato and R.H. Abeles (1972) J.Biol.Chem. <u>247</u>, 4270-4276.

Cardinale, G.J., T.J. Carty & R.H. Abeles (1970) J.Biol.Chem. <u>245</u>, 3771 - 3775.

Frenkel, E.P. (1973) J.Clin.Invest. <u>52</u>, 1237 - 1245.

Green, R., S.V. van Tonder, G.J. Oettle, G. Cole & J. Metz. Nature <u>254</u>, 148 - 150 (1975).

ALKANE BIOSYNTHESIS BY THE PERIPHERAL NERVOUS SYSTEM

C.Cassagne, D.Darriet and J.M.Bourre

Département de Biochimie, Université de Bordeaux II, 33405 Talence, FRANCE
and Laboratoire de Neurochimie, Hôpital de la Salpêtrière,75634 Paris, FRANCE

The myelin accumulates specifically high amounts of alkanes (Bourre et al., 1977), while mitochondria, synaptosomes and microsomes contain only minor amounts of them (Darriet et al.,1978). A widely accepted idea is that alkanes found in mammalian tissues are almost certainly exogenous. An alternative possibility however is that these very long aliphatic chains are synthesized in situ by the animal cells ; supporting that hypothesis, the alkane concentration in the brain myelin was reduced by 70 % in the Quaking mutant, which is a recessive mutant characterized by a defective myelination (Bourre et al.,1977; Cassagne et al.,1977a).

The excised sciatic nerve of the rabbit is able to synthesize alkanes from 1-^{14}C stearate (Cassagne et al.,1977b).

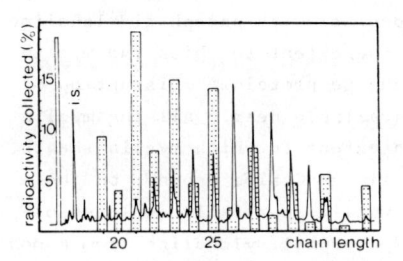

Fig.1. Gas-liquid chromatography of alkanes obtained from a homogenate incubated with [1-^{14}C]stearate.

Table 1
Alkane synthesis by rabbit siatie nerve fractions

	Specific activity (nmol/mg protein/h)	%
Homogenate	0.19	100
20 000 × g pellet	0.09	5.4
150 000 × g pellet	2.4	78.4
150 000 × g supernatant	0.05	16.2

Incorporation of [1-^{14}C]stearate into alkanes. For experimental details see Materials and methods. Results are given as nmol alkane synthesized/mg protein/h. % represents the total activity recovered in each fraction

From stearate as the labelled substrate, the synthesis of alkanes in the C19-C33 range is observed (Fig 1). The synthesis occurs chiefly in the microsomal fraction (Table 1),which contains ca 80% of the total activity; the specific activity in that fraction is the highest ever reported,whatever the enzyme source. The study of the various cofactors involved in that synthesis showed that the malonylCoA is the elongating agent and NADPH is the preferred reductant. When (1-^{14}C)stearate was the substrate, CoA and ATP were also needed, suggesting that alkane synthesis requires stearoylCoA.

Bourre,J.M., Cassagne,C.,Larrouquère-Régnier,S. and Darriet,D.(1977)J.Neurochem. 29:645-648

Cassagne,C.,Bourre,J.M.,Larrouquère-Régnier,S.,and Darriet,D.(1977a) Proc.Int.Soc. Neurochem.6:552

Cassagne,C.,Darriet,D. and Bourre,J.M.(1977b) FEBS Letters 82:51-54

Darriet,D.,Cassagne,C. and Bourre,J.M.(1978) Neuroscience Letters- in press

THE ROLE OF SOLUBLE, CYTOPLASMIC PHOSPHOLIPID EXCHANGE PROTEINS IN MYELIN SHEATH FORMATION AND TURNOVER OF MYELIN PHOSPHOLIPIDS

E. M. CAREY
Department of Biochemistry, University of Sheffield, Sheffield S10 2TN U.K.

Brain cytoplasmic proteins catalyse the transfer of phospholipids between membranes. Their possible involvement in myelin membrane formation when large amounts of phospho- and glycolipids are incorporated into the proliferating plasma membrane of the myelinating cell, and in the replacement of existing myelin lipids, has been studied.

Isolated myelin has a limited capacity to participate in phospholipid exchange, but the exchange is enhanced when myelin is osmotically shocked (Carey and Foster, 1977). The ability of myelin phosphatidylcholine to participate in exchange processes with phosphatidylcholine in other membrane pools is related to the extent to which the myelin membrane is exposed to the soluble exchange proteins. Disruption of the multilamellar organisation makes available nearly all the myelin phosphatidylcholine for exchange. The extent to which myelin sheath lipids are readily degraded by phospholipase C also depends on the extent to which myelin is disrupted. Myelin, partially depleted of phospholipid showed increased uptake of phosphatidylcholine from a donor membrane by a supernatant protein dependent net transfer process, until the amount of phosphatidylcholine in the myelin sheath was restored to that of the untreated myelin. _In vivo_ exchange of phospholipid could be envisaged as occurring where the myelin sheath is in contact with the glial cell cytoplasm.

There was an increase in phospholipid exchange activity of rat brain supernatant between the late foetal/early neonatal period and the period of maximum myelination. In the adult brain, the phospholipid exchange activity is higher in white than in grey matter but as a fundamental cellular activity it is not confined solely to the myelinating cell. All phospholipids except ethanolamine-containing phospholipids are exchanged by soluble, cytoplasmic proteins from brain, whereas with liver supernatant phosphatidylethanolamine is also exchanged. Although the ethanolamine phospholipids of myelin are replaced _in vivo_ it could be by a process other than through the functioning of specific phospholipid exchange proteins

Carey, E.M. and Foster, P.C. (1977) Biochem. Soc. Trans. 5: 1412-1414

THE MYELINATING EXPLANT CULTURE AS A MODEL FOR MYELIN MATURATION IN VIVO

G.E. Fagg, H.I. Schipper[*] and V. Neuhoff

Max-Planck-Institut für experimentelle Medizin, Forschungsstelle Neurochemie, and [*]Neurologische Klinik der Universität, 3400 Göttingen, FRG

Cultures of explanted central neural tissue have been employed extensively for morphological studies of myelination and demyelination, although fewer biochemical investigations have been reported. In the present study, characterisation of the process of myelination in culture is extended to an analysis of myelin protein composition during development.

Fragments of foetal rat spinal cord were cultured in Maximow assemblies essentially as described by Bornstein and Murray (1958). Myelin fractions were prepared (Norton and Poduslo, 1973) from pooled (100-140) explants after 12, 18, 24 and 30 days in vitro (DIV) and, for comparison, from the spinal cords of rats at equivalent developmental ages (5, 11, 17 and 23 days old). Proteins were fractionated by micro-SDS-polyacrylamide gradient gel electrophoresis (Rüchel et al., 1974).

Qualitatively similar protein profiles were obtained for myelin isolated from either cultures or from spinal cords. Quantitatively, myelin from cultures contained a higher proportion of high molecular weight proteins (suggesting greater contamination of the fraction by non-myelin structures) than in that from spinal cords and, in the 12 DIV fraction, the typical low molecular weight myelin proteins were barely detectable (consistent with the light microscopic observation that myelination began after 10-12 DIV). At all later ages studied, however, myelin protein development in culture closely resembled that occurring in vivo; parallelism was particularly apparent in the case of the basic proteins. These data confirm the value of tissue culture systems for studies of both myelin maturation and myelin diseases. (Supported by a fellowship from the Royal Society (GEF) and a grant (SFB 33) from the Deutsche Forschungsgemeinschaft).

Bornstein, M.B. and Murray, M.R. (1958) J. biophys. biochem. Cytol. 4: 499-504
Norton, W.T. and Poduslo, S.E. (1973) J. Neurochem. 21: 749-757
Rüchel, R. et al. (1974) Hoppe-Seyler's Z. Physiol. Chem. 355: 997-1020

STUDIES ON MYELIN ASSOCIATED GLYCOLIPIDS IN CULTURES OF C_6 GLIAL AND DISSOCIATED BRAIN CELLS.

Louis L. Sarlieve, G. Subba Rao, and Ronald A. Pieringer
Temple Univ. School of Medicine, Phila., Pa. 19140

Previous studies from our lab (J.B.C., 252-5884 (1977) showed that triiodothyronine (T_3) caused a precocious and specific accumulation and biosynthesis of the myelin associated galactosyldiacylglycerols (gal DG) in brains of young rats. The interaction of T_3 and other effectors with cells of nerve origin is perhaps best studied in tissue culture. In pursuit of this idea our initial experiments demonstrated the absence of gal. DG and sulfogal. DG in C_6 cells. However, C_6 cells possess a very active sulfotransferase which synthesizes sulfogal. DG from gal DG and PAPS in vitro. T_3 in the C_6 growth media did not stimulate the sulfotransferase activity nor the uptake of $^{35}SO_4$ from the media into lipids. Cells from dissociated brains of one day old mice grown in the presence of fetal calf serum converted more $^{35}SO_4$ into lipid at 8 days in culture better than any other culture age. However, cells derived from the cerebrum of the 15 day mouse embryo took up ^{35}S into sulfolipids at a much faster rate than the 1 day old new born with increasing days in culture (the longest time measured was 19 DIC). The ^{35}S-lipids were characterized as sulfogalactosyl ceramide (80%), and sulfogalactosyl(diacyl and monoalkylmonoacyl) glycerol (20%). There was 70% diacyl and 30% of alkylacyl forms of the latter lipid. T_3 added to these primary cultures had no effect on the synthesis of ^{35}S-lipid nor on the amount of protein synthesized. When rat serum was substituted for fetal calf serum there was a slight augmentation of both of these parameters. However, when grown in the presence of serum derived from rats treated with T_3 both the protein content and the conversion of $^{35}SO_4$ to sulfolipid in the cells increased more than 2-fold. We have encountered some variation in the extent of stimulation in different experiments. The latter may be caused by a near optimum endogenous concentration of T_4 and T_3 in the serum. We are led to this conclusion by the fact that we can observe a stimulation with T_3 more readily with rat serum than with calf serum which has a higher amount (5x) of T_4 than rat serum. Also serum from a hypothyroid rat did not support the development of the cells or their ability to take up ^{35}S into lipid nearly as well as the control.

Symposium

Factors Influencing Neuronal Development

Chairman:
D. Biesold

THE DEVELOPMENT OF ORDER IN NEURAL CONNECTIONS

G. Székely
Department of Anatomy, University Medical School,
4012 Debrecen /Hungary/

Abstract

The "neuronal specificity theory", as formulated by Sperry, attributes differential chemical properties to neurons and mutual chemoaffinities control the development of specific neuronal interconnections. Theoretical considerations reveal a number of problems inherent in the theory. Experimental results which are at variance with the theory are shown. It is concluded that a number of other mechanisms, including the Hebbian coincidence principle in synapse formation, a sequential order in neurogenesis, the morphological characters of neurons, may play an instrumental role in the development of neural organization.

1. Introduction

From the restoration of normal vision, and from rotated vision in the case of eye rotation, Sperry /1944, 1945/ inferred "that the nerve cells of the retina and tectum must acquire cell-unique cytochemical tags" that serve as a kind of identification marker, and control the development of ordered interconnections between related points of the retina and of the tectal visual center. Similar results were obtained from experiments on other systems. Rotated wiping reflexes were evoked in frogs in which the skin on one side of the body was cut out and replaced with a 180° rotation in tadpoles /Miner, 1956/. Stimulation of a supernumerary limb grafted into the back elicited characteristic reflexes of the normal limb /Miner, 1956/. From an extra eye grafted into the head region corneal reflexes of the normal eye could be evoked in salamanders /Weiss, 1942/. These and similar results were interpreted in the sense that neurons in the sensory ganglia acquire from the innervated periphery "cell-unique cytochemical tags" which control the establishment of specific connections with cytochemically matching neurons. The apparent consistency of results obtained with this experimental paradigm warranted a claim to a general application of the theory to the formation of ordered nerve connections. Indeed, a formidable body of data has been collected in favour of the "chemoaffinity theory", or "neuronal specificity theory", as it became called since the first experiment of Sperry /for ref: Gaze, 1970; Jacobson, 1971/. Nevertheless theoretical considerations as well as experimental results reveal difficulties and contradictions inherent in the theory.

2. Theoretical considerations

The first problem is caused by a kind of queer dualism in neuronal specificity. In the case of the retina and tectum the neuronal specifity seems to be firmly established well before the functional stage of the visual system. In other words, eye rotation done at the earliest possible embryonic stage results in rotated visual fields /Székely, 1954/. One may, therefore, say that the specific connections of retina and tectum are genetically determined. The other cases, for example skin rotation, in which the neuronal specificity is supposedly imposed upon the sensory neurons by the innervated area, leave us in a very ambiguous situation. We do not know, and perhaps never shall, whether sensory neurons had ever been determined functionally, or whether they must make contact with the periphery in order to acquire the appropriate specificity. If there are sensory neurons with, "back" and "belly" /or "limb", or "cornea"/ characters, there must be "matching" neurons that receive the "specific" sensory fibers in the spinal cord and brain stem as well. Are their characters genetically determined or is their specificity imposed upon them by the specified sensory fibers? In the former case where is the interface on which genetic determination and peripheral influences meet in the center? By what mechanisms is the periphery able to tune the genetic machinery for the production of specific receptor proteins for cell-cell recognition? If this tuning is achieved by axonal transport to the perikarya of various compounds which are capable of interfering with chromosomal activities, then an incredibly refined chemical specification must be attributed to the periphery /skin, limb, etc,/. The number of questions of this kind is limited only by the time one is willing to spend on them.

A second problem emerges as one comes to consider convergence and divergence in neuronal interconnections, a basic principle of central organization. There is virtually not a single place in the central nervous system where self-contained reflex arcs could be found at work. Instead each individual neuron is involved in a variety of activities, receiving and giving connections from, and to, hundreds or thoudands of other neurons. It hardly seems conceivable how to make a rational blue print which could cope with the formidable task of wiring such a structure.

Finally, it is often forgotten that the alleged "cytochemical tag" and the refined peripheral chemical specification, though looked for, were never found /Brackenbury et al., 1977; Thiery et al., 1977/. That is, not a single bit of direct evidence is available in favour of the assumed chemical basis of connection specificity. Therefore the theory of "neuronal specificity", as proposed by Sperry, is no less conjectural than any other hypothesis in explaining the development of order in the nervous system.

3. Experimental controversies

Within the frame of this short review it is not possible to give a complete list of all experiments which, in some way or other, do not comply with the "neuronal specificity" theory. Only two examples will be chosen, one for the case when the "cytochemical tags" are allegedly stamped upon the sensory fibers by the innervated periphery, and the second for the case when the genetic machinery determines the specificity of neurons.

3.1. The "apparent" corneal specificity

A lucky accident in an experiment has actually raised the first doubt against the concept of "neuronal specificity". I have repeated Weiss' /1942/ original experiment in which he grafted an eye in the place of the ear capsule in salamander larvae and could evoke a lid-closure reflex of the host's own eye by touching the cornea of the grafted eye after metamorphosis. The lid-closure /corneal/ reflex consists of the withdrawal of the eye bulb with a consequent passive closure of the eye lids. The efferent limb of the reflex is the abducens nerve which innervates the eye retractor muscles. In his interpretation Weiss suggested that the sensory nerves which came to innervate the grafted cornea changed their old central connections, and established new connections with the abducens nucleus.

The result could easily be reproduced. In a second group of animals a supernumerary limb was grafted into the head region in order to investigate whether "limb specific" reflexes could be evoked from an extra limb with cranial nerve sensory innervation /Székely, 1959/. No such reflexes could be evoked . It happened, however, that the foot of a grafted limb was bitten off by another animal kept in the same dish, and the regeneration blastema which developed on the end of the limb graft proved to be as effective in evoking a blink reflex as an eye graft. As regeneration proceeded the thereshold increased, and with complete regeneration of the foot the area was no longer effective in evoking the reflex. Reamputation of the foot enabled a blink reflex to be elicited again from the new blastema.

A detailed comment on this phenomenon can be found elsewhere /Székely, 1974/. Let it be only mentioned that it is very improbable that the cornea and the blastema have any biochemical properties in common with respect to the control of synaptic formation. Another way of interpretation is using neurophysiological terms. The corneal reflex may be looked upon as a withdrawal reflex in the terminology of Sherringtonian reflexology. It is known that a regeneration blastema is innervated by free epithelial terminals /Hay, 1960/ like the cornea, and both are supplied by nerves much more richly than the surrounding skin /Singer, 1952/. The cornea and the blastema being thus more sensitive than the skin, the corneal reflex elicited from eye and limb grafts may be regarded as a reflex irradiation phenomenon. Another interpretation may be that the similar types of nerve terminals in the cornea and blastema encode similar kinds of sensory messages in the impulse series conveyed by the respective sensory nerves. We may note that neither of these interpretations are more conjectural, in fact they are better supported, than the "neuronal specificity" theory.

3.2. Controversial data on retino-tectal connections

With modern techniques using unit recordings from the optic tectum, Sperry's /1944, 1945/ early prediction has been repeatedly corroborated; that is, optic fibers exclusively established tectal connections according to the retinal positions of their parent cell /Fig. 1 a, b/, irrespective of whether the eye was in normal or in variously rotated position /for ref: Gaze, 1970; Jacobson 1971/. Eye transplantation experiments suggested that the cytochemical specificity of the retina became determined in early embryos /Székely, 1954/; and this determination disclosed a close similarity with the morphogenetic fields in the determination of organ primordia, including independent functional determination along different axes /Székely, 1954; Jacobson, 1971/, and the functional regulation of mutilated eye primordia /Székely, 1955; Feldman and Gaze, 1975/. Experiments in which rotation of the tectum resulted in a corresponding rotation in the retino-tectal projection, suggested the presence of "cytochemical tags" in the tectum as well /Sharma and Gaze, 1971/.

The first doubt that the "cell-unique cytochemical tags" could identify optic fibers and tectal cells came from experiments in which the tectal connections of so called compound eyes were studied. In frog embryos double nasal /NN/ and double temporal /TT/ compound eyes were prepared by removing one half of the eye primordium and replacing it by the remaining half /N or T/ taken from another embryo /Gaze et al., 1963/. Electrophysiological recordings indicated that fibers from both poles of the eye grew uniformly in the corresponding part /caudal in case of NN eyes, and rostral in case of TT eyes/ of the tectum, while fibers from the middle part of the eye occupied the vacant part, that is, either the rostral or the caudal part, of the tectum /Fig. 1 c/. In other words the tectal projection of each hemiretina expanded uniformly to cover the

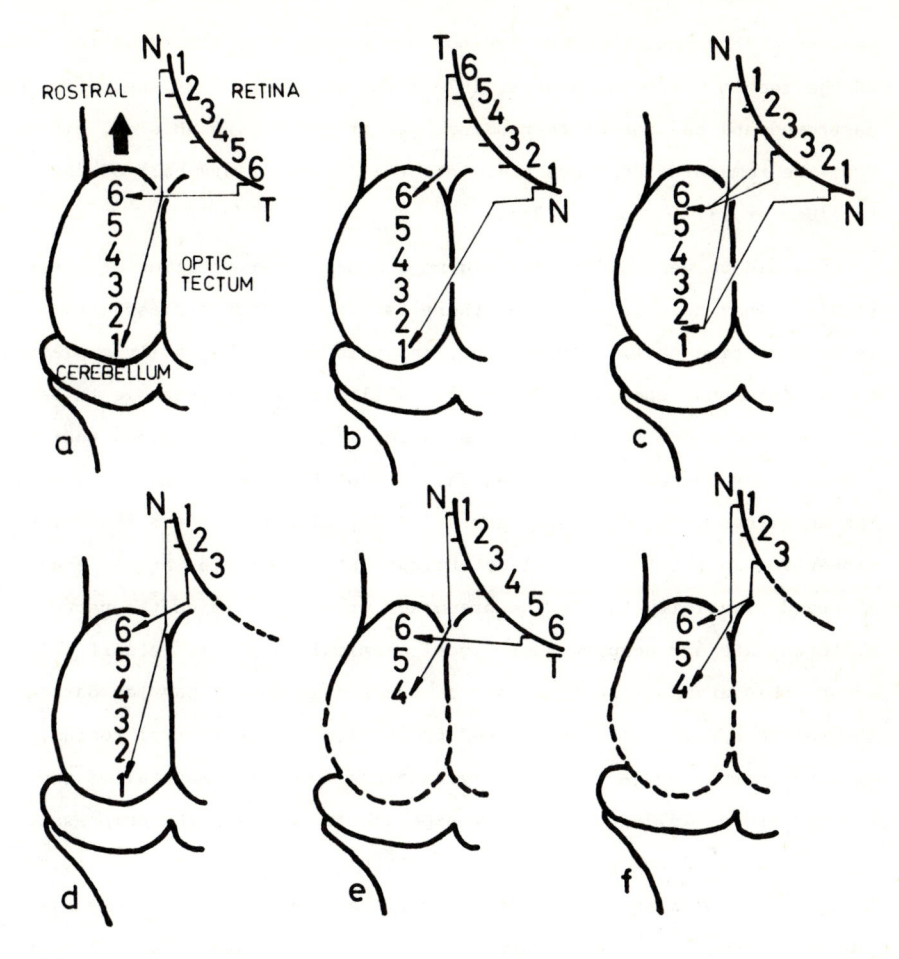

Fig. 1. Diagramatic representation of the different retino-tectal projections. a: the normal projection pattern, the corresponding retinal and tectal sites are labeled with arabic numbers; b: projection pattern of a rotated retina; c: expanded projection from a NN compound eye; d: expanded projection of a half eye; e: compressed projection of a whole eye onto a half tectum; f: projection of a half eye onto the inappropriate half tectum.

entire surface of the tectum. In the expanded projections, fibers gradually shifted in one direction and terminated in foreign sites of the tectum, while maintaining their exact retinotopical relations to each other. Comparable results were obtained from experiments in which the operations were made in adult animals /Fig. 1 d, e, f/. An expansion was recorded in the projection of a half eye onto the tectum /Horder, 1971; Yoon, 1972b/. Conversely, if one half of the tectum was removed the projection of the entire retina was now compressed onto the remaining half of the tectum /Gaze and Sharma, 1970/; and if one half of the retina was destroyed and the non-corresponding part of the retina ablated, fibers from the remaining half eye grew into the inappropriate half of the tectum /Horder,1971; Yoon, 1972b/.

It would be very difficult to give a short interpretation of these results. They clearly show that there is a considerable plasticity even in adult retino-tectal connections, and this cannot be easily explained by the "neuronal specificity" theory. While it is not possible to review all the literature related to this issue, it must be mentoined that some attempts try to save the general validity of the theory in the face of the above results. Meyer and Sperry /1974/, alluding to the field-like properties in the "functional determination" of the eye, argue that the observed plasticity "is in the precursor process by which the nerve cells differentiate and acquire their local chemical tags". It is difficult to accept this argument in the light of the result that a barrier put across the tectum results in a compressed projection in the rostral tectum, but that the normal projection quickly recovers following removal of the barrier /Yoon, 1972a/. In the few experiments in which the compression phenomenon could not be reproduced, it seems that the experimental technique rather than the lack of plasticity could be blamed for the failure /Meyer and Sperry, 1974; Strazniczky, 1973/. To suggest that the optic fibers do not actually form contacts in inappropriate places /Jacobson, 1974/ merely takes the argument towards the irrational, but is of little help in saving the theory. Recently, a high degree of plasticity has been demonstrated in the mammalian optic system as well /Schneider and Jhaveri, 1974/. However fragmentary this review may be the selected experiments bear a strong warning against the oversimplification by appeal to "neuronal specifictiy" of the problem of organization in the visual system.

4. Mechanisms alternative to "neuronal specificity"

By dismissing the "neuronal specificity" theory we certainly do not want to replace order by disorder in the CNS. But what do we mean by order? The question is more relevant than it may look at first sight, because with its definition, however vague, we form a kind of conceptual nervous system. Since the epoch-making work of Cajal and Sherrington it is generally accepted that the functional unit of the CNS is the reflex arc. It can hardly be doubted that the "neuronal specificity" theory is the best possible hypothesis which can explain the establishment of this kind of structural organization. But, perhaps because of its apparent clarity, the theory seriously curtails our immagination in developing new ideas about the nervous system.

An alternative view regards groups of neurons with complex interactions as basic units of nervous function. According to Hebb's /1949/ original idea, the structure of these groups is not determined by preaddressed connectivity patterns, however a regular structure may develop on the basis of the Hebbian coincidence principle. As he proposed, coinciding activities of converging axons strenghthen their relation to the target neuron resulting in the development of well defined pathways within, and between, groups. The principle demands two kinds of coincidence: the synchronization of activity in converging axons and the anatomical convergence of the fibers themselves. It is clear that these properties provide the neuron groups with a high degree of dynamism and plasticity.

This view of nervous function has gained support during the years. Works on sensory cortices provided very good evidence that cell groups arranged in columns in the cortex may be viewed as functional units /Mountcastle, 1957; Hubel and Wiesel, 1962/. Increasingly accumulating evidence is suggestive of the assumption that various forms of locomotory movements are controlled by groups of neurons arranged specifically in the spinal cord and brain stem /Herman et al., 1976/. On the basis of a whealth of experimental data, Szentágothai and Arbib /1974/ consider a variety of conceptual models in which structural "modules" constitute the basis of neural organization. The flow of nervous impulses within and between the modules cannot be defined with the aid of the familiar blue-prints of reflex chains. Order is expressed in the form of general tendencies of making connections in one direction or another, and in an apparently disordered lattice of neural elements very specific pathways may take shape on the basis of morphological and physiological properties of the components, and on the basis of past and present functional states of the network. It is clear that not a single, but only a number of various mechanisms can account for the establishment of this kind of order. In the following pages we shall briefly show three examples, each of which may be viewed as contributing mechanisms to the establishment of order. It will be seen that they are not mutually exclusive, but rather that they are interdependent and conditional upon one another in the process of bringing order in the CNS. It will also be clear that there must be many more mechanisms which supplement one another. For this reason a synthesis of the three mechanisms shown will not be attempted.

4.1. Simulation studies

Since the invention of "learning machines" and mathematical models to simulate neural functions, several attempts have been made to study the development of specific pathways in structurally non-determined systems. Only a few examples will be shown here, they deal with the visual system and are therefore most relevant to the present subject.

One of them is von der Malsburg's /1973/ model of the striate cortex. The model was made with the intent of simulating the organization of cortical neurons into "functional columns" and the differential sensitivity of their neurons to light bars and edges presented in a certain orientation /Hubel and Wiesel 1962, 1963/. The basic assumption of the model can be traced back to the Hebbian concept of synapse formation.

The elements of a "retinal plate" have acess to all elements of a "cortical plate", and if the activity of a fiber coincides with the firing of neurons to which it is connected, then these connections will be strengthened. After twenty presentations of nine differently orientated light bars, individual neurons in the "cortical plate" showed differential sensitivity, and neurons with the same or similar orientation sensitivity tended to appear in clusters. The model demonstrates that with this simple learning process an organization can be achieved without the assumption of any predetermined circuitry. In fact, it has been shown that in contrast to the original hypothesis of Hubel and Wiesel /1963/, the visual cortex of young kittens discloses a high degree of plasticity in this respect /Blakemore and Cooper, 1970; Hirsch and Spinelli, 1970/.

The complement of this model by the "functional verification hypothesis" /Willshaw and von der Malsburg, 1976/ leads to the simulation of an ordered projection of one neuron layer /retina/ onto another /visual center/. The hypothesis is that postsynaptic elements tend to maintain connections with fibers having similar patterns of activity. Thus the positional information of retinal elements present in spatial activity patterns can yield an ordered projection, but fails to specify the size or orientation of the projection. Adding, however, the action of "polarity markers" to the process /the orientation of a few initial contacts is specified/, the model is capable of simulating the ordered retino-tectal projection in amphibia, and adjusts for enlargement of one or both layers /expansion and compression phenomena in retino-tectal maps/.

The "arrow model" of Hope et al. /1976/ proposes another mechanism for orientation information. It assumes that any two retinal fibers which are close to each other on the tectum must be able to tell the direction /but not the distance/ of their cells of origin from each other in the retina; while at any point on the tectum a retinal fiber must be able to tell the directions rostral and caudal from local information. Computer simulation has indicated that the model can account for the results of almost all experimental findings, including rotation of eye or tectum and compression and expansion of maps; it cannot, however, account for the result obtained from experiments in which two tectal grafts were reciprocally translocated.

The arrow-model does not require a point-to-point specification of the retina and tectum, but it still requires a considerable amount of information to be available to the optic fibers. It is known that the retina also has an ordered projection onto the diencephalic optic centers, the lateral geniculate complex and the pretectal nucleus, and has a direct ipsilateral projection from the temporal half on the same centers /Lázár 1971, Scalia and Fite 1974/. There seem to be strong reciprocal interconnections between tectum and diencephalon /Lázár 1969, Trachtenberg and Ingle 1974/. The differentiation of diencephalon proceeds from the caudal to the rostral direction /Picouet, 1975/, while the optic tectum differentiates from the rostral to the caudal direction /Lázár, 1973/. The sequential differentiation in the opposite directions of tectum and diencephalon suggests that reciprocal tecto-diencephalic connections may be also ordered. Each retinal locus, therefore, sends fibers both to diencephalic and tectal centers, and these centers are mutually interconnected probably in an ordered fashion. It would be interesting to investigate whether these series of interconnections could replace the polarity information in the Willshow-Malsburg model and in the arrow model.

4.2. Sequential patterns in neurogenesis

In almost all neural centers investigated, the differentation of the various neurons follows a definite time sequence. In the cerebellum, for example, which is one of the most investigated structures, neurons of the deep nuclei are the first to differentiate, they are followed by Purkinje cells and then by the large Golgi cells. The smaller cells of the cerebellar cortex, the basket cells and stellate cells differentiate later, while granule cells come last in the order.

Following the exact time course of cerebellar development in the rat, Altman /1972/ has observed that neurons of the molecular layer are generated sequentially from the bottom upward in the cerebellar primordium. The granule cells are generated in the external germinal layer, and before migrating downward to settle in the granular layer, they extend their axons, the parallel fibers, in the molecular layer. Due to the sequential differentation of granule cells, the parallel fibers are stacked from the bottom upward and form synapses in the same order in the molecular layer. Altman and Bayer /1978/ have extended these investigations to the brain stem precerebellar nuclei. They found that of the neurons which make contact with granule cells, the lateral reticular nucleus forms first, it is followed by the nucleus reticularis tegmenti pontis, and pontine neurons are the last to differentiate. Since the growth of mossy fibers is arrested upon making contact with granule cells /Altman, 1973/, this chronological relation implies a stratification of the mossy terminals; the first complement of descending granule cells should contact mossy fibers from early differentiating precerebellar nuclei, whereas later descending granule cells meet mossy fibers from late forming precerebellar nuclei. Since the corresponding parallel fibers are stacked in a bottom-upward order in the molecular layer, this chronology has the further implication that the lateral reticular nucleus should influence the proximal domain of the dendrites of Purkinje cells, while pontine nuclei should influence the distal domain.

This hypothesis of Altman, which is directly testable with physiological techniques, offers an alternative mechanism of how a delicately ordered connectivity pattern can be established without assuming any sort of "neuronal specificity". This possibility suggests that further investigations in this direction may be very fruitful and instructive in studying the development of specific neural functions and specific neuronal interconnections. In fact, this hypothesis was advanced at the end of the previous section suggesting order in the reciprocal tecto-diencephalic connections on the basis of a developmental time-table of mesencephalon and diencephalon.

4.3. Specificity of neuronal morphology

As an alternative to "neuronal specificity", it has been proposed that the gene-determined morphological characters of nerve cells may play an instrumental role in the establishment of order in the CNS /Székely 1966, 1974/. This idea can be traced back to Sholl's /1956/ work in which with the determination of cell density and geometry of dendritic and axonal arborization, he tried to give a quantitative description of neuronal connectivity in the cortex. Similar form-function relationship has been considered in the retina /Lettvin et al., 1961/, in the dorsal horn of the spinal cord /Scheibel and Scheibel, 1968/ and in a number of different centers /Szentágothai and Arbib, 1974/. To close the present survey, a simple example will be shown in which some characteristic morphological features of motoneurons can be associated with their specific function in the spinal cord.

This observation was rendered possible by the adaptation of the cobalt staining technique to vertebrates /Székely and Gallyas, 1975/. The technique provides a selective and probably quantitative staining of spinal motoneurons /Székely, 1976/, of which three distinctly different types can be discerned. The first type has an oval or spindle form body and a lateral and a medial dendritic stem. The former arborizes in the lateral and ventral funiculi, the medial dendrite crosses to the opposite side and sends branches to both the white and gray matter. These neurons innervate the axial muscles. The second type has a polygonal body which lies laterally and dorsally to the first type. Two or more dendrites leave the perikaryon, and they assemble into a lateral group which spreads over the lateral funiculus, and into a medial group which turns toward the central canal. The ventral part of this group decussates to the other side. These neurons innervate ventral trunk muscles. The third type of neuron can be found exclusively at limb levels, and they inner-

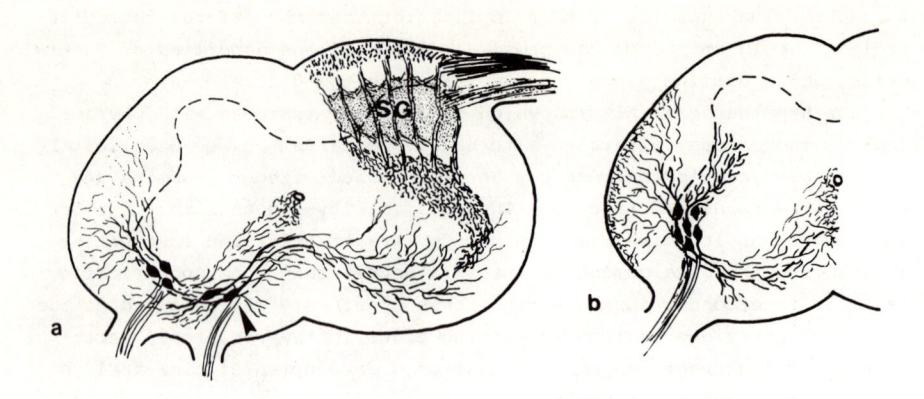

Fig. 2. Crossectional diagrams of the spinal cord. **a:** on the left side motoneurons to the axial musculature /arrow head/ and to ventral trunk muscles are shown, on the right side the ventral projections of dorsal root fibers are depicted. SG=substantia gelatinosa; **b:** motoneurons to limb muscles. Compare the sites of dorsal root terminals with the dendritic arbor.

vate limb muscles. Their polygonal bodies generate three groups of dendrites: a lateral group to the lateral funiculus, a medial group to the central gray matter and a dorsomedial group extending in the direction of the substantia gelatinosa /Fig. 2 a, b/. Staining the dorsal root fibers with cobalt reveals that only this type of neuron receives dorsal root terminals and they do it through the dorsomedial dendrites /Székely, 1976/; and only these neurons can be monosynaptically activated through the dorsal root /Czéh, 1972; Cruce, 1974/.

Embryological experiments /Székely, 1963/ indicate that the functional capacity of the spinal cord segments to move a limb is determined very early in the embryonic life, and no other spinal sections are capable of this function. The third type of motoneuron is, therefore, indispensable for the control of coordinated limb movements. The first type of motoneuron, which innervates axial muscles, is apparently involved in swimming movement, and its large contralateral dendrite may be essential in the control of alterning contractions of segmental muscles in the act of swimming. It seems, therefore, that the functional specificity is associated with a morphological specificity of the different sections of the spinal cord. This coincidence is suggestive of a closer form-function relationship in the sense that the differential morphological characters of neurons, controlled very probably genetically, may be responsible for the inherent functional differences in the spinal cord.

References

Altman, J. /1972/ Postnatal development of the cerebellar cortex in the rat, I,II,III. J. Comp. Neurol. 145: 353-514.

Altman, J. /1973/ Experimental reorganization of the cerebellar cortex, III. J. Comp. Neurol. 149: 153-180.

Altman, J. and Bayer, S.A. /1978/ Prenatal development of the cerebellar system in the rat. I,II. J. Comp. Neurol. 179: 23-75.

Blakemore, C. and Cooper, G.F. /1970/ Development of the brain depends on the visual environment, Nature 228: 477-478.

Brackenbury, R., Thiery, J.-P., Rutishauser, U. and Edelman, G.M. /1977/ Adhesion among neural cells of the chick embryo, I.J. Biol. Chem. 252: 6835-6840.

Cruce, W.L.R. /1974/ A supraspinal monosynaptic input to hindlimb motoneurons in lumbar spinal cord at the frog, Rana catesbeiana, J. Neurophysiol. 37: 691-704.

Czéh, G. /1972/ The role of dendritic events in the initiation of monosynaptic spikes in the frog motoneurons, Brain. Res. 39: 505-509.

Feldman, J.D. and Gaze, R.M. /1975/ The development of half-eyes in Xenopus tadpoles, J. Comp. Neurol. 162: 13-22.

Gaze, R.M. /1970/ The Formation of Nerve Connections, Acad. P., London.

Gaze, R.M., Jacobson, M. and Székely, G. /1963/ The retino-tectal projection in Xenopus with compound eyes. J. Physiol. /Lond./ 165: 484-499.

Gaze, R.M. and Sharma, S.C. /1970/ Axial differences in the reinnervation of the goldfish optic tectum by regenerating optic nerve fibers, Exp. Brain. Res. 10: 171-181.

Hay, E.D. /1960/ The fine structure of nerves in the epidermis of regenerating salamander limbs, Exp. Cell. Res. 19: 299-317.

Hebb, D. /1949/ The Organization of Behavior, J. Wiley and Sons, Inc., New York.

Herman, R.M., Grillner, S., Stein, P.S.G. and Stuart, D.G. /1976/ Neural Control of Locomotion. Advances in Behavioral Biology, Vol. 18. pp. 822, Plenum P., New York.

Hirsch, H.V.B. and Spinelli, D.N. /1970/ Visual experience modifies distribution of horizontally and vertically oriented receptive fields in cats, Science 168: 869-871.

Hope, R.A., Hammond, B.J. and Gaze, R.M. /1976/ The arrow model: retinotectal specificity and map formation in the goldfish visual system, Proc. Roy. Soc. B 194: 447-466.

Horder, T.J. /1971/ Retention by fish optic nerve fibers regenerating to new terminal sites in the tectum, of "chemospecific" affinity for their original sites. J. Physiol. /Lond./ 216: 53P-55P.

Hubel, D.H. and Wiesel, T.N. /1962/ Receptive fields, binocular interaction and functional architecture in the cat's visual cortex. J. Physiol. /Lond./ 160: 106-154.

Hubel, D.H. and Wiesel, T.N. /1963/ Shape and arrangement of columns in cat's striate cortex, J. Physiol. /Lond./ 165: 559-568.

Jacobson, M. /1967/ Retinal ganglion cells: Specification of central connections in larval Xenopus laevis, Science 155: 1106-1108.

Jacobson, M. /1971/ Developmental Neurobiology, Holt, Rinehart and Winston, New York.

Jacobson, M. /1974/ Neuronal plasticity: Concepts in pursuit of cellular mechanisms. In: D.G. Stein, J.J. Rosen and N. Butters /eds/: Plasticity and Recovery of Function in the Central Nervous System, 31-43, Acad. Press. Inc., New York.

Lázár, G. /1969/ Efferent pathways of the optic tectum in the frog. Acta Biol. Acad. Sci. Hung. 20: 171-183.

Lázár, G. /1971/ The projection of the retinal quadrants on the optic centres in the frog; a terminal degeneration study. Acta Morph. Acad. Sci. Hung. 19: 325-334.

Lázár, G. /1973/ The development of the optic tectum in Xenopus laevis: a Golgi study, J. Anat. 116: 347-355.

Lettvin, J. Y., Maturana, H.R., Pitts, W.H. and McCulloch, W.S. /1961/ Two remarks on the visual system of the frog. In: W.A. Rosenblith /ed/: Sensory Communication, 757-776, J. Wiley and Sons, New York.

Malsburg, von der, Chr. /1973/ Self-organization of orientation sensitive cells in the striate cortex, Kybernetik 14: 85-100.

Meyer, R.L. and Sperry, R.W. /1974/ Explanatory models for neuroplasticity in retinotectal connections. In: D.G. Stein, J.J. Rosen and N. Butters /eds/: Plasticity and Recovery of Function in the Central Nervous System, 45-63. Acad. Press, Inc., New York.

Miner, N. /1956/ Integumental specification of sensory fibres in the development of cutaneous local sign, J. Comp. Neurol. 105: 161-170.

Mountcastle, V.B. /1957/ Modality and topographic properties of single neurons of cat's somatic sensory cortex. J. Neurophysiol. 20: 4o8-434.

Picouet, M.J. /1975/ Le developpement du thalamus des Anoures au course de le métamorphose, J. Embryol. Exp. Morph. 33: 313-333.

Scalia, F. and Fite, K. /1974/ A retinotopic analysis of the central connections of the optic nerve in the frog. J. Comp. Neurol. 158: 455-478.

Scheibel, M.E. and Scheibel, A.B. /1968/ Terminal axonal pattern in the cat spinal cord, II, Brain Res. 9: 32-58.

Schneider, G.E. and Jhaveri, S.R. /1974/ Neuroanatomical correlates of spared or altered function after brain lesions in the newborn hamster. In: D.G. Stein, J.J. Rosen and N. Butters /eds/: Plasticity and Recovery of Function in the Central Nervous System, 65-109, Acad. Press. Inc., New York.

Sharma, S.C. and Gaze, R.M. /1971/ The retinotopic organization of visual responses from tectal reimplants in adult goldfish. Arch. Ital. Biol. 109: 357-366.

Sholl, D.A. /1956/ The Organization of the Cerebral Cortex, Methuen, Co., London.

Singer, M. /1952/ The influence of the nerve in regeneration of the amphibian extremity, Quart. Rev. Biol. 27: 169-200.

Sperry, R.W. /1944/ Optic nerve regeneration with return of vision in anurans, J. Neurophysiol. 7: 57-69.

Sperry, R.W. /1945/ Restoration of vision after crossing of optic nerves and after contralateral transposition of the eye, J. Neurophysiol. 8: 15-28.

Straznicky, K. /1973/ The formation of the optic fibre projection after partial tectal removal in Xenopus, J. Embryol. Exp. Morph. 29: 397-409.

Székely, G. /1954/ Zur Ausbildung der lokalen funktionellen Spezifität der Retina, Acta Biol. Acad. Sci. Hung. 5: 157-167.

Székely, G. /1957/ Regulationstendenzen in der Ausbildung der "funktionellen Spezifität" der Retinaanlage bei Triturus vulgaris, Roux Arch.

Entw.-Mech. 150: 48-60.

Székely, G. /1959/ The apparent "corneal specificity" of sensory neurons, J. Embryol. Exp. Morph. 7: 375-379.

Székely, G. /1963/ Functional specificity of spinal cord segments in the control of limb movements. J. Embryol. Exp. Morph. 11: 431-444.

Székely, G. /1966/ Embryonic determination of neural connections, Advanc. Morphogen. 5: 181-219.

Székely, G. /1974/ Problems of neuronal specificity in the development of some behavioral patterns in amphibia, In: G. Gottlieb /ed./: Studies on the Development of Behavior and the Nervous System, Vol. 2. 115-150. Acad. Press. New York - London.

Székely, G. /1976/ The morphology of motoneurons and dorsal root fibers in the frog's spinal cord. Brain. Res. 103: 275-290.

Székely, G. and Gallyas, F. /1975/ Intensification of cobaltous sulphide precipitate in frog nervous tissue, Acta Biol. Acad. Sci. Hung. 26: 175-188.

Szentágothai, J. and Arbib, M.A. /1974/ Conceptual Models of Neural Organization. NRP Bulletin Vol. 12, No. 3, MIT Press, Cambridge, Mass.

Thiery, J.-P., Brackenbury, R., Ruitshauser, U. and Edelman, G.M. /1977/ Adhesion among neural cells of the chick embryo, II, J. Biol. Chem. 252: 6841-6845.

Trachtenberg, M.C. and Ingle, D. /1974/ Thalamo-tectal projections in the frog, Brain Res. 79: 419-430.

Weiss, P. /1942/ Lid-closure reflex from eyes transplanted to atypical locations in *Tritorus torosus*: Evidence of a peripheral origin of sensory specificity, J. Comp. Neurol. 77: 131-169.

Willshaw, D.J. and Malsburg, von der, Chr. /1976/ How patterned neural connections can be set up by self-organization, Proc. Roy. Soc. B 194: 431-445.

Yoon, M. /1972a/ Reversibility of the reorganization of retino-tectal projection in goldfish, Exp. Neurol. 35: 565-577.

Yoon, M. /1972b/ Transposition of the visual projection from the nasal hemiretina onto the foreign rostral zone of the optic tectum in goldfish. Exp. Neurol. 37: 451-462.

MORPHOGENETIC RELATIONS BETWEEN CELL MIGRATION AND SYNAPTOGENESIS IN
THE NEOCORTEX OF RAT

J.R. Wolff, B.M. Chronwall and M. Rickmann
Dept. of Neurobiology, Neuroanatomy, Max-Planck-Institute for biophysical Chemistry, Am Faßberg, 3400 Göttingen, GFR

Introduction

This paper tries to point out possible interrelations between
different morphogenetic events during the development of the neocortex.
For practical reasons, we had to choose an unconventional form of
presenting a synopsis of the results of various types of experiments
and data from the literature. As far as our own experiments are
concerned, data on material and methods will be given in each chapter
together with the results.

The nervous system can be described as an assembly of neurons which
is differentiated into various types showing different biochemical and
biophysical properties (e.g. excitatory and inhibitory neurones) and
which communicate in a non-uniform and non-random manner. Such a system
could be defined by the total number of neurons involved, by the
relative number of neurons belonging to each type, and by the quantitative and qualitative characteristics of inter-neuronal communication.
This concept allows to search for decisive steps during the development
of neuronal networks.

The amount of cells in a nervous system depends on a sufficient
number of cell cycles to be induced in ventricular cells or other types
of undifferentiated matrix cells ((1) in fig. 1; for ref. see: Sidman,
1970, Sidman and Rakic, 1973, Berry, 1974, Rakic, 1975). A reasonable
fraction of these cells is lost by cell death which has been excluded
from fig. 1, because it occurs in various stages of differentiation
(Chu-Wang a. Oppenheim, 1978). The exponential proliferation of undifferentiated matrix cells terminates when mitotic divisions produce non-
proliferating proneurons and/or proliferative glioblasts (e.g. Angevine,
1970). After these decisive mitoses both determined classes of cells
leave the proliferation zone migrating along certain guiding structures
(2.1 in fig. 1). The cell migration is obviously involved in a number
of important decisions determining such properties as the definite
position of each neuron and the number of neurons reaching a certain
part of the nervous system at a certain time. Long axons show a directed

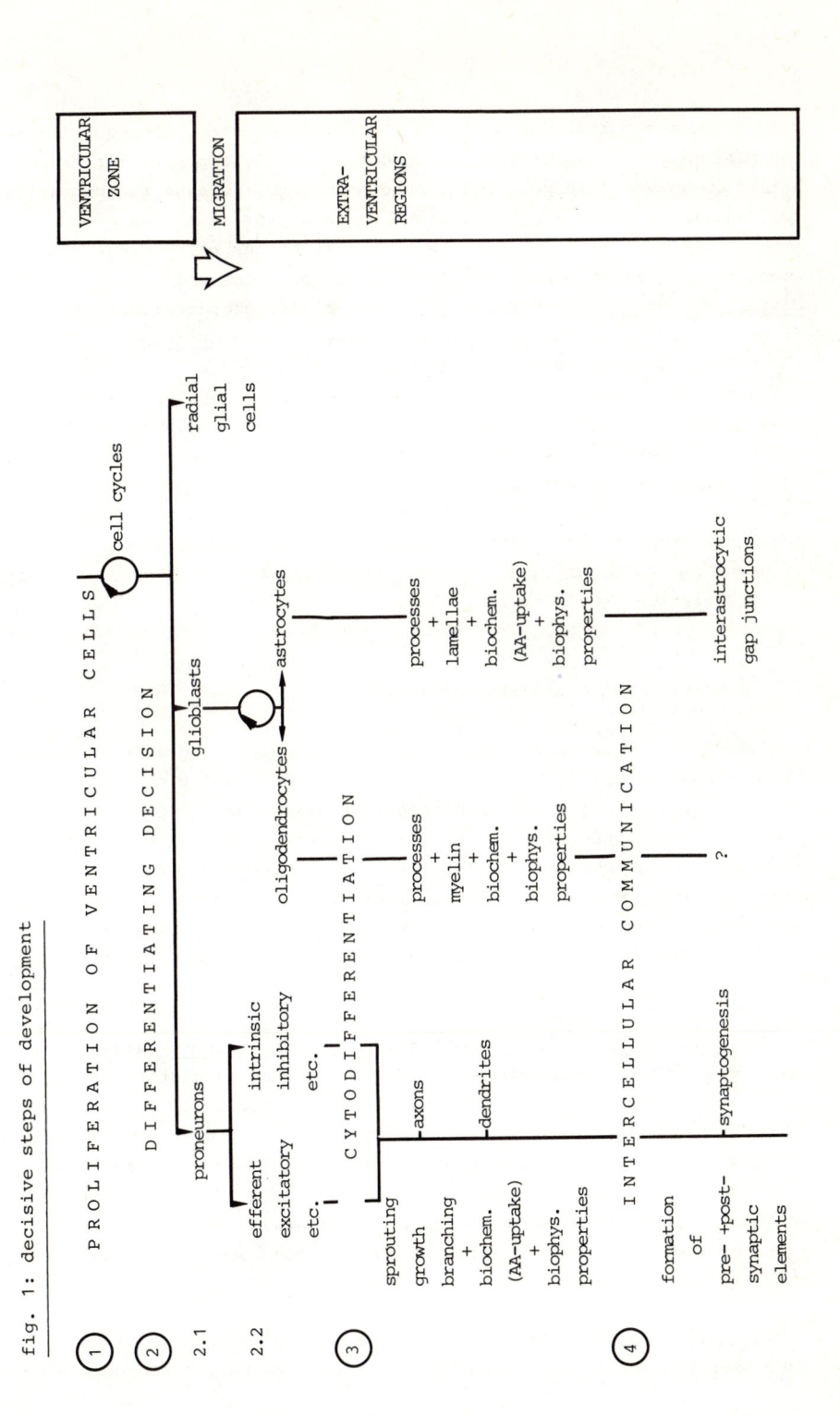

fig. 1: decisive steps of development

① PROLIFERATION OF VENTRICULAR CELLS — cell cycles — radial glial cells

② DIFFERENTIATING DECISION

2.1 proneurons { intrinsic inhibitory etc. / efferent excitatory etc. }

2.2 glioblasts → astrocytes — processes + lamellae + biochem. (AA-uptake) + biophys. properties
           ↔ oligodendrocytes — processes + myelin + biochem. + biophys. properties

③ CYTODIFFERENTIATION — axons / dendrites — sprouting, growth, branching + biochem. (AA-uptake) + biophys. properties

④ INTERCELLULAR COMMUNICATION — synaptogenesis — formation of pre- +post-synaptic elements — ? — interastrocytic gap junctions

VENTRICULAR ZONE ⇒ MIGRATION ⇒ EXTRA-VENTRICULAR REGIONS

growth and reach specific parts of the nervous system. Consequently, the migration and deposition of neurons limit their possibilities to establish random interneuronal connections, but increase the probability to realize contacts with those axons which appear in the same locus. In the following we will concentrate on the relations which might exist among the migration mode, the localization of axons and the <u>differentiation of neurons</u> and <u>glial cells</u> into characteristic types (e.g. GABA accumulating cells) on one side and the <u>synaptogenesis</u> on the other. Synaptic interactions were selected, because these seem to represent at present that kind of interneuronal communication which is best defined in terms of biochemistry as well as of biophysics and morphology. However, evidence will be presented that interactions with extrasynaptic receptors might be important during synaptogenesis. As far as possible we will restrict the considerations to the neocortex, sometimes even to its occipital region, to avoid the high complexity of developmental processes caused by the temporo-spatial gradients of differentiation (see Angevine, 1970).

Cell migration and cell differentiation

Postmitotic neurons are generally displaced from the site of their last mitosis before they start their final differentiation including the formation of axons and dendrites. The displacement of neurones was observed in various parts of the CNS of several species, irrespective of their origin from the ventricular zone, subventricular mitoses or from the extragranular layer of the cerebellum. Since the evidence for this general phenomenon has been reviewed for several times, it will not be repeated here (e.g. Sidman and Rakic, 1973, Berry, 1974, Rakic, 1975).

The <u>temporo-spatial distribution of migrating neurons</u> varies from one part of the nervous system to another. In the neocortex the deposition of neurons follows a characteristic gradient, the so-called "inside-out layering" which exists similarly in the cortices of various species (mouse: Angevine and Sidman, 1961; rats: Berry and Rogers, 1965; monkey: Rakic, 1972; human: Sidman and Rakic, 1973). According to this gradient neurons of the deep layers are formed earlier than those of the middle layers while neurons of the superficial layers are the last to be formed.

To analyse further the mode of migration and deposition of neurons we prepared a series of autoradiograms from rats which had received one injection of ^{3}H-thymidine between embryonic day 13 and 22. The survival time varied between one hour and one month. In some cases a second injection of ^{14}C-labelled thymidine followed the first by 12 hours. Two emulsion autoradiograms prepared according to the method of Schultze et al. (1976) allow to separate neurons being labelled during their last mitosis from those which underwent further mitoses after the first labelling.

These studies have demonstrated that the inside-out principle although being valid for most pyramidal neurons in Lamina II to VI does obviously not hold true for non-pyramidal neurons. The so-called Cajal-Retzius neurons of lamina I are known to reach the pallial anlage, before the first cortical plate neurons are formed, yet they are situated more superficially (Raedler and Sievers, 1975, König et al., 1977). Our autoradiograms show additionally that neurons of different age are added to the most superficial layer during the whole period of neuron proliferation (Rickmann et al., 1977). Similarly, non-pyramidal neurons are permanently added to all layers as soon as these have been formed by the initial wave of neurons which follow the inside-out layering (fig. 2, Chronwall and Wolff, 1978). In the original descriptions the inside-out gradient was noted to be accompanied by scattered neurons which were partly situated in the subcortical white matter, but were also scattered in the layers surrounding the peak of the inside-out gradient (e.g. Berry and Rogers, 1965). These scattered neurons were mainly explained as a fraction of slowly migrating neurons which follow the major wave of rapidly and synchroneously moving neurons (see Rakic, 1974, 1975). In contrast to this explanation, not only old neurons were observed in more superficial positions, but late labelled neurons were also found in deep layers.

In autoradiograms counterstained with cresylviolet many cortical neurons could not be classified as pyramidal or non-pyramidal neurons. On the other hand, a significant number of pyramidal neurons could unequivocally be recognized as some fusiform and stellate neurons in various layers. Only relying on these identified neurons it was found that all labelled pyramidal neurons were located within the band of major labelling, i.e. within the population of neurons following the inside-out principle. Surprisingly enough, no identified labelled pyramid has yet been detected among the scattered labelled neurons in any layer. It should be added that pyramidal neurons according to our definition possess a vertically ascending apical dendrite excluding all improperly oriented pyramids (see van der Loos, 1965). In contrast, labelled fusiform and stellate neurons could be identified in all

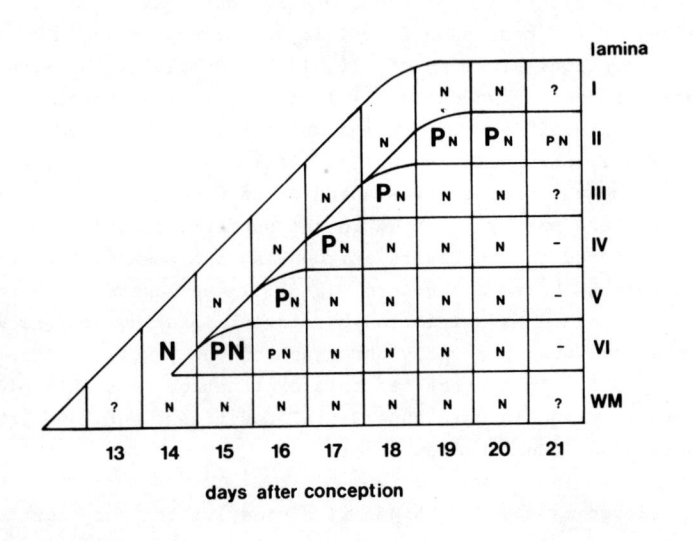

fig. 2:
The diagram demonstrates the days of origin of pyramidal (P) and non-pyramidal (N) neurons and their definite laminar position after migration. The size of the letters indicates the relative amount of cells.

layers, though the majority of them seemed to be located outside the major band of labelled neurons (see Chronwall and Wolff, 1978).

Thus, in the neocortex of rats postmitotic neurons seem to follow at least two different modes of deposition: (1) the inside-out layering which during a limited period of time places all pyramidal neurons into a certain layer and (2) a diffuse mode of deposition which contributes to all layers permanently single neurons, after the layer has been preformed by the deposition of pyramidal neurons (fig. 2). The second mode is less conspicious, as it does not produce local accumulations of neurons which are labelled by ^{3}H-thymidine at the same time. Its contribution to a local population of neurons may yet be significant, especially in the deeper layers, because it adds permanently neurons of different age to the same level.

From these observations the question arises what is the mechanism of migration and are there two different types of migration? Several mechanisms of displacement have been proposed. For several years the nucleus and the perikaryon were thought to move through the primitive processes reaching the external limiting zone of the cortex (Berry and Rogers, 1965, Morrest, 1970). This mode of displacement has not yet been ruled out as a mechanism for the rapid and synchroneous movement of neurons for short distances between the ventricular zone and the early cortical anlage (see Rakic, 1975). However, in the cerebellar cortex as well as in the neocortex Rakic (1971, 1972) has shown that migrating neuroblasts are submicroscopically separated from, but adapted to the surface of elongated processes of the so-called radial glial cells in the pallium or Bergmann glial fibers in the cerebellum. The neurons can migrate either from or towards the pial surface along the radial glial processes as in the cerebellar and cerebral cortex, respectively (see Rakic, 1975). Hence, the glial processes might provide a more or less passive guidance from the site of origin (ventricular zone or external granular layer) to the site of definite position of the neurons.

As Cajal's epithelial cells (1960), radial glial cells usually extend from the ventricular surface to the pial surface to which they are connected by one to several terminal branches and end-foot-like thickenings (compare Rakic 1972, 1975). These long radial glial cells can guide migrating neurons from the ventricular zone directly to the surface of the cortical anlage. Such a mode of migration deposing the younger neurons always on top of the older ones fits well the inside-out layering.

In Golgi preparations another type of radially oriented cells can be seen additionally. It resembles the radial glial cell except for the fact that the radial process is shorter and does not reach the pial surface. These cells have been described as freely arborizing spongioblasts (Stensaas, 1967). A careful light microscopical analysis revealed, however, that these processes showing a variable length form regularly a distinct contact with a blood vessel before the terminal branches appear (fig. 3). Sometimes these freely arborizing terminal branches can be missing, but the vascular contact has always been observed when the terminal part of the process was included in the section. Although, in our Golgi preparations migrating neuroblasts were not regularly stained, some cells with the shape and size of migrating cells (compare Rakic, 1972) were seen to be adapted to these <u>short radial glial cells</u>. If electronmicroscopic studies which are still to be undertaken would confirm that neurons migrate also along short radial glial cells, then the second migration mode would depend on the distribution of radial glial cells terminating at blood vessels, i.e. it should be related to the distribution of vascular branching along the radial vessels (Wolff, 1976, 1978).

The autoradiograms show that there are cells which dilute the labelling by further mitotic divisions after they have been deposed in extraventricular levels. These cells probably represent glioblasts which can be observed from E14 onwards on all parts of the pallial (see Rickmann et al., 1977) and cortical anlage suggesting that glioblasts migrate also along both, the long and the short radial glial cells.

Our knowledge about the decisive steps of the development of the cortical structure is largely incomplete. It is, for instance, unknown what terminates the cell migration. Most neurons migrating along the long radial glial cells do not move to the marginal end of the radial processes, but stop at the lower border of Lamina I. The induction of the apical dendrite of pyramidal neurons is probably not simply related to a certain type of axon located in lamina I, because non-pyramidal neurons arrive and differentiate also in this layer during the whole period of neuron production (see Rickmann et al., 1977). Hence, we do not know what might induce the differentiation of pyramidal and non-pyramidal neurons. Therefore, we had to rely on the Golgi method to demonstrate the earliest developmental stages of neurons. Non-pyramidal neurons being characterized by horizontally orientated primary dendrites can be distinguished from primitive pyramidal neurons which develop the vertically oriented, apical dendrite, first (Berry and Rogers, 1965). Using this criterion non-pyramidal or horizontal neurons can be first

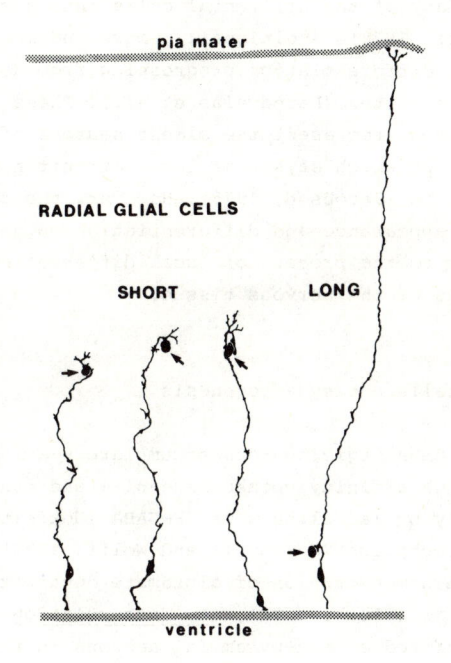

fig. 3:
Camera lucida drawing of long and short radial glial cells (Golgi impregnation). Long processes terminate at the pia mater, though vascular contacts occur with side branches (arrow). Short processes terminate in a variable arborization beyond a vascular contact (arrows).

visualized in lamina I during early stages of embryonic development
(embryonic day = E15; Raedler and Sievers, 1976, König et al., 1977,
Rickmann et al., 1977). During the following week non-pyramidal neurons
start their differentiation sequentially in several discrete cortical
levels, such as the lower border of lamina VI, laminae V, III and II
(Wolff, 1978). After birth, no further levels containing horizontal
neurons are established. By this time, secondary dendrites and
dendritic branches expanding in other than the horizontal direction
seem to transform many of the horizontal cells into stellate neurons
(Chronwall and Wolff, 1978). Additionally, more and more non-pyramidal
neurons start their differentiation progressing from deep to the superficial levels of the cortex (Parnavelas et al.). These results showing
that horizontal neurons represent the oldest neurons of the cortex
contradict the concept which says that local circuit neurons appear
and differentiate late (Jacobson, 1975). However, the sequential
character of their appearance and differentiation suggests that they
are closely related to the process of local differentiation during all
developmental stages of the nervous tissue.

GABA accumulating cells and synaptogenesis

Glial cells and GABAnergic neurons accumulate y-amino-butyric
acid (GABA) by a high affinity uptake mechanism and can be localized
autoradiographically by labelling with ^{3}H-GABA (Hökfelt and Ljungdahl,
1972). Using this technique (Chronwall and Wolff, 1978) as well as the
immunohistochemical demonstration of glutamate decarboxylase, i.e. the
GABA producing enzyme (Ribak, 1978), GABA accumulation and production
was selectively confined to non-pyramidal neurons in the neocortex of
adult rats.

In a series of developmental stages varying between embryonic day
14 and adults we applied ^{3}H-GABA by superfusion or microinjection to
the prospective or definite visual cortex of albino rats. Autoradiograms of semithin sections (1 to 5 μm) of the Epon embedded brains
were used to determine: the time and the site at which GABA-accumulating
cells appear first and their distribution within the cortex. In some
cases, resectioned autoradiograms were analysed by electron microscopy
to ascertain whether the accumulation of GABA took place in a neuron or
a glial cell.

Already one day after the first neurons had been deposed in the presumptive visual neocortex, i.e. at 15 days p.c. (E15), single GABA-

accumulating cells can be seen in lamina I and in the intermediate zone (see fig. 4). At E 17 these have increased in number and exist also within the cortical plate. During the perinatal and early postnatal period GABA-accumulating cells show roughly a laminated distribution with maxima in lamina I, II/III, V/VI and in the subcortical level near the lower border of the cortex. Our attempts to identify cells by resectioning revealed as yet that by E 17 there are neurons and glial cells in lamina I which can accumulate GABA. The neurons are postsynaptic to a number of symmetric and asymmetric synapses and possess axons. GABA-accumulating glial cells represent only a small fraction of all glial precursors in Lamina I. They are more numerous near the border of the bipolar cortical plate and are at least in part innervated by axons. The laminar distribution of GABA-accumulating glial cells and neurons resembling each other is very similar to the branching pattern of intracortical radial vessels (Wolff, 1976) as well as to the distribution of early horizontal neurons (see above). This distribution pattern is not only coincident with the appearance and aggregation of the first symmetrical synapses (Wolff and Wolff, 1977), but also with the peaks of early synaptogenesis (Molliver et al., 1973, Wolff, 1976). Therefore, the question has arisen whether the similarity is coincidental or might indicate a causal relationship between the appearance of releasable GABA-pools and the local synaptogenetic activity.

To test this hypothesis we applied GABA to certain neurons and checked their structural changes by electron microscopy. We selected the superior cervical ganglion of adult rats, because a limited number of afferent axons enters the ganglion and no direct neuronal feedback exists to the site of their origin. Thus, one could hope that postsynaptic changes would not result in the formation of new synapses which could not be distinguished from the preexisting ones. Although GABA is not a physiological transmitter in this ganglion, it had been shown before to exert inhibitory effects on the ganglion cells. Cold or ^{3}H-GABA was infused for 3 to 7 days into the ganglion by means of thin glass capillaries connected to small glass bulbs which were fixed in the M. longus colli by a tissue glue. Autoradiograms showed that GABA being constantly released from the applicator accumulated in the glial cells as well as in the capsular cells. Besides, other cytological changes in the dendrites of the ganglion cells filamentous material aggregated locally along the inner aspect of the plasma membrane and produced structural complexes at many sites which resembled the postsynaptic membrane thickenings of Gray's type I synapses. Since presynaptic elements were lacking at these sites, we called them free postsynaptic thickenings (F-Post) developing in the presence of normal

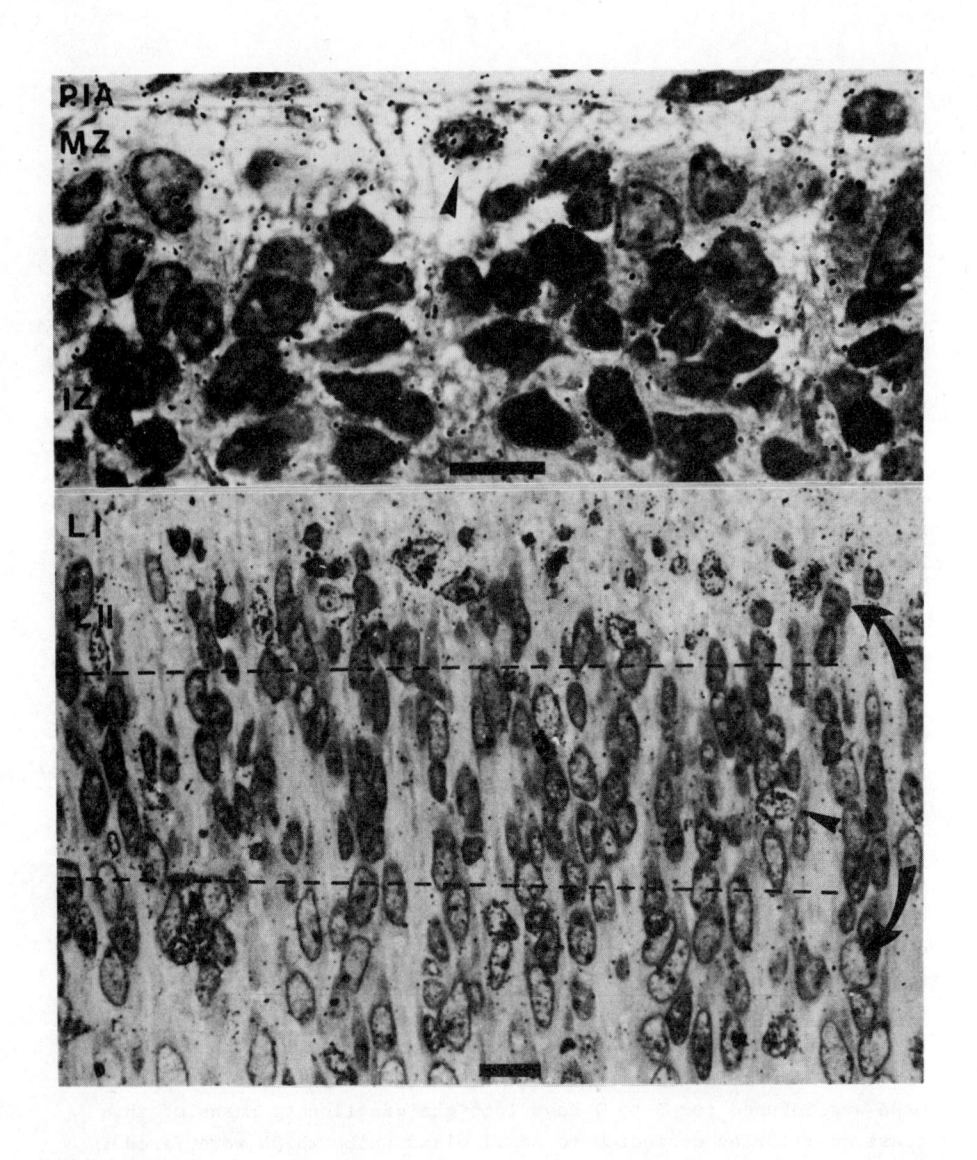

fig. 4:
Autoradiograms after superfusion of the occipital cortex with ^{3}H-GABA.
Top: Labelled cell (arrow head) in the marginal zone (MZ) at E 15.
IZ = intermediate zone. Bottom: labelled cells accumulate at the border
of L I and II as well as between the bipolar and multipolar cortical
plate (arrows). In between only few isolated cells are labelled (arrow
head); bars = 10 µm.

afferent synapses (for further details see Wolff et al., 1978).

Structural complexes similar to F-Post have been seen to occur in the brain under various conditions. They obviously can be produced in many parts of the brain by degeneration of afferent nerves. This led to the assumption that F-Post persist after the degenerating presynaptic element has vanished (e.g. Pinching, 1969, Matthews et al., 1976, Raisman et al., 1974). The present results show, however, that F-Post can be newly formed under the unfluence of GABA without interfering with the afferent axons and their synapses. F-Post have often been reported to occur at least transitorily during normal synaptogenesis (for ref. see Hinds and Hinds, 1976). In the neocortex F-Post have been observed in about the same temporal and laminar distribution as the GABA-accumulating cells (König et al., 1975, Wolff, 1978). This suggests that F-Post might be formed under the influence of GABA or any substance with a similar action not only under the experimental conditions in the superior cervical ganglion, but also during the normal development of the cortex. This condition might not only be necessary for the formation of F-Post, but also for their maintenance. This could explain why F-Post persist for a long time on the dendrites of Purkinje cells of mutant mice in the absence of granule cell axons, but in the presence of the GABAnergic basket cell (Sotelo, 1973).

To elucidate the molecular mechanisms of the promoting effect of GABA on the synaptogenesis, experiments are being carried out in our laboratory to check whether the F-Post represent membrane sites at which receptors either of excitatory or inhibitory nature are accumulated.

Morphogenetic interrelations

The synopsis of data as described above suggest that relations exist among different morphogenetic events which can be summarized in the following hypothetical sequence:

Intracerebral radial blood vessels appearing in a sequence between embryonic day 12 to 13 and postnatal day 10 to 14 terminate and branch in discrete levels of the cortical anlage. Their maturation induces the appearance of short radial glial cells terminating with endfeet in the perivascular glial sheaths. In contrast to the pyramidal neurons, postmitotic neurons and glioblasts migrating along the short radial glial cells do not reach the lower border of lamina I and for unknown reasons differentiate into non-pyramidal neurons. GABAnergic neurons differentiating selectively from the group of non-pyramidal neurons promote synaptogenesis directly or indirectly via GABA-accumulating glial cells. The heterogeneous activation of synaptogenesis by influencing the spatial distribution of metabolic demands probably determines the final distribution of capillaries which arise from the early formed pre- and postcapillary intracerebral vessels. Such a sequence of morphogenetic interrelations could explain the surprising coincidence between the levels of branching of radial vessels which are partly formed long before the synaptogenesis reaches a significant level and the distribution of synapses in the adult cortex.

Summary

Autoradiographic studies after ^{3}H-thymidine labelling revealed that in the neocortex of rats glioblasts and postmitotic proneurons find their definite positions by two different modes of deposition: (1) the well known inside-out layering (Berry and Rogers, 1965) and (2) a diffuse mode of deposing permanently single cells in all layers.

As revealed by Golgi preparations, two types of radial glial cells exist. The long ones reaching the meningeal surface have been shown to serve as guidances for the migration of neurons according to the inside-out principle (Rakic, 1972). Short radial glial cells terminating at intracerebral blood vessels (4 sets of different length, see Wolff, 1978) seemingly guide migrating cells to various levels of the pallial and cortical anlage. The first mode apparently supplies the cortical anlage with pyramidal neurons, while the second mode regulates the deposition of non-pyramidal neurons.

In certain glial cells and in non-pyramidal neurons the capacity of accumulating ^{3}H-GABA was demonstrated autoradiographically to appear very early during the prenatal development ($\geq$ E 15). Both types of cells are first aggregated in the same discrete levels of the cortex (Lamina I, II, III, V/VI and lower cortical border) in which the rate of synaptogenesis is high during the embryonic and early postnatal period (see Wolff, 1978).

In a series of experiments long lasting infusions of GABA into the superior cervical ganglion of adult rats produced free postsynaptic thickenings (F-Post) on ganglion cells without interfering with the afferent nerves (Wolff et al., 1978). The relation of these F-Post to the normal synaptogenesis is discussed. It is suggested that GABA and perhaps other inhibitory agents may regulate synaptogenesis by increasing the number of postsynaptic offerings for making synaptic contacts.

<u>Acknowledgements</u>: The authors are indebted to Dr. Ferenc Joó for valuable comments improving the manuscripts, to Miss Gisela Kotte for preparing the manuscript and to Mrs. B. Pirouzmandi and A. Wolff for their skilful technical assistance.
The study was supported by a grant from the Deutsche Forschungsgemeinschaft, SFB 33, Proj. E 3.

References

Angevine, J.B. (1970) Critical cellular events in the shaping of neural centers. In: The Neurosciences: Second Study Program, Ed. F.O. Schmitt, Rockefeller University Press, New York, pp. 62-72.

Berry, M. and Rogers, A.W. (1965) The migration of neuroblasts in the developing cerebral cortex, J. Anat. 99: 691-709.

Berry, M. (1974) Development of the cerebral neocortex of the rat. In: Studies of the development of behavior and the nervous system. Aspects of Neurogenesis 2, Academic Press, New York, pp. 8-67.

Cajal, S. Ramón y (1960) Studies on vertebrate neurogenesis (revised and translated by Lloyd Guth), Charles C. Thomas, Springfield, Illinois.

Chronwall, B.M. and Wolff, J.R. (1978a) Classification and location of neurons taking up ^{3}H-GABA in the visual cortex of rats. In: Amino acids as chemical transmitters, Plenum Press, New York, pp. 297-303.

Chronwall, B.M. and Wolff, J.R. (1978b) Aspects on the development of non-pyramidal neurons in the neocortex of rat, Zoon, in press.

Chronwall, B.M. and Wolff, J.R. (1978c) Non-pyramidal neurons in early developmental stages of the rat neocortex, Acta Anat., in press.

Chu-Wang, J. and Oppenheim, R.W. (1978) Cell death of motoneurons in the chick embryo spinal cord. II. A quantitative and qualitative analysis of degeneration in the ventral root, including evidence for axon outgrowth and limb innervation prior to cell death, J. Comp. Neurol. 177: 59-86.

Hinds, J.W. and Hinds, P.L. (1976) Synapse formation in the mouse olfactory bulb, II. Morphogenesis, J. Comp. Neurol. 169: 41-62.

Hökfelt, T. and Ljungdahl, Å. (1972) Autoradiographic identification of cerebral and cerebellar cortical neurons accumulating labelled gamma-aminobutyric acid (^{3}H-GABA), Exp. Brain Res. 14: 354-362.

Jacobson, M. (1975) Development and evolution of type II neurons: Conjectures a century after Golgi. In: Golgi Centennial Symposium. Perspectives in Neurobiology, M. Santini, ed., Raven Press, New York, pp. 147-151.

König, N., Roch, G. and Marty, R. (1975) The onset of synaptogenesis in rat temporal cortex, Anat. Embryol. 148: 73-87.

König, N., Valat, J., Fulcrand, J. and Marty, R. (1977) The time of origin of Cajal-Retzius cells in the rat temporal cortex. An autoradiographic study, Neurosci. Lett. 4: 21-26.

Matthews, D.E., Cotman, C. and Lynch, G. (1976) An electron microscopic study of lesion-induced synaptogenesis in the dentate gyrus of the adult rat. II. Reappearance of morphologically normal synaptic contacts, Brain Res. 115: 23-41.

Molliver, M.E., Kostoviç, I. and van der Loos, H. (1973) The development of synapses in the cerebral cortex of the human fetus, Brain Res. 50: 403-407.

Morrest, D.K. (1970) A study of neurogenesis in the forebrain of opossum pouch young, Z. Anat. Entwickl.-Gesch. 130: 265-305.

Parnavelas, J.G., Bradford, R., Mounty, E.J. and Lieberman, A.R. The development of non-pyramidal neurons in the visual cortex of the rat, submitted for publication.

Pinching, A.J. (1969) Persistance of postsynaptic membrane thickenings after degeneration of olfactory nerves, Brain Res. 16: 277-281.

Raedler, A. and Sievers, J. (1975) The development of the visual system of the albino rat. Advances in Anatomy, Embryology and Cell Biology, Vol. 50, Fasc. 3, Springer-Verlag, Berlin-Heidelberg-New York.

Raisman, G., Field, P.M., Ostberg, A.J.C., Iversen, L.L. and Zigmond, R.E.(1974) A quantitative ultrastructural and biochemical analysis of the process of reinnervation of the superior cervical ganglion in the adult rat, Brain Res. 71: 1-16.

Rakic, P. (1971) Neuron-glia relationship during granule cell migration in developing cerebellar cortex. A Golgi and electron microscopic study in Macacus rhesus, J. Comp. Neurol. 141: 283-312.

Rakic, P. (1972) Mode of cell migration to the superficial layers of fetal monkey neocortex, J. Comp. Neurol. 145: 61-84.

Rakic, P. (1974) Neurons in rhesus monkey visual cortex: systematic relation between time of origin and eventual disposition, Science 183: 425-427.

Rakic, P. (1975) Timing of major ontogenetic events in the visual cortex of the rhesus monkey. In: Brain mechanisms in mental retardation, Eds. M. Buchwald and M. Brazier, Academic Press, New York, pp. 3-40.

Ribak (1978) Spinous and sparsely-spinous stellate neurons contain glutamic acid decarboxylase in the visual cortex of rats, J. Neurocytol., in press.

Rickmann, M., Chronwall, B.M. and Wolff, J.R. (1977) On the development of non-pyramidal neurons and axons outside the cortical plate: The early marginal zone as a pallial anlage, Anat. Embryol. 151: 285-307.

Sidman, R.L. (1970) Cell proliferation, migration, and interaction in the developing mammalian central nervous system. In: The Neurosciences: Second Study Program, Ed. F.O. Schmitt, Rockefeller University Press, New York, pp. 100-107.

Sidman, R.L. and Rakic, P. (1973) Neuronal migration, with special reference to developing human brain: a review, Brain Res. 62: 1-35.

Sotelo, C. (1973) Permanence and fate of paramembraneous synaptic specializations in mutants and experimental animals, Brain Res. 62: 345-351.

Stensaas, L.J. (1967) The development of hippocampal and dorsolateral pallial regions of the cerebral hemisphere in fetal rabbits. V. Sixty millimeter stage, glial cell morphology, J. Comp. Neurol. 131: 423-436.

Van der Loos, H. (1965) The "improperly" oriented pyramidal cell in the cerebral cortex and its possible bearing on problems of neuronal growth and cell orientation, Bull. Johns Hopkins Hosp. 117: 228-250.

Wolff, J.R. (1976) Quantitative analysis of topography and development of synapses in the visual cortex, Exp. Brain Res. Suppl. I: 259-263.

Wolff, J.R. (1976) An ontogenetically defined angioarchitecture of the neocortex, Drug Res. 26: 1246 (Abstract).

Wolff, J.R. and Wolff, A. (1977) Development of inhibitory synapses in the neocortex, Acta Anat. 99(3): 355 (Abstract).

Wolff, J.R. (1978) Ontogenetic aspects of cortical architecture: Lamination. In: Architectonics of the Cerebral Cortex, Eds. M.A.B. Brazier and H. Petsche, Raven Press, New York, pp. 159-173.

Wolff, J.R., Joó, F. and Dames, W. (1978) Plasticity of dendrites shown by continuous GABA administration in superior cervical ganglion of adult rat, Nature, in press.

MEMBRANE GLYCOPROTEINS IN SYNAPTOGENESIS

G. Gombos, G. Vincendon, A. Reeber, M.S. Ghandour and J-P. Zanetta
Centre de Neurochimie du CNRS et Institut de Chimie Biologique de la Faculté de Médecine de l'Université Louis Pasteur, Strasbourg, France.

One of the central problems in developmental neurobiology today is that of the molecular mechanism by which the different neurons make specific synapses and thus form reproducible and specific circuits. Put in extremely simplified terms, the problem of synapse specification is to answer the following schematic question : why and how nerve endings from neuron A make synapses with neuron A' and not with neuron B' (while nerve endings from neuron B make synapses with neuron B' and not with neuron A'). However before an axon can make a synapse with its partner cell, this axon must make "contact" with the cell. If axonal growth is not oriented, "contact" is probable only in the case of neighbouring short-fiber neurons where the areas covered by the processes of one cell overlap with that of another. But when neuronal perikarya are far from each other, oriented axonal growth is required for "contact" to be made. Evidently when the axon branches out in the proximity of a cell, the situation is similar to that of the neighbouring neurons. Oriented axonal growth is one link in a long chain of events : neuroblasts multiply, migrate toward their final position and then differentiate. Their axons grow toward their target, ignoring some cells, making contacts which are transient with some cells and stable with others. These stable contacts develop into synapses. Some synapses disappear after being formed. A series of modifications occur in the cells when they receive their innervation.

1. EVENTS WHICH LEAD TO SYNAPSE FORMATION

1.1. Neuroblast migration and neurite growth

In most cases the type of neuron in which the neuroblast is going to differentiate is already determined before migration occurs and its immediate environment does not modify its basic morphology. The mechanisms that start, stop and orient neuronal migration are not known. Possibly they are in principle similar to those orienting axonal growth (see § 3).
In in vitro cultures, axonal growth occurs at the axonal tip, made of a particular structure called the growth-cone which consists of a "varicosity" from which thin philopodia expanding in all directions, continuously emerge and retract and seem to explore the environment. The growth cone appears to be the only structure of the axon that adheres to the support by means of transient, continuously changing adhesive contacts made by the philopodia. It appears that axonal direction is established by the path followed by the growth-cone of the axon. Possibly the same mechanism of axonal growth takes place in the nervous system in vivo. The following observations suggest that the point of emergence of the axon is determined by factors in-

trinsic to the neuron examined, but that the final direction taken by the axon depends on factors extrinsic to the cell examined. In fact the axon of pyramidal cells of cerebral cortex (Van der Loos, 1965) and of Purkinje cells in cerebellum emerge from the usual point even if these cells are "disoriented" as seen occasionally or in neurological mutant mice (Sidman et al., 1965) and in their phenocopies (Altman and Anderson, 1972; 1973). But these axons then, bend and go toward their usual target. In transplantation experiments in fish embryos (Hibbard, 1965) in which the "polarity" of the transplanted medulla is opposite to that of the host medulla, the Mauthner cell axon grows in the direction determined by the polarity of the transplant but once it enters the host medulla, makes a hairpin bend and follows the "polarity" of the host medulla. This could also indicate that the signals guiding axonal direction come from the immediate environment of the growth-cone. Other examples however suggest that "guiding signals" come also from long distance. Axons diverted from their normal path by surgical manipulation not only can return to this path by a short detour (Arora, 1963; Arora and Sperry, 1962) but sometimes they find their target cell by following other paths (i.e. optic fibers could be made to follow either the controlateral pathway or the oculomotor nerve and they find their normal partner cells in the correct tectum) (Hibbard, 1967; Gaze, 1959; Sharma, 1972). Similarly the Mauthner cell axon finds its target even if obstacles (transplant of other tissue) are in its path. Optic nerve fiber cross the mid-line to meet the cells in the controlateral colliculus if, the homolateral colliculus containing their target cells is removed (Schneider, 1970). It appears however that the "long distance" signals do not come uniquely from the target cells. In fact fibers continue to grow in the same direction even if the target is removed (see mossy fibers in rat cerebella and Mauthner cell axons in Xenopus in § 1.5).

Signals directing axonal growth are differently perceived by the growing axons since different axons growing in the same tract can selectively distribute themselves along different pathways. Sensory and motory fibers of peripheral nerve or optic nerve fibers are an example of this.

1.2. Growth-migration of cerebellar granule cells

In at least one case, neuroblasts migrate and grow processes at the same time : the round shaped neuroblast in cerebellum which differentiates into a granule cell at first becomes fusiform, then T shaped, then while growth-cones at the tip of two transverse arms of the "T" grow in the classical manner to give the "parallel fiber", the third arm (which includes at its tip the granule cell perikaryon), grows as the perikaryon migrates (Altman, 1972a,b,c). This growth-migration is in the direction of the presynaptic partner fiber : the mossy fibers. Bergmann glia processes have been suggested as a "guide" for this migration(Rakic, 1971) (see § 3.1).

1.3. Axons growing in the target cells area

When the axon enters the target area, the many retracting and exploring philopodia probably contact many cells in the area. Some cells reject contact, others accept it. Possibly synapses are formed where philopodia remain in stable contacts. Electrophysiological experiments show that during development of Xenopus levis tectum, the incoming fibers migrate by forming transient functional contacts with many cells until they find the correct target cells (Gaze et al., 1972; Straznicky and Gaze, 1972).

A great variety of situations (concerning type, number and the state of differentiation of the cells in the area as well as type, number and chronology of arrival of the incoming fibers) exists when the growing axon reaches the "target area". Synapse specification appears not to require a precise state of differentiation common to all neurons, since in some cases axons make contacts, which evolve into synapses, on apparently undifferentiated neuroblasts (like the Daphnia neurons of the lamina (Levinthal et al., 1976; Lo Presti, 1975) or the neurons of the medial trapezoid nucleus which receive fibers from the neurons of the cochlear nucleus (Morest, 1968; 1969) in which "contact" appears to trigger differentiation (Lo Presti et al., 1973) while in other cases post-synaptic densities are already in place before the arrival of the fiber. The best example for this is the post-synaptic density on spiny processes of Purkinje cell dendrites which is not only detectable by electron microscopy before "contact" with parallel fibers but also is present in agranular cerebella (Altman and Anderson, 1973; Llinas et al., 1973) or in some neurological mutant mice (Hirano and Dembitzer, 1973).

It should be pointed out however that the detection of post-synaptic density depends on specific staining techniques and thus it does not mean that all constituents of such density are already present but only those responsible for the staining. Thus for example, there is no indication for the early presence of neurotransmitters receptor in the post-synaptic density of Purkinje cells, and conversely the lack of detection of post-synaptic specialization in other cases, could simply be due to the absence of the specific molecules responsible for the staining while other components of the post-synaptic portion of the junction could be absent.

1.4. <u>Selection of the site of synapses with the post-synaptic cell</u>

In the adult, synapses do not cover the whole surface of the neuronal perikaryon and dendrites but large areas of it are free of synapses and are covered by glial cell processes. In cells receiving synapses from recognizable fibers of different type, these synapses have a specific distribution on the receiving cell surface. This has been shown for the Mauthner cell of the Goldfish (Stefanelli, 1951), granule cells of dentate girus (Alksne et al., 1966; Blackstad, 1956; Gottlieb and Cowan, 1972; Mosko et al., 1973; Nafstad, 1967; Raisman et al., 1965; Raisman, 1966) and Purkinje cells of cerebellum (Palay and Chan Palay, 1974). In the last case, climbing fibers terminate on smooth dendrites, the parallel fibers, axons of granule cells, make synapses with the spiny processes of these dendrites, while the basket cell axons terminate on the Purkinje cell perikaryon and axonal hillock. It would appear that specific sites on cell surface and specific fibers recognize each other, but a simple point to point "recognition" is unlikely. For example, lesions made during development to one type of fibers afferent to the granule cells of the dentate girus is followed by a redistribution of the other synapses (Cotman et al., 1973; Lynch et al., 1973a,b; Steward et al., 1973).

1.5. "Stop" signal for axonal growth

In most cases, axonal growth stops after the contact is made. If the region containing the target cells of Mauthner cell axons in Xenopus is removed, the fiber continues to grow, even into another spinal cord grafted "tail to tail" (Swisher and Hibbard, 1967). Monolateral removal of the anterior quadrigeminal body of new born hamster, results in continued growth of the incoming optical fibers until they reach and make synapses with cells in the controlateral colliculus (Schneider, 1970). Removal during development of large portions of projection fields cause "sprouting" of the incoming fibers, with formation of synapses in other cells (Raisman and Field, 1973; Linch et al, 1973a; Zimmer, 1973). In agranular rat cerebella, mossy fibers keep growing beyond the Purkinje cell layer (Altman, 1973). The same is true for cerebella in which granule cell migration is delayed (Altman and Anderson, 1972; 1973). In this case synapses with granules are made even in the molecular layer. The contact of mossy fibers with granule cells appears to be the stop-growth signal. On the contrary, mossy fibers of human cerebellum reach their adult position and remain there to wait for granules (Rakic and Sidman, 1970). We suggest that each cell has the capacity to grow axons of a certain maximum length and size but in some cases stop-growth signals intervene before (rat mossy fibers) and in other cases after (man mossy fibers) that this capacity is fully realized.

2. EVOLUTION OF NEWLY FORMED SYNAPSES

Space limitation does not allow us to summarize what is known on synapse construction and we will limit ourselves to briefly consider the following two points : (i) synapses, apparently in excess, can degenerate during development (ii) synapses can migrate to a different point on the cell surface. In fetal (d17) rat diaphragm, axons develop multiple branches and make complex network and the muscle fibers show several synaptic profiles at the end-plate but by the 3rd-5th postnatal week, there is only one nerve terminal per muscle fiber (Bennet and Pettigrew, 1974). Purkinje cells in developing cerebella receive functional contacts by more than one climbing fiber but in the adult, only one climbing fiber innervates one Purkinje cell (Eccles et al., 1966; Delhaye-Bouchaud et al., 1975). The first contact between climbing fibers and Purkinje cells is followed by "early synapses" on the Purkinje cell perikaryon (Altman, 1972b). At a later date these synapses disappear but the connection of climbing fibers is with Purkinje cell dendrites.

3. THEORIES ON SYNAPSE SPECIFICATION

3.1. Mechanical and mechanochemical guidance or restriction

This theory deals only with axonal orientation which would be determined by mechanical interaction with preexisting structures. If selective adhesion to preexisting axons or glial cells is added, we have the "contact guidance" or "selective fasciculation" theories. The growth of optic nerve axons in "Daphnia" (Lopresti et al., 1973; Macagno et al., 1973) and the migration of granule cells along Bergmann glia processes (Rakic, 1971) in cerebellum could support these views.

3.2. Chemical guidance : chemiotaxis, chemiotropism, neurotropism

Already in 1905, Ramon y Cajal suggested that axonal growth could be directed by specific diffusible substances. We can imagine that receptors, at different points of the growth cone could guide growth all the way to the target by perceiving different concentrations of a gradient of chemiotactic substances.

The chemiotaxis theory cannot explain (i) how and why a philopodium attracted by neurotropic factors is sometimes rejected and (ii) how synapses are specifically formed on precise areas of a given cell surface. In addition to this, an unbelievably high number of neurotropic substances would be needed, in order to account for all nerve fibers.

Substance such as the lectin-like molecules described in slime molds, electric organ, muscle, cultured myoblasts, neuroblastoma cells and, probably, chick brain (Barondes and Rosen, 1976; Nowak et al., 1976; Teichberg et al., 1975) could be neurotropic as well as adhesion substances. The transient presence of lectin-like molecules in developing rat brain has been recently shown (Simpson et al., 1977).

3.3. "Cytodifferentiation" or "(chemioaffinity)" theory

In this hypothesis (Sperry, 1963), all nerve cells acquire early in development some specific "label" (specific cell surface molecules ?) that make them different from all other nerve cells. This "label" is complementary to the "label" of the target cell (receptor ?). "Complementarity" implies site recognition and interaction between one molecule in one cell and its receptor on the other. Recent adhesion studies with dissociated retinal cells and the tectum surface indeed show preferential adhesion of cells of one retinal region to the corresponding quadrant of the tectum. However, it is too premature to conclude that these studies have any relation with synapse specificity. The limitation of the cytodifferentiation theory is that the axon recognizes its partner cells only when cell surfaces contact one another and thus axon oriented growth and choice of the target cell before such contact is made remains unexplained. Several authors have introduced the concepts of "graded affinity" (Barondes, 1970; Roberts and Flexner, 1966) between axon and post-synaptic target and of "competition" (Gottlieb and Cowan, 1972; Guillery, 1972; Schneider, 1971;

Mark, 1969) between ingrowing nerve terminals for post-synaptic sites to explain the final approach of the axon to the target cell and its precise distribution on precise cells. Cotman has proposed a mechanism (Cotman and Banker, 1974) for a guided approach of optic nerve fibers to the target cells of the tectum by a grid determined by two perpendicular systems of adhesion-recognition molecules which lead the probing philopodia of the growth cone to the cell surface of maximum adhesion for both adhesion systems ("graded change of adhesive properties").

The chemioaffinity hypothesis and all the others derived from it do not explain in great detail the axonal directional growth over long distances. But also in this case graded affinity along the path of axonal growth could be postulated.

The major limitation of the chemioaffinity hypothesis and of elaborations of it is the extremely high number of specific molecules, and of molecules complementary to them, necessary for specification of all synapses. This number will be even greater if we assume that recognition processes also guide axonal migration.

3.4. Timing theory

This theory (Gaze, 1970; Jacobson, 1969) deals only with synapse specification without discussing the mechanism of axonal oriented growth. The post-synaptic cell surface is not complementary but only compatible with the surface of the incoming axons and the "timing" of fiber arrival and of cell migration is the only factor that specifies connections. Thus no "recognition" is needed. The best known example of such a possible mechanism is that in the Daphnia visual (Levinthal et al., 1976; Lo Presti, 1975; Lo Presti et al., 1973) system in which the axons from the 8 photoreceptor cells of each ommatidium arrive at the optic ganglion at the moment at which a single set of 5 undifferentiated lamina neuroblasts migrates to the proper position to receive them. In addition one of the eight axons of each ommatidium leads the other seven (contact guidance ?) and appears to trigger differentiation of the lamina cells. The second set of 8 fibers, after arrival, can make synapses only with the freshly migrated second set of 5 cells, and the process goes on until all 22 sets of cells and fibers of the 22 ommatidia are properly connected. In other examples however, timing alone cannot be sufficient for synapse specification : in fact of the two sets of fibers that simultaneously reach the ciliary ganglion one forms selective synapses with the ciliary cell and the other with the choroid cells (Landmesser and Pilar, 1972). Transient cross connections are never seen. Axons in the optic tectum of Xenopus appear to migrate to their target cells by establishing transient adhesive contacts with many cells in the area (Gaze et al., 1972).

3.5. Selective stabilization of synapses

This theory, described in detail recently (Changeux and Danchin, 1976), is based on observations that redundant fiber terminals, contacts and even synapses present during development disappear in the adult (see § 2). Since neurotransmission appears to be already present before synapses are morphologically evident, function could determine which synpases or contacts survive and which will degenerate.

4. CRITICAL EVALUATION OF ALL THE THEORIES ON SYNAPSE SPECIFICATION

Each of the reported theories does not account for all the phenomena which lead to synapse specification. Probably the experiments on which they are based just show a single facet of the many concurrent processes all of which lead to specific synapse formation. Many factors such as the orientation, speed and distances covered by the growing axon, and before it by the migrating cells, determine the appropriate "place" and "time" for a fiber to meet its partner cells. Hormones (i.e. thyroid hormones) and other factors (Balazs et al., 1975) which influence the multiplication and differentiation of neurons also affect synapse specification by modifying "timing". Mechanical guidance as well as molecular interaction with preexisting surfaces or with the intercellular medium also must play a role, admittedly minor, in determining orientation of axonal growth and cell migration. The orientation of fibers growing at a later date can be facilitated by following the path opened by the early fibers. All these factors should work in the proper order, if the fiber is to meet the partner cell. "Timing" makes the encounter possible and, by limiting the number of cells and fibers meeting at any given place, it facilitates the sorting out of fibers and cells but in most cases is not sufficient alone for synapse specification. Also the timing of the arrival of glial processes which cover neuronal surface can help to specify synapse location in preventing the formation of synapses with nerve fibers arriving later.

Synapse stabilization (which certainly occurs in some cases and may possibly be a general phenomenon) appears to be a mechanism for perfecting the circuitry already established by other mechanisms. Such mechanisms imply at least surface compatibility.

It appears then that in most cases, some mechanism of "recognition" is needed. However how is this possible without an unbelievable high number of molecules needed to "code" for each complementary neuronal surface ? One hypothesis is that many cells and fibers in different areas of the nervous system have the same recognition molecules but most of them never meet because of spatial segregation. This hypothesis would also overcome another objection to "cell specificity" theories : that axons, in the absence of the normal target cells, can make connections with cells with which they are normally not connected. For example, the synapses of mossy fibers with Purkinje cells present in agranular rat cerebella (Altman and Anderson, 1973; Llinas et al., 1973; Sotelo, 1975), are absent in normal cerebella. Purkinje cells could have a "recognition" molecule (similar to that of parallel fibers) which reacts with mossy fibers. In normal cerebella spatial segregation (see § 1.5) does not allow contact between mossy fibers and Purkinje cells. On the contrary the continued growth of mossy fibers in agranular cerebella allows this contact. Another possible mechanism of coding of the cell surface with a limited number of molecules is that the coding elements are stable molecular aggregates specific to the cell but assembled from molecules each of which is common to many cells.

Also in the case of the chemiotaxis hypothesis, the high number of substance can be reduced if the mechanism is not simply that of one fiber-one neurotropic agent, but is that of coordinate systems of different neurotropic agents diffusing from properly spaced points, which, thus, could act on many axon types at the same time particularly if each axon can perceive these agents in a specific way because of its specific combination of "receptors".

In conclusion, it appears that in one way or another a "recognition" mechanism exists. Thus at present the problem of synapse specification is not that of the existence of such mechanisms but is that of defining its nature, modality of function and limits of its specification.

5. POTENTIAL "RECOGNITION" MOLECULES

Three main approaches have been followed in the search for recognition molecules involved in synapse specification. In one approach the possible role of post-synaptic receptors for neurotransmitters in the selection of the site of synapses has been examined with opposite conclusions (Changeux and Danchin, 1976; Diamond and Miledi, 1962; Sytokowski et al., 1973). In another approach aggregation and adhesion phenomena were investigated in a search for surface specialization molecules (Moscona and Hausman, 1977; Glaser et al., 1977; Rutishauser et al., 1976). The limitation of space does not allow us to deal with these studies here.

The third approach is based on the idea that, (as suggested in other systems) in the nervous system, surface glycoproteins could be informational molecules.

5.1. Glycoproteins as informational molecules

Some authors (Barondes, 1970; Brunngraber, 1969; Sharon, 1975) have suggested that the sugar moiety (glycan) of glycoproteins can act as informational molecules : the letters in the "code" are the constituent sugars of the glycans and the information is contained in the overall structure of the glycan.

Few sugar types are systematically found in glycoproteins. In mammalian central nervous system, they are 6 or 7 (Fuc, L-fucose ; Gal, D-galactose ; GalNAc, N-acetyl-D-galactosamine ; GlcNAc, N-acetyl-D-glucosamine ; Man, D-mannose ; NeuNac, N-acetyl-neuraminic acid and possibly Glc, D-glucose). Other sugars have been shown in glycoproteins of other organs and species.

In all organs glycoprotein sugars can be linked in highly branched glycans, very different from the linear arrangement of sugars in GAG (Glycosamino-glycans = mucopolysaccharides). Thus in spite of the limited number of constituent sugar types, a glycan of a glycoprotein, can have highly individual properties because of the particular sugar sequence, and branching pattern. Recent systematic studies show that glycoprotein glycans can be classified according to few general "designs". The best studied "design" is that of the glycans linked N-glycosidically to asparagine (ASN). (Montreuil, 1975). The sugar sequence ("core") immediately linked to the asparagine shows little variations in many glycoproteins (with the possible addition of fucose as a lateral branch of one of the first two GlcNAc) and is probably synthetized by the dolychol phosphate pathway and transferred as a whole on the polypeptide backbone.

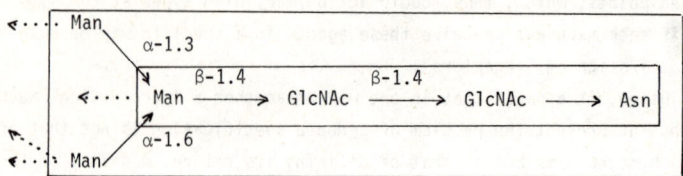

```
      ←··· Man
              ↘ α-1.3
                β-1.4        β-1.4
      ←··· Man ────→ GlcNAc ────→ GlcNAc ────→ Asn
         ↗ α-1.6
      ←···Man
```

Specific glycosyltransferases add different sugars (including Man and GlcNAc) to this "core". The only general rule is that AcNeu and Fuc are only in terminal positions and Gal or GalNAc are penultimate when AcNeu and Fuc are terminal. A large variety of glycans can thus be constructed from a similar "core" by variation of sugar chain length, number and type of sugars added, glycosidic linkages between sugar carbons, length and composition of each sugar branch etc. Although the overall design of each complex glycan is distinctive of a given glycoprotein, errors in glycosidase activity and/or the action of glycosidase, can result in the presence of "incomplete glycan" in glycoprotein molecules with the same polypeptide backbone (microheterogeneity). Whether microheterogeneity can have functional implication is not known.

Glycans with other designs and different linkages to the polypeptide backbone have been also described. In addition to the large possible choice of glycans, it should be emphasized that several glycans are usually on the same polypeptide backbone of a glycoprotein and in many cases they can be of completely different structures.

The most peripheral cell surface components are glycans of glycoproteins or glycolipids (Rambourg and Leblond, 1967). This localization and their alleged coding properties (Sharon, 1975) have stimulated hypothesis that they play a role (for review see Cook and Stoddart, 1973; Sharon, 1975) in different cell systems as determinants of cell surface specificity and as signal donors and/or receptors, particularly in contact phenomena. These hypotheses, applied to the nervous system (Barondes, 1970; Brunngraber, 1969) would make glycans good candidates for molecules which specify the surface of neurons of different type either as individual molecules or as "patterns" of molecules.

In addition if surface glycans are "informational molecules" they can play a role in tissue morphogenesis : as surface molecules specific for a given cell type and thus as donors or receptors of specific messages, they could affect processes of cell differentiation and migration, of axonal growth and intercellular recognition. Moreover their possible developmental modification in the same cells would make cell surface interactions or sensitivity to diffusible macromolecules different at different developmental ages.

The data in the literature tend to indicate that in developing nerve cells, glycoprotein composition and distribution indeed change. In fact carbohydrate material accumulates at the surface of the growth cone of some neuroblasts (James and Tresman, 1972; Pfenninger and Maylie-Pfenninger, 1975). The composition of cell surface glycoproteins of neuroblastoma cells is modified by neurite sprouting (Brown, 1971; Glick et al., 1973; Truding et al., 1974) and that of the bulk of brain glycoproteins changes throughout development (Di Benedetta and Cioffi, 1972; Krusius et al., 1974; Margolis and Gomez, 1974; Margolis et al., 1976). The use of labelled lectins (sugar-binding proteins) of different specificity, has shown not only that the distribution of glycans which bind to one lectin is not uniform on the whole surface of one type of neuroblast but is different in different neuroblasts of the same type and also that neuroblasts of different types have different patterns of lectin binding sites on their surface (Pfenninger and Maylie-Pfenninger, 1975; Pfenninger and Rees, 1976; Vaughn et al., 1976). However, direct experimental evidence that glycans of cell surface glycoproteins play a role in nervous tissue morphogenesis, has not yet been obtained.

The coding capacity of glycolipids has been emphasized less, probably because they have simpler glycans than most glycoproteins and thus have a more limited "coding" possibility, probably limited to cell type (neurons versus glial cells). In addition glycolipids (polysialogangliosides, specific to neurons, cerebrosides and sulphatides specific to glial cells and myelin) appear during nervous tissue development later than the phenomena of cell migration and differentiation and thus seem to follow rather than to "code" for these phenomena.

5.2. Transient Con A binding glycoprotein in parallel cell surface as recognition molecules

Studies in our laboratory (Zanetta et al., 1978; Gombos et al., 1977 and Reeber et al., this meeting) have shown that the glycans of 4 Con A* binding-glycopolypeptides transiently present on the surface of parallel fibers are recognized and specifically bound by some molecule on the post-synaptic partner cells (Purkinje, stellate, basket and Golgi neurons). These glycans are of small molecular weight and simple sugar composition (virtually only mannose and N-acetyl-glucosamine). The similarity of these glycans with the "core" of most glycans of glycoproteins (core-like glycans) raised the possibility that they were "incomplete" glycans to which other sugars were added later in development and thus lost affinity for Con A. However incorporation studies with labelled sugar precursors of glycoproteins, binding of other lectins (specific for more peripheral sugars) to cerebella slices at different ages,

---

*Concanavalin A : a lectin specific for manno- and gluco-pyranosides provided that their carbons 3, 4 and 6 are not involved in glycosidic linkages hence these sugars will bind Con A only if they are terminal or if linked through carbons 1 and 2 when internal to a glycan.

developmental curves of glycosidases all indicated that the transient character of the Con A binding sites was due to the disappearance of either the glycan alone or of the whole glycoproteins molecule.

Two points are of great interest, one is that both Con A binding glycoproteins on parallel fibers and their receptors are redundant. In fact the former appears to be on the whole surface of parallel fibers, the latter on the whole surface of partner cells, including their cell body. The second point is that both groups of molecules are transient and both are present only between the 10th and the 16th postnatal day. Before this date and between the 16th and the 24th day, only one groups of molecules is present and both are absent after the 24th day. In effect, the period during which each parallel fiber carries Con A-binding sites is much shorter since fiber formed earlier looses its Con A binding glycans before the 16th day and the last formed fibers start to grow very close to the 16th day.

We suggest that the transient Con A-binding glycoproteins on parallel fibers and their binding sites on the post-synaptic neurons are "recognition" molecules which, by their interaction, possibly form a specific "early connection" between complementary neuronal surfaces, when neuronal processes contact each other. The redundant presence of the two types of molecules increases the chances for molecular interaction when cell surfaces contact one another (contact being determined by the direction of growth of parallel fibers and of dendrites of their partner cells). These molecules apparently are not involved in synapse maintenance and stabilization since they disappear simultaneously with synapse formation but the establishment of the early connection could constitute the anterograde and retrograde signals that initiate the implementation of the program of synapse formation. Both groups of molecules, the Con A binding glycoproteins and their binding sites, are transient and the presence of each group is partially out of phase with that of the other. Obviously connection can occur only during the phase of simultaneous presence of these complementary molecules. One fundamental problem is that of the mechanism that triggers the disappearance of these molecules from cell surface. The signal could be the establishment of "the early connection"; in this case the response is differently delayed in the granule cells on one hand and on the partner cells of parallel fibers on the other. Alternatively the signal could be endogenous and thus disappearance will occur also in absence of early connection or be simply accelerated by them.

If this last hypothesis is correct, the fact that complementary recognition molecules are not present exactly at the same time has large implications with regard to the mechanism of synapse specification because the presence of complementary molecules on the two cell surfaces becomes a necessary but not sufficient condition for the establishment of the "early connection". This should occur only if both molecules are present at the appropriate time. Thus the two cells can be compatible and even complementary at one stage of development and not compatible at another. In fact acceleration or slowing down of differentiation of one of the two cells which usually form synapses, can result in an encounter of the two cell processes when the presence of one recognition molecule is out of phase with the presence of the complementary molecule on the other cells. Synapses will not be formed if the complementary molecule appears after the earlier molecule has disappeared. Desynchronization of Purkinje cell maturation and granule cell formation by interfering with thyroid hormone levels or by X-ray (Lewis et al., 1976; Nicholson and Altman, 1972; for other references see Altman, 1976; Balazs et al., 1975) should confirm or reject these hypotheses. Similarly the genetically determined absence of either group of complementary molecules could be the basis for the absence of synapses between parallel fibers and Purkinje cell dendrites and of the trans-synaptic degeneration "en cascade" of granule cells in staggerer mutant mice (Sotelo and Changeux, 1974).

6. CONCLUSION

If the transient presence of recognition molecules on the neuronal surface is a general phenomenon and not limited only to the (parallel fibers)-(Purkinje, stellate, basket, Golgi neurons) system, it would mean that the chemiospecificity theory should be modified by suggesting that the "label" of each cell with regard to synapse specification does not last for the whole cell lifetime, but is transient.

A limited number of recognition molecules transiently present at successive times on different cell types becomes sufficient for the establishment of even complex circuitry. We have already suggested before (§ 4) that several cells and fibers could have the same recognition molecules and never connect because of spatial segregation, but if "recognition molecules" appear at different times on the surface of different cells, also cells and fibers in the same areas having the same recognition molecules could be connected in different circuits. The presence of the same "recognition" molecules on more than one cell type appears to be feasible, and thus, as discussed before (§ 4) the formation of alternative pathways when the target cell is missing can be explained. We can imagine that in the example given at the § 4 , the mossy fiber growth cone contains the same Con A binding glycoproteins as parallel fibers.

The hypothesis that transient molecules determine synapse specification is consistent with the requirement for a limited number of recognition molecules and, (in conjunction with all other factors which determine cell/fiber contact) permits the great plasticity in the system demanded by many experimental results.

The presence at the same time or at different times of different "recognition" molecules on the same cell could answer many other questions. For example specific connections of different fibers, arriving at the same cell could be due to the presence, and possibly sequential appearance and disappearance of different recognition molecules in the same cell.

REFERENCES

Altman, J. (1972a)J. comp. Neurol. 145: 353-398.
Altman, J. (1972b)J. comp. Neurol. 145: 399-464.

Altman, J. (1972c) J. comp. Neurol. 145: 465-514

Altman, J. and Anderson, W.J. (1972) J. comp. Neurol. 146: 355-406

Altman, J. and Anderson, W.J. (1973) J. comp. Neurol. 149: 123-152

Altman, J. (1976) J. comp. Neurol. 165: 65-76

Alksne, J.F., Blackstad, T.W., Walberg, F. and White, L.E. Jr. (1966) Erg. Anat. Entw. Gesch. 39: 1

Arora, H.L. and Sperry, R.W. (1962) Amer. Zoologist, 2: 389

Arora, H.L. (1963) Anat. Record 145: 202

Balasz, R. Lewis, P.D. and Patel, A.J. (1975) In Brazier Growth and development of the brain, Raven Press, New York, pp. 83-115.

Barondes, S.H. (1970) In Neurosciences Second Study Program (F.O. Schmitt, ed.) Rockefeller University Press, New York, pp. 747-760

Barondes, S.H. and Rosen, S.D. (1976) In Barondes Neuronal recognition, Plenum Press, New York, pp. 331-356.

Bennett, M.R. and Pettigrew, A.G. (1974) J. Physiol. 241: 515-545.

Blackstad, T.W. (1956) J. Comp. Neurol. 105: 417

Brown, J.C. (1971) Expl. Cell Res. 69:440-442.

Brunngraber, E.G. (1969) Perspect. Biol. Med. 12: 467-470

Changeux, J-P and Danchin, A. (1976) Nature 264: 705-712

Cook, G.M.W. and Stoddart, R.W. (1973) In Surface carbohydrates of the eukaryotic cell, Academic Press, London, pp. 257-270

Cotman, C.W., Matthews, D.A., Taylor, D. and Lynch, G. (1973) Proc. Nat. Acad. Sci. U.S. 70: 3473

Cotman, C.W. and Banker, G.A. (1974) In Reviews of Neuroscience, Vol. 1 (S. Ehrenpreis and I.J. Kopin eds.) Raven Press, New York, pp. 1-62

Delhaye-Bouchaud, N., Crepel, F. and Mariani, J. (1975) C.R. Acad. Sci. Paris 281: 909-912

Diamond, J. and Miledi, R. (1962) J. Physiol. (London) 162:393.

Di Benedetta, C. and Cioffi, I.A. (1972) In Glycolipids, glycoproteins and mucopolysaccharides of the nervous system (Zambotti, Tettamanti and Arrigoni eds) Plenum Press, New York, pp. 115-124

Eccles, J.L., Llinas, R. and Sasaki, K. (1966) J. Physiol. Lond. 182: 268-296

Gaze, R.M. (1959) Quart. J. Exp. Physiol. 44:209

Gaze, R.M. (1970)   The formation of nerve connections, Academic press, London.

Gaze, R.M., Chung, S.H. and Keating, M.J. (1972) Nature New Biol., 236: 133-135

Glaser, L., Santala, R. Gottlieb, D.I. and Merrell, R. (1977) In Cell and Tissue Interactions (J.W. Lash and M.M. Burger eds.) Raven Press, New York, pp. 197-208.

Glick, M.C., Kimhi, Y. and Littauer, U.Z. (1973) Proc. Nat. Acad. Sci. USA 70: 1682-1687

Gombos, G., Ghandour, M.S., Vincendon, G., Reeber, A. and Zanetta, J-P. (1977) In Maturation of Neurotransmission, Vol. 3 (E. Giacobini and A. Vernadakis eds.) S. Karger, A.G., Basel, Switzerland, in press.

Gottlieb, D. and Cowan, W. (1972) Brain Res., 41: 452-456

Guillery, R.W. (1972) J. Comp. Neurol. 144: 117

Hibbard, E. (1965) Exp. Neurol. 13: 289

Hibbard, E. (1967) Exp. Neurol. 19: 350

Hirano, A. and Dembitzer, H.M. (1973) J. Cell Biol. 56: 478-486

Jacobson, M. (1969) Science 163: 543-547

James, D.W. and Tresman, R.L. (1972) J. Neurocytol. 1: 383-395

Krusius, T., Finne, J. Karkkainen, J. and Jarnefelt, J. (1974) Biochim. biophys. Acta, 365: 80-92

Landmesser, L. and Pilar, G. (1972) J. Physiol. 222: 691

Levinthal, F., Macagno, E.R. and Levinthal, C. (1976) Cold Spring Harbor Symp. Quant. Biol. 40: 321-331

Lewis, P.D., Patel, A.J., Johnson, A.L. and Balazs, R. (1976) Brain Res. 104: 49-62

Llinas, R., Hillman, D. and Prescht, W. (1973) J. Neurobiol. 4: 69-94

Lo Presti, V. (1975) The development of the eye-optic lamina neuroconnections in Daphnia magna: A serial section electron micrsocope study. Ph.D. dissertation, Columbia University, New York.

Lo Presti, V., Macagno, E.R. and Levinthal, C. (1973) Proc. Natl. Acad. Sci. USA 70: 433-437

Lynch, G., Deadwyler, S. and Cotman, C. (1973a) Science 180: 1364-1366

Lynch, G. Mosko, S., Parks, T. and Cotman, C. (1973b) Brain Res. 50: 174-178

Macagno, E.R., Lopresti, V. and Levinthal, C. (1973) Proc. Nat. Acad. Sci. U.S. 70: 57-61

Margolis, R.K. and Gomez, Z. (1974) Brain Res. 74: 370-372

Margolis, R.K., Preti, C. Lai, D. and Margolis, R.U. (1976) Brain Res. 112: 363-369

Mark, R.F. (1969) Brain Res. 14: 245-254

Moscona, A.A. and Hausman, R.E. (1977) In Cell and Tissue Interactions (J.W. Lash and M.M. Burger eds.) Raven Press, New York, pp. 173-185

Mosko, S., Lynch, G.S. and Cotman, C.W. (1973) J. Comp. Neurol. 152: 163

Montreuil, J. (1975) In Pure and Applied Chemistry, Vol. 42 pp. 431-477

Morest, D.K. (1968) Z. Anat. Entwickl. 127: 201

Morest, D.K. (1969) Z. Anat. Entwickl. 128: 271

Nafstad, P.M.J. (1967) Z. Zellforsch. 76: 532

Nicholson, J.L. and Altman, J. (1972) Brain Res. 44: 25-36

Nowak, T.P., Haywood, P.L. and Barondes, S.H. (1976) Biochem.biophys.Res.Commun. 68: 650-657

Palay, S.L. and Chan-Palay, V., (1974) In Cerebellar Cortex Cytology and Organization, Springer-Verlag, Berlin.

Pfenninger, K.H. and Maylie-Pfenninger, M.R. (1975) J. Cell Biol. 67: 332a

Pfenninger, K.H. and Rees, R.P. (1976) In Barondes Neuronal recognition, Plenum Press, New York, pp. 131-173

Rambourg, A. and Leblond, C.P. (1967) J. Cell Biol. 32: 27-53

Raisman, G., Cowan, W.M. and Powell, T.P.S. (1965) Brain 88:963

Raisman, G. (1966) Brain 89: 317

Raisman, G. and Field, P.M. (1973) Brain Res. 50: 241-264

Rakic, P. (1971) J. Comp. Neurol. 141: 283

Rakic, P. and Sidman, R.L. (1970) J. Comp. Neurol. 139: 473

Roberts, R.B. and Flexner, L.B. (1966) Amer. Sci. 54: 174

Rutishauser, U., Thiery, J-P., Brackenbury, R., Sela, B.A. and Edelman, G.M. (1976) Proc. Natl. Acad. Sci. USA, 73: 577-581

Schneider, G.E. (1970) Brain Behav. Evol. 3: 295

Schneider, G.E. (1971) Anat. Rec. 169: 420

Sharma, S.C. (1972) Nature New Biol. 238: 286-287

Sharon, N. (1975) In Complex carbohydrates. Their chemistry, biosynthesis and function, Addison-Wesley, Reading, Mass, pp. 26-29

Sidman, R.L., Green, M.C. and Appel, S.H. (1965) Catalog of the Neurological Mutants of the Mouse, Harvard Univ. Press, Cambridge, Mass., pp. 66-67

Simpson, D.L., Thorne, D.R. and Loh, H.H. (1977) Nature, Lond. 266: 367-369

Sotelo, C. and Changeux, J.P. (1974) Brain Res. 67: 519-526

Sotelo, C. (1975) Brain Res. 94, 19-44

Stefanelli, A. (1951) Quart. Rev. Biol. 26: 17

Steward, O., Cotman, C.W. and Lynch, G.S. (1973) Exptl. Brain Res. 18: 396-414

Straznicky, K and Gaze R.M. (1972) J. Embryol. Exp. Morph. 28: 87

Sytkowski, A.J. Vogel, Z. and Nirenberg, I.M.W. (1973) Proc. Nat. Acad. Sci. U.S. 70: 270-274

Swisher, J.E. and Hibbard, E. (1967) J. Exp. Zool., 165: 433

Teichberg, V.I., Silman, I. Beitsch, D.D. and Resheff, G. (1975) Proc.Natl. Acad. Sci. USA, 72: 1383-1387

Truding, R., Shelanski, M.L., Daniels, M.P. and Morell, P. (1974) J. biol. Chem. 249:3973-3982

Van der Loos, H. (1965) Bull. John Hopkins Hosp. 117: 228

Vaughn, J.E., Henrikson, C.K. and Wood, J.G. (1976) Brain Res. 110: 431-445

Zanetta, J-P., Roussel, G., Ghandour, M.S., Vincendon, G. and Gombos, G. (1978) Brain Res. 142: 301-319

Zimmer, J. (1973) Brain Res. 64: 293-311

EFFECT OF MALNUTRITION ON GANGLIOSIDE DEVELOPMENT

I. Karlsson

Psychiatric Research Centre, Department of Neurochemistry, S:t Jörgen Hospital, S-422 03 Hisings Backa, Sweden.

Introduction

The brain tissue is highly complex, due to the different cellular elements, the wide extension of the neuronal cells and the presence of synaptic connexions. The study of the development of this tissue provides many obstacles. With histologic methods it is possible to make qualitative descriptions but quantitative studies are laborious and difficult. Biochemical studies of brain tissue can be made exact and reproducible. The difficulty comes in interprating the results. To facilitate this, compounds only bound to certain structures have been searched. The DNA, which is found only in the cell nucleus and in a constant amount in each cell, could be used for estimation of the cell number. The cerebrosides are a part of the myelin membrane and only small amounts are found in other structures than myelin or the oligodendroglia cell membrane. The cerebroside concentration is therefore used to estimate the amount of myelin in a brain tissue specimen.

It has been suggested for many years that the gangliosides could be used as marker substances for neuronal membranes (Svennerholm, 1957). They were given their name by Klenk (1942) who isolated them from brain gray matter. Gangliosides have later been found in several membrane structures in the brain which have made their topografic localisation uncertain. However, they have shown a strong association with the neuronal cell membranes.

Chemical structure of the major brain gangliosides

Gangliosides are sphingolipids with a carbohydrate chain to which sialic acid is bound. The mammalian brain contains four major gangliosides:

$$GM1 \quad Gal\beta \rightarrow 3GalNAc\beta \rightarrow 4Gal\beta \rightarrow 4Glc\beta \rightarrow Cer$$
$$3$$
$$\uparrow$$
$$\alpha NeuAc$$

$$GD1a \quad Gal\beta \rightarrow 3GalNAc\beta \rightarrow 4Gal\beta \rightarrow 4Glc\beta \rightarrow Cer$$
$$3 \qquad\qquad\qquad 3$$
$$\uparrow \qquad\qquad\qquad \uparrow$$
$$\alpha NeuAc \qquad\quad \alpha NeuAc$$

$$GD1b \quad Gal\beta \rightarrow 3GalNAc\beta \rightarrow 4Gal\beta \rightarrow 4Glc\beta \rightarrow Cer$$
$$3$$
$$\uparrow$$
$$\alpha NeuAc8 \leftarrow \alpha NeuAc$$

$$GT1 \quad Gal\beta - 3GalNAc\beta - 4Gal\beta - 4Glc\beta - Cer$$
$$3 \qquad\qquad\qquad 3$$
$$\uparrow \qquad\qquad\qquad \uparrow$$
$$\alpha NeuAc \qquad\quad \alpha NeuAc8 \leftarrow \alpha NeuAc$$

Many other gangliosides have been isolated from brain tissue but in lower concentrations, for review see Leeden & Yu (1976).

Studies of the topographic localisation

Klenk (1942) could only demonstrate gangliosides in gray matter of the brain. Svennerholm (1957) showed them to be present also in white matter, but in a much lower concentration. Later studies have shown the concentration to be 2.5, measured as μmol lipid-NeuAc/g wet weight, in gray matter and slightly below 1 in white matter of adult human brain (Svennerholm & Vanier, 1972). Different mammalian species have similar values (for review see Leeden & Yu, 1976).

The uneven distribution between gray and white matter suggested a distribution in the neuronal cell. It was pointed out, by Svennerholm (1957) that the gangliosides, for quantitative reasons, must be constituents not only of the cell body but also of the axon and dendrites. In the unmyelinated human brain white matter, which consists mainly of unmyelinated axons, the ganglioside concentration is higher than in the cortex of the same immature brain (Svennerholm & Vanier, 1972). Synaptosomal plasma membranes have given the highest ganglioside concentration of brain subcellular fractions (Wiegand, 1967; Lapetina et al., 1967; Spence & Wolfe, 1967; Breckenridge et al., 1972). Small amounts of gangliosides were isolated in purified myelin by Norton & Autilio (1966) and subsequent studies have strongly supported the assumption that gangliosides are a part of the myelin membrane (Suzuki et al., 1968; Suzuki, 1970; Leeden et al., 1973) showing that gangliosides are not only associated with the neuronal cell.

Bovine myelin has a ganglioside content of 1.35 μmol/g dry weight (Suzuki et al., 1968). The concentration is 3.2 in bovine dry white matter (Yu & Leeden, 1970) constituting 50% of the white matter (Norton & Autilio, 1966). The calculated concentration in the non myelin material will be almost four times that in myelin, indicating a high concentration in the axonal membrane.

Myelin can probably be more purely prepared than other brain membrane subfractions. The yield of synaptosomal plasma membranes is low and when great attempts are made to get a "pure" fraction it will be very low (Morgan et al., 1971). The representativity of such fractions is still to be shown.

A further possibility for evaluating the topographic distribution of the gangliosides was offered by the technique of separating neurons and astroglia cells. The values in the glia cell fractions have been consistently larger than in the neuronal cell fractions (Hamberger & Svennerholm, 1971; Norton & Poduslo, 1971; Abe & Norton, 1974). The ganglioside pattern has been the same in the two fractions (Hamberger & Svennerholm, 1971; Abe & Norton, 1974). However, the two cell populations are not comparable. The axons and dendrites are lost from the neurons and only the cell body remains, while the astroglia cells are better preserved, and the amount of plasma membranes will, therefore, be much higher in this fraction. The studies show the presence of gangliosides in the astroglia cells, but do not give the relative concentration in these plasma membranes. The highest ganglioside level in isolated cell fractions was reported by Derry & Wolfe

(1967) who hand dissected Deiter's neurons from ox brain stem. They reported a high concentration in the neurons, compared with dissected clumps of glia cells, the value, however, was lower than that found in whole gray matter of ox.

A new and elegant technique for microscopic visualization of gangliosides was recently developed independently by Manuelidis & Manuelidid (1976) and Hansson, Hamberger & Svennerholm (1977). It used the ability of cholera toxin to bind GM2 specifically in equimolar proportions (Holmgren et al., 1973). Hansson et al. (1977) incubated the specimen with cholera toxin and then, after rinsing, incubated with cholera toxin antibodies to which peroxidase was conjugated. Manuelidis & Manuelidid used peroxidase bound to cholera toxin. In this way peroxidase was bound to GM1 and could be visualized for electron microscopic studies. The other major brain gangliosides were visualized by Hansson et al. (1977) by exposing them to sialidase which converts them all to GM1.

Hansson et al. (1977) showed that the gangliosides were situated in the plasma membranes of the neurons and in the glia cells, some precipitates were found in the myelin. There was a considerably higher ganglioside concentration in the synaptic junctions than in other membranes. Pre-treatment with sialidase greatly increased the precipitates, especially in the synaptic junctions.

The studies of the topografic distribution have thus revealed that:

> the highest ganglioside concentration is found in the synaptic junctions
> gangliosides are distributed in the plasma membranes of the neuron and astroglia cells
> The myelin has only a low concentration of gangliosides.

The majority of the gangliosides are bound to the neuronal cells, either in the plasma membranes or in the synaptic junctions. Changes in the amounts of the neuronal membranes or synaptic junctions will be reflected in the ganglioside concentration. During development the ganglioside concentration will reflect the outgrowth of axon and dendrites, as well as the formation of interneuronal connextions. The maturation of the neurons occurs simultaniously with that of the astroglia cells (Caley & Maxwell, 1968) and therefore changes which occur in the latter will not disturb the use of gangliosides (lipid-NeuAc) as a neuronal cell marker substance.

Rat material

A highly inbred Sprague-Dawley rat strain was used in the experiments. The litters were reduced to six on the day of birth. Housing and breeding conditions were previously described by Alling et al. 1974. Normal development was analysed in rats fed ordinary laboratory pellets.

Nutritional design

The protein-calorie deficiency was induced by feeding a diet deficient in protein. Caloric

TABLE 1. CALORIC COMPOSITION OF THE DIETS

Diet	Percentages of total calories	
	Control	Deficient
Fish protein	16	6
Starch (wheat)	58	68
Sucrose	5	5
Total fat	21	21

composition of the diets is given in Table 1. This diet was fed to 3-month-old females from one week before mating. Previous studies had shown that a reduction of protein content from 16 to 8 calorie-% reduced growth up to 21 days of age, to half that found in the rats fed the control diet (Alling et al., 1974). The effect on forebrain growth was small. A further reduction of the protein content to 6 calorie-% aggrevated body growth reduction and at 21 days of age the weight was 1/3 of that of the control group. Mortality increased considerably and was 60-80% from birth to 21 days of age. The deficient rats did not survive weaning at 21 days of age, and both groups were, therefore, weaned at 30 days of age. All diets were fed at libitum.

Brain material

Normal development was analysed in pooled litters taken at frequent intervals up to 30 days of age. In the nutritional experiment, one sample of newborn rats and four pooled litters at 10, 15, 21, 30 and 120 days of age were taken from both diets. The same number of males and females were taken in the pooled litters at 30 and 120 days of age.

Lipid extraction and quantitative methods

The pooled brains were homogenized with an equal volume of water in an all-glass homogenizer. Portions of the homogenate were taken for lipid extraction with chloroform-methanol (Svennerholm & Vanier, 1972). Lipid NeuAc and ganglioside patterns were determined with the resorcinol method (Vanier et al., 1971), and cerebrosides were quantified with a thin layer scanning method (Karlsson & Svennerholm, 1978).

Students t-test were used for the statistical interpretations.

Normal development

Ganglioside concentration is low at birth (Fig. 1) indicating the immaturity of the brain in the newborn rat. During postnatal development, up to 21 days of age, the concentration increases to values near those found in adult animals. The syntheses is slow during the first days of life but later increases to a distinct phase of rapid increase between 10 and 15 days of age. Cerebrosides start to increase after 10 days of age, and the most rapid formation is seen at about 20 days of

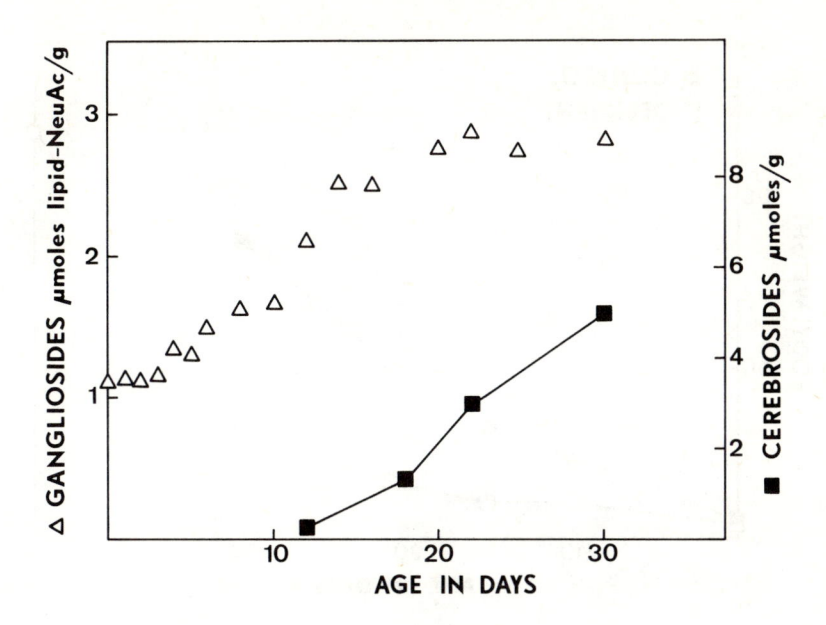

Fig. 1. Postnatal increase of gangliosides and cerebrosides in rat forebrain

age (Fig. 1). It is evident that the neuronal growth and differentiation preceeds myelination.

Similar developmental curves for gangliosides have been shown for whole rat brain by Spence & Wolfe (1967) and Wells & Dittmer (1967). Merat & Dickerson (1973) found two peaks in the ganglioside development for rat and pig forebrain. This two peak curve has not been found by any other investigator, and is, from a developmental point of view, difficult to explain. The values they report from 30 and 60 days of age are only half of those found in our laboratory, which suggests difficulties in their ganglioside determinations.

In human cerebral cortex a rapid phase of ganglioside increase starts during the last trimester, and levels off a few months after birth (Svennerholm & Vanier, 1972). The developmental curve shows similarities with that of rat forebrain when the different time of brain maturation is taken into account.

GD1a increases rapidly from birth to 10 days of age in the rat forebrain (Vanier et al., 1971) and this change preceeds the most rapid phase of increase of total gangliosides. GM1 and GT1 decrease concomitant with the changes in GD1a. The same changes are also seen in human cerebral cortex during the last half of the gestation (Vanier et al., 1971). During further development GD1a decreases slowly while the other three major gangliosides increase in both rat and human.

Human white matter differs little from cerebral cortex at birth (Vanier et al., 1971), however, the ganglioside pattern changes during myelination from that in the cortex with increases of GM1

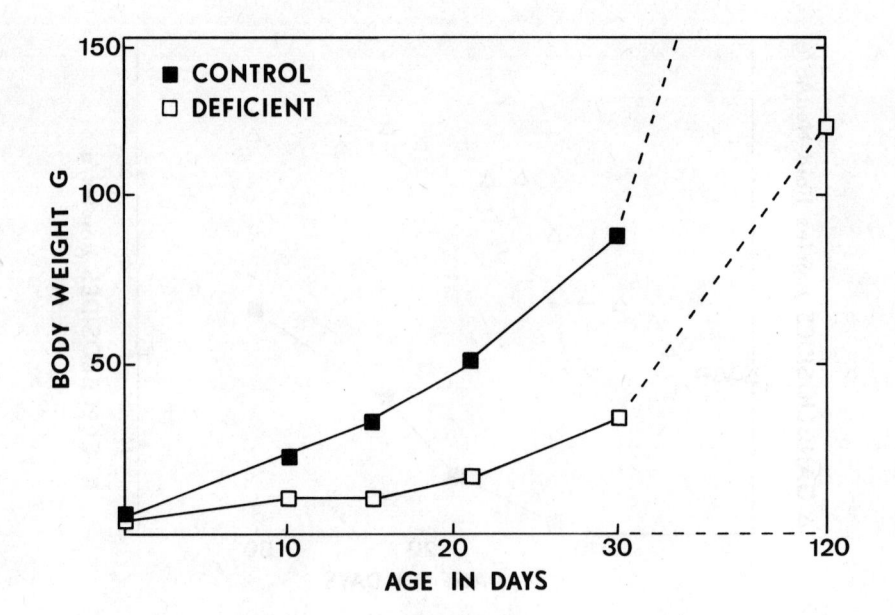

Fig. 2. Body growth during postnatal development in protein-calorie deficiency.

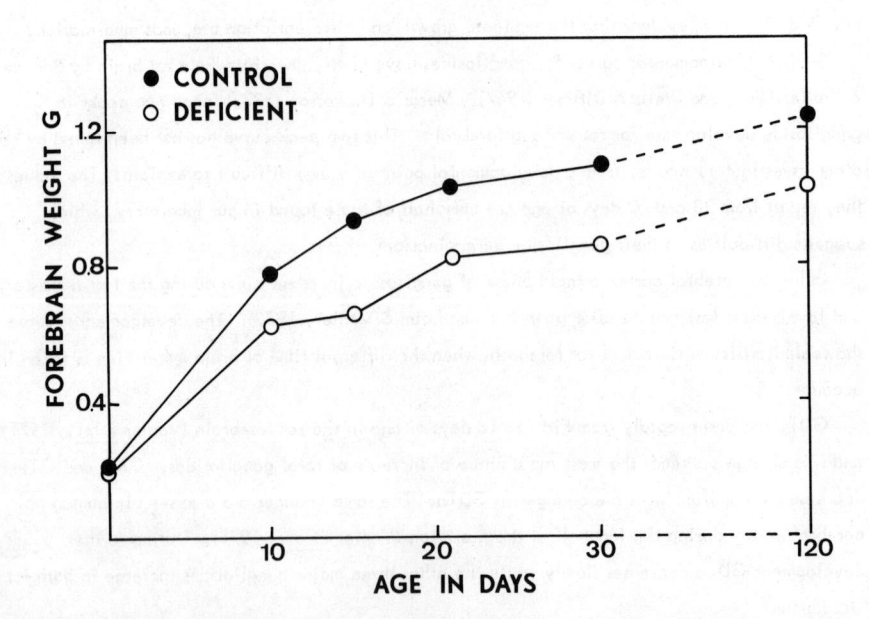

Fig. 3. Postnatal forebrain growth in protein-calorie deficient rats.

which thus becomes the dominating ganglioside, while the other three major gangliosides are reduced to smaller proportions.

Effects of malnutrition on forebrain development and body growth

Body weight was only a little altered by malnutrition during gestation (Fig. 2) but growth was considerably reduced during lactation. The body weight difference increased during the whole period studied.   We could not demonstrate a sure difference in the forebrain weight at birth (Fig. 3). Postnatal growth was reduced in the protein-calorie deficient group especially between 10 and 15 days of age. The difference in weight at 15 days of age was reduced during the further development.

Effect of malnutrition on the gangliosides

The severe protein-calorie deficiency did not alter ganglioside formation more than very slightly during gestation or the 10 first days of postnatal life (Fig. 4). The rapid increase seen in control rats thereafter was reduced in the deficient group resulting in a lower concentration at 15 days of age ($p<0.01$). However, between 15 and 21 days of age the concentration in the deficient group increased considerably and from 21 days of age it was higher than in the control rats. The higher value in the deficient rats was significant at 30 and 120 days of age ($p<0.01$).

Cerebroside concentration was already lower in the deficient rats at 10 days of age (Fig. 5). In contrast to the ganglioside concentration there was no decrease of the difference seen during development and the malnourished rats had a considerable deficit at 120 days of age.

Previous studies have given conflicting results of the effect of protein-calorie deficiency on the ganglioside concentration during development. Geison & Waisman (1970) found no effects of undernutrition during the lactation period while *ad libitum* feeding to rats who had been undernourished to 21 days of age resulted in a higher ganglioside concentration at 8 weeks of age compared with control rats.

Bass *et al.*, (1970) reported dramatic effects of protein-calorie deficiency during the lactation period on the gangliosides. They analysed, with a microchemical method, small samples of both cerebral cortex and white matter. However, they give a ganglioside concentration in their control rats 2-3 times higher than generally accepted for the gangliosides in mammalian brains.

Ghittoni & Faryna de Raveglia (1972) undernourished rats during the lactation period and found a 50% reduction in the ganglioside concentration at 21 days of age. Merat & Dickerson (1974) undernourished one group of rats during the lactation period. In a second group, a protein deficient diet was given to the dam during the gestation and lactation periods, as in this study. They found in this latter group a drastically reduced ganglioside concentration, at 21 days of age. Their values for the control rats are also low only 2/3 of those found in this study.

The ganglioside pattern was identical in the control and deficient rats (Table 2). At 120 days

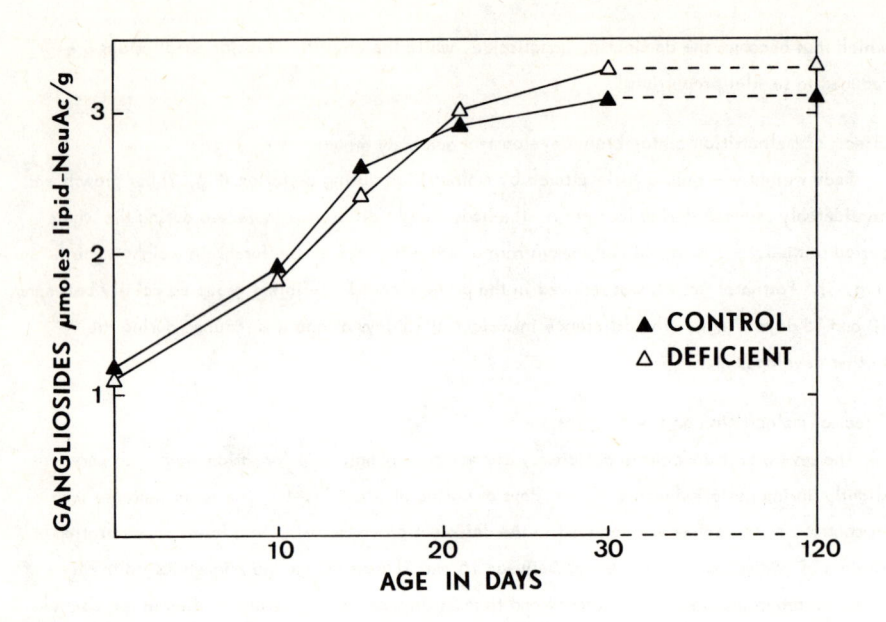

Fig. 4. Forebrain gangliosides during postnatal development in protein-calorie deficient rats.

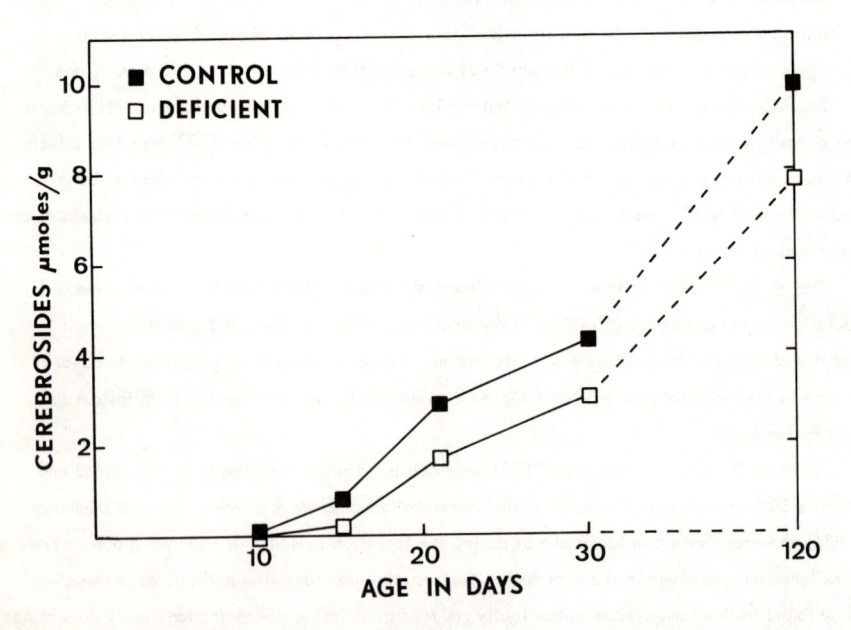

Fig. 5. Forebrain cerebrosides during the postnatal development in protein-calorie deficient rats.

TABLE 2. PATTERN OF MAJOR GANGLIOSIDES

The composition is expressed in molar percentage of gangliosides

	30 day-old rats		120 day-old rats	
	Control	Deficient	Control	Deficient
GM1	30	29	37	33
GD1a	36	38	31	32
GD1b	13	13	14	15
GT1	11	12	11	13

of age slightly lower values were found for GM1 in the deficient rats. This difference is probably due to the lower proportion of myelin in these rats.

In the experiment a protein-calorie deficiency was induced from early gestation to adult age. The effect during the intrauterine period was small compared with that seen during postnatal life. Further pregnancies in females continuously fed the deficient diet did not aggrevate intrauterine malnutrition. The greatest effects of protein-calorie deficiency were seen during the lactation period, illustrated both by the great mortality and inhibition of body growth between 10 and 15 days of age. The ganglioside formation, which in the rat forebrain is normally most rapid between 10 and 15 days of age, was astonishingly little affected. The phase of most rapid increase was delayed but the concentration later caught up and indeed later became higher in the deficient animals. The results indicate that gangliosides, localised dominantly in the neuronal membranes and synapses, were selectively resistant to malnutrition. This suggests that the neuronal cell growth and differentiation continue to reach a normal degree of maturation, even in severe protein-calorie deficiency.

References

Abe T. and Norton W.T. (1974) J. Neurochem. 23, 1025-1036.

Alling, C., Bruce, Å., Karlsson, I. and Svennerholm, L. (1974) Nutr. Metabol. 16, 38-50.

Bass, N.H., Netsky, M.G. and Young, E. (1970) Arch. Neurol. 23, 289-302.

Breckenridge, W.C., Gombos, G. and Morgan, I.G. (1972) Biochem. Biophys. Acta 249, 695-707.

Caley, D.W. and Maxwell, D.S. (1968) J. comp. Neurol. 133, 17-44.

Derry, D.M. and Wolfe, L.S. (1967) Science 158, 1450-1452.

Geison, R.L. and Waisman, H.A. (1970) J. Nutr. 100, 315-324.

Ghittoni, N.E. and Faryna de Raveglia, I. (1972) Neurobiology 2, 41-48.

Hamberger, A. and Svennerholm, L. (1971) J. Neurochem. 18, 1821-1829.

Hansson, H.-A., Holmgren, J. and Svennerholm, L. (1977) Proc. Natl. Acad. Sci. 74, 3782-3786.

Holmgren, J., Lönnroth, I., Månsson, J.-E. and Svennerholm, L. (1973) Infect. Immun. 8, 208-214.

Karlsson, I. and Svennerholm, L. (1978) J. Neurochem. in press.

Klenk, E. (1942) Z. Physiol. Chem. 273, 76-86.

Lapetina, E.G., Soto, E.F. and De Robertis, E. (1967) Biochim. Biophys. Acta 135, 33-43.

Leeden, R.W. and Yu, R.K. (1976) in Glycolipid Methodology (Witting, L.A. ed.) pp. 187-214. The American Oil Chemists' Society, Champain III.

Leeden, R.W., Yu, R.K. and Eng, L.F. (1973) J. Neurochem. 21, 829-839.

Manuelidis, L. and Manuelidis, E.E. (1976) J. Neurocytol. 5, 575-589.

Merat, A. and Dickerson, J.W.T. (1973) J. Neurochem. 20, 873-880.

Merat, A. and Dickerson, J.W.T. (1974) Biol. Neonate 25, 158-170.

Morgan, I.G., Wolfe, L.S., Mandel, P. and Gombos, G. (1971) Biochim. Biophys. Acta 241, 737-751.

Norton, W.T. and Autilio, L.A. (1966) J. Neurochem. 13, 213-222.

Norton, W.T. and Poduslo, S.E. (1971) J. Lipid Res. 12, 84-90.

Spence, M.W. and Wolfe, L.S. (1967) Can. J. Biochem. 45, 671-688.

Suzuki, K. (1970) J. Neurochem. 17, 209-213.

Suzuki, K., Poduslo, S.E. and Poduslo, J.F. (1968) Biochim. Biophys. Acta 152, 576-586.

Svennerholm, L. (1957) Acta Soc. Med. Upsalien. 62, 1-16.

Svennerholm, L. and Vanier, M.-T. (1972) Brain Res. 47, 457-468.

Vanier, M.-T., Holm, M., Öhman, R. and Svennerholm, L. (1970) J. Neurochem. 18, 581-592.

Wells, M.A. and Dittmer, J.C. (1967) Biochemistry 6, 3169-3174.

Wiegandt, H. (1967) J. Neurochem. 14, 671-674.

Yu, R.K. and Leeden, R.W. (1970) J. Lipid Res. 11, 506-516.

SENSORY DEPRIVATION AND BRAIN DEVELOPMENT

V. Bigl and D. Biesold
Paul Flechsig Institute of Brain Research, Department of Neurochemistry, Karl Marx University, Leipzig (GDR)

1. Introduction and scope of the article

The postnatal development of the brain is prevailed by two
opposing principles which in their interaction shape the irritating
complexity of adult brain structure, the precise arrangement of
individual neuronal cell bodies as well as the accuracy of their
interconnections: genetically determined specifity and environment-
mediated plasticity. The different answers to the question, to what
extent the genetic Anlage of the brain might be modified by adequate
stimulation from environment and experience in terms of behaviour,
intellectual capabilities etc. of the adult have already today an
impact upon society which cannot be unterestimated. What might have
been a mere academic question at the turn of the century when Von
Gudden (1870) published his experimental work on the influence of
visual deprivation upon the morphology of the developing brain is of
profound practical importance today. The growing concern on these
issue is reflected best in the increasing number of relevant papers
devoted to the study of the influence of environmental manipulation
on the development of the brain or certain brain structures.

For experimental studies the complex influence of the environment
has been reduced to specific forms of sensory input, which can be
both readily altered in quantity and quality and relatively easy
measured. The visual system, in addition, offers the advantage that
its relay stations are easily accessible and the functional consequences of the lack of adequate stimulation can be followed up by
testing visually guided behaviour in animals. Most of the work on
sensory deprivation is confined, therefore, to visual deprivation.
Several comprehensive reviews have appeared in the last years summarizing morphological aspects (Globus, 1975) and electrophysiological
as well as functional implications (Chow, 1973) of visual deprivation
but a detailed biochemical analysis is still lacking. The available
data reveal a marked discrepancy between the striking changes in
morphological as well as physiological developmental pattern of the
brain and the minor changes in brain biochemistry. In the present

paper, therefore, an attempt is made to correlate available biochemical data to morphological and physiological findings. It was not intended to give a complete review but instead to assess critically some problems concerning the biochemical aspects of visual deprivation and brain development.

2. Functional-structural correlates

The increasing number of papers on the influence of visual deprivation upon brain development provoked by the observation of Hubel and Wiesel (1963) that after monocular visual deprivation in kitten most cells of the visual cortex lose their effective input from the deprived eye and become monocularly driven, also stimulated research on the physiological as well as morphological basis of normal visual function. Thus the last ten years have witnessed successful efforts to unravel the morphological and functional heterogeneity and complexity of the visual pathway and to deduce from microelectrophysiological recordings on single cells how the visual system might work as a whole. The original observations by Hubel and Wiesel on the electrophysiological affects of sensory deprivation were confirmed, partly corrected and in challenging the different aspects and their interpretations step by step extended (Barlow, 1975). Morphological studies supplemented the electrophysiological findings (Hubel et al., 1975). Biochemistry of the central visual pathway, in contrast, is still at best fragmentary: nothing is known on the chemical specificity which might characterize functionally and morphologically different classes of cells within the visual pathway. Another obstacle for the integration of biochemistry into the present day picture of visual information processing and its development in dependence on visual input are the different species used by the different disciplines. Whereas most of the electrophysiological studies were done in cats and partially in monkeys biochemical data are mainly reported for rat, pigeon and rabbit. As remarkable differences exist in the organization of the visual pathway between the different species the result of similar studies can hardly be correlated. Before reviewing biochemical data on visually deprived animals it seems, therefore, appropriate to give a general account on the morpho-functional organization of the visual pathway and on some recent findings on electrophysiological changes following light deprivation. For a more detailed account the reader is referred to reviews of Barlow (1975), Globus (1975), Guillery (1974), Rose and Haywood (1976), Rose (1977) and others.

On the basis of cell size and electrophysiological properties three different classes of ganglion cells can be distinguished in the retina of the cat: large and medium-sized cells classified as Y- and X-neurons and small neurons called W-cells. Visual information is transmitted by the axons of these cells to three similar populations of cells in the lateral geniculate body (LGN) and from there in parallel separate channels to the visual cortex (Wilson et al., 1976). The different P-cells in the LGN projecting towards the visual cortex comprise in the rat about 87 %, 13 % being interneurons (Burke and Sefton, 1966).

Monocular deprivation reduces greatly the number of electrophysiologically identifiable Y-cells in the LGN which is consistent with a reduced labelling of Y-cells by the horseradish peroxidase method. X-cells seem to be unaffected (Sherman et al., 1972; Garey and Blakemore, 1977). Reduction in cell size and density were also described for mono- and binocular deprived kitten (Wiesel and Hubel, 1963 b), rat (Fifkova, 1968) as well as dark raised mice (Gyllensten et al., 1965). Changes in the size and number of synapses have been described repeatedly (for ref. see Globus, 1975). In contrast, it has also been argued that monocular deprivation has no arresting influence on LGN cell growth but leads to a hypertrophic growth of the cells receiving input from the undeprived eye (see Guillery, 1974).

The geniculo-cortical fibers terminate mainly in layer IV of the visual cortex. The pyramidal cells of layer V and VI, in turn, give rise to the cortico-geniculate or cortico-collicular projection. The visual input is not only modulated by the cortical afferents but at the level of the LGN also by projections from other thalamic nuclei, the reticular formation and aminergic pathways (Singer, 1977), but the integration of these fiber systems into structural models of the information processing within the LGN is still missing.

The electrophysiological and morphological changes in monocular deprived kittens are fully reversible within few days when the deprived eye is brought to forced usage by opening the deprived and suturing the formerly open eye if the operation is done within the critical period (Blakemore and van Slyters, 1974). After the critical period changes are irreversible (Hubel and Wiesel, 1970).

As in very young, visually inexperienced kitten receptive field properties as well as single cell responses to stimulation of the eye are comparable to those of normal adult cats it was suggested that the main fiber connections within the visual pathway are specified from birth and develop without experience (Hubel and Wiesel, 1963). From studies of receptive field properties and neuronal connectivity in the visual cortex of conture-deprived rats Singer and Trettner (1976) also concluded "that most if not all afferent, intrinsinc and efferent connections of area 17 and 18 are specified from birth and depend only little from visual experience". This supports the notion by Jacobson (1970) that "there is no evidence that functional activity is necessary for the formation of any neuronal connections". Where, then, comes experience into play? What factors are responsible for the deleterious effects of visual deprivation leading to almost complete blindness?

The critical period in kitten and rat parallels the main period of synaptic development in the LGN resp. superior colliculus (Winfield et al., 1976; Lund and Lund, 1972). Most likely, sensory input interferes somehow with this process of synaptic development and allows for some freedom in the domain of orientation tuning, binocular correspondence and retinotopy which might be specified within the predetermined structural plan by visual experience (Singer and Tretter, 1976). Three obvious possibilities are conceivable for this process: (1) physical withdrawal of existing synaptic connections or re-directing or arresting growth of newly developing afferents; (2) modification of existing synapses in the sense of increasing or decreasing their efficiency by changing, for example, the number of quanta of transmitter released per impulse or the number or density of receptor sites on the postsynaptic membrane (Blakemore and Hillman, 1977); and (3) excessive growth of synaptic terminals which by the influence of sensory deprivation might become released of some inhibitory control or gain some dominance over others. The causative factors which trigger these events are largely unknown. Binocular competition seems, however, to play a major role in these processes (Guillery, 1974). In monocular deprived animals the synapses related to the open eye might gain some advantage and develop better than the "deprived" synapses. The dominance of the open eye is not related to interocular differences in light intensity or temporal pattern (Wilson et al., 1977) nor can it be prevented by electrical stimulation of the deprived eye (Blakemore and Hillman, 1977) but seems related to spatial pattern.

(1) The suggestion of a withdrawal of already developed fiber connections or arresting the growth of newly developing fibers fits with the bulk of morphological data on the reduction of cell size, number of dendritic spines, size and number of synaptic terminals and dendritic branching (see Globus, 1975; Vrenssen and De Groot, 1975). But as pointed out previously most of the studies did not characterize the synaptic terminals involved in view of their origin and this easy-to-go correlation seems too simplified.

(2) In agreement with a suggested change in the efficiency of synapse it was found that intravienous injection of the GABA antagonist bicuculline in monocular deprived kittens restores the normal binocular responsiveness of the cortical neurons instantaneously (Duffy et al., 1976) and 1 hour after enucleation of the dominating eye in monocular deprived cats the number of cortical cells responding to stimulation of the deprived eye increases significantly. The lack of responsiveness of cortical neurons in monocular deprived cats might therefore be due to tonic inhibition of the normally developed excitatory input. The highly significant decrease in the number of synaptic vesicles in cortical synapses (Vrensen and De Groot, 1974) point to an additional involvement of excitatory synapses otherwise normally developed in size and number.

(3) An increase in the number of afferents was shown by Hubel et al., (1975) in deprived monkeys and related to the geniculo-cortical pathway. The increased number of synaptic profiles in the visual cortex of monocular deprived rabbits shown by Vrensen and De Groot (1975) was, in contrast, restricted to cortical laminae I - III indicating an increase in unspecific pathways.

From this shortly outlined structural-functional aspects of visual deprivation three main lines for a biochemical approach evolve.
(1) characterization of number and functional state of synaptic terminals; (2) macromolecular synthesis as equivalent of brain growth and development and (3) postnatal differentiation of cells which could account for the reduction in the volume of visual brain areas in deprived animals. A point which will not be considered here concerns possible hormonal changes brought about by complete visual deprivation. Available biochemical evidence on these problems will be reviewed in the following paragraph.

3.1 Changes in the synaptic compartment following dark rearing and monocular deprivation

One tool used to characterize the synaptic compartment in deprivation studies is the determination of transmitter-specific enzyme activities. Table I summarize results of relevant studies. Not all of the enzymes listed are strictly specific markers for a certain type of neurotransmission. Na/K-ATPase was included as possible marker for the development of the general bioelectric activity (Meisami, 1975). AChE and ChAc activity in general seems to be reduced in visually deprived animals. Changes were reported by our group for all enzymes of the amine metabolism. No significant influence of dark rearing was found on GAD activity, the key enzyme of GABA metabolism.

In view of the important role the reticular formation has on visual information processing which seems to involve cholinergic neurons, and the conflicting results on AChE activity in table I we reinvestigated the development of cholinergic enzyme systems in the visual structure of dark reared rats. Fig. 1 and 2 show the development of AChE and ChAc activities in the visual cortex and the LGN, respectively. ChAc activity in the visual cortex (VC) increases steeply during the second and third postnatal week reaching adult levels. In the LGN enzyme activity develops more slowly adult values being reached at the end of the 7th week. AChE activity approaches adult levels during the 7th and 9th week of life. Only minor differences are to be observed between dark raised animals and their appropriate controls.

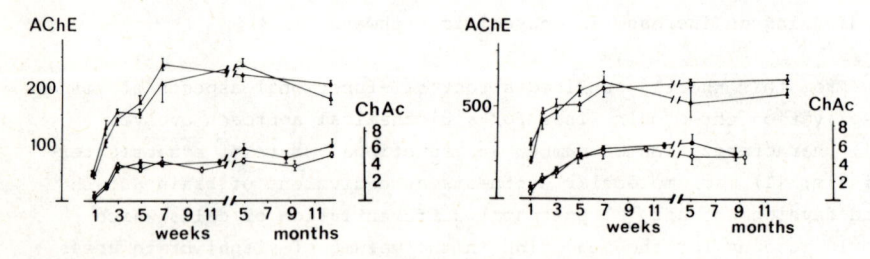

Fig. 1 (left) and Fig. 2 (right): Development of ChAc and AChE activity in the visual cortex (fig. 1) and lateral geniculate nucles of normal and dark raised rats.

Rats were reared in complete darkness or under normal laboratory conditions. At the age indicated in the figure they were killed by decapitation and the visual structures as previously described prepared (Bigl et al., 1974 a). Enzyme activities were assayed by microradiochemical methods (Bigl and Schober, 1977).

Table I: Changes in the activity of transmitter-related enzymes in visual and non-visual areas of the brain of visually deprived or dark-reared animals

Enzyme	\multicolumn{4}{c}{Changes in enzyme activity in per cent of control values or undeprived (ipsilateral) site}	Experimental conditions	Author
	VC	EVC	LGN	CS		
Na/K-ATPase	-14	-35	not studied	-20	rat, 40 d dark rearing	Meisami (1975)
Na/K-Mg ATPase	n.s.	n.s.	not studied	not studied	rat, 50 d dark rearing	Sinha and Rose (1976)
AChE	-30 (n.s.)	-7 (n.s.)	-65	-27	rat, 21 d dark rearing	Maletta and Timiras (1967)
AChE	+15 (n.s.)	not studied	-25	-3 (n.s.)	rat monocular depriv. 9-11 m	Maraini et al. (1969)
AChE	-11 (cerebral hemisphere)		-11 (optic lobe)		chicken, dark reared 4 d after hatch	Bondy and Margolis (1969)
AChE	+9 (n.s.)	+8 (n.s.)	not studied	not studied	rat, 50 d dark reared	Sinha and Rose (1976)
ChAc	n.s.	n.s.	-52	-64	rat, 21 d dark rearing	Maletta and Timiras (1968)
ChAc	-16	-14	not studied	not studied	rat, 50 d dark reared	Sinha and Rose (1976)
GAD	n.s.	n.s.	not studied	not studied	rat, 50 d dark reared	Sinha and Rose (1976)
MAO	+39	+17 (n.s.)	+35	+50	rat, 7 weeks dark reared	Bigl et al. (1974 a)
COMT	-7 (n.s.)	-66	-25	-19	same	same
TyrOH	-7 (n.s.)	-28	-33	+14 (n.s.)	same	Bigl et al. (1974 b)
TrpOH	-30	-43	-39	-62	same	Usbekow et al. in prep.

Data were recalculated from the papers quoted.

<u>Abbreviations used</u>: VC visual cortex; EVC extra-visual cortical area; LGN lateral geniculate body; CS superior colliculus; n.s. non significant

We have shown earlier that ACh is unlikely to be involved in the direct optic transmission (Bigl and Schober, 1977), instead almost all cholinergic afferents to all neocortical areas including the VC seem to derive from a small group of cells in the basal forebrain of the rat (Wenk et al., in preparation) which play a role in many higher brain functions (for ref. see Divac, 1975). In addition, we could recently obtain evidence that there is a direct cholinergic link between the mesencephalic reticular formation and the LGN, but most of the AChE activity in this structure is to be localized in intrinsic neurons (Bigl et al., in preparation). Hence our results do not give evidence that sensory deprivation acts upon the unspecific cholinergic systems which facilitate transmission within the visual pathway. The results are in agreement with data from Rose's group that in the VC of dark reared rats there is no significant difference in muscarinic receptor binding (Rose, 1977).

All enzymes involved in amine metabolism are sensitive to dark rearing (table I). In order to evaluate whether the changes in enzyme activity in dark reared animals reflect transient functional-adaptive mechanisms or point to a structural basis within the aminergic fiber systems, uptake activities for noradrenaline (NA), dopamine (DA) and serotonin (5-HT) were studied. The results summarized in table II show a significant reduction ($P < 0.05$) in DA uptake in the LGN only; NA and 5-HT uptake are not affected by dark rearing. Concentrations of all three amines in the visual structures were also unchanged (Bigl, unpublished).

In contrast to NA and 5-HT which are known to occur in the LGN there is no evidence available for DA as a physiological transmitter in the LGN (Tebēcis, 1974). Possibly DA is localized partly in retinal fibers of the nervus opticus.

A third transmission system which can be correlated to certain neuronal structures within the visual pathway is the GABA-system which most probably is involved in the (inhibitory) transmission of interneurones in the LGN and superior colliculus (CS) as well as the cortex (Tebēcis, 1974). Developmental pattern of GAD activity showed no differences in dark raised rats as compared with normal controls (Kunert, unpublished). Preliminary results on the effect of monocular deprivation on some parameters of the GABA system show that there is no difference in GAD activity between deprived and non-deprived structures but enzyme activity in both sites of CS and VC tends to be higher

Table II: In vitro uptake of [³H] noradrenaline, [³H] dopamine and [³H] serotonin in the visual structure of normal and dark reared rats.

Uptake (mole x 10^{-12}/g wet weight/10 min)

		Noradrenaline	Dopamine	Serotonin
LGN	control	86.97 ± 15.58	118.65 ± 11.14	510.73 ± 39.85
	dark	65.37 ± 15.26	74.66 ± 5.61	545.39 ± 42.85
CS	control	14.68 ± 3.60	53.57 ± 3.24	429.91 ± 43.26
	dark	9.70 ± 2.26	67.56 ± 13.96	500.16 ± 49.10
VC	control	22.87 ± 6.44	31.96 ± 13.63	128.55 ± 12.91
	dark	30.55 ± 2.92	50.97 ± 17.32	129.02 ± 24.60

Rats were raised and the visual structures prepared and homogenized as described previously (Bigl et al., 1974). Noradrenaline and dopamine uptake studies were performed at amine concentrations of 1x10^{-8} m. Uptake at 37° was corrected for uptake at 0°C. Values are means ± s.e.m. from 4 or 5 independent experiments.

than in unoperated littermates (fig. 3).

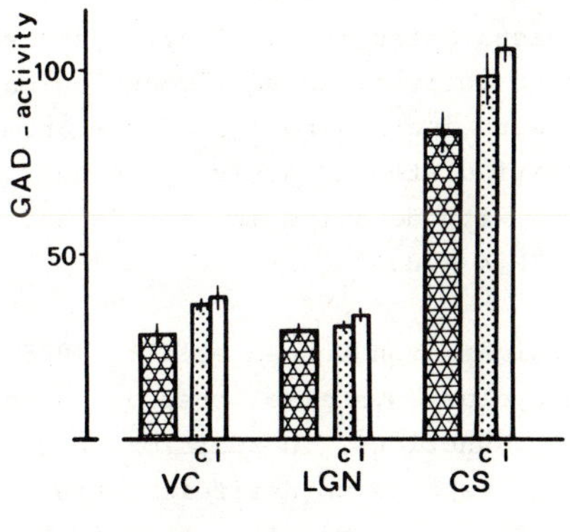

Fig. 3: GAD activity in monocular deprived rats and normal controls.

The left eye lid was sutured in nembutal anaesthesia at the age of ten days. Animals were raised until three months and processed as described in legend to fig. 1. GAD activity was assayed using [1-^{14}C] glutamic acid (Kunert, 1977). Results are expressed as μmole glutamic acid decarboxylated/ g wet weight/h ± S.D. from n = 3.
Cross hatched columns: controls, dotted columns and empty columns: contra- and ipsilateral structures of monocular deprived rats.

Table III: Uptake of [^{3}H] GABA in homogenates of visual structures in normal and dark reared rats.

GABA uptake (mole × 10^{-9}/g wet weight/h)

Experimental condition	VC	LGN	CS
normal control	6.7 ± 0.7	7.1 ± 0.6	10.4 ± 0.7
dark reared	6.1 ± 0.9	8.1 ± 0.5	5.5 ± 0.5

Adult animals (3 months old) were used. Rearing conditions and preparation of samples as fig. 1 [^{3}H] GABA uptake was studied at substrate concentration of 1 × 10^{-8} m [^{3}H] GABA. Values are corrected for uptake and absorption at 0°C. Mean ± S.E.M. from n = 4.

GABA uptake in dark raised rats (table III) is reduced to almost 50% in the CS and increased to about 15% in the LGN the latter difference not being significant. No change could be found in the visual cortex. Although our data on the GABAergic system need further confirmation they strongly support electrophysiological evidence for an important role of inhibitory circuitry in visual deprivation.

The factors which link the synaptic events to the metabolic machinery of the cells which finally realize the changes in cell growth, enzyme activities etc. are largely unknown. One possible candidate for a "second messenger" in synaptic transmission is the cyclic AMP system. In a recent study (Ruschke et al., submitted for publication) we investigated the effect of dark rearing on the development of the transmitter-specific stimulation of adenylate cyclase (AC) activity in cell-free preparations of three visual structures. Fig. 4 demonstrates that all three amines stimulate AC activity in very young animals. The degree of stimulation decreases with increasing age of the animals. Significant increase in transmitter-stimulated AC activity in dark reared vs. normal rats was only found with dopamine in the LGN and 5-HT in the CS of 15 days old rats. ($P < 0.05$).

As it was shown in other brain regions (Von Hungen et al., 1975) that the transmitter-sensitive adenylate cyclase system(s) precede in their development the formation of synaptic junctions. In addition, receptor hypersensitivity might develop in response to deafferentation as well as denervation (see Globus, 1975). Our results, therefore, lend support to the speculation that also in the visual system the transmitter-sensitive AC may be involved in the establishment of synaptic function (for ref. see von Hungen et al., 1975) and sensory deprivation possibly interferes partly with that process.

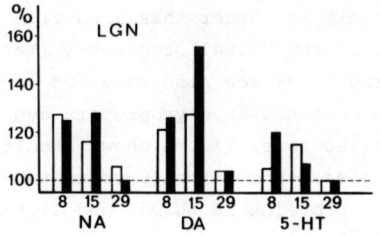

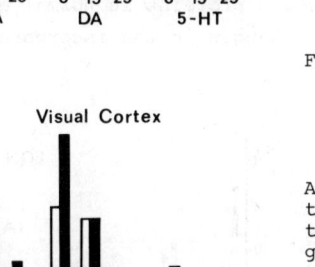

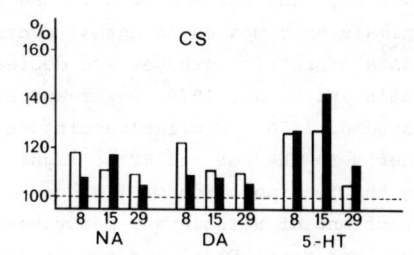

Fig. 4: Stimulatory influence of biogenic amines on the activity of adenylate cyclase activity in vitro.

AC activity of a particulate fraction from the isolated visual structure was assayed according to Maguire and Gilman (1974) with and without the amines indicated (final concentration 50 μm). Results are expressed as per cent of unstimulated (basal) activity. Numbers indicate the age of the animals. Black columns: dark raised, empty columns: normal control. n = 5.

3.2 Visual deprivation and macromolecular synthesis

There can be little doubt that the brain does not only react to increased stimulation by changes in its electrophysiological properties but also quickly adapts its macromolecular synthesis to the momentary functional demands. The effects of muscular activity, electrical stimulation and other stimuli upon synstesis of protein and RNA are well known and extensively studied (for review see Jakoubek, 1974). Studies on the influence of visual deprivation or dark rearing on macromolecular synthesis are carried out under two different aspects. In the first approach previously dark reared animals are subjected to visual stimulation. Under this condition, transient changes in RNA and protein synthesis restricted to the visual structure were observed. As several recent reviews (Rose, 1977; Rose and Haywood, 1976) cover this field we will not dwell on this subject. The short, transient nature of the increase in protein synthesis makes it unlikely that is represents a biochemical correlate of functional recovery but rather an adaptive response to the altered functional activity. The second approach, more directly related to the impact of sensory deprivation upon brain development, deals with more or less persistent, long-term changes in RNA and protein turnover in dark reared animals as compared to normal control animals. Under this condition RNA and protein turnover are decreased in the visual structures (Margolis and Bondy, 1970, Dewar and Reading 1973; see also Rose and Haywood, 1976). Similar results can be obtained in glycoprotein synthesis in the LGN and VC of light deprived rats. Fig. 5 shows results on the incorporation of [^{3}H]fucose in electrophoretically separated fractions of soluble and membrane-bound proteins in normal and light deprived rats. Small but consistent changes in the incorporation of

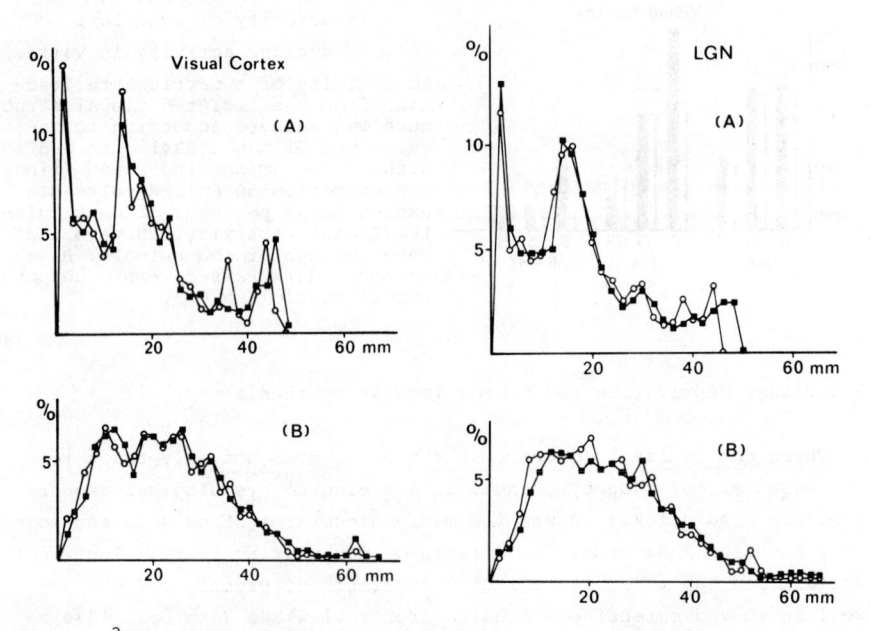

Fig. 5: [^{3}H]Fucose incorporation profiles into water soluble (A) and membrane-bound (B) proteins from the visual cortex and the lateral geniculate body of normal and dark reared rats.

Rats reared in darkness for 8 weeks and normal controls of the same age received an intracisternal injection of 150 µCi [^{3}H]L-fucose and were killed 4 hours later. Water-soluble and membrane-bound proteins of the two structures were isolated and separated by disc-electrophoresis. Radioactivity was measured in segments of the gel by liquid scintillation counting. Data are given as radioactivity per gel segment in per cent of the total activity of the gel.

label in some protein fraction can easily be seen. The interpretation
of this type of experience, however, is much more difficult. It is
not the place to delve more deeply into the argumentation on the
methodological aspects of in vivo studies of protein synthesis (Jakoubek, 1974; Rose and Haywood, 1976) but in the context of this review
the functional interpretation has to be questioned. Only when specific
proteins can be accurately assayed some structural-functional correlations seem to be possible. For example, tubulin synthesis in the
visual cortex was demonstrated to be in a certain extent dependent
on the visual input in a critical period of the postnatal development
of the rat: incorporation of [^{14}C]leucine into tubulin is greatly enhanced in normal rats following eye opening but no increase can be
observed in dark reared rats at the same time (Perry and Cronly-
Dillon, 1978). The increased tubulin synthesis coincides well with
the "critical period" in the rat. This adds more direct proof of
disturbances in axonal transport processes in visual deprivation. As
there are only a few functionally characterized brain proteins known
so far, most of the observed changes in protein turnover can not be
explained in functional terms.

3.3. Influence of visual deprivation on postnatal DNA synthesis and cell division

The reduction in thickness of the visual cortex and in the volume
of the LGN in dark reared mice (Gyllensten et al., 1965) points to
the involvement of glial cells in the morphological changes observed
in light deprived animals.

In similar experimental situations involving complex sensory stimulation the glia/neuron ratio of the cortex was found to be higher
than in control animals (Diamond et al., 1964). We studied, therefore,
the postnatal genesis of non-neuronal cells in the VC and the LGN.
In rats the normal decline in the rate of glia proliferation which
occurs during development proceeds slower than during visual deprivation. Whereas in postnatal week 4 there is no difference in the
frequency of non-neuronal cells labelled by a pulse of [^{3}H] thymidine
(labelling index) between dark raised and normal control animals, in
the 11th postnatal week we found a significant reduction in the labelling index in the LGN. Labelled cells in the visual cortex and the
medial geniculate body, a non-visual thalamic area, are also reduced
but statistically not significant (fig. 6). The results provide

evidence that sensory deprivation controls also maturation of non-neuronal cells.

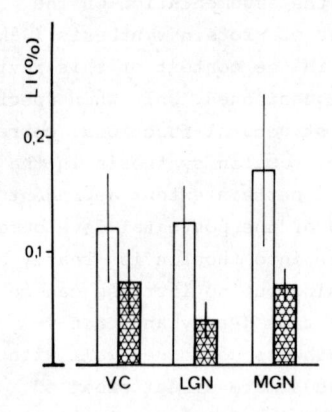

Fig. 6: Percentage of labelled non-neuronal cells in the visual cortex (VC), lateral geniculate (LGN) and medial geniculate (MGN) of 11-week-old rats.

White columns: controls; crosshatched columns: dark reared (modified from Mareš et al., 1978).

In this review some functional and morphological data on the influence of visual deprivation upon the development of the visual system were outlined with the intention to emphasize the need for a closer correlation of biochemical data to structural and functional parameters of normal and altered development of the visual pathway. The complexity of the functional and structural organization of the visual structures with respect to the different nerve cell populations, different in their chemical specificity, their developmental pattern and their morphological characteristics, might be further unravelled in biochemical terms with the application of micromethods allowing, for example, quantitative assessment of parameters of the transmitter metabolism in the monocular and binocular segments of LGN and VC. With this approach the gap should be narrowed between electrophysiological, morphological, and biochemical research in visual deprivation.

Acknowledgement: This study was supported by a grant from the Ministry of Science and Technology of the G.D.R. We wish to thank all our colleagues from the Paul Flechsig Institute for Brain Research for stimulating discussions and comments, especially Dr. L. Müller, Mrs. Martina Brückner, Mrs. Irene Ruschke and Mr. E. Kunert for allowing their unpublished results to be included. Thanks are also due to Mrs. Agnes Greif for typing the manuscript and Mrs. Brigitte Bigl for technical collaboration and drawing of the graphs.

References

Barlowe, H.B. (1975) Visual experience and cortical development. Nature 258: 199-204.

Bigl, V., Biesold, D. and Weisz, K. (1974 a) The influence of functional alteration on monoamine oxidase and catechol-O-methyl transferase in the visual pathway of rats. J. Neurochem. 22: 505-509.

Bigl, V., Brückner, G. and Steudel, A. (1974 b) Alterations in catecholamine metabolism in the visual system of the rat induced by different light stimuli, In: Neurobiological basis of memory formation (Matthies, H., ed.) VEB Verlag Volk und Gesundheit, Berlin, pp. 212-225.

Bigl, V. and Schober, W. (1977) Cholinergic transmission in subcortical and cortical visual centers of rats: no evidence for the involvement of primary optic system. Exp. Brain Res. 27: 211-219.

Blakemore, C. and van Sluyters, C. (1974) Reversal of the physiological effects of monocular deprivation in kittens: further evidence for a sensitive period. J. Physiol. (Lond.) 237: 195-216.

Blakemore, C. and Hillman, P. (1977) An attempt to assess the effects of monocular deprivation and strabismus on synaptic efficiency in the kitten's visual cortex. Exp. Brain Res. 30: 187-202.

Bondy, S.C. and Margolis, F.L. (1969) Effects of unilateral visual deprivation on the developing avian brain. Exp. Neurol. 25: 447-459.

Burke, W. and Sefton, A.J. (1966) Inhibitory mechanisms in lateral geniculate nucleus of rat. J. Physiol. (Lond.) 187: 231-246.

Chow, K.L. (1973) Neuronal changes in the visual system following visual deprivation, In: Handbook of sensory physiology, Vol. VII/3 (Jung, R. ed.) Springer Verlag, Berlin, pp. 599-627.

Dewar, A.J. and Reading, H.W. (1970) Nervous activity and RNA metabolism in the visual cortex of rat brain. Nature 225: 869-870.

Diamond, M.C., Krech, D. and Rosenzweig, M.R. (1964) The effects of an enriched environment on the histology of the rat cerebral cortex. J. comp. Neurol. 123: 111-119.

Divac, J. (1975) Magnocellular nuclei of the basal forebrain project to neocortex, brain stem and olfatory bulb. Review of some functional correlates. Brain Res. 93: 385-398.

Duffy, F.H., Snodgrass, S.R., Burchfield, J.L. and Conway, J.L. (1976) Bicuculline reversal of deprivation amblyopia in the cat. Nature 260: 256-257.

Fifkova, E. (1968) Changes in the visual cortex of rats after unilateral deprivation. Nature 220: 379-381.

Garey, L.J. and Blakemore, C. (1977) Monocular deprivation: Morphological effects on different classes of neurones in the lateral geniculate nucleus. Science 195: 414-416.

Globus, A. (1975) Brain morphology as a function of presynaptic morphology and activity, In: The developmental neuropsychology of sensory deprivation (Riesen, A.H., ed.) Academic Press, New York, pp. 9-91.

Gudden, von B. (1870) Experimentaluntersuchungen über das peripherische und zentrale Nervensystem. Arch. Psych. Nervenkr. 2: 693-723.

Guillery, R.W. (1974) On structural changes that can be produced experimentally in the mammalian visual pathways. In: Essays on the nervous system (Bellairs, R. and Gray, E.G., eds.), Clarandon Press, Oxford, pp. 299-326.

Gyllensten, L., Malmfors, T. and Norrlin, M.L. (1965) Effect of visual deprivation on the optic centers of growing and adult mice. J. comp. Neurol. 124: 149-160.

Hubel, D.H. and Wiesel, T.N. (1963) Receptive fields of cells in striate cortex of very young visually inexperienced kittens. J. Neurophysiol. 26: 994-1002.

Hubel, D.H. and Wiesel, T.N. (1970) The period of susceptibility to the physiological effects of unilateral eye closure in kittens. J. Physiol. (Lond.) 206: 419-436.

Hubel, D.H. , Wiesel, T.N. and LeVay, S. (1975) Functional architecture of area 17 in normal and monocular deprived macaque monkeys. Cold. Spr. Harb. quant. Biol. 40: 581-589.

Hungen, K. von, Roberts, S. and Hill, D.F. (1974) Developmental and regional variation in neurotransmitter-sensitive adenylate cyclase systems in cell-free preparations from rat brain. J. Neurochem. 22: 811-819.

Jacobson, M. (1970) Development, specification and diversification of neuronal connections. In: The Neurosciences, Second study program (Schmitt, F.O. ed.) The Rockefeller University Press, New York, p. 126.

Jakoubek, B. (1974) Brain function and macromolecular synthesis, Dion Limited, London.

Kunert, E. (1977) Estimation of GABA metabolism enzymes GAD and GABA-T in rat brain homogenates. In: Neurochemical methods for the study of putative transmitter metabolism in the nervous system. Manual of an IBRO-UNESCO training course (Bigl. V., ed.) Leipzig.

Lund, J.S. and Lund, R.D. (1972) The effects of varying periods of visual deprivation on synaptogenesis in the superior colliculus of the rat. Brain Res. 42: 21-32.

Maletta, G.J. and Timiras, P.S. (1967) Acetylcholinesterase activity in optic structures after complete light deprivation from birth, Exp. Neurol. 19: 513-518.

Maletta, G.J. and Timiras, P.S. (1968) Choline acetyltransferase activity and total protein content in selected optic areas of the rat after complete light deprivation during CNS development. J. Neurochm. 15: 787-793.

Macquire, M.E. and Gilman, A.G. (1974) Adenylate cyclase assay with adenylyl imidodiphosphate and product detection by competitive protein binding. Biochim. biophys. Acta 358: 154-163.

Maraini, G., Corta, F. and Franquelli, F. (1969) Metabolic changes in the retina and the optic centres following monocular light deprivation in the new-born rat. Exp. Eye Res. 8: 55-89.

Mareš, V., Brückner, G., Narovec, T. and Biesold, D. (1978) DNA synthesis and cell division in the rat visual centres. An autoradiography study. Life Sci. in press.

Margolis, F.L. and Bondy, S.C. (1970) Effect of unilateral visual deprivation by eyelid suturing on protein and ribonucleic acid metabolism of avian brain. Exp. Neurol. 27: 353-358.

Meisami, E. (1975) Early sensory influences on regional activity of brain ATPase in developing rats. In: Growth and development of the brain (Brazier, M.A.G., ed.) International Brain Research Organization Monograph Series, Vol. 1, Raven Press, New York, pp. 51-74.

Perry, G.W. and Cronly-Dillon, J.R. (1978) Tubulin synthesis during a critical period in visual cortex development. Brain Res. 142: 374-378.

Rose, S.P.R. and Haywood, J. (1976) Experience, learning and brain metabolism, In: Biochemical Correlates of Brain Structure and Function (Davisson, A.W., ed.) Academic Press, New York.

Rose, S.P.R. (1977) Transient and lasting biochemical responses to visual deprivation and experience in the rat visual cortex. IBRO/ICPS Satellite Symposium "Brain mechanisms in memory and learning", London.

Sherman, S.M., Hoffmann, K.-P. and Stone, J. (1972) Loss of a specific cell type from dorsal lateral geniculate nucleus in visually deprived cats. J. Neurophysiol. 35: 532-541.

Singer, W. and Tretter, F. (1976) Receptive-field properties and neuronal connectivity in striate and parastriate cortex of contour-deprived cats. J. Neurophysiol. 39: 613-630.

Singer, W. (1977) Control of thalamic transmission by corticofugal and ascending reticular pathways in the visual system. Physiol. Rev. 57: 386-420.

Sinha, A.K. and Rose, S.P.R. (1976) Dark rearing and visual stimulation in the rat: effect on brain enzymes. J. Neurochem. 27: 921-926.

Tebēcis, A.K. (1974) Transmitters and identified neurons in the mammalian central nervous system. Scientechnica (Publ.) LtD., Bristol.

Vrensen, G. and Groot, de D. (1974) The effect of dark rearing and its recovery on synaptic terminals in the visual cortex of rabbits: a quantitative electron microscopy study. Brain Res. 78: 263-278.

Vrensen, G. and Groot, de D. (1975) The effect of monocular deprivation on synaptic terminals in the visual cortex of rabbits. A quantitative electron microscopy study. Brain Res. 93: 15-24.

Wiesel, T.N. and Hubel, D.H. (1963 a) Single cell responses in striate cortex of kittens deprived of vision in one eye. J. Neurophysiol. 26: 1003-1017.

Wiesel, T.N. and Hubel, D.H. (1963 b) Effects of visual deprivation on morphology and physiology of cells in the cat's lateral geniculate body. J. Neurophysiol. 26: 978-993.

Wilson, P.D., Rowe, M.H. and Stone, J. (1976) Properties of relay cells in cat's lateral geniculate nucleus: a comparison of W-cells with X- and Y-cells. J. Neurophysiol. 39: 1193-1209.

Wilson, J.R., Webbs, S.V. and Sherman, S.M. (1977) Conditions for dominance on one eye during competitive development of central connections in visually deprived cats. Brain Res. 136: 277-287.

Winfield, D.A., Headon, M.D. and Powell, T.P.S. (1976) Postnatal development of the synaptic organisation of the lateral geniculate nucleus in the kitten with unilateral eyelid closure. Nature 263: 591-594.

GLIAL CELLS IN DEVELOPING RAT CEREBELLUM : BIOCHEMICAL AND IMMUNOHISTOCHEMICAL STUDY
M.S. Ghandour, G. Vincendon, E. Bock, D. Filippi, G. Laurent, J-P. Zanetta and G. Gombos. Institut de Chimie Biologique, Faculté de Médecine de l'Université Louis Pasteur and Centre de Neurochimie du CNRS, Strasbourg, France. Laboratoire de Chimie Biologique, Faculté de Médecine, Marseille, France. The Protein Laboratory, University of Copenhagen, Copenhagen, Denmark.

In contrast to the detailed information concerning nerve cells during the postnatal development of rat cerebellum, those concerning glial cells are very scarce and confined to Bergmann glia. In order to study the development of glial cells, we have followed in cerebellar slices by immunohistochemical techniques, and in cerebellar homogenates by immunological and biochemical techniques the appearance and accumulation of the four general or specific markers of glial cells shown in the accompanying communication (Vincendon et al. this meeting). All four markers, with the possible exception of butyrylcholinesterase, remain strictly localized in the same cell as in the adult and keep the same subcellular localization throughout development. The developmental curves of both the soluble and membrane bound form of each marker are of similar shape and chronology.

S100 and CA II, although localized in different glial cell types, have similar developmental curves : both are at low constant levels from birth to the 7-10th postnatal day, both start increasing after the 7-10th day with maximum accumulation rate between the 20th and the 25th day. Both continue to slowly accumulate after the 25th day, thus oligodendrocytes and astrocytes develop and differentiate simultaneously.

Cells containing either marker are present at birth. The number of astrocytes estimated by cells containing S100 and that of oligodendrocytes estimated by cells containing CA II, increase between the 4th and the 10th day, until by the 12th day both cell types are present apparently in the same number as in the adult. Thus it appears that although S100 and CA II are present early in newformed glial cells, the content per cell of S100 in astrocytes and that of CA II in oligodendrocytes increases greatly after cell multiplication, that is during differentiation (myelination being a sign of oligodendrocyte differentiation).

BuChE, present in both oligodendrocytes and astrocytes accumulates earlier than CA II and S100, and it does not appear to increase with glial cell differentiation as do CA II and S100. A transient appearance of BuChE has been observed in Purkinje cells of the nodulus. This is probably due to the extremely high level of acetylcholinesterase in these cells which could not be inhibited (without inhibition of butyrylcholinesterase) by the concentration of B.W. 284 C51 used in the biochemical method to reveal butyrylcholinesterase.

Astrocyte markers show that Bergmann glia is present at birth, but the number of these cells increases until the 4th-7th day. This confirms the perinatal formation of these cells.

GLIAL CELLS MARKERS IN ADULT RAT CEREBELLUM : BIOCHEMICAL AND IMMUNOHISTOCHEMICAL STUDY
G. Vincendon, M.S. Ghandour, E. Bock, D. Filippi, G. Laurent, J-P. Zanetta and G. Gombos
Institut de Chimie Biologique, Faculté de Médecine de l'Université Louis Pasteur and Centre de Neurochimie du CNRS, Strasbourg, France. Laboratoire de Chimie Biologique, Faculté de Médecine, Marseille, France. The Protein Laboratory, University of Copenhagen, Copenhagen, Denmark.

Cerebellar cortex (folium) is a good experimental model for obtaining unambiguous results concerning the localization of specific cell markers because of its layered structure of very regular architecture and distinctive cells. We have examined by immunofluorescence and immunoperoxidase methods with specific antisera, the cellular localization in adult rat cerebellum of : S100 protein, gliofibrillar acidic protein (GFA) and of the form II of Carbonic anhydrase (CA II). Distribution between cytosol and particulate fractions was also measured by immunological methods.

GFA, as described in the literature, is specifically localized in all astrocytes (Bergmann glia, fibrous, protoplasmic and velate astrocytes). A very important feature of our immunohistological preparations is that the GFA fluorescent antiserum clearly reveals the pattern of glial fibrils bundles. This is probably due to the fixation method used.

S100 protein is also almost exclusively present in astrocytes but minor staining of a few oligodendrocytes cannot be excluded. Neurons, neuronal membranes or nuclei never bound the specific antiserum. Similar to the distribution of lactate dehydrogenase, 90% or more of S100 protein is in the cytosol.

CA II is exclusively localized in oligodendrocytes and myelin. 35% of it is membrane-bound, the rest being in the cytosol.

Thus the three markers are not only exclusively glial markers but specific of different glial cell types.

These studies were completed by a histochemical study on another putative glial cell marker : butyrylcholinesterase (BuChE). The distribution in the cytosol and membranes was also followed in this case. BuChE is indeed a marker of glial cells but is present in all glial cells, in myelin and probably also in the capillary endothelium. In contrast with current ideas, BuChE is largely a membrane bound enzyme in cerebellum.

Our study also shows that : (1) the honey-comb like network of glial processes which separate the granular layer into groups of granule cell and glomeruli is made not only of astrocytic but also of oligodendrocytic processes (2) Purkinje cell perikarya are enveloped by glial processes which derive not only from Bergmann glia but also from other satellite cells of oligodendrocytic type. This oligodendrocytic envelope is particularly abundant around the basket cell "pinceau". (3) oligodendrocytes in the molecular layer appear to contribute to the glial envelope of Purkinje cell dendrites.

EFFECT OF NEONATAL HYPOTHYROIDISM ON RAT CEREBELLUM ONTOGENESIS
F. Vitiello, M.S. Ghandour, J. Clos, J. Legrand, G. Vincendon, G. Gombos
Institut de Chimie Biologique, Faculté de Médecine de l'Université Louis Pasteur and Centre de Neurochimie du CNRS, 11 rue Humann, Strasbourg, France
and Laboratoire de Physiologie Comparée, Université des Sciences et Techniques du Languedoc, Montpellier, France.

It is well known that in mammals neonatal thyroid deficiency has a great influence on postnatal development; it produces a marked and frequently irreversible damage to the histological, biochemical and behavioral maturation of the brain.

In order to better define the role of thyroid hormone in regard to both neuronal growth and myelination, we have followed, from birth to the 40th day of age, the developmental curves of some parameters of cell multiplication (DNA and RNA), plasma membrane formation and myelination (Na^+, K^+-ATPase, 5'-nucleotidase, gangliosides, phospholipids, cholesterol, galactolipids and CNPase).

We focused our attention on the cerebellum because its development is mostly postnatal and its morphology is well known.

Rats were rendered hypothyroid by administring propylthiouracil (50 mg/day) to the mothers from the 18th day of gestation to the day of sacrifice of the pups.

Each of the parameters studied presents a characteristic sigmoid-shaped curve. The shape and the time course of each curve is roughly similar in control and experimental animals. But, after the 8th-10th day of age, values in hypothyroid group are significantly lower than those of the control. Results on DNA, RNA and proteins are in good agreement with published data.

a) DNA synthesis is retarded and, at the 40th day, the cell number per cerebellum is 85 % of controls, as estimated from DNA values. RNA and proteins values are always lower than control.

b) Specific activity of Na^+, K^+-ATPase and 5'-nucleotidase, marker enzymes of plasma membrane, are significantly lower than controls from the 20th day after birth. From this age to 40 days, control values continue to increase, while hypothyroid values remain nearly at the same level. The developmental curve of gangliosides, compounds highly concentrated in neuronal plasma membrane, is significantly different from control after the 10th-12th day. Of the various parameters of membrane formation only phospholipids reach the control values after the 30th day.

c) Myelin is greatly affected in its deposition and probably in its composition. Cholesterol and galactolipids show a deficit which increases with age from the 12th to the 40th day (at 40 days they are respectively 20% and 25% less than in control), while CNPase specific activity is, at 40 days, about 50% of control.

These results are in good agreement with morphological data on cerebellar development in hypothyroid rat. Moreover, the poor development of the markers of plasma membrane testifies to a damage of membrane deposition and the strong and unequal decrease of myelin markers reflects the great influence of hypothyroidism on myelination.

REGIONAL PATTERNS OF GLYCOLYTIC ENZYMES IN DEVELOPING RAT BRAIN

H.H. Gustke, D. Schiffer[*] and S.L. Kowaleski[*], Max-Planck-Institut für experimentelle Medizin, Forschungsstelle Neurochemie, Göttingen, and
*Universitäts-Kinderklinik, Bonn, FRG

A considerable fraction of the total activity of several glycolytic enzymes in rat brain homogenate is recovered in the particulate matter, the fractionation of which by sucrose gradient centrifugation showed that hexokinase, pyruvate kinase and lactate dehydrogenase are associated mainly with synaptosomes and not with mitochondria (MacDonnell and Greengard, 1974). Therefore, monitoring of glycolytic enzyme activity in subcellular fractions of rat brain during development is a convenient procedure for the evaluation of biochemical as well as morphological maturation under normal and experimentally altered conditions.

Developmental patterns of hexokinase (HK), phophofructokinase (PFK) and pyruvate kinase (PK) have been examined in homogenates, soluble fraction and particulate matter of rat cerebral cortex, cerebellum and brain stem. HK and PFK total activity showed in all three brain regions a moderate increase in the first postnatal week, followed by a steep rise until day 21, whereas PK activity decreased from birth to a minimum at day 7 and thereafter steadily increased to about four-fold of this value at day 21. PFK and PK activities in U/g wet tissue were about two and four times as high as hexokinase activity. The activities of all three enzymes were lowest in cerebellum and nearly equal in cortex and brain stem.

The increase in hexokinase activity was completely confined to the particulate fraction, the soluble activity being constant over the whole period, whereas PFK and PK particulate and soluble activity both rose in parallel to the homogenate activity, the particulate activity always being higher than the soluble one. No substantial regional differences of these patterns were found. Protein analysis of the three cell fractions showed that the increase in enzyme activity is only to a minor part due to cell multiplication and growth, and mainly indicates the developmental completin of the enzymatic equipment of the cell organelles.

MacDonell, P.C. and Greengard, O. (1974) Arch. Biochem. Biophys. 163: 644-655

STUDY OF GLYCOSIDASES DURING THE DEVELOPMENT OF HUMAN BRAIN

P.ANNUNZIATA,G.C.GUAZZI and A.FEDERICO

Clinica Neurologica,Clinica Neurologica (R),1^Facoltà di Medicina,Università degli Studi di Napoli;Clinica Neurologica,Facoltà di Medicina,Università degli Studi di Siena.

It is known that during development great changes occur in glycoconjiugates in rat (Krusius et al.,1974;Margolis et al.,1975) as well in human brain (Federico and Di Benedetta,in press).In order to better understand these phenomena,several authors analyzed the activities of glycosidases during rat brain development (Quarles and Brady, 1970;Traurig et al.,1973).None report about similar changes in human brain is present until now in the literature.

We report the results of a study of seven glycosidase activities during the development of human brain.

The enzyme activity has been determined according to Quarles and Brady (1970) and Van Hoof and Hers (1968) on white and gray frontal matter of human brains obtained after general autopsies of subjects deceased after non neurological diseases and stored at -80°C until the analyses.The age span was from 1 day to 64 years old.

N-acetyl-β-glucosaminidase and N-acetyl-β-galactosaminidase have a maximal activity in newborn age.Afterward they decrease.In adult and old age their activities linearly increase to reach values that are little lower than those found in newborn age.

Maximal β-glucuronidase activity was found in newborn age.It afterward decreases and at 24 years old age has similar values to those found in 64 years old brain.

β-galactosidase and β-glucosidase activities increase,the first to have maximal values in the first days of life,the second in childhood.They afterward decrease to have in 64 years old brain values similar to those found in newborn brain.

α-fucosidase and α-mannosidase have their maximal activity in newborn age and linearly decrease during development.

These data,compared to analogous results on rat brain development,will be discussed in relationship to the well known glycoconjugates changes reported during the maturation of the brain.

ACID PHOSPHATASE DURING DEVELOPMENT OF CHICKEN SPINAL CORD
ANTERIOR HORN CELLS UNDER NORMAL AND EXPERIMENTAL CONDITIONS

R.A.van Welsum and J.Drukker

Dept.of Anatomy and Embryology, University of Amsterdam

Neuronal development is described as occurring in several phases. The phase of the neuroblasts is followed by a resting phase and a phase of rapid dendritic growth which is followed by a gradual transition to the adult situation. The so-called resting phase is characterized by the lack of considerable morphological alteration and only slight changes in biochemical parameters. During this phase axotomy was performed by sectioning the sciatic nerve of chick embryos at 8-12 days of incubation. The anterior horn cells in the lumbar spinal cord then showed a very strong morphological reaction,(Houthoff & Drukker, 1977). Enzyme histochemistry (a.o. acid phosphatase) showed a strong increase after axotomy during the resting phase. In similar experiments Drukker & Vos (1974) could not find any differences between experimental and control sides. This led to the introduction of ultramicrochemical methods according to Lowry (Lowry & Passonneau, 1972).

Acid phosphatase of spinal cord anterior horn cells with 4-methylumbelliferyl phosphate as substrate was studied. A characterization of the enzyme, normal control series and the results after section of the sciatic nerve in chick embryos will be presented. The differences between the experimental sides and normal controls will be discussed.

Drukker,J and Vos,J (1974)Acta morph.neerl.-scand.12:47-60
Houthoff,H.J. and Drukker,J (1977) Neuropathol.and appl.Neurobiol.
    3:441-451
Lowry,O.H. and Passonneau,J.V. (1972) A flexible system of
    enzymatic analysis.Academic Press, New York

GLUTAMATE AND KAINIC ACID BINDING TO SYNAPTIC MEMBRANES OF CEREBRAL CORTEX FROM DEVELOPING RATS

Cath Sanderson and Sean Murphy
Brain Research Group, Open University,
Milton Keynes, U.K.

L-glutamate has been shown to be a potent excitant when applied iontophoretically to neurons in the central nervous system, and has been suggested as a putative excitatory neurotransmitter in the mammalian brain. We are investigating this hypothesis using labelled L-glutamate and the proposed glutamate agonist, kainic acid as ligands in binding studies. The binding assay employs a rapid, post-incubation centrifugation method and non-specific binding is estimated in parallel samples containing in addition to labelled substrate, excess unlabelled ligand (0.1 m$\underline{M}$). We are particularly interested in characterising binding sites for glutamate on a developmental basis and in distinguishing between glutamate binding to postsynaptic receptors from binding to high (reuptake) and low (transport) affinity sites.

"Crude synaptic membrane" was prepared from cerebral cortex of 50 day old rat and exhibited high affinity, Na^+-independent specific binding of labelled glutamate. This binding was inhibited if the tissue was pre-incubated with cold kainic acid. Lower affinity, Na^+-dependent binding was also observed but kainate had no affinity for these sites. Specific binding of kainic acid was not observed in rats less than 30 days of age in either crude synaptic membrane or a purified synaptic membrane preparation.

From these studies, kainic acid appears to be specific to the binding sites of the glutamate receptor: the apparent late postnatal development of these receptors is currently under investigation.

AVAILABILITY OF AMINO ACIDS TO THE DEVELOPING BRAIN: CHANGES IN THE KINETICS AND SPECIFICITY OF MEMBRANE TRANSPORT SYSTEMS IN SEPARATED CELLS FROM THE RAT CEREBRAL CORTEX DURING MATURATION

Sean Murphy and Arun Sinha,
Brain Research Group, Open University,
Milton Keynes, U.K.

Protein synthesis rates are high in the neonatal rat brain, declining to adult levels by 7 weeks of age. As the blood brain barrier is not fully functional until this time, the availability of amino acids to the developing brain may be governed simply by plasma levels and the presence of membrane transport systems on the cells themselves.

In a continuing investigation into the interactions between protein synthetic activity and amino acid transport into cells, we have been following the development of membrane transport systems both in terms of their specificity and the kinetics of transport. Employing a cell separation technique for neurons and neuropil from neonatal rat cerebral cortex developed in our laboratories (J. Neurochem. in press), we have looked at the transport of three amino acids (L-leucine, L-alanine and L-lysine) from birth to three weeks of age.

Cell suspensions and neuronal and neuropil fractions were incubated in ^{14}C-labelled substrates and then rapidly filtered under vacuum. Specific transport systems for all three amino acids were present from birth, that for L-lysine being Na^+-independent while transport of L-alanine and L-leucine was Na^+-dependent (uptake inhibited in the presence of ouabain or in low Na^+ conditions).

Kinetic parameters (K_M, K_D, Vmax) for all three transport systems changed developmentally. These changes may be related to the rates of specific protein synthesis in the various cellular "compartments" during maturation of the cerebral cortex.

BRAIN AND PLASMA AMINO ACID POOLS IN EARLY AND ADULT PROTEIN-CALORIE UNDERNUTRITION.

Rodés, M., Sabater, J. and González-Sastre, F.
Dept. Neurochem. Instituto Provincial de Bioquímica Clínica,
Universidad Autónoma. Barcelona. Spain.

Early undernutrition (group I) was obtained on suckling rats by intermitent separation from the mother. Analysis were performed at the 21st day on plasma and brain samples, and repeated at 30 days after a 9 days period of nutritional rehabilitation. Undernutrition was assesed by measuring the differences in body and organ weight between control and undernourished animals.

Male adults rats (∼290gr) were undernourished by decreasing to one half (group II) and to one third (group III) the verified normal food intake. The undernutrition was assesed as with the young animals. The analysis were performed when the decrease in body weight with respect to controls was of 22% (II) and 35% (III).

Brain samples were extracted with 5% perchloric acid according to Gaitonde, M.K. (1974). Plasma samples were deproteinized with 10% sulphosalicylic acid. Amino acid analysis were carried out in a "Cromaspek" Amino acid Analizer.

Undernutrition causes marked alterations in the free amino acid pools of both young and adult animals. The changes in the concentration of individual amino acids are considerably higher in the young animals than in the adult ones and the plasma modifications are always larger than those detected in brain. Of all changes detected in plasma amino acids levels, some are found at both ages: high levels of glycine and lysine and decreased levels of glutamine, proline, tyrosine and ornithine. In the adult these changes occours with a decrease in the concentration of aspartic and glutamic acids, threonine and leucine. In the young animals the additional changes consists in significant increases of taurine, aspartic acid, glutamic acid as well as of the essential amino acids threonine, valine, isoleucine, leucine and phenilalanine. Alanine, citrulline and arginine are found decreased. The changes in brain tissue except for a few amino acids follow a pattern similar than in plasma. On the contrary, in the adult the changes in amino acid levels verified in plasma have not been found in the brain tissue pool.

The effects of nutritional rehabilitation after early undernutrition have been investigated after 9 days of feeding ad libitum. The amino acid levels in plasma either attain the control values or become altered in the opposite sens than under nutritional deprivation. Similar changes, although not identical, were observed in the brain tissue pool.

SOME ASPECTS OF THE REGULATION OF GABA LEVEL IN DEVELOPING RAT BRAIN

L. Ossola, M. Maitre, J. M. Blindermann and P. Mandel
Centre de Neurochimie du CNRS,
11 Rue Humann, 67085 Strasbourg Cedex, France

Gamma aminobutyric acid (GABA) is the major inhibitory neurotransmitter in the central nervous system. Previous studies in various species have indicated that cerebral GABA levels are relatively high at early stages of development and particularly in the newborn (Roberts et al., 1951; Roberts, 1961; Vernadakis and Woodbury, 1961; Van Den Berg et al., 1965). However, glutamic acid decarboxylase (GAD) activity in the newborn is only about 10 % that of the adult (Coyle and Enna, 1976). This discrepancy between the amounts of GABA and GAD might be explained by regulation at the level of GABA transport or degradation. We have investigated this latter possibility by determining the molecular activities of the GABA catabolizing enzyme, GABA aminotransferase (GABA-T), in the developing rat brain, in parallel with GABA levels and GAD activity.

Our data show that the specific activities of GAD and GABA-T increase during postnatal development in rat brain, but that the level of GABA remains constant in brain hemispheres and in cerebellum. A radioimmunoassay, allowing the measurement of the protein apoenzyme of GABA-T, gave more precise information than measurement of enzymatic activity. Actually the quantity of enzyme per mg of protein remained constant during cerebral development. Therefore, the observed increase in specific activities of GABA-T in brain regions seems to indicate the presence of an effector of GABA-T activity. Thus, it appears that modulation of the level of this effector, but not of the level of GABA-T protein, modifies total GABA-T activity during growth. This mechanism appears to contribute to the regulation of GABA level in the developing rat brain.

Coyle, J.T. and Enna, S.J. (1976) Brain Res.
    111: 119-133.
Roberts, E. (1961) in "Regional Neurochemistry", Kety, S.S. and Elkes, J., Eds.,
    Pergamon Press, Oxford - New York, pp. 324-347.
Roberts, E., Harman, P.J. and Frankel, S. (1951) Proc. Soc. Exp. Biol.
    78: 799-803.
Van Den Berg, C.J., Van Kempen, G.M.J., Schade, J.P. and Veldstra, M. (1965)
    J. Neurochem. 12: 863-869.
Vernadakis, S.A. and Woodbury, D.M. (1962) Am. J. Physiol.
    203: 748-752.

GLYCOCONJUGATES AND OTHER BIOCHEMICAL CHANGES IN AGING RAT AND HUMAN BRAIN
A. FEDERICO, R.M. CORONA, P. ANNUNZIATA

Clinica Neurologica e Clinica Neurologica (R), 1°Facoltà di Medicina, Università degli Studi di Napoli

In recent years, efforts have intensified to learn more about aging processes at molecular level, in particular to know whether age associated changes in the brain are related to neuron loss or to the presence, in the nerve cell, of age-associated constituents that are responsable of the impairement of the cell function. It was hypothesized that glycoconjugates, for their localization on the external surface of plasma membrane, could have an important role on ageing mechanism (Sullivan and Debusk, 1974). There are reported many data about changes in lipid (Horrocks et al, 1972, 1977), DNA, RNA, proteins and neurotransmitter substances (Samorajski and Rolstein, 1972), while no data are present in the literature about membrane specific constituents (glycoproteins, gangliosides, glycosaminoglycans).

The present research reports results of biochemical analyses of aging human (65-80 years old) and rat (18 months) brains. Human brains were choosed from brain of subjects deceased after non neurological diseases.

Gangliosides did not change.

Glycoprotein sugar composition did not change in the "soluble" fraction, while a trend to increase was found in the "insoluble" fraction.

"Soluble" glycosaminoglycans are decreased. On the contrary, an increase was found in the "insoluble" fraction. Changes were also found in some glycosidase activities. Comparatively, lipids, phospholipids, cholesterol, sulfatides, cerebrosides and proteins have been analyzed. Total phospholipids are unchanged, while a t.l.c. showed a decrease of phophatidylethanolamine. These changes are similar to those reported by Horrocks et al. (1977).

These biochemical data on human and rat brain will be discussed on the basis of the other results of the literature and in relationship to the well known histological findings of this age.

Bibliography

Horrocks L.A., in Neurobiological Aspects of maturation and aging (D.N.Ford Ed), pag. 393, Elsevier, Amsterdam, 1972

Horrocks L.A., Proceed. Int. Soc. Neurochem., Vol.6, pag.176, 1977

Samorajski T. and C. Rolstein, in Neurobiological Aspects of maturation and aging (D.N. Ford Ed.) pag.253, Elsevier, Amsterdam, 1972

Sullivan J.L. and Debusk A.G., J.Theor.Biol., 46, 291, 1974

EFFECT OF CORTISOL ON CHOROIDAL SODIUM-POTASSIUM ADENOSINE TRIPHOSPHATASE AND CISTERNAL CEREBROSPINAL FLUID PRESSURE IN DEVELOPING CHICK EMBRYO

F. Šťastný[1], Z. Rychter[2], R. Jelínek[3]
Inst. of Physiology[1]), Inst. of Histology[2]), Fac. of Med., Charles Univ., Inst. of Exp. Med.[3]), Teratol. Lab., Prague, Czechoslovakia

In embryonic avian brain, as in mammalian target organs, the effects of ubiquitous glucocorticoids are mediated by a transient presence of specific cytoplasmic receptors. After translocation into a nucleus the glucocorticoid-receptor complex induces a prolongation of cell mitotic cycle and initiates synthesis of specific mRNA that codes certain proteins which express own effects of glucocorticoids on the morphological, biochemical and functional differentiation of cells.

The involvement of choroidal cell membrane Na^+-K^+-ATP phosphohydrolase (Na^+-K^+-ATPase; EC 3.6.1.3) was studied to determine whether the enzyme system might be, at least partially, responsible for the developmental changes in the osmolarity and composition of cerebrospinal fluid (CSF) in normal and cortisol-treated chick embryos. The enzyme activity was assayed separately in three main parts of the telencephalic choroid plexus (base, body, apex) of the chick embryos between days 9 and 19 of incubation. We have found a rapid increase of the Na^+-K^+-ATPase activity after day 13, mainly in the apex (13-times) containing the most differentiated cells. On the contrary, the enzyme activity increase was low in the base (5-times) which represents the proliferation center of this structure. The ATPase activity increase was followed by an increase in the cisternal CSF pressure (5-fold) which is one of the main morphogenetic factors in the brain enlargement. Exogenous cortisol, administered in a dose of 10 µg (on days 7 or 9), 20 µg (on days 11 or 13) and 40 µg (on days 15 or 17) onto the chorio-allantoic membrane 48 hours before the assay of choroidal Na^+-K^+-ATPase activity, stimulated the enzyme activity in the base, but an inhibitory effect of the steroid was observed in the body and apex after day 13. On day 19 cortisol had the inhibitory effect only, which has been observed under in vitro conditions as well. The inhibitory effect of cortisol in a concentration of 3×10^{-4} M was slightly weaker than that of ouabain in the same concentration. Contemporarily, on day 19 cortisol caused a decrease of the CSF pressure and water content in the cerebral hemispheres of chick embryos.

The results suggest that the Na^+-K^+-ATPase system is implicated in the regulation of CSF production and may participate on the beneficial effect of glucocorticoids in the therapy of brain edema and intracranial hypertension in the postnatal period of development.

PRENATAL MATERNAL PHENOBARBITAL ALTERS PLASMA CONCENTRATION OF CORTICOSTERONE IN DEVELOPING OFFSPRING

J.W. Zemp, L.D. Middaugh, and W.O. Boggan
Dept. of Biochemistry and of Psychiatry and Behavioral Sciences, MUSC
Charleston, S.C., USA

We have previously reported that phenobarbital injected into mice for the last third of pregnancy increased neonatal mortality and reduced body weight of surviving offspring (Zemp and Middaugh, 1976). Young (21 day) offspring prenatally exposed to phenobarbital have reduced conversion of ^{3}H-tyrosine to dopamine and norepinephrine (Grover et al., 1978). The effects of early exposure to phenobarbital appear to persist into adulthood since offspring of animals injected with the drug differ from control animals on a number of behavioral tasks after maturity (Zemp and Middaugh, 1976). The mechanisms mediating the long term behavioral effects are unknown. It is known that manipulation of the pituitary-adrenal system, which is partially under catecholaminergic control, can alter performance on behavioral tasks similar to those reported in our previous studies. In the present study potential alterations in the pituitary-adrenal system from prenatal exposure to phenobarbital was assessed by determining plasma corticosterone concentration in newborn and 21-day-old offspring of mice injected daily with either saline or phenobarbital in the last third of pregnancy. Offspring of mice treated with phenobarbital exhibited elevated plasma concentrations of corticosterone on the day of birth and reduced concentration 21 days after birth. The high concentrations at birth are consistent with reported compensatory increases in fetal corticosterone due to phenobarbital induced reductions in maternal corticosterone (Milković et al., 1973). The lower concentrations at 21 days are consistent with reported effects of neonatal injections of corticosterone on plasma concentrations of the glucocoricoid in young rats (Nyakas and Endröcze, 1972). (Supported by grant #DA01624 (JWZ) from the National Institute on Drug Abuse and by grant #AA01865 (WOB) from the National Institute on Alcohol Abuse and Alcoholism.)

Grover, T.A., Middaugh, L.D., Simpson, L.W., and Zemp, J.W. (1978) submitted to J. Neurochem. May, 1978.
Milković, K., Paunović, J., Kniewald, Z., Milković, S. (1973) Endocrinol. 93:115-118
Nyakas, C. and Endröczi, E. (1972) Acta Physiol. Acad. Sci. Hung. 42:231-241
Zemp, J.W. and Middaugh, L.D. (1977) Perinatal Addiction, Harbison, R.D. ed. p. 307-331, Spectrum, N.Y.

USE OF ARRHENIUS PLOTS OF Na-K-ATPase AND ACETYLCHOLINESTERASE AS A TOOL FOR STUDYING CHANGES IN LIPID-PROTEIN INTERACTIONS IN NEURONAL MEMBRANES DURING BRAIN DEVELOPMENT.

M. Gorgani and E. Meisami
The Institute of Biochemistry and Biophysics, University of Tehran, Tehran, Iran

In order to study the changes in the mode of interaction of neuronal membrane enzymes with the lipid components during postnatal development, the activities of Na-K-ATPase and acetylcholinesterase in the rat brain cortex were measured at different postnatal ages as a function of temperature. Arrhenius plots of the data were prepared and the apparent energies of activation were computed for each plot. It was observed that all plots were biphasic except that for Na-K-ATPase of the immature (5 day old) brain which showed no trasnition temperature, with an apparent energy of activation of 15.5 Kcal/mole. The enzyme from the mature brain (25 day old) showed an average transition temperature of 22.6 C, with average apparent energies of activation of 15.3 and 27.2 Kcal/mole above and below the transition temperature respectively. The cortex of 1 day old rat showed no Na-K-ATPase activity. Arrhenius plots of acetylcholinesterase studied at ages 1,5 and 25 days postnatally, all showed transition temperature which increased from an average of 16.1 C for 1 day old to 17 and 21.5 C for 5 and 25 day old animals respectively. The average apparent energies of activation for acetylcholinesterase below the transition temperature changed from 8.3 Kcal/mole at day 1 to 8.7 and 7.2 Kcal/mole at days 5 and 25, while above the transition temperature they were 4.3, 5.2 and 4.1 at days 1,5 and 25 respectively.

These results indicate that Na-K-ATPase becomes a truly "integral" membrane enzyme only at a relatively late postnatal age, while acetylcholinesterase appears to have already established its "peripheral" association with the neuronal membranes at birth. It may further be generalized that lipid-protein interaction for membrane-associated enzymes changes with development, and the extent of the change depends on the enzyme and the degree of its association with the membranes.

AchE and Ach in rat olfactory bulb during development and under olfactory deprivation

E. Meisami, R. Mousavi and R. Safaii
Institute of Biochemistry and Biophysics,
University of Tehran, P.O. Box 314-1700
Tehran, IRAN

The activity of acetylcholinesterase (AchE) and the amount of acetylcholine (Ach) was measured in the olfactory bulbs (OB) of the rat during postnatal development. The results indicated that at birth AchE is present at a mean activity level of 12.8 nmoles Ach hydrolyzed per min per mg protein. This value increases by 2.5 times until day 5 when it remains constant up to day 10. Then it increases extremly sharply, reaching a value of 76.3 by day 25, and 101.2 by day 60 when adult values are obtained. Measurement of Ach by bioassay of leech dorsal muscle revealed a 10 fold increase in the amount of this transmitter between days 1 and 25 (9.9 nmoles/g wet wt. at day 1 and 86.6 at day 25). It therefore appears that during postnatal development, the activity of AchE and the amount of Ach increase by the same amount (ca.10X), indicating a very significant increase in the number of cholinergic synapses in OB, particularly after day 10. A further study on the distribution of AchE activity in central olfactory strucutres, using finely separated samples of regions from the 25-day-old freeze-dried brain, revealed that while the main OB and the anterior olfactory nucleus show the same activity, the accessory OB shows half as much activity.

Neonatal unilateral olfactory deprivation, produced by closure of a nare, which has been shown to cause retarded growth of the OB (1), resulted in no significant change in the specific activity of AchE in OB, measured at days 25,60 and 120 postnatally. These results indicate that bulbar cholinergic synapses are probabely not involved in the primary olfactory pathways, and that olfactory stimulation is not necessary for development and maintainance of these synapses.

1) Meisami, E. (1976) Brain Res.
   107: 437-444

LOCALIZATION AND ACTIVITY OF CHOLINESTERASES IN THE RAT DEVELOPING DIAPHRAGM

T. Kiauta and M. Brzin
Institute of Pathophysiology, University of Ljubljana, 61105, Yugoslavia

Ultrastructural localization of AChE and BuChE activities in developing rat diaphragm motor endplates was investigated. The method used was the CNS⁻ or CN⁻ modification of the one-step Cu-thiocholine procedure. The precipitates obtained after a short incubation period are finely granulated and make a high resolution possible.

Concomitant quantitative measurements of AChE and BuChE activities in the same muscles were made by using a radiometric method.

Cytochemically, the activity of both enzymes was found to be localized at the same sites as in adult endplates, i.e., along the presynaptic and the postsynaptic membranes, in the primary and in the rather poorly developed secondary synaptic clefts, and in the Schwann cell-nerve ending interspace. In addition, the reaction product was found at the perinuclear membrane of Schwann cells where it was never observed in adult motor endplates.

Quantitatively, BuChE activity, which is some four times lower than AChE activity in adult diaphragms, was found to be somewhat higher than AChE activity in developing muscles. The high BuChE persisted relatively unchanged until the 12th day after birth. Even in 40-day-old rats, BuChE activity was higher than in adult animals.

The high BuChE activity in developing rat diaphragms does not disagree with the hypothesis according to which BuChE serves as a precursor for AChE at cholinergic synapses.

MICRO-ANALYSIS OF LIPIDS IN PERIPHERAL NERVE BIOPSIES IN TWO AGE GROUPS

S. Pollet, J. J. Hauw[+], J. C. Turpin, F. Le Saux, M. Monge and N. Baumann
Laboratoire de Neurochimie INSERM U. 134, CNRS ERA 421, and [+] Laboratoire de Neuropathologie Charles Foix, Hôpital de la Salpêtrière, 75634 Paris Cédex 13, France

Lipid analysis were performed on superficial peroneal nerve biopsies after elimination of epineurium. Sonication was required for lipid extraction in chloroform-methanol-water 70:30:4. Separation of the lipid classes was obtained on TLC glass plates (10 x 10 cm) coated with silica gel HPTLC 60 F 254 Merck using one multiple two dimensional chromatography (1). Most of the methods used for quantitation were standard ones which had been scaled down. This technique allows the ponderal study of cholesterol, cerebrosides, sulfatides, ethanolamine phospholipids, phosphatidyl-choline, -serine, -inositol, and gangliosides on 500 µg of lipid.

All the nerve specimens were normal as confirmed by neuromorphological examinations using light microscopic study of transverse sections of adjacent fragment of the biopsy after either epon or paraffin embedding and section. Samples were divided in two groups : one consisted of 6 specimens from birth to age 16, the other comprised biopsies of 7 patients from 36 to 77 years old. A significant increase in cholesterol was found in the older age group (30 µg $\pm$ 3 instead of 18 µg $\pm$ 5 per mg fresh tissue) ; the same was true for glycolipids (hexose content 8 µg $\pm$ 4 instead of 2 µg $\pm$ 0.5 per mg fresh tissue). These variations must be taken into account for studies of pathological samples.

(1) Pollet, S., Ermidou, S., Le Saux, F., Monge, M. and Baumann, N. Micro-analysis of brain lipids employing multiple two-dimensional thin-layer chromatography. J. Lipid Res. (1978) In press.

THE EFFECT OF THYROID HORMONE ON FATTY ACID ACTIVATION DURING MYELINATION.

R.C.CANTRILL*, L.KERR* AND E.M.CAREY**

* Department of Medical Biochemistry
  University of the Witwatersrand Medical School
  Johannesburg, South Africa.
** Department of Biochemistry
  University of Sheffield
  Sheffield, England.

The effects of thyroid hormone and its specific inhibitor propylthiouracil on normal brain development have been studied. T_3 administration causes a permanent decrease in the number of glial cells (Balazs et al, 1971a) but increases the rate at which the enzymes of glucose and amino acid metabolism attain adult activities (Cocks et al, 1970). Neonatal thyroidectomy causes a retardation in the rate of DNA increase although normal cell numbers are attained after some delay (Balazs et al, 1968 ; Balazs et al, 1971b). There is also a reduction in the amount of myelin deposited and lower levels of synaptosomal enzyme activity and glucose metabolism; this may be due to a decrease in axonal density (Balazs et al, 1968; Eayrs, 1968). The peak incorporation of cerebrosides, sulphatides and cholesterol into myelin is delayed by 10 days in the thyroidectomised rat (Tsujimura, et al).

In this present study we have investigated the effect of hypo- and hyper-thyroidism on the fatty acid activating system of the developing rat brain. Palmitoyl-CoA synthetase activity was assayed in brain subfractions obtained during the first 20 days after birth. Hypothyroid animals were shown to have a lower enzyme activity in the mitochondria-enriched fraction than control and thyroxine-treated rats.

Balazs,R., Kovacs,S., Teichgraber,P., Cocks,W.A. and Eayrs,J.T.(1968) J. Neurochem. 15: 1335-1349
Balazs,R., Kovacs,S., Cocks,W.A., Johnson,A.L. and Eayrs,J.T.(1971a) Brain Res. 25: 555-570.
Balazs,R., Cocks, W.A., Eayrs,J.T. and Kovacs,S. (1971b) in Hormones in Development (Hamburgh,M. and Barrington, E.J.W. Appleton-Century Crofts, pp.357-379)
Cocks,W.A., Balazs,R., Johnson,A.L. and Eayrs,J.T. (1970) J. Neurochem. 17: 1275-1285
Eayrs,J.T. (1968) in Endocrinology and Human Behaviour (Michael,R.P. Oxford University Press, London. pp.238-255)
Tsujimura,R., Kariyama,N. and Hatotani,N. (1973) in Hormones and Brain Function (Lissak,K. Plenum Press, N.Y. pp.69-78)

Metobolism of 1-alkyl-and 1-alkenyl-sn-glycero-3-phosphoethanolamine in subcellular fractions of myelinating rat brains.

J.Gunawan, M.Vierbuchen and H.Debuch
Institute for physiological chemistry, Lehrstuhl II
University of Cologne, Germany

At least two enzyme-activities compete with 1-monoradylphospholipids resulting in either 1-radylglycerol or 1-radyl, 2-acyl phospholipid. Therefore we prepared: 1-$[^{14}C]$-alkyl GPE (A_1), 1-$[^{3}H]$-alkyl GPE (A_2), 1-$[^{14}C]$-alkenyl GPE (B) and 1-$[^{14}C]$-alkyl-glycerophospho$[^{3}H]$-ethanolamine (C) and incubated these substrates in the presence of Mg^{2+} with different subcellular fractions of 14 days old rats.

Using substrate A_2 we found the highest hydrolysis rate in the microsomal fraction, which could be inhibited by sulfhydryl blocking reagents. The lipid soluble hydrolysis products were identified as 1-alkyl-, or 1-alkenyl-glycerol respectively through all the experiments with microsomal fractions.

Comparing the catabolism of A_1 an B, we observed a much lower affinity of the enzyme system to the lysoplasmalogen (in contrast to Wykle and Schremmer, 1974) which also inhibited the hydrolysis of the saturated ether compound.

To get more information on this reaction, we identified also the water soluble products by paper chromatography after incubation of substrate C. It turned out, that ethanolamine was the main product formed. At any incubation time (15 - 180 min) however, about 10% of $[^{3}H]$-activity of the upper phase was found in phosphorylethanolamine. Therefore we assume, that the hydrolysis of the substrate occurs at two different sites. On the other hand, the acylated compounds of A_1 and B did not serve as substrates for the degradation enzymes.

Furthermore, we found a slightly higher acylation rate for the lysoplamalogen compared with 1-alkyl-GPE in the lysosomal fractions.

In order to investigate a correlation between the total hydrolysing enzyme activities of rat brain and the age of the animals, we incubated brain homogenates of 7, 14, 21 and 365 days old rats with substrate A_2. The youngest animals had twice the specific hydrolysing enzyme activity compared with the adults.

Wykle, R.L. and Schremmer, J.M. (1974) J. Biol. Chem. 249: 1742 - 1746

MUSCARINIC RECEPTOR IN DEVELOPING RAT CEREBELLUM

J. Mallol, C. Sarraga, M. Bartolomé, J-P. Zanetta, G. Vincendon and G. Gombos
Departamento de Farmacologia, Facultad de Medicina, Universitad de Badajoz, Spain
and Institut de Chimie Biologique, Faculté de Médecine de l'Université Louis Pasteur
and Centre de Neurochimie du CNRS, Strasbourg, France.

Cerebellum is a good model for developmental studies aiming to correlate morphological to biochemical events because of (i) its architecture in layers each containing neurons and structures of distinct morphology and (ii) the well-known chronology of development (postnatal in rat) of each of its components. The small amounts of markers of cholinergic transmission in adult rat cerebellum indicate that only a small portion of cerebellar synapses are cholinergic. However also in this case, cerebellum remains a good model since these synapses are confined to precise types of cells and fibers (although possibly not to all individual cells or fibers of a given morphology).

In this paper, we describe some studies on the development of the acetylcholine receptors in the cerebellum of inbred Wistar rats grown under standardized conditions. These receptors are only of the muscarinic type (ther is no α-bungarotoxin binding). Their amount was assessed by the specific binding of quinuclidinyl benzilate (QNB) to whole homogenate by the filtration technique (Yamamura and Snyder (1974) PNAS, 71 : 1725). Specificity of binding was also controlled by displacement for acetylcholine, buscapine, atropine and dimethyl aminoethyl benzoic acid. Optimal conditions of pH (7.4) temperature (37°) incubation time (60 min) were found. Particularly critical for quantitative estimate of the binding is the type of filter used. Gelman filters (0.45 μ) and Whatman GF/F filters retain only 50 and 80 %, respectively, of the material which specifically binds QNB as compared with Millipore filters (0.45μ). The results obtained with the Millipore filters indicate that during development, if the characteristics of the muscarinic receptor remain constant, its amount increases continuously from birth to the 20th postnatal day and remains constant thereafter (about 25 pmole of QNB/g of wet tissue) when expressed on a per cerebellum basis, but shows two peaks with maxima, one at the 8th, the other at the 20th postnatal day, when expressed per mg protein. These data suggest at least two sets of cholinergic synapses each localized on cholinoceptive cells formed at different times. Correlations with the chronology of morphological development, with the developmental curves of acetylcholinesterase, and with data concerning transmitters localization, indicate that the peak at the 20th postnatal day corresponds to the accumulation of receptors necessary for the synapses between the afferent mossy fibers and granule cells in glomeruli (even though it is possible that not all mossy fibers are cholinergic). The peak at the 8th day could correspond to the formation of receptors for (i) the "early synapses" on Purkinje cells if acetylcholine is the transmitter of climbing fibers, (ii) cholinergic synapses on deep nuclei cells (the only cholinergic-cholinoceptive cells in rat cerebellum).

IMPAIREMENT OF BRAIN DEVELOPMENT IN MAN: NEUROCHEMICAL STUDY OF TWO DISEASES

A. FEDERICO, R.M. CORONA, I.D'AMORE, P. ANNUNZIATA and G.C. GUAZZI
Clinica Neurologica e Clinica Neurologica (R), 1^Facoltà di Medicina dell'Università di Napoli e Clinica Neurologica dell'Università di Siena

It is known that thyroid deficiency determines a marked change in normal maturational processes of the brain.So pre- and post-natal thyroid hormon deficiency has been used as an experimental model to study brain development in animals.

Few neurochemical data are reported about analogous changes in man.

Here we report a neurochemical study of two human pathological conditions that are characterized by impairment of the brain development:
- a case of congenital athyroidism,dead at 18 years old age;
- a case of liss-enkephaly,deceased at the age of 18 months.Liss-enkephaly is characterized by a smooth cerebral cortex as in a 12 weeks embryo,in relationship to defect of cell migration processes.

We analyzed lipids,phospholipids,gangliosides,cerebrosides,sulfatides,mucopolysaccharides,glycoproteins and several glycosidases.All results were compared to analogous data obtained from control brain of same age and sex.

In the case of athyroid patient a decrease of lipids,a marked change in gangliosi de and phospholipid patterns,an increase in "insoluble" mucopolysaccharides and a decrease in "insoluble" glycoproteins in gray matter have been found.

Less dramatic changes have been found in the case of Liss-enkephaly.

These neurochemical data give some indications,in humans,about the biochemical basis of brain maturation impairement.

Symposium

Neurochemistry of Hypoxia

Chairman:
H. S. Bachelard

CLINICAL ASPECTS OF ANOXIA IN THE NERVOUS SYSTEM

Lindsay Symon, TD, FRCS, Professor of Neurological Surgery,
London University. Gough Cooper Department of Neurological Surgery,
Institute of Neurology,
London WC1N 3BG.

Difficulties in the analysis of hypoxic brain damage have been that present brain damage is often consequent upon accidental and therefore incompletely documented events. Thus a drowning, an accidental exposure to toxic gas or an anaesthetic accident, are by their accidental nature frequently imperfectly monitored. The relative contribution of pure oxygen lack, of secondary cardio-vascular damage or pulmonary disturbance are hard to ascertain and the pathology is often clouded by artefactual change consequent upon the passage of hours or days post mortem before detailed neuropathological examination or even fixation, has been attempted.

The nervous system depends for its function on oxidative metabolism and complete withdrawal of either oxygen or of substrate will result in a rapid cessation of neuronal function; indeed total withdrawal of the oxygen supply of the brain as for example by circulatory arrest will lead to the first evidence of brain dysfunction, a complete loss of consciousness, within about ten seconds. This interval will be prolonged if the circulation to the brain continues in the absence of oxygen, as for example in exposure to an atmosphere of inert gas, or to lowered pressure as in explosive decompression in an aircraft, but even so the interval will be scarcely doubled and loss of consciousness soon supervenes. From this time on, cellular damage in the brain is a possibility. The actual anoxic damage to each neurone may very well be similar in a wide variety of circumstances, but the total clinical and neuropathological picture will depend very much upon the character of the initial incident, and upon the secondary or sequential changes which ensue in the organism as a whole or within the head. Thus, a severe head injury may be complicated by micro-vascular ruptures and by the development of extensive secondary brain swelling with impaired cerebral perfusion consequent upon raised intracranial pressure. A severe hypoxic incident during anaesthesia may similarly result in secondary brain swelling and the clinical picture in the two circumstances at the end of several years may be almost indistinguishable.

In order to codify our thinking it is necessary to consider first the various classifications of anoxia which pertain as much to the nervous system as any other. Barcroft's original (1925) classification may now be modified as suggested by Brierley (1977) into the following main types:-

Ischaemic/oligaemic anoxia

Here the initiation of anoxia depends on reduction or arrest of blood flow in the brain as a whole or in the territory of a single cerebral vessel. A variety of disorders of the cardiovasculature may contribute, and the picture may be dominated by disease of the cerebral vessels themselves.

Anoxic/hypoxic anoxia

Here the initiating phenomenon is a reduction of the arterial PO_2 to zero (anoxia) or to vastly reduced levels (hypoxia) - and the primary phenomena concerned are respiratory obstruction or disease, inert gas inhalation, and acute decompression as for example in aircraft accidents.

Anaemic anoxia

Here the problem centres upon reduction of the oxygen carrying capacity of the blood - it may occasionally occur in severe haemorrhage or in anaemia though this must be rare these days, but its commonest type in clinical practice is carbon-monoxide poisoning where the haemoglobin molecule itself becomes incapable of transporting oxygen.

Histotoxic anoxia

Here the oxidative mechanisms at cellular level are poisoned and the classical example is cyanide intoxication.

Hypoglycaemia

Strangely, it has come to be realised that this belongs most properly in the group of anoxic problems in the nervous system although the deficiency is of the substrate (glucose) and the oxidative mechanisms are disturbed at cellular level secondary to this lack.

This pure classification of anoxic types is necessary for an adequate scientific understanding of the problem, but it is helpful I think to realise that from the clinical point of view clinical anoxic brain damage may be codified in much simpler terms. Indeed one may regard anoxia from the clinical point of view a diffuse, generalised phenomenon, or as a predominantly focal abnormality.

Table 1 characterises the common clinical phenomena forming these two groups and it is upon these two clinical varities of anoxic damage to the brain that I will concentrate in the course of this discussion.

THE CLINICAL FEATURES OF DIFFUSE HYPOXIC DAMAGE TO THE NERVOUS SYSTEM

Table 1 specifies the circumstances in which diffuse hypoxic damage to the nervous system may occur. The dominant circumstances in clinical practice are either diminution in the oxygen supply from embarrassment of respiration such as occurs in anaesthetic accidents, drowning, gassing or acute decompression at high altitude, or circulatory insufficiency such as occurs in cardiac arrest during surgery, diffuse microvascular embolisation as in air embolism or fat embolism following trauma, and possibly the circumstances of caisson disease or decompression sickness in which microvascular embolisation by released nitrogen bubbles is believed to be a major dominant pathological cause. There is a community of influence of all these various aetiological factors on the central nervous system and that is generally a reduction in cerebral perfusion often on a basis of central circulatory failure.

Table 1

Diffuse or generalised

Lack of oxygen	-	anaesthetic accidents
	-	drowning
	-	gassing
	-	decompression
Circulatory insufficiency	-	cardiac arrest
	-	air embolism
	-	decompression sickness
	-	severe brain injury e.g. brain swelling

Focal

Occlusive vascular disease	-	Embolic
	-	Thrombolic
	-	Vasospastic
	-	Haemorrhagic

In all these types of diffuse anoxic damage to the nervous system the distribution of neuro-pathological damage is in the boundary zone between cerebral vessels. The control of the cerebral circulation is characterised by the phenomenon of auto-regulation that is the capacity of each cerebral vessel to respond by vaso-dilatation to falls in perfusion pressure, thus maintaining tissue blood flow in the face of changing perfusion pressure. Generalised falls in systemic perfusion however result in maximal vaso-dilatation and the circulation then becomes pressure-passive. We find that impairment of perfusion occurs clasically not in the centre of arterial fields, but as Zulch and Behrend (1961), were first to point out, the last field or border zone where the territories of perfusing arteries met. Classically this occurs in the boundary zones between the middle cerebral, posterior cerebral and anterior cerebral vessels along the fronto-parietal cortex, and in the deep surface of the temporal lobe including the hippocampus, in the corpus striatem, and in the boundary zone between the two major supplying arteries of the cerebellar convexity.

The clinical feature therefore, depends on the intensity with which the circulation is impaired and the mode of development of symptoms depends more on the secondary sequel induced by the hypoxic brain damage than on the distribution of the hypoxic damage itself. All these diffuse phenomena are characterised by damage to the permeability characteristics of the nervous system either at vascular or cellular membrane level, and the production of secondary brain swelling which I will deal with in greater detail when we come to discuss focal cerebral ischaemia.

The clinical picture of severe cerebral swelling

The key to the understanding of the clinical features of severe cerebral swelling in the anatomy of the tentorial hiatus. We may regard the brain as somewhat similar to a gel, although it is not a gel, it is traversed by numerous elastic structures and there is a fine fibre network within it, but it will respond to deformation by plastic creep, and the tentorial hiatus is a focus upon which this deformation concentrates. It has long been known clinically that the development of cerebral swelling results in herniation of the uncinate portion of the temporal lobes through the tentorial hiatus with deformation and compression of the upper brain stem and direct compression of one or both ocular-motor nerves which pass across the edge of the tentorium towards the back of the cavernous sinus from their origin in the interpeduncular fissure. The diencephalon and upper midbrain can be crudely regarded in clinical terms as the seat of consciousness, and their compression results in impairment of consciousness. Associated with this is a variable degree of loss of influence of cerebral function on the brain stem control of motor function, characterised by the gradual development of isolated brain stem responses originally described by Magnus.

Associated with the development of this picture is loss of pupillary reactivity, the pupils become fixed and dilated, and the loss of the oculo-cephalic and oculo-vestibular reflexes which are commonly present in states of unconsciousness unaccompanied by severe midbrain compression.

Abundant clinical evidence has been adduced particularly in the management of head injuries, to show that the development of decerebration and loss of the brain stem reflexes are uniformly disastrous prognostic signs. Once again I must emphasize to you however, that these are not by themselves the product of hypoxic, anoxic or ischaemic brain damage, but in general the product of the extensive brain swelling which follows such cerebral insult.

PHENOMENA ASSOCIATED WITH FOCAL ISCHAEMIA IN THE CENTRAL NERVOUS SYSTEM

Occlusive vascular disease is the occasion of focal brain ischaemia. Vascular disease may either be embolic in which case it is usually associated with disintegrating vegetations or cardiac valves, thrombi on myocardial ischaemic areas, micro-emboli associated with such recently described phenomena as mitral leaflet prolapse or more directly concerned with cerebral vessels and emanating from disintegrating plaques at the bifurcation of the terminal carotid artery or elsewhere in the afferent cerebral vascular tree. It may be primarily thrombotic and occasioned by development of coagulation of blood elements on top of an athero-sclerotic vascular plaque in one of the major cerebral arteries or their branches and perhaps in small vessels, associated with the focal micro-vascular dilatations originally described by Charcot and Bouchard and beautifully demonstrated more recently by Ross Russell, occasioning thrombosis and sometimes haemorrhage in vessels in the deep nuclei and characteristically occurring in hypertension. It may be due to a more dynamic obstruction occasioned by the curious narrowing of cerebral vessels associated with disordered vascular re-activity which is seen in subarachnoid haemorrhage. Whatever the primary nature of the vascular obstruction, the effect is the same, blood supply to the area supplied by the vessel in question is reduced, and a number of pathophysiological phenomena are set in train both in the area of ischaemia and in the surrounding regions where an attempt is made to replace the lost blood flow from collateral vessels. It is an everyday clinical experience that a dense neurological deficit soon after a cerebral ischaemic episode may gradually resolve and finally even disappear. There has long been intrigued clinical speculation as to whether such resolution of neurological deficit could relate to the recovery of neurones which, having been so damaged following the ischaemic episode as to cease functioning, yet subsequently recovered, or whether this recovery of function represents the assumption by associated areas outwith the zone of the immediate ischaemia, of some of the activities of the irreversibly damaged cells. It now seems more likely that cells may survive in a state of structural integrity yet with paralysis of function. Possible explanations for subsequent recovery may either be improvement in the residual circulation with expansion of collateral vessels from neighbouring cerebro-vascular beds, or modification of the neuronal metabolism itself so that function may be resumed at a lower basal level of blood flow.

The experimental model

We have described a stroke model (Symon et al. 1971, Symon 1975) in which the middle cerebral artery is occluded either by an intracranial approach along the sphenoidal wing, or by a transorbital approach after removal of the contents of the orbit. The vessel may be occluded temporarily by a light spring clip, but if a prolonged study of ischaemia is required, the artery is divided between two small haemostatic clips. The model permits variation in that division of the artery in the first millimetre of its course spares the perforating vessels as Shellshear (1921) and Abbie (1934) have indicated. One may increase the intensity of ischaemia in the basal ganglia by placing the peripheral clip further laterally, thus partially occluding the blood supply of the basal ganglia.

Perfusion studies following acute middle cerebral occlusion Symon (1961), and studies of the brains of animals which have been allowed to survive up to three years following middle cerebral occlusion Brierley and Symon (1977) indicate that the regional distribution of the middle cerebral artery in the experimental primate is very similar to the descriptions of Shellshear and Abbie, in man.

Experimental occlusion of the middle cerebral artery in a baboon, produces a clinically recognisable stroke with facial weakness, a semi-flexed resting position of the elbow wrist and fingers, and impairment of joint position sense in the arm which is obvious in placing reactions but which varies considerably in extent from animal to animal. The rapidity of recovery of elbow and finger movements is much greater in animals where perforating vessels have been spared and in fact the deficit may vary from virtually undetectable weakness to a quite evident stroke. The infarct involves the basal ganglia, the lips of the sylvian fissure (opercular and insular) while the cortex of the lateral aspect of the hemisphere apart from the immediate area of the temporal pole, appears normal.

The relationship between blood flow and function in an experimental stroke model

Throughout our correlations of clinical and experimental ischaemia, we have used methods of determination of brain blood flow which employ the same basic principle, the clearance of inert gas from saturated tissue. In man, the technique has been the use of Xenon 133 with external collimation, as described by Lassen and Ingvar (1961) and in animals, the technique has been the study of clearance of hydrogen gas Pasztor et al. (1973), where v CBF may be assessed over very small areas of brain with platinum electrodes of a diameter of 300 . or less, the area of response being within a radius of 2 millimetres, whereas even with the best external collimation available, inert radioactive gas sampling technique involves very much larger areas impractical in the small primate brain. With middle cerebral occlusion, there is a contour of reduction of blood flow as indicated in Figure 1. the ischaemia being densest in the region of the sylvian opercula and in the basal ganglia where flow is reduced to about 20% of the basal control flow of around 55 mls. per 100 grams per minute, i.e. to 10-12 mls per 100 grams per minute. There is a graduated reduction in flow decreasing in intensity as the midline and parasagittal areas are approached. During acute occlusion, tissue blood flow determination shows no actual increase in flow in collateral zones, but such may be detected (Symon, Ishikawa and Keyer 1963) in the pia assessing flow directly from collateral vessels, and it is clear that the production of a middle cerebral stroke evokes rapidly an increase in flow in collateral vessels from anterior and posterior cerebral circulation.

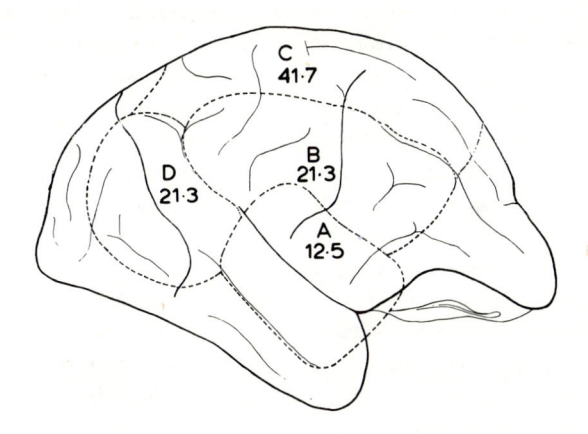

Figure 1.   Averaged regional blood flow (ml/100Gm/min) in baboon cortex within 1 hour of middle cerebral artery occlusion.

The relation between reduced blood flow determined in this way and the electrical function of cortical neurones can be assessed by the reaction of a Somato-sensory evoked response recorded by a plate electrode on the post Rolandic Cortex. Stimulation is applied to the contra-lateral mandibular division of the 5th nerve, and the relationship of cortical electrical activity to local perfusion can be determined since the relay's immediately preceeding the cortex, in the ventro-lateral nucleus of the thalamus, lie, in primates, in an area supplied by choroidal vessels and is unaffected by middle cerebral occlusion. The cortical evoked response normally lies in an area of intermediate ischaemia and the density of ischaemia in this intermediate zone can be varied at will by the induction of systemic hypotension, since autoregulation is lost following middle cerebral occlusion (Symon et al. 1976). We have demonstrated (1974) a threshold relationship between regional blood flow and the somato-sensory evoked response, the response being maintained to levels of blood flow of 20 mls. per 100 grams per minute, while at levels below 12 mls per 100 grams per minute the response is absent. In the area of 14 - 16 mls. per 100 grams per minute there is a very sharp decline in the evoked response, the 50% reduction of response being at 16 mls. per 100 grams per minute. Similar findings have been reported by Boysen et al. (1973) in Scandinavia, looking at the relationship between EEG frequency and carotid perfusion during endarterectomy. Heiss (1977) and his colleagues have also found a similar relationship between r CBF and neuronal activity in the cat.

The relationship between blood flow and structural integrity

In many animals the somato-sensory evoked response will be abolished by middle cerebral occlusion without added hypotension. Figure 2 shows the approximate area of what we might term functional suppression in such a hemisphere. It is particularly interesting to note that this area of cortex is very much wider than the area of cortex which will proceed to infarction in chronic experiments, a group of comparable animals having been maintained over three years while the characteristics of their clinical stroke was studied. Detailed pathological studies after perfusion fixation have been performed Brierley and Symon (1977) and indicate that the ultimate area of infarction is confined to areas where blood flow reduction in the acute stage of the stroke is certainly below values of 10 mls. per 100 grams per minute, indicating that in the acute stage of infarction, loss of function will affect neurones in a much wider distribution than the ultimate structural loss. In accordance with this view, Morawetz et al. (1978) found that histopathological signs of structural infarction following a 2 to 3 hour period of focal ischaemia in the monkey were only obtained at sites where local blood flow was below 10 - 12 mls. per 100 grams per minute.

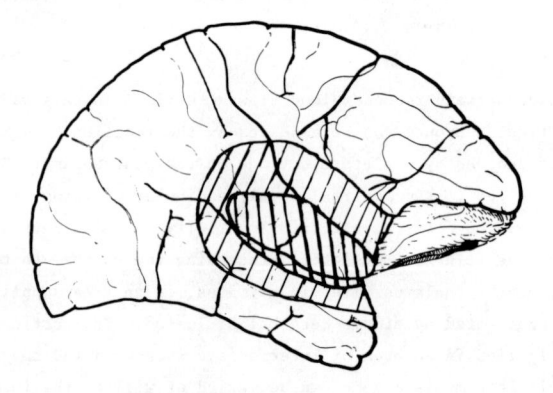

Figure 2   Schematic representation of the "ischaemic penumbra" following acute middle cerebral artery occlusion in the baboon. The inner hatched area represents brain which may infarct, the outer, very approximately, the area of functional loss which will remain structurally intact.

From these observations, we developed a concept that an area of structural loss in stroke was probably surrounded for sometime in the acute phase, by an area of functional neuronal suppression, in which the structural integrity of the neurones was immediately, and even permanently, preserved. This we termed the "ischaemic penumbra".

We now have some evidence that in the penumbral areas, certain basic physiological mechanisms of the neurone remain intact, particularly concerned with ionic homoeostasis at the cell membrane level. The development of potassium sensitive micro-electrodes by Walker (1971) and their modification by Astrup (1977) and ourselves (1977) have enabled us to determine movements of extracellular space potassium during ischaemia and to relate regional cerebral blood flow changes to the development of failure of the ionic pump and the flux of potassium into the extracellular space. Potassium has long been known to accumulate in the extra-cellular space during hypoxia but in our experiments, control levels of extra-cellular potassium ranging from 3 - 9 millimoles (mean 5.7 $\pm$ 1.5) were maintained with only minor changes at about the level of the threshold for electrical function, significant and massive movements of potassium ion into the extracellular space occurring only when flow fell to between 7 and 11 mls. per 100 grams per minute, a level significantly lower than the levels for failure of electrical function. The assumption must be therefore, that there is a differential failure of neuronal metabolic processes so that in the penumbral area, synaptic transmission is impaired but the energy state and the ionic balance are maintained at normal levels. Presumably the metabolic rate of oxygen is reduced to a minimal level by the arrest of function. Some support for this view has come from studies of Siesjo and his group in bicuculline induced status epilepticus, where it has been found that during progressive ischaemia and failure of seizural activity, the tissue content of ATP is maintained at about normal levels until the point at which potassium is massively released from the cells.

Evidence of the relationship between neuro-transmitter function and ischaemia

With Professor Alan Davison and Dr. David Bowen, (Bowen et al. 1976) we attempted to link the degree of ischaemia to the function of subcellular components particularly neuro-transmitter uptake by synaptosomes. The evidence so far suggests that the synaptosound uptake of the neuro-transmittor GABA shows a linear decline in relation to flow which probably commences at a level of flow higher than that required to interfere with electrical function. Cholinergic neurones however, show no such sensitivity, uptake remaining unimpaired to very much lower levels.

The development of oedema

Work from our own laboratory Symon et al. (in press) and from Hossmann and Shuier (1978) have indicated a definite relationship between the development of cerebral swelling and the intensity of brain ischaemia. In our experimental model, we assessed brain water content in life by brain impedance measurements and intracranial pressure changes, and following the sacrifice of the animal by the use of graded density kerosene bromobenzine columns.

We find that significant ischaemia is associated at 1½ hours, with an increase in water content in the most densely ischaemic zones and in the area of the penumbra; a significant relationship between the increase in water content and blood flow is evident. Movement of water occurs when flow falls below 20 mls. per 100 grams per minute and it appears to be the initial depth of ischaemia which acts as a trigger mechanism releasing water which thereafter advances through the hemisphere as described for cold oedema fronts by Klatzo and Reulen. Experiments in which ischaemic areas have been re-perfused indicate that after 1½ hours of ischaemia, re-perfusion is associated with an increase rather than a decrease in ischaemic oedema. Restoration of blood flow after a significant period of ischaemia may thus compound the problem of brain swelling. The evident levels of flow necessary to evoke brain swelling are significantly higher than those concerned with evident impairment of cell membrane permeability. Impairment of vascular permeability to large molecules is unlikely to be the cause, since there is clear evidence that such vasogenic brain oedema generally develops only after 4 - 6 hours and reaches its maximum after a few days. Leakage of large molecules such as Technetium or R.I.H.S.A. is maximum in an ischaemic stroke only after ten days to three weeks. O'Brien (1974) and his associates have indicated an increase in water content of cortex a few days after an experimental stroke of up to 90%, and it seems probable that such oedema is an accompaniment to tissue necrosis since it does not seem to be evoked by vascular occlusion of less than 6 hours duration, which from the work of Morawetz would be expected to produce significant tissue loss.

The early phase of brain oedema we must therefore regard as a cytotoxic rather than vasogenic. Our observations and those of Hossmann indicate that the degree of this type of oedema is determined by the initial flow reduction after vascular occlusion. Hossmann has also shown that following one hour of complete ischaemia, there is a significant increase in tissue osmolality, from 308 to 353 milliosmoles, creating a gradient of about 50 milliosomoles between brain and blood Hossmann and Takagi (1976). It may well be that the correspondence of the level of ischaemia necessary to evoke brain swelling, and the approximate level necessary to abolish brain function, may indicate an increase in osmolality associated with failure of synthesis of large neuro-transmitter molecules, their smaller percursors being allowed to remain free, significantly increasing osmolality as a result.

Possible approaches to the management of the consequences of hypoxia
in the nervous system

It will be clear that throughout this review the evidence indicates that significant ischaemia in the brain will result in an irreversible loss of function in certain finite areas of tissue, surrounded by a penumbral zone in which though tissue function is lost, tissue structure is maintained. This prompts a search for methods of re-perfusion though under certain circumstances this may be dangerous. We may consider a clinical case in point. A patient, a man of 53 years old had a giant right posterior communicating aneurysm occasioning sub-arachnoid haemorrhage. Operation was technically difficult and involves dissection of the anterior choroidal artery from the fundus of the aneurysm and temporary occlusion of the middle cerebral and distal carotid arteries for four minutes. CBF on the first post operative day was 38 mls. per 100 grams per minute and mild pyramidal signs were evident on the left. On the third day he became drowsy and developed a frank left hemiparesis and further lowering of CBF. CT Scan showed a right capsular low density area. CBF was raised by elevation of the systemic blood pressure using intravenous metaraminol, and for two days the focal neurological deficit could be titrated against the systemic blood pressure, as the blood pressure fell the paresis returned, but thereafter the Aramine could be discontinued and the patient went on to make a full recovery, accompanied by disappearance of the right capsular low density area. It is clear therefore that this patient suffered from pre-infarct associated with low cerebral blood flow, but that absolute loss of structure was nowhere present since increasing the blood flow by systemic hypertension in the presence of presumed failure of autoregulation in the ischaemic region, resulted in a sufficient increase in flow to resolve the deficit and convert the penumbral zone back to normality. Whether in fact a very small area of structural loss could be present in such a case it would be impossible to say. It is clear, however, that induced hypertension could be harmful if vasogenic brain oedema were already present and the detailed relationship between re-perfusion increase in oedema and the time necessary to evoke such a potentially dangerous complication, has yet to be further investigated.

A further approach to the solution of brain ischaemia would be to reduce the metabolic demand for oxygen of the brain to such a low level that brain structure could be maintained in the face of vastly lower blood flow. Such a suggestion was put forward by Michenfelder and his group (1976) in the use of deep barbiturate anaesthesia. They suggested that under deep barbiturate anaesthesia, animals would survive with ischaemic lesions which would otherwise prove fatal. Flamm and his co-workers have indicated that a free radical scavenging effect of barbiruates might explain their beneficial effect in reducing ischaemic brain lesions, since free radical chain reactions may have a deleterious effect on a variety of cell membranes.

Flamm (1977) showed that ascorbic acid and unsaturated fatty acids decreased in the area of an experimental stroke, and that this change could be reduced by barbiturates. Ascorbic acid is a naturally occuring antioxidant which decreased in the presence of free radical reactions. Fatty acids in turn can be lost due to paraoxidation in membrane lipids Strosznajder and Dabrowiecki (1975). The free radicals are believed to cause paraoxidation of unsaturated fatty acids and thus may damage the membranes of mitochondria which in turn would adversely effect the function of the ischaemic neurone and might lead to brain oedema. In our own experiments in ischaemic baboon cortex, abolition of the evoked response occurred at identical flow levels with and without barbiturate, although a significant slowing of the rate of change in relation to a finite degree of ischaemia could be demonstrated. Other metabolic inhibitors could well be investigated in this respect.

The clinician viewing the field of cerebral ischaemia, particularly from a focal aspect, which after all constitutes the major portion of ischaemic brain disease, is driven therefore to ask his biochemical colleagues where the keys might be to the protection of already suggested critical ischaemic areas. How could one prevent a potential build up of small molecules associated with a failure of neuro-transmitter generation? How could one prevent accumulation of free radicals presumably in the penumbral state - leading on to cell membrane and mitochondrial membrane damage? How could one more effectively ensure dispersal of brain oedema in the cytotoxic phase and finally in what way might it be possible to preserve the permeability characteristics of vascular membranes to large molecules which being lost, lead to the complicating and lethal vasogenic phase of brain oedema?

References

Abbie, A.A. (1934) The morphology of the forebrain arteries with special reference to the evolution of the basal ganglia. Journal of Anatomy, 68: 433-470.

Astrup, J., Symon, L., Branston, N.M. and Lassen, N. (1977) Cortical evoked potential and extra cellular K+ and H+ at critical levels of brain ischaemia. Stroke 8: 51-57.

Barcroft, J. (1925) The respiratory function of the blood Part I: Lessons from high altitudes. Cambridge University Press, Cambridge.

Bowen, M.D., Goodhard, N.J., Strong, A.J., Smith, S.B., White, B., Branston, N.M. Symon, L. and Davison, A.N. (1976) Biochemical indices of brain structure, function, and hypoxia in cortex from baboons with middle cerebral artery occlusion. Brain Research, 117: 503-507.

Boysen, G., Engell, H.C. and Trojaborg, W. (1975) Effect of mechanical r CBF reduction on EEG in man. In: Cerebral Circulation and Metabolism. Ed. Langfitt, T.W., McHenry, L.C. Jr., Reivich, M. and Wollman, H. Springer Verlag. 378-379.

Branston, N.M., Symon, L., Crockard, H.A. and Pasztor, E. (1974) Relationship between the cortical evoked potential blood flow and local cortical floow flow following acute middle cerebral artery occlusion in the baboon Experimental Neurology, 45: 195-208.

Branston, N.M., Strong, A.J. and Symon, L. (1977) Extracellular potassium activity evoked potential and tissue blood flow; relationships during progressive ischaemia in baboon cerebral cortex. J. Neurol. Sci., 532: 305-321.

Brierley, J.B. and Symon, L. (1977a) The extent of infarcts in baboon brains three years after division of the middle cerebral artery. J. Neuropath. and Appl. Neurobiol. 3: 217-218.

Brierley, J.B. (1977b) Experimental hypoxic brain damage. J. Clin. Path., 30 Suppl. (Roy. Coll. Path.) 11: 181-187.

Charcot, J.M. and Bouchard, C. (1868) Nouvelles Recherches sur le pathogenie de l'hemorrhagie cerebral. Archive de Physiologie Normale et Pathologique 110: 643-725.

Heiss, W.D., Hayakawa, T. and Waltz, A.G. (1976) Cortical neuronal function during ischaemia. Effects of occlusion of one middle cerebral artery on single unit activity in cats. Arch. Neurol. 33: 813-820.

Flamm, E.S., Demopoulos, H.B., Seligman, M.L. and Ransohoff, J. (1977) Possible molecular mechanisms of barbiturate mediated protection in regional cerebral ischaemia. Acta. Neurol. Scand. 56, Suppl. 64: 150-151.

Hossman, K.A. and Takagi, S. (1976) Osmolality of brain in cerebral ischaemia. Exp. Neurology 51: 124-131.

Hossman, K.A. and Schuier, F.J. (1978) The metabolic (cytotoxic) type of brain oedema following middle cerebral artery occlusion in cats. Proc. of the Princeton Conference 1978) (In Press).

Lassen, N.A. and Ingvar, D.H. (1971) The blood flow of the cerebral cortex determined by radioactive Krypton 85. Experientia (Basel) 17: 42-43.

Magnus, R. (1924) 'Korperstellung'. Springer.

Michenfelder, J.D., Milde, J.H. and Stundt, E.M. (1976) Cerebral protection of barbiturate anaesthesia. Use after middle cerebral artery occlusion in Java monkeys. Arch. Neurol. 33: 345-350.

Morawetz, R.B., de Girolami, Ojemann, R.G. Marcoux, F.W. and Crowell, R.M. (1978) Cerebral blood flow determined by hydrogen clearance during middle artery occlusion in unanaesthetised monkeys. Stroke, 9: 143-149.

O'Brien, M.D., Waltz, A.G. and Jordan, M.M. (1974) Ischaemic cerebral oedema. Distribution of water in brains in cats after occlusion of the middle cerebral artery. Arch. Neurol. 30: 456-460.

Pasztor, E., Symon, L., Dorsch, N.W.C. and Branston, N.M. (1973) The hydrogen clearance method in assessment of blood flow in cortex white matter and deep nuclei of baboons. Stroke 4: 556-557.

Russell, R.W.R. (1963) Observations on intracerebral aneurysms. Brain 86: 425.

Shellshear, J.L. (1921) The basal arteries of the forebrain and their functional significance. Journal of Anatomy, 55: 27-35.

Siesjo, (1978) Personal communication.

Strosznajder, J. and Dabrowiecki, Z. (1975) Peroxidation of fatty acids in microsomes prepared from normal and ischaemic guinea pig brains. Bull Acad. Sci. (Biol.) 23: 647-653.

Symon, L. (1961) Studies of leptomeningeal collateral circulation in macacus rhesus. Journal of Physiology, 159: 68-86.

Symon, L., Ishikawa, S. and Meyer, J.S. (1963) Cerebral arterial pressure changes and development of leptomeningeal collateral circulation. Neurology, 13: 237-250.

Symon, L., Khodadad, M.D. and Montoya, G. (1971) The effect of carbon dioxide inhalation on the pattern of gaseous metabolism in ischaemic zones of the primate cortex. An experimental study of the 'intracerebral steal' phenomen in baboons. Journal of Neurology, Neurosurgery and Psychiatry, 34: 481-486.

Symon, L. (1975) Experimental model of stroke in the baboon. Advances in Neurology, 10: 199-212.

Symon, L., Branston, N.M. and Strong, A.J. (1976) Autoregulation in acute focal ischaemia. Stroke 7: 547-554.

Symon, L. and Branston, N.M. (1978) Water movement in Brain ischaemia. Proc. 4th Cologne Symposium (In Press).

Walker, J.L. Jr. (1971) Ion specific liquid ion exchanger mocro electrodes. Analytical chemistry. 43: 83a-93a.

Zulch, K.J. and Behrend, R.C.H. (1961) The pathogenesis and topography of anoxia, hypoxia and ischaemia of the brain in man. In Cerebral Anoxia and the Electrocephalogram, edited by H. Gastaut and J.S. Meyer. 144-163. Thomas Springfield, Illinois.

BRAIN ENERGY METABOLISM AND CIRCULATION IN HYPOXIA

L. Berntman and B.K. Siesjö
Laboratory of Experimental Brain Research, E-blocket, and the Department of
Anaesthesia, University of Lund, Lund, Sweden

Even moderate degrees of hypoxia affect brain function and, when severe, cerebral hypoxia leads to gross functional derangement, and to neuronal death. In spite of considerable effort, the mechanisms that cause cerebral dysfunction and cell injury have been only partially clarified. In this review, we will summarize current information on the effects of hypoxia on cerebral energy metabolism and blood flow, emphasizing three main issues. First, we will recall the evidence indicating that neuronal dysfunction in moderate hypoxia shows a poor relationship to energy failure. Second, we will be concerned with the influence of hypoxia on cellular redox state and on carbohydrate and amino acid metabolism. Third, we will pose the question whether maintainance of cerebral energy state in hypoxia is solely due to the increase in cerebral blood flow, or if decreased energy use rate contributes. In that context, we will discuss mechanisms responsible for the increase in cerebral blood flow (CBF). The discussion will be confined to situations in which tissue hypoxia is due to a lowering of arterial oxygen content (C_{O_2}), secondary to a fall in oxygen tension ("hypoxic hypoxia") or a reduction in hemoglobin concentraion ("anemic hypoxia"). In many respects, cerebral metabolic changes of a similar nature accompany cerebral ischemia but, in that condition, interpretation of results are often complicated by the inhomogenous nature of changes in flow and metabolism.

A. Cerebral energy state

In man, the acuity of the dark-adapted eye and short-term memory are affected at arterial P_{O_2} values exceeding 50 mm Hg. With more severe degrees of hypoxia (at P_{O_2} values of about 35-40 mm Hg) symptoms include dimming of vision, lightheadedness and nausea. Finally, at arterial P_{O_2} values of about 30 mm Hg, consciousness is lost (see Luft 1965, Cohen et al. 1967). Small animals like rats and mice are more resistant since they usually show preserved consciousness when their arterial P_{O_2} falls to about 20 mm Hg during exposure to 5% O_2 (see Duffy et al. 1972, Lewis et al. 1973). Many studies of metabolic and circulatory effects of hypoxia have been conducted in anaesthetized and artificially ventilated animals. In these, functional changes must be evaluated from the EEG. A clear slowing of EEG frequences is observed at Pa_{O_2} values (about 25 mm Hg) exceeding those leading to unconsciousness in unanaesthetized and spontaneously breathing animals. Presumably, cessation of spontaneous EEG activity in the former roughly corresponds to loss of consciousness in the latter.

In the paralyzed rat, reduction in Pa_{O_2} to a level of about 25 mm Hg (equivalent to inhalation of about 6% O_2) gives rise to dissociated changes in labile metabolites.

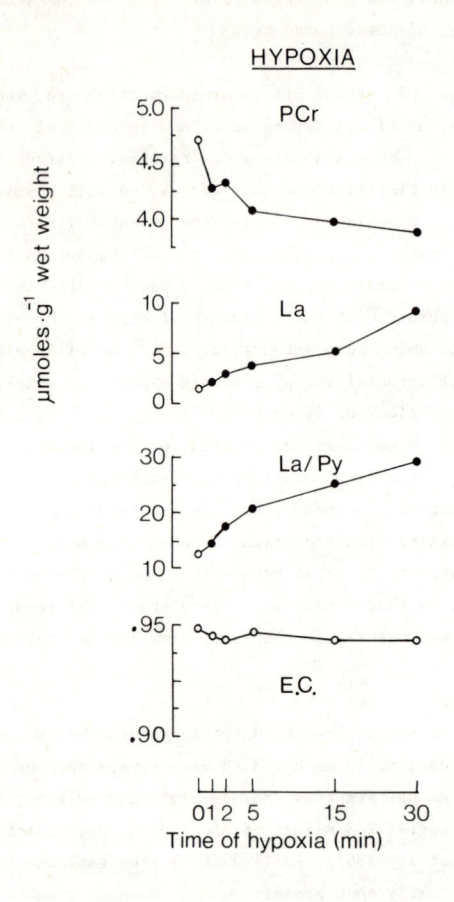

Fig. 1. Cortical concentrations ($\mu mol \cdot g^{-1}$) of phosphocreatine (PCr), lactate (La), and lactate-pyruvate ratio, as well as the energy charge of the adenine nucleotide pool (EC) (Atkinson, Biochemistry 7: 4030-4034, 1968) after 1 to 30 min of hypoxia (Pa_{O_2} 25 mm Hg). Values are taken from Norberg and Siesjö (1975a). Filled circles denote statistical significant differences from normoxic control animals ($p<0.05$).

HYPOXIA

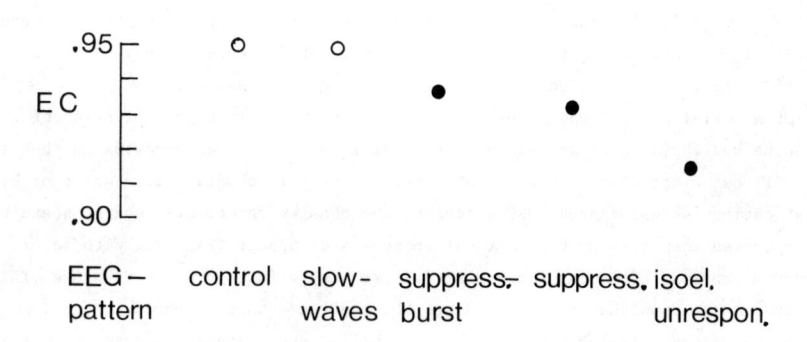

EC
.95
.90

EEG– pattern: control | slow-waves | suppress.- burst | suppress. | isoel. unrespon.

Fig. 2. Relationship between deranged EEG, patterns induced by hypoxia and the adenylate energy charge (EC) in the cerebral cortex. Filled circles indicate statistical significant changes from normoxic control values ($p<0.05$).

As fig. 1 demonstrates there is a fall in PCr concentration and rises in lactate concentration and the lactate/pyruvate ratio but adenine nucleotide concentrations are strikingly unaltered (Siesjö and Nilsson 1971, Duffy et al. 1972, MacMillan and Siesjö 1972, Bachelard et al. 1974). Current information suggests that there is a moderate perturbation of cerebral energy state only during the first minute of hypoxia, presumably at a time when the circulatory response has not yet become maximal (Norberg and Siesjö 1975a). Later, energy state seems well maintained. Thus, the small increase in ADP concentration that is occasionally observed could well be due to a change in the apparent equilibrium constant of the adenylate kinase reaction, and the fall in PCr concentration should largely reflect an influence of pH on the creatine kinase equilibrium (e.g. Norberg and Siesjö 1975a). In summary, if one allows for an initial and transient maladjustment of ATP production to the energetic needs of the cells changes in cerebral energy metabolism are largely confined to the lactic acidosis, and the accompanying redox change (see below).

If arterial P_{O_2} is reduced below 20 mm Hg the phosphorylation potential of the adenine nucleotide pool is reduced, and additional lactate accumulates in the tissue. A small but clear change in adenylate energy charge is observed at degrees of hypoxia that extinguish spontaneous EEG activity, the changes in cerebral energy state being exaggerated when also evoked cortical responses disappear (Fig. 2). With severe hypoxia there is often cardiovascular failure with a fall in blood pressure. This, or indeed any complication that reduces the hyperemia that otherwise occurs, exaggerates the cellular hypoxia. Experimentally, such situations have been mimicked by inducing hypoxia in animals in which one carotid artery has been clamped (Salford et al. 1973a,b, Salford and Siesjö 1974, Levy et al. 1975). Since blood flow is maintained, albeit at a lower level than on the contralateral side, one observes a severely deranged cerebral energy state with excessive accumulation of lactic acid. From an energetic point of view such conditions form a transition to complete anoxia, e.g. as occurs during respiratory standstill. However, a severe fall in P_{O_2} usually leads to cardiovascular collapse and, although energy failure rapidly occurs, interruption of glucose supply usually limits the lactic acidosis to values determined by the preanoxic tissue stores of glucose and glycogen. To take two examples, complete ischemia is normally accompanied by increases in tissue lactate content to 12-15 $\mu mol \cdot g^{-1}$ (Ljunggren et al. 1974) whereas severe hypoxia with some remaining circulation can cause lactate concentration to rise above 35-40 $\mu mol \cdot g^{-1}$ (Salford and Siesjö 1974). In view of the fact that some tissue acidosis has been incriminated as one important mechanism of cell damage circulatory alterations should provide an important modulator of the cerebral metabolic response to hypoxia.

Qualitatively, cerebral metabolic and circulatory changes in anemic hypoxia are similar (Borgström et al. 1975, Jóhannsson and Siesjö 1975). However, for a comparable decrease in arterial oxygen content, metabolic changes are much less pronounced than in hypoxic hypoxia. Presumably, part of this difference is due to the fact that, anemia, by causing a decrease in blood viscosity, creates favourable conditions for the microcirculation. Another factor of possible importance is that, in anemia,

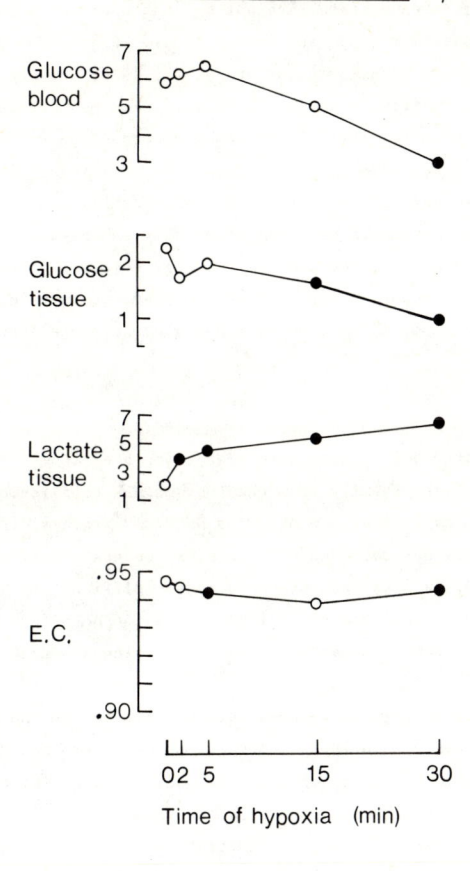

Fig. 3. Concentrations ($\mu mol \cdot g^{-1}$) of glucose in arterial blood and of glucose and lactate in cortical tissue, and the adenylate energy charge (E.C.) obtained after 2 to 30 min of hypoxia (Pa_{O_2} 25-30 mm Hg) in rats, starved for 24 hours.

P_{O_2} is not reduced on the inflow side of the microcirculation.

It has been shown both in man (Cohen et al. 1967), in the perfused dog's head (Drewes and Gilboe 1973a) and in the rat (Borgström et al. 1976) that moderate to marked hypoxia is accompanied by increased glucose consumption. The observations in man were carried out at semi-steady state, indicating that there is a sustained increase in (anaerobic) glucose flux (see also Drewes et al. 1973). However, results on mice (Duffy et al. 1972) and on rats (Bachelard et al. 1974, Norberg and Siesjö 1975a) show that lactate production is largest initially, and measurements of glucose consumption indicate that, after an initial increase, glucose flux may return towards control values (Borgström et al. 1976).

Recent results on starved animals indicate that, under semi-steady state conditions of hypoxia, glucose requirements are only moderately enhanced (Berntman and Siesjö 1978). Thus, in spite of the fact that arterial glucose concentrations fell to 3 $\mu mol \cdot g^{-1}$ after 30 min of hypoxia, cerebral energy state was upheld (Fig. 3) at values usually encountered in fed animals whose blood glucose concentrations increase during hypoxia.

B. Redox changes

In most studies, cellular redox changes have been calculated from the equilibrium of NAD-linked substrate couples, usually from the LDH and MDH equilibria. A disadvantage with such calculations is that, in deriving NADH/NAD$^+$ ratios, it is not only necessary to estimate lactate/pyruvate and/or malate/oxaloacetate ratios but also changes in intracellular pH. Such changes are notoriously difficult to quantitate, especially in non-steady state conditions. Indirect estimates on rats suggested that cytoplasmic NADH/NAD$^+$ ratios rose when arterial P_{O_2} fell below about 50 mm Hg (Siesjö et al. 1975).

In starved rats, derivation of redox changes were facilitated by the facts that there was only a moderate increase in lactate concentrations (Berntman and Siesjö 1978). Since P_{CO_2} fell, within 0.1 units of control, facilitating derivation of NADH/NAD$^+$ ratios. As table 1 shows, at a Pa_{O_2} of about 25 mm Hg, ratios derived from the lactate/pyruvate, malate/OAA, and β-hydroxybutyrate/acetoacetate couples all increased, indicating that both cytoplasmic and mitochondrial redox systems become reduced. Results on mitochondrial redox ratios corroborate those reported by Rosenthal et al. (1976) who used noninvasive optical techniques to monitor the redox state of cyt-a-a₃. In fact, these results suggest that there is a direct relationship between tissue P_{O_2} and the redox state of cyt-a-a₃, i.e. that even very moderate reductions in P_{O_2} are accompanied by mitochondrial redox changes. In this connection, it should be remarked that derivation of mitochondrial redox changes from the glutamate dehydrogenase system shows a paradoxical decrease in NADH/NAD$^+$ during hypoxia probably indicating that this system is unsuitable for estimating mitochondrial redox changes in the brain (Berntman and Siesjö 1978).

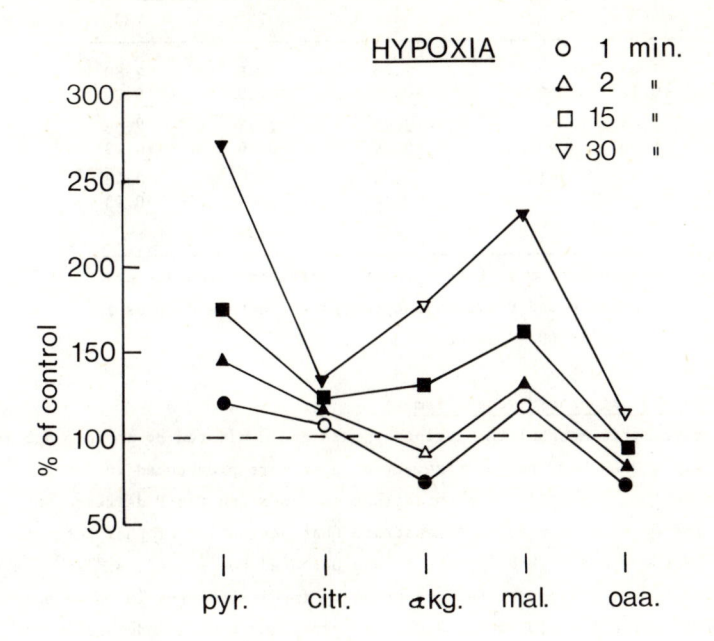

Fig. 4. Changes in pyruvate and some citric acid cycle intermediates in the rat brain after 1 to 30 min of hypoxia (Pa$_{O_2}$ 25 mm Hg), expressed as percent of normoxic control values. Statistical significant changes are indicated by filled symbols. Control values (μmol·g^{-1}): pyruvate 0.119 ±0.003, citrate 0.309 ±0.005, α-ketoglutarate 0.139 ±0.006, malate 0.341 ±0.010 and oxaloacetate 0.0056 ±0.0002. From Norberg and Siesjö (1975b).

Table 1.

NADH/NAD$^+$	Control	Hypoxia (min)			
		2	5	15	30
LDH·10^{-3}	1.45 ±0.12	2.16** ±0.09	2.46** ±0.19	2.46*** ±0.10	2.80** ±0.24
MDH·10^{-3}	1.44 ±0.04	2.53*** ±0.14	2.79*** ±0.09	2.75*** ±0.16	2.93*** ±0.12
GDH	1.42 ±0.11	1.26 ±0.16	1.04 ±0.14	0.79** ±0.04	0.43*** ±0.03
β HBDH	0.12	0.33	–	0.33	–

· NADH/NAD$^+$ ratios derived from lactate/pyruvate, malate/oxalocatate and glutamate/α-ketoglutarate and β-hydroxybutyrate/acetoacetate couples in animals starved for 24 or 48 (β HB/AcAc) hours.

C. Carbohydrate and amino acid metabolism

Changes in glycolytic metabolites during hypoxia are dominated by an increase in pyruvate and, since NADH/NAD$^+$ ratios rise, by an even more pronounced increase in lactate concentration. Initially, concentrations of G-6-P and F-6-P decrease and, since flux is increased, the results demonstrate that activation of phosphofructo- kinase occurs (Norberg and Siesjö 1975a). In the perfused dog's head, and at severe degrees of hypoxia, the triggering factors probably involve changes in the concentra- tions of adenine nucleotides (Drewes and Gilboe 1973b) but during moderately pro- nounced hypoxia in the rat, such changes are so small that the triggering mechanisms remain elusive.

In the initial phase of hypoxia changes in citric acid cycle intermediates are dominated by reductions in the concentrations of α-KG and OAA (Fig. 4). Later α-KG increases to normal or supranormal concentrations, and there is an increase in the citric acid cycle pool (Norberg and Siesjö 1975b). Amino acid changes include a rise in alanine and a fall in aspartate concentrations (Duffy et al. 1972, Norberg and Siesjö 1975b). At least in unanaesthetized animals and with prolonged hypoxia, GABA concentration increases (Wood et al. 1968, Duffy et al. 1972).

Most of the changes in citric acid cycle intermediates and associated amino acids can be explained by the redox change, and the increase in pyruvate concentration (Duffy et al. 1972, Norberg and Siesjö 1975b). Thus, a rise in malate/OAA ratio may, by reducing OAA concentration, give rise to a shift in the aspartate aminotransferase reaction, explaining the (initial) decreases in α-KG and aspartate concentrations. Furthermore, the increase in pyruvate concentration will favour anaplerotic reactions. One such reaction, catalyzed by alanine aminotransferase will replenish α-KG and give rise to accumulation of alanine. Presumably, there is also increased CO_2 fixa- tion by an acceleration of the reaction catalyzed by pyruvate carboxylase (cf. Mahan et al. 1975).

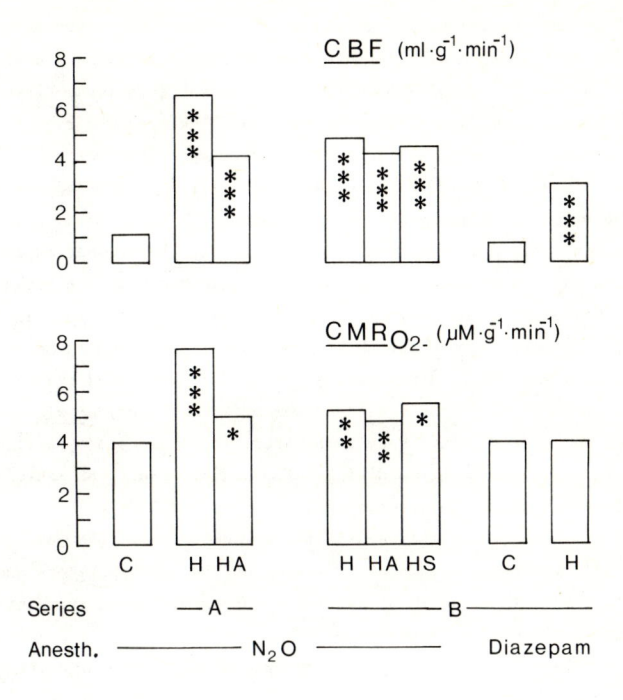

Fig. 5. Cerebral blood flow and oxygen consumption in hypoxic rats (Pa$_{O_2}$ 25-30 mm Hg). C = normoxic control animals, H = hypoxia, A = adrenalectomized rats, S = 24 hours of starvation.

D. Metabolic rate and blood flow

The results discussed demonstrate that relatively moderate degrees of hypoxia are accompanied by metabolic signs of an insufficient cellular supply to oxygen, mainly in terms of a redox change and an acceleration of anaerobic glycolysis. However, the data indicate that gross energy failure does not occur until oxygen availability is severely curtailed. The results lead to two pertinent questions. First, which are the molecular mechanisms underlying the functional changes observed at even mild degrees of hypoxia? Second, what mechanisms are responsible for maintaining energy homeostasis? Since the first of these questions will be discussed elsewhere in this symposium we will be concerned with the second.

Theoretically, brain tissues could compensate for the reduced oxygen availability by decreasing the energy demands. It has been suggested that one mechanisms for curtailing neuronal firing and metabolic rate is provided by the increased GABA levels (Wood 1967). More direct evidence that a purposeful reduction in metabolic rate may occur in hypoxia was provided by Duffy et al. (1972) who noted that cerebral energy use rate, as measured with the closed system method of Lowry et al. (1964), fell in mice exposed to 5-3% O_2. Since body temperature fell, and since it is not known if the method is valid under the conditions of hypoxia, results of measurements of oxygen consumption should be scrutinized.

Previous measurements in man have failed to demonstrate that moderate hypoxia in man (Pa_{O_2} 35-40 mm Hg) is accompanied by a reduction in CMR_{O_2} (Kety and Schmidt 1948, Cohen et al. 1967). Thus, at these degrees of hypoxia, oxygen availability (and consumption) is upheld by the increase in CBF.

Results obtained in rats during more severe degrees of hypoxia (Pa_{O_2} 22-25 mm Hg) were essentially similar, i.e. there was an increase in CBF with no significant change in CMR_{O_2} (Jóhannsson and Siesjö 1975). Since the CBF method used in these studies were not well suited to measure the excessive flow rates encountered (5-6 $ml \cdot g^{-1} \cdot min^{-1}$) the experiments were repeated, using an improved CBF technique (Berntman et al. 1978a). The results gave unexpected results, indicating that CMR_{O_2} may in fact rise during hypoxia and that the mechanisms involve "extrinsic" and "intrinsic" catecholamines (Berntman et al. 1978a). Fig. 5 shows that in one rat strain studied (series A), a reduction in Pa_{O_2} to 25-27 mm Hg gave rise to a 6- to 7-fold increase in CMR_{O_2} and to a rise in CMR_{O_2} to 180% of control. Since most (but not all) of this increase was prevented by prior removal of the adrenal glands we conclude that under certain circumstances circulating catecholamines penetrate the blood-brain barrier and provoke a marked increase in oxygen consumption. This conclusion is supported by previous results obtained in "immobilization stress" (Carlsson et al. 1975, 1977) and by results obtained following systemic or intracisternal administration of β-adrenoceptor agonists (MacKenzie et al. 1976a,b, Berntman et al. 1978b).

Fig. 5 also shows that in another rat strain (series B) the excessive increase in CMR_{O_2} did not occur, indicating that in these animals catecholamines were either not released in the same amounts or did not penetrate the blood-brain barrier. However, in all groups studied, whether adrenalectomized or not, CMR_{O2} increased by

20-30% (cf. also the corresponding increase in CMR_{O_2} in the adrenalectomized animals of series A). Animals starved for 24 hrs showed the same rise in oxygen consumption (Berntman and Siesjö 1978), indicating that glucose levels of about 1 $\mu mol \cdot g^{-1}$ in cortical tissue is sufficient to maintain the elevated metabolic rate. Suspecting that intrinsic catecholamines were involved, animals were given propranolol or diazepam prior to exposure to hypoxia. Due to cardiovascular complications the experiments with propranolol failed, but diazepam completely prevented the increase in CMR_{O_2} during hypoxia. In view of the fact that this drug has been reported to prevent a stress-induced increase in the activity of locus coeruleus neurons (Taylor and Laverty 1969, Corrodi et al. 1971, Lidbrink and Farnebo 1973) we tentatively conclude that in animals maintained on 70% N_2O the increase in CMR_{O_2} is related to increased activity in noradrenergic neurons. This conclusion is supported by the observation that a similar increase in CMR_{O_2} occurs during hypercapnia and that this is prevented both by diazepam and by propranolol (Berntman et al. 1978c).

Obviously, measurements of CMR_{O_2} fail to corroborate the assumption that cerebral metabolic rate falls at degrees of hypoxia that do not affect cerebral energy state. On the contrary, like other stressful situations hypoxia may cause activation of extrinsic and intrinsic catecholaminergic systems that tend to enhance metabolic rate. Current information suggests that a fall in metabolic rate only occurs when oxygen is virtually completely extracted from cerebral blood (Blennow et al. 1978). This occurs if Pa_{O_2} is reduced below 20 mm Hg, or if the rise in CBF is curtailed by a reduction in cerebral perfusion pressure.

The results quoted indicate that the main mechanism maintaining energy homoestasis in the hypoxic brain is the rise in CBF, but they do give no clue to the mechanisms regulating CBF. It has been commonly held that the decrease in extracellular pH (pH_e) is responsible (e.g. Lassen 1968). However, recent results show that CBF increases before pH_e falls, also that other putative vasodilating agents like extracellular potassium and adenosine cannot be made responsible for the vasodilatation (Nilsson et al. 1978). Results shown in fig. 5 demonstrate that even if the rise in CMR_{O_2} is prevented by diazepam there is an appreciable increase in CBF during hypoxia. Although other transmitters than catecholamines could contribute to the vasodilatation, it cannot be excluded that hypoxia directly affects metabolism and function in the resistance vessels themselves.

References

Bachelard, H.S., Lewis, L.D., Pontén, U. and Siesjö, B.K. (1974) Mechanisms activating glycolysis in the brain in arterial hypoxia, J. Neurochem. 22: 395-401.

Berntman, L. and Siesjö, B.K. (1978) Cerebral metabolic and circulatory changes induced by hypoxia in starved rats, J. Neurochem. In press.

Berntman, L., Carlsson, C. and Siesjö, B.K. (1978a) Cerebral oxygen consumption and blood flow in hypoxia: Influence of sympathoadrenal activation. Submitted to Stroke.

Berntman, L., Dahlgren, N. and Siesjö, B.K. (1978b) Influence of intravenously administered catecholamines on cerebral oxygen consumption and blood flow in the rat, Acta physiol. scand. In press.

Berntman, L., Dahlgren, N. and Siesjö, B.K. (1978c) Influence of intravenously administered catecholamines on cerebral oxygen consumption and blood flow in the rat, Acta physiol. scand. In press.

Berntman, L., Dahlgren, N. and Siesjö, B.K. (1978d) Influence of extreme hypercapnia on cerebral blood flow and oxygen consumption in the rat brain, Anesthesiology. Accepted for publication.

Blennow, G., Nilsson, B. and Siesjö, B.K. (1978) Cerebral functional and metabolic changes provoked by hypoxia during bicuculline-induced seizures in rats, Ann. Neurol. Submitted.

Borgström, L., Jóhannsson, H. and Siesjö, B.K. (1975) The relationship between arterial P_{O_2} and cerebral blood flow in hypoxic hypoxia, Acta physiol. scand. 93: 423-432.

Borgström, L., Norberg, K. and Siesjö, B.K. (1976) Glucose consumption in rat cerebral cortex in normoxia, hypoxia and hypercapnia, Acta physiol. scand. 96: 569-574.

Carlsson, C., Hägerdal, M. and Siesjö, B.K. (1975) Influence of amphetamine sulphate on cerebral blood flow and metabolism, Acta physiol. scand. 94: 128-129.

Carlsson, C., Hägerdal, M., Kaasik A.E. and Siesjö, B.K. (1977) A catecholamine-mediated increase in cerebral oxygen uptake during immobilization stress in rats, Brain Res. 119: 223-231.

Cohen, P.J., Alexander, F.C., Reivich, M. and Wollman, H. (1967) Effects of hypoxia and normocarbia on cerebral blood flow and metabolism in conscious man, J appl. Physiol. 23: 183-189.

Corrodi, H., Fuxe, K., Lidbrink, P. and Olsson, L. (1971) Minor tranquilizers, stress and central catecholamine neurons, Brain Res. 29: 1-16.

Drewes, L.R. and Gilboe, D.D. (1973a) Glycolysis and the permeation of glucose and lactate in the isolated, perfused dog brain during anoxia and postanoxic recovery, J Biol. Chem. 218: 2489-2496.

Drewes, L.R. and Gilboe, D.D. (1973b) Cerebral metabolite and adenylate energy charge recovery following 10 min of anoxia, Biochim. Biophys. Acta 320: 701-707.

Drewes, L.R., Gilboe, D.D. and Betz, A.L. (1973) Metabolic alterations in brain during anoxic-anoxia and subsequent recovery, Arch. Neurol. 29: 385-390.

Duffy, T.E., Nelson, S.R. and Lowry, O.H. (1972) Cerebral carbohydrate metabolism during acute hypoxia and recovery, J. Neurochem. 19: 959-977.

Jóhannsson, H. and Siesjö, B.K. (1975) Cerebral blood flow and oxygen consumption in the rat in hypoxic hypoxia, Acta physiol. scand. 93: 269-276.

Kety, S.S. and Schmidt, C.F. (1948) The effects of altered arterial tensions of carbon dioxide and oxygen on cerebral blood flow and cerebral oxygen consumption in normal young men, J. clin. Invest. 27: 484-491.

Lassen, N.A. (1968) Brain extracellular pH: the main factor controlling cerebral blood flow, Scand. J. clin. Lab. Invest. 22: 247-251.

Levy, D.E., Brierley, J.B., Silverman, D.G. and Plum, F. (1975) Brain hypoxia initially damages cerebral neurons, Arch. Neurol. (Chic.) 32: 450-455.

Lewis, L.D., Pontén, U. and Siesjö, B.K. (1973) Arterial acid-base changes in unanesthetized rats in acute hypoxia, Resp. Physiol. 19: 312-321.

Lidbrink, P. and Farnebo, L:O. (1973) Uptake and release of noradrenaline in rat cerebral cortex in vitro: No effect of benzoediazepines and barbiturates, Neuropharm. 12: 1087-1095.

Ljunggren, B., Ratcheson, R.A. and Siesjö, B.K. (1974) Cerebral metabolic state following complete compression ischemia, Brain Res. 73: 291-307.

Lowry, O.H., Passonneau, J.V., Hasselberger, F.X. and Schulz, D.W. (1964) Effect of ischemia on known substrates and cofactors of the glycolytic pathway in brain, J. Biol. Chem. 239: 18-30.

Luft, U.C. (1965) Aviation physiology - the effect of altitude. In: Handbook of Physiology (eds. W.O.Fenn and H.Rahn) Section 3: Respiration. Vol. II. pp. 1099-1145. American Physiological Society, Washington.

MacKenzie, E.T., McCulloch, J., O'Kean, M., Pickard, J.D. and Harper, A.M. (1976a) Cerebral circulation and norepinephrine: Relevance of the blood-brain-barrier, Amer. J. Physiol. 231: 483-488.

MacKenzie, E.T., McCulloch, J. and Harper, A.M. (1976b) Influence of endogenous norepinephrine on cerebral blood flow and metabolism, Amer. J. Physiol. 231: 489-494.

MacMillan, V. and Siesjö, B.K. (1972) Brain energy metabolism in hypoxemia, Scand. J. Clin. Lab. Invest. 30: 126-136.

Mahan, D.E., Mushahwar, I.K. and Koeppe, R.E. (1975) Purification of rat brain carboxylase, Biochem. J. 145: 25-35.

Nilsson, B., Rehncrona, S. and Siesjö, B.K. (1978) Coupling of cerebral metabolism and blood flow in epileptic seizures, hypoxia and hypoglycaemia. In: Cerebral Vascular Smooth Muscle and its Control. Ciba Foundation Symposium 56. pp. 199-218. Elsevier/Excerpta Medica, Amsterdam-Oxford-New York.

Norberg, K. and Siesjö, B.K. (1975a) Cerebral metabolism in hypoxic hypoxia. I. Pattern of activation of glycolysis, a re-evaluation, Brain Res. 86: 31-44.

Norberg, K. and Siesjö, B.K. (1975b) Cerebral metabolism in hypoxic hypoxia. II. Citric acid cycle intermediates and associated amino acids, Brain Res. 86: 45-54.

Rosenthal, M., LaManna, J., Jöbsis, F.F., LeVasseur, J., Kontos, H. and Patterson, J. (1976) Effects of respiratory gases on cytchrome $\underline{a}$ in intact cerebral cortex: Is there a critical P_{O_2}? Brain Res. 108: 143-154.

Salford, L.G., Plum, F. and Siesjö, B.K. (1973a) Graded hypoxia-oligemia in rat brain. I. Biochemical alterations and their implications, Arch. Neurol. 29: 227-233.

Salford, L.G., Plum, F. and Brierley, J.B. (1973b) Graded hypoxia-oligemia in rat brain. II. Neuropathological alterations and their implications, Arch Neurol. 29: 234-238.

Salford, L.G. and Siesjö, B.K. (1974) The influence of arterial hypoxia and unilateral carotid artery occlusion upon regional blood flow and metabolism in the rat brain, Acta physiol. scand. 92: 130-141.

Siesjö, B.K., Jöhannsson, H., Norberg, K. and Salford, L.G. (1975) Brain function, metabolism and blood flow in moderate and severe arterial hypoxia. In: Brain Work, Alfred Benzon Symposium VIII (eds. D.H.Ingvar, N.A.Lassen) pp. 101-125. Munksgaard, Copenhagen.

Siesjö, B.K. and Nilsson, L. (1971) The influence of arterial hypoxemia upon labile phosphates and upon extracellular and intracellular lactate and pyruvate concentrations in the rat brain, Scand. J. Clin. Lab. Invest. 27: 83-96.

Taylor, K.M. and Laverty, R. (1969) The effect of chlordiazepoxide, diazepam and nitrazepam on catecholamine metabolism in regions of the rat brain, Europ. J. Pharmacol. 8: 296-301.

Wood, J.D. (1967) A possible role of gamma-aminobutyric acid in the homeostatic control of brain metabolism under conditions of brain hypoxia, Exp. Brain Res. 4: 81-84.

Wood, J.D., Watson, W.J. and Ducker, A.J. (1968) The effect of hypoxia on brain γ-aminobutyric acid levels, J. Neurochem. 15: 603-608.

EFFECTS OF LOW OXYGEN ON BRAIN MONOAMINE METABOLISM

A. Carlsson
Department of Pharmacology, University of Gothenburg, Gothenburg, Sweden

Introduction

It is well established that certain steps in the synthesis and metabolism of monoamines require molecular oxygen. They may thus be expected to slow down in hypoxia. These steps are the aromatic hydroxylation of tryptophan to 5-hydroxytryptophan (5-HTP) and of tyrosine to dopa (3,4-dihydroxyphenylalanine), the side-chain hydroxylation of dopamine to noradrenaline, and the oxidative deamination of the monoamines to the corresponding aldehydes. The enzymes involved in these reactions are tryptophan hydroxylase, tyrosine hydroxylase, dopamine β-hydroxylase (DBH), and monoamine oxidase, respectively (Kaufman 1974, Goldstein, 1966, Pletscher et al., 1960).

This oxygen dependence formed the starting-point of our studies on the influence of hypoxia on brain monoamine metabolism several years ago. Our primary aim was to find out if the oxygen dependence led to diminished rates of the conversions mentioned above, as measured in brain in vivo in hypoxic but otherwise intact animals. Our studies showed that the expected changes do indeed occur. However, hypoxia will not only influence monoamine synthesis and utilization in this direct manner. More complex, apparently regulatory mechanisms also appear to influence these processes. The different types of hypoxia-induced changes in monoamine metabolism and their possible functional significance will be briefly discussed in the present paper.

Aromatic amino-acid hydroxylases

When rats were exposed to a low oxygen atmosphere, a marked decrease in the in vivo rates of tryptophan and tyrosine hydroxylation respectively, was found in all brain regions investigated (Davis and Carlsson, 1973a and b). The method used for measuring hydroxylation rates was to determine the rate of accumulation of 5-HTP and dopa following inhibition of the aromatic amino-acid decarboxylase by NSD 1015 (3-hydroxybenzylhydrazine 100 mg/kg i.p.). Similar results were obtained with other more indirect techniques. A close correlation was found between the rate of tryptophan hydroxylation and cerebral venous P_{O_2} (Davis et al., 1973). Hydroxylation rates were reduced already at low hypoxia degrees that left the adenylate energy charge uninfluenced. In fact, the decrease in 5-HTP formation preceded the accumulation of lactate in the brain cells.

Hypocapnia and hypercapnia, induced by hyperventilation and addition of CO_2 to the inhaled air, respectively, caused changes in the P_{O_2} of cerebral venous blood, from about half to about twice the normal value, with concomitant marked changes in the rate of 5-HTP formation. The dependence of 5-HTP formation on P_{O_2} seemed to be unchanged by the variations in P_{CO_2} (Carlsson et al., 1977, Garcia de Yebenes Prous et al., 1977). These observations suggest that tryptophan hydroxylase is normally about 40% saturated with oxygen. Keeping in mind that this enzyme is only about half saturated with its other substrate, i.e. tryptophan (see Carlsson and Lindqvist, 1978), it appears that the rate of 5-HT synthesis can easily be varied by an order of magnitude, simply by manipulating the two substrate levels! The regulatory mechanisms possibly linked to this phenomenon present an intriguing area of investigation (cf. Wurtman and Fernstrom, 1976).

The hydroxylation of tyrosine is somewhat less strikingly influenced by varying P_{O_2} than that of tryptophan. On the other hand, tyrosine hydroxylation, in contrast to tryptophan hydroxylation, is markedly dependent on P_{CO_2} (apart from the indirect action of P_{CO_2} mediated via P_{O_2} changes). Thus, while hypercapnia (about 8% CO_2 in the inspired gas mixture) enhanced the hydroxylation of both amino-acids by about 25%, additional moderate O_2 reduction to normalize venous P_{O_2} (about 15% O_2 in the inspired gas mixture) restored the hydroxylation of tryptophan to normal but, if anything, further enhanced tyrosine hydroxylation (Carlsson et al., 1977). These observations suggest that the smaller sensitivity of tyrosine hydroxylase than tryptophan hydroxylase to hypoxia is at least partly due to the stimulation of the former but not the latter enzyme by acidosis (cf. Brown et al., 1974).

Monoamine oxidase

Despite the decrease in the first ratelimiting steps of 5-HT and catecholamine synthesis in hypoxia, the levels of these neurotransmitters in the brain are but slightly influenced. The logical conclusion must be that also the metabolic degradation of the monoamines is retarded in hypoxia. As is well known, monoamine oxidase also requires molecular oxygen and should thus be inhibited in hypoxia. Accordingly we found a marked decrease in the concentrations of 5-hydroxyindoleacetic acid and homovanillic acid in the brains of hypoxic rats, with or without probenecid pretreatment. Moreover, if monoamine oxidase is inhibited beforehand, the accumulation of monoamines is retarded by hypoxia (Davis and Carlsson, 1973 b).

The apparently equal dependence of monoamine-synthesizing enzymes and of monoamine oxidase on P_{O_2} may be looked upon as purposeful way of making some important transmitter systems less dependent of variations in oxygen availability.

Other monoamine-synthesizing and degradating enzymes

The other enzymes involved in the synthesis and metabolism of monoamines have been studied less extensively. However, available data indicate that dopamine-β-hydroxylase, which requires molecular oxygen, is inhibited by hypoxia in vivo, whereas the other enzymes, i.e. the aromatic amino-acid decarboxylase and catechol-O-methyltransferase (COMT), are not influenced (Brown et al., 1975).

Indirect actions of hypoxia on monoamine metabolism

The accumulation of 3-methoxytyramine, i.e. the 3-O-methylated metabolite of dopamine, was found to be markedly retarded by hypoxia (Brown et al., 1975). As mentioned, COMT activity appears to be intact in hypoxia, and thus reduced substrate availability is the most likely explanation of this finding. The dopamine level in brain is not markedly reduced in hypoxia. However, dopamine appears to become available for COMT only after release into the synaptic cleft. Hypoxia thus seems to inhibit this release, possibly by reducing the firing by dopaminergic neurons. Accordingly, the depletion of dopamine following inhibition of tyrosine hydroxylase is retarded by hypoxia, whereas noradrenaline shows the opposite change. In other words, hypoxia appears to inhibit firing by dopaminergic neurons and to enhance firing by noradrenergic neurons.

The mechanism by means of which hypoxia influences the physiological activity of catecholamine neurons remains to be clarified. Possibly GABA is involved; its degradation rather than synthesis is dependent on oxygen availability, and it induces changes in monoamine metabolism similar to those seen in hypoxia (cf. Mršulja et al., 1976, Biswas and Carlsson, 1977).

Role of catecholamines for hypoxia-induced behavioural changes

Hypoxia has been found to disrupt reversibly conditioned avoidance response (CAR) in rats. This disruption is reversed by treating the animals with dopa or apomorphine, a synthetic dopamine-receptor agonist (Brown and Engel, 1973, Brown et al., 1973, 1975). Since CAR is known to depend on intact catecholamine systems in the brain, these observations suggest that inibition of catecholaminergic, especially dopaminergic activity is involved in hypoxia-induced CAR disruption. Perhaps we are dealing with a regulatory mechanism which serves to inhibit in order to preserve energy in hypoxia.

Interaction between stress and hypoxia on tyrosine hydroxylase

Foot-shock or immobilization stress enhances the rate of tyrosine hydroxylation in the brain and in the adrenal medulla (Snider et al., 1974, Brown et al., 1974, Davis, 1976). As mentioned, hypoxia has the opposite effect. Simultaneous exposure to stress and hypoxia results in a remarkable interaction: the hydroxylation of tyrosine is enhanced and the dependence on P_{O_2} is hardly demonstrable any more. In contrast, the hydroxylation of tryptophan retained its sensitivity to hypoxia under stress. These observations suggest that tyrosine hydroxylase may undergo a conformational change in stressed animals, resulting in increased affinity for oxygen.

Effects of neonatal oxygen deprivation

In neonatal, like in adult rats hypoxia or anoxia were found to cause marked inhibition of brain-monoamine synthesis of short duration. However, exposure of rats to hypoxia (6% O_2) for 4.5 h at 1 day of age also caused certain persistent changes. At the age of 28 days these rats were found to acquire a conditioned avoidance response at a markedly reduced rate. Moreover, the activities of brain tyrosine and tryptophan hydroxylases, measured in vivo, were markedly reduced at this age. Hypoxia at a critical phase of neuronal development thus appears to give rise to persistent changes in the biochemistry and function of monoaminergic neurons (Hedner et al., 1978).

Summary

The overall effect of hypoxia on central monoaminergic mechanisms is complex and consists of a) direct actions on oxygen dependent enzymatic mechanisms and b) indirect actions.

The direct inhibitions of tryptophan hydroxylase, tyrosine hydroxylase, dopamine-β--hydroxylase and monoamine oxidase occur already in moderate hypoxia before any change in energy metabolism in the brain is detectable. The functional significance of these inhibitions is difficult to evaluate; the effects on the synthetic enzymes and on monoamine oxidase appear to balance each other, leaving brain-monoamine levels largely unaffected.

Among indirect actions an inhibitory effect on dopamine turnover is particularly prominent. It is probably due to inhibition of nerve impulse-induced transmitter release and may be important for the behavioural inhibition and preservation of energy in hypoxia.

Neonatal hypoxia has been shown to produce long-lasting changes in monoaminergic activity as well as a learning deficit.

References

Biswas, B. and Carlsson, A. (1977) The effect of intracerebroventricularly administered GABA on brain monoamine metabolism, Naunyn-Schmiedeberg's Arch. Pharmacol. 299: 47-51.

Brown, R.M., Davis, J.N. and Carlsson, A. (1973) Dopa reversal of hypoxia-induced disruption of the conditioned avoidance response, J. Pharm. Pharmacol. 25: 412-414.

Brown, R.M. and Engel, J. (1973) Evidence for catecholamine involvement in the suppression of locomotor activity due to hypoxia, J. Pharm. Pharmacol. 25: 815-819.

Brown, R.M., Kehr, W. and Carlsson, A. (1975) Functional and biochemical aspects of catecholamine metabolism in brain under hypoxia, Brain Res. 85: 491-509.

Brown, R.M., Snider, S.R. and Carlsson, A. (1974) Changes in biogenic amine synthesis and turnover induced by hypoxia and/or foot shock stress. II The central nervous system, J. Neural Transm. 35: 293-305.

Carlsson, A., Holmin, T., Lindqvist, M. and Siesjö, B. (1977) Effect of hypercapnia and hypocapnia on tryptophan and tyrosine hydroxylation in rat brain, Acta physiol. scand. 99: 503-509.

Carlsson, A. and Lindqvist, M. (1978) Dependence of 5-HT and catecholamine synthesis on concentrations of precursor amino-acids in rat brain, Naunyn-Schmiedeberg's Arch. Pharmacol. (in press).

Davis, J.N. (1976) Brain tyrosine hydroxylation: alteration of oxygen affinity in vivo by immobilization or electroshock in the rat, J. Neurochem. 27: 211-215.

Davis, J.N. and Carlsson, A. (1973 a) Effect of hypoxia on tyrosine and tryptophan hydroxylation in unanaesthetized rat brain, J. Neurochem. 20: 913-915.

Davis, J.N. and Carlsson, A. (1973 b) The effect of hypoxia on monoamine synthesis, levels and metabolism in rat brain, J. Neurochem. 21: 783-790.

Davis, J.N., Carlsson, A., MacMillan, V. and Siesjö, B.K. (1973) Brain tryptophan hydroxylation: Dependence on arterial oxygen tension, Science 182: 72-74.

Garcia de Yebenes Prous, J., Carlsson, A. and Mena Gomez, M.A. (1977) The effect of CO_2 on monoamine metabolism in rat brain, Naunyn-Schmiedeberg's Arch. Pharmacol. 301: 11-15.

Goldstein, M. (1966) Inhibition of norepinephrine biosynthesis at the dopamine-β-hydroxylation stage, Pharm. Revs. 18: 77-82.

Hedner, T., Lundborg, P. and Engel, J. (1978) Biochemical and behavioral changes in young adult rat brain after neonatal oxygen deprivation. To be published.

Kaufman, S. (1974) Properties of the pterin-dependent aromatic amino acid hydroxylases. In Wolstenholme, G.E.W. and Fitzsimors (eds.) Aromatic amino acids in the brain. Ciba Foundation Symp. 22 (new series). Elsevier, Amsterdam, pp. 85-115.

Mršulja, B.B., Lust, W.D., Mršulja, B.J., Passionneau, J.V. and Klatzo, I. (1976) Post-ischemic changes in certain metabolites following prolonged ischemia in the gerbil cerebral cortex, J. Neurochem. 26: 1099-1103.

Pletscher, A., Gey, K.F. and Zeller, P. (1960) Monoaminoxydase-hemmer, Progr. in Drug Res. 2: 417-590.

Snider, S.R., Brown, R.M. and Carlsson, A. (1974) Changes in biogenic amine synthesis and turnover induced by hypoxia and/or foot shock stress. I. The adrenal medulla, J. Neur. Transm. 35: 283-291.

Wurtman, R.J. and Fernstrom, J.D. (1976) Control of brain neurotransmitter synthesis by precursor availability and nutritional state, Biochem. Pharmacol. 25: 1691-1696.

ARACHIDONIC ACID METABOLISM AND CYCLIC NUCLEOTIDES IN THE CNS DURING
HYPOXIA

C. Galli, C. Spagnuolo, L. Sautebin and G. Galli
Institute of Pharmacology and Pharmacognosy, University of Milan, Italy

I. Introduction

Formation of metabolic derivatives of arachidonic acid endowed of biological activity has been described in most tissues. The original attention of investigators was centered on the prostaglandins E_2 and $F_{2\alpha}$ (PGE_2 and $PGF_{2\alpha}$), but the discovery of short-lived potent intermediates (the prostaglandin endoperoxides PGG_2 and PGH_2)(Samuelsson, 1965) and of new unstable products, thromboxane A_2 (Hamberg et al., 1974) and prostacyclin (Bunting et al., 1976), derived from arachidonic acid through the activity of the cycloxygenase, has considerably broaden the field of investigation. Simultaneous formation of hydroxy fatty acids, mainly derived through the activity of a lipoxygenase, has also been described (Hamberg and Samuelsson, 1974).

The conversion of the endoperoxides - which appear to be the key intermediates in the prostaglandin forming system - to the primary prostaglandins PGE_2 and $PGF_{2\alpha}$ or to the thromboxanes and to prostacyclin (PGI_2), differs from one tissue to the other. Also, since prostaglandins are not stored, but are formed and either released through the circulation or metabolized, investigations have centered more on the metabolic capacity of tissue to convert the precursor fatty acid to the various products, rather than on the determination of prostaglandin levels. However, the rapid release of endogenous fatty acid precursors from the tissue phospholipid pool occurring after death and during manipulation of tissues, overloads the PG synthetase system thus affecting the rate of free arachidonic acid metabolism in "in vitro" studies. Also the presence and relative concentrations of cofactors such as glutathione influences the relative metabolic conversion of the precursor to the various products "in vitro" (Christ-Hazelhof et al., 1976). Cyclic nucleotides (cyclic AMP and cyclic GMP) are ubiquitous endogenous compounds which are responsible of the regulation of cellular responses to stimulation. They are involved in promoting different cellular events that in most instances appear to be strikingly contrasting. The complex biological roles of these compounds are under study in a large number of laboratories and several reviews have attempted to describe the specific functions in various system (e.g. Goldberg et al., 1975; Nathanson, 1977).

Cyclic AMP may be the factor responsible of prostaglandin formation in tissues.

II A. Prostaglandins and the CNS

The presence of prostaglandins in the CNS has been described many years ago (Samuelsson, 1964) and formation of PGE_2 and $PGF_{2\alpha}$ in brain tissue from endogenous substrate has been described in "in vitro" incubations (Nicosia and Galli, 1975; Wolfe et al., 1976a). More recently, formation of thromboxane B_2, the relatively stable compound derived from thromboxane A_2 (Wolfe et al., 1976b) and of Prostaglandin D_2 (Abdel-Halim et al., 1977) have also been observed in "in vitro" studies. These products accumulate progressively in the tissue up to a certain period of time because of the low prostaglandin metabolizing activity of most areas in the adult animals brain (Nakano et al., 1972), except for Purkinje cells in cerebellum where a high 15 PG Dehydrogenase activity is present (Siggins et al., 1971).

II B. Prostaglandin precursors in the CNS.

The CNS contains high levels of long chain polyunsaturated fatty acids derived from the dietary essential fatty acids (EFA) linoleic and linolenic acids. Arachidonic acid, precursor of prostaglandins of the PG_2 series, is the major fatty acid of the linoleic acid family in brain, it accumulates during the period of most active neuronal growth (Sinclair and Crawford, 1972) and is mainly located in the 2 position of glycerophospholipids. Changes of the fatty acid composition of the diet and especially of the relative balance of EFA of the linoleic and linolenic acid series, modify the levels of arachidonic acid in the growing brain (Galli et al., 1970; Alling et al., 1972; Galli et al., 1971; Galli et al., 1972). The concentration of di-homo-γ-linoleic acid (20:3 n-6), precursor of PG_1 series, in all tissues, is negligible, supporting the generally accepted concept that this type of prostaglandins does not occur naturally in mammals (Dawes, 1978).

Conversion of the precursor polyunsaturated fatty acid to the various products of the prostaglandin synthetase is considered to depend upon release of these compounds from tissue phospholipids through activation of lipolytic enzymes, mainly of the type of phospholipases (Vogt 1978). In fact, stimulation of prostaglandin formation also stimulates arachidonic acid release (Isakson et al., 1977), thus rising tissue levels of this fatty acid in free form. In brain, concentrations of free arachidonic acid are raised much above the "in vivo" values by ischemia following decapitation and by administration of convulsant drugs (Bazan, 1970).

II C. Formation of Prostaglandins in the CNS

1. In vitro studies. Prostaglandins are formed in incubated brain slices (Nicosia and Galli, 1975; Wolfe et al., 1976a) as well as in

homogenates (Wolfe, 1976a) from endogenous arachidonic acid. Attempts to study the conversion of exogenous labelled substrate were unsuccessful (Wolfe et al., 1976a) because exogenous arachidonic acid does not equilibrate with the endogenous compound and does not reach the active sites of prostaglandin synthetase.

Formation of $PGF_{2\alpha}$ exceeds that of PGE_2, and is much greater in brain cortex than in cerebellum or other brain areas (Wolfe et al., 1976a; Bosisio et al., 1976). Formation of Thromboxane B_2 has also been described in rat brain (Wolfe et al., 1976b) and of Prostaglandin D_2 in brain homogenates (Habdel-Halim, 1977).

Studies of prostaglandin biosynthesis "in vitro" give useful information on the metabolic activity of the tissue, but have limited quantitative correlations with the "in vivo" situation. In fact, the free arachidonic acid pool formed during ischemia following death provides a large excess of substrate for prostaglandin formation. It seems reasonable to assume that stimulation of PG formation "in vivo" is obtained by a more precise coupling between release of the substrate and conversion by the cycloxygenase activity. The enzymatic release of arachidonic acid "in vitro" is not affected by the presence of neuromediators, such as catecholamines, in the incubation medium, but formation of $PGF_{2\alpha}$ (and not of PGE_2) is increased (Wolfe et al., 1976a).

2. In vivo studies. Determinations of "in vivo" levels of prostaglandins in brain may lead to erroneous and variable results unless precautions are taken in order to prevent the increase of free arachidonic acid levels occurring after decapitation, which leads to unphysiological formation of prostaglandins. Our approach has been to sacrifice small laboratory animals by use of a microwave oven (Cenedella et al., 1975), whereas alternative methods are based on rapid freezing of cerebral tissue (Veech et al., 1973). Basal levels of prostaglandins in brain of animals sacrificed with the above precautions are very low, of the order of few ng/g fresh tissue, but are greatly raised by several experimental conditions. Administration of convulsant drugs, surgical trauma of nervous tissue or ischemia are able to greatly stimulate prostaglandin formation especially in brain cortex. $PGF_{2\alpha}$ formation exceeds that of PGE_2. Recently "in vivo" formation of endoperoxides after intraventricular injection of corbachol or pentamethylenetetrazole has been shown to exceed that of $PGF_{2\alpha}$ or Thromboxane B_2 (Wolfe, 1978).

II D. Biological role of prostaglandins in the CNS.

Little is known on the biological role of Prostaglandins in the CNS, since the available informations have been obtained by studying the effects of the stable products of arachidonic acid metabolism used in pharmacological doses. Most of the data have been obtained by using

PGE$_1$, which does not appear to occur naturally in mammals. It is generally assumed that prostaglandins are involved in cell function as modulators of responses to stimulation. The hypothesis has been formulated that prostaglandins of the E type act by modulating the response of adenylcyclase to norepinephrine either presinaptically (Hedqvist, 1973) or postsynaptically (Bloom et al., 1975). However the inability to confirm this hypothesis in several regions of the CNS and the discovery of new potent derivatives of arachidonic acid metabolism, seriously challenge the general validity of the proposed scheene (For a recent review on Prostaglandins in the CNS, see Coceani and Pace-Asciak, 1976).

The most accepted role of Prostaglandins in the CNS is the implication of E type of prostaglandins in the pyrogen-induced fever at the hypothalamic level, although it is questionable that they are essential in all kinds of fever (Veale et al., 1978).

II E. Changes of the prostaglandin forming system in brain during development.

The ability of the CNS to respond to the administration of convulsant drugs with enhanced formation of PGF$_{2\alpha}$ is quite low in the newborn rat and increases up to the adult values within three weeks of life (Paoletti et al., 1978). Since seizures are not induced by convulsant drugs in the newborn, it is possible that the reduced stimulation of brain Prostaglandin formation under these experimental conditions depends upon insensitivity of the CNS to the pharmacological treatment more than upon inactivity of PG synthetase. In addition the observation of a reduced stimulation of free fatty acid and free arachidonic acid release in immature brain during post-mortem ischemia or administration of convulsant drugs (Bazan, 1971a) may provide the explanation for the reduced prostaglandin formation under the same conditions. The reported initial inability of newborns to produce fever following pyrogen injection has also been interpreted as a result of inability of the immature hypothalamus to elaborate or to be sensitive to prostaglandins (Blatteis, 1976; Cooper et al., 1975).

It has been also reported that predominant formation of PGE$_2$ rather than of PGF$_{2\alpha}$ occurs in fetal lamb brain, associated with high prostaglandin metabolizing activity, which, insted is very low in the adult brain (Pace-Asciak and Rangaraj, 1976).

Since prostaglandins and cyclic nucleotides are likked together in the control of cell functions (Kuehl, 1974), it is of interest that the responsiveness of the enzymes involved in the formation of cyclic nucleotides to neuromediators in very low at birth and develops during brain maturation (Cheney et al., 1976).

Thus it appears that full responsiveness of both Prostaglandin synthetase and nucleotide cyclases in the CNS is reached only after the critical stages of brain maturation. It is possible that the stage of development of the adenylcyclase system is the limiting factor in the series of events leading also to prostaglandin formation.

III A. Anoxia in the CNS

Arterial hypoxia occurs as a consequence of a decrease in arterial pO_2 (hypoxic hypoxia) or in hemoglobin content (anemic hypoxia).

Under these conditions the cerebral perfusion pressure and the availability of energetic substrates to the brain are not reduced. The above hypoxic conditions, which have been termed "uncomplicated anoxia" (Sjesjo et al., 1977), result in complex alterations both of energy metabolism and of neurotransmitter turnover, depending upon the degree of fall of arterial pO_2. Other possible situations responsible for a progressively reduced oxygen supply to the brain are a) a state of "hypoxia with relative ischemia" (relative ischemia) occurring when reduced arterial pO_2 values are associated with reduced cerebral blood flow (CBF), b) a state of "incomplete ischemia", and finally c) a state of "complete ischemia". Complete ischemia differs from relative and incomplete ischemias because oxygen and substrate supplies are both completely excluded: thus rapid depletion of energy stores (glycogen, glucose), of pyruvate, ketoglutarate, oxaloacetate and ATP, and accumulation of lactate, succinate, alanine, GABA and ammonia have been described (Siesjo et al., 1977).

Our work on the prostaglandin forming system and on cyclic nucleotides has been carried out in anoxia following complete ischemia, obtained either by decapitation, or, more recently, by ligation of both common carotid arteries in the gerbil.

III B. The prostaglandin forming system during ischemic anoxia in brain.

1. Precursors. In ischemic brain a rapid rise of the levels of free fatty acids and especially free arachidonic acid is observed (Bazan, 1970; Bazan, 1971b). This appears to be an enzymatic process since is prevented when the animals are killed by using a microwave oven (Cenedella et al., 1975) technique which rapidly inactivates brain enzymes (Schmidt et al., 1972).

The composition of the released free fatty acid pool is characterized by high levels of arachidonic and stearic acid and closely resembles that of brain diglycerides and phosphatidyl inositol. This suggests that activation of lipolytic enzymes specific for the 1 and 2 positions of the above lipids may be responsible for the observed release of free fatty acids during ischemia. Studies performed by prelabelling brain lipids "in vivo" by intracerebral injection of labelled arachidonic acid have shown high incorporation of radioactivity mainly in choline and inositol phospholipids, with the highest relative specific activity in the latter (Galli et al., 1977). Again the lipid classes undergoing greater changes during ischemia were those of diacylglycerols and phosphatidylinositol. Another possible explanation of the increased size of the free fatty acid and the free arachidonic acid pools in the ischemic tissue, besides the activation of tissue phospholipases, is an unbalancement between phospholipid breakdown and resynthesis, under these conditions (Rodriguez de Turco et al., 1977). Ischemic release of free fatty acids in brain is prevented by pretreatment with anticonvulsant drugs (Bazan, 1978), whereas compounds known to interfere with arachidonic acid mobilization in other tissues such as mepacrine (Vargaftig and Dao Hai, 1972) and corticosteroids (Kantrowitz et al., 1975) are ineffective (Wolfe, 1978). The rate of release of free arachidonic acid during ischemia is much greater in brain cortex than in cerebellum (Bosisio et al., 1976) and is affected by changes of dietary levels of the precursor linoleic acid (Galli and Spagnuolo, 1976). Further release of arachidonic acid is observed also from incubated cerebral tissue; however the rate of release is much lower and is less specific for this fatty acid in comparison with the release occurring "in vivo" during ischemia (Bosisio et al., 1976).

2. Prostaglandins. Ischemic anoxia has been shown to release prostaglandins in several organs including kidneys (McGiff et al., 1970; Jaffe et al., 1972) and heart (Wennmalm et al., 1974). Hypoxic lung has also been shown to release prostaglandin-like compounds (Said et al., 1974). The release of the newly discovered very potent metabolites of arachidonic acid, the thromboxanes and prostacyclin (PGI_2) in hypoxic tissues has not yet been fully investigated.

The possible significance of prostaglandin liberation by hypoxia is that the released compounds might contribute to improve blood flow in the tissue through vasodilatation. Nothing is known about the mechanism(s) responsible for the stimulation of prostaglandin formation in hypoxia tissues, although it has been suggested that the rise of cyclic nucleotide levels followed by stimulation of phospholipases may be the factor involved. Prostaglandin release from the CNS is greatly enhanced under several conditions, some of which are not physiologic, and the rate of release appears generally to correlate with neuronal activity. The sites of release are not defined but may vary with the kind of stimulus applied: during ischemia probably several sites of release may be involved, neurons, glial cells, vessels and platelets. PGF compounds were reported to be preferentially released when neuronal activity is pathologically enhanced and as a consequence of inflammation or brain damage following ischemia (Ramwell et al., 1966; Bradley et al., 1969;

Coceani et al., 1971), although the absolute identification and quantitative determinations prior to the development of mass fragmentographic techniques were not reliable. The predominance of $PGF_{2\alpha}$ formation during ischemia may reflect a change in the synthesis of prostaglandins towards the formation of compounds which have been shown to be pharmacologically inactive with respect to most neuronal systems to be cleared into the CSF and general circulation (Wolfe, 1978). High levels of $PGF_{2\alpha}$ have been found, in fact, in the CSF of patients with altered cerebral haemodynamics (La Torre et al., 1974). $PGF_{2\alpha}$ formation during brain ischemia in the rat is much greater in cortex than in cerebellum, correlating also with a differential release of arachidonic acid in these two brain areas (Bosisio et al., 1976). Differential formation of Prostaglandin in cerebral cortex or cerebellum has been described also "in vitro" (Wolfe et al., 1975); however arachidonic acid release was not measured under these conditions. No data is available in the literature on the formation of thromboxanes in brain during ischemia, although it has been in "in vitro" studies (Wolfe et al., 1976b) and "in vivo" during convulsions (Wolfe, 1978). Also, no information is so far available in the literature on the synthesis in the CNS of the hydroxy fatty acids which have been described in the metabolic conversion of arachidonic acid in platelets (Hamberg and Samuelsson, 1974), and on the metabolic balance between the lipoxygenase and cycloxygenase systems.
In experiments carried out in our Institute we have studied the metabolic balance of the lipoxygenase and cycloxygenase systems during anoxia in brain. The data so far obtained show, for the first time, that hydroxy fatty acids derived from arachidonic acid through the activity of the lipoxygenase system are formed in rat brain (Sautebin L., Spagnuolo C., Galli G. and Galli C., unpublished). The presence of thromboxane B_2 has also been detected "in vivo" prior and after ischemia (Folco G.C. and Spagnuolo C., unpublished). In the last condition, a considerable increase of Thromboxane B_2 levels is observed in rat brain cortex.

IV. Cyclic nucleotides and ischemia.

A vast literature has described the important biological role of cyclic nucleotides in the Central Nervous System (Nathanson, 1977).

Changes of brain levels of cyclic nucleotides during ischemia have been reported many years ago (Breckenridge, 1964) and subsequently confirmed by a number of laboratories (Schmidt et al., 1972; Kimura et al., 1974). Lust et al. (1975) have shown in the gerbil model of ischemia (Levine and Payan, 1966), a considerable increase of cyclic AMP in the ischemic cerebral cortex, associated with a decrease of cyclic GMP. Further work has described the changes of cyclic nucleotides during ischemia also in cerebellum and spinal cord (Lust et al., 1977). The complex changes of cyclic nucleotides in cerebral cortex during and after an ischemic episode are considered to reflect functional alterations of the cortex, whereas the almost opposite changes of cyclic nucleotide levels in cerebellum and spinal cord could be a consequence of a decreased output from cortical areas. The possibility that the changes in the non ischemic regions (cerebellum and spinal cord) during cortical ischemia are neuronally mediated or, less likely, are due to extra-neuronal factors has been discussed (Lust et al., 1977). The same authors also proposed that the increase of cyclic AMP, which inhibits excitability, and the decrease of cyclic GMP, an excitatory compound, in brain cortex are responsible of the loss of recordable electrical activity following the onset of ischemia (Hossmann and Sato, 1971; Swaab and Boer, 1972). From the specific tissue-course of cyclic AMP and GMP changes, it has been suggested that the synthesis and degradation of the two cyclic nucleotides are regulated independently (Kimura et al., 1974). The rise of cGMP in cerebellum of gerbils after long lasting ischemia following bilateral occlusion of common carotids is similar to that found in cerebellum after administration of convulsant drugs (Goldberg et al., 1978; Mao et al., 1975) and this observation, on the basis of the pharmacological activity of this cyclic nucleotide, may explain the onset of seizures observed in the ischemic animals.

The mechanisms involved in the changes of cyclic nucleotides during ischemia are not understood, but it has been suggested that the rise of cAMP may be due to the release of endogenous substances, such as adenosine (Berne et al., 1974), which causes activation of adenyl cyclase (Sattin and Rall, 1970).

V. Interactions of the prostaglandin forming system with cyclic nucleotides.

Cyclic nucleotides and prostaglandins are linked together in the control of cell function (Kuehl, 1974).

The mediatory role of cyclic nucleotides in the response of cellular systems to external stimuli, suggests a possible effect of these endogenous compounds on prostaglandin formation. Limited data in few systems have shown that cyclic AMP stimulates phospholipase A_2, thus providing the free substrate for prostaglandin synthesis. TSH, for instance, increases cAMP levels and the synthesis of prostaglandins in the thyroid preceded by stimulation of phospholipase A_2 (Haye et al., 1973). Stimulation of phospholipase A by cyclic nucleotides has been observed also in fat cell homogenates (van den Bosch and van den Besselaar, 1978), whereas phospholipase A in brain synaptosomes is inhibited by cyclic AMP (Kunze, 1973).

Prostaglandin are considered to modulate the cyclase system, which, in turn, provides the second messenger of external stimuli to intracellular receptor sites. Most data, obtained in extraneuronal tissues, show that prostaglandins of the E type generally raise cAMP levels. PGE_1 and E_2 stimulate cAMP formation "in vitro" in cultured neuronal tissue (Gilman and Niremberg, 1971) and in slices of cerebral cortex (Berti et al., 1973). Species differences were observed in the response to PGE_1 and E_2, whereas PGF_s were inactive. However, PGE_1, although widely used as an active agent, is not considered a physiological compound (Dawes, 1978) and PGE_2 is not the major metabolite of arachidonic acid in most tissues and in brain.

Elevation of cyclic AMP levels induced by arachidonic acid in cell-culture systems (Cohen and Jaffe, 1973; Hong et al., 1976; Burstein et al., 1977) suggested that other compounds, besides PGE_2 may stimulate the cyclase system. The discovery of new products of arachidonic acid metabolism such as thromboxane A_2 and prostacyclin, endowed of potent and antagonistic biological activities, suggests a more active role of these compounds in the modulation of the cyclase systems. However studies of the effects of these metabolites on cyclic nucleotides, are few, due to the high instability of these compounds, limiting the general availability to the investigator.

The effects of the labile thromboxane A_2 on platelet adenyl cyclase has been studied by coupling a thromboxane generating system (endoperoxide plus platelet microsomes) with platelets: thromboxane A_2 inhibits the cAMP stimulating activity of PGE_1, whereas the endoperoxides, thromboxane B_2 and the products of arachidonate lipoxygenase are inactive (Miller et al., 1977). Prostacyclin in contrast has been shown to increase platelet cAMP levels (Gorman et al., 1977). Furthermore, peroxides derived from unsaturated fatty acids through lipoxidase activity have been shown to stimulate human platelet guanylate cyclase (Hidaka and Asano, 1977). As previously mentioned, thromboxane B_2 is formed in brain tissue, whereas formation of prostacyclin does not seem to occur under the same conditions (Wolfe, 1978). It would appear from the available data that, at least in certain systems, stimulation of arachidonic acid metabolism is activated by cyclic AMP and that arachidonic acid metabolites may, in turn, modulate the formation of cyclic nucleotides. Pharmacological blockade of the cycloxygenase system, however, does not affect the rise of cAMP induced by TSH in the thyroid (Haye et al., 1973) and the increase of cGMP occurring "in vivo" in the cerebellum after stress (Dinnendahl and Gumulka, 1977), suggesting that the proposed modulatory role of prostaglandins on cyclic nucleotides may not be functionally important in all systems.

The effects of coupling a thromboxane generating system with homogenates of brain cortex assayed for basal and noradrenaline-stimulated adenylcyclase activity are also under study. The physiological significance of the formation of endoperoxides, thromboxanes and hydroxy fatty acids in the CNS remains elusive. It appears however, that their functions may be not absolutely essential, since pharmacological blockade of brain cycloxygenase "in vivo" does not appreciably alter neurological activities.

Acknowledgement

Supported by contract number 76.012.37 by CNR. We wish to thank the Fidia S.p.A. for financial and technical assistance.

References

Abdel-Halim, M.S., Hamberg, M., Sjöquist, B. and Änggard, E. (1977) Identification of Prostaglandin D_2 as a major Prostaglandin in Homogenates of Rat Brain. Prostaglandins 14, 633-643.

Alling, C., Bruce, A., Karlsson, I., Sapia, O. and Svennerholm, L. (1972) Effect of maternal essential fatty acid supply on fatty acid composition of brain, liver, muscle and serum in 21-day-old rats. J. Nutrition, 102, 773-782.

Bazan, N.G. (1970) Effects of ischemia and electroconvulsive shock on free fatty acid pool in the brain. Biochim. Biophys. Acta 218, 1-10.

Bazan, N.G. (1971a) Modifications in the free fatty acids of developing rat brain. Acta Physiol. Latino-Americana, 21, 15-20.

Bazan, N.G. (1971b) Changes in free fatty acids of brain by drug-induced convulsions, electroshock and anaesthesia. J. Neurochem. 18, 1379-1385.

Berne, R.M, Rubio, R. and Curnish, R.R. (1974) Release of adenosine from ischemic brain. Circulation Res. 35, 262-271.

Berti, F., Trabucchi, M., Bernareggi, V. and Fumagalli, R. (1973). Prostaglandins on cyclic AMP formation in cerebral cortex of different mammalian species. Adv. Biosc. 9, 475-480.

Blatteis, C.M. (1976) Fever. Exchange of shivering by non shivering pyrogenesis in cold-acclimated guinea pigs. J. Appl. Physiol. 40, 29-34.

Bloom, F.E., Siggins, G.R., Hoffer, B.J., Segal, M. and Oliver, A.P. (1975) Cyclic nucleotides in the central synaptic actions of catecholamines, in Adv. Cyclic Nucleotide Res., 5, 603-618.

Bosisio, E., Galli, C., Galli, G., Nicosia, S., Spagnuolo, C. and Tosi, L. (1976) Correlation between release of free arachidonic acid and prostaglandin formation in brain cortex and cerebellum. Prostaglandins 11, 773-781.

Bradley, P.B., Samuels, G.M.R. and Shaw, J.E. (1969) Correlation of prostaglandin release from the cerebral cortex of cats with electrocorticogram, following stimulation of the reticular formation. Br. J. Pharmac. 37, 151-157.

Breckenridge, B. McL. (1964) The measurement of cyclic adenylate in tissue, Proc. Nat. Acad. Sci. USA, 52, 1580-1586.

Bunting, S., Gryglewski, R., Moncada, S. and Vane, J.R. (1976) Arterial walls generate from prostaglandin endoperoxides a substance (prostaglandin X) which relaxes strips of mesenteric and coeliac arteries and inhibits platelet aggregation. Prostaglandins 2, 897-914.

Burstein, S., Gagnon, G., Hunter, S.A. and Maudsley, D.V. (1977) Elevation of Prostaglandin and cyclic AMP levels by Arachidonic Acid in Primary Epithelial Cell Cultures of C3H Mouse Mammary Tumors. Prostaglandins 13, 41-53.

Cenedella, R.A.J., Galli, C. and Paoletti, R. (1975) Brain free fatty acid levels in rats sacrificed by decapitation versus focused micro wave irradiation. Lipids, 10, 290-293.

Cheney, L., Costa, E., Racagni, G. and Szilla, G. (1976) in "Brain Disfunction in Infantile Febrile Convulsions", edited by M.A.B. Brazier and F. Coceani, Raven Press, pag. 41-53.

Christ-Hazelhof, E., Nugteren, D.H. and Van Dorp, D.A. (1976) Conversions of prostaglandin endoperoxides by glutathione-S-transferases and serum albumins. Biochim. Biophys. Acta, 450, 450-461.

Coceani, F. and Pace-Asciak, C.R. (1976) Prostaglandins and the Central Nervous System in "Prostaglandins: Physiological, Pharmacological and Pathological Aspects", Edited by S.M.M. Karim, MTP Press, Ltd. pag. 1-37.

Coceani, F., Puglisi, L. and Lavers, B. (1971) Prostaglandins and neuronal activity in spinal cord and cuneate nucleus. Ann. N.Y. Acad.Sci. USA, 180, 289-301.

Cohen, F. and Jaffe, B. (1973) Production of prostaglandins by cells in vitro: Radioimmunoassay measurement of the conversion of arachidonic acid to PGE_2 and $PGF_{2\alpha}$. Biochim. Biophys. Res. Comm. 55, 724-729.

Cooper, K.E., Pittman, Q.T. and Veale, W.L. (1975) Observation on the development of the "fever" mechanism in the fetus and newborn. In: Temperature Regulation and Drug Action, edited by P. Lomax, E. Schonbaum and J. Jacob, Karger, Basel, pp. 43-50.

Dawes, G.S. (1978)"Conclusion" in "Prostaglandin and Perinatal Medicine" Edited by F. Coceani and P.M. Olley. Raven Press, pag. 387-393.

Dinnendahl, V. and Gumulka, S.W. (1977) Stress induced alterations of cyclic nucleotide levels in brain: effects of centrally acting drug. Psychopharmacology, 52, 243-249.

Galli, C., Galli, G., Spagnuolo, C., Bosisio, E., Tosi, L., Folco, G.C. and Longiave, D. (1977)"Dietary essential fatty acids, brain polyunsaturated fatty acids and prostaglandin biosynthesis" in "Function and biosynthesis of lipids" Edited by N.G. Bazan, R.R. Brenner and N.M. Giusto, Plenum Press, pag. 561-575.

Galli, C. and Spagnuolo, C. (1976) The release of brain free fatty acids during ischemia in essential fatty acid deficient rats. J. Neurochem. 26, 401-404.

Galli, C., Trzeciak, H. and Paoletti, R. (1971) Effects of dietary fatty acids on the fatty acid composition of brain ethanol amine phosphoglyceride. Reciprocal replacement of n-6 and n-3 polyunsaturated fatty acids. Biochim. Biophys. Acta, 248, 449-459.

Galli, C., Trzeciak, H.I. and Paoletti, R. (1972) Effects of essential fatty acid deficiency on myelin and various subcellular structures in rat brain. J. Neurochem., 19, 1863-1867.

Galli, C., White, H.B.Jr. and Paoletti, R. (1970) Brain lipid modification induced by essential fatty acid deficiency in growing male and female rats. J. Neurochem., 17, 347-355.

Gilman, A.G. and Niremberg, (1971) Regulation of adenosine 3',5' cyclic monophosphate metabolism in cultured neuroblastoma cells. Nature (Lond.) 234, 356-358.

Goldberg, N.D., Haddox, M.K., Hartle, D.K. and Hadden, J.W. (1973) The biological role of cyclic 3',5'-guanosine monophosphate. In

"Pharmacology and the future of man, Proc. 5th Intern. Conf. Pharmacology, Basel, Karger, vol. 5, pp. 146-169.

Goldberg, N.D., Haddox, M.K., Nicol, S.E., Glass, D.B., Sanford, C.H., Kuehl, F.A.Jr. and Estensen, R. (1975) Biological regulation through opposing influences of cyclic GMP and cyclic AMP. The Yin Yang Hypothesis, in "Adv. in Cyclic Nucleotides Res." vol. 5, Edited by G.I. Drummond, P. Greengard and G.A. Robison, Raven Press, pp. 307-330.

Gorman, R.R., Bunting, S. and Miller, O.V. (1977) Modulation of human platelet adenylate cyclase by Prostacyclin (PGx).Prostaglandins, 13, 377-388.

Hamberg, M. and Samuelsson, B. (1974) Prostaglandin endoperoxides. Novel transformations of arachidonic acid in human platelets. Proc. Natl. Acad. Sci. USA, 71, 3400-3404.

Hamberg, M., Svensson, J. and Samuelsson, B. (1975) Thromboxanes: a new group of biologically active compounds derived from prostaglandin endoperoxides. Proc. Natl. Acad. Sci. USA, 72, 2994-2998.

Haye, B., Champion, S. and Jacquemin, C. (1973) Control by TSH of a phospholipase A_2 activity, a limiting factor in the biosynthesis of prostaglandins in the thyroid. Febs Lett. 30, 253-259.

Hedqvist, P. (1973) Autonomic neurotransmission. In: The Prostaglandins vol. 1, Edited by P.W. Ramwell, Plenum Press, pp. 101-131.

Hidaka, H. and Asano, T. (1977) Stimulation of human platelet guanylate cyclase by unsaturated fatty acid peroxides. Proc. Natl. Acad. Sci. USA, 74, 3657-3661.

Hong, S.-C., L., Polsky-Cykin, R. and Levine, L. (1976) Stimulation of prostaglandin biosynthesis by vasoactive substances in methylcholantrene-transformed mouse BALB/373. J. Biol. Chem., 251, 776-780.

Hossmann, K.A. and Sato, K. (1971) Effect of ischemia on the function of the sensorimotor cortex in cat. Electroenceph. Clin. Neurophysiol. 30, 535-545.

Isakson, P.C., Raz, A., Denny, S.E., Wyche, A. and Needleman, P. (1977) Hormonal stimulation of arachidonate release from isolated perfused organs, relationship to prostaglandin biosynthesis. Prostaglandins 14, 853-873.

Jaffe, B.M., Parker, C.W., Marshall, G.R. and Needleman, P. (1970) Renal concentration of prostaglandin E in acute and chronic renal ischemia. Biochim. Biophys. Res. Comm. 49, 799-806.

Kantrowitz, F., Robinson, D.R. and McGuire, M.B. (1975) Corticosteroids inhibit prostaglandin production by rheumatoid synovia. Nature 258, 737-739.

Kimura, H., Thomas, E. and Murad, F. (1974) Effects of decapitation, ether and pentobarbital on guanosine 3',5'-phosphate and adenosine 3',5'-phosphate levels in rat tissues. Biochim. Biophys. Acta 343, 519-528.

Kuehl, F.A.Jr. (1974) Prostaglandins, cyclic nucleotides and cell function. Prostaglandins, 5, 325-340.

Kunze, H. (1973) Stimulation of phospholipases in synaptosomes of guinea pig brain. Naunyn Schmiedebergs Arch. Pharmacol. 277, R41.

La Torre, E., Patrono, C., Fortuna, A. and Grossi-Belloni, D. (1974) Role of prostaglandin $F_{2\alpha}$ in human cerebral vasospasm. J. Neurosurg. 41, 293-298.

Levine, S. and Payan, H. (1966) Effects of ischemia and other procedures on the brain and retina of the gerbil (Meriones Ungulatus). Exptl. Neurol., 16, 255-262.

Lust, W.D., Kobayashi, M., Mrsulja, B.B., Wheaton, A. and Passonneau, J.V. (1977) Cyclic nucleotide levels in the gerbil cerebral cortex, cerebellum and spinal cord following bilateral ischemia in "Tissue hypoxia and ischemia" Edited by M. Reivich, R. Coburn, S. Lahiri and B. Chance, Plenum Press, pag. 287-299.

Lust, W.D., Mrsulja, B.B., Mrsulja, B.J., Passonneau, J.V. and Klatzo, I. (1975) Putative neurotransmitters and cyclic nucleotides in prolonged ischemia of the cerebral cortex. Brain Res., 98, 394-399.

Mao, C.C., Guidotti, A. and Costa, E. (1975) Evidence for an involvement of GABA in the mediation of the cerebellar cyclic GMP decrease and the anticonvulsant action of diazepam. Naunyn-Schmiedeberg's Arch. Pharmacol. 289, 369-378.

McGiff, J.C., Crowshaw, K., Terragno, N.A. and Lonigro, A.J. (1970) Release of prostaglandin like substance into renal blood flow in response to angiotensin. II. Circ. Res. Suppl. 26, 121-127.

Miller, O.V., Johnson, R.A. and Gorman, R.R. (1977) Inhibition of PGE_1 stimulated cAMP accumulation in human platelets by thromboxane A_2. Prostaglandins 13, 599-689.

Nakano, J., Prancan, A.V. and Moore, S.E. (1972) Metabolism of prostaglandin E_1 in the cerebral cortex and cerebellum of the dog and the rat. Brain Res. 39, 545-548.

Nathanson, J.A. (1977) Cyclic Nucleotides and Nervous System Function. Physiological Reviews, 57, 157-256.

Nicosia, S. and Galli, G. (1975) A mass fragmentographic method for the quantitative evaluation of brain prostaglandin biosynthesis. Prostaglandins, 9, 397-403.

Pace-Asciak, C.R. and Rangaraj, G. (1976) Prostaglandin biosynthesis and catabolism in the developing fetal sheep brain. J. Biol. Chem. 251, 3381-3385.

Paoletti, R., Folco, G.C., Spagnuolo, C. and Terzi, C. (1978) Relationship between the prostaglandin system and responsiveness to epileptogenic agents in mature and immature rats. in "Prostaglandins and Perinatal Medicine" Edited by F. Coceani and P.M. Olley, Raven Press, pag. 191-199.

Ramwell, P.W., Shaw, J.E. and Jessup, R. (1966) Spontaneous and evoked release of prostaglandins from frog spinal cord. Am. J. Physiol. 211, 998-1004.

Rodriguez De Turco, E.B., Cascone, G.D., Pediconi, M.F. and Bazan, N.G. (1977) Phosphatidate, phosphatidylinositol, diacylglycerols and free fatty acids in the brain following electroshock, anoxia or ischemia, in "Function and Biosynthesis of Lipids", Edited by N.G. Bazan, R.R. Brenner and N.M. Giusto, Plenum Press, pag. 389-397.

Said, S.I., Yoshida, T., Kitamura, S. and Wreim, C. (1974) Pulmonary alveolar hypoxia: Release of prostaglandins and humoral mediators. Science, 185, 1181-1183.

Samuelsson, B. (1964) Identification of a smooth muscle-stimulating factor in bovine brain. Prostaglandin and related factors 25. Biochim. Biophys. Acta, 84, 218-219.

Samuelsson, B. (1965) On the incorporation of oxygen in the conversion of 8,11,14-eicosatrienoic acid to prostaglandin E_1. J. Am. Chem. Soc., 87, 3011-3013.

Sattin, A. and Rall, T.W. (1970) The effect of adenosine and adenine nucleotides on the cyclic adenosine 3',5'phosphate content of guinea pig cerebral cortex slices. Mol. Pharmacol., 6, 13-2 .

Schmidt, M.J., Schmidt, D.E. and Robison, G.A. (1972) Cyclic AMP in the rat brain: microwave irradiation as a means of tissue fixation. In "Advances in Cyclic Nucleotide Research" Edited by Greengard,P., Paoletti, R. and Robison, G.A., Raven Press, pp. 425-434.

Siesjo, B.K., Nordstrom, C.H. and Rehncrona, S. (1977) in "Tissue Hypoxia and Ischemia" Edited by M. Reivich, R. Coburn, S. Lahiri and B. Chance, Plenum Press, pag. 261-271.

Siggins, G., Hoffer, B. and Bloom, F. (1971) Prostaglandin-norepinephrine interactions in brain: microelectrophoretic and histochemical correlates. Ann. N.Y. Acad. Sci. 180, 302-323.

Sinclair, A.J. and Crawford, M.A. (1972) The accumulation of arachidonate and docosahexaenoate in the developing rat brain. J. Neurochem. 19, 1753-1758.

Swaab, D.F. and Boer, K. (1972) The presence of biologically labile compounds during ischemia and their relationship to the EEG in rat cerebral cortex and hypothalamus. J. Neurochem. 19, 2843-2853.

van den Bosch, H. and van den Besselaar, A.M.H.P. (1978) Intracellular formation and removal of lysophospholipids in "Phospholipases and Prostaglandins" Edited by C. Galli, G. Galli and G. Porcellati, Raven Press, pag. 69-75.

Vargaftig, B.B. and Dao Hai, N. (1972) Selective inhibition by mepacrine of the release of "rabbit aorta contracting substance" evoked by the administration of bradykinin. J. Pharm. Pharmacol. 24, 159-161.

Veale, W.L., Cooper, K.E. and Pittman, Q.I. (1978) Prostaglandin action on central thermoregulation and fever: ontogenetic aspects, in "Prostaglandins and Perinatal Medicine" Edited by F. Coceani and P.M. Olley, Raven Press, pag. 199-215.

Veech, R.L., Harris, R.L., Veloso, D. and Veech, E.H. (1973) Freeze-blowing: a new technique for the study of brain in vivo. J. Neurochem. 20, 183-188.

Vogt, W. (1978) Role of phospholipase A_2 in prostaglandin formation in "Phospholipases and Prostaglandins" Edited by C. Galli, G. Galli and G. Porcellati, Raven Press, pag. 89-97.

Wennmalm, A., Phan-Hou-Chank and Junstad, M. (1974) Hypoxia causes prostaglandin release from perfused rabbit hearts. Acta Physiol. Scand. 91, 133-135.

Wolfe, L.S. (1975) Possible role of prostaglandins in the nervous system. In "Advances in Neurochemistry" Edited by Agranoff B.W. and Aprison M.H., Plenum Press, p. 1-49.

Wolfe, L.S., Pappius, H.M. and Marion, J. (1976a) The biosynthesis of prostaglandin by brain tissue in vitro. In "Advances in Prostaglandin and Thromboxane Research", vol. 1, Edited by B. Samuelsson, and R. Paoletti, Raven Press, N.Y., pp. 345-356.

Wolfe, L.S., Rostworowski, K. and Marion, J. (1976b) Endogenous formation of the prostaglandin endoperoxide metabolite, thromboxane B_2, by brain tissue. Biochem. Biophys. Res. Comm. 70, 907-913.

Wolfe, L.S. (1978) Some facts and thoughts on the biosynthesis of prostaglandins and thromboxanes in brain. In "Prostaglandins and Perinatal Medicine" vol. 4, Edited by F. Coceani and P.M. Olley, Raven Press, p. 215-221.

PHOSPHOLIPID AND ITS METABOLISM IN ISCHEMIA

G. Porcellati[1], G.E. De Medio[1], C. Fini[1], A. Floridi[1], G. Goracci[1], L.A. Horrocks[2], J.W. Lazarewicz[3], C.A. Palmerini[1], J. Strosznajder[3], and G. Trovarelli[1]

[1]Department of Biochemistry, University of Perugia, Italy, [2]Department of Physiological Chemistry, The Ohio State University, Columbus, Ohio, U.S.A. and [3]Department of Neurochemistry, Medical Research Center, Polish Academy of Sciences, Warsaw, Poland

INTRODUCTION

Several models are known to produce ischemic lesions in brain, including those of the anoxic-schemic type. Siesjö et al. (1977) have recently reviewed the subject. One of the techniques to study ischemia is the artery occlusion approach, which has been improved either by combining ischemia with anoxic treatments in normal rat or by adopting, as the animal source, the desert rodent, the Mongolian gerbil (meriones unguiculatus) as reported by Levine and Payan (1966). The basis and advantages for using this last procedure, which will be here classified as the "gerbil model", are well known (Levine and Payan, 1966 ; Levy and Duffy, 1977 ; Kobayashi et al., 1977). Almost all of bilaterally ligated gerbils display the characteristic biochemical changes and positive neurological signs of ischemia (see Kobayashi et al., 1977).

Decapitation, another model of ischemia, converts the brain into a closed biological system, and has been widely used in the past for studying various biochemical parameters. However, the model which will be here classified as that of postdecapitative ischemia, is not known whether is really comparable to ischemic processes, and already differences for some biochemical changes have been found between this and the gerbil model (Kobayashi et al., 1977).

Various contributions on structural and metabolic changes of brain lipid have been reported in the past following postdecapitative ischemia (see Bazàn, 1976 and Rodrìguez de Turco et al., 1977), but no reports on these aspects have been reported for the gerbil model. The present work, therefore, deals with some studies carried out on lipid changes occurring in brain during bilateral ischemia induced in the gerbil by ligation of both common carotid arteries, with the aim of comparing the results with those obtained with the postdecapitative model. In addition, extension will be made of our previous works (Lazarewicz et al., 1972) on some aspects of phospholipid metabolism and functional impairment of rat brain mitochondria during postdecapitative ischemia. Brief observations on the mechanisms of the metabolic changes during general ischemia will finally be presented.

Abbreviations : FFA, free fatty acids ; DG, diacyl glycerols ; TG, triacyl glycerols ; CPG, choline phosphoglycerides ; EPG, ethanolamine phosphoglycerides ; GPC, glycero-3-phosphorylcholine ; GPE, glycero-3-phosphorylethanolamine ; GPI, glycero-3-phosphorylinositol ; GPS, glycero-3-phosphorylserine ; IPG, inositol phosphoglycerides ; PC, phosphorylcholine ; PE, phosphorylethanolamine ; CDPC, cytidine-5'-diphosphate choline ; CDPE, cytidine-5'-diphosphate ethanolamine ; CMP, cytidine-5'-monophosphate.

A BRIEF REVIEW OF THE PAST WORK

This review will deal chiefly with changes of lipid metabolism during postdecapitative ischemia, with exceptional reference to other models, if necessary.

Lunt and Rowe (1968) and Bazàn (1970) observed an increase of the FFA pool in the mouse and rat brain during the first few minutes from decapitation. A two- to four-fold increase was found in the rat brain 4-5 min from decapitation by others (Bazàn et al., 1971 ; Cenedella et al., 1975 ; Galli and Spagnuolo, 1976), the most pronounced changes occurring always in the first few minutes. The increase was considered a sign of enzymic nature occurring rapidly and linearly during the early phase of the ischemic process, rather than post mortem autolytic degradation. Increases were chiefly observed in stearic (18:0) and arachidonic (20:4) acids, followed by palmitic (16:0), oleic (18:1) and docosahexaenoic (22:6) acids (Bazàn, 1976), although some different results were obtained in other laboratories. All these findings were extended by Majewska et al. (1974), who found a three-fold increase of the FFA content in ischemic rat brain mitochondria after 1 min from decapitation ; the authors were also able to correlate these changes with functional mitochondrial impairements occurring during the same time (Lazarewicz et al., 1972, 1975). The increase of the FFA levels was also observed in brain microsomal particles of rats undergoing postdecapitative ischemia (Rossowska et al., 1977) or hypoxia (Strosznajder et al., 1978 a), being accompanied by a significant decrease of their CPG and EPG content (Rossowska and Dabrowiecki, 1978 ; Strosznajder et al., 1978 b) and by modification of their Na^+-K^+-ATPase and glucose-6-phosphate phosphohydrolase activities (Rossowska et al., 1977 ; Rossowska and Dabrowiecki, 1978).

Banschbach and Geison (1974) and Aveldaño and Bazàn (1975 b) have reported noticeable increase of the DG pool in the rat and mouse brain following decapitation, with much smaller changes of TG content (Bazàn, 1976). Moreover, the concentrations of alkenylacyl-sn-GPE and -GPC were reported to decrease in whole brain or brain microsomes during postdecapitative ischemia and hypoxia (Strosznajder et al., 1978 a, 1978 b). This decrease, which occurs at brief time intervals from the ischemic treatment, has been recently related to the increased plasmalogenase activity of brain tissue in the gerbils after bilateral ligation of common carotid arteries (Horrocks et al., 1978).

Ischemic processes inhibit also lipid synthesis from labelled acetate, glucose or appropriate phospholipid precursors (Strosznajder et al., 1978 a, 1978 b), a finding already reported for hypoxic brain by Kosow et al. (1966). Noticeable inhibition of plasmalogen synthesis has been found in brain microsomes at short time intervals from decapitation (Strosznajder and Dabrowiecki, 1977). Ischemia produces also a rapid inhibition of arachidonate incorporation into brain lipids (Rodrìguez de Turco et al., 1977).

The decrease of phospholipid concentration during brain ischemia is therefore thought to be due not only to activation of catabolic processes, but also to inhibition of synthetic mechanisms. The phospholipid classes indicated to be the most likely candidates for rapid breakdown upon ischemia are IPG and the plasmalogens, although no direct proof exists as yet for this assumption. Possible involvements of phospholipases A (see Bazàn, 1976) and plasmalogenase (Horrocks et al., 1978) are also at play.

THE EXPERIMENTAL MODEL

Ligation has been carried out in adult gerbils (50-60 g body wt.), obtained from Donald Robinson (Tumble Brook, Mass., U.S.A.), for both common carotid arteries with the aid of surgical threads (Kobayashi et al., 1977) in order to produce bilateral ischemia without ambiguity. As known, in fact, there is still a low percentage of successful unilateral ischemia in the same animals (Levine and Payan, 1966 ; Kobayashi et al., 1977). Only short time intervals from ligation were investigated (30", 60", 180" and 300 "), since the main interest of the present work was to examine real and rapid biochemical changes, readily reversible after appropriate recirculation (see Kobayashi et al., 1977 ; Mazzari and Finesso, 1978) and which would certainly not compromise membrane integrity in an irreversible way. Mortality after ligation was never observed, neither tonic-clonic seizures, which otherwise would have affected severely the results. Barbiturate anaesthesia was not used during the surgical intervention to avoid misleading results (Levy and Duffy, 1977 ; Majewska et al., 1974, 1977) and subcutaneously injected 1% novocaine (50 µl) was adopted. This procedure might not have avoided, however, stress-conditioned phenomena. Sham surgery was carried out for all controls. Anaesthesia was completely avoided to inject intraventricularly labelled lipid precursors (Yau and Sun, 1974), two hours before ligation.

Liquid nitrogen was used to inactivate the tissue. This and/or microwave treatment were found similarly to have no effect on the concentration of brain hydrosoluble compounds and FFA (see Medina et al., 1975 and Cenedella et al., 1975). Freezing the excised gerbil head gave reliable results : the cooling was rapid (0.1-0.2 min), due also to the small size of the animals (see Banschbach and Geison, 1974 and Levy and Duffy, 1975). To minimize thawing, all manipulations of the frozen brain were done with the sample kept in contact with liquid nitrogen, following reported reccomendations (Bazàn et al., 1971). The brains, weighed frozen, were immediately homogenized in lipid solvents. As a check of our technical procedure, estimation of DG in the normal rat brain (GLC procedure), removed and treated as described or inactivated by focused microwave irradiation (model 70-50 microwave oven, Litton Microwave Co., 2450 Mhz, 1330 W, 3 sec) gave values, respectively of 173 (3) and 158 (3) nmol/g wet wt., which agreed rather well with the levels reported by Sun (1970) and Banschbach and Geison (1974).

Estimations have been made of the chief high-energy phosphates and related compounds in the brain of the gerbils at different time intervals from ligation (Mazzari and Finesso, 1978) to monitor each time the changes due to ischemia[*]. The data were in substantial agreement with previous results (Kobayashi et al., 1977) and provided evidence that the energy status of the brain was severely compromised. Data for the sham-operated gerbils were not different from already published values (Levy and Duffy, 1975).

GENERAL METHODOLOGY

Lipid extraction and analysis. Lipid from frozen brains (two for each experimental point) was extracted according to Folch-Pi et al. (1957). The chloroform contained 0.1%

---

[*] Thanks are given to Dr.S.Mazzari and Mr.M.Finesso for controlling these parameters.

2,6-di-tert-butyl-p-cresol, as antioxidant. The lipids were dissolved in 5 ml of chloroform-methanol (98:2). 50 µl of this solution was used for radioactivity determinations and the remainder passed through a Bio-Sil A (Bio-Rad Labs., Richmond, Ca., U.S. A.) column (5 g) equilibrated with the above mentioned solvent system. The neutral and polar lipids were eluted respectively with 100 ml of chloroform-methanol (98:2) and 150 ml of methanol.

The neutral lipid-containing fraction was concentrated in vacuo, dissolved in 2 ml of chloroform and, after removal of 50 µl for radioactivity measurements, resolved by TLC on silica gel G plates (0.5 mm) activated at 120°C for 30 min, with petroleum ether (60°-70°)-diethyl ether-acetic acid (80:20:1) as the solvent (Mangold, 1969). Stearic acid was used as standard. The FFA-, DG- and TG-containing spots were eluted with 40 ml each of antioxidant-containing chloroform-methanol (98:2). After removal of the solvent in vacuo, extracts were kept under N_2 in n-hexane at -20°C until use.

FFA fraction was dissolved in 5 ml of chloroform. 0.5 ml were used for radioactivity determinations and 2 ml colorimetrically for mass estimation (Cenedella et al., 1975), by using stearic acid as the standard. Extraction and chromatographic procedures for estimating lipid components do not generate FFA from phospholipids (Bazàn et al., 1971 ; Cenedella et al., 1975). The remaining 2.5 ml were methylated with diazomethane, after addition of 17:0 as internal standard, and gas-chromatographed on glass column (180 x 3 mm) with the Fractovap GV model (Carlo Erba, Milan, Italy) using 10 % SP-2330, as stationary phase on Cromosorb WAW (Supelco, Inc., Bellefonte, Penn.,U.S.A.).

DG or TG fractions were supplemented with appropriate amounts of 17:0 and dissolved in 2 ml of chloroform. 0.5 ml were used for radioactivity measurements and the remaining 1.5 ml saponified under controlled conditions (Kates, 1972). Glycerol from DG or TG was assayed according to Wieland (1974), using purified distearin or tripalmitin, as standards, whereas fatty acids from the same lipids were determined, as previously reported.

The polar lipid fraction was dissolved with 5 ml of chloroform-methanol (2:1), as above. Total P was determined (Ernster et al., 1950) on 2 µl, total radioactivity on 10 µl and individual phospholipid classes or subclasses on 50 µl (Horrocks and Sun, 1972). TLC was carried out by including always purified monoacyl-sn-GPC and monoacyl-sn-GPE standards. For each experimental point three plates were run, often in duplicate, for determinations of radioactivity content (Porcellati and Goracci, 1976), P content (Binaglia et al., 1973) and fatty acid constitution. For this last purpose, the spots were scraped off into stoppered test-tubes, to which 1 ml of 3% H_2SO_4-containing benzene-methanol (1:1) and appropriate amounts of 20:0 were added. After methanolysis for 60 min at 80°C, the tubes were cooled, the content extracted three times with 3 vol of petroleum ether (50°-70°), concentrated in vacuo, and analyzed by GLC, as reported previously.

Radioactivity measurements were carried out as reported elsewhere (Porcellati and Goracci, 1976) by setting the 3330 Packard liquid scintillation spectrometer for doubly labelled compound estimation with the use of external standard standardization.

Analysis of water-soluble compounds. Analysis of the water-soluble compounds was carried out on the residue and washings from the first Folch extraction (see previous paragraph), by treating them with 10 ml each of 0.5 N perchloric acid (PA) and centrifuging. The solution was adsorbed on a 1 x 3 cm alumina column (Woelm, 1st grade) equilibrated with 0.5 N PA. After washing (50 ml of H_2O), elution was performed with 0.5 N NH_4OH (about 30 ml). Eluate was directly chromatographed on a 0.7 x 17 cm AG 1 x 4 (formiate form) column equilibrated with H_2O. After adsorption and washing (50 ml of H_2O) elution was carried out with 0.1 N formic acid (40 ml). Eluate was dried in vacuo at 37°C, taken up in 1.5 ml of 2-methyl-2-amino-1-propanol (MAP) - 0.02 N NaCl puffer (pH 10.5). Aliquots were then used for determining water-soluble, labelled or unlabelled phospholipid precursors and CMP (Floridi et al., 1978) by high resolution liquid liquid chromatography. Analysis was carried out on Aminex A 14 resin, 20 $\pm$ 3 μ (Cl^- form), as stationary phase (Bio-Rad Labs., Richmond, Ca., U.S.A.) with the Pye-Unicam LC 20 Liquid Liquid Chromatography apparatus equipped with detectors. Estimation was carried out as reported elsewhere (Floridi et al., 1978).

Isotope administration. Animals were injected intraventricularly and simultaneously without anaesthesia with 8 μCi of [methyl-^{3}H] choline (27.8 nmoles, S.A. of 287.7) and 0.4 μCi of [1-^{14}C] arachidonic acid (6.64 nmoles, S.A. of 60.2), by following previously published procedures (Sun, 1977). After 120 min from injection, gerbils were ligated or sham-operated, and analyses carried out as reported before. Two similarly-treated gerbils (either controls or not) were constantly used for each experimental point. Each analysis was normally carried out in quadruplicate.

Experiments with brain isolated mitochondria. Ischemia was produced by incubating the decapitated rat heads for 5 min at 37°C. Mitochondria, isolated and purified from from cerebral cortex (Lazarewicz et al., 1974), were used either for FFA determination or phospholipid assay or ^{45}Ca uptake measurements. FFA were extracted and determined according to Dole and Meinertz (1960) and Itaya and Ui (1965), respectively. Methyl esters were formed according to McGee (1974) and separated by GLC with a Pye-Unicam S-104 model (Rossowska et al., 1977). Phospholipids were extracted according to Folch-Pi et al. (1957), separated according to Horrocks and Sun (1972) and their P content determined according to Bartlett (1959). In the experiments describing ^{45}Ca uptake, mitochondria (0.3 mg of protein/ml) were incubated at 37°C in 0.3 M sucrose, 5 mM Tris-HCl puffer (pH 7.4), 2 mM Na-phosphate (pH 7.4), 5 mM KCl, 5 mM $MgCl_2$, 5 mM succinate, 2.5 μM rotenone, 3 mM ATP and 1 mM $^{45}CaCl_2$ (0.5 μCi/μmol). Calcium accumulation was determined according to Carafoli and Lehninger (1967) and the values corrected for passively bound ^{45}Ca.

RESULTS AND DISCUSSION

Lipid Metabolism in the Gerbil

Fatty acids. Fig.1 shows the noticeable increase of the FFA pool which takes place in gerbil brain after bilateral carotid ligation. A 2.3-fold increase is shown at 5 min (2.2-fold in the colorimetric estimation), which agrees rather well with values found in postdecapitative ischemia (Bazàn et al., 1971 ; Cenedella et al., 1975 ; Avel-

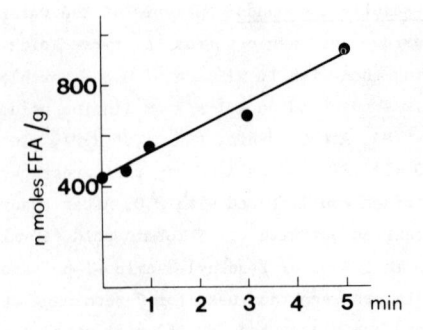

Fig.1. - FFA content of gerbil brain at various times after ligation. Data expressed as nmoles/g wet wt., relative to stearic acid, as standard. Each point represents data from six to ten brains. The values obtained by GLC (present figure) were similar to those calculated colorimetrically. Standard deviation values between 10-15 %.

daño and Bazàn, 1975 a ; Galli and Spagnuolo, 1976). Detectable changes are already evident at 30 sec from ligation. The starting level of the FFA pool, however, is higher than that found by the mentioned authors, although very rapid freezing were used in our studies. Stress-conditioned phenomena or species differences may be at play in explaining this variation.

The major components of FFA were palmitic (25%), stearic (27%), oleic (21%), arachidonic (15%) and docosahexaenoic (2-3%) acids. The composition of this pool is not different from published values, exception being made of the higher 20:4 and the lower 22:6 amounts. Bilateral ischemia has shown to produce a noticeable increase of the relative 20:4 and 22:6 concentrations (increases by 50 and 62 %, respectively) at 5 min from ligation, while those of 16:0, 18:0 and 18:1 did not change. These variations do not seem different from those obtained after postdecapitative ischemia, although, on the whole, gerbils appear to behave during ischemia more similarly to mouse (Aveldaño and Bazàn, 1975 a) than to rat (Cenedella et al., 1975 ; Galli and Spagnuolo, 1976). When expressed in absolute amounts, the change observed for the 20:4 and 22:6 concentrations are more evident. Fig.2 shows that the concentrations of these two fatty acids are tripled at 5 min from ligation. Moreover, at 5 min that of 18:0 is doubled and those of 16:0 and 18:1 increase by 73 % and 85 %, respectively (results not shown in the figure).

FFA have been determined in whole blood of gerbils, and were found to possess the following main composition : 16:0 (37.2 %), 18:0 (22 %), 18:1 (24.5 %), 18:2 (12 %), 20:4 (3.8 %) and 22:6 (traces). The noticeable difference between this composition and that of brain would exclude that the FFA obtained from the brain samples could have been substantially contaminated by the blood.

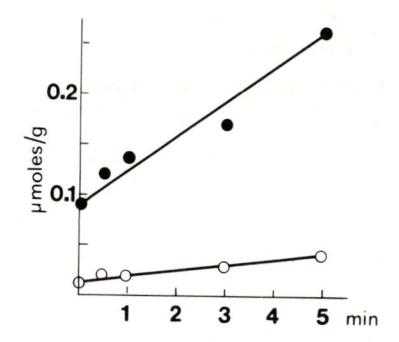

Fig.2.- Brain content (μmoles/g wet wt.) of arachidonic acid (20:4, ●-●-●-●) and docosahexaenoic acid (22:6, O-O-O-O) at various times after ligation in the gerbil. The per cent of 20:4 over total lipid pool at the indicated times was 1, 1.2, 1.4, 2.3 and 2.7. Each point represents data from six brains. Standard deviation values within 10 %.

Neutral glycerides. Table I indicates that DG concentration does not change within the interval examined or slightly increases at very short time intervals. The starting levels vary from those reported by Sun (1970) and Banschbach and Geison (1974), being two- to three-fold higher. The difference is not due to methodology (see the section on "The Experimental Model") and might be due either to species differences or to stress-conditioned phenomena. The presence of a very effective lipase acting on endogenous DG (Cabot and Gatt, 1976), presumably highly stimulated in gerbil brain at short times from ligation, and which superimposes its catabolic effect upon DG production at very early times of damage, might also explain the relatively small increases which take place in the DG pool during ischemia. The fatty acid composition of the DG pool at 0-30'''' was similar to that observed at 3 min of postdecapitative ischemia (Aveldaño and Bazàn, 1975 b) or noticed at "zero time" by Sun (1970) who did not use particular freezing procedures. This point would support the above mentioned assumptions.

When the 20:4 content of DG is reported as the per cent of the 20:4 content of total lipid, a significant increase (from 10 to 30 %) is seen at early stages from ligation (Table II). Again, the 20:4 content of DG, when reported in absolute values, increases from mean starting levels of 0.16 to 0.23 and 0.20 μmoles/g wet wt., at 30'''' and 1', respectively (increases of 43 % and 25 %), and that of 18:0 from 0.198 to 0.220 and 0.222, respectively. Therefore, a small enrichment in DG of arachidonate and probably of stearate takes place during ischemia in the gerbil model, but this could have been probably higher without the concomitant effect of very active lipase activities. It is worth to say that the noticeable increase of the 20:4 in the FFA pool (see legend of figure 2) follows to this initial enrichment of the DG pool indicating that the two events are linked together.

TABLE I

The Change of Diglyceride and Triglyceride Content of Gerbil Brain at various Times after Ligation

Time after ligation (sec)	Diglyceride	Triglyceride
0	0.36	0.24
30	0.42	0.24
60	0.35	-
180	0.37	0.28
300	0.39	0.25

Determinations carried out by GLC, as reported in the text. The values were similar to those obtained by glycerol assay. Data expressed as μmoles/g wet wt. Mean values from six to ten brains. Standard deviation values within 10 - 15 %.

The content of brain TG does not change during the time intervals investigated, as reported in Table I, nor does its fatty acid constitution. A relative stability of the brain TG pool during postdecapitative ischemia has also been found by Aveldaño and Bazàn (1975 a).

TABLE II

Changes of Arachidonic Acid Content in some Neutral and Polar Lipids of Gerbil Brain at various Times after Ligation

Lipid	\multicolumn{5}{c}{Time after ligation (sec)}
	0	30	60	180	300
Diacyl-GPC	25.9	25.8	26.6	20.2	20.6
Alkenylacyl-GPE	25.8	28.4	21.0	20.1	21.4
Diacyl-GPI *	2.5	1.9	2.1	1.9	2.5
Diacyl-glycerol	2.2	2.8	2.4	2.1	2.0

Estimation of 20:4 was carried out, as explained in the text. Data expressed as per cent of the total 20:4 occurring in polar and neutral lipid fractions (excluding the FFA fraction). Data were excluded for lipids not undergoing significant variation. Mean value (six brains) of total 20:4 for control gerbils (excluding FFA) : 7.09 μmol/g wet wt.

Polar lipids. Bilateral carotid ligation produces noticeable changes in the concentration of polar lipids in brain (De Medio et al., 1978), and decreases slightly the total phospholipid content. The most evident changes are shown in Fig.3. Alkenylacyl-GPE and alkenylacyl-GPC content decreases noticeably at 1 min from ligation reaching thereafter more stationary levels ; monoacyl-GPC levels increase within similar time intervals (Fig.3). The decrease in alkenylacyl-GPC is not accounted for by the increase in monoacyl-GPC, and this could be due to lysophospholipase activity which further deacylates the lysocompound. The data are in favour of an active plasmalogen hydrolysis during ischemia ; similar results have been reported during hypoxia or post-decapitative ischemia (Strosznajder et al., 1978 a, 1978 b). Plasmalogenase activity, on the other hand, has been found noticeably increased in the brain of gerbils during carotid ligation (Horrocks et al., 1978), and might be effectively involved in the breakdown of phospholipids during early ischemia (Horrocks et al., 1978). Other mechanisms may be, however, at play and will be discussed in the section which deals with water-soluble compounds. Sphingomyelin content also decreases during ischemia (De Medio et al., 1978). Diacyl-GPI levels did not change, but most probably they were on the contrary sensibly decreased, since the fraction runs together with monoacyl-GPE which is supposed, on the contrary, to increase (De Medio et al., 1978), due to the ethanolamine plasmalogen breakdown.

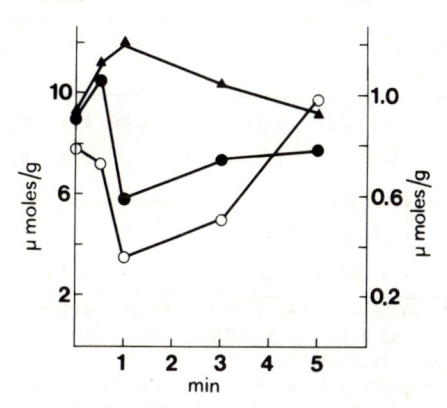

Fig.3.- Changes of alkenylacyl-GPE (o-o-o-o), alkenylacyl-GPC (●-●-●-●) and monoacyl-GPC (▲-▲-▲-) content of gerbil brain after ligation. Values on the left refers to alkenylacyl-GPE and those on the right to other two lipids. Data from six brains for each experimental point. Standard deviation values within 10%.

Table II indicates a decrease of 20:4 in diacyl-GPI, diacyl-GPC and alkenylacyl-GPC, while no change occurs (not shown) in the levels of this fatty acid in diacyl-GPS, diacyl-GPE, PA, TG and sphingomyelin. The finding of a not remarkable loss of 20:4 diacyl-GPI, contrary to previous suggestions (see Bazàn, 1976 and Rodrìguez de Turco, 1977), may be explained in view of the monoacyl-GPE impurity (whose concentrations on the contrary are supposed to increase) mentioned above.

On the whole, the results of this section would point to noticeable catabolic activities of brain tissue during ischemia, particularly on both plasmalogens, diacyl-GPI and probably diacyl-GPC.

Incorporation of labelled 20:4 into brain lipids. Table III shows the percentage distribution of 20:4 radioactivity of different lipids (two hours from isotope administration) at different time intervals from ligation. With preliminary experiments it has been observed that FFA and lipid radioactivity was due only to 20:4, as checked by argentation TLC and successive GLC of the separated fatty acids. Labelled species were only those containing 20:4. No characteristic changes take place in diacyl-GPS, PA, alkenylacyl-GPC and diacyl-GPE, with more or less evident decreases in diacyl-GPC and alkenylacyl-GPE and increases in monoacyl-GPC, FFA, DG and, unespectedly, in TG. The levels of 20:4 in diacyl-GPI appear rather stable, but presumably they decrease due to the contamination by radioactive monoacyl-GPE (see previous section).

TABLE III

The Percentage Distribution of Radioactivity (20:4) among Gerbil Brain Lipids at various Times from Ligation

Time from ligation (sec)	PS	PA	PC	CP	LPC	PE	EP	PI*	FFA	DG	TG
0	2.1	1.2	52.0	4.3	7.2	5.7	2.4	8.1	1.5	2.1	6.7
30	2.3	1.1	52.1	4.4	–	5.5	2.0	9.2	2.3	2.1	7.5
60	1.6	1.0	46.2	3.7	9.3	4.9	–	8.9	4.0	2.9	11.3
180	2.8	1.4	39.9	–	8.6	6.7	2.1	–	5.5	3.0	9.3
300	2.2	1.2	46.8	4.2	5.1	5.5	2.2	8.6	4.4	2.6	8.3

Ligation was performed two hours after isotope administration (see text). Data are expressed as per cent of total lipid 20:4 radioactivity. Mean values from six brains. Standard deviation values within 10 %.
PS = diacyl-GPS ; PA = phosphatidic acid ; PC = diacyl-GPC ; CP = choline plasmalogen ; LPC = monoacyl-GPC ; PE = diacyl-GPE ; EP = ethanolamine plasmalogen ; PI = diacyl-GPI ; FFA = free fatty acids ; DG = diglycerides ; TG = triglycerides.

* Contains monoacyl-sn-GPE.

When values of incorporation of 20:4 are being worked out in terms of relative specific activities (i.e., % distribution of 20:4 radioactivity/ % of 20:4 of total phospholipid) for each lipid, then no particular lipid seemed to acquire different value of R.S.A. from 0 to 5 min of ischemia, except FFA (Fig.4). The variation in R.S.A. of FFA along with time, unlike that of other lipids, would indicate the probable derivation of the free 20:4 from different lipid sources, of higher specific activities than that of FFA at first times from ligation and of lower at later times. Small active pools of TG (specific activity at zero time of about 40), diacyl-GPI (specific activity of more than 4) and probably of diacyl-GPC are most likely candidates for the first event, while pools of plasmalogens (very low specific activities at zero time) might be involved for relatively later changes.

<u>Water-soluble metabolites</u>. Fig.5 shows the concentration of the main CPG and EPG metabolites, together with CMP, at different times from ligation. It is clear that CDP-choline levels increase rather linearly up to 3 min, being then accompanied by a rapid production of PC. Levels of CDP-ethanolamine, on the contrary, remain stable, while those of PE increase very rapidly at early stages of ischemia. CMP content inter-

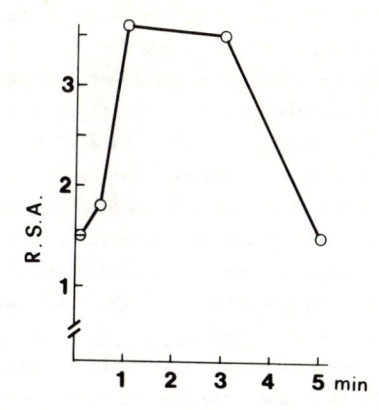

Fig.4.- Relative specific activity (R.S.A.) of the brain FFA at various time intervals from ligation in the gerbil. R.S.A. = per cent of distribution of radioactivity (see Table III) over per cent of total lipid 20:4 content. Single experiment with duplicate estimations. Comparable results obtained in similarly performed experiments.

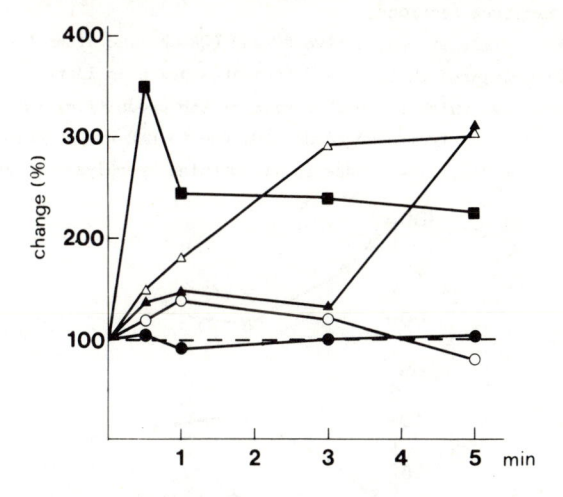

Fig.5.- Changes of concentration of water-soluble components in the gerbil brains at different times from ligation. ■-■-■-■-, PE ; -△-△-△-, CDP-choline ; ▲-▲-▲-▲-, PC ; O-O-O-O-, CMP ; -●--●-, CDP-ethanolamine. Data presented as change per cent of controls (= 100), shown by dotted line. Mean data from six brains. See the text for additional information and technical data.

estingly increases at first time intervals from ligation and then slowly returns towards control values. The values at 3 and 5 min for PE are influenced by the presence of a phosphorus-containing, UV-insensitive non-identified contaminating peak.

The data of Fig.5 point clearly to a probable effect of a phospholipase C activity upon EPG, whereas, due to the difference in the rate of appearance of extra CDP-choline and PC, this does not seem to be the important factor for the cleavage of CPG. Rather, in this last connection, brain tissue is able to display, like liver, a "back reaction" mechanism (Goracci et al., 1977), which is responsible, in the presence of excess of CMP, for the cleavage of endogenous diacyl-GPC (and probably also of alkenylacyl-GPC) back to CDP-choline and DG. On the other hand, the "back reaction" is much less active on EPG degradation (Goracci et al., unpublished observations). The decrease of the initially higher CMP concentration (Fig.5), which would obviously decrease the rate of the "back reaction", might be responsible for the decrease of plasmalogen breakdown shown to occur after 1 min (Fig.3). The successive increase of PC concentration shown in Fig.5 might be due to CDP-choline hydrolysis to PC and pyrophosphate. Phospholipase C may act also on sphingomyelin, whose levels decrease in ischemia (see previous sections), yielding PC. Phospholipases C acting on EPG (Williams et al., 1973) and sphingomyelin (Barenholz et al., 1966) have been described in brain tissue. It seems that the combined action of phospholipase C and of the "back reaction" mechanism is responsible for production of DG, which however do not seem to increase noticeably during ischemia, as a whole, owing probably to the very active neutral lipase influence, as reported in the previous sections.

Fig.6 shows the levels of radioactive PC and CDP-choline after the intraventricular administration of tritiated choline, and indicates again an increasing appearance of label into CDP-choline, which is another sign of the production, through "back reaction", of this nucleotide from endogenous CPG, due to CMP accumulation. The later decrease of CDP-choline labelling is due to its partial hydrolysis to PC and pyrophosphate.

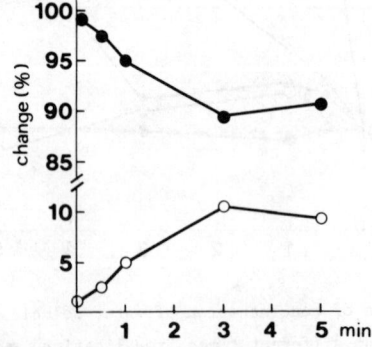

Fig.6.- Changes in radioactivity content (expressed as change % of the sum of labelled PC and CDP-choline) of PC (●-●-●-) and CDP-choline (o-o-o-) in gerbil brain after ligation. Choline was eluted from column before analysis ; GPC was not examined. Data from 4 brains.

Lipids in Mitochondria during Ischemia

Table IV shows that post-ischemic degradation of brain phospholipid characterizes not only whole brain and microsomal fraction after decapitation (Rossowska and Dabrowiecki, 1978), but also brain mitochondria. The most pronounced hydrolysis of CPG and EPG was observed. It is known that proper structural composition of lipids, which constitute the backbone of biological membranes, is a prerequisite for functional integrity of membranes and of certain phospholipid-requiring metabolic processes. The results of Table IV might explain the close relationships which might exist between lipid composition and impaired functional activity of mitochondria (Lazarewicz et al., 1972) during ischemia.

TABLE IV

The Phospholipid Content of Rat Brain Mitochondrial Fraction during Postdecapitative Ischemia

Phospholipid	Data (μg P/mg protein) in Control rats	Ischemic rats
Total phospholipids	17.88	13.10
Choline phosphoglycerides	7.73 (43.2)	5.56 (42.4)
Diacyl-GPE	2.65 (14.8)	2.03 (15.5)
Alkenylacyl-GPE	4.19 (23.4)	2.77 (21.2)
Diacyl-GPS + Diacyl-GPI	1.43 (8.0)	1.16 (8.9)
Sphingomyelin	1.27 (7.1)	1.00 (7.6)
Diphosphatidylglycerol	0.61 (3.4)	0.58 (4.4)

Mean values from three experiments. Per cent of total phospholipid P is shown in brackets. See the text for experimental details.

Table V shows a two-fold increase of FFA content in brain mitochondria after 5 min of postdecapitative ischemia. The increase consists mainly of arachidonic and oleic acids, but the increase of palmitic and stearic acid content was also significant. These results suggest the probable participation of both phospholipases, A_1 and A_2, in the release of FFA. This phenomenon might be at least partially responsible for the observed disfunctions of intracellular brain organelles and mitochondria (Lazarewicz et al., 1972, 1975).

One of the above mentioned effects could be reflected in the decreased ability of brain ischemic mitochondria for massive calcium accumulation, which is inhibited by about 40 %. in the conditions reported in Fig.7. In order to distinguish between effects of phospholipid hydrolysis and FFA increase on deranged mitochondrial calcium transport, additional experiments were performed. More precisely, before calcium accumulation, control brain mitochondria were pretreated with phospholipase A_2 and phospholipase C with subsequent removal of products of degradation with bovine serum albumin. It appeared that the treatment of mitochondria with both phospholipases inhibits calcium accumulation by 50 %. This effect was surprisingly independent within some limits on the amount of the enzyme added, which might suggest the existence of specific phospholipid pools responsible for calcium transport.

TABLE V

The Level of Fatty Acids in Rat Brain Mitochondrial Fraction during Postdecapitative Ischemia

Free Fatty Acids	Content in Control rats	Ischemic rats
16:10	7.5	13.6
16:1	0.4	0.8
18:0	7.1	11.0
18:1	5.5	12.6
20:4	2.7	10.8
Total	23.2	48.8

Mean values from three experiments, expressed as nmoles/mg protein. See the text for experimental details.

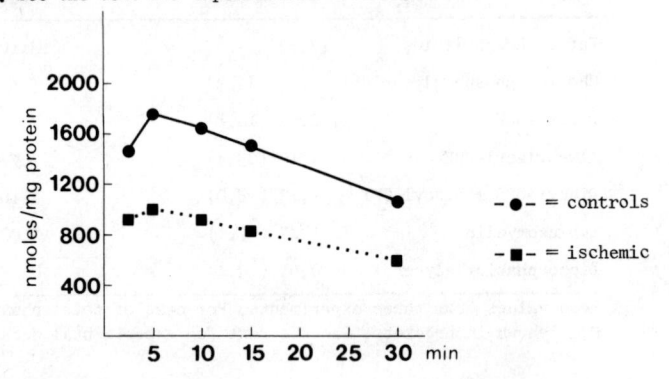

Fig.7.- Calcium accumulation in rat brain mitochondrial fraction during postdecapitative ischemia. See the text, for details.

The increase of FFA seems to play an important role in the injury of brain mitochondrial membranes induced by ischemia, and the present results substantiate this hypothesis (Lazarewicz et al., 1972, 1975). It is worth mentioning that membrane phospholipid structure, which is altered during ischemia, is also of great significance for this aspect. The answer to this problem must wait the elucidation of which phospholipid molecular species are the most affected by the ischemic process.

The relationship between FFA release and calcium seems to be of great interest (Lazarewicz et al., 1972, 1975, 1978) and ischemia alters mitochondrial function linked to a correct calcium metabolism (Ozawa et al., 1967). It is not known whether FFA released in ischemia probably under the effect of cyclic-AMP, which increases during this condition (see Lust et al., 1977), plays a causative role in calcium alterations and therefore contribute, together with energy imbalance, to calcium redistribution, or calcium changes, due to energy disturbances, are the primary factor responsible for phospholipid degradation (Majewska et al., 1977 ; Lazarewicz et al., 1978).

CONCLUSIONS

The use of gerbils is of interest in studying biochemical parameters undergoing changes during ischemia. Bilateral ligation is preferred to unilateral approach, and gives more reliable results. Although similarities have been reported as regard to changes observed during postdecapitative ischemia and the gerbil model (Levy and Duffy, 1975), others have described substantial differences (Kobayashi et al., 1977 ; Mayevski, 1978). In the present study, changes such as FFA production and phospholipid degradation, have been found in the gerbil model which can at least be compared with those involved in postdecapitative model. However, differences have been reported in the course of the experiments described.

A variety of enzymic mechanisms seems to be involved in generating FFA at the brain level at the very early stages of bilateral ischemia. Phospholipase C-mediated effects, "back-reaction" mechanisms and phospholipases A have certainly their role in elevating the level of FFA above normal, and probably they all are participating in the overall phenomenon. The phospholipid most involved in the process are the plasmalogens, the diacyl-GPI and the diacyl-GPC, although sphingomyelin seems also to be interested. A very active neutral lipase partially masks the DG increase in the ischemic brain of the gerbil ; however, the increase of the arachidonate content in DG always preceeds the elevation of free 20:4, being probably the most efficient source of this acid in the gerbil brain after ligation. The significant decrease in the plasmalogen content and the lack of production of stoichiometric amounts of lysocompounds points also to the occurrence in the ischemic brain of very active plasmalogenase and lysophospholipase activities, which almost consecutively degrade endogenous phospholipid to water-soluble products and FFA (Horrocks et al., 1978).

The activation of phospholipases might be due to effects brought about by cyclic-AMP (Bazàn et al., 1971 ; Bazàn, 1976 ; Cenedella et al., 1975). The levels of this nucleotide are certainly elevated either during postdecapitative ischemia or in the gerbil model (see Lust et al., 1977 for review). However, this hypothesis is not yet clearly defined, and several other considerations might be advanced in explaining the degradative events that take place in ischemic models. Phospholipase activities might undergo variations due to calcium redistribution caused by energy imbalance (Lazarewicz et al., 1977, 1978) and might also be dependent probably at later stages of ischemia by the plasmalogen content which is certainly decreasing, as stated elsewhere (see Strosznajder et al., 1978 a, 1978 b). Probably more than one mechanism plays primary roles for production of events in the ischemic brain.

ACKNOWLEDGEMENTS

Thanks are given to Mr. Antonio Boila for skilled technical assistance throughout the work of lipid analyses. Partial financial aid was possible through funds from the Consiglio Nazionale delle Ricerche, Rome (contribute no.77.01455.04).

REFERENCES

Aveldaño, M.I. and Bazàn, N.G. (1975 a) Differential lipid deacylation during brain ischemia in a homeotherm and a poikilotherm. Content and composition of free fatty acids and triacylglycerols, Brain Research 100: 99-110.

Aveldaño, M.I. and Bazàn, N.G. (1975 b) Rapid production of diacylglycerols enriched in arachidonate and stearate during early brain ischemia, J.Neurochem. 25: 919-920.

Banschbach, M.W. and Geison, R.L. (1974) Post-mortem increase in rat cerebral hemisphere diglyceride pool size, J.Neurochem. 23: 875-877.

Barenholz, Y., Roitman, A. and Gatt, S. (1966) Enzymatic hydrolysis of sphingolipids. II. Hydrolysis of sphingomyelin by an enzyme from rat brain, J.Biol.Chem. 241: 3731-3737.

Bartlett, G.R. (1959) Phosphorus assay in column chromatography,J.Biol.Chem.234:466-468.

Bazàn, N.G. (1970) Effects of ischemia and electroconvulsive shock on free fatty acid pool in the brain, Biochim.Biophys.Acta 218: 1-10.

Bazàn, N.G. (1976) Free arachidonic acid and other lipids in the nervous system during early ischemia and after electroshock, in Function and Metabolism of Phospholipids in Central and Peripheral Nervous Systems (Porcellati, G., Amaducci, L. and Galli, C., eds.) 317-335, Plenum Press, New York.

Bazàn, N.G., de Bazàn, H.E.P., Kennedy, W.G. and Joel, C.D. (1971) Regional distribution and rate of production of free fatty acids in rat brain, J.Neurochem. 18: 1387-1393.

Binaglia,L., Goracci, G., Porcellati, G., Roberti, R. and Woelk, H. (1973) The synthesys of choline and ethanolamine phosphoglycerides in neuronal and glial cells of rabbit in vitro, J.Neurochem. 21: 1067-1082.

Cabot, M.C. and Gatt, S. (1976) Hydrolysis of neutral glycerides by lipases of rat brain microsomes, Biochim.Biophys.Acta 431: 105-115.

Carafoli, E. and Lehninger, A.L. (1967) Energy-linked transport of Ca^{2+}, phosphate and adenine nucleotides, in Methods in Enzymology, vol.X (Estabrook, R.W. and Pullman, M.E., eds.) 745-749, Academic Press, New York.

Cenedella, R.J., Galli, C. and Paoletti, R. (1975) Brain free fatty acid levels in rats sacrified by decapitation versus focused microwave irradiation, Lipids 10: 290-293.

De Medio, G.E., Trovarelli, G., Goracci, G., Palmerini, C.A., Floridi, A., Fini, C., Mazzari, S., Finesso, M. and Porcellati, G., This Congress.

Dole, V.P. and Meinertz, H. (1960) Microdetermination of long-chain fatty acids in plasma and tissues, J.Biol.Chem. 235: 2595-2599.

Ernster, L., Zetterström, R. and Lindberg, O. (1950) Method for the determination of tracer phosphate in biological material, Acta Chem.Scand. 4: 942-947.

Floridi, A., Palmerini, C.A., Fini, C., Goracci, G. and Trovarelli, G. (1978) Analysis of hydrosoluble precursors of choline and ethanolamine glycerophosphatides by high performance liquid chromatography, It.J.Biochem. in the press.

Folch-Pi, J., Lees, M. and Sloane-Stanley, S.H. (1957) A simple method for the isolation and purification of total lipids from animal tissues, J.Biol.Chem.226: 497-509.

Galli, C. and Spagnuolo, C. (1976) The release of brain free fatty acids during ischemia in essential fatty acid-deficient rats, J.Neurochem. 26: 401-404.

Goracci, G., Horrocks, L.A. and Porcellati, G. (1977) Reversibility of ethanolamine and choline phosphotransferases (E.C. 2.7.8.1 and 2.7.8.2) in rat brain microsomes with labelled alkylacylglycerols, Febs Letters 80: 41-44.

Horrocks, L.A. and Sun, G.Y. (1972) Ethanolamine plasmalogens, in Research Methods in Neurochemistry (Rodnight, R. and Marks, N., eds.) 223-231, Plenum Press, New York.

Horrocks, L.A., Spanner, S., Mozzi, R., Fu, S.C., D'Amato, R.A. and Krakowka, S. (1978) Plasmalogenase is elevated in early demyelinating lesions, in Myelination and Demyelination:Recent Chemical Advances (Palo, J., ed.) 342-347, Plenum Press, New York.

Itaya, K. and Ui, M. (1965) Colorimetric determination of free fatty acids in biological fluids, J.Lipid Res. 6: 16-21.

Kates, M. (1972) Techniques of Lipidology, in Laboratory Techniques in Biochemistry and Molecular Biology (Work, T.S. and Work, E., eds.) Vol.III, 269-600, North Holland Publishing Company, Amsterdam.

Kobayashi, M., Lust, W.D. and Passonneau, J.V. (1977) Concentrations of energy metabolites and cyclic nucleotides during and after bilateral ischemia in the gerbil cerebral cortex, J.Neurochem. 29: 53-59.

Kosow, D.P., Schwarz, H.P. and Marmolejo, A. (1966) Lipid biosynthesis in anoxic-ischemic rat brain, J.Neurochem. 13: 1139-1142.

Lazarewicz, J.W., Strosznajder, J. and Gromek, A. (1972) Effects of ischemia and exogenous fatty acids on the energy metabolism in brain mitochondria, Bull.Acad.Polon. Sci., Ser.Sci.Biol. 20: 599-606.

Lazarewicz, J.W., Haljamae, H. and Hamberger, A. (1974) Calcium metabolism in isolated brain cells and subcellular fractions, J.Neurochem. 22: 33-45.

Lazarewicz, J.W., Strosznajder, J. and Dabrowiecki, Z. (1975) Effect of cerebral ischemia on calcium transport in isolated brain mitochondria, in Proc.VIIth Intern.Congr. Neuropath., Vol.II, 605-608, Excerpta Medica, Amsterdam.

Lazarewicz, J.W., Majewska, M.D. and Wròblewski, J.T. (1978) Possible participation of calcium in the pathomechanism of ischemic brain damage, in Pathophysiological, Biochemical and Morphological Aspects of Cerebral Ischemia and Arterial Hypertension (Mossakowski, M.J., Zelman, I.B. and Kroh, H., eds.) 79-86, Pol.Med.Publ., Warsaw.

Levine, S. and Payan, H. (1966) Effects of ischemia and other procedures on the brain and retina of the gerbil (Meriones unguiculatus), Exptl.Neurol. 16: 255-262.

Levy, D.E. and Duffy, T.E. (1975) Effect of ischemia on energy metabolism in the gerbil cerebral cortex, J.Neurochem. 24: 1287-1289.

Levy, D.E. and Duffy, T.E. (1977) Cerebral energy metabolism during transient ischemia and recovery in the gerbil, J.Neurochem. 28: 63-70

Lunt, G.G. and Rowe, C.E. (1968) The production of unesterified fatty acids in brain, Biochim.Biophys.Acta 152: 681-693.

Lust, W.D., Kobayashi, M., Mrsulja, B.B., Wheaton, A. and Passonneau, J.V. (1977) Cyclic nucleotide levels in the gerbil cerebral cortex, cerebellum and spinal cord following bilateral ischemia, in Tissue Hypoxia and Ischemia (Reivich, M., Coburn, R., Lahiri, S. and Chance, B., eds.) 287-298, Plenum Press, New York.

Majewska, D., Gromek, A. and Strosznajder, J. (1974) Properties of brain mitochondria in conditions of ischemia and nembutal anaesthesia in guinea-pigs, Bull.Acad.Polon. Sci., Ser.Sci.Biol. 22: 267-273.

Majewska, D., Lazarewicz, J.W. and Strosznajder, J. (1977) Catabolism of mitochondrial membrane phospholipids in conditions of ischemia and barbiturate anaesthesia, Bull. Acad.Polon.Sci., Ser.Sci.Biol. 25: 125-131.

Mayevsky, A. (1978) Ischemia in the brain : the effects of carotidal artery ligation and decapitation on the energy state of the awake and anaesthesized rat, Brain Res. 140: 217-230.

Mangold, M.K. (1969) Aliphatic lipids, in Thin-layer Chromatography (Stahl, E., ed.) 363-421, Springer-Verlag, Berlin.

Mazzari, S. and Finesso, M. (1978), This Congress.

McGee, J. (1974) Preparation of methyl esters from the saponifiable fatty acids (micro methyl ester preparation), J.Chromat. 100: 35-42.

Medina, M.A., Jones, D.J., Stavinoha, W.B. and Ross., D.H. (1975) The levels of labile intermediary metabolites in mouse brain following rapid tissue fixation with microwave irradiation, J.Neurochem. 24: 223-227.

Ozawa, K., Seta, K., Araki, G. and Handa, H. (1967) The effect of ischemia on mitochondrial metabolism, J.Biochem. (Tokyo) 61: 512-514.

Porcellati, G. and Goracci, G. (1976) Phospholipid composition and turnover in neuronal and glial membranes, in Lipids, Vol.I (Paoletti, R., Porcellati, G. and Jacini, G., eds.) 203-214, Raven Press, New York.

Rodrìguez de Turco, E.B., Cascone, G.D., Pediconi, M.F. and Bazàn, N.G. (1977) Phosphatidate, phosphatidylinositol, diacylglycerols, and free fatty acids in the brain following electroshock, anoxia, or ischemia, in Function and Biosynthesis of Lipids (Bazàn, N.G., Brenner, R.R. and Giusto, N.M., eds.) 389-396, Plenum Press, New York.

Rossowska, M., Lewandowski W. and Dabrowiecki Z. (1977) Effect of ischemia on the activity of Na^+-K^+-ATPase in the microsomal fraction of guinea-pig brain, Bull.Soc.Polon.Sci., Ser.Sci.Biol. 24: 691-696.

Rossowska, M. and Dabrowiecki, Z. (1978) Effect of hypoxia and ischemia on the activity of glucose-6-phosphatase in the guinea-pig brain, J.Neurochem., in the press.

Siesjö, B.K., Nordström, C.-H. and Rehncrona, S. (1977) Metabolic aspects of cerebral hypoxia-ischemia, in Tissue Hypoxia and Ischemia (Reivich, M., Coburn, R., Lahiri, S. and Chance, B., eds.) 261-269, Plenum Press, New York.

Strosznajder, J. and Dabrowiecki, Z. (1977) Enzymic synthesis of ethanolamine plasmalogens in the microsomal fraction of rat brain under oxygen deficiency, Bull.Soc. Polon.Sci., Ser.Sci.Biol. 25: 133-139.

Strosznajder, J., Dabrowiecki, Z. and Radomińska-Pyrek, A. (1978 a) Effect of hypoxia on enzymic synthesis of diacyl and ether types of choline and ethanolamine phosphoglycerides in rat brain microsomes, in Cerebral Ischemia and Arterial Hypertension (Mossakowski, M.J., Zelman, I.B. and Kroh, M., eds.) 103-109, Pol.Med.Publ., Warsaw.

Strosznajder, J., Dabrowiecki Z. and Domańska-Janik, K. (1978 b) Effect of anoxia and hypoxia on the lipid metabolism in rat brain, submitted to Brain Res.

Sun, G.Y. (1970) Composition of acyl groups in the neutral glycerides from mouse brain, J.Neurochem. 17: 445-446.

Sun, G.Y. (1977) Metabolism of arachidonate and stearate injected simultaneously into the mouse brain, Lipids 12: 661-665.

Yau, T.M. and Sun, G.Y. (1974) The metabolism of [$1-^{14}C-$]arachidonic acid in the neutral glycerides and phosphoglycerides of mouse brain, J.Neurochem. 23: 99-104.

Wieland, O. (1974) Glycerol-UV-Method, in Methods of Enzymatic Analysis (Bergmeyer,H. U., ed.) Vol.III, 1404-1409, Verlag Chemie, Weinheim.

Williams, D.J., Spanner, S. and Ansell, G.B. (1973) A phospholipase C in brain tissue active towards phosphatidylethanolamine, Biochem.Soc.Trans. 1: 466-467.

RESPIRATION OF SYNAPTOSOMES
H. Wise, H.S. Bachelard and G.G. Lunt,
Department of Biochemistry
University of Bath, Bath, U.K.

In order to study the dependence of brain function on its oxygen supply, rat cerebral cortex synaptosomes were prepared by the method of Bradford et al (1975), and their respiration was followed at 30°C in Krebs phosphate using an oxygen electrode in a closed system. Optimal conditions (glucose and oxygen) gave rates of 63μmol O_2/hr/100mg.protein. If the synaptosomes were added to oxygen-saturated glucose-free medium, respiration rates decreased slightly with decreasing oxygen to about 14μM-O_2, when they fell sharply. In the presence of 10mM glucose, rates remained constant until 14μM-O_2, when they also fell sharply.

Respiration in media with oxygen initially low

The synaptosomes were pre-incubated in the presence or absence of 10mM glucose and their respiration was measured at oxygen concentrations of 14μM and below. Respiration rates were steady until 6μM oxygen when they fell off sharply.

Pre-incubation (30°, 35min)	Respiration (no glucose)	(with glucose)
None	100 (control)	172*
Non-shaking, no glucose	90	113
Shaking, no glucose	48*	-
Shaking, plus glucose	130	174*

Results as percent of control. *Significantly different from control.

The results show that glucose protects against damage caused to the synaptosomes during pre-incubation and that oxygen becomes limiting for their respiration below approx. 14μM.

Bradford, H.F., Jones, D.G., Ward, H.K. and Booher, J. (1975). Brain Res. 90: 245-249

EFFECT OF HYPOXIA ON THE IN VIVO INCORPORATION OF $[2-^3H]$ GLYCEROL AND $[1-^{14}C]$ PALMITATE INTO LIPIDS OF VARIOUS SUBCELLULAR FRACTIONS PURIFIED FROM DIFFERENT REGIONS OF GUINEA PIG BRAIN.

M. Alberghina, I. Serra, E. Geremia, A.M. Giuffrida
Institute of Biochemistry, University of Catania, Italy

The effect of hypoxia on the in vivo incorporation of labelled precursors into lipids (total, neutral and phospholipids) extracted from subcellular fractions of guinea pig brain regions was investigated.

Intermittent normobaric hypoxia was accomplished keeping the animals in a glass chamber flushed with nitrogen atmosphere containing 9% of oxygen, for 80 hrs. $[2-^3H]$ glycerol and $[1-^{14}C]$ palmitate were injected either intraperitoneally (75 µCi and 30 µCi, respectively, per 100 g b.w.) or in the lateral ventricle (0.9 µCi and 2.7 µCi, respectively, per g brain), dissolved in saline containing BSA. The labelling time was 2 hrs for the intraperitoneal and 40 mins for the intraventricular injections. Mitochondria, myelin, synaptosomes and microsomes were purified from cerebral hemispheres, cerebellum and brain stem; the purity of subcellular fractions was checked by measuring marker enzyme activities. The lipids extracted from the various subcellular fractions were separated by TLC; the amount and the radioactivity of the various lipid classes were measured; the labelling of individual phospholipids was evaluated, after elution of the corresponding TLC spots, by phosphorus determination and radioactivity measurement. The distribution of ^3H and ^{14}C label in the phospholipids and sphingolipids was checked by chemical hydrolysis and by phospholipase A_2 degradation. To evaluate the dilution of the labelled precursors, after the intraperitoneal injection, the concentration and the specific radioactivity of total lipids, FFA, glycerides and phospholipids were measured in blood samples from both control and hypoxic animals, at the end of labelling time.

In the hypoxic animals compared to the controls the serum concentration of FFA and glycerides increased, on the contrary their specific radioactivity greatly decreased.

The incorporation of labelled precursors into lipids and particularly into phospholipids of purified mitochondria from the various brain regions examined markedly decreased in hypoxic animals compared to the controls. These results were evident for phospholipid species which are synthesized by the mitochondria as well as for those synthesized at the microsomal level. The effect of hypoxia on lipid metabolism in the other subcellular fractions investigated differed among the three brain regions examined.

THE EFFECT OF HYPOXIA ON THE METABOLISM OF LABELED GLUCOSE AND ACETATE IN THE RAT BRAIN

K. Domańska-Janik, T. Zalewska
Medical Research Centre, Pol. Acad. Sci., 00-784 Warsaw, Poland.

The effect of mild hypoxia on the metabolism of 6-^{3}H-glucose and 2-^{14}C-acetate in the rat brain was studied. The animals were kept in the mixture of 7% O_2 in N_2 for 2 hours. Normal body temperature was maintained. Gasometric parameters of blood and concentrations of metabolities in blood and brain were measured in separate experiments throughout the hypoxic period. Between 15 min and 2 hours of hypoxia all these parameters were relatively stable and their values were as follow:

Blood /μmole/ml/

	pH	pO_2	pCO_2	Lact.	Pyruv.	Glucose
Control /12/	7,34	90	39	1,31	0,109	7,1
Hypoxia /7/	7,35	39	32	7,88	0,281	9,8

Brain /μmole/g wet weight/

	ATP	CrP	ADP	Lact.	Pyruv.	Glucose
Control /12/	3,17	2,60	0,56	3,17	0,099	0,98
Hypoxia /7/	2,58	2,60	0,42	8,96	0,168	1,29

After injection of isotopes at 10, 20, 30 min before the end of hypoxic period, the conversion of 6-^{3}H-glucose and 2-^{14}C-acetate label into amino acid fraction was reduced by about 20%. Differences in the availability of both substrates to the brain /calculated from the ratio of the brain acid soluble radioactivity to blood specific radio activity /indicate in spite of similar reduction of the flux of ^{14}C and ^{3}H labels into amino acids different effects of hypoxia on basal metabolism in these two compartments. In case of glucose, the retention of the label in non-amino acid fraction seems to be connected with acceleration of glycolysis and accumulation of lactate from additional glucose delivered to the brain. In the acetate compartment the observed reduction of conversion of ^{14}C into amino acid fraction is connected probably with inhibition of the Krebs cycle activity.

In some experiments the incorporation of ^{14}C from U-^{14}C-glucose and 2-^{14}C-acetate into macromolecular fractions /lipid, protein and nucleic acid/ was also measured under hypoxic conditions. The incorporation of the label from U-^{14}C-glucose was much lower than from ^{14}C-acetate. The possible effect of hypoxia on synthetic processes in both compartments will be discussed.

ENERGY UTILIZATION AND CHANGES IN SOME INTERMEDIATES OF GLUCOSE METABOLISM IN NORMAL AND HYPOXIC RAT BRAIN AFTER DECAPITATION

T. Zalewska, K. Domańska-Janik
Medical Research Centre, Pol. Acad. Sci., 00-784 Warsaw, Poland.

Energy metabolism was studied in the cerebral cortex of rats during and following hypoxia induced by breathing gas mixture of 7% O_2 in N_2 by 2 hours.
Cortical energy stores /2 ATP + ADP + Pcreatine/ remained unchanged after hypoxic treatment. Lactate rose over 3-fold. Pyruvate, glucose and G-6-P level also increased significantly. Metabolic activity in the cortex expressed as the utilization of high-energy phosphates was markedly decreased /-30%/ after 30 sec. following decapitation of hypoxic animals and remained lowered also after 3 hours of recovery. After 6 hours following hypoxic episode the high-energy phosphates utilization was closest to the control values suggesting the restoration of normal metabolic activity. Immediately after hypoxic period the utilization of high-energy compounds was connected almost exclusively with glycolytic activity of the brain tissue. Other energy-dependent processes inhibited by hypoxia were gradually normalised, reaching 70% of control value after 6 hours of recovery period.

REGULATION OF THE LEVEL OF CYCLIC AMP IN THE BRAIN DURING HYPOXIA
AND ISCHEMIA

L.Khatchatrian and K.Domańska-Janik
Department of Neurochemistry, Medical Research Centre, Pol.Acad.Sci.,
Warsaw, Poland

Cyclic AMP plays an important role in the nervous system: it participates in the mediation of postsynaptic action of certain neurotransmitters and in regulation of their biosynthesis. The role of cAMP in some destructive processes in ischemic-anoxic brain has also been postulated.

The aim of the present studies was to investigate the levels of cyclic AMP and the activity of adenylate cyclase which is responsible for its biosynthesis from ATP. The studies were performed on rat brain in postdecapitative ischemia and in hypoxic hypoxia in 7 % O_2 in nitrogen.

The level of cyclic AMP was increased in ischemia. The most pronounced increase was observed 30 sec after decapitation / 46 % /. A rapid decrease of this level was then observed and in the 3rd min after decapitation it reached values similar to control ones. In animals submitted 3 and 6 hours before to 2 hr hypoxia the cAMP after 30 sec from decapitation was increased by 60 %. This may suggest an increased susceptibility of adenylate cyclase system due to previous hypoxia.

In order to find whether the changes in the activity of adenylate cyclase can be responsible for the observed cAMP levels the activity of this enzyme was determined in rat brain after hypoxia / 7 % O_2 for 2 hr / and after 5 min of postdecapitative ischemia. It has been found that the activity of adenylate cyclase in cortex and corpus striatum was decreased by 15 % after hypoxia and by 20 % after ischemia. This decrease was most pronounced in synaptosomal fraction obtained from ischemic brain where it was found to be only 50 % of the control

The short-term changes in cAMP levels found in the brains after ischemia are in agreement with Lust and Passonneau /1976/ and suggest that cAMP may play a role of a trigger initiating some postischemic metabolic disturbances. It seems that apart from other mechanisms regulating the metabolism of cAMP the found decrease of adenylate cyclase activity may be responsible for the rapid drop of cAMP level observed already after 1 min of ischemia.

Lust W.D., Passonneau J.V. /1976/ J. Neurochem. 26: 11-16

MECHANISMS OF DAMAGE OF MITOCHONDRIAL LIPID CHEMICAL STRUCTURE AFTER ISCHEMIC-ANOXIA

M.D.Majewska, L.Khatchatrian, J.Strosznajder, J.W.Łazarewicz, Medical Research Centre, Pol.Acad.Sci., Warsaw, Poland.

Short time postdecapitative ischemic-anoxia /till 5 min./ leads to pronounced damage of mitochondrial lipid architecture, manifested by a decrease of phospholipid content as well as by increase of free fatty acid pool. Phospholipid degradation starts rapidly at 0.5 min of postdecapitative ischemia.

The question arises, which is the mechanism of this post-ischemic destruction of brain mitochondrial phospholipids. Possible participation of two potential agents of ischemic-anoxia in the activation of endogenous mitochondrial phospholipase /c.AMP and Ca^{2+} ions/ was investigated. Antiradical activity of brain tissues and process of lipid peroxidation were also examined.

In in vitro experiments c.AMP, especially its dibutyryl-derivative in concentrations 10^{-6} M significantly stimulates hydrolysis of phospholipids of purified mitochondrial fraction in rat and guinea pig brains. Activation of adenyl-cyclase in homogenates of rat brain cerebral cortex by noradrenaline or NaF is connected with the stimulation of phospholipase activity. In in vivo experiments pretreatment of rats with β-adrenergic receptor antagonists /propranolol/, in the dose /0.3 mg per kg of body weight/, decreasing twice the cardiac action, had a protecting effect on post-ischemic hydrolysis of mitochondrial lipids, what also suggest participation of c.AMP in brain phospholipid hydrolysis. Calcium ions, the other candidate for phospholipase activation, in concentrations 2-10 μmoles/mg mit. prot. have also stimulating effect on phospholipid hydrolysis of brain mitochondrial fraction in vitro. Postdecapitative ischemic anoxia significantly decreases the antiradical activity of brain tissue, which is closely connected with the acceleration of lipoperoxidation in mitochondrial membranes.

Mechanism of ischemic activation of hydrolysis of brain mitochondrial lipids is complex. There are various related agents. First of them, being probably catecholamine activation of adenyl-cyclase, leading to the increase of c.AMP level and phospholipase stimulation. Further activation of mitochondrial phospholipase may be due to increased level of Ca^{2+} ions. In the post-ischemic period, as a result of lowered antiradical activity of brain tissue process of lipoperoxidation rapidly developes.

THE ROLE OF CMP ON THE REGULATION OF BRAIN CHOLINE PHOSPHOTRANSFERASE

E.Francescangeli, G.Goracci, L.A.Horrocks[*] and G.Porcellati
Department of Biochemistry, University of Perugia, Italy and [*]Department of Physiological Chemistry, The Ohio State University, Columbus, Ohio, U.S.A.

Recently, some indications of the reversibility of brain choline phosphotransferase (CPT), which catalyzes the reaction : DG + CDP-choline $\rightleftharpoons$ CPG + CMP, have been reported indicating that CMP could react with microsomal choline phosphoglycerides (CPG) producing diglycerides (DG) and CDP-choline (Goracci et al., 1977). This hypothesis has been confirmed by the following experiments. [Methyl-^{3}H] choline was injected intracerebrally into young rats. The animals were sacrificed after two hours. Microsomes containing labeled CPG were incubated in the presence of 4 mM CMP under optimal conditions for CPT activity and the radioactivity of CDP-choline measured after different time intervals of incubation. CMP was omitted in controls. The addition of CMP increased the radioactivity of CDP-choline up to 10 % after 60 min of incubation. A parallel decrease of the radioactivity was observed in CPG.

Incubation Time		0	2'	4'	15'	30'	60'
CPG	Control	87.8	85.5	84.9	90.1	75.3	88.6
	+ CMP	87.8	82.4	80.9	76.5	70.7	70.9
CDP-choline	Control	0.3	0.4	0.9	0.9	0.4	0.4
	+ CMP	1.0	2.4	3.6	7.6	8.4	10.3

Results are presented as the percentage of total radioactivity, and are the mean of several experiments.

Experiments carried out with microsomes labeled with [^{3}H] oleic acid by the same procedure have shown that the addition of 4 mM CMP enhanced the radioactivity of free fatty acids (FFA) and, to a lesser extent, that of DG with a simultaneous decrease of labeling in CPG. Since DG and monoglyceride lipases are present in brain microsomes (Cabot and Gatt, 1977), in other experiments labeled microsomes were incubated with DFP and then with CMP. Under these conditions the labeling of FFA was reduced and that of DG increased following the addition of CMP.

These results strongly support the reversibility of brain CPT which could be involved in the degradation of endogenous CPG when the concentration of CMP increases, as for instance in brain ischemia (Porcellati et al., 1978).

Cabot, M.C. and Gatt, S. (1977) Biochemistry
    16 : 2330-2334

Goracci, G., Horrocks, L.A. and Porcellati, G. (1977) Febs Letters
    80 : 41-44

Porcellati, G., De Medio, G.E., Fini, C., Floridi, A., Goracci, G., Horrocks, L.A., Lazarewicz, H.W., Palmerini, C.A., Strosznajder, J. and Trovarelli, G., This Congress

EFFECT OF ISCHEMIA ON ENERGY METABOLISM AND CATECHOLAMINE LEVELS IN THE GERBIL BRAIN IN VIVO

S. Mazzari and M. Finesso
Fidia Research Laboratories, Abano Terme - Italy

Adult Mongolian Gerbils (Meriones unguiculatus) had both common carotid arteries ligated under local novocaine anesthesia. Animals were sacrified by decapitation and the head allowed to fall immediately in liquid nitrogen 1.5 and 10 min later.

In these experimental conditions there occurred an extensive increase of lactate and a parallel decrease of both glucose and glycogen in the brain of operated animals respect to the sham-operated ones. Similarly ATP and P-creatine levels decreased, reaching a maximum decrease after 1 min of ischemia and persisting until the end of the experiment; 5'-AMP increased. Brain catecholamines were only slight reduced after 10 min.

In the recovery studies, both common carotid arteries were open after 5 min of ischemia and recirculation re-established for 5 min. Glucose, ATP and P-creatine reached normal values after 5 min of recirculation. The glycogen resynthesis was slow and lactate concentration remained elevated. Obtained data reproduced previous results (Kobayashi, M., 1977) using a different short-lasting operating procedure, suggesting the validity of this experimental model.

The modification in energy metabolism is a more sensitive parameter respect to catecholamine modification. Furthermore the low values of P-creatine and the increase of blood glucose suggest a condition of induced stress which may affect the observed modifications in a manner unrelated with the induced anoxia.

Kohayashi, M., Lust, W.D. and Passonneau, J.V. (1977) J. Neurochem. 29: 53-59

CHANGES IN GERBIL BRAIN PHOSPHOGLYCERIDES DURING BILATERAL ISCHEMIA

G.E.De Medio, G.Trovarelli, G.Goracci, C.A.Palmerini, A.Floridi, C.Fini, S.Mazzari*, M.Finesso* and G.Porcellati.
Department of Biochemistry, University of Perugia, Italy.
*FIDIA Research Laboratories, Padua, Italy.

It is well known that decapitation enhances the levels of free fatty acids in brain. Their origin and the mechanisms involved are not yet well established. Since phosphoglycerides are probably their source, the phospholipid content and their distribution in Gerbil brains, at different time intervals after bilateral ligation of common carotide arteries, have been studied. Sham-operated animals were used, as controls. The phospholipid phosphorus decreased by about 13%, 5 min after ligation. Data, expressed as percentages of total phospholipid P, indicate that the major changes were located in choline and ethanolamine plasmalogens. During the first minute after ligation normal choline plasmalogen content (2.4%) decreased to 1.5%. At longer intervals no further decreases were observed. A similar decrease in ethanolamine plasmalogens was found (from 17.5% to 8.9% at 1 min from ligation). An increase of monoacyl-glycerophosphorylcholine(-GPC) and monoacyl-glycerophosphorylethanolamine(-GPE) occurred, althoug not in a parallel way. Sphingomyelin content decreased from 7.9% to 5.4% after 5 minutes. No significant changes of the other major phospholipid classes were found.

Other experiments were carried out by injecting [methyl-^{3}H]-choline in the lateral ventricle of Gerbil brains. After 2 hours, both carotide arteries were ligated and the animals sacrified at different time intervals. The labeling of choline-containing lipids was determined during 5 min period from ligation. A significant increase of radioactivity in monoacyl-GPC was found and this agrees with the above mentioned data. In the same period changes of labeling of water-soluble precursors of phosphoglycerides were observed, mainly of CDP-choline which greatly increases. Our results indicate that phospholipids undergo degradation for which more than one mechanism could be involved.

Fatty acid composition of the main phospholipid classes have been determined up to 5 min from ligation of carotide arteries and changes have been found, especially of phosphatidic acid, phosphatidylinositol and phosphatidylcholine.

PHOSPHOLIPASES IN ISCHEMIC GERBIL BRAIN

L. A. Horrocks, W. R. Snyder, A. D. Edgar, M. E. Nesham and J. N. Allen
Depts. of Physiological Chemistry and Medicine (Neurology), The Ohio State University, Columbus, Ohio 43210 USA

Bilateral ligation of the common carotid arteries in the Mongolian gerbil, Meriones unguiculatus, for 30 min produces severe cerebral infarction and death of 80% of gerbils whereas infarction seldom occurs with less than 20 min of ligation (1). Cerebral edema is evident after 10 min of ligation and reaches a maximum after 30 min of ligation (1). Coincident with the edema is a rapid 2.3-fold increase in plasmalogenase activity up to 20 min of ligation. The plasmalogenase activity then decreases markedly (2). From these results we have formulated the following working hypothesis: a. Plasmalogenase is located primarily in oligodendroglia (3), probably on the plasma membrane. b. Ischemia produces an activator of oligodendroglial plasmalogenase. c. Prostaglandin precursors and lysophosphatidyl ethanolamines are formed. d. These products disrupt the oligodendroglial plasma membranes and produce inflammation. e. Myelin that was maintained by the necrotic oligodendroglia is then removed.

Since activation of other phospholipases could also produce similar effects, we have assayed the acid phospholipases A_1 and A_2 in gerbil brains after bilateral ligation. The A_1 activity assayed at pH 4.0 decreased and the A_2 activity assayed at pH 4.6 did not change. Both of these activities are much lower than plasmalogenase activities. The acid phospholipases A_1 and A_2 apparently are not involved in the pathological changes during the first hour. Lysophospholipase activity at 15, 30 and 60 min of ligation was about 20% above the control value. The relatively high activity of the lysophospholipase suggests that lyso compounds do not accumulate appreciably. Since most of the prostaglandin precursors in white matter are at the 2-position of ethanolamine plasmalogens (4), further hydrolysis of the lyso compounds produced by plasmalogenase will lead to a large increase in 20:4 (n-6) and 22:4 (n-6) acids which can be converted to prostaglandins.

1 Wise, G., Stevens, M., Shuttleworth, E.C. and Allen, N. (1973) Abst. Soc. Neurosci. 375.

2 Horrocks, L.A., Spanner, S., Mozzi, R., Fu, S.C., D'Amato, R.A. and Krakowka, S. (1978) In: "Myelination and Demyelination: Recent Chemical Advances", ed. by J. Palo, Plenum Press, New York, in press.

3 Dorman, R.V., Toews, A.D. and Horrocks, L.A. (1977) J. Lipid Res. 18: 115-117.

4 Horrocks, L.A. and Fu, S.C. (1978) In: "Enzymes of Lipid Metabolism", ed. by S. Gatt, L. Freysz and P. Mandel, Plenum Press, New York, pp. 397-406.

ROLE OF PHOSPHOLIPIDS IN CALCIUM ACCUMULATION IN BRAIN MITOCHONDRIA FROM ADULT RAT AFTER ISCHEMIC ANOXIA AND HYPOXIC HYPOXIA

J. Strosznajder, Dept. of Neurochemistry, Medical Research Centre, Polish Academy of Sciences, Warsaw, Poland

The accumulation of calcium ions was measured in purified mitochondria (0.3 mg protein) incubated with 0.3 M sucrose, 5 mM succinate, 2.5 μM rotenone, 3 mM ATP, and 1 mM $CaCl_2$ in a final volume of 1 ml. Ischemic anoxia was produced by incubation for 5 min at 37°C of the heads after decapitation. For hypoxic hypoxia, rats were kept in a chamber through which 7% oxygen in nitrogen was passed at a rate of 2 $l(min)^{-1}$ for 2 hours. Body temperature of the rats was maintained at 37°C.

After ischemic anoxia the accumulation of calcium ions was decreased by about 35%. This process was also inhibited by about 28% after hypoxic hypoxia. At the same time the content of phospholipids decreased in ischemic and hypoxic brain mitochondria by about 27% and 16% respectively. In particular cholineglycerophospholipids and ethanolamine plasmalogens were diminished. Simultaneously the content of free fatty acids increased. Additions of low concentrations of fatty acids, lysoglycerophospholipids, deoxycholate or sodium dodecyl sulfate inhibited calcium accumulation in control mitochondria. In another experiment mitochondria were incubated with phospholipase A_2 (EC 3.1.1.4) or phospholipase C (EC 3.1.4.3) in the presence of albumin. After washing and reisolation, there was a large decrease in the rate of calcium accumulation. These results suggest that the integrity of the membrane phospholipids is important for calcium transport.

The decrease of ethanolamine plasmalogens may be a factor responsible for the inhibition of calcium accumulation in mitochondria after ischemic anoxia and hypoxic hypoxia. Another factor may be the nonspecific detergent action of fatty acids and lysoglycerophospholipids.

EFFECT OF ISCHEMIC ANOXIA AND HYPOXIC HYPOXIA ON ACYLATION OF LYSOGLYCEROPHOSPHOLIPIDS IN RAT BRAIN SUBCELLULAR FRACTION

J. Strosznajder, Department of Neurochemistry, Medical Research Centre, Polish Academy of Sciences, Warsaw, Poland

In the presence of ATP, Mg^{2+} and CoA, brain subcellular fractions (crude mitochondria, microsomal and synaptosomal fractions) actively catalyze the incorporation of [1-^{14}C]linoleic acid into endogenous phospholipids. Among the different endogenous glycerophospholipids investigated, the relative order for linoleic incorporation was phosphatidylcholine > phosphatidylethanolamine > phosphatidylinositol and phosphatidylserine > ethanolamine plasmalogen > sphingomyelin.

The incorporation of linoleate into brain subcellular fraction was affected by ischemic anoxia and hypoxia. Ischemic anoxia was produced by incubation for 5 min at 37°C of the heads after decapitation. For hypoxic hypoxia rats were kept in a chamber through which 7% oxygen in nitrogen was passed at a rate of 2 l(min)$^{-1}$ for 2 hours. Body temperature of the rats was maintained at 37°C. Both conditions inhibited by about 20-25% the incorporation of [1-^{14}C]linoleic acid into ethanolamine plasmalogens and into inositol and serine glycerophospholipids. The pattern of linoleate incorporation into crude mitochondrial lipids changed during development (4, 8, 16 and 90 day old rats). For young rats oxygen deficiency mostly inhibited the incorporation of linoleate into inositol and serine glycerophospholipids.

Acyl CoA formation may have been the rate limiting reaction. The incorporation of [1-^{14}C]oleyl CoA into the glycerophospholipids of microsomal and synaptosomal fractions isolated from ischemic brain was stimulated. Thus the acylation reactions of endogenous lysoglycerophospholipids are not disturbed. The stimulation of acylation with oleyl CoA indicates that the endogenous lysoglycerophospholipid substrate levels are increased by about 20-30% after ischemic anoxia.

Symposium

Neurochemistry of Addictive Drugs

Chairman:
A. Herz

BIOCHEMICAL ASPECTS OF TOLERANCE TO, AND PHYSICAL DEPENDENCE ON, CENTRAL DEPRESSANTS

H. Kalant
Department of Pharmacology, University of Toronto, and Addiction Research Foundation of Ontario, Toronto, Canada   M5S 1A8

What Are "Central Depressants"?

Very many drugs of quite diverse types can depress various functions of the central nervous system, when given in appropriate dosage. For example, substances as dissimilar as digitalis glycosides, cocaine, magnesium chloride and insulin are all capable, under certain conditions, of producing central depressant phenomena, e.g. drowsiness, hyporeflexia, coma and respiratory depression and arrest. Among pharmacologists and clinicians, however, it is traditional to reserve the term "central depressants" for a group of drugs which meet the following criteria:

(1) their *primary* actions at low and intermediate dose ranges are exerted upon the central nervous system (CNS);
(2) though very low doses produce neuroexcitatory phenomena, the major effects through most of the dose range involve graded degrees of neuronal hypoexcitability (Smith, 1977);
(3) these effects are exerted rather indiscriminately upon all types of neuron, which differ only in threshold concentration required for the drug effect (Staiman & Seeman, 1974).

The drugs which meet these criteria are a chemically heterogeneous group of general anesthetic gases and liquids, alcohols, barbiturates, benzodiazepines, and other hypnotics and sedatives (Smith, 1977). Neuroleptics and opiates are not usually included, even though their neuropharmacological actions bear many similarities to, and are often synergistic with, those of the previously named drugs. The reason is a traditional but increasingly questionable one. Neuroleptics and opiates have well-characterized stereospecific receptors which are not homogeneously distributed in the CNS (Seeman et al., 1975; Höllt et al., 1977; Snyder, 1977). Accordingly, they have selective effects (mainly depressant) on certain groups of neurons bearing these receptors, while other neurones are affected only by very much higher (and often lethal) concentrations of these drugs, at which nonspecific interactions with cell membranes can occur. Though stereospecific CNS receptors for benzodiazepines have recently been described (Squires & Braestrup, 1977), the resemblances between *in vivo* effects of diazepam and those of barbiturates and ethanol suggest either that diazepam receptors must be very widely distributed throughout the brain, or that non-receptor-mediated effects of these drugs must be clinically important. The different spectra of central effects produced by central depressants, neuroleptics and opiates may therefore depend more on the distribution of cells affected by each, than on differences in types of effect produced in cells which are affected (Kalant, 1977). This is an important question in relation to the development of tolerance.

Characteristics of Tolerance to Central Depressants

If one is to search, in a rationally directed way, for the neurochemical mechanisms underlying tolerance and physical dependence, it is necessary to note the empirically determined features of tolerance and dependence which a postulated mechanism must be capable of explaining.

The first such feature is the characteristic parallel shift of the log-dose/response (LDR) curve. For most drugs acting on the CNS system, a plot of the logarithm of the dose against the measured effect yields a sigmoid curve, the middle section of which shows a roughly linear relation between the log-dose and the effect. Tolerance to ethanol, barbiturates and other central depressants is characterized by a parallel shift of the LDR curve, so that the middle section retains the same slope, and all points on the curve are moved farther along the dose axis by a constant logarithmic increment (Kalant et al., 1971). However, the maximum response attainable (R_{max}) is unchanged, as long as the dose is increased appropriately. This is a point of difference from opiate tolerance, which is characterized at first by a parallel shift, and then (on treatment with much larger doses) a progressive decrease in R_{max} but no further shift of the midpoint of the curve (Mucha et al., 1978). It is conceivable that the reduction in R_{max} corresponds to some change in opiate receptor binding, and that the lack of it in ethanol tolerance reflects the absence of specific receptors for ethanol and similar depressants.

The second feature to be accounted for is the quantitative relationship between the degree of tolerance attained (degree of parallel shift of LDR curve, in log units), the rate at which it develops, and the size and frequency of drug dose used to induce it. Surprisingly, there have been few systematic studies of this relationship, but such as there are indicate a fairly good proportionality (Kalant et al., 1971; Smith, 1977). This means that any neurochemical process underlying tolerance cannot be all-or-none in type, but must be continuously variable in rate and degree.

A third feature is that the process is reversible, with a character-

istic half-life of its own (LeBlanc et al., 1976a), and with some sort of
residual "memory" left in the nervous system after the tolerance is gone.
The rate of loss of tolerance, when drug treatment is discontinued, differs substantially from one species to another, but is largely independent of the treatment schedule used to produce tolerance (LeBlanc et al.,
1976a). There are insufficient data to permit valid comparisons among the
numerous central depressant drugs, but the rate of loss of tolerance is
quite similar for ethanol and pentobarbital. These observations imply
that tolerance involves some change in neuronal function and/or composition
which obeys the same type of kinetics as the biosynthesis and decay of other
known cell constituents. Once the tolerance has been completely lost, a
new period of drug treatment re-evokes it more rapidly than in the first
treatment period (Kalant et al., 1978a). The cell appears to retain a long-
lasting "trace" of this process, analogous to the ability of an immuno-
competent cell to re-synthesize antibodies more rapidly on re-exposure
to an antigen than during its first exposure.

A fourth feature is that the rate of tolerance development can be
markedly influenced by behavioral and environmental influences. For
example, a rat which is required to perform a certain task (e.g. maze
test or moving belt test) every day under the influence of a small dose
of ethanol, develops tolerance much more rapidly than one which receives
the same dose of ethanol <u>after</u> the daily task performance (LeBlanc et
al., 1976a). This "behavioral augmentation" of tolerance suggests strongly that the stimulus to development of tolerance is not the drug itself,
but the functional disturbance produced by the drug, which varies according to the external influences acting upon the subject during the time
in which the drug is present in the brain. According to this view, the
neurons would be adapting in the same way as to any comparable functional
disturbance, regardless of whether it is produced by a drug or by a physiological stimulus of any other kind. This is an important consideration, because it makes it easier to understand the development of cross-
tolerance between drugs with quite dissimilar chemical structures, such
as ethanol and barbiturates, and between these drugs which do not have
specific receptors and those which may or do have them (Smith, 1977).

A final feature which must be accounted for is the relationship between tolerance and physical dependence. By physical dependence is meant
a state produced by chronic administration of the drug, such that sudden
withdrawal of the drug gives rise to functional disturbances which can
be corrected promptly by renewed drug administration. Most hypotheses
treat tolerance and physical dependence as complementary manifestations
of the same adaptive changes in the CNS. For example, typical acute ef-

fects of central depressants in the intermediate and high dosage range include elevation of sensory thresholds, hyporeflexia, muscular flaccidity, decreased respiratory rate and depth, and elevated threshold for electrically or chemically induced seizures (Wallgren & Barry, 1970). These can all be viewed as manifestations of decreased neuronal excitability, and tolerance is viewed as the result of compensatory changes leading to increased excitability which neutralizes the drug effects. In the absence of the drug, these changes are unmasked and can give rise to the converse phenomena: lowered sensory thresholds, hyperreflexia, increased muscle tone and tremor, hyperventilation and spontaneous or easily evoked seizures (Freund, 1971; Majchrowicz, 1975; Gildea & Bourn, 1977; Boisse & Okamoto, 1978).

However, this view of physical dependence and tolerance as reflections of the same underlying process is not rigorously proven. Most studies have shown the two to appear and disappear roughly in parallel. However, methods for demonstrating tolerance are usually more sensitive than tests for dependence. Therefore, tolerance is generally found sooner and disappears later than signs of dependence, so that it is possible, indeed common, to have some degree of tolerance without obvious physical dependence. The opposite has also been claimed, i.e. the presence of physical dependence under conditions in which tolerance is not demonstrable (Tabakoff & Ritzmann, 1977), while others have reported that physical dependence can reach a maximum while tolerance continues to intensify (Majchrowicz & Hunt, 1976). These apparent contradictions may well prove to reflect differences in techniques used to demonstrate and measure the two phenomena. But the point is not yet settled, and it is therefore highly desirable to study possible neurochemical alterations under conditions both of tolerance (i.e. drug present) and of dependence (i.e. drug withdrawn).

Is Tolerance a Single or a Multiple Process?

If tolerance to central depressants is really a specific example of a more general family of CNS adaptations, there should be several types of tolerance with different cellular mechanisms. This appears to be true. Rapid short-term adaptations, such as sensory habituation or short-term memory, find their parallel in acute (within-session) drug tolerance which begins to appear within minutes of the start of drug exposure and disappears within hours (Smith, 1977). Although this might rest on a rapidly adaptable biosynthetic process, a functional synaptic modification, comparable to post-tetanic facilitation or change in presynaptic inhibition, seems more plausible. Intermediate-range changes such as temperature adaptation or long-term memory may be analogous to chronic (between-sessions) tolerance, developing over a period of days or weeks and lasting for a similar time after drug withdrawal. Such processes are known to involve neurochemical mechanisms, which will be discussed below. Finally, the long-range carry-over of ability for rapid re-evocation of tolerance mentioned above, like the long-range ability to re-evoke antibody formation, almost certainly implies a biochemical change in nuclear or other regulatory mechanisms within the neurons.

At the same time, there is evidence that tolerance does not develop at the same rate to all the effects of the same drug. For example, CNS tolerance to alcohol and barbiturates can be demonstrated on motor coordination and body temperature tests, after treatment with doses which do not produce tolerance in tests of "sleeping time" (J.M. Khanna et al., unpublished results). It is not yet clear whether these results reflect different neuronal pathways having different thresholds for initiation of tolerance, or whether the degree of tolerance, like the susceptibility to acute drug effects, represents summation of changes at different numbers of synapses in simpler and more complex pathways.

Possible Neurochemical Mechanisms of Tolerance to Central Depressants

Over the past half-century, many possible mechanisms have been investigated. Not surprisingly, they reflect the growing sophistication of neurochemical knowledge and techniques. No attempt will be made here to review exhaustively all the possibilities that have been explored. However, it is worth while to mention some of the major groups of related phenomena that have been studied.

General metabolic processes. Some of the earliest studies involved acute effects of central depressants on basic aspects of cell metabolism, such as oxygen consumption, glucose utilization and oxidative phosphorylation, and changes in these effects during chronic drug administration. More recently there have been comparable studies of the acute and chronic effects on other substances central to cellular respiration, such as brain levels of coenzyme A, adenine nucleotides, tricarboxylic acid cycle intermediates, and free amino acids (Brunner et al., 1975, Rawat, 1975). In general, the drug concentrations needed to affect processes of neuronal respiration and intermediary metabolism are in the general anesthesia range, and are quite high relative to those which are capable of inducing tolerance on behavioral tests.

Neurotransmitter turnover. A more promising approach, therefore, appeared to be in relation to processes more directly related to neuronal transmission. Ethanol, barbiturates and other central depressants have been shown to affect acutely the release, and indirectly the synthesis and degradation, of most transmitter substances (Wallgren & Barry, 1970; Kalant, 1975; Smith, 1977). Comparable results have been found with a range of different techniques, including cortical superfusion cups (Phillis & Jhamandas, 1971) and push-pull cannula perfusion methods (Erickson & Graham, 1973), in vivo turnover rate measurements (Frankel et al., 1975; Ngai et al., 1978), and release of endogenous or isotopically labelled exogenous transmitters from isolated cortical slices in vitro (Kalant et al., 1977).

At low concentrations of "depressant" drugs, at which increased neuronal activity and overt behavior are commonly seen, there appears to be a small increase in release of acetylcholine (ACh), while at higher concentrations there is a progressive reduction, especially in the release stimulated by applied electrical pulses or endogenous nerve impulses (Clark et al., 1977). Other transmitters, including norepinephrine (NE), dopamine (DA), serotonin (5-HT) and GABA, are similarly affected, but generally at considerably higher drug concentrations (Carmichael & Israel, 1975). It is not surprising, therefore, that in tissue from animals made tolerant by chronic drug administration, only ACh release (Clark et al., 1977) and re-uptake (Carson et al., 1975) have been shown to be increased in the drug-withdrawn state.

Nevertheless, when ethanol, barbiturates or benzodiazepines are administered in vivo, marked and complex patterns of changes in levels and turnover rates of 5-HT, DA, NE and GABA are seen, both during drug administration and in the withdrawal state. These changes are reviewed by Engel elsewhere in this symposium, and will not be detailed here. The point to be made, however, is that these alterations vary in different parts of the brain, with different drug regimens, and at different stages, in a way which cannot be related to a simple mechanism of direct drug action. Therefore most of these changes probably reflect altered impulse flow, secondary to the drug actions elsewhere in the CNS.

Cyclic nucleotides. Since various neurotransmitters have been shown to exert at least some of their actions via changes in the levels of c-AMP and c-GMP in the post-synaptic neurons, it is not surprising that the foregoing complex effects of alcohol and other central depressants on transmitter release in vivo are accompanied by equally complex patterns of change in cyclic nucleotide metabolism (Hess et al., 1975; Hunt et al., 1977; Volicer & Hurter, 1977) and in the enzymes related to c-AMP and c-GMP (Kuriyama, 1977). It is also not surprising that corresponding changes in receptor sensitivity have been found (Engel & Liljequist, 1976; French et al., 1977). The problem of interpretation, however, is exactly the same for the transmitter-related changes as for the changes in turnover of the transmitters themselves.

Cation transport. Ethanol at rather high concentrations has been found to inhibit active transport of K^+ in brain slices, and to inhibit allosterically the membrane-bound (Na + K)-dependent adenosine triphosphatase (ATPase) which is responsible for this transport (Kalant et al., 1978c). After chronic ethanol treatment, the ATPase activity is increased in synaptosomal membranes isolated from physically dependent animals at times corresponding to maximum intensity of withdrawal reactions (Roach et al., 1973; Wallgren et al., 1975, Rangaraj & Kalant, 1978). However, it is not yet possible to say whether it is increased in the tolerant animal that has not undergone a withdrawal reaction. The problem is that the enzyme becomes extremely sensitive to inhibition by ethanol in the presence of NE or DA at the concentrations found in normal brain. Thus, during isolation of the enzyme in the presence of ethanol, the increased activity may be masked.

Membrane structure. All the central depressants are postulated to share the common feature of inserting themselves into either the lipid bilayer or the lipid-protein interface of the cell membrane, according to their membrane:water partition coefficients and their various polar and non-polar groups. The common effect at relatively low concentrations is a "fluidization" of membrane lipids, and an associated expansion of the membrane, preventing the conformational changes necessary for opening of the Na^+ channels during the initial depolarization phase of the action potential (Seeman, 1972; Hunt, 1975). There is indeed a substantial body of evidence to support this explanation of the local anesthetic action of these drugs. However, it is not yet technically possible to demonstrate whether or not such an action underlies the milder levels of intoxication produced by these drugs in the CNS.

If this is a basic mechanism of drug action, one might expect that tolerance would be accompanied by a change in neuronal membrane structure such as to favor easier opening of the Na^+ channels, to offset the membrane packing effect of the drugs. In the drug-free state, greater Na^+ flux should be evident, as a basic mechanism of neuronal hyperexcitability. So far, this has been investigated only indirectly by measurement of N-ethyl maleimide reactivity as an index of membrane conformational change (Gruber et al., 1977), and by use of electron spin resonance probes to test the degree of membrane fluidity (Chin & Goldstein, 1977a,b). Ethanol, added in vitro, was found to increase synaptosomal membrane fluidity as predicted. However, membranes from tolerant animals were not less fluid than normal membranes, but were merely less affected by addition of ethanol in vitro. These findings therefore do not yet suggest a mechanism which can account for both tolerance and physical dependence.

Protein synthesis. Regardless of the underlying mechanism, if tolerance is accompanied by increased activity of certain enzymes, increased turnover of certain neurotransmitters, or increased synthesis of certain constituents of cell membranes or organelles, then obviously it must involve increased activity of at least some sequences of DNA-RNA-protein synthesis. One might reasonably predict that addition of central depressants acutely would inhibit protein synthesis, and that tolerance and physical dependence would be accompanied by compensatory increases. To date this has been studied mainly by incorporation of labelled amino acids into protein by brain cortex slices (Jarlstedt, 1977) or by cerebral ribosomes (Tewari et al., 1977; Khawaja et al., 1978) in vitro, and the results are not consistent with this simple hypothesis. Ethanol, added in vitro to ribosomes from normal brain, does reduce amino acid incorporation. However, ribosomes from chronically ethanol-treated animals show reduced rather than increased incorporation in the absence of ethanol in vitro. In vivo incorporation studies have given contradictory results: alcohol-withdrawn animals have shown decreased incorporation in some studies, and increased in others (Mørland & Sjetnan, 1976). The results are difficult to interpret because proper paired-feeding has not been used in all cases, and also because some of the findings may represent the effects of withdrawal reaction rather than of tolerance per se.

General Problems of Interpretation

All of the neurochemical studies summarized above share various major difficulties of interpretation. The first is that the great majority have not included behavioral measures of tolerance or physical dependence. It has often been assumed that chronic administration of a drug is sufficient to ensure tolerance. The recent work on behavioral modification of tolerance shows this assumption to be simplistic.

A second problem is that the use of whole brain for preparing ribosomes for protein synthesis studies, or of cortical slices for transmitter release, may mask important regional differences in response to a drug, acutely or chronically. For example, ACh turnover was found to be differentially affected in different parts of the brain by various anesthetic agents (Nordberg & Sundwall, 1977; Ngai et al., 1978). There is suggestive evidence of differential ethanol effects on amino acid incorporation into different brain structures in vivo (Jarlstedt, 1977).

A third problem is that it is impossible to tell whether any neurochemical change found after chronic drug administration is a possible mechanism of tolerance, an effect of tolerance or withdrawal, an incidental or parallel disturbance, or even a manifestation of drug-induced pathology rather than tolerance. Even if the changes disappear over the same time course as the disappearance of tolerance after the drug is stopped, this still does not answer the basic question of cause vs. effect.

For example if one examines cerebral cortical ATPase activity, or ACh turnover, or c-AMP synthesis after administration of a drug in vivo, it is impossible to be certain whether any changes found are due to direct action of the drug on the cortex, or are secondary to reduced neuronal activity because of reduced impulse flow from subcortical sites where the drug may be acting. Conversely, changes found during tolerance or withdrawal may represent adaptations to altered impulse flow from other parts of the brain, rather than adaptations to direct local action of the drug (Kalant, 1975). For these reasons, it is advantageous to use the opposite approach, i.e. to produce specific identifiable neurochemical lesions or dysfunctions and see how these affect the acute action of a drug and the ability of the organism to develop tolerance to it under defined conditions. There have been relatively few such studies.

Biochemical Interventions Affecting Tolerance Development

In view of the resemblances noted earlier, between tolerance and learning, or tolerance and physiological adaptations, it is not surprising that biochemical interventions that modify these latter processes have also been tested for their effects on development of tolerance. Cycloheximide, an inhibitor of cerebral protein synthesis, which had previously been shown to block the consolidation of new learning, has also been found to block completely the development of tolerance in rats treated chronically with ethanol (LeBlanc et al., 1976b). This is not surprising, since tolerance development, as noted above, must involve new synthesis of at least some proteins. However, its action is so broad that it really provides little help in localizing or identifying the basic mechanism(s) of tolerance.

Depletion of specific neurotransmitters provides a more selective approach, provided that the transmitter in question is not essential for the expression of the acute effects of the drug under study. For example, 5-HT depletion by p-chlorophenylalanine (p-CPA) was shown to inhibit the development of tolerance to morphine analgesia, but it also antagonized the acute analgesic effect. Therefore it was not clear whether p-CPA was in fact blocking the mechanism of tolerance development or simply removing the stimulus to tolerance. Fortunately the situation with central depressants has proved somewhat simpler. Pre-administration of p-CPA had no effect on the acute impairment of motor performance by ethanol or pentobarbital, but clearly retarded the development of tolerance to both these drugs (Frankel et al., 1975), and of cross-tolerance between them and barbital. Conversely, elevating the level of brain 5-HT by administration of tryptophan accelerated the development of tolerance. Further, p-CPA also retarded the development of tolerance to ethanol and barbiturates as measured by changes in drug-induced hypothermia and in drug-induced sleep time. Comparable effects have been obtained by pre-treatment with 5,7-dihydroxytryptamine, another inhibitor of 5-HT synthesis (J.M. Khanna et al., in preparation).

The effects of p-CPA on ethanol and barbiturate tolerance have been examined in greater detail. When animals have been made maximally tolerant before the start of p-CPA treatment, and treatment with the central depressant is continued, p-CPA has no effect on the maintenance of tolerance. However, if the drug is stopped, p-CPA accelerates the loss of tolerance (Frankel et al., 1978). Therefore it appears that the effect of 5-HT on acquisition of tolerance may be to facilitate the retention of short-term tolerance from one session to another. Thus, consolidation and incremental build-up of the adaptive changes may proceed more rapidly. With very slowly developing tolerance, such as that produced in time by many repetitions of the test doses used to see if tolerance is occurring, 5-HT appears to have little effect on the rate.

NE may also be involved in the development of tolerance in a way which does not involve alteration of the acute drug effects. Destruction of noradrenergic axon terminals by injection of 6-hydroxydopamine has been found to block the development of tolerance to the hypothermia-producing effect of phenobarbital and ethanol in mice (Tabakoff & Ritzman, 1977; Tabakoff et al., 1978). It was also claimed that tolerance and physical dependence were separable, since physical dependence was demonstrable in the 6-hydroxydopamine-treated mice while tolerance was not. However, physical dependence was assessed 12 hrs after ethanol withdrawal, while tolerance was tested the next day. If NE depletion, like 5-HT depletion, accelerates the loss of tolerance, it could have altered in that way the apparent relation between tolerance and physical dependence.

A very recent method of biochemical intervention in the study of drug tolerance has been the use of various endogenous <u>brain and hypophyseal peptides or fragments</u> prepared from them. This intervention also grows out of the study of the neurochemistry of learning. Over the past 15 years, much information has accumulated on the ability of vasopressin and ACTH, and hormonally inactive fragments of these, to facilitate the acquisition, or prevent the extinction, of newly learned behaviors of several types (DeWied, 1977). Radioactively labelled peptides were found to bind preferentially in the region of the parafascicular nuclei, and lesioning of the binding site gave rise to degeneration of 5-HT terminals in the cortex. In view of the effects of 5-HT on tolerance development mentioned above, it was inevitable that effects of these peptides on tolerance would be tested. Acceleration of morphine tolerance and physical dependence by DGVP and by Pro-Leu-GlyNH$_2$ (PLG, the side chain of oxytocin) has been reported (Van Ree & DeWied, 1977). However, the results to date with ethanol tolerance have been unclear. Both DGVP and PLG have been found to inhibit the development of "behaviorally augmented" tolerance when given in doses of 2 µg/kg or more (Kalant et al., 1978b), but smaller doses in the range of 0.2 µg/kg appear to accelerate it. So far, there has not been a systematic investigation of the effects of endorphins and related opioid peptides on development of tolerance to central depressants.

This brief sampling of biochemical interventions perhaps gives an overly optimistic picture of their utility in the analysis of mechanisms of tolerance and dependence on central depressants. As more such interventions are tried, it may well be found that major disturbance of the metabolism of <u>any</u> neurotransmitter will inhibit tolerance development, and that the real requirement is for an optimal flexible balance of all transmitters. That would not be surprising, in view of the numerous reciprocal modulatory loops that have already been found among known transmitters. In that case, the study of tolerance will have to move to a higher level of complexity, involving adaptive changes in neuron systems rather than in single transmitters or intracellular messengers.

Conclusions
(1) To date, neurochemical investigations of central depressant tolerance and physical dependence have concentrated chiefly on changes in whole brain, following chronic administration of the drugs in question. There is a great need of much more selective studies involving localized change in identifiable functional units, ranging from single neurons or subcellular fractions to specific brain nuclei or pathways.

(2) There is a clear need for much more careful correlation of neurochemical changes with behavioral evidence of tolerance and dependence. Since the same dosage regimen of a drug can produce quite different degrees of tolerance, depending on the behavioral and environmental circumstances, it is no longer justifiable to assume that chronic administration of a drug automatically implies tolerance and dependence.

(3) Whenever possible, comparisons should be made between different drugs, including various central depressants and opiates, and between different dosage regimens. This may help to clarify which neurochemical changes are specific for tolerance to a given drug, and which are secondary to behavioral changes produced by all of them.

(4) It will likely prove more useful in future to study the effects of known specific biochemical interventions on the ability to produce tolerance to a drug, rather than the effects of drug tolerance on a limitless range of biochemical processes which go on in living neurons.

References

Boisse, N.R. and Okamoto, M. (1978) Physical dependence to barbital compared to pentobarbital. III. Withdrawal characteristics. J. Pharmacol. Exp. Ther. 204: 514-525.

Brunner, E.A., Chang, S.C. and Berman, M.F. (1975) Effects of anesthesia on intermediary metabolism. Ann. Rev. Med. 26: 391-401.

Carmichael, F.J. and Israel, Y. (1975) Effects of ethanol on neurotransmitter release by rat brain cortical slices. J. Pharmacol. Exp. Ther. 193: 824-834.

Carson, V.G., Jenden, D.J. and Noble, E.P. (1975) Acetylcholine (AcCh) uptake in cortical slabs taken from $C_{57}Bl$ mice after acute and/or chronic ethanol treatment. Proc. West. Pharmacol. Soc. 19: 341-345.

Chin, J.H. and Goldstein, D.B. (1977a) Effects of low concentrations of ethanol on the fluidity of spin-labeled erythrocyte and brain membranes. Molec. Pharmacol. 13: 435-441.

Chin, J.H. and Goldstein, D.B. (1977b) Drug tolerance in biomembranes: A spin label study of the effects of ethanol. Science 196: 684-685.

Clark, J.W., Kalant, H. and Carmichael, F.J. (1977) Effect of ethanol tolerance on release of acetylcholine and norepinephrine by rat cerebral cortex slices. Can. J. Physiol. Pharmacol. 55: 758-768.

Engel, J. and Liljequist, S. (1976) The effect of long-term ethanol treatment on the sensitivity of the dopamine receptors in the nucleus accumbens. Psychopharmacol. 49: 253-257.

Erickson, C.K. and Graham, D.T. (1973) Alteration of cortical and reticular acetylcholine release by ethanol in vivo. J. Pharmacol. Exp. Ther. 185: 583-593.

Frankel, D., Khanna, J.M., LeBlanc, A.E. and Kalant, H. (1975) Effect of p-chlorophenylalanine on the acquisition of tolerance to ethanol and pentobarbital. Psychopharmacol. 44: 247-252.

Frankel, D., Khanna, J.M., Kalant, H. and LeBlanc, A.E. (1978) Effect of p-chlorophenylalanine on the loss and maintenance of tolerance to ethanol. Psychopharmacol. 56: 139-143.

French, S.W., Palmer, D.S. and Wiggers, K.D. (1977) Changes in receptor sensitivity of the cerebral cortex and liver during chronic ethanol ingestion and withdrawal. In: M.M. Gross (ed.), Alcohol Intoxication and Withdrawal, Vol. IIIA, pp. 515-538. New York: Plenum Press.

Freund, G. (1971) Alcohol, barbiturate and bromide withdrawal syndrome in mice. In: N.K. Mello and J.H. Mendelson (eds.), Recent Advances in Studies of Alcoholism, pp. 453-471. Washington, D.C.: U.S. Dept. H.E. & W.

Gildea, M.L. and Bourn, W.M. (1977) Effect of barbiturate withdrawal on pentylenetetrazol seizure threshold. Comm. Psychopharmacol. 1: 123-129.

Gruber, B., Dinovo, E.C., Noble, E.P. and Tewari, S. (1977) Ethanol-induced conformational changes in rat brain microsomal membranes. Biochem. Pharmacol. 26: 2181-2185.

Hess, S.M., Chasin, M., Free, C.A. and Harris, D.N. (1975) Modulators of cyclic AMP systems. In: E. Costa and P. Greengard (eds.), Mechanisms of Action of Benzodiazepines, pp. 153-167. New York: Raven Press.

Höllt, V., Członkowski, A. and Herz, A. (1977) The demonstration in vivo of specific binding sites for neuroleptic drugs in mouse brain. Brain Res. 130: 176-183.

Hunt, W.A. (1975) The effects of aliphatic alcohols on the biophysical and biochemical correlates of membrane function. In: E. Majchrowicz (ed.), Biochemical Pharmacology of Ethanol, pp. 195-210. New York: Plenum Press.

Hunt, W.A., Redos, J.D., Dalton, T.K. and Catravas, G.N. (1977) Alterations in brain cyclic guanosine 3':5'-monophosphate levels after acute and chronic treatment with ethanol. J. Pharmacol. Exp. Ther. 200: 103-109.

Jarlstedt, J. (1977) Alcohol and brain protein synthesis. In: M.M. Gross (ed.), Alcohol Intoxication and Withdrawal, Vol. IIIA, pp. 155-171. New York: Plenum Press.

Kalant, H. (1975) Direct effects of ethanol on the nervous system. Fed. Proc. 34: 1930-1941.

Kalant, H. (1977) Comparative aspects of tolerance to, and dependence on, alcohol, barbiturates, and opiates. In: M.M. Gross (ed.), Alcohol Intoxication and Withdrawal, Vol. IIIB, pp. 169-186. New York: Plenum Press.

Kalant, H., LeBlanc, A.E. and Gibbins, R.J. (1971) Tolerance to, and dependence on, some non-opiate psychotropic drugs. Pharmacol. Rev. 23: 135-191.

Kalant, H., LeBlanc, A.E., Gibbins, R.J. and Wilson, A.P. (1978a) Acceleration of development of tolerance during repeated cycles of alcohol exposure. Psychopharmacology, in press.

Kalant, H., Mucha, R.F. and Niesink, R. (1978b) Effects of vasopressin and oxytocin fragments on ethanol tolerance. Proc. Can. Fed. Biol. Soc. 21: 71.

Kalant, H., Woo, N. and Endrenyi, L. (1978c) Effect of ethanol on the kinetics of rat brain ($Na^+ + K^+$)ATPase and K^+-dependent phosphatase with different alkali ions. Biochem. Pharmacol. 27(8): in press.

Khawaja, J.A., Lindholm, D.B. and Niittylä, J. (1978) Selective inhibition of protein synthetic activity of cerebral membrane-bound ribosomes as a consequence of ethanol ingestion. Res. Comm. Chem. Pathol. Pharmacol. 19: 185-188.

Kuriyama, K. (1977) Ethanol-induced changes in activities of adenylate cyclase, guanylate cyclase and cyclic adenosine 3',5'-monophosphate dependent protein kinase in the brain and liver. Drug Alc. Depend. 2: 335-348.

LeBlanc, A.E., Kalant, H. and Gibbins, R.J. (1976a) Acquisition and loss of behaviorally augmented tolerance to ethanol in the rat. Psychopharmacol. 48: 153-158.

LeBlanc, A.E., Matsunaga, M. and Kalant, H. (1976b) Effects of frontal polar cortical ablation and cycloheximide on ethanol tolerance in rats. Pharmacol. Biochem. Behav. 4: 175-179.

Majchrowicz, E. (1975) Induction of physical dependence upon ethanol and the associated behavioral changes in rats. Psychopharmacol. 43: 245-254.

Majchrowicz, E. and Hunt, W.A. (1976) Temporal relationship of the induction of tolerance and physical dependence after continuous intoxication with maximum tolerable doses of ethanol in rats. Psychopharmacol. 50: 107-112.

Mørland, J. and Sjetnan, A.E. (1976) Reduced incorporation of [^{3}H]leucine into cerebral proteins after long-term ethanol treatment. Biochem. Pharmacol. 25: 220-221.

Mucha, R.F., Niesink, R. and Kalant, H. (1978) Tolerance to morphine analgesia and immobility measured in rats by changes in log-dose-response curves. Life Sci., in press.

Ngai, S.H., Cheney, D.L. and Finck, A.D. (1978) Acetylcholine concentrations and turnover in rat brain structures during anesthesia with halothane, enflurane and ketamine. Anesthesiol. 48: 4-10.

Nordberg, A. and Sundwall, A. (1977) Effect of sodium pentobarbital on the apparent turnover of acetylcholine in different brain regions. Acta Physiol. Scand. 99: 336-344 (1977).

Phillis, J.W. and Jhamandas, K. (1971) The effects of chlorpromazine and ethanol on in vivo release of acetylcholine from the cerebral cortex. Comp. Gen. Pharmacol. 2: 306.

Rangaraj, N. and Kalant, H. (1978) Effects of ethanol withdrawal, stress and amphetamine on rat brain ($Na^+ + K^+$)-ATPase. Biochem. Pharmacol., in press.

Rawat, A.K. (1975) Effects of ethanol on brain metabolism. In: E. Majchrowicz (ed.), Biochemical Pharmacology of Ethanol, pp. 165-177. New York: Plenum Press.

Roach, M.K., Khan, M.M., Coffman, R., Pennington, W. and Davis, D.L. (1973) Brain ($Na^+ + K^+$)-activated adenosine triphosphatase activity and neurotransmitter uptake in alcohol-dependent rats. Brain Res. 63: 323-329.

Seeman, P. (1972) The membrane actions of anesthetics and tranquilizers. Pharmacol. Rev. 24: 583-655.

Seeman, P., Chau-Wong, M., Tedesco, J. and Wong, K. (1975) Brain receptors for antipsychotic drugs and dopamine: Direct binding assays. Proc. Nat. Acad. Sci. (Wash.) 72: 4376-4380.

Smith, C.M. (1977) The pharmacology of sedative/hypnotics, alcohol, and anesthetics: sites and mechanisms of action. In: W.R. Martin (ed.), Handbuch der experimentellen Pharmakologie, Vol. 45/1, pp. 413-587. Berlin: Springer-Verlag.

Snyder, S.H. (1977) Opiate receptors and internal opiates. Scientific American 236(3): 44-56.

Squires, R.F. and Braestrup, C. (1977) Benzodiazepine receptors in rat brain. Nature 266: 732-734.

Staiman, A. and Seeman, P. (1974) The impulse-blocking concentrations of anesthetics, alcohols, anticonvulsants, barbiturates, and narcotics on phrenic and sciatic nerves. Can. J. Physiol. Pharmacol. 52: 535-550.

Tabakoff, B., Hoffman, P. and Moses, F. (1977) Neurochemical correlates of ethanol withdrawal: alterations in serotoninergic function. J. Pharm. Pharmacol. 29: 471-476.

Tabakoff, B. and Ritzmann, R.F. (1977) The effects of 6-hydroxydopamine on tolerance to and dependence on ethanol. J. Pharmacol. Exp. Ther. 203: 319-331.

Tabakoff, B., Yanai, J. and Ritzmann, R.F. (1978) Brain noradrenergic systems as a prerequisite for developing tolerance to barbiturates. Science 200: 449-451.

Van Ree, J. and DeWied, D. (1977) Effect of neurohypophyseal hormones on morphine dependence. Psychoneuroendocrinol. 2: 35-41.

Volicer, L. and Hurter, B.P. (1977) Effects of acute and chronic ethanol administration and withdrawal on adenosine 3':5'-monophosphate and guanosine 3':5'-monophosphate levels in the rat brain. J. Pharmacol. Exp. Ther. 200: 298-305.

Wallgren, H. and Barry, H., III. (1970) The Actions of Alcohol. Amsterdam: Elsevier.

Wallgren, H., Nikander, P. and Virtanen, P. (1975) Ethanol-induced changes in cation-stimulated adenosine triphosphatase activity and lipid-proteolipid labeling of brain microsomes. In: M.M. Gross (ed.), Alcohol Intoxication and Withdrawal, II., pp. 23-36. New York: Plenum Press.

PATHWAYS OF ALCOHOL METABOLISM

K.F. Tipton, A.J. Rivett and I.L. Smith
Department of Biochemistry, Trinity College, Dublin 2, Ireland.

I. Introduction

The diverse effects of acute and chronic ethanol ingestion make it difficult to identify the primary effects of this drug and it is unlikely that there will be a single common mechanism underlying the various aspects of intoxication and dependence. In this review we will consider the pathways involved in ethanol breakdown and some of the theories that might explain its neurochemical effects.

II. Ethanol Metabolism in Liver

The liver is by far the most important site of metabolism of ingested ethanol, accounting for some 75% of the total (see Li, 1977). The metabolism involves its oxidation first to acetaldehyde and then to acetate. Minor pathways for ethanol include excretion unchanged and as acetaldehyde, condensation of acetaldehyde with amines, conjugation to form sulphate or glucuronide esters and formation of fatty acid esters, but these have little quantitative significance (Hawkins and Kalant, 1972).

A. Oxidation by Alcohol Dehydrogenase

Alcohol dehydrogenase (ADH), which is the most important enzyme involved in the formation of acetaldehyde, catalyses the reaction:

$$C_2H_5OH + NAD^+ \rightleftharpoons CH_3CHO + NADH + H^+$$

It has a wide substrate specificity and its normal physiological functions may include the oxidation of steroids and retinol and the omega-oxidation of fatty acids as well as the oxidation of small amounts of endogenously produced ethanol. Despite the relatively long history of studies on this enzyme, it is only recently that the full complexity of its constituent isoenzymes in human liver is beginning to become clear. In some livers as many as 3-10 isoenzymes may be separated by electrophoresis and the genetic basis of their formation has been considered (see Li, 1977, for review). The majority of these isoenzymes show quite similar K_m values for ethanol (Pietruszko et al., 1972) although they may differ in their substrate-specificities, specific activities and inhibitor sensitivities. Two specific isoenzymes have, however, attracted special interest because of unusual features in their properties that may have direct relevance to the metabolic effects of ethanol.

The "atypical" ADH isoenzyme, that was first detected by von Wartburg et al., 1974), has been shown to be present in about 5-10% of the British and American white population, 20% of the Swiss population and 85% of the Japanese population. It differs from the other isoenzymes in having an optimum pH for the conversion of ethanol to acetaldehyde of about 8.5 whereas the optimum for the normal ADH isoenzymes is greater than pH 10. This results in the atpical isoenzyme having a higher activity at physiological pH values which can cause a rapid rise in the concentration of acetaldehyde following ethanol ingestion and it has been speculated that this may explain the phenomenon of alcohol sensitivity. This condition, which is common amongst the Japanese, involves vasodilation, flushing and tachycardia following the ingestion of ethanol. Similar responses can be induced in many normal individuals by the administration of the drug disulphiram (Antabuse), which inhibits the metabolism of acetaldehyde, and this has led to the suggestion that alcohol sensitivity may result from the production of higher-than-normal acetaldehyde concentrations in response to ethanol.

An additional isoenzyme, π-ADH, purified from human liver (Li et al., 1977), differs from the other isoenzymes in having a considerably higher K_m for ethanol (about 20 mM at pH 7.5 as compared with about 0.4 mM for the others) which suggests that it might be of particular importance in the metabolism of high doses of ethanol above those that are sufficient to saturate the other isoenzymes. Li et al. (1977) have, calculated that this isoenzyme might account for less than 15% of the ethanol metabolism at 5 mM ethanol but as much as 40% at intoxicating doses of 50 mM. The activity of this isoenzyme showed considerable variations between individual human livers leading to the speculation that this may provide a basis for the observed genetic variation in alcohol tolerance.

B. Other Pathways of Ethanol Oxidation

It appears that the capacity of liver ADH to deal rapidly with intoxicating doses of ethanol in vivo is limited by the availability of NAD$^+$ (Grunnet et al., 1973; Hawkins and Kalant, 1972) and this is supported by the observation that an increased rate of ethanol oxidation results from administration of metabolites such as fructose, pyruvate or glyceraldehyde (Grunnet et al., 1973) that can increase the rate of NADH re-oxidation.

A proportion of hepatic ethanol oxidation is insensitive to the ADH inhibitor pyrazole (see e.g. Lieber 1977a; Grunnet et al., 1973). The ADH isoenzymes differ in pyrazole sensitivity and in particular the π-isoenzyme is relatively insensitive (Li et al., 1977) but there is certainly a proportion of ethanol metabolism not involving NAD^+ which is shown by its failure to alter the cellular NAD^+/NADH ratio (Grunnet and Theiden, 1972) and by isotope labelling which gives the value 10 - 11% for the ADH-independent pathway (Harve et al., 1977).

Possible alternative pathways are the reaction catalysed by catalase:
$$C_2H_5OH + H_2O_2 \longrightarrow CH_3CHO + 2H_2O$$
and a microsomal cytochrome P-450-containing ethanol oxidizing system (MEOS) that catalyses:
$$C_2H_5OH + NADPH + H^+ + O_2 \longrightarrow CH_3CHO + NADP^+ + 2H_2O$$
The relative importance of the two is the subject of considerable controversy (see Li, 1977; Lieber, 1977a).

The concentration of the enzyme-substrate complex between catalase and H_2O_2 in perfused liver is found to be reduced in the presence of ethanol oxidation (Oshino et al., 1973). Thurman et al, (1972) have suggested that this is the sole alternative pathway since they found that treatment with pyrazole and the catalase inhibitor 3-amino-1,2,4-triazole completely inhibits ethanol oxidation in rat liver. They suggested that MEOS activity was solely due to contamination by catalase. Lieber (1977a) has, however, calculated that catalase can accoung for only 2% of the total ethanol oxidation under normal conditions when the reaction is limited by H_2O_2 availability.

In addition the MEOS has recently been purified to a state that is free from both alcohol dehydrogenase and catalase activities and activity has been reconstituted with cytochrome P-450, NADPH-cytochrome-c reductase and phospholipid (Ohnishi and Lieber, 1977; Miwa et al., 1978) but the doubts about its importance in ethanol metabolism in vivo have not yet been resolved.

C. Oxidation of Acetaldehyde

Over 90% of the acetaldehyde produced is normally oxidised to acetate in the liver (Lindross, 1975; Eriksson and Sippel, 1977). Mitochondrial NAD^+-dependent aldehyde dehydrogenases play the predominant role (Lindross, 1975) with the cytoplasmic enzyme only becoming significant at high aldehyde concentrations (> 400 µM). Most of the acetate produced is released into the blood stream and then taken up by other tissues for further oxidation.

D. Enzyme Induction by Ethanol

Chronic investion of ethanol results in an increased ability to metabolise it (see e.g. Cederbaum et al., 1978). Since the factor limiting the oxidation of ethanol by liver ADH appears to be the rate of reoxidation of NADH, changes in its activity for which there are conflicting reports (see Lieber, 1977a), may have little effect.

Several enzymes may be involved in the increased metabolism of ethanol during chronic ingestion. Increased activity of the mitochondrial sodium and potassium-dependent ATPase leads to a hypermetabolic state involving increased NADH oxidation (see e.g. Bernstein et al., 1975). Catalase activity is probably little affected (Ishii et al., 1973) but there is an increase in hydrogen peroxide formation due to NADPH oxidase activity (Lieber and De Carli, 1970). The activity of the MEOS has been found to be increased (Ishii et al., 1973; Ohnishi and Lieber, 1977) but, on withdrawal, the raised activity of this system persists after ethanol oxidation rates fall to normal (Mezey, 1972). The increase in MEOS activity is associated with a proliferation of the smooth endoplasmic reticulum and increases in the activities of a number of drug metabolising enzymes assocated with it, which may account for the resistance of many alcoholics to drugs such as sedatives. The effects of ethanol ingestion on the activities of liver aldehyde dehydrogenase are relatively small.

III. Ethanol Metabolism in Brain and Other Organs

A number of other tissues including brain, heart, kidney, lung, skeletal muscle, stomach, gut and blood contain ADH activity but their individual contributions to the total ethanol metabolism are small. ADH activity in brain is only 0.025% of that in liver but has been found to be increased by chronic ethanol consumption (Raskin and Sokoloff, 1972). The properties of the enzymes from rat brain and liver have been found to be very similar (Duncan et al., 1976). The observation that acute and chronic ethanol treatment result in an increase in the brain NADH/NAD$^+$ ratio (Rawat and Kuriyama, 1972) suggests that this organ may indeed be involved in metabolising ingested ethanol though Mukherji et al. (1975) failed to detect ethanol oxidation in perfused brain. Aldehyde dehydrogenase activity in brain is higher than that of ADH and the enzyme from this source has been studied in some detail. (see. Tipton et al., 1977; Tabakoff, 1977).

IV. The Effects of Ethanol Ingestion

A. Release of Acetaldehyde

Some of the effects of ethanol, for example the sweating, headache, nausea and vomiting that contribute to a severe hangover, have been attributed to the actions of acetaldehyde and are enhanced by the presence of aldehydes in alcoholic drinks (see e.g. Magrinat et al., 1973). The possible role of acetaldehyde in the other effects of ethanol is, however, less clear.

Ethanol consumption causes elevated acetaldehyde levels in blood as well as in liver. Earlier determinations of the blood acetaldehyde levels may have given overestimates because of methodological problems and its non-enzymic production but recent determinations indicate an upper value of less than 150 µM (Lindros, 1975; Eriksson and Sippel, 1977). The values are reduced by prior treatment with an alcohol dehydrogenase inhibitor and elevated with the aldehyde dehydrogenase inhibitor disulfiram.

The acetaldehyde concentrations in brain have been found to be negligable after correction for the presence of blood except at extremely high concentrations of administered ethanol (Eriksson and Sippel, 1977). In contrast Pettersson and Kiessling (1977) have reported a steady-state concentration of about 30 µM acetaldehyde in cerebrospinal fluid (CSF) when the blood concentration resulting from ethanol ingestion was 20-60 µM and also increased values when brain aldehyde dehydrogenase was inhibited. There is no known barrier between the CSF and brain tissue (Fenstermacker et al., 1974) and one might expect the concentrations in these two compartments to be similar. If the equilibration between the two pools is not very rapid, however, brain aldehyde dehydrogenase with a very low K_m for acetaldehyde (Duncan and Tipton, 1971) might cause a lower concentration to result in brain.

B. Effects on Neurotransmitters

The effects of ethanol upon the metabolism (Fig. 1), release and reuptake of biogenic amines have been extensively studied.

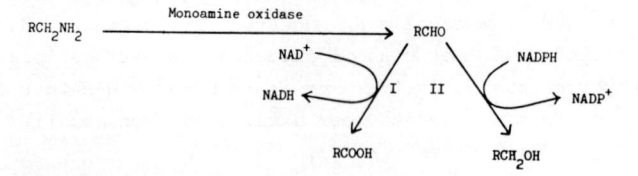

$$RCH_2NH_2 \xrightarrow{\text{Monoamine oxidase}} RCHO$$

I   Aldehyde Dehydrogenase
II Aldehyde Reductase

Fig.1 The Metabolism of Biogenic Amines. In liver ADH may also be involved in the production of the alcoholic metabolite, but this enzyme does not appear to play a significant role in brain.

Ethanol ingestion results in a decreased urinary excretion of the acid metabolite and increased formation of the alcoholic metabolite (see e.g. Huff et al., 1971). The dominant factor is acetaldehyde acting as a competitive substrate for aldehyde dehydrogenase and inhibiting oxidation of the biogenic amine-derived aldehydes (Lahti and Majchrowicz, 1969) rather than the fall in NAD^+ levels resulting from ethanol oxidation. This effect, which entails an increase in the steady-state levels of the biogenic aldehydes (Turner et al., 1974b) is mainly confined to the liver and has not been observed in the central nervous system (Eccleston et al., 1969).

Ethanol consumption appears to have little effect on the biogenic amine concentration in brain (see Noble and Tewari, 1977) although there may be a short-term stimulation of noradrenaline turnover. Thadain and Truitt (1977) have shown that ingested ethanol initially stimulates release of noradrenaline in rat brain and inhibits its neuronal uptake whereas intracranially injected acetaldehyde only increases its release.

Although ethanol ingestion appears to have no consistent effect on the levels of γ-aminobutyric acid (GABA) in brain (see e.g. Noble and Tewari, 1977; Häkkinen and Kulonen, 1976) drugs that tend to increase its concentration by inhibiting its breakdown increase the effects of ethanol intoxication whereas the GABA receptor antagonist bicuculline reduces it (Häkkinen and Kulonen (1976). These treatments were found to have the opposite effect on the hyperexcitability associated with alcohol withdrawal (Goldstein, 1975).

Ethanol ingestion inhibits the release of acetyl choline and thus lowers its concentration in brain (see Feldstein, 1971 for review). The levels return to normal during chronic ethanol consumption and this might be associated with the development of alcohol tolerance (Kalant and Grose, 1963; but see Rawat, 1974) but drugs that effect the cholinergic system appear to have no effect on withdrawal (Goldstein, 1975). The mechanism of the effect on release is unclear but action on the Na^+/K^+ balance has been suggested (Eriksson and Graham, 1973; Israel et al., 1975). Ethanol inhibits changes of cellular Na^+ and K^+ levels that result from electrical stimulation suggesting that it may inhibit the action potential in brain. In addition the effects of ethanol in inhibiting acetylcholine release from brain cortex slices are overcome by a high K^+ concentration in the medium (Kalant and Grose, 1968).

Calcium ions, which are known to have an important role in neurotransmitter release, have been shown to be significantly decreased in brain by ethanol (Ross et al., 1974). A similar effect could be produced by the administration of morphine or salsolinol, and the morphine antagonist naloxone reversed the effect in all cases.

C. Conjugation Reactions

Several condensation reactions between biogenic amines and aldehydes leading to the formation of tetrahydroisoquinoline alkaloids may occur. Dopamine and the aldehyde derived from it can react to form tetrahydropapaveroline (Holtz et al., 1964) which is structually similar to morphine (Fig.2). Dopamine and acetaldehyde react in a similar way to form salsolinol and condensation between tryptamine or 5-hydroxyptamine and aldehydes could occur as shown in Fig. 3 (Holman et al., 1975).

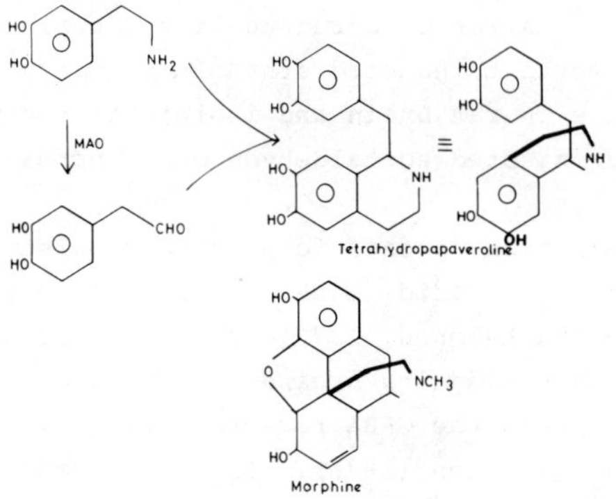

Fig.2 The Formation of Tetrahydropapaveroline from Dopamine and Dihydroxyphenylacetaldehyde, and its structural similarity to Morphine

These compounds are pharmacologically active (see e.g. Deitrich, 1977; Cohen, 1975) and it has been suggested that they may cause some of the effects of ethanol. However elevated brain concentrations of aldehydes are required for formation of the condensation products (Walsh, 1973) and there is little evidence that substantial amounts of acetaldehyde reach the brain. They could, however be formed peripherally and it has been suggested that a derivative of salsolinol can cross the blood-brain barrier (Marshall and Hirst, 1976). Cohen (1975) has demonstrated the formation of salsolinol when adrenal glands are perfused with acetaldehyde, and salsolinol and tetrahydropapaveroline have been detected in the urine of Parkinsonian patients treated with DOPA (Sandler et al., 1973) and in rat brain following the administration of L-DOPA and ethanol (Turner et al., 1974a; Collins and Bigdeli, 1975). Thus it appears that these compounds can be formed in vivo, but it remains to be shown whether they have any importance in determining the effects of ethanol.

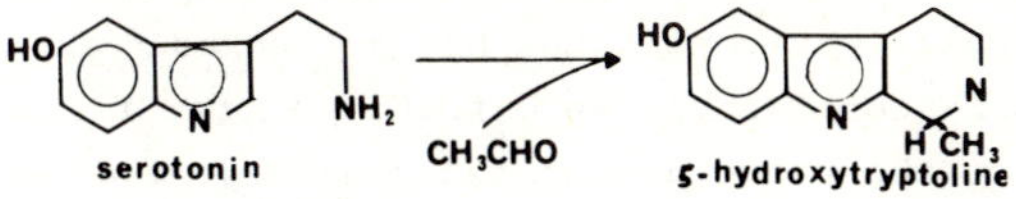

Fig.3 The Formation of 5-Hydroxytryptoline.

D. Effects on Membranes

Ethanol causes an expansion and fluidisation of membranes (Seeman, 1972 and 1974) which is belived to be related to its anaesthetic actions. These effects, which appear to involve alterations in the membrane lipid (Miceli and Ferrel, 1973; Hosein et al., 1977) and protein (Seeman, 1974; Hoffman and Hosein, 1978), have been observed in mitochondrial membranes, where they are probably responsible for the impairment of mitochondrial function that results from chronic ethanol consumption (Hosein et al., 1977), and in neural membranes where they affect neuro-transmitter release and cation permeability (Seeman, 1974). In addition to direct effects of ethanol, aldehydes may also affect these systems since they have been shown to bind to neural membranes (Alivisatos and Ungar, 1968). In contrast to these effects the development of ethanol tolerance in experimental animals has been reported to be associated with a rigidifying of the membranes (Curran and Seeman, 1977) which renders them less sensitive to the effects of ethanol and anaesthetics.

E. Sleep

The relationships between the effects discussed above and the sedative effects of ethanol and the sleep disturbances that occur in alcoholic psychosis are unclear. Several workers have shown that the metabolites of the biogenic amines have sleep-inducing properties (Sabelli and Giardina, 1973; Blum et al., 1973) and that they have a synegistic effect with ethanol. The idea that the ethanol-produced alterations in the metabolism of the biogenic amines, are responsible for the sleep-producing effects of ethanol is attractive but far from proven, since, as discussed earlier, the central effects of ethanol do not appear to include any detectable alteration of amine metabolite patterns.

The pathway involved in the metabolism of GABA is similar to that of amines since the succinate semialdehyde produced by transamination can be converted by a specific dehydrogenase to succinate or reduced by an aldehyde reductase to form γ-hydroxybutyrate (GHB) (see Tipton et al., 1977 for review). Since GHB or its parent aldehyde have a powerful hypnotic action (Laborit, 1973) the possibility that ethanol may affect this metabolic process was considered, but at "normal" concentrations neither ethanol nor acetaldehyde were found to have any appreciable effects upon either the reductase or the dehydrogenase (unpublished observations). It is probably an oversimplification to view these effects solely in terms of disturbances in aminergic systems since cholinergic systems are implicated by the observation that physostigmine has an antagonistic effect in mice (Eriksson and Burnam, 1971).

F. Interactions with Drugs

The alcohol dehydrogenase inhibitor pyrazole has been shown to potentiate a number of the acute effects of ethanol and also to delay the onset of the withdrawal reaction in alcohol-dependent mice, although it has a convulsant action if administered during ethanol withdrawal (see Goldstein, 1975). In addition pyrazole alone has been reported to be capable of producing a withdrawal reaction in some conditions (Thurman and Pathman, 1975). In contrast disulfiram appears to affect neither the development of ethanol-dependence nor the withdrawal reaction which has been interpreted as indicating that acetaldehyde is not directly involved in these phenomona (Thurman and Pathman, 1975). The acute interaction between disulfiram and ethanol has already been mentioned and this has led to its use in the aversion-treatment of alcoholics, the intensity of the reaction in different individuals is, however, very variable. Interpretation of the effects of this drug are complicated by its lack of specificity since it inhibits dopamine β-hydroxylase at higher concentrations, but both these effects have been incorporated into a proposed mechanism for the induced ethanol sensitivity which involves the elevated acetaldehyde levels causing release of catecholamines (Truit and Walsh, 1971).

Chloral hydrate and paraldehyde also act as inhibitors of aldehyde dehydrogenase producing characteristic shifts in amine metabolism (Huff et al., 1971). Their action, however, is more complicated than that of disulfiram since they produce dependence. There are similarities between the effects of these drugs and that of ethanol and any of these compounds will substitute for another in preventing withdrawal symptoms. There is also cross-tolerance between ethanol and the barbiturates and the latter drugs have been shown to affect the mechanism of the biogenic amines by inhibiting the aldehyde reductases. Differences between the specificities of the aldehyde reductases and dehydrogenases will result in the effects of barbiturates being different from those of acetaldehyde (see

Tipton et al., 1977). The relationship, if any, between this effect and the pharmacological effects of barbiturates is unclear particularly since the sedative effects of barbiturates are sterospecific whereas the ability to inhibit the aldehyde reductases is not, and this has led to the suggestion that the inhibition is involved in the anticonvulsant rather than the sedative effects of these compounds (Erwin and Deitrich, 1973). If at least some of the symptoms of ethanol dependence were due to the formation of morphine-like alkaloids one might expect some cross-tolerance between ethanol and morphine but none has been demonstrated. In addition the withdrawal symptoms from morphine are very different from those from ethanol and the administration of the morphine antagonist naloxone did not precipitate withdrawal symptoms in alcohol-dependent rats (Goldstein and Judson, 1971).

E. Other Effects

In this review we have only been able to deal with selected aspects of the effects of ethanol and the reader is referred to the recent reviews edited by Lieber (1977b) for detailed treatments of other aspects including the effects on cellular metabolism, lipid, protein and nucleic acid synthesis and the tissue damage that can result from chronic ethanol consumption.

References

Alivisatos, S.G.A. and Ungar, F. (1968) Incorporation of radioactivity from labelled serotonin and tryptamine into acid-insoluble material from subcellular fractions of brain. I. The nature of the substrate, Biochemistry 7: 285-292.

Bernstein, J., Videla, L. and Israel, Y. (1975) Hormonal influences in the development of the hypermetabolic state of the liver produced by chronic administration of ethanol, J. Pharmac. Exp. Ther. 192:583-591.

Blum, D., Calhoun, Merritt, J.H. and Wallace, J.H. (1973) Synergy of ethanol and alcohol-like metabolites:tryptophol and 3,4-dihydroxyphenylethanol, Pharmac. 9: 294-299.

Cederbaum, A.I., Dicker, E., Lieber, C.S. and Rubin, E. (1978) Ethanol oxidation by isolated hepatocytes from ethanol-treated and control rats; factors contributing to the metabolic adaption after chronic ethanol consumption, Biochem. Pharmac. 27: 7-15.

Cohen, G. (1975) Some neuropharmacologic properties of tetrahydroisoquinolines derived from the condensation of aldehydes with catecholamines, Finnish Foundation for Alcohol Studies 23: 187-195.

Collins, M.A. and Bigdeli, M.G. (1975) Tetrahydroisoquinolines in vivo. I. Rat brain formation of salsolinol, a condensation product of dopamine and acetaldehyde, under certain conditions during ethanol intoxication, Life Sci. 16: 585-602.

Curran, M. and Seeman, P. (1977) Alcohol tolerance in a cholinergic nerve terminal:relation to the membrane expansion-fluidization theory of ethanol action, Science 197: 910-911.

Deitrich, R.A. (1977) Ethanol and biogenic aldehyde metabolism, in Structure and Function of Monoamine Enzymes (ed. Usdin, E., Weiner, N. and Youdim, M.B.H.) Marcel Dekker, New York, pp 651-674.

Duncan, R.J.S. and Tipton, K.F. (1971) The kinetics of pig brain aldehyde dehydrogenase, Eur. J. Biochem. 22: 538-543.

Duncan, R.J.S., Kline, J.E. and Sokoloff, L. (1976) Identity of brain alcohol dehydrogenase with liver alcohol dehydrogenase, Biochem. J. 153: 561-566.

Eccleston, D., Reading, H.W. and Ritchie (1969) 5-Hydroxytryptamine metabolism in brain and liver slices and the effects of ethanol, J. Neurochem. 16: 274-276.

Ericksson, C.K. and Burnam, W.L. (1971) Cholinergic alteration of ethanol-induced sleep and death in mice, Agents and Actions 2: 8-13.

Ericksson, C.K., and Graham, D.T. (1973) Alteration of cortical and reticular acetylcholine release by ethanol in vivo, J. Pharmac. Exp. Ther. 185: 583-593.

Eriksson, C.J.P. and Sippel, H.W. (1977) The distribution and metabolism of acetaldehyde in rats during ethanol oxidation-I. The distribution of acetaldehyde in liver, brain, blood and breath, Biochem. Pharmac. 26: 241-247.

Erwin, V.G. and Deitrich, R.A. (1973) Inhibition of bovine brain aldehyde reductase by anticonvulsant compounds in vitro, Biochem. Pharmac. 22: 2615-2624.

Feldstein, A. (1971) Effect of Ethanol on Neurohumeral Amine Metabolism, in The Biology of Alcoholism (ed. Kissin, B and Begleiter) vol 1, Plenum Press, New York, pp 127-159.

Fenstermacher, J.D., Pattak, C.S. and Blasberg, R.G. (1974) Transport of material between brain extracellular fluid, brain cells and blood. Fed. Proc. 33: 2070-2074.

Goldstein, D.B. (1975) Physical dependence on alcohol in mice, Fed. Proc. 34: 1953-1961.

Goldstein, A. and Judson, B.A. (1971) Alcohol dependence and opiate dependence: lack of relationship in mice, Science, 172: 290-292.

Grunnet, N. and Thieden, H.I.D. (1972) The effect of ethanol concentration upon in vivo metabolite levels of rat liver, Life Sci. 11: 983-993.

Grunnet, N., Quistorff, B. and Thieden, H.I. (1973) Rate-limiting factors in ethanol oxidation by isolated rat liver parenchymal cells, Eur. J. Biochem. 40: 275-282.

Häkkinen, H-M. and Kulonen, E. (1976) Ethanol intoxication and γ-aminobutyric acid, J. Neurochem. 27: 631-633.

Havre, P. Abrams, M.A., Corrall, R.J.M., Yu, L.C., Szczepanik, P.A., Feldman, H.B., Klein, P., Kong, M.S., Margolis, J.M. and Landau, B.R. (1977) Quantitation of pathways of ethanol metabolism, Arch. Biochem. Biophys. 182: 14-23.

Hawkins, R.D. and Kalant, H. (1972) The metabolism of ethanol and its metabolic effects, Pharmac. Rev. 24: 67-157.

Hofmann, I and Hosein, E.A. (1978) Effects of chronic ethanol consumption on the rate of rat liver protein turnover and synthesis, Biochem. Pharmac. 27: 457-463.

Holman, R.B., Elliott, G.R., Seagraves, E., DoAmaral, J.R., Vernikos-Danellis, J.,Kellar, K.J. and Barchas, J.D. (1975) Tryptolines: their potential role in the effects of ethanol, Finnish Foundation for Alcohol Studies 23: 207-216.

Holtz, P., Stock, K. and Westerman, E. (1964) Formation of tetrahydropapaveroline from dopamine in vitro, Nature 203: 656-658.

Hosein, E.A., Hofmann, I. and Linder, E. (1977) The influence of chronic ethanol feeding to rats on the integrity of liver mitochondrial membrane as assessed with the Mg^{2+}-stimulated ATPase enzyme, Arch. Biochem. Biophys. 183: 64-72.

Huff, J.A., Davis, V.E., Brown, H. and Clay, M.M. (1971) Effects of chloral hydrate, paraldehyde and ethanol on the metabolism of [^{14}C]-serotonin in the rat, Biochem. Pharmac. 20: 476-482.

Ishii, H., Joly, J.G. and Lieber, C.S. (1973) Effect of ethanol on the amount and enzyme activities of hepatic rough and smooth microsomal membranes, Biochim. Biophys. Acta 291: 411-420.

Israel, Y., Carmichael, F.J. and Macdonald, J.A. (1975) Effects of ethanol on electrolyte metabolism and neurotransmitter release in the CNS, Adv. Exp. Med. Biol. 59: 55-68.

Kalant, H. and Grose, W. (1963) Effects of ethanol and pentobarbital on release of acetylcholine from cerebral cortex slices. J. Pharmac. Exp. Ther. 158: 386-393.

Laborit, H. (1973) Gamma-hydroxybutyrate, succinic semialdehyde and sleep, Prog. Neurobiol. 1: 255-274.

Lahti, R.A. and Majchrowicz, E. (1969) Acetaldehyde - an inhibitor of the enzymatic oxidation of 5-hydroxyindoleacetaldehyde, Biochem. Pharmac. 18: 535-538.

Li, T-K. (1977) Enzymology of Human Alcohol Metabolism, Adv. Enzymol. 45: 427-483.

Li, T-K., Bosron, W.F., Dafeldecker, W.P., Lange, L.G. and Valee, B.L. (1977). Isolation of π-alcohol dehydrogenase of human liver: Is it a determinant of alcoholism? Proc. Nat. Acad. Sci. U.S.A. 74: 4378-4381.

Lieber, C.S. (1977a) Metabolism of Ethanol, in Metabolic Aspects of Alcoholism (ed. Lieber, C.S.) MTP Press, Lancaster, pp 1-29.

Lieber, C.S. (1977b) (Editor) Metabolic Aspects of Alcoholism. MTP Press, Lancaster.

Lieber, C.S. and DeCarli, L.M. (1970) Reduced nicotinamide-adenine dinucleotide phosphate oxidase :activity enhanced by ethanol consumption, Science 170: 78-80.

Lindros, K.O. (1975) Regulatory factors in hepatic acetaldehyde metabolism during ethanol oxixation, Finnish Foundation for Alcohol Studies 23: 67-81.

Magrinat, G., Dolan, J.P., Biddy, R.L., Miller, L.D. and Korol, B. (1973) Ethanol and methanol metabolites in alcohol withdrawal, Nature 244: 234-235.

Marshall, A. and Hirst, M. (1976) Potentiation of ethanol narcosis by dopamine- and L-DOPA-based isoquinolines, Exper. 32: 201-203.

Mezey, G. (1972) Duration of the enhanced activity of the microsomal ethanol oxidising enzyme system and rate of ethanol degradation in ethanol-fed rats after withdrawal. Biochem. Pharmac. 21: 137-142.

Miceli, J.M. and Ferrel, W.J. (1973) Effects of ethanol on membrane lipids III. Quantitative changes in lipid and fatty acid composition of non-polar and polar lipids of mouse total liver mitochondria and microsomes following ethanol feeding. Lipids 8: 722-727.

Miwa, G.T., Levin, W., Thomas, P.E. and Lu, A.Y.H. (1978) The direct oxidation of ethanol by a catalase- and alcohol dehydrogenase-free reconstituted system containing cytochrome P-450, Arch. Biochem. Biophys. 187: 464-475.

Mukherji, B., Kashiki, Y., Ohyanagi, H. and Sloviter, H.A. (1975) Metabolism of ethanol and acetaldehyde by the isolated perfused rat brain. J. Neurochem. 24: 841-843.

Noble, E.P. and Tewari (1977) Metabolic Aspects of Alcoholism in the brain, in Metabolic Aspects of Alcoholism (ed. Leiber, C.S.) M.T.P. Press, Lancster, pp 149-185.

Ohnishi, K. and Lieber, C.S. (1977) Reconstitution of the microsomal ethanol-oxidising system. Qualitative and quantitative changes of cytochrome P-450 after chronic ethanol consumption, J. Biol. Chem. 252: 7124-7131.

Oshino, N., Oshino, R. and Chance, B. (1973) The characteristic of the "peroxidatic" reaction of catalase in ethanol oxidation, Biochem. J. 131: 555-567.

Petersson, H. and Kiessling, K-H. (1977) Acetaldehyde occurance in cerebrospinal fluid during ethanol oxidation in rats and its dependence on the blood level and on dietary factors, Biochem. Pharmac. 26: 237-240.

Pietruszko, R., Theorell, H. and de Zalenski, C. (1972) Heterogeneity of alcohol dehydrogenase from human liver. Arch. Biochem. Biophys. 153: 279-293.

Rawat, A.K. (1974) Brain levels and turnover rates of presumptive neurotransmitters as influenced by administration and withdrawal of ethanol in mice, J.Neurochem. 22: 915-922.

Rawat, A.K. and Kuriyama, K. (1972) Ethanol oxidation: effects on the redox state of brain in mouse, Science 176: 1133-1135.

Raskin, N.H. and Sokoloff, L. (1972) Ethanol-induced adaptation of alcohol dehydrogenase activity in rat brain, Nature 236: 138-140.

Ross, D.H., Medina, M.A. and Cardenas, H.L. (1974) Morphine and ethanol: selective depletion of regional brain calcium, Science 186: 63-64.

Sabelli, H.C. and Giardina, W.A. (1973) Amine modulation of affective behaviour, in Chemical Modulation of Brain Function (ed Sabelli, H.C.) Raven Press, New York, pp 2-23.

Sandler, M., Bonham-Carter, S., Hunter, K.R. and Stern, G.M. (1973) Tetrahydroisoquinoline alkaloids: in vivo metabolites of L-Dopa in man, Nature 241: 439-443.

Seeman, P. (1972) The membrane actions of anaesthetics and tranquilizers, Pharmac. Rev. 24: 583-665.

Seeman, P. (1974) The membrane expansion theory of anesthesia: direct evidence using ethanol and a high-precision density meter. Exper. 30: 759.

Tabakoff, B. (1977) Brain aldehyde dehydrogenases and reductases, in Structure and Function of Monoamine Enzymes (ed. Usdin, E., Weiner, N and Youdim, M.B.H.) Marcel Dekker, New York pp 629-649.

Thadani, P.V. and Truitt, G.B. (1977) Effect of acute ethanol or acetaldehyde on the uptake, release, metabolism and turnover rate of norepinephrine in rat brain, Biochem. Pharmac. 26: 1147-1150.

Thurman, R.G. and Pathman, D.E. (1975) Withdrawal symptoms from ethanol evidence against the involvement of acetaldehyde, Finnish Foundation for Alcohol Studies 23: 217-231.

Thurman, R.G., Ley, H.G. and Scholz, R. (1972) Hepatic microsomal ethanol oxidation: hydrogen peroxide formation and the role of catalase, Eur. J. Biochem. 25: 420-430.

Tipton, K.F., Houslay, M.D. and Turner, A.J. (1977) Metabolism of Aldehydes in Brain, Essays in Neurochem. Neurobiol. 1, 103-138.

Truitt, E.B. and Walsh, M.J. (1971) The role of acetaldehyde in the actions of ethanol, in The Biology of Alcoholism (ed. Kissin, B. and Begleiter) vol 1, Plenum Press, New York pp 161-195.

Turner, A.J., Baker, K.M., Algeri, S. and Frigerio, A. (1974a) The identification and quantitation of tetrahydropapaveroline in rat brain by mass fragmentography, in Mass Spectrometry in Biochemistry and Medicine (ed. Frigerio, A. and Castagnoli, N.), Raven Press, New York, pp 99-109.

Turner, A.J., Illingworth, J.A. and Tipton, K.F. (1974b) Simulation of biogenic amine metabolism in brain, Biochem. J. $\underline{144}$: 353-360.

von Wartburg, J.P., Berger, D., Buhlmann, Ch. Dubdied, A. and Ris, M.M. (1974) Heterogeneity of pyridine nucleotide dependent alcohol and aldehyde metabolising enzymes, in Alcohol and Aldehyde Metabolising Systems (ed. Thurman, R.G., Yonetani, T., Williamson, J.R. and Chance, B.)Academic Press, New York, pp 33-44.

Walsh, M.J. (1973) Biogenesis of biologically active alkaloids from amines by alcohol and acetaldehyde, Ann. N.Y. Acad. Sci. $\underline{215}$: 98-110.

MEMBRANE EFFECTS OF ALCOHOL IN THE NERVOUS SYSTEM

Henrik Wallgren and Pekka Virtanen
Division of Physiology, Department of Zoology, University of Helsinki
and Research Laboratories of the Finnish State Alcohol Monopoly (Alko),
Helsinki, Finland

Study of the direct effects of addictive drugs, as well as of the causation for dependence upon them, is a central topic in pharmacological research. The results have obvious applications in the social and clinical fields. The fact that drugs of abuse modify mind and behaviour also has important implications for neurochemical research. As pointed out by Loh et al., (1977) we may "by the careful simultaneous evaluation of a number of biochemical, behavioral, and pharmacological parameters ... learn something about not only the mechanism of action of these drugs but also the neurochemical mechanisms underlying behavior". Few research institutes are equipped to carry through all the steps required in such an analysis. However, the aggregate effect of research in this area constitutes an important contribution to our knowledge about the processes regulating behaviour. One advantage provided by drugs is the production of predictable and reproducible modifications of behaviour.

The aliphatic alcohols are usually classified among general depressants. Presumably, they act in a manner similar to that of volatile anaesthetics and inert gases, by a rather unspecific physicochemical interaction with hydrophobic moieties in nerve cell membranes. Most other drugs of abuse seem to react with specific receptors in the nervous system. Superficially, some aspects of the direct effects and of physical dependence may appear similar, but presumably the underlying mechanisms differ for various types of drugs. For an explanation of these phenomena, it thus seems necessary to have an analysis extending from the molecular machinery at the subcellular level via electrophysiological study of individual cells and complex networks to the level of overt behaviour.

General reviews of the effects of alcohol on the nervous system can be found in Israel (1970), Wallgren and Barry (1970), Seixas and Eggleston (1973), Kalant (1974, 1975) and Smith (1977). Reviews of the mechanism of general anaesthesia have been written by, e.g., Seeman (1972), Aston (1972), Halsey et al. (1974) and Fink (1975). A recent, important review of the neurochemical and behavioural mechanisms of ethanol and opiate actions has been edited by Blum (1977). Membrane effects of alcohols have been reviewed and discussed by, for instance, Kalant (1971), Grenell (1972), Seeman (1972) and Hunt (1975). In the follow-wing, we shall attempt a review of the neurochemical aspects of membrane-related effects of alcohol in the nervous system. It seems useful to begin with a brief survey of the evidence for the membrane actions of aliphatic alcohols.

Evidence for Direct Membrane Actions of Aliphatic Alcohols

As pointed out earlier (Wallgren, 1970, 1977), a satisfactory theory of the mechanism by which aliphatic alcohols act should account for the relationship between their carbon chain length, physicochemical properties, and pharmacological potency, as formulated in the classical studies by Meyer (1901), Overton (1901), Brink and Posternak (1948) and Ferguson (1939, 1951). By physicochemical techniques, it has, for instance, been demonstrated that alcohols including ethanol cause lateral expansion of brain lipid films (Skou, 1958). This is one of the pieces of experimental evidence giving support to the membrane expansion theory of anaesthesia (Seeman, 1972). The concept has, on the other hand recently been challenged because theoretical analysis and experiments on membrane structure indicate that the extent of expansion is far less than that assumed in Seeman's theory (Trudell, 1977). On the basis on thermodynamic calculations and experimental data from model systems such as liposomes, Bangham (1975) and Hill and Bangham (1975) propose that alcohols may act by increasing membrane fluidity. Usually, rather high concentrations of the drug are needed for measurable effects to occur (e.g., Paterson et al., 1972). However, Chin and Goldstein (1977) applied electron paramagnetic resonance spin label techniques to membrane lipid preparations from mouse brain and succeeded in obtaining definite effects with concentrations compatible with reversible intoxication _in vivo_. The fluorescence probe technique has also yielded results supporting increased membrane disordering (Spero and Roth, 1970, and acc. to Seeman, 1972).

Voltage clamp studies on squid giant axons (Armstrong and Binstock, 1964, Moore, Ulbricht and Takata, 1964) as well as more indirect measurements of ion movements in rat brain cortex _in vitro_ (Wallgren, Nikander, von Boguslawsky and Linkola, 1974, Israel, Carmichael and Macdonald, 1975) indicate that an important aspect of the action of ethanol is depression of the increase in sodium inward current forming the rising part of the spike potential. This interpretation is supported by the results of the most thorough electrophysiological study on effects of ethanol so far performed (Bergman et al., 1974, Faber and Klee, 1977). The object was _Aplysia_ neurons. In these, ethanol relatively specifically inhibited the early inward sodium (or calcium) current. Furthermore, ethanol was found to decrease the postsynaptic increase in ionic current on application of transmitter substance both in excitatory and inhibitory synapses. Finally, synaptic transmission was considerably more sensitive than axonal conduction (Faber and Klee, 1977). In all instances in this invertebrate preparation, the aspect of nerve function affected was ionic currents associated with nerve activity. In goldfish Mauthner cells, on the other hand, synaptic transmission seemed to be inhibited through a decrease in transmitter release (Faber and Klee, 1977). In vertebrate preparations, it is somewhat controversial whether synaptic transmission is more sensitive to aliphatic alcohols than axonal conduction (Larrabee and Posternak, 1952; reviews by Wallgren and Barry, 1970, Smith, 1977). A conclusion emerging particularly from the meticulous studies of Faber, Klee and associates is, however, that ethanol always seems to interact with membrane structures, but that the sensitivity of the structure and its type of response depend on special characteristics of the actual membrane region involved.

It thus seems possible that ethanol interacts in a variety of ways with structures involved in the control of the excitation cycle of nerve cells both along the cell body and axon and in the synaptic region. Faber and Klee (1977) assume that in goldfish, the behavioural excitation seen with relatively low concentrations of ethanol may be due to inhibition of synaptic function whereas deeper intoxication may be associated with direct effects on axonal excitability.

Tolerance to and physical dependence on alcohol may be due to functionally compensatory changes in the hydrophobic membrane components assumed to be the primary target for the effects of alcohol. This provides an analogy to the temperature effects as seen in the alterations of membrane composition and viscosity occurring in acclimation to new temperature levels in poikilothermic organisms. A purely physical change in the membrane sets off a biochemical response, adjusting membrane composition to the altered conditions. This has been reported in unicellular organisms (Melchior et al., 1970), as well as the nervous system of fish and other poikilothermic animals (review by Hazel and Prosser 1974). Study of such changes may aid in the identification and chemical characterization of the structures involved in the control of the excitation cycle.

Effects on Active Transport Processes

Most of the studies in this area related to nerve tissue concern microsomal Na^+,K^+-ATPase preparations in vitro. There are also some studies on electrolyte transport in cerebral slice preparations, as well as on the uptake of transmitters. The subject has been reviewed by e.g. Kalant (1971), Sun, Seeman and Middleton (1977) and Smith

(1977). Inhibition of Na^+, K^+-ATPase has been reported among others by Järnefelt (1961), Israel, Kalant and Laufer (1965), Kalant and Israel (1967), Sun and Samorajski (1970) and Lin (1976). There are several reasons for suspecting that this inhibition of cation-stimulated ATPase is not responsible for the actions of ethanol. First, the concentrations of ethanol required for affecting the ATPase are generally higher than those producing physiological effects, second, partial inhibition of Na^+, K^+-ATPase causes depolarization and hyperexcitability (Hunt, 1975). Ethanol, although having biphasic effects in most nerve and muscle preparations, produces primarily depolarization and depression at higher concentrations. In *Aplysia* neurons effects on membrane potentials are quite variable, ranging from depolarization to hyperpolarization; functional depression is typical at the cellular level (Faber and Klee, 1977). Third, the higher aliphatic alcohols do not affect membrane potentials or cause hyperpolarization, but nevertheless produce behavioural effects resembling those of ethanol.

An interesting aspect is the apparent competition between ethanol and potassium in inhibition of the enzyme (Israel et al., 1965). Also in cerebral cortex slices, depleted of potassium, K^+-transport is more strongly inhibited than that of Na^+-ions (Israel-Jacard and Kalant, 1965; Israel, Kalant and LeBlanc, 1966; Wallgren, Nikander, von Boguslawsky and Linkola, 1974).

The reported effects of chronic exposure to ethanol on Na^+, K^+-ATPase are controversial. An "adaptive" increase has been described by Israel et al. (1970), Knox et al. (1972), Roach et al. (1973b), Sun et al. (1976) and Sun et al. (1977). In contrast, Israel and Kuriyama (1971), Goldstein and Israel (1972), Akera et al. (1973) as well as Nikander (in Wallgren et al., 1975) failed to find such an increase. Wallgren et al. (1975), however, detected a relative increase in Na^+, K^+-ATPase and decrease in Mg^{2+}-ATPase that correlated with the intensity of the withdrawal excitability in rats. The fact that a decrease rather than an increase in ion transport activity should cause increased excitability makes the interpretation of the significance of these findings even more difficult than the discrepancies as such. However, it is of interest in this connection that von Boguslawsky and Nikander (1972) found inhibition of K^+-transport in cerebral cortex slices from a low drinking, ethanol-sensitive strain of rats (ANA), but no inhibition in the high-drinking, ethanol-tolerant AA-strain. - According to Tabakoff (1977), active Ca^{2+} transport in the choroid plexus is inhibited by ethanol.

High concentrations of ethanol inhibit active accumulation of noradrenaline, glutamate, and 5-hydroxytryptamine (Israel et al., 1973; Roach et al., 1973a; Sun et al. 1977). With synaptosomal preparations isolated from rats given various acute doses of ethanol, noradrenaline uptake was found to be decreased with high doses whereas hypothalamic content of noradrenaline was elevated with low doses and depressed with high doses (Sun et al., 1977). The same authors report that chronic administration of ethanol to rats raises both hypothalamic noradrenaline content and the uptake of noradrenaline in isolated synaptosomes.

Effects of Ethanol on Membrane Phospholipids, Ca^{2+}-Binding, and Membrane Composition

Calcium is known to be a key factor in the control of axonal excitability (Frankenhaeuser and Hodgkin, 1957; Rothstein, 1968, Triggle, 1973). Ca^{2+}-ions stabilize the membrane and calcium removal raises the excitability. Calcium is also involved in transmitter release and uptake (Katz and Miledi, 1967; Rubin, 1970). The sodium gates in the axonal membrane are probably lined by protein molecules, but Ca^{2+}-ions controlling access to the channels for sodium ions seem to be bound to phospholipids, perhaps at the phosphate group (D'Arrigo, 1975). On the other hand, Low and Cook (1972) and Hauser, Darke and Phillips (1976) have shown that at physiological pH, a likely site of Ca^{2+}-binding is the carboxyl groups of phosphatidylserine. Membrane-bound phosphatidylserine in the nervous system shows a Ca^{2+}-dependent exchange reaction with free serine (Porcellati et al., 1971). For these reasons, it seems to be of interest to explore whether the changes in nerve function evident in drug tolerance and withdrawal excitability might be related to changes in phospholipid metabolism and Ca^{2+}-binding.

According to Seeman (1972) and Ross (1977), alcohols increase binding of Ca^{2+} to membranes, an action compatible with decreased excitability. Chronic treatment with ethanol causes decreased Ca^{2+}-binding (Ross, 1977), a finding that fits with the hyperexcitability. This effect is inhibited by naloxone, a parallel to observations with morphine. Calcium content of synaptosomes on the other hand decreases in acute intoxication and increases as a consequence of chronic intoxication (Ross, 1977).

In rat brain microsomal preparations, lipid labelling with ^{14}C-serine is inhibited in acute intoxication and the hangover-phase following intoxication (Wallgren et al., 1975). Labelling of synaptosomal phosphatidylserine with ^{14}C-serine in the brains of rats in vivo was decreased in acute hangover and further depressed in the withdrawal period after prolonged intoxication (Virtanen, 1977). This observation seems to tie in quite interestingly with Ross' (1977) observations on the behaviour of Ca^{2+}-ions in acute and chronic intoxication.

Whereas the observations on altered Ca^{2+}-binding and phospholipid labelling suggest a possible basis for the increase in excitability seen after prolonged action of ethanol, very little is known about the possible alterations in membrane composition. Ross (1977) isolated polypeptides from rat brain synaptic membranes and found decreased amounts of a peptide with a molecular weight close to 210,000 with no change in other fractions. Noble et al. (1976) report that surface sialic acid of cultured hamster astroblasts is released in greater quantity by neuraminidase after chronic incubation with ethanol (19 - 110 days) than in controls. They suggest that such a change may alter the membrane uptake of organic cations such as choline. Ross et al. (1977) report an increase in sialic acid exposure on synaptic membranes of rats receiving 15 % (v/v) ethanol as their sole drinking fluid for 14 days. This also was seen as indicating that Ca^{2+}-ions may be bound to sialic acid in addition to phospholipids.

Dinovo et al. (1976; Biochem. Biophys. Res. Commun. 68,1975) found that rats drinking ethanol chronically showed an increased number of microsomal SH groups reacting with N-ethyl maleimide and 5,5'dithiobin (2-nitrobenzoic acid), but not of total SH groups. According to Khawaja, Lindholm and Niittylä (1978), ethanol ingestion causes selective inhibition of protein synthesis by cerebral membrane-bound ribosomes, possibly indicating damage to the endoplasmic reticulum.

Isoquinoline Alkaloids and Alcohol Effects

A basic difference between the action of narcotic analgesics of the opiate type and of alcohols is that opiates are bound to specific receptor sites (Pert and Snyder, 1973) whereas the alcohols interact in an unspecific manner with the membranes, as already discussed in this paper. Two independent proposals have been made that acetaldehyde formed in the oxidation of ethanol by condensation with monoamine metabolites might give rise to tetrahydroisoquinoline (TIQ) alkaloids acting either as false transmitters (Cohen and Collins, 1970) or binding directly to receptors (Davis et al., 1970). These proposals have the appeal of affording a common basis for dependence on opiates and ethanol and have led to extensive research efforts. Although formation of such compounds can be easily demonstrated in vitro (Davis et al., 1975), in vivo production has not been conclusively shown to occur during ethanol intoxication. After pre-treatment with pyrogallol, salsolinol was detected in the brains of ethanol-intoxicated rats (Collins and Bigdeli, 1975). Turner et al. (1974) detected small amounts of tetrahydropapaveroline in rat brain, but the levels were not changed during ethanol intoxication. Direct infusion of TIQ-alkaloids and tetrahydropapaveroline into the cerebral ventricles of rats induces strong preference for ethanol (Myers and Melchior, 1977a,b). In these experiments, withdrawal symptoms were also seen after the animals had been on the experimental regimen for about a week. However, possible effects of the alkaloids alone were not checked separately from those of the combination alkaloid infusion-ethanol ingestion, and the withdrawal symptoms described seemed more similar to those characteristic of opiate than of ethanol dependence.

Although it may appear attractive to assume a common basis for ethanol and opiate dependence, there are a number of difficulties with the hypothesis. The careful comparisons between opiates and ethanol made by Eidelberg (1977) and Herz et al., (1977) (also reviews by Smith, 1977, Blum, 1977) show important pharmacological neurophysiological differences in their mode of action. Acetaldehyde does not penetrate in detectable amounts into the brain of rats until the acetaldehyde concentrations in the blood are above approximately 0.25 mM, which seldom occurs during ethanol intoxication (Eriksson and Sippel, 1976). Thus conditions in the brain do not seem to permit formation of the assumed alkaloids. Furthermore, tert.butanol which does not give rise to aldehyde in the organism induces acute symptoms and effects on ion movements in brain tissue similar to ethanol (Wallgren et al., 1973; Wallgren et al., 1974). Tert.butanol also gives rise to withdrawal symptoms quite similar to those produced by ethanol in the rat (Wallgren et al., 1973; Thurman and Patham, 1975; Bellin and Edmonds, 1975; Wood and Laverty, 1976). Thus, acetaldehyde and TIQ-formation cannot be necessary for development of alcohol dependence and the withdrawal syndrome. It is too early to conclude that alkaloids are not involved at all in ethanol dependence, but at least they can explain only part of the syndrome, a part which remains to be defined. However, it is perhaps more likely that the molecular mechanisms and cellular sites affected by opiates and alcohols are different. The partial similarity of the symptoms and the interactions in voluntary consumption (Sinclair et al., 1973; Sinclair, 1974; Myers and Melchior, 1977a,b) may arise because the structures involved in the central nervous system are to some extent shared or because of the relationships with calcium (Ross, 1977). This area of research has been progressing rapidly but is also hampered by a difficulty common in biological research - that of obtaining clean-cut, unambiguous results.

Conclusions

Electrophysiological analysis on the cellular level indicates that interaction with membrane structures is the primary reason for the disturbance of nerve function caused by ethanol. The sensitivity and type of effect seems to depend on the special characteristics of the membrane structure involved. Important types of effect are changes in ionic currents during impulse generation and alterations in transmitter release. Physicochemical data suggest that ethanol and other alcohols increase membrane disorder by unspecific interaction with hydrophobic membrane components.

Evidence is emerging that chronic exposure to alcohols may cause changes in the organization of membrane macromolecules and their turnover, perhaps as part of an adaptive process. It has been proposed that tetrahydroisoquinoline alkaloids formed by condensation of ethanol-derived acetaldehyde with monoamines would explain much of the phenomena of alcohol addiction and dependence. The hypothesis is controversial. Direct evidence such as the similarity of physical dependence on ethanol and on tert.butanol argues against the hypothesis. Behavioral interactions between opiates and ethanol as well as the observation that reduction in calcium binding caused by chronic exposure to ethanol is inhibited by naloxone are evidence that support a connection between opiate and ethanol dependence. Much more detailed knowledge is needed of how these drugs interact with the molecular machinery of nerve cells before safe conclusions can be made about to what extent their mechanisms of action are similar.

Acknowledgement

This paper was prepared while one of us (Pekka Virtanen) was supported by a grant from the Finnish Foundation for Alcohol Studies.

References

Akera, T., Rech, R.H., Marquis, W.J., Tobin, T. and Brody, T.M. (1973) Lack of relationship between brain $(Na^+ + K^+)$-activated adenosine triphosphatase and the development of tolerance to ethanol in rats, J. Pharmacol. Exp. Ther. 185: 594-601.

Armstrong, C.M. and Binstock, L. (1964) The effects of several alcohols on the properties of the squid giant axon, J. Gen. Physiol. 48: 265-277.

Aston, R. (1972) Barbiturates, alcohol, and tranquilizers, In Chemical and Biological Aspects of Drug Dependence, CRC Press, Cleveland, Ohio.

Bangham, A.D. (1975) Alcohol, anaesthetics & membranes, Proc. 6th Int. Congr. Pharmacol., Helsinki, Vol. 3, 33-39, Finnish Pharmacological Society.

Bergmann, M.C., Klee, M.R. and Faber, D.S. (1974) Different sensitivities to ethanol of three early transient voltage clamp currents of Aplysia neurons, Pflügers Arch. Ges. Physiol. 348: 139-153.

Bellin, S.I. and Edmonds, H.L. (1976) The use of tert.butanol in alcohol dependence studies, Proc. West. Pharmacol. Soc. 19: 351-354.

Blum, K., Ed. (1977) Alcohol and Opiates. Neurochemical and behavioral mechanisms, Academic Press, New York-San Francisco-London, 403 pp.

Boguslawsky, P. v. and Nikander, P. (1972) A difference between high- and low-drinking rats in effects of ethanol on ion movements in cerebral tissue, Acta Physiol. Scand. 84: 12A-13A.

Brink, F. and Posternak, J.M. (1948) Thermodynamic analysis of the relative effectiveness of narcotics. J. Cell. Comp. Physiol. 32: 211-233.

Chin, J.H. and Goldstein, D.B. (1977) Electron paramagnetic resonance studies of ethanol on membrane fluidity, in M.M. Gross, Ed., Alcohol Intoxication and Withdrawal - IIIa: 111-122, Plenum Press, New York and London.

Cohen, G. and Collins, M. (1970) Alkaloids from catecholamines in adrenal tissue: possible role in alcoholism, Science 170: 1749-1751.

Collins, M.A. and Bigdeli, M.G. (1975) Biosynthesis of tetrahydroisoquinoline alkaloids in brain and other tissues of ethanol-intoxicated rats, in M.M. Gross, Ed., Alcohol Intoxication and Withdrawal - II: 79-91, Plenum Press, New York and London.

D'Arrigo, J.S. (1975) Axonal surface charges: evidence for phosphate structure, J. Membr. Biol. 22: 255-263.

Davis, V.E., Cashaw, J.L. and McMurtrey, K.D. (1975) Disposition of catecholamine-derived alkaloids in mammalian systems, in M.M. Gross, Ed., Alcohol Intoxication and Withdrawal - II: 65-78, Plenum Press, New York and London.

Davis, V.E., Walsh, M.J. and Yamanaka, Y. (1970) Augmentation of alkaloid formation from dopamine by alcohol and acetaldehyde in vitro, J. Pharmacol. Exp. Ther. 174: 401-412.

Dinovo, E.C., Gruber, B. and Noble, E.P. (1976) Alterations of fast-reacting sulfhydryl groups of rat brain microsomes by ethanol, Biochem. Biophys. Res. Commun. 68: 975-981.

Eidelberg, E. (1977) On the possibility that opiate and ethanol actions are mediated by similar mechanisms, in M.M. Gross. Ed., Alcohol Intoxication and Withdrawal - IIIb: 87-94, Plenum Press, New York and London.

Eriksson, C.J.P. and Sippel, H.W. (1977) The distribution and metabolism of acetaldehyde in rats during ethanol intoxication. I. The distribution of acetaldehyde in liver, brain, blood and breath, Biochem. Pharmacol. 26: 241-247.

Faber, D.S. and Klee, M.R. (1977) Actions of ethanol on neuronal membrane properties and synaptic transmission, in K. Blum, Ed., Alcohol and Opiates, p. 41-63, Academic Press, New York-San Francisco-London.

Ferguson, J. (1939) The use of chemical potentials as indices of toxicity, Proc. Roy. Soc. (London) B 127: 387-404.

Ferguson, J. (1951) Relations between thermodynamic indices of narcotic potency and the molecular structure of narcotics, in Mécanisme de la narcose, Colloq Int. Cent. Natl. Réch. Sci. (Paris) 26: 25-39.

Fink, B.R. (1975) Progress in Anesthesiology, Vol. 1: Molecular Mechanisms of Anesthesia, Raven Press, New York.

Frankenhaeuser, B. and Hodgkin, A.L. (1957) The action of calcium on the electrical properties of squid axons, J. Physiol. (London) 137: 245-260.

Goldstein, D.B. and Israel, Y. (1972) Effects of ethanol on mouse brain (Na + K)-activated adenosine triphosphatase, Life Sci. 11: 957-963.

Grenell, R.G. (1972) Effects of alcohol on the neuron, in B. Kissin and H. Begleiter, Eds., The Biology of Alcoholism, Vol. 2: Physiology and Behavior, p. 1-19, Plenum Press, New York and London.

Halsey, M.J., Millar, R.A. and Sutton, J.A. (1974) Molecular Mechanisms in General Anaesthesia, Churchill Livingstone, Edinburgh-London-New York.

Hauser, H., Darke, A. and Phillips, M.C. (1976) Ion binding to phospholipids: interactions of calcium with phosphatidyl serine, Eur. J. Biochem. 62: 335-344.

Hazel, J.R. and Prosser, C. Ladd (1974) Molecular mechanisms of temperature compensation in poikilotherms, Physiol. Rev. 54: 620-677.

Herz, A., Zieglgänsberger, W., Schulz, R., Fry, J.P. and Satoh, M. (1977) Neuronal aspects of opiate dependence and tolerance in comparison to central depressants, in M.M. Gross, Ed., Alcohol Intoxication and Withdrawal - IIIb: 117-140, Plenum Press, New York and London.

Hill, M.W. and Bangham, A.D. (1975) General depressant drug dependency: a biophysical hypothesis, in M.M. Gross, Ed., Alcohol Intoxication and Withdrawal - II: 1-9, Plenum Press, New York and London.

Hunt, W.A. (1975) The effects of aliphatic alcohols on the biophysical and biochemical correlates of membrane function, in E. Majchrowicz, Ed., Biochemical Pharmacology of Ethanol, p. 195-210, Plenum Press, New York and London.

Israel, Y. (1970) Cellular effects of alcohol. A review, Quart. J. Stud. Alc. 31: 293-316.

Israel, Y. and Kuriyama, K. (1971) Effect of *in vivo* ethanol administration on adenosinetriphosphatase, Life Sci. 10: 591-599.

Israel, Y., Kalant, H. and Laufer, I. (1965) Effects of ethanol on Na, K, Mg-stimulated microsomal ATPase activity, Biochem. Pharmacol. 14: 1803-1814.

Israel, Y., Kalant, H. and LeBlanc, E. (1966) Effects of lower alcohols on potassium transport and microsomal adenosine triphosphatase activity of rat cerebral cortex, Biochem. J. 100: 27-33.

Israel, Y., Kalant, H., LeBlanc, E., Bernstein, J.C. and Salazar, I. (1970) Changes in cation transport and (Na + K)-activated adenosine triphosphatase produced by chronic administration of ethanol, J. Pharmacol. Exp. Ther. 174: 330-336.

Israel, Y., Rosemann, E., Hein, S., Colombo, G. and Canessa-Fischer, M. (1971) Effects of alcohol on the nerve cell, in Y. Israel and J. Mardones, Eds., Biological Basis of Alcoholism, p. 53-72, Wiley-Intersciences, New York.

Israel, Y., Carmichael, F.J. and Macdonald, J.A. (1975) Effects of ethanol on electrolyte metabolism and neurotransmitter release in the CNS, in M.M. Gross, Ed., Alcohol Intoxication and Withdrawal - II: 55-64, Plenum Press, New York and London.

Israel-Jacard, Y. and Kalant, H. (1965) Effects of ethanol on electrolyte transport and electrogenesis in animal tissues, J. Cell. Comp. Physiol. 65: 127-132.

Järnefelt, J. (1961) Inhibition of the brain microsomal adenosine-triphosphatase by depolarizing agents, Biochim. Biophys. Acta 48: 111-116.

Kalant, H. (1971) Absorption, diffusion, distribution, and elimination of ethanol: Effects on biological membranes, in B. Kissin and H. Begleiter, Eds., The Biology of Alcoholism Vol. 1, p. 1-62, Plenum Press, New York and London.

Kalant, H. (1974) Ethanol and the nervous system: experimental neurophysiological aspects, Int. J. Neurol. 9: 111-120.

Kalant, H. (1975) Direct effects of ethanol on the nervous system, Fed. Proc., Fed. Amer. Soc. Exp. Biol. 34: 1930-1941.

Kalant, H. and Israel, Y. (1967) Effects of ethanol on active transport of cations, in R.D. Maickel, Ed., Biochemical Factors in Alcoholism, p. 25-37, Pergamon Press, Oxford.

Katz, B. and Miledi, R. (1965) The effect of calcium in acetylcholine release from motor nerve terminals, Proc. Roy. Soc. Biol. 161: 496-503.

Khawaja, J.A., Lindholm. D.B. and Niittylä, J. (1978) Selective inhibition of protein synthetic activity of cerebral membrane-bound ribosomes as a consequence of ethanol ingestion, Res. Commun. Chem. Pathol. Pharmacol., in press.

Knox, W.H., Perrin, R.G. and Sen, A.K. (1972) Effect of chronic administration of ethanol on (Na^+, K^+)-activated ATPase activity in six areas of the cat brain, J. Neurochem. 19: 2881-2884.

Larrabee, M.G. and Posternak, J.M. (1952) Selective action of anesthetics on synapses and axons in mammalian sympathetic ganglia, J. Neurophysiol. 15: 91-114.

Lin, D.C. (1976) Effect of ethanol on the kinetic parameters of brain (Na^+ + K^+)-activated adenosine triphosphatase, Ann. N. Y. Acad. Sci. 273: 331-337.

Loh, H.H., Lee, N.M. and Harris, R.A. (1977) Alterations of macromolecule biosynthesis after chronic administration of opiates and ethanol, in M.M. Gross, Ed., Alcohol Intoxication and Withdrawal - IIIb: 65-84, Plenum Press, New York and London.

Low, E. and Cook, A.M. (1972) The role of phosphatidyl serine in neural excitation and conduction, Biomed. Sci. Instrum. 9: 115-121.

Melchior, D.L., Morowitz, H.J., Sturtevant, J.M. and Tsong, T.Y. (1970) Characterization of the plasma membrane of *Mycoplasma laidlawii*, Biochim. Biophys. Acta 219: 114-122.

Meyer, H.H. (1901) Zur Theorie der Alkoholnarkose. III Mitt. Der Einfluss wechselnder Temperatur auf Wirkungsstärke und Teilungskoefficient der Narkotika, Arch. Exp. Pathol. Pharmakol. 46: 338-350.

Moore, J.W., Ulbricht, W. and Takata, M. (1964) Effect of ethanol on the sodium and potassium conductances of the squid axon membrane. J. Gen. Physiol. 48: 279-295.

Myers, R.D. and Melchior, C.L. (1977a) Alcohol drinking: abnormal intake caused by tetrahydropapaveroline in brain, Science 196: 554-556.

Myers, R.D. and Melchior, C.L. (1977b) Differential actions on voluntary alcohol intake of tetrahydroisoquinolines or a β-carboline infused chronically in the ventricle of the rat, Pharmacol. Biochem. Behav. 7: 381-392.

Noble, E.P., Syapin, P.J., Vigran, R. and Rosenberg, A. (1976) Neuraminidase-releasable surface sialic acid of cultured astroblasts exposed to ethanol, J. Neurochem. 27: 217-221.

Overton, E., (1901) Studien über die Narkose: zugleich ein Beitrag zur allgemeinen Pharmakologie, G. Fischer, Jena.

Paterson, S.J., Butler, K.W., Huang, P., Labelle, J., Smith, I.C.P. and Schneider, H. (1972) The effects of alcohols on lipid bilayers: A spin label study, Biochim. Biophys. Acta 266: 597-602.

Pert, C.P. and Snyder, S.H. (1973) Opiate receptor: Demonstration in nervous tissue, Science 179: 1011-1014.

Porcellati, G., Arienti, G., Pirotta, M. and Giorgini, D. (1971) Baseexchange reactions for the synthesis of phospholipids in nervous tissue: the incorporation of serine and ethanolamine into the phospholipids of isolated brain microsomes, J. Neurochem. 18: 1395-1417.

Roach, M.K., Davis, D.L., Pennington, W. and Nordyke, E. (1973a) Effect of ethanol on the uptake by rat brain synaptosomes of ^{3}H-DL norepinephrine, ^{3}H-5-hydroxytryptamine, ^{3}H-GABA, and ^{3}H-glutamate, Life Sci. 12: 433-441.

Roach, M.K., Khan, M.M., Coffman, R., Pennington, W. and Davis, D.L. (1973b) Brain (Na$^+$, K$^+$)-activated adenosine triphosphatase activity and neurotransmitter uptake in alcohol-dependent rats, Brain Res. 63: 323-329.

Ross, D.H. (1977) Adaptive changes in Ca^{++}-membrane interactions following chronic ethanol exposure, in M.M. Gross, Ed., Alcohol Intoxication and Withdrawal - IIIa: 459-471, Plenum Press, New York and London.

Ross, D.H., Kibler, B.C. and Cardenas, H.L. (1977) Modification of glycoprotein residues as Ca^{2+} receptor sites after chronic ethanol exposure, Drug Alc. Dep. 2: 305-315.

Rothstein, A. (1968) Membrane phenomena, Annu. Rev. Physiol. 30: 15-72.

Rubin, R.P. (1970) The role of calcium in the release of neurotransmitter substances and hormones, Pharmacol. Rev. 22: 389-428.

Seeman, P. (1972) The membrane actions of anesthetics and tranquilizers, Pharmacol. Rev. 22: 389-428.

Seixas, F.A. and Eggleston, S., Eds. (1973) Alcoholism and the central nervous system, Ann. N.Y. Acad. Sci. 215: 1-389.

Sinclair, J.D., Adkins, J. and Walker, S. (1973) Morphine-induced suppression of voluntary alcohol drinking in rats, Nature (London) 246: 425-427.

Sinclair, J.D. (1974) Morphine suppresses alcohol drinking regardless of prior alcohol access duration, Pharmacol. Biochem. Behav. 2: 409-412.

Skou, J.C. (1958) Relation between the ability of various compounds to block nervous conduction and their penetration into a monomolecular layer of nerve-tissue lipid, Biochim. Biophys. Acta 30: 625-629.

Smith, C.M. (1977) The pharmacology of sedative/hypnotics, alcohol and anesthetics, in G.V.R. Born, O. Eichler, A. Farah, H. Herken and A.D. Welch, Eds., Handbook of Experimental Pharmacology, Vol. 45/1 (W.R. Martin, Ed.), p. 413-587, Springer, Berlin-Heidelberg-New York.

Spero, L. and Roth, S. (1970) Fluorescent hydrophobic probe study of the interaction of local anesthetics and red cell ghosts, Fed. Proc., Fed. Amer. Soc. Exp. Biol. 29: 474.

Sun, A.Y. and Samorajski, T. (1970) Effect of ethanol on adenosine triphosphatase and acetylcholinesterase activity in isolated synaptosomes of guinea pig brain, J. Neurochem. 17: 1365-1372.

Sun, A.Y., Seaman, R.N. and Middleton, C.C. (1977) Effects of acute and chronic alcohol administration on brain membrane transport systems, in M.M. Gross, Ed., Alcohol Intoxication and Withdrawal - IIIa: 123-138, Plenum Press, New York and London.

Sun, A.Y., Sun, G.Y. and Middleton, C.C. (1976) Alcoholmemrane interaction in the brain. II. Effect of chronic ethanol administration. Ann. N. Y. Acad. Sci. in press.

Tabakoff, B. (1977) Neurochemical aspects of ethanol dependence, in K. Blum, Ed., Alcohol and Opiates. Neurochemical and behavioral mechanisms, p. 21-39, Academic Press, New York-San Francisco-London.

Thurman, R.G. and Pathman, D.E. (1975) Withdrawal symptoms from ethanol: evidence against the involvement of acetaldehyde, in K. O. Lindros and C.J.P. Eriksson, Eds., The Role of Acetaldehyde in the Actions of Ethanol, p. 217-231, The Finnish Foundation for Alcohol Studies, Vol, 23, Helsinki.

Triggle, D.J. (1973) Effects of calcium on excitable membranes and neurotransmitter action, Progr. Surf. Membr. Sci. 5: 267-331.

Trudell, J.R. (1977) The membrane volume occupied by anesthetic molecules: a reinterpretation of the erythrocyte expansion data, Biochim. Biophys. Acta 470: 509-510.

Turner, A.J., Baker, K.M., Algeri, S., Frigerio, A. and Garattini, S. (1974) Tetrahydropapaveroline: Formation in vivo and in vitro in rat brain, Life Sci. 14: 2247-2257.

Virtanen, P. (1977) Labeling of rat brain synaptosomal phosphatidyl serine in the after state of acute intoxication and in the withdrawal state, in M.M. Gross, Ed., Alcohol Intoxication and Withdrawal - IIIa: 193-202, Plenum Press, New York and London.

Wallgren, H. (1970) Absorption, diffusion, distribution and elimination of ethanol. Effect on biological membranes, in J. Trémolières, Ed., Int. Encyclopedia Pharmacol. Ther., Section 20, Vol. 1, Alcohol and Derivatives, p. 161-188, Pergamon Press, Oxford and New York.

Wallgren, H. (1977) The mechanisms for actions of ethanol on the nervous system, in C.-M. Ideström, Ed., Recent Advances in the Study of Alcoholism, p. 1-7, Excerpta Medica International Congress Series No. 407, Excerpta Medica, Amsterdam.

Wallgren, H. and Barry, III, H. (1970) Actions of Alcohol, Vol. 1, Elsevier, Amsterdam-London-New York.

Wallgren, H., Kosunen, A.-L. and Ahtee, L. (1973) Technique for producing an alcohol withdrawal syndrome in rats, Isr. J. Med. Sci. 9: Suppl., 63-71.

Wallgren, H., Nikander, P., von Boguslawsky, P. and Linkola, J. (1974) Effects of ethanol, tert.butanol and clomethiazole on net movements of sodium and potassium in electrically stimulated cerebral tissue, Acta Physiol. Scand. 91: 83-93.

Wallgren, H., Nikander, P. and Virtanen, P. (1975) Ethanol-induced changes in cation-stimulated adenosine triphosphatase activity and lipid-proteolipid labeling of brain microsomes, in M.M. Gross, Ed., Alcohol Intoxication and Withdrawal - II: 23-36, Plenum Press, New York and London.

Wood, J. and Laverty, R. (1976) Alcohol withdrawal syndrome following prolonged t-butanol administration to rats, Proc. Univ. Otago Med. Sch. 54: 86-87.

EFFECTS OF DEPENDENCE-PRODUCING DRUGS ON NEUROTRANSMITTERS AND NEURONAL EXCITABILITY

S. Liljequist

Department of Pharmacology, University of Göteborg, Fack, S-400 33 Göteborg, Sweden

Introduction

The central effects of dependence-producing drugs on brain monoamines are well documented both in man and experimental animals (Wallgren and Barry, 1970). Evidence has accumulated over the past decade indicating that the ability of drugs to produce stimulatory effects and euphoric sensations and thus a rewarding experience - which has been considered as an important factor in the dependence liability of drugs (see e.g. Seevers, 1968) - may be related to their capacity to alter certain neuronal functions in the central nervous system (see Engel, 1977). Consequently, specific neurochemical events might be connected to the development of drug dependence. Naturally, it should be kept in mind that in addition many different psychological and sociological factors most certainly contribute to the initial excessive consumption of various drugs which gradually may result in increasing dependence to the drugs. The purpose of the present contribution is to briefly discuss in which way dependence-producing drugs, especially ethanol, influence central neuronal functioning and to which degree the drug-induced changes might be related to the development of dependence to the drug.

Effects of Acute Administration of Ethanol on Neurotransmitters

Today it is generally accepted that central monoaminergic neurons play an important role in the regulation of mental processes. Furthermore, the ability of small doses of dependence-producing drugs, e.g. ethanol, amphetamine and morphine to induce behavioural stimulation has been clearly demonstrated in man as well as in experimental animals. In view of such observations it is not surprising that several attempts have been made to find a relationship between the mood-elevating (i.e. euphoric, stimulatory) properties of those drugs and their ability to alter the activity of central monoaminergic neurons (see Table 1). As a matter of fact, several studies seem to support such a relationship. Ethanol, amphetamine, and morphine have thus by various techniques been shown to stimulate the metabolism of central catecholamines (for rev., see Engel and Carlsson, 1977). That the stimulatory properties of these drugs are mediated via central catecholamine mechanisms is supported by the observation that the drug-induced stimulation can be blocked by administration of the specific inhibitor of catecholamine synthesis α-methyltyrosine in doses which per se have no behavioural effects (Figure 1). Furthermore, this blockade by α-methyltyrosine is reversed by the catecholamine

Table 1. Catecholamines and dependence-producing drugs (from Engel, 1977)

	Opiates (morphine)	Psychomotor stimulants (amphetamine)	Ethanol
Animal experiments:			
Catecholamine turnover	↑	↑	↑
Central stimulation	+	+	+
Effect of α-methyltyrosine (α-MT) on the central stimulation	↓	↓	↓
L-DOPA reversal of α-MT effect	+	+	+
Effect of dopamine (DA)-agonists on the central stimulation	↓	↓	↓
Human studies:			
Euphoria and excitement	+	+	+
Antagonism of euphoria and excitement (= removing primary reinforcing properties)			
by α-MT	n.p.	+	+
by DA-agonists	+	+	+

↑ denotes increase; ↓ denotes antagonism; and + denotes that the effect is present. n.p. = not performed.

precursor dopa in small doses which per se are behaviourally inactive (Engel et al., 1974). Similar observations have been obtained also in human studies, where it has been shown that small doses of α-methyltyrosine are able to reduce ethanol- and amphetamine-induced euphoric sensations in man (Jönsson et al., 1969; Ahlenius et al., 1973). Additional support for the involvement of similar neuronal mechanisms is given by the findings that α-methyltyrosine suppresses the self-administration of ethanol, amphetamine as well as morphine (for ref. see Engel and Carlsson, 1977) and the preference for ethanol and morphine (Amit and Levitan, 1975).

Another interesting observation is that small doses of dopamine agonists have been shown to block the central stimulatory effects of ethanol (Carlsson et al., 1974) as well as those of morphine (Kuschinsky and Hornykiewicz, 1974). A reasonable explanation for these paradoxical results comes from recent studies where is was demonstrated that small doses of various dopamine agonists were able to suppress the activity of central dopamine neurons, probably by activating presynaptic inhibitory receptors, so called autoreceptors (Carlsson, 1975). To which degree this phenomenon may be related to the decreased craving for alcohol reported to occur in alcoholics treated with the dopamine agonist apomorphine (Schlatter and Lal, 1972) remains to be clarified.

An increasing number of recent studies indicate that also the neurotransmitter GABA may be involved in the alteration of neuronal functioning following acute administration of ethanol (e.g. Sutton and Simmonds, 1973). As a matter of fact the stimulatory effects of

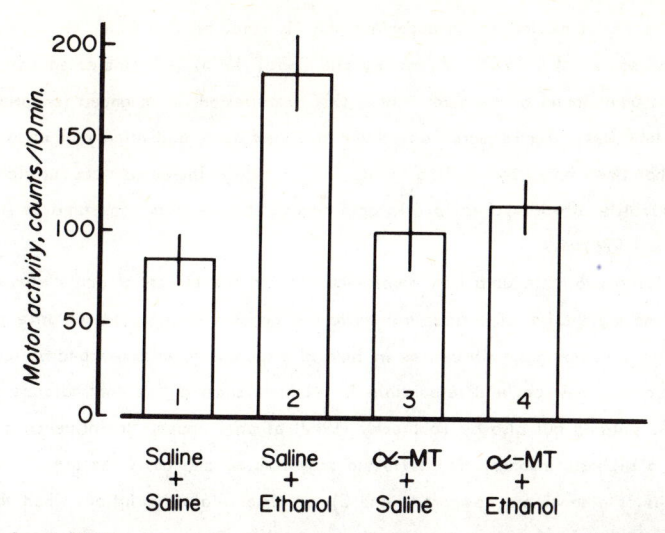

Figure 1. Inhibition by subthreshold doses of α-methyltyrosine (α-MT) of ethanol-induced locomotor stimulation in rats. α-MT (40 mg/kg i.p.) was given 2 h 35 min and ethanol (1 g/kg 20% w/v i.p.) 35 min before the recording of the locomotor activity started. Shown are the means ± s.e.m. of 5-9 rats. 1-2: $p < 0.01$; 2-3 and 2-4: $p < 0.05$; 1-3, 1-4 and 3-4: not significant. (From Carlsson et al., 1972)

ethanol appear to be blocked following pretreatment with GABA-like drugs (Cott et al., 1976; Carlsson and Biswas, 1978).

However, despite the rather good agreement between the pharmacological manipulations in the studies above, little is known about the mechanisms by which central stimulants affect central catecholamine metabolism. When concentrating on ethanol a series of recent studies suggest that ethanol, in doses which functionally have stimulatory effects on the behaviour of experimental animals (i.e. rats and mice) appears to influence the metabolism of central catecholamines in a much more complex manner than previously assumed. This may explain some conflicting observations, such as 1) that ethanol induces a dose-dependent increase in the catecholamine synthesis (measured as the accumulation of dopa following inhibition of central aromatic amino acid decarboxylase) preferentially in dopamine-rich brain areas in the rat (Carlsson and Lindqvist, 1973), 2) that the catecholamine synthesis remains unchanged after ethanol in both mice and rats (Waldeck, 1974; Bustos and Roth, 1976), and 3) that there is a dose-related decrease of the dopamine synthesis in striatal synaptosomes in rats (Pohorecky and Newman, 1977). Corrodi et al. (1966) and Hunt and Majchrowicz (1974 a) reported that ethanol enhances the depletion of noradrenaline after inhibition of the catecholamine synthesis with α-methyltyrosine in rat brain. An accelerated disappearance of both dopamine and noradrenaline in mouse brain has been reported by Cott (1975). In several studies an increased net formation of ^{3}H-dopamine and ^{3}H-noradrenaline from ^{3}H-tyrosine has been reported (Carlsson et al., 1973; Pohorecky and Jaffe, 1975). An increased catecholamine turnover has been found by Karoum et al. (1976) manifested as enhanced formation of catecholamine metabolites. Furthermore, an ethanol-induced decreased release of dopamine in striatal slices has been noted by Gysling et al. (1976), while increased accumulation of the dopamine metabolite dihydroxyphenylacetic acid (dopac) has been demonstrated in striatum by Bustos and Roth (1976).

When summarizing available data from studies concerning the effects of acute ethanol administration on the metabolism of various transmitters it appears that procedural differences (e.g. large variations in dosages, differences in time of exposure to ethanol, and various routes of administration) must be held responsible for at least some of the contradictory results reported. As pointed out also by Pohorecky (1977) ethanol appears to influence outcoming results in a biphasic manner, that is 1) via dose-related changes when the effects of low and high doses of ethanol are compared, and 2) via time-related variations when the effects of the same dose of ethanol are compared at two different occasions after administration of the drug.

Some recent series of experiments (Liljequist and Carlsson, 1978) in which we were dealing with the stimulatory effects of ethanol were carried out at the Department of Pharmacology in Gothenburg in order to further investigate possible mechanisms by which the metabolism of central catecholamines may be altered by acute administration of ethanol. Thus, we studied the effects of a small stimulatory dose of ethanol on the accumulation of the O-methylated dopamine metabolite 3-methoxytyramine and on the O-methylated noradrenaline metabolite normetanephrine, respectively. As will be seen from Table 2 the administration of a small dose of ethanol markedly retarded the formation of 3-methoxytyramine ($p < 0.005$). No effect of ethanol on the accumulation of normetanephrine could be observed.

Table 2. Effect of acute ethanol administration on the formation of O-methylated metabolites in whole brain of mice. (From Liljequist and Carlsson, 1978).

Ethanol (2.36 g/kg, 15% w/v) was given i.p. followed in 30 min by 100 mg/kg pargyline. Sixty minutes later the animals were decapitated. Concentrations are given in nanogram/g ± s.e.m. (n of determinations). Statistics by t-test.

	3-Methoxytyramine	Normetanephrine
Saline	137 ± 20 (9)	105 ± 3 (4)
Ethanol	65 ± 8 (9)	104 ± 5 (4)
	$p < 0.005$	N.S.

Table 3. Catechols and O-methylated metabolites in whole brain of mouse after dopa loading: effect of acute ethanol administration. (From Liljequist and Carlsson, 1978)

Ethanol, 2.36 g/kg or saline was given i.p. followed in 25 min by 50 mg/kg benserazid and 100 mg/kg L-dopa. After a further period of 5 min 400 mg/kg α-methyltyrosine and 100 mg/kg pargyline were given. Sixty minutes later the animals were decapitated. Mean values ± s.e.m. of concentrations (nanogram/g) from 8 determinations are given. Statistics by t-test.

	Dopa	Dopamine	3-Methoxy-tyramine	Noradrenaline	Normeta-nephrine
Saline	25 986 ± 1618	2 678 ± 40	1 527 ± 79	846 ± 54	95 ± 16
Ethanol	30 442 ± 1316	1 573 ± 124	432 ± 32	807 ± 12	76 ± 17
	$p < 0.05$	$p < 0.001$	$p < 0.001$	N.S.	N.S.

These results might be interpreted to mean that acute administration of small doses of ethanol inhibit the release of dopamine without affecting the release of noradrenaline. In order to further analyze this possibility the effects of dopa loading on the accumulation of dopa, dopamine, noradrenaline, 3-methoxytyramine, and normetanephrine in ethanol- and saline-pretreated animals were studied. As will be seen from Table 3 ethanol significantly ($p < 0.05$) increased the accumulation of dopa in ethanol-pretreated animals following dopa loading. On the other hand, the levels of dopamine were significantly ($p < 0.001$) lower in animals treated with ethanol as compared to animals pretreated with saline. No effect of the ethanol pretreatment on the levels of noradrenaline could be observed. As in the first series of experiments the formation of the dopamine metabolite 3-methoxytyramine was markedly retarded in ethanol-treated animals ($p < 0.001$). No difference in the accumulation of normetanephrine after dopa loading was observed.

These results indicate that small doses of ethanol (2.36 g/kg) which functionally have stimulatory effects on the behavioural activity of mice (see above), appear to influence the metabolism of central catecholamines in a complex manner. The conversion of dopa into dopamine is retarded, possibly due to inhibition of dopa decarboxylase, as indicated by increased levels of dopa simultaneously with decreased levels of dopamine. Moreover, the formation of 3-methoxytyramine from dopamine is markedly retarded. Whether this retardation is entirely due to lowered substrate availability, or whether the COMT activity is also inhibited, cannot be decided from the present data.

A puzzling phenomenon is the inability of ethanol to inhibit the accumulation of noradrenaline from dopa and its further conversion into normetanephrine, whereas the formation of dopamine and 3-methoxytyramine is markedly inhibited. These observations suggest that we are dealing with a selective action on dopamine neurons leaving the noradrenaline neurons unaffected. Since the dopa decarboxylase and COMT of the two types of neurons are believed to be very similar if not identical, it is at present difficult to envisage the mechanism involved. Theoretically a selective action on the cell membrane of dopaminergic neurons might account for the following sequence of effects, 1) reduced uptake of dopa by the dopaminergic neuron, leading to reduced conversion of dopa to dopamine, 2) reduced release of dopamine from dopaminergic neurons leading to a more marked inhibition of 3-methoxytyramine formation than would be expected from the moderately lowered dopamine levels. It should be pointed out that the dopa level of the brain after a loading dose of dopa does not give any information about the dopa level in the cytoplasm of dopaminergic neurons. The latter may very well be lowered in spite of an increase of the former level.

Furthermore, it should be pointed out that an interaction between ethanol and the other drugs used may have influenced our results. Experiments are underway to clarify this point.

Effects of Chronic Ethanol Administration on Central Neurotransmitters

Since increasing evidence suggests a causal relationship between central catecholamines and the central stimulant actions of ethanol it is not surprising that much attention has been focussed also on the influence of chronic ethanol administration on these transmitter systems. However, as in the case of acute ethanol effects, a more precise evaluation of existing data is complicated by large variations in the experimental situation (i.e., variating periods of exposure to ethanol, disparate levels of ethanol in blood, different ways of ethanol administration).

So far, no consistent change in catecholamine metabolism has been detected during and after chronic ethanol administration. However, during very prolonged ethanol administration (one year) increased levels of dopamine and noradrenaline have been found by Post and Sun (1973) in the rat caudate. Also Griffiths et al. (1974) have observed increased levels of dopamine and noradrenaline in the whole brain of mice during chronic inhalation of ethanol. Similar observations have been made at our Department in that an increased level of dopamine has been found in the limbic forebrain of the rat during chronic ethanol treatment (9 months) (Figure 2). Since the catecholamine synthesis in these animals is not changed, our data might be interpreted to mean that the release of dopamine is decreased, which can be in accordance with the findings of a decreased rate of dopamine turnover in ethanol dependent rats reported by Hunt and Majchrowicz (1974). It should, however, be noted that both time of ethanol exposure and routes of administration were different in the two series of experiments. Furthermore, Ahtee and Svartström-Fraser (1975) have not been able to find any changes in the accumulation of homovanillic acid during ethanol intoxication.

Withdrawal of chronic ethanol treatment has been reported both to enhance the net accumulation of ^{3}H-noradrenaline after ^{3}H-tyrosine (Pohorecky et al., 1974) and to accelerate the rate of disappearance after α-methyltyrosine (Hunt and Majchrowicz, 1974; Ahtee and Svartström-Fraser, 1975). Such results are in substantial agreement with findings from our Department in that we after ethanol withdrawal have found an increased catecholamine synthesis both in limbic and striatal brain structures accompanied by decreased levels of both

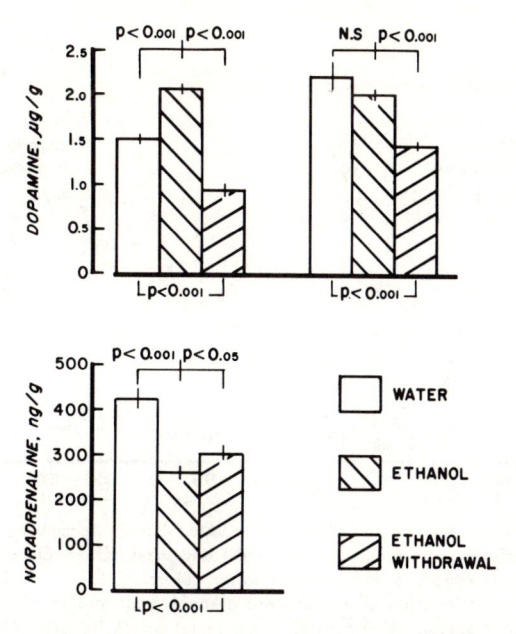

Figure 2. Levels of dopamine (upper part) and noradrenaline (lower part) in limbic forebrain and dopamine in striatum. The withdrawal analyses were performed 4 days after discontinuation of the 270 days lasting chronic ethanol treatment at a time when the animals showed a marked increase in their behavioural activity. Shown are the means ± s.e.m. of 6 determinations. (From Liljequist et al., 1977).

dopamine and noradrenaline in the limbic part of the brain (see Figure 2). Thus, following ethanol withdrawal there appears to be an increased activity in both central dopamine and noradrenaline neurons.

Additional support for the contention that central catecholamine mechanisms are affected by chronic ethanol administration comes from our recent studies (Engel and Liljequist, 1976) where it was shown that direct application of dopamine into the nucleus accumbens causes a more pronounced increase in the locomotor activity of animals chronically treated with ethanol than in untreated control animals (see Figure 3). This effect of dopamine was antagonized by haloperidol indicating a specific effect on dopamine receptors. These observations have been interpreted to mean that prolonged ethanol administration produces an increased sensitivity to locally applied dopamine at or beyond dopamine receptors at least in the nucleus accumbens. Furthermore, the increased receptor sensitivity appears to be mainly related to postsynaptic structures since it could be demonstrated also after excluding the influence of presynaptic mechanisms (Liljequist, 1978). Moreover, it has been shown (Liljequist, 1978) that the phenomenon of increased receptor sensitivity is first observed after 5 months of ethanol treatment and lasts for about 4 weeks after cessation of the chro-

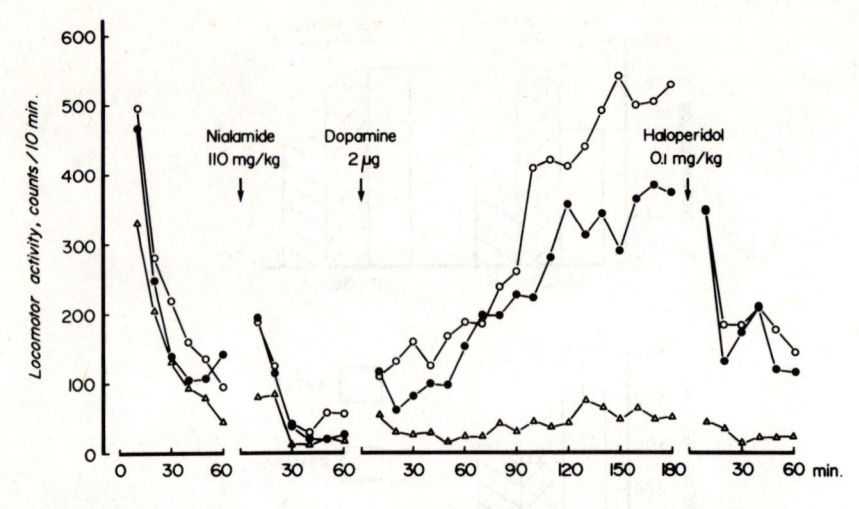

Figure 3. Effect of local application of dopamine into nucleus accumbens (2 μg to each side) on locomotor activity of rats during (●---●) and after (○----○) a chronic ethanol treatment of 380 days as compared to untreated water controls (△——△). Test was performed 4 days after discontinuation of the chronic ethanol treatment in withdrawal rats. Each value is mean activity ± s.e.m. of 4-6 rats. (From Engel and Liljequist, 1976)

nic ethanol administration.

When testing the neurotransmitter specificity of the observed receptor supersensitivity it was shown that the effects of the central noradrenaline receptor agent clonidine on blood pressure, heart rate, and flexor reflex activity were similar in animals chronically treated with ethanol and in animals never subjected to ethanol (Liljequist et al., 1978). Also oxotremorine-induced tremor activity appears to be of the same magnitude in both groups of treatment (Liljequist, 1978) suggesting that chronic ethanol treatment does not alter the sensitivity of cholinergic mechanisms. Taken together, these results indicate that the phenomenon of increased receptor sensitivity is mainly related to dopamine-sensitive neuronal functions.

In order to further test the hypothesis that chronic ethanol treatment especially affects central dopamine-sensitive mechanisms we studied the influence of gamma-butyrolactone (GBL) on the central catecholamine synthesis (Liljequist and Engel, in preparation). It has been claimed that GBL specifically interferes with the activity of dopamine neurons (for discussion of the effects of GBL, see Roth et al., 1974). Thus it has been shown that the administration of GBL induces a markedly increased accumulation of dopa in the striatum and that this effect of GBL appears to be related to its ability to inhibit the nerve impulse flow in dopamine neurons of the nigro-striatal pathway (Walters and Roth, 1974). Due to the GBL-induced inhibition of neuronal firing there will be both an increased synthesis of dopamine as well as greatly elevated levels of endogenous dopamine. Roth and his associates have argued that the GBL-induced cessation of impulse flow in all probability disturbs normally acting homeostatic mechanisms by altering the sensitivity of tyrosine hydroxylase to inhibition by endogenous levels of dopamine (for discussion, see Roth et al., 1974). A possible explanation for the altered susceptibility to end-product inhibition by dopamine might be that the GBL-induced cessation of impulse flow leads to a diminished influx of calcium into the cellular space. The attenuated influx of calcium would therefore decrease the level of intracellular calcium causing a change in the kinetics of striatal tyrosine hydroxylase. Actually, this assumption was confirmed in in-vitro experiments where it was shown that GBL altered the kinetic properties of tyrosine hydroxylase both by reducing the K_m for tyrosine and cofactor and by increasing the K_i for dopamine. Furthermore, it was demonstrated that the dopamine agonist, apomorphine, in doses which preferentially activate presynaptic mechanisms was able to reverse the GBL-induced alterations in the kinetic properties of tyrosine hydroxylase as well as the GBL-induced increased accumulation of dopa in the striatum. These results might be interpreted to mean that the tyrosine hydroxylase activity is regulated via presynaptic mechanisms in dopamine neurons.

The effect of GBL (750 mg/kg i.p.) on the rate of tyrosine hydroxylation was studied in animals subjected to our ethanol regimen (for details, see Liljequist et al., 1977) for 150 days and compared to age-matched animals never subjected to ethanol. As can be seen from Table 4 the administration of GBL produced a marked, significant increase in the dopa accumulation in the striatum of both groups of treatment. However, the GBL-induced dopa accumulation was significantly lower ($p < 0.05$) in ethanol-treated animals. Furthermore, it should be noted that the enhanced dopa accumulation after administration of GBL in the present experiments seems to be of the same magnitude as that observed by Walters and Roth (1974).

Table 4. Effect of GBL on the accumulation of dopa after inhibition of dopa decarboxylase by means of NSD-1015 in the striatum of rats chronically treated with ethanol for 150 days. (From Liljequist and Engel, in preparation.)

GBL (750 mg/kg was given i.p. followed in 10 min by apomorphine 0.125 mg/kg i.p. After a further period of 5 min 100 mg/kg NSD-1015 was given i.p. Thirty minutes later the animals were decapitated. Mean values ± s.e.m. of concentrations (µg/g) from 10 determinations are given. Statistics by t-test.

	Controls		Ethanol
NSD-1015	0.672 ± 0.035	N.S.	0.650 ± 0.037
	$p < 0.001$		$p < 0.001$
GBL + NSD-1015	1.989 ± 0.224	$p < 0.05$	1.386 ± 0.115
	$p < 0.01$		N.S.
GBL + Apo + NSD-1015	1.308 ± 0.064	N.S.	1.227 ± 0.079

According to the above hypothesis concerning the underlying mechanism of GBL-induced changes in neuronal functioning a plausible explanation for the results obtained in the present study would be that chronic ethanol administration alter the capacity of GBL to induce membrane polarization. Thus it may be suggested that GBL is not able to fully prevent depolarization of dopamine neurons (which would be predicted to follow a blockage of impulse flow) in rats chronically treated with ethanol. A natural consequence would be that a partial depolarization of the dopamine neurons involved causes leakage of calcium into the cell thereby resulting in a weaker influence of GBL on the tyrosine activating mechanisms. If such a hypothesis is correct, it would be predicted that administration of a presynaptic dose of the dopamine receptor agonist apomorphine would cause a less pronounced reversal of the GBL-induced activation of tyrosine hydroxylase in ethanol-treated rats than in water control rats, since the intracellular level of calcium would already be elevated in ethanol rats due to depolarization of the membrane.

In order to test this possibility, apomorphine (0.125 mg/kg i.p., the smallest dose of apomorphine able to cause a significant reversal of GBL-induced increased accumulation of dopa in normal rats) was given to GBL-pretreated animals. As can be seen from Table 4 apomorphine significantly reversed the dopa accumulation in water control rats without affecting the tyrosine activity in GBL-pretreated ethanol rats. Thus it may be suggested that chronic ethanol administration induces a subsensitivity of presynaptic mechanisms in central dopamine neurons. Whether this decreased response of dopamine-sensitive mechanisms is due to the proposed changes of calcium flux or is caused by alternative effects on cell membrane properties needs further clarification.

Effect of Dependence-Producing Drugs on Neuronal Excitability

Electrophysiological studies suggest that ethanol and opiates may have common sites of action both in brain and in spinal cord (for references, see Eidelberg, 1977). Interestingly enough, Eidelberg and his associates have reported that ethanol modifies neuronal firing in cerebellum while morphine seems to be without effect in the same brain area (Eidelberg, 1977). To which degree the inability of morphine to affect neuronal firing is related to the reported absence of opiate receptors in cerebellum (Pert and Snyder, 1978) remains to be clarified.

However, some regional similarities between the effects of ethanol and morphine seem to exist. Thus it has been observed that both ethanol (at doses producing ataxia) and morphine exert a depressive influence upon the activity of spinal interneurons (Eidelberg and Wooley, 1970; Le Bars et al., 1975). Furthermore, a stimulatory action of ethanol upon Renshaw cells in spinal cord has been reported by Meyer-Lohman et al. (1972). However, with regard to morphine both depression (Felpel et al., 1970) and excitation (Davies and Duggan, 1974) of Renshaw cells have been observed.

Studies, in which the microelectrophoretic technique has been utilized, indicate that opiates might be able to inhibit firing in both cortical and striatal neurons (Herz et al., 1977). Furthermore, this effect of morphine is antagonized by administration of the morphine antagonist naloxone, suggesting a specific effect on opiate receptors. Recently, Pohorecky and Brick (1977) have reported that systemic administration of ethanol causes inhibition of neuronal firing in the locus coeruleus. These findings could, however, not be confirmed in experiments at our Department (Svensson, 1978).

However, an interesting observation has been made in connection with studies concerning the firing rate of various neurons. Thus it has been reported by Eidelberg and Wooley (1970) concerning ethanol, and by Satoh et al. (1975) as well as by Zieglgänsberger et al. (1976) concerning morphine, that these agents exert non-specific actions on neuronal membranes very similar to those seen after administration of local anesthetic drugs. Moreover, it has been proposed (Eidelberg, 1977) that these local anesthetic-like effects of ethanol and morphine might be due to interference with the sodium current occurring during the conduction of action potentials (for reference, see Israel et al., 1975) and that calcium would be involved in this process.

As a matter of fact, further indications for a possible role of calcium in these phenomena are coming from the investigations of Ross et al. (1974) who have shown that both ethanol and morphine deplete the content of brain calcium in a dose-dependent manner. Furthermore, the calcium-depleting influence of ethanol and morphine can be antagonized by administration of naloxone.

To which degree the non-specific, local anesthetic-like membrane effects of ethanol and morphine are related to an altered state of calcium is not known. However, several possibilities seem to exist. Today it is for example rather generally accepted that calcium ions influence the membrane permeability to Na^+ and $K+$ (Brismar, 1977). Calcium ions are probably also of importance for the release of transmitter substances via their effect on the process underlying the fusion between the vesicles containing transmitter substances and the nerve terminal plasma membrane (Papahadjopoulos et al., 1975).

Summary

No neurochemical or neurophysiological studies of the effects of dependence-producing drugs have so far managed to identify specific primary mechanisms of action that may explain the dependence-producing properties of these agents. The demonstration of opiate receptors in the brain (Pert and Snyder, 1973) will certainly be of great importance in future attempts to solve the problem of drug tolerance and dependence. Concerning ethanol, it will be much more difficult to analyze its primary actions on central nervous mechanisms since it is generally assumed that ethanol exerts a non-specific influence on most neurons.

However, the present brief survey suggests that it perhaps would be possible to identify some key effects of ethanol which may be of importance in the regulation of central neurotransmitter function. Thus it was demonstrated that acute administration of ethanol after loading with dopa alters the content of dopamine in the brain and probably also changes the release of dopamine. The possibility that the observed effects could be due to a selective action of ethanol on membranes in dopamine neurons was discussed. Furthermore, in the experiments concerning chronic effects of ethanol it was shown that the capacity of GBL to activate tyrosine hydroxylase in dopamine-rich brain areas was impaired, which perhaps can be explained by an altered ability of GBL to inhibit depolarization of dopamine neurons. Furthermore, data from electrophysiological studies indicate that both opiates and ethanol produce non-specific, local anesthetic-like effects on cell membranes.

Taken together, it would appear that the above described actions of ethanol are in some way related to alterations of neuronal membrane functions in which calcium-dependent mechanisms might be involved. As pointed out above the importance of calcium in synaptic transmission and neurotransmitter release is today well-recognized. Thus is may be hypothesized that ethanol via alteration of calcium-dependent processes secondarily can affect the metabolism of neurotransmitters thereby producing its known influence on behaviour in man and experimental animals.

Acknowledgements. This study was supported by the Swedish Medical Research Council (projects Nos. 2157, 4247), Ferrosans Jubileumsstiftelse, Åke Wibergs Stiftelse and the Faculty of Medicine, University of Göteborg. The technical assistance of Mrs. Barbro Aldäng, Miss Christel Carlsson and Mr. Kenn Johannessen is gratefully acknowledged.

References

Ahlenius, S., Carlsson, A., Engel, J., Svensson, T.H. and Södersten, P. (1973) Antagonism by alpha methyltyrosine of the ethanol-induced stimulation and euphoria in man. Clin. Pharmacol. Ther. 14: 586-591.

Ahtee, L. and Svartström-Fraser, M. (1975) Effect of ethanol dependence and withdrawal on the catecholamines in rat brain and heart. Acta pharmacol. et toxicol. 36: 289-298.

Amit, Z. and Levitan, D.E. (1975) In: J.D. Sinclair and K. Kiianmaa (Eds.), The effects of centrally active drugs on voluntary alcohol consumption. Finnish Foundation for Alcohol Studies 24: 85-100.

Brismar, T. (1977) Slow action of Ca on myelinated nerve fibres of Xenopus laevis. Acta physiol. scand. 99: 361-367.

Bustos, G. and Roth, R.H. (1976) Effect of acute ethanol treatment on transmitter synthesis and metabolism in central dopaminergic neurons. J. Pharm. Pharmac. 28: 580-582.

Carlsson, A. (1975) Receptor-mediated control of dopamine metabolism. In: E. Usdin and W.E. Bunney Jr. (Eds.), Pre- and Postsynaptic Receptors, pp. 49-65. M. Dekker Inc., New York.

Carlsson, A. and Biswas, B. (1978) Effect of GABA administered intracerebro-ventricularly or intraperitoneally on brain monoamines and locomotor activity. In: S. Garattini, J.F. Pujal, and R. Sananin (Eds.), Interactions Between Putative Neurotransmitters in the Brain, pp. 305-315. Raven Press, New York.

Carlsson, A., Engel, J., Strömbom, U., Svensson, T.H. and Waldeck, B. (1974) Suppression by dopamine-agonists of the ethanol-induced stimulation of locomotor activity and brain dopamine synthesis. Naunyn-Schmiedeberg's Arch. Pharmacol. 283: 117-128.

Carlsson, A., Engel, J. and Svensson, T.H. (1972) Inhibition of ethanol-induced excitation in mice and rats by α-methyl-p-tyrosine. Psychopharmacologia (Berl.) 26: 307-312.

Carlsson, A. and Lindqvist, M. (1973) Effect of ethanol on the hydroxylation of tyrosine and tryptophan in rat brain in vivo. J. Pharm. Pharmac. 25: 437-440.

Carlsson, A., Magnusson, T., Svensson, T.H. and Waldeck, B. (1973) Effect of ethanol on the metabolism of brain catecholamines. Psychopharmacologia (Berl.) 30: 27-36.

Corrodi, H., Fuxe, K. and Hökfelt, T. (1966) The effect of ethanol on the activity of central catecholamine neurons in rat brain. J. Pharm. Pharmac. 18: 821-823.

Cott, J.M. (1975) Investigations into the mechanisms of reduction of ethanol-induced sleep by thyrotropin-releasing hormone (TRH). Ph.D. Thesis, Department of Pharmacology, University of North Carolina at Chapel Hill.

Cott, J., Carlsson, A., Engel, J. and Lindqvist, M. (1976) Suppression of ethanol-induced locomotor stimulation by GABA-like drugs. Naunyn-Schmiedeberg's Arch. Pharmacol. 295: 203-209.

Davies, J. and Duggan, A.W. (1974) Opiate agonist-antagonist effects on Renshaw cells and spinal inter-neurones. Nature 250: 70-71.

Eidelberg, E. (1977) On the possibility that opiate and ethanol actions are mediated by similar mechanisms. In: M.M. Gross (Ed.) Alcohol Intoxication and Withdrawal III b. Studies in Alcohol Dependence. Adv. Exp. Med. Biol. 85B: 87-94. Plenum Press, New York.

Eidelberg, E. and Wooley, D. (1970) Effects of ethyl alcohol upon spinal cord neurones. Arch. int. Pharmacodyn. Therap. 185: 388-396.

Engel, J. (1977) Neurochemical aspects of the euphoria induced by dependence-producing drugs. In: C.-M. Ideström (Ed.) Recent Advances in the Study of Alcoholism, pp. 16-22. Excerpta Medica, Amsterdam.

Engel, J. and Carlsson, A. (1977) Catecholamines and behavior. In: L. Valzelli and W.B. Essman (Eds.) Current Developments in Psychopharmacology, 4: 1-32.

Engel, J. and Liljequist, S. (1976) The effect of long-term ethanol treatment on the sensitivity of the dopamine receptors in the nucleus accumbens. Psychopharmacology 49: 253-257.

Engel, J., Strömbom, U., Svensson, T.H. and Waldeck, B. (1974) Suppression by α-methyl-tyrosine of ethanol-induced locomotor stimulation: partial reversal by L-dopa. Psychopharmacologia (Berl.) 37: 275-279.

Felpel, L.P., Sinclair, J.G. and Yim, G.K.W. (1970) Effects of morphine on Renshaw cell activity. Neuropharmacology 9: 203-210.

Griffiths, P.J., Littleton, J.M. and Ortiz, A. (1974) Changes in monoamine concentrations in mouse brain associated with ethanol dependence and withdrawal. Br. J. Pharmac. 50: 489-498.

Gysling, K., Bustos, G., Coucha, I. and Martinez, G. (1976) Effect of ethanol on dopamine synthesis and release from rat corpus striatum. Biochem. Pharmac. 25: 157-162.

Herz, A., Zieglgänsberger, W., Schulz, R., Fry, J.P. and Satoh, M. (1977) Neuronal aspects of opiate dependence and tolerance in comparison to central depressants. In: M.M. Gross (Ed.), Alcohol Intoxication and Withdrawal III b. Studies in Alcohol Dependence. Adv. Exp. Med. Biol. 85 B: 117-140. Plenum Press, New York.

Hunt, W.A. and Majchrowicz, E. (1974) Alterations in the turnover of brain norepinephrine and dopamine in alcohol-dependent rat. J. Neurochem. 23: 549-552.

Israel, Y., Carmichael, F.I. and Macdonald, J.A. (1975) Effects of ethanol on electrolyte metabolism and neurotransmitter release in the CNS. In: M.M. Gross (Ed.) Alcohol Intoxication and Withdrawal. Exp. Stud. II 59: 55-64. Plenum Press, New York.

Jönsson, L.-E., Gunne, L.-M. and Änggård, E. (1969) Effects of α-methyl-tyrosine in amphetamine-dependent subjects. Pharmacol. Clin. 2: 27-29.

Karoum, F., Wyatt, R.J. and Majchrowicz, E. (1976) Brain concentrations of biogenic amine metabolites in acutely treated and ethanol-dependent rats. Br. J. Pharmac. 56: 403-411.

Kuschinsky, K. and Hornykiewicz, O. (1974) Effects of morphine on striatal dopamine metabolism: possible mechanism of its opposite effect on locomotor activity in rats and mice. European J. Pharmacol. 26: 41-50.

Le Bars, D., Menetrey, D., Conseiller, C. and Besson, J.M. (1975) Depressive effects of morphine upon lamina V cells activities in the dorsal horn of the spinal cat. Brain Res. 98: 261-277.

Liljequist, S. (1978) Changes in the sensitivity of dopamine receptors in the nucleus accumbens and in the striatum induced by chronic ethanol administration. Acta pharmacol. et toxicol., in press.

Liljequist, S., Ahlenius, S. and Engel, J. (1977) The effect of chronic ethanol treatment on behaviour and central monoamines in the rat. Naunyn-Schmiedeberg's Arch. Pharmacol. 300: 205-216.

Liljequist, S., Andén, N.-E., Engel, J. and Henning, M. (1978) Noradrenaline receptor sensitivity after chronic ethanol administration. J. Neural Transm. 42: 000-000, in press.

Liljequist, S. and Carlsson, A. (1978) Alteration of central catecholamine metabolism following acute administration of ethanol. J. Pharm. Pharmac., in press.

Meyer-Lohmann, J., Hagenah, R., Hellweg, C. and Benecke, R. (1972) The action of ethyl alcohol on the activity of individual Renshaw cells. Naunyn-Schmiedeberg's Arch. Pharmacol. 272: 131-142.

Papahadjopoulos, D. and Poste, G. (1975) Calcium-induced phase separation and fusion in phospholipid membranes. Biophys. J. 15: 945-948.

Pert, C.B. and Snyder, S.H. (1973) Opiate receptor: demonstration in nervous tissue. Science 179: 1011-1014.

Pohorecky, L.A. (1977) Biphasic action of ethanol. Biobehav. Rev. 1: 231-240.

Pohorecky, L.A. and Brick, J. (1977) Activity of neurons in the locus coeruleus of the rat: inhibition by ethanol. Brain Res. 131: 174-179.

Pohorecky, L.A. and Jaffe, L.S. (1975) Noradrenergic involvement in the acute effects of ethanol. Res. Comm. Chem. Path. Pharmacol. 12: 433-445.

Pohorecky, L.A., Jaffe, L.S. and Berkeley, H.A. (1974) Ethanol withdrawal in the rat: involvement of noradrenergic neurons. Life Sci. 15: 427-437.

Pohorecky, L.A. and Newman, B. (1977) Effect of ethanol on dopamine synthesis in rat striatal synaptosomes. Drug and Alcohol Dependence 2: 329-334.

Post, M.E. and Sun, A.Y. (1973) The effect of chronic ethanol administration on the levels of catecholamines in different regions of the rat brain. Res. Comm. Chem. Path. Pharmacol. 6: 887-894.

Ross, D.H., Medina, M.A. and Cardenas, H.L. (1974) Morphine and ethanol. Selective depletion of regional brain calcium. Science 186: 63-64.

Roth, R.H., Walters, J.R. and Morgenroth III, U.H. (1974) Effects of alterations in impulse flow on transmitter metabolism in central dopaminergic neurons. In: E. Usdin (Ed.), Neuropsychopharmacology of Monoamines and Their Regulatory Enzymes, pp. 364-384. Raven Press, New York.

Satoh, M., Zieglgänsberger, W. and Herz, A. (1975) Interaction between morphine and putative excitatory neurotransmitters in cortical neurones in naive and tolerant rats. Life Sci. 17: 75-80.

Schlatter, E.K.E. and Lal, S. (1972) Treatment of alcoholism with Dent's oral apomorphine method. Quart. J. Stud. Alcohol 33: 430-436.

Seevers, M.H. (1968) Psychopharmacological elements of drug dependence. J. Amer. Med. Assoc. 206: 1263-1266.

Sutton, I. and Simmonds, M.A. (1973) Effects of acute and chronic ethanol on the γ-aminobutyric acid system in rat brain. Biochemical Pharmacology 22: 1685-1692.

Svensson, T.H. (1978) Personal communication.

Waldeck, B. (1974) Ethanol and caffeine: a complex interaction with respect to locomotor activity and central catecholamines. Psychopharmacologia (Berl.) 36: 209-220.

Wallgren, H. and Barry III, H. (1970) Actions of Alcohol, Vol. 1-2. Elsevier Publ. Comp. Amsterdam.

Walters, J.R. and Roth, R.H. (1974) Dopaminergic neurons: drug-induced antagonism of the increase in tyrosine hydroxylase activity produced by cessation of impulse flow. J. Pharmacol. exp. Ther. 191: 82-91.

Zieglgänsberger, W., Fry, J.P., Herz, A., Moroder, L. and Wünsch, E. (1976) Enkephalin-induced inhibition of cortical neurones and the lack of this effect in morphine tolerant/dependent rats. Brain Res. 115: 160-164.

BIOCHEMICAL THEORIES OF OPIOID DEPENDENCE: AN ANALYSIS

H.O.J. Collier
Miles Laboratories Limited, Stoke Poges, Slough SL2 4LY, England.

Use of Words

Those who discuss theories must take care how they use words, because theories are mainly expressed in words. Let me therefore begin by describing how I shall be using some words in this discussion.

Whereas the term addiction has various usages, including an everyday one, with several shades of meaning, the terms dependence and tolerance can be restricted to more defined usage. I shall therefore speak of drug dependence, which may be regarded as a state arising from exposure to drug such that withdrawal of that drug causes an observable disturbance, of behavioural, biochemical or other type. Tolerance can be regarded as a state of reduced responsiveness to drug, arising from exposure to that drug, such that a given dose produces a lesser effect and a larger dose is needed to produce the same effect. Since, among opioids, tolerance accompanies dependence, theories of opioid dependence should also account for tolerance.

Whereas an opiate may be regarded as a derivative of opium having specific morphine-like action, the term opioid, introduced by Acheson (Martin, 1967) includes any substance, whether derived from opium, or of synthetic or endogenous origin, that has such an action. Morphine and many other opioids have mixed agonist and antagonist actions; but naloxone is a pure antagonist (Kosterlitz & Watt, 1968). Naloxone therefore provides a test and a definition of opioid -- as any substance having morphine-like actions that are competitively antagonized by naloxone. Such actions may be described as specific.

The specific actions of opioids occur at different rates, one measurable in minutes and one in hours or longer. The faster are the acute or immediate actions; the slower are the subacute or delayed, expressed as tolerance and dependence. Since naloxone also antagonizes the delayed actions of opioids, these, too, can be regarded as specific.

Only nerve cells possessing opioid receptors can be expected to respond specifically to opioids. Such nerve cells may be termed opioid-sensitive neurones. They provide a starting point in the analysis of opioid dependence.

Dependence and the Opioid-sensitive Neurone

Distinctions between theories

Biochemical theories of opioid dependence can be classified in various ways. One way is to divide the theories into two types -- one stemming from the theory of Tatum et al (1929) and the other from that of Himmelsbach (1943). Tatum et al proposed that dependence is an exaggeration of the stimulatory relative to the inhibitory actions of morphine; whereas Himmelsbach proposed that dependence is a state of neuronal excitation arising in homeostatic compensation for the inhibitory action of morphine. Most of the biochemical mechanisms discussed are of the Himmelsbach type; but at least one belongs to that of Tatum et al.

Alternatively, biochemical theories of opioid tolerance and dependence can be divided into two other types, according to whether they apply to events outside the opioid-sensitive neurone (extra-cellular theories) or to events in (including on) this neurone (intra-cellular theories). Extra-cellular theories are usually humoral or neurological. Most humoral theories suppose that a change in the supply of a hormone, transmitter or other regulatory substance is responsible for dependence. For example, Kosterlitz & Hughes (1975) have recently proposed that dependence arises because, in response to treatment with exogenous opioid, the natural production of endogenous opioid diminishes, leaving the total supply of opioid in the brain inadequate after exogenous drug is withdrawn. Unless it is supposed that the opioid-sensitive neurone itself produces enkephalin, this is an extra-cellular theory.

An example of a neurological extra-cellular theory is the redundant pathway theory of Martin (1968). Martin supposes that, in response to blockade by drug of one neuronal pathway, a second pathway is opened up. The extra pathway would lessen response to drug, and this would be expressed as tolerance. When the drug was withdrawn, the extra and the hitherto blocked pathways would both function, giving additional activity, expressed as withdrawal effects.

Intra-cellular theories suppose that a change in or on the opioid-sensitive neurone is responsible for dependence. Such a change might be in a receptor responding to a messenger from another cell, in an enzyme of the cell, in the liberation of a transmitter or in some other mechanism.

The division of theories into extra- and intra-cellular leads to the question -- can opioid dependence develop within a single cell? According to the answer, one group of theories can be discarded as of secondary importance.

Tolerance and dependence within one cell

Experiments on three preparations provide evidence that opioid tolerance and dependence can occur within a single cell. One preparation is the neuroblastoma x glioma (NG) hybrid cell possessing opioid receptors (strain 108-15), which may be regarded as a model of the opioid-sensitive neurone (Klee & Nirenberg, 1974). This cell specifically responds to opioids by a reduced production of cyclic AMP (Sharma et al, 1975a; Traber et al, 1975a). Exposure of NG 108-15 cells in culture for more than 12h to morphine induces both tolerance and dependence. Tolerance takes the form of a lessened inhibition by opioid of cyclic AMP production. Dependence is expressed as a higher production of cyclic AMP on withdrawal of morphine, especially in the presence of stimulants of adenylate cyclase, such as E prostaglandin (PGE) or adenosine (Sharma et al, 1975b, Traber et al, 1975b).

The conclusion that opioid tolerance and dependence can occur within one cell is reinforced by experiments on the myenteric plexus of the isolated guinea-pig ileum. In this preparation, opioids specifically inhibit the neurally-evoked contraction of the longitudinal smooth muscle, by lessening the release of the excitatory transmitter, acetylcholine, from a post-ganglionic motor neurone. Tolerance and dependence can be generated in the myenteric plexus of the ileum by incubation at $4^{\circ}C$ with morphine or other opioid (Hammond et al, 1976; Villarreal et al, 1977). Tolerance, in this preparation, takes the form of a lessened inhibition by morphine, and dependence of an intense contraction to naloxone. Tolerance and dependence can be induced in this way in the presence of the ganglion-blocking drug, hexamethonium, and the withdrawal response to naloxone can be prevented by the cholinergic blocker, hyoscine. These findings suggest that tolerance/dependence has developed in a single type of motor neurone. This suggestion is supported by the observation of North & Karras (1978) that, after exposure to morphine for 24h, at room temperature, single neurones of the myenteric plexus can exhibit tolerance and dependence towards opioids.

Satoh et al (1976) have reported another experiment indicating that tolerance and dependence can occur within one cell. These authors recorded the discharge of nerve impulses in comparable single cortical neurones of naive and morphine-dependent rats. In naive, but not in dependent animals, morphine, administered by microelectrophoresis, inhibited the elicitation of nerve impulses by acetyl-choline or l-glutamate, administered in the same way. In naive animals, naloxone had no effect on spontaneous discharge of nerve impulses; but, in dependent animals, naloxone increased impulse discharge. These experiments reveal, in the sensorimotor cortex of morphine-dependent rats, the existence of single neurones exhibiting tolerance and dependence.

Molecular mechanisms of opioid action

   Since opioid tolerance and dependence can occur within a single cell possessing opioid receptors, we can discard extra-cellular theories of opioid dependence as, at most, of secondary importance, and concentrate on intra-cellular theories. There is by now much evidence that opioid tolerance/dependence expresses an increased capacity for the formation and possibly also increased effectiveness of cyclic AMP in appropriate neurones (Mehta & Johnson, 1974; Collier & Francis, 1975, 1978; Collier et al 1974, 1975, 1978; Francis et al, 1975; Sharma et al, 1975b, 1977; Traber et al, 1975b; Hosein & Lau, 1977). Several theories have been proposed for the biochemical mechanism(s) whereby this heightened activity of the cyclic AMP mechanism is brought about within one cell. To survey these theories, it is convenient to consider in turn the separate elements in opioid interaction with the cyclic AMP mechanism, illustrated in Fig.1. These elements are -- (1) the opioid receptor, (2) receptors for excitatory messengers, (3) coupling mechanisms between receptors and (4) the catalytic unit of adenylate cyclase, (5) phosphodiesterase, (6) protein kinase and associated phosphorylase, (7) transmitter formation and (8) release.

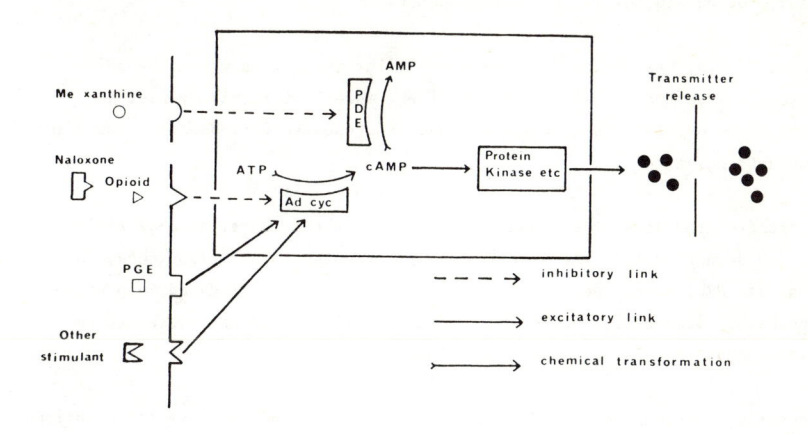

Fig.1. Molecular mechanism of opioid action within the opioid-sensitive cell (Collier, 1979). The box encloses enzymatic parts of the cyclic AMP mechanism.

Analysis of Dependence Mechanisms

Opioid receptor

A decrease in the available number of opioid receptors on a cell could produce tolerance (Axelrod, 1956; Collier, 1966) but this would not be sufficient to account for dependence. Dependence induction could, however, be explained in terms of two different opioid receptors, or of two forms of one receptor. Such explanations seem to be particular cases of the theory of Tatum et al (1929).

Let us suppose that a separate opioid receptor were involved in the induction of tolerance and dependence. To distinguish this from the normal opioid receptor, let us term it the dependence receptor. The dependence receptor would react to the opioid molecule in a way roughly opposite to the normal receptor. Thus, the dependence receptor would respond to opioid with excitation of the neurone, and this excitation would antagonize the normal inhibitory response. It would respond to naloxone also with excitation. Lastly, let us suppose that, in the opioid-naive cell, the opioid receptors are overwhelmingly of normal type, and that exposure to opioid induces the formation of dependence receptors at the expense of normal receptors. Such a theory would seem to offer a workable mechanism of opioid tolerance and dependence.

Snyder (1975) has sketched a somewhat similar theory, in which the opioid receptor is supposed to exist in distinct but interconvertible agonist and antagonist forms, and exposure to opioid is supposed to increase the proportion of the antagonist form.

Apart from the lack of evidence for theories of this type, an objection is that they do not explain the increased responsiveness to stimulant messenger substances (PGE, adenosine, 5-hydroxytryptamine) nor the decreased responsiveness to inhibitory messengers (adrenaline, noradrenaline, dopamine), observed in opioid-dependent cells _in vitro_ (see below).

Recently, Jacquet et al (1977) have suggested that, because the (+) enantiomer of morphine, injected into the rat brain, had stimulatory non-specific actions, "another important mechanism (of dependence) may be the selective stimulation by opiates of the naloxone-insensitive receptors following blockade by naloxone of the naloxone-sensitive receptors which normally act to inhibit the former". There is no evidence to suggest, however, that this mechanism applies to a single opioid-sensitive neurone.

Receptors for excitatory messengers

A second type of theory based on receptor change concerns not the opioid receptor, but receptors for the humoral messengers, such as PGE, that, coupled to adenylate cyclase, stimulate cyclic AMP formation, which opioids antagonize. This theory postulates that these receptors increase in number (or efficiency) during dependence, thus making it harder for the opioid to inhibit their stimulation (Collier, 1966, 1969, 1972).

The main objection to this theory is that, although increased responsiveness to excitatory messenger substances is observed in opioid-dependent preparations, the effect is non-specific. For example, the dependent NG 108-15 cell, after withdrawal, responds more intensely to both PGE_1 and adenosine (Sharma et al, 1975b). Again, the withdrawn dependent ileum is supersensitive to both PGE and 5-hydroxytryptamine, which may be supposed to act on different receptors (Schulz & Goldstein, 1973; Schulz & Herz, 1976).

Coupling mechanisms

Opioids do not inhibit fluoride-stimulated cyclic AMP formation in rat brain homogenate nor in NG 108-15 cells, although they do inhibit formation of cyclic AMP stimulated by PGE (Collier & Roy 1974a,b; Sharma et al, 1975a, 1977; Traber et al, 1975a; Havemann & Kuschinsky, 1978). Likewise, in withdrawn opioid-dependent NG 108-15 cells, the response to fluoride is not exaggerated, whereas that to PGE_1 is (Sharma et al, 1977).

A possible explanation of this difference between stimulation by fluoride and by PGE_1 is that fluoride may act on the catalytic unit without the mediation of a receptor, whereas the action of PGE_1 is mediated via an E prostaglandin receptor. If so, the coupling between receptor and catalytic unit may be important in the mechanism of tolerance and dependence.

Although we do not know the nature of the coupling mechanism, the "floating-receptor" hypothesis (Cuatracasas, 1974, 1975; de Haën, 1976) proposes that the receptors for adenylate cyclase are flexibly related to the catalytic unit and separable from it. Consistently with this, receptors from one cell can be made to stimulate the adenylate cyclase of another cell (Schramm et al, 1977).

The possible importance of the coupling mechanism rather than the receptors themselves in the induction of opioid dependence is indicated by experiments on the sensitivity of dependent guinea-pig ileum, some of which were mentioned above. Not only is the dependent ileum supersensitive to excitation by PGE and

5-HT, but it is also subsensitive to inhibition by noradrenaline, adrenaline and dopamine (Goldstein & Schulz, 1973). It is, of course, subsensitive also to inhibition by morphine, as tolerance implies.

Let us suppose that the coupling between the receptors and the catalytic unit of adenylate cyclase is non-specific and can only be either excitatory or inhibitory, according to the type of receptor activated and that, if both excitatory and inhibitory receptors are activated at the same time, the message conveyed to the catalytic unit is a resultant of the two influences. Then the non-specific supersensitivity and subsensitivity of dependence could be explained as a change, in favour of excitation, in the coupling between the catalytic unit of adenylate cyclase and the receptors for inhibitory and stimulant substances. This explanation would also provide a workable mechanism for opioid tolerance and dependence.

Catalytic unit of adenylate cyclase

On the basis of their thorough and revealing experiments on NG 108-15 cells, Sharma et al (1977) have put forward the hypothesis that "opiates, by inhibiting adenylate cyclase, alter the relative abundance of low- and high-activity forms of the enzyme". This wording seems to apply primarily to the catalytic unit of the enzyme, but if it is applied to the whole enzyme system, then the altered coupling hypothesis would presumably be included.

The hypothesis of Sharma et al comes close to the general theory of tolerance and dependence proposed by Shuster (1961) and by Goldstein & Goldstein (1961, 1968), known as the enzyme expansion theory. This theory, however, did not propose that one form of an enzyme replaced another.

Phosphodiesterase

There is good evidence that inhibitors of phosphodiesterase (PDE), such as methylxanthines, intensify withdrawal responses in morphine-dependent animals (Collier & Francis, 1975; Aceto & Harris, 1978). Moreover, methylxanthines and some other PDE inhibitors, such as ICI-63197 and RO-201724, induce in opioid-naive rats a pattern of behaviour closely resembling that of withdrawal in opioid-dependent animals (Collier et al, 1974; Francis et al, 1975; Collier et al 1978). The behaviour pattern induced by phosphodiesterase inhibitors in naive rats has been termed the quasi-morphine withdrawal (or abstinence) syndrome. This syndrome is intensified by naloxone and specifically and stereospecifically suppressed by opioids with a potency correlated with their analgesic potencies.

Whereas opioids readily suppress the behavioural excitation induced by methylxanthines, they quickly induce at the same time a heightened responsiveness to naloxone (Francis et al, 1976). Likewise, when isolated ileum is incubated with caffeine as well as morphine, greater tolerance is produced than with morphine alone (Hammond et al, 1976). These observations seem to have parallels in dependence itself. Thus, after challenge of morphine-dependent mice with naloxone, the responsiveness to naloxone declines, but this can be restored with a single dose of morphine (Way et al, 1976). Likewise, in the dependent ileum, normorphine quickly restores a fading responsiveness to naloxone (Schulz & Herz, 1976). These observations, coupled with those of Cox et al (1975) on the offset of tolerance, suggest a duality in the mechanism of opioid tolerance/dependence. There is a slow phase both in induction of dependence and likewise in the decline of tolerance/dependence after withdrawal. From the evidence of methylxanthines, this seems to correspond with the development (or decline) of an increased potential for forming cyclic AMP. There is also a faster phase, induced by opioids and revealed by rapid increases and decreases of responsiveness to naloxone.

It follows from these findings and from the analysis of opioid action embodied in Fig.1, that a lessening of PDE activity, through inhibition of enzyme, would provide another mechanism of dependence. We have not been able, however, to detect a difference between the high affinity phosphodiesterase activities of normal and dependent brains (N.M. Butt, H.O.J. Collier, N.J. Cuthbert, D.L. Francis, J.R. Lovely & S.A. Saeed, unpublished).

Protein kinase

Fig.1 indicates that an increased amount or activity of protein kinase would offer another mechanism of opioid dependence. Such a change has, in fact, been found by Clark et al (1972) in homogenate of dependent rat brain. If protein kinase is regarded as a receptor for cyclic AMP within the cell, then the increase of protein kinase in dependent brain may be interpreted as a special case of the theory that dependence arises from a drug-induced change in number (or efficiency) of receptors for an endogenous messenger substance (Collier, 1966,1972).

Transmitter release

An early theory of opioid tolerance and dependence is that this is due to accumulation of unreleased transmitter at the nerve terminal (Crossland & Slater, 1968; Paton, 1969). In this theory, termed the "surfeit" theory, tolerance would be due to increased leakage of transmitter, resulting from the overfilling of a reservoir within the terminal. Withdrawal effects would result from the release of the unusually large amount of accumulated transmitter. This theory remains a possible component of the molecular mechanism of dependence. If so, however, it seems likely, from experiments with specific antagonists in the whole animal, that several transmitters would be involved (Collier et al, 1972). Inhibition of transmitter release might also lead to post-synaptic super-sensitivity, which, too, could contribute to tolerance and dependence (Paton, 1969; Collier, 1969).

Kosterlitz & Hughes (1975) have proposed a converse of the surfeit theory, which would apply to this discussion if enkephalin were produced and liberated as a feed-back inhibitor by the opioid-sensitive neurone itself. They have suggested that exposure to exogenous opioid would lessen production of enkephalin by the cell and, on withdrawal of drug, a deprivation of enkephalin would result, which would be intensified by naloxone. At present there seems to be neither objection to this theory nor evidence for it.

Summary and Conclusions

Biochemical theories of opioid tolerance and dependence fall into two classes, according to whether they apply to events outside or inside (including on) the opioid-sensitive cell. That opioid tolerance and dependence can be induced in neuroblastoma x glioma cells in culture, in post-ganglionic neurones of the myenteric plexus of the guinea-pig ileum _in vitro_ and in single cortical neurones of the rat brain argues that this state can arise in the opioid-sensitive cell. Attention is therefore directed towards intra-cellular mechanisms of tolerance/dependence induction.

There is strong evidence that opioid tolerance/dependence expresses an increased potential or actual activity of a cyclic AMP mechanism of the cell. The involvement is therefore discussed of the following elements in the interaction of opioids with cyclic AMP -- receptors, receptor coupling mechanisms, adenylate cyclase, phosphodiesterase, protein kinase and transmitter release.

The evidence now available points to one or more of the following processes as likely to be involved in the induction of opioid tolerance and dependence in an opioid-sensitive cell.
1. A change, in favour of stimulation, in the coupling between the receptors for inhibitory and stimulant substances and the catalytic unit of adenylate cyclase.
2. A change, in favour of high activity, of the balance between high- and low-activity forms of adenylate cyclase.
3. An increase in the amount or activity of cyclic AMP dependent protein kinase. All of these mechanisms would represent an increase in the activity of the

cyclic AMP system of opioid-sensitive cells in response to the inhibition of adenylate cyclase by opioids.

Since the induction of tolerance/dependence appears to consist of two elements -- a slow change and a faster one that can only be expressed after the slow change has happened. There is thus room for more than one of these mechanisms to participate. Possibly the fast mechanism is of another type.

References

Aceto, M.D. & Harris, L.S. (1978). 3-Isobutyl-1-methylxanthine (IBMX) elicited withdrawal signs in morphine-dependent rhesus monkeys. Fedn.Proc., 37, 764 (Abs.2893).

Axelrod, J. (1956). Possible mechanism of tolerance to narcotic drugs. Science, 124, 263-264.

Clark, A.G., Jovic, R., Ornellas, M.R. & Weller, M. (1972). Brain microsomal protein kinase in the morphinized rat. Biochem. Pharmacol., 21, 1989-1990.

Collier, H.O.J. (1966). Tolerance, physical dependence and receptors. Adv.Drug. Res., 3, 171-188.

Collier, H.O.J. (1969). Humoral transmitters, supersensitivity and dependence. In Scientific Basis of Drug Dependence, edit. H. Steinberg, pp.49-66. Churchill, London.

Collier, H.O.J. (1972). Drug dependence: a pharmacological analysis. Br.J.Addict. 67, 277-286.

Collier, H.O.J. (1979). Consequences of interaction between opioid molecule and specific receptor. In Mechanisms of Pain and Analgesic Compounds, edit. R.F. Beers & E.G. Bassett, Raven, New York, in press.

Collier, H.O.J. & Francis, D.L. (1975). Morphine abstinence is associated with increased brain cyclic AMP. Nature (Lond.), 255, 159-162.

Collier, H.O.J. & Francis, D.L. (1978). A pharmacological analysis of opiate tolerance/dependence. In The Bases of Addiction, edit. J. Fishman. Dahlem Konferenzen, Berlin, in press.

Collier, H.O.J. & Roy, A.C. (1974a). Morphine-like drugs inhibit the stimulation by E prostaglandins of cyclic AMP formation by rat brain homogenate. Nature (Lond.), 248, 24-27.

Collier, H.O.J. & Roy, A.C. (1974b). Hypothesis: inhibition of E prostaglandin-sensitive adenyl cyclase as the mechanism of morphine analgesia. Prostaglandins, 7, 361-376.

Collier, H.O.J., Francis, D.L. & Schneider, C. (1972). Modification of morphine withdrawal by drugs interacting with humoral mechanisms: some contradictions and their interpretation. Nature (Lond.), 237, 220-223.

Collier, H.O.J., Francis, D.L., Henderson, G. & Schneider, C. (1974). Quasi-morphine-abstinence syndrome. Nature (Lond.), 249, 471-473.

Collier, H.O.J., Francis, D.L., McDonald-Gibson, W.J., Roy, A.C. & Saeed, S.A. (1975). Prostaglandins, cyclic AMP and the mechanism of opiate dependence. Life Sci., 17, 85-90.

Collier, H.O.J., Butt, N.M., Francis, D.L., Roy, A.C. & Schneider, C. (1978). Mechanism of opiate dependence elucidated by analysis of the interaction between opiates and methylxanthines. In Proc. 10th Congress Collegium Internationale Psychopharmacologicum, edit. P. Deniker, C. Radouco-Thomas &

A. Villeneuve, pp. 1331-1338. Pergamon, Oxford.

Cox, B.M., Ginsburg, M., Willis, J. & Davies, J.R. (1975). The offset of morphine tolerance in rats and mice. Br.J.Pharmac. 53, 383-391.

Crossland, J. & Slater, P. (1968). The effect of some drugs on the "free" and "bound" acetylcholine content of rat brain. Br.J.Pharmac. Chemother., 33, 42-47.

Cuatracasas, P. (1974). Membrane receptors. Ann.Rev.Biochem., 45, 169-214.

Cuatracasas, P. (1975). Hormone receptors -- their function in cell membranes and some problems related to methodology. Advan.Cyclic Nucleotide Res., 5, 79-104.

de Haën, C. (1976). The mon-stoichimetric floating receptor model for hormone sensitive adenylyl cyclase. J.Theor.Biol., 58, 383-400.

Francis, D.L., Roy, A.C. & Collier, H.O.J. (1975). Morphine abstinence and quasi-abstinence effects after phosphodiesterase inhibitors and naloxone. Life Sci., 16, 1901-1906.

Francis, D.L., Cuthbert, N.J., Dinneen, L.C., Schneider, C. & Collier, H.O.J.(1976). Methylxanthine-accelerated opiate dependence in the rat. In Opiates and Endogenous Opioid Peptides, edit. H.W. Kosterlitz. pp. 177-184. Elsevier/North Holland, Amsterdam.

Goldstein, A. & Schulz, R. (1973). Morphine-tolerant longitudinal muscle strip from guinea-pig ileum. Br.J.Pharmac., 48, 655-666.

Goldstein, D.B. & Goldstein, A. (1961). Possible role of enzyme inhibition and repression in drug tolerance and addiction. Biochem.Pharmacol., 8, 48.

Goldstein, D.B. & Goldstein, A. (1968). Enzyme expansion theory of drug tolerance and physical dependence. In The Addictive States, edit. A. Wikler, pp 265-267. Williams & Wilkins, Baltimore.

Hammond, M.D., Schneider, C. & Collier, H.O.J. (1976). Induction of opiate tolerance in isolated guinea-pig ileum and its modification by drugs. In Opiates and Endogenous Opioid Peptides edit. H.W. Kosterlitz, pp 169-176. Elsevier/North Holland, Amsterdam.

Havemann, U. & Kuschinsky, K. (1978). Interactions of opiates and prostaglandins E with regard to cyclic AMP in striatal tissue of rats in vitro. Naunyn-Schmiedebergs Arch.Pharmacol., 302, 103-106.

Himmelsbach, C.K. (1943). With reference to physical dependence. Fedn.Proc., 2, 201-203.

Hosein, E.A. & Lau, A. (1977). Adenyl cyclase activity during tolerance and dependence to morphine. Trans.Amer.Soc.Neurochem. 8, 83 (Abs. 39).

Jacquet, Y.F., Klee, W.A., Rice, K.C., Iijima, I. & Minamikawa, J. (1977). Stereospecific and non-stereospecific effects of (+) and (-) morphine: evidence for a new class of receptors? Science, 198, 842-845.

Klee, W.A. & Nirenberg, M. (1974). A neuroblastoma x glioma hybrid cell line with morphine receptors. Proc.Nat.Acad.Sci.U.S.A., 71, 3474-3477.

Kosterlitz, H.W. & Hughes, J. (1975). Some thoughts on the significance of enkephalin, the endogenous ligand. Life Sci, 17, 91-96.

Kosterlitz, H.W. & Watt, A.J. (1968). Kinetic parameters of narcotic agonists and antagonists, with particular reference to n-allylnoroxymorphine (naloxone). Br.J.Pharmac. Chemother., 33, 266-276.

Martin, W.R. (1967). Opioid antagonists. Pharmac.Rev., 19, 463-521.

Martin, W.R. (1968). A homeostatic and redundancy theory of tolerance to and dependence on narcotic analgesics. In The Addictive States, edit. A. Wikler, pp. 206-225. Williams & Wilkins, Baltimore.

Mehta, C.S. & Johnson, W. (1974). Elevation of brain adenosine 3':5' monophosphate during naloxone precipitated withdrawal in morphine dependent rats. Fedn.Proc,

33, 493 (Abs. 1594).

North, R.A. & Karras, P.J. (1978). Opiate tolerance and dependence induced in vitro in single myenteric neurones. Nature (Lond.), 272, 73-75.

Paton, W.D.M. (1969). A pharmacological approach to drug dependence and drug tolerance. In Scientific Basis of Drug Dependence, edit. H. Steinberg, pp. 31-41. Churchill, London.

Satoh, M., Zieglgänsberger, W. & Herz, A. (1976). Actions of opiates upon single unit activity in the cortex of naive and tolerant rats. Brain Res., 115, 99-110.

Sharma, S.K., Nirenberg, M. & Klee, W.A. (1975a). Morphine receptors as regulators of adenylate cyclase activity. Proc.Nat.Acad.Sci.U.S.A., 72, 590-594.

Sharma, S.K., Klee, W.A. & Nirenberg, M. (1975b). Dual regulation of adenylate cyclase accounts for narcotic dependence and tolerance. Proc.Nat.Acad.Sci. U.S.A., 72, 3092-3096.

Sharma, S.K., Klee, W.A. & Nirenberg, M. (1977). Opiate-dependent modulation of adenylate cyclase. Proc.Nat.Acad.Sci.U.S.A., 74, 3365-3369.

Schulz, R. & Goldstein, A. (1973). Morphine tolerance and supersensitivity to 5-hydroxytryptamine in the myenteric plexus of the guinea-pig. Nature (Lond.), 244, 168-170.

Schulz, R. & Herz, A. (1976). Aspects of opiate dependence in the myenteric plexus of the guinea-pig. Life Sci., 19, 1117-1128.

Schramm, M., Orly, J., Eimerl, S. & Korner, M. (1977). Coupling of hormone receptors to adenylate cyclase of different cells by cell fusion, Nature (Lond.), 268, 310-313.

Shuster, L. (1961). Repression and de-repression of enzyme synthesis as a possible explanation of some aspects of drug action. Nature (Lond.), 189, 314-315.

Snyder, S.H. (1975). VIII. A model of opiate receptor function with implications for a theory of addiction. In Opiate Receptor Mechanisms, edit. S.H. Snyder & S. Matthysse, pp. 137-141. MIT Press, Cambridge, Massachusetts.

Tatum, A.L., Seevers, M.H. & Collins, K.H. (1929). Morphine addiction and its physiological interpretation based on experimental evidences. J.Pharmac. Exp.Ther., 36, 447-475.

Traber, J., Fischer, K., Latzin, S. & Hamprecht, B. (1975a). Morphine antagonizes action of prostaglandin in neuroblastoma x glioma hybrid cells. Nature (Lond.), 253, 120-122.

Traber, J., Gullis, R. & Hamprecht, B. (1975b). Influence of opiates on the levels of adenosine 3':5'-cyclic monophosphate in neuroblastoma x glioma hybrid cells. Life Sci., 16, 1863-1868.

Villarreal, J.E., Martinez, J.N. & Castro, A. (1977). Validation of a new procedure to study narcotic dependence in the isolated guinea-pig ileum. Proc.Committee on Problems of Drug Dependence, Boston, 1977, in press.

Way, E.L., Brase, D.A., Iwamoto, E.T., Shen, J. & Loh, H.H. (1976). In Opiates and Endogenous Opioid Peptides, edit. H.W. Kosterlitz, pp. 311-318. Elsevier/ North Holland, Amsterdam.

OPIATE RECEPTORS AND ENDORPHINS IN OPIATE ADDICTION

V. Höllt, J. Bläsig, J. Dum, R. Przewłocki and A. Herz
Department of Neuropharmacology, Max-Planck-Institut für Psychiatrie,
Kraepelinstrasse 2, D-8000 München 40, G.F.R.

Introduction

The identification of opiate receptors five years ago (for review see Höllt and Wüster, 1978) opened an area of explosively rapid progress in the opiate field, culminating in the discovery of the naturally occurring endogenous ligands for these receptors - the so-called endorphins (for review see Teschemacher, 1978). Since then, speculation has arisen as to the possible roles of opiate receptors and endorphins in opiate addiction, a phenomenon characterized by tolerance, physical dependence and compulsive drug abuse. Opiate tolerance and dependence are generally thought to reflect the activity of adaptive processes responding to the acute effects of these drugs (Himmelsbach, 1943). Tolerance implies that repeated drug administration decreases the sensitivity to opiates, with the consequence that higher doses are required to elicit the same effect. Physical dependence refers to a state in which repeated administration of opiate is necessary to maintain an apparently normal functioning of the organism. This state being manifested by the withdrawal signs that occur when the continued exposure to the drug is interrupted. There is agreement that the adaptive changes of opiate tolerance and dependence take place within the nervous system itself, general alterations in metabolic disposition appear to be of minor, if any importance (for review see Kuschinski, 1977). The present report focus on two questions: firstly, are the adaptive changes observed after chronic exposure to opiates due to quantitative or qualitative changes of the opiate receptors? Secondly, is the development of opiate tolerance/dependence accompanied by alterations of the endorphin system(s)?

Opiate receptors in opiate addiction

On theoretical grounds, chronic treatment with opiates might be expected to cause a change of either the number or the affinity of the receptors for these drugs. To explain tolerance to narcotics one would most likely expect a decrease in the number of receptors and/or a decrease in the affinity of the receptors for the opiate agonists. In addition, to explain the increased efficacy of the opiate antagonists to elicit withdrawal during an increase in opiate tolerance/dependence, one might also expect an increased affinity of the receptors for opiate antagonists.

Both in vitro and in vivo approaches have been undertaken to evaluate differences in the properties of opiate binding between naive animals and animals chronically treated with opiates.

Binding studies in vitro

The most direct approach is to measure opiate receptor interaction by binding studies in vitro. Pert and Snyder (1973), Simon et al. (1973) and Terenius (1973) independently advanced a technique which made it possible to study opiate/receptor interactions under test tube conditions. By employing opiate ligands of high specific radioactivity and high receptor affinity these authors elegantly demonstrated a high affinity binding to a particulate fraction of animal brain. The fact that the binding affinities of a wide variety of opiates were closely related to their in vivo potency provided a strong evidence that this binding reflected binding to pharmacologically effective receptors. Receptor binding of an opiate is characterized by two main parameters: the number of sites and the affinity constant, both parameters which can be easily derived from a saturation analysis of the binding data. When interpreting such parameters, however, one has to keep in mind that receptors demonstrated in binding studies reflect the recognition function of the receptors only. Some investigators use the term "receptor" only to refer to the combination of the binding or recognition component with the component involved in the biological response, the so-called effector system. Numerous efforts have been undertaken to obtain evidence for an alteration of the recognition properties to opiate receptors during the development of opiate tolerance/dependence.

Table 1 summarizes the data reported in literature. As can be seen, the majority of the binding data show no change in either the affinity or the number of binding sites in the brain of rats or mice chronically treated with morphine. Several investigators found an increase in the receptor binding of opiate agonists and antagonists, which was exclusively due to an increase in the number but not in the affinity of opiate receptor sites. These findings are surprising since opiate tolerance/dependence might be more readily explained by a decrease rather than an increase in the number of opiate receptor sites. This increase in the number of specific opiate binding sites, however, was also observed as early as 5 minutes after acute injection of morphine, indicating that the time course of this effect does not parallel that for the development of opiate tolerance/dependence (Pert et al., 1973; Pert and Snyder, 1976). Moreover, an enhancement of opiate receptor binding was also found following the acute application of opiate antagonists, which do not induce opiate tolerance/dependence (Pert et al., 1973).

The mechanism by which this increase in opiate binding sites occurs remains unclear, although it appears to be unrelated to the phenomenon of tolerance/dependence. Pert and Snyder (1976) suggested that the increase in the number of receptor sites following the administration of opiates _in vivo_ might be due to displacement of endorphins from their receptors, which then remain more accessible for labelling by the tritiated opiates _in vitro_. Another possible explanation for this effect could be that the receptors occupied by opiates after administration _in vivo_ are less susceptible than unoccupied receptors to possible destruction caused during the homogenizing procedure carried out prior to the _in vitro_ binding studies. The fact that bound opiates can protect opiate receptors from destruction has been reported by Simon et al. (1975). These authors were able to solubilize an opiate-receptor complex but not free receptors.

A disadvantage inherent in all binding studies _in vitro_ is that possible alterations of the binding properties of opiate receptors which occur during opiate tolerance/dependence might be lost during tissue processing and remain undetected. In line with this suggestion are experiments reported by Davis et al. (1975). These investigators found a decreased affinity of morphine binding to receptors in brainstem slices from tolerant rats, but no change in binding to brainstem homogenates of the same animals (see Table 1). They suggested that the use of more intact tissue preparations, such as brain slices, might be more adequate to detect possible alterations of opiate receptor binding during opiate tolerance/dependence than the use of crude homogenates or subcellular fractions of brain tissue. Very recently, Davis et al. (1978) also reported a decrease in the affinity of receptors for the enkephalin analog d-Ala-enkephalin amide in brain slices of morphine tolerant rats. In contrast, no difference between naive and morphine tolerant/dependent rats with respect to number and binding sites of opiate receptors in rat brain slices has been reported by Huang and Takemori (1975). Takemori's group (Kitano and Takemori, 1977), however, recently found evidence for a decreased affinity for morphine and an increased affinity for naloxone of the receptors in striatal slices of tolerant mice. The authors performed experiments in which they found that naloxone was somewhat more effective in displacing morphine from the receptors in striatal slices of morphine tolerant mice than from slices of naive mice.

In view of the possibility that alterations of the binding properties of opiate receptors can be detected in intact tissues, it seems worthwhile to discuss the data which characterize receptor binding properties _in vivo_ in naive and tolerant animals.

In vivo approaches to characterize receptor binding

Procedures undertaken to compare the properties of opiate receptor binding in naive and tolerant/dependent animals in vivo have been necessarily indirect. Initially, the apparent pA_2 method was applied (Schild, 1957; Cox and Weinstock, 1964), a method which relies upon behavioral measurements to trace opiate receptor interactions. The pA_2 is calculated from the molar dose of an opiate antagonist necessary to reduce the potency of an opiate agonist by one half. The negative logarithm of this dose, the so-called pA_2 value, is assumed to be proportional to the receptor affinity constant of the antagonist. Using this method Takemori and co-workers (see Takemori, 1974) found an 8-fold increase in the apparent affinity constant of the antagonist naloxone in morphine tolerant/dependent mice. However, later attempts (Höllt et al., 1975; Dum et al., 1978, in preparation) have been unable to confirm these results. The authors (Dum et al., 1978) suggest that the apparent pA_2 values may appear increased in some studies because either measurements were made during withdrawal, which seems to interfere with the behavioral measurements, or because investigations in morphinized animals failed to consider the level of pretreatment morphine in the brains of the tolerant/dependent animals. Such morphine would artificially inflate the apparent pA_2 values.

A second approach to trace opiate receptor interaction in vivo relies upon the measurement of opiate concentrations in the brain of naive and tolerant/dependent animals. Shen and Way (1975a,b) observed a remarkable decrease in the morphine concentrations (measured by a radioimmunoassay) in the brain of chronically morphinized mice after the injection of naloxone. Since no changes in the morphine brain levels were observed when naloxone was injected into naive mice, the authors proposed that the result reflected an increased affinity of the receptors for the opiate antagonist, naloxone, or a decreased affinity for the agonist, morphine, in the tolerant/dependent mice. Although these findings are in agreement with the aforementioned in vitro data of Kitano and Takemori (1977), they could not be confirmed by other groups (Don Catlin et al., 1977; Dum et al., 1977). On the contrary, Don Catlin et al. (1977) found that naloxone tended to increase the brain concentrations of morphine (measured by a radioimmunoassay) when injected into tolerant/dependent mice. These authors hypothesized that this discrepancy might be due to antibody specificity and that the compound displaced in the Shen and Way study might not be morphine, but a chemically similar endogenous compound which has been recognized by the antibodies. Interestingly, such an endogenous compound which binds to morphine antibodies has recently been detected by Gintzler et al. (1976). Dum et al. (1977) also found no evidence for a decrease in the

TABLE 1

Animal	Morphine treatment	radio-labelled ligand	receptor preparation (brain)	tolerant versus naive binding affinity number of sites

	Animal	Morphine treatment		radio-labelled ligand	receptor preparation (brain)	binding	affinity	number of sites
Terenius et al., 1973	rats	20-35 mg/kg 20-35 mg/kg	4 d 17 d	dihydro-morphine dihydro-morphine	synaptic plasma membranes	= =		
Pert et al., 1973	mice	10-15 mg/kg 75 mg pellet	5-8 min 2-108 hrs	dihydro-morphine	washed homogenates	+ +	= =	+ +
Klee and Streaty, 1974	rats	75 mg pellet	72 hrs	dihydro-morphine	crude mito-chondrial fraction	=	=	=
Frederickson et al., 1974	rats	slow release suspension 300 mg/kg	36 hrs 36 hrs	dihydro-morphine naloxone	striatal homogenates washed	= +	= =	= +
Hitzemann et al., 1974	mice	75 mg pellet	24 48 72	naloxone	crude mito-chondrial fraction	= + =	= = =	= + =
Harris and Kazmierowski, 1975	rats	20 mg/kg 5-20 mg/kg	5 min 2 hrs - 28 d	naloxone	washed homogenates	+		
Höllt et al., 1975	rats	6x75 mg pellets	10 d	naloxone etorphine	washed homogenates	=	=	=
Davis et al., 1975	rats	2x75 mg pellets	72 hrs	morphine	brainstem homogenates brainstem slices	= -	= =	= -

Table 1 continued

Huang and Takemori, 1975	rats	2x75 mg pellets	72 hrs	etorphine	cerebral and striatal slices	=	=
Bonnet et al., 1976	rats	75 mg pellet	72 hrs	naltrexone	crude mitochondrial fraction of medial hypothalamus	=	=
Pert et al., 1976	mice	75 mg pellet	2–108 hrs	dihydromorphine	washed homogenates	+	=	+
Davis et al., 1978	rats	–		D-Ala-enkephalinamide	slices	–	–	=

Table 1: Opiate receptor binding studies in vitro.

= no change
+ increase in binding
– decrease in binding
during development of opiate tolerance/dependence.

brain concentrations of morphine after the injection of naloxone in chronically morphinized mice, using both radioimmunoassay and gas liquid chromatography for the measurement of morphine. They suggested that morphine might not be a suitable compound for measurement of opiate receptor interactions in vivo, since it has a low affinity for the receptors and would cause only a small fraction of the total drug to be bound to the receptors in the brain. A displacement of morphine from the receptors by naloxone would, therefore, probably be unable to significantly affect the total brain concentration of this drug.

Consistent with the above hypothesis is the fact that opiates which have a much higher receptor affinity than morphine can be displaced by antagonists from the brain of intact animals (Dobbs, 1968; Cerletti et al., 1974; Pert et al., 1975; Höllt et al., 1975). The principle of this technique is to inject very small doses of radiolabelled opiate into animals either alone or together with high doses of unlabelled opiates. The nonlabelled opiates displace the tracer opiate from its receptor sites in the brain, leading to a decrease in radioactive content, an effect most apparent in those areas of the brain that contain a high concentration of receptors (Höllt and Herz, 1978). Using this technique Höllt et al. (1975) studied the displacement of tracer amounts of tritiated naltrexone from the brains of naive and of morphine tolerant mice by increasing doses of nonlabelled naltrexone, a potent opiate antagonist. An analysis of the displacement data revealed no evidence for a substantial change in the number of opiate receptor sites in the brains of morphine tolerant/dependent mice. An apparent decrease in affinity for naltrexone in the brains of these animals was attributed to the residual morphine present following chronic morphinization. Pert et al. (1976) showed no change in the receptor binding of the potent antagonist diprenorphine in vivo in chronically morphinized rats. This problem has been reinvestigated recently, using a similar but improved method (Dum to be published). The new technique for measurement receptor binding in vivo (Höllt and Herz, 1978) uses the concentration of an opiate in the cerebellum as an "internal standard", since the cerebellum contains no opiate receptors. Figure 1 shows an experiment in which the increasing occupation of opiate receptors by the opiate antagonist is indicated by a dose dependent decrease in the brain/cerebellum ratio of ^{3}H-naloxone, when increasing doses of unlabelled naloxone are simultaneously applied. In naive mice the brain/cerebellum ratio declines from 2.8 to 1.3 when a tracer dose of ^{3}H-naloxone is injected together with increasing doses of unlabelled naloxone ranging from 0.01 to 1.0 mg/kg. The decrease of the ratio reflects the increasing displacement of ^{3}H-naloxone by unlabelled naloxone from specific receptors in the brain. At doses of about 1 mg/kg naloxone all

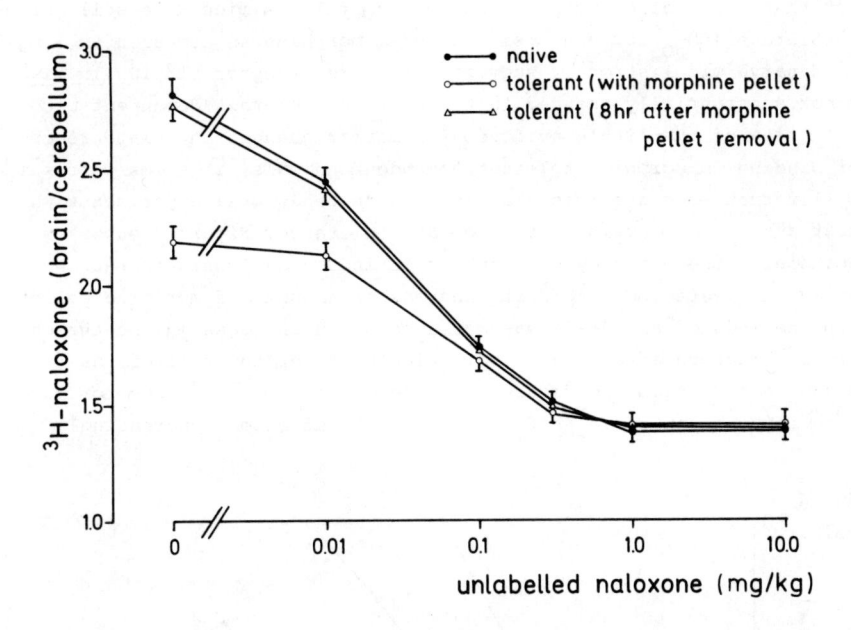

Fig. 1: Displacement of ^{3}H-naloxone by unlabelled naloxone in mouse brain. 20 μCi/kg (0.3 °g/kg) ^{3}H-naloxone was injected i.v. into mice alone or together with increasing doses of unlabelled naloxone. The animals were killed 20 min after injection and their radioactivity in the brain (without cerebellum) and in the cerebellum measured. Mice were rendered tolerant/dependent by the s.c. implantation of one pellet containing 75 mg morphine. Means ± S. E. of the means of 8 animals. The brain/cerebellum ratio is taken as a measure for receptor occupation (see text).

receptor sites are saturated, the higher dose of 10 mg/kg causing no further reduction in the brain/cerebellum ratio of ^{3}H-naloxone. In mice rendered morphine tolerant/dependent by the s.c. implantation of a morphine pellet containing 75 mg morphine base 72 hrs previously, the receptor binding of ^{3}H-naloxone was less, as indicated by the decrease of the brain/cerebellum ratio from 2.8 to 2.1 when the tracer dose of ^{3}H-naloxone was injected alone. Again, unlabelled naloxone was able to displace ^{3}H-naloxone in about the same dose range in naive and in tolerant animals. The decrease in the receptor binding of naloxone in the brains of morphine tolerant/dependent mice appears to be due to morphine released from the s.c. pellet competing with ^{3}H-naloxone for common receptor sites, since 8 hrs after pellet removal the dose response curve for the brain/cerebellum ratio of ^{3}H-naloxone is not longer significantly different to that seen in naive mice (see figure 1). The fact that 8 hrs after morphine pellet removal the mice were still highly tolerant (ED$_{50}$ for analgesia 80 mg/kg morphine as compared to 5 mg/kg in naive mice) strongly suggests that the receptor binding in vivo is not substantially changed in the morphine tolerant/dependent mice.

In view of the little evidence for marked changes in opiate receptor binding in morphine tolerant/dependent animals, it seems worthwhile to illustrate the dramatic changes seen in behavioral experiments which occur during the development of opiate tolerance. Figure 2 shows the decrease of the efficacy of morphine to induce analgesia in rats when the animals were implanted with increasing amounts of morphine pellets. Morphine induced analgesia was measured by an increase in the threshold for vocalization after electrical stimulation of the tail of the rats. It can be seen from the figure that the dose response curves for morphine induced analgesia shift to the right but become increasingly

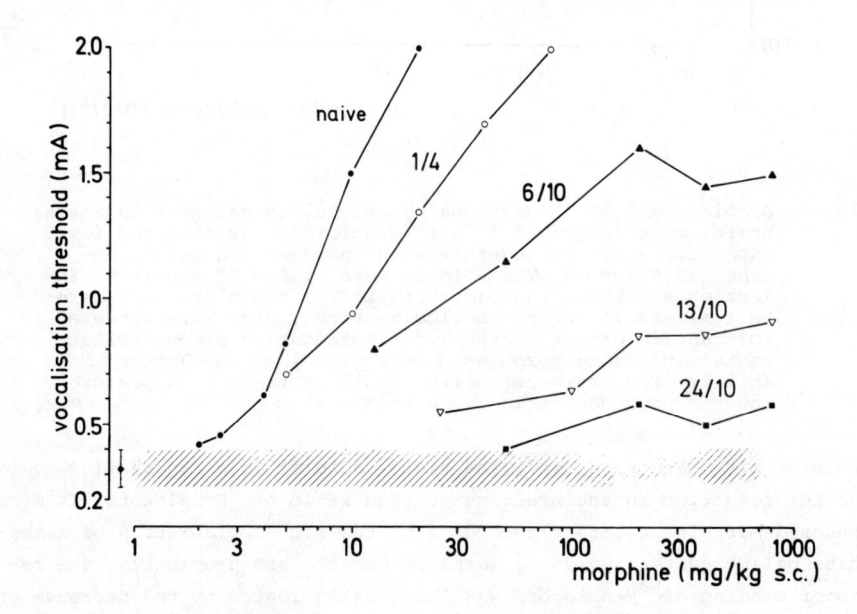

Fig. 2: Dose-response curves for morphine induced analgesia. Analgesia was measured 45 min after s.c. administration morphine in rats. The first number of the codes gives the number of pellets (75 mg morphine) implanted, the second number the duration of the implantation period in days. The ordinate indicates stimulation threshold in mA, the abscissa the dose of morphine in mg/kg. Each point represents the mean of at least 8 rats (for details see Bläsig et al., 1976).

flatter with increasing degrees of tolerance. In the highly tolerant animals morphine was no longer able to completely prevent the stimulation evoked vocalization response. Moreover, at morphine doses of 300 mg/kg the analgesic effect appears to be maximal, higher doses having no more effect. This finding cannot be explained solely by a decrease in the affinity of morphine receptors. In this case a parallel shift of the dose response curve to the right would have been expected. The decrease in the efficacy of morphine could be much more easily explained by a decrease in the number of opiate receptor sites with increasing degrees of tolerance. As discussed above, however, there is little evidence from either <u>in vitro</u> or <u>in vivo</u> binding studies that the number of opiate recognition sites changes during tolerance and dependence. It appears more likely that the decreased efficacy of morphine in morphine tolerant/dependent rats is due not to changes in the recognition function of opiate receptors, but reflects rather a desenzitation of their effector function.

Endorphins in opiate addiction

The recent discovery of the endorphins, the naturally occurring ligands for the opiate receptors, opened a series of questions concerning their involvement in opiate addiction. It has been proposed by Kosterlitz and Hughes (1975) that in opiate addiction the exogenously applied opiates increase the physiological effects which the endorphins have in naive animals. As a consequence, the organism becomes wholly dependent on the exogenous narcotics. A possible negative feedback from the constantly stimulated receptors may then result in a decreased rate of enkephalin synthesis.

Until now, the few experiments undertaken to test this working hypothesis rely predominantly upon concentration measurements of structurally identified endorphins (e.g. enkephalin, ß-endorphin) in different areas of brain and pituitary of naive and chronically morphinized rats.

Enkephalins

Methionine- and leucine-enkephalin, two pentapeptides, were the first endorphins to have been isolated and structurally identified by Hughes et al. (1975). The first attempt to investigate possible changes of the enkephalin brain concentration in morphine tolerant/dependent rats was performed by Simantov and Snyder (1976). In this study the enkephalins were measured by their ability to bind to opiate receptors <u>in vitro</u>, a method which is called radioreceptor assay. The authors found an approximately 100% increase in the enkephalin content of whole rat brain 5 days after the s.c. implantation of a 75 mg morphine pellet into the animals. This increase in enkephalin concentrations was reversed within 1 hour following the injection of naloxone. In order to explain their findings, the authors hypothesized that prolonged exposure to morphine might have led to a diminution of the enkephalin release via feedback regulation, thereby increasing the amount of stored enekphalin in the brain. The application of the opiate antagonist naloxone was thought to reverse this effect and restore brain enkephalin concentrations to normal.

Unfortunately, attempts to confirm these results were unsuccessful. Using a specific radioimmunoassay for determination of methionine-enkephalin, Fratta et al. (1977) found no alteration in the concentration of this peptide in striatum, hypothalamus and the remainder of the brain of morphine tolerant/dependent rats. Wesche et al. (1977) studied the distribution of immunoreactive methionine-enkephalin and leucine-enkephalin in various rat brain areas. These authors were also unable to find significant differences in the content of the enkephalins between naive and chronically morphinized rats. Similar data have now been obtained by Snyder's group (Childers et al., 1977): Using a specific radioimmunoassay for methionine- and leucine-enkephalin Childers et al. (1977) found no consistent differences between naive and chronically morphinized rats in respect to the enkephalin content in whole brain, striatum and hypothalamus of the animals.

There is present agreement that enkephalin concentrations in various brain areas are not markedly changed by chronic morphine administration. This, however, does not exclude the possibility that morphine tolerance/dependence is associated with changes in the turnover of the peptides. Presently, there exists only one preliminary report concerning the turnover of the enkephalins in morphine tolerant rats. Clouet and Ratner (1976) studied the incorporation of ^{3}H-glycine into enkephalins after intracysternal injection of the aminoacid into rats. The authors found a 50% decrease of the concentrations of enkephalins in the brain of morphine tolerant/dependent rats, which was accompanied by an increased specific radioactivity of the peptides, two hours after ^{3}H-glycine application, indicating an increased formation of enkephalins from the free aminoacid. The enkephalin levels reported by the authors, however, were more than 100-fold higher than those reported by others (Fratta et al., 1977; Wesche et al., 1977). It seems, therefore, likely that the changes in the enkephalin content and turnover found by the authors in chronically morphinized rats might be due to changes of other compounds which had been codetermined.

ß-Endorphin

The detection of endogenous opiate-like acting compounds in the pituitary (Teschemacher et al., 1975) and, in particular, the structurally identification of ß-endorphin, the 61-91 fragment of the pituitary peptide ß-lipotropin, raised many questions about the relationship between ß-endorphin and methionine-enkephalin, since the latter pentapeptide shares its full aminoacid sequence with the N-terminus of ß-endorphin. There appears to be now agreement that the enkephalins in the brain do not derive from the pituitary ß-endorphin, since hypophysectomy does not change the enkephalin content in several brain areas (Wesche et al., 1977; Kobayashi et al., 1978). The fact that ß-endorphin is predominantly localized in the pituitary (Guillemin et al., 1977a) and released into blood stream in response to stress (Guillemin et al., 1977b; Rossier et al., 1977; Höllt et al., 1978) suggests a hormonal function of this peptide. Recently we tested the influence of chronic opiate treatment on the ß-endorphin concentrations in pituitaries, plasma and hypothalamus of rats, measured by a radioimmunoassay (Höllt et al., 1978b). Table 2 shows some of the results. As can be seen, there is no significant alteration in the levels of ß-endorphin-like immunoreactivity in plasma and pituitary of rats rendered tolerant/dependent by the implantation of 6 morphine pellets within 10 days. In these animals, however, the administration of naloxone elicits a severe abstinence syndrome which is accompanied by a marked increase in plasma levels of ß-endorphin and by a significant decrease of ß-endorphin in the anterior lobe of the pituitary. It is difficult to distinguish whether this release of ß-endorphin from the adenohypophysis into blood is due to specific (receptor coupled) release mechanism or only due to a nonspecific stress effect of the morphine withdrawal. Whereas the chronic treatment with morphine for 10 days did not change pituitary concentrations of ß-endorphin, a long term treatment with morphine by implanting morphine pellets continuously for one month caused a significant reduction in the concentration of this peptide in the anterior and in the intermediate/posterior lobe of the pituitary (see Table 2).

Discussion

From the data discussed above, there is no evidence that the brain levels of the enkephalins are significantly altered in morphine tolerant/dependent animals. It might be possible, however, that the chronic administration of opiates can be accompanied by a change in the turnover of the pentapeptides. Until now a thorough analysis of the turnover of the enkephalins in morphine tolerant/dependent animals has not been reported.

The content of ß-endorphin in the pituitary also remained unchanged

ß-endorphin-like immunoreactivity

	plasma (ng/ml)	pituitary anterior lobe ng/mg	pituitary intermediate/ posterior lobe ng/mg
control (6 placebo pellets within 10 days) + saline	0.26 ± 0.15	80.4 ± 19.6	313.2 ± 106.5
chronic morphine (6 morphine pellets (75 mg) within 10 days) + saline	0.35 ± 0.16	89.9 ± 20.9	296.4 ± 54.6
withdrawal (6 morphine pellets (75 mg) within 10 days) + 10 mg/kg naloxone	1.05 ± 0.32	62.6 ± 8.8 *	345.8 ± 100.2 *
control (21 placebo pellets within 30 days)	0.19 ± 0.11	83.1 ± 5.2	486.5 ± 120.5
chronic morphine (21 morphine pellets (75 mg) within 30 days)	0.24 ± 0.13	21.6 ± 14.2 **	129.3 ± 30.0 **

* $p < 0.01$ withdrawal versus chronic morphine

** $p < 0.005$ chronic morphine versus control (placebo)

Table 2: Effect of chronic morphine treatment and withdrawal on the concentration of ß-endorphin in plasma and pituitaries of rats measured by a radioimmunoassay (for details see Höllt et al., 1978a,b). Mean ± S. D. of 6-8 rats. Saline + naloxone were administered i.p. 45 min before sacrifice. Morphine pellets contained 75 mg morphine and were s.c. implanted.

in rats chronically morphinized for ten days. Interestingly, however, long term treatment with morphine caused a marked decrease in the ß-endorphin concentration in the pituitary (Table 2). It is possible that long term exposure of opiate receptors to morphine leads to a decrease in the synthesis of ß-endorphin in the pituitary, an assumption which is in line with the aforementioned hypothesis of Kosterlitz and Hughes (1975). Perhaps the decreased content of ß-endorphin reflects some sort of deficiency of the ß-endorphin system in the pituitary. It has been suggested that such a deficiency of an endorphinergic system may play part in the protracted abstinence syndrome that can be observed long after the dramatic withdrawal signs have ceased (Goldstein, 1976). Moreover, a deficiency of an endorphinergic system after long term administration of opiates might at least partly explain the high rate of recidivism after abstinence in heroin addicts. In this context it is interesting to note that narcotic addiction has been assumed to be some sort of "metabolic" disease (Dole and Nyswander, 1967). It has been speculated that this metabolic disease might represent a deficiency of an endorphinergic system (Goldstein, 1976). It would be most interesting if such a deficiency of the ß-endorphin system in heroin addicts could be detected by a decreased ability of the patients to release ß-endorphin from the pituitary into the blood.

The fact that the decrease in the levels of ß-endorphin in the pituitary can only be detected after long term opiate treatment, indicates that this effect is not directly related to the marked adaptive changes responsible for opiate tolerance/dependence.

These adaptive changes also appear not to be due to changes in the binding function of opiate receptors, since the *in vitro* and *in vivo* binding studies outlined above provide no evidence for marked changes in the binding characteristics of opiates in chronically morphinized animals. This lack of a significant alteration of opiate receptor binding is surprising, since in other membrane receptor systems such changes in the binding function have been reported. For instance, a desentization in ß-adrenergic receptor activity is accompanied by a reduced number of the recognition sites of the ß-adrenergic receptors (Mukherjee et al., 1975). In addition, the supersensitivity of the dopamine receptors after chronic treatment with neuroleptics has been shown to be associated with an increase of the number of dopamine receptor sites (Burt et al., 1977). Since there are many similarities between different membrane receptor systems, a change of the binding function of opiate receptors after chronic exposure to opiates is not unlikely. Perhaps, our methods to measure opiate receptor binding *in vitro* and *in vivo* are not sensitive enough to detect subtle changes in the receptor binding in tolerant/dependent animals.

It seems rather questionable, however, whether such subtle changes in the binding properties can explain the marked tolerance of the receptors to the pharmacological action of opiates (see figure 2). Moreover, on theoretical grounds, an alteration of the binding function of opiate receptor can explain tolerance, but hardly dependence. To explain the adaptive mechanisms underlying the development of opiate tolerance/dependence in an unitary manner it has been proposed that opiates inhibit the action of an "endogenous substance" (e.g. a "neurohormone") in the central nervous system. Chronic treatment with opiates may result in an increased synthesis of the "neurohormone" (Shuster, 1961; Goldstein and Goldstein, 1961) or in an increase of the receptors for the "neurohormone" (Collier, 1966). Tolerance to the action of opiates develops because more opiates are now required to elicit the same effect. If opiates are withdrawn the "neurohormone" not inhibited by opiates causes the typical withdrawal response.

Putative candidates for such "endogenous substances" are several neurotransmitters. In fact, chronic opiate treatment leads to a supersensitivity for dopamine, noradrenaline and serotonin which becomes manifest during morphine withdrawal (Herz and Schulz, 1978).

Direct experimental evidence for the involvement of such an "endogenous substance" in the mechanism of opiate tolerance/dependence has been provided by experiments investigating the inhibition of cAMP formation by opiates in cultured neuroblastoma-glioma hybrid cells (Klee et al., 1975; Traber et al., 1975). It has been found that these cells continuously exposed to opiates show adaptation to the inhibitory action of morphine by increasing their adenylate cyclase activity. If the exposure of the cells to the opiates was terminated, the increased adenylate cyclase activity was unmasked and the cAMP content of the cells increased. This elevation of cAMP might be a biochemical correlate of the abstinence syndrome.

References

Bläsig, J., Höllt, V., Meyer, G. and Herz, A. (1976) Relationship between tolerance to morphine and occupation of specific binding sites by morphine. In: Opiates and Endogenous Opioid Peptides, H.W. Kosterlitz (ed.), Elsevier/North-Holland Biomedical Press, Amsterdam, pp. 391-394.

Bonnet, K.A., Hiller, J.M. and Simon, E.J. (1976) The effects of chronic opiate treatment and social isolation on opiate receptors in the rodent brain. In: Opiates and Endogenous Opioid Peptides, H.W. Kosterlitz (ed.), Elsevier/North-Holland Biomedical Press, Amsterdam, pp. 335-343.

Burt, D.R., Creese, J. and Snyder, S.H. (1977) Antischizophrenic drugs: Chronic treatment elevates dopamine receptor binding in brain, Science 196: 326-328.

Catlin, Don H., Liewen, M.B. and Schaeffer, I.C. (1977) Brain levels of morphine in mice following removal of a morphine pellet and naloxone challenge: No evidence for displacement, Life Sci. 20: 133-140.

Cerletti, C., Manara, L. and Mennini, T. (1974) Brain levels of the potent analgesic etorphine in rats and their functional significance, Brit. J. Pharmacol. 52: 440 P.

Childers, S.R., Simantov, R. and Snyder, S.H. (1977) Enkephalin: Radioimmunoassay and radioreceptor assay in morphine dependent rats, Europ. J. Pharmacol. 46: 289-293.

Clouet, D.H. and Ratner, M. (1976) The incorporation of ^{3}H-glycine into enkephalins in the brains of morphine treated rats. In: Opiates and Endogenous Opioid Peptides, H.W. Kosterlitz (ed.), Elsevier/North-Holland Biomedical Press, Amsterdam, pp. 335-343.

Collier, H.O.J. (1966) Tolerance, physical dependence and receptors, Advances in Drug Research 3: 171-188.

Cox, B.M. and Weinstock, M. (1964) Quantitative studies of the antagonism by nalorphine of some of the actions of morphine-like analgesic drugs, Brit. J. Pharmacol. 22: 289-300.

Davis, M.E., Akera, T. and Brody, T.M. (1975) Saturable binding of morphine to rat brain-stem slices and the effect of chronic morphine treatment, Res. Commun. Chem. Pathol. Pharmacol. 12: 409-418.

Davis, M.E., Brody, T.M. and Akera, T. (1978) Reduction of (D-ALA)^2-methionine enkephalinamide binding in morphine tolerant rats and differences of receptor binding between D-ALA and opiate alkaloids, Fed. Proceed. 37: 132.

Dobbs, H.E. (1968) Effect of cyprenorphine (M 285), a morphine antagonist, on the distribution and excretion of etorphine (M 99) a potent morphine-like drug, J. Pharmacol. exp. Ther. 169: 406-414.

Dole, V.P. and Nyswander, M. (1967) Heroin addiction - a metabolic disease, Arch. Intern. Med. 120: 19-24.

Dum, J., Meyer, G., Höllt, V., Herz, A. and Catlin, Don H. (1977) Inability of naloxone to change brain morphine levels in tolerant mice, Europ. J. Pharmacol. 46: 165-170.

Dum, J., Bläsig, J., Meyer, G. and Herz, A. (1978) Lack of alteration in the analgesic receptor-antagonist interaction during the development of tolerance to morphine, submitted to Europ. J. Pharmacol.

Fratta, W., Jang, H.Y., Hong, J. and Costa, E. (1977) Stability of met-enkephalin content in brain structures of morphine dependent or foot-shock-stressed rats, Nature 268: 452-453.

Frederickson, R.C.A., Horng, J.S., Burgis, V. and Wong, D.T. (1974) Alteration of opiate receptors in physically dependent rodents. In: Problems of Drug Dependence, National Academy of Sciences, Washington, pp. 411-434.

Gintzler, A.R., Leong, A. and Spector, S. (1976) Antibodies as a means of isolating and characterizing biologically active substances: Presence of a non-peptide, morphine-like compound in the central nervous system, Proc. Natl. Acad. Sci. USA 73: 2132-2136.

Goldstein, D.B. and Goldstein, A. (1961) Possible role of enzyme inhibition and repression in drug tolerance and addiction, Biochem. Pharmacol. 8: 48.

Goldstein, A. (1976) Opioid peptides (endorphins) in pituitary and brain, Science 193: 1081-1086.

Guillemin, R., Ling, N., Lazarus, L., Burgus, R., Minick, S., Bloom, F., Nicoll, R., Siggins, G., Segal, D. (1977a) The endorphins, novel peptides of brain and hypophyseal origin, with opiate-like activity: Biomedical and biological studies. In: ACTH and Related Peptides: Structure, Regulation and Action, D.T. Krieger and W.F. Young (eds.), The New York Academy of Sciences, New York, pp. 131-157.

Guillemin, R., Vargo, T., Rossier, J., Minick, S., Ling, N., Rivier, C., Vale, W., Bloom, F. (1977b) ß-Endorphin and adreno-corticotropin are secreted concomitantly by the pituitary gland, Science 197: 1367-1369.

Harris, J. and Kazmierowski, D.T. (1975) Morphine tolerance and naloxone receptor binding, Life Sci. 16: 79-84.

Herz, A. and Schulz, R. (1978) Changes in neuronal sensitivity during addictive processes. In: The Basis of Addiction, J. Fishman (Ed.), Dahlem Konferenzen, Berlin, in press.

Himmelsbach, C.K. (1943) Symposium: Can the europhoric, analgetic and physical dependence effects of drugs be separated? IV. With reference to physical dependence, Fed. Proceed. 2: 201-203.

Hitzemann, R.J., Hitzemann, B.A. and Loh, H.H. (1974) Binding of ions and tolerance development, Life Sci. 14: 2393-2404.

Höllt, V., Dum, J., Bläsig, J., Schubert, P. and Herz, A. (1975) Comparison of in vivo and in vitro parameters of opiate receptor binding in naive and tolerant/dependent rodents, Life Sci. 16: 1823-1828.

Höllt, V. and Herz, A. (1978) In vivo receptor occupation by opiates and correlation to the pharmacological effect, Fed. Proc. 37: 158-161.

Höllt, V. and Wüster, M. (1978) The opiate receptors. In: Developments in Opiate Research, A. Herz (ed.), Marcel Dekker, New York, pp. 1-65.

Höllt, V., Przewłocki, R. and Herz, A. (1978a) Radioimmunoassay of ß-endorphin: Basal and stimulated levels in extracted rat plasma, Naunyn-Schmiedeberg's Arch. Pharmacol., in press.

Höllt, V., Przewłocki, R. and Herz, A. (1978b) ß-Endorphin-like immunoreactivity in plasma, pituitaries and hypothalamus of rats following treatment with opiates, submitted to Life Sci.

Huang, J.T. and Takemori, A.E. (1975) Accumulation of etorphine by slices of cerebral cortex and corpus striatum of rats, Biochem. Pharmacol. 25: 47-51.

Hughes, J., Smith, T.W., Kosterlitz, H.W., Fothergill, L.A., Morgan, B.A. and Morris, H.R. (1975) Identification of two related pentapeptides from the brain with potent opiate agonist activity, Nature 258: 577-579.

Kitano, T. and Takemori, A.E. (1977) Enhanced affinity of opiate receptors for naloxone in striatal slices of morphine-dependent mice, Res. Commun. Chem. Pathol. Pharmacol. 18: 341-351.

Klee, W.A. and Streaty, R.A. (1974) Narcotic receptor sites in morphine-dependent rats, Nature 248: 61-63.

Klee, W.A., Sharma, S.K. and Nirenberg, M. (1975) Opiate receptors as regulators of adenylate cyclase, Life Sci. 16: 1869-1874.

Kobayashi, R.M., Palkovits, M., Miller, R.J., Chang, K.J. and Cuatrecasas, P. (1978) Brain enkephalin distribution is unaltered by hypophysectomy, Life Sci. 22: 527-550.

Kosterlitz, H.W. and Hughes, J. (1975) Some thoughts on the significance of enkephalin, the endogenous ligand, Life Sci. 17: 91-96.

Kuschinsky,K. (1977) Opiate dependence, Progr. in Pharmacol. Vol. 1: No. 2.

Mukherjee, C., Caron, M.G. and Lefkowitz, R.J. (1975) Catecholamine-induced subsensitivity of adenylate cyclase associated with loss of ß-adrenergic binding sites, Proc. Nat. Acad. Sci. USA 72: 1945-1949.

Pert, C.B. and Snyder, S.H. (1973) Opiate receptor: Demonstration in nervous tissue, Science 179: 1011-1014.

Pert, C.B., Pasternak, G.W. and Snyder, S.H. (1973) Opiate agonists and antagonists discriminated by receptor binding in brain, Science 182: 1359-1361.

Pert, C.B., Kuhar, M.J. and Snyder, S.H. (1975) Autoradiographic localization of the opiate receptor in rat brain, Life Sci. 16: 1849-1854.

Pert, C.B. and Snyder, S.H. (1976) Opiate receptor binding - enhancement by opiate administration in vivo, Biochem. Pharmacol. 25: 847-853.

Pert, C.B., Snyder, S.H. and Kuhar, M.J. (1976) Opiate receptor binding in intact animals. In: Tissue Responses to Addictive Drugs, D.H. Ford and D.H. Clouet (eds.), Spectrum Publications, Inc., New York, pp. 89-101.

Rossier, J., French, E.D., Rivier, C., Ling, N., Guillemin, R. and Bloom, F.E. (1977) Footshock induced stress increases ß-endorphin levels in blood but not brain, Nature 170: 618-620.

Schild, H.O. (1957) Drug antagonism and pA, Pharmacol. Rev. 9: 242-246.

Shen, J.W. and Way, E.L. (1975a) Displacement of morphine from the brain during antagonist precipitated abstinence, Comm. on Problems of Drug Dependence (NAS-NRC), 37th Ann. Scientific Meeting, pp. 635-644.

Shen, J.W. and Way, E.L. (1975b) Antagonist displacement of brain morphine during precipitated abstinence, Life Sci. 16: 1829-1830.

Simantov, R. and Snyder, S.H. (1976) Elevated levels of enkephalin in morphine-dependent rats, Nature 262: 505-507.

Simon, E.J., Hiller, J.M. and Edelman, J. (1973) Stereospecific binding of the potent narcotic analgesic ^{3}H-etorphine to rat brain homogenates, Proc. Natl. Acad. Sci. USA 70: 1947-1949.

Simon, E.J., Hiller, J.M. and Edelman, J. (1975) Solubilization of a stereospecific opiate-macro molecular complex from rat brain, Science 190: 389-390.

Shuster, L. (1961) Repression and de-repression of enzyme synthesis as a possible explanation of some aspects of drug action, Nature 189: 314-315.

Takemori, A.E. (1974) Biochemistry of drug dependence, Ann. Rev. Biochem. 43: 15-32.

Terenius, L. (1973) Stereospecific interaction between narcotic analgesics and a synaptic plasma membrane fraction of rat cerebral cortex, Acta pharmacol. et toxicol. 32: 317-320.

Teschemacher, H., Opheim, K.E., Cox, B.M. and Goldstein, A. (1975) A peptide-like substance from pituitary that acts like morphine. 1. Isolation, Life Sci. 16: 1771-1776.

Teschemacher, Hj. (1978) Endogenous ligands of opiate receptors (endorphins). In: Developments in Opiate Research, A. Herz (ed.), Marcel Dekker, New York, pp. 67-151.

Traber, J., Gullis, R. and Hamprecht, B. (1975) Influences of opiates on the levels of adenosine 3',5'-cyclic monophosphate in neuroblastoma x glioma hybrid cells, Life Sci. 16: 1863-1868.

Wesche, D., Höllt, V. and Herz, A. (1977) Radioimmunoassay of enkephalins. Regional distribution in rat brain after morphine treatment and hypophysectomy, Naunyn-Schmiedeberg's Arch. Pharmacol. 301: 79-82.

Acknowledgements

The authors thank Mrs. U. Bäuerle and Mrs. F. Sailer for excellent technical assistance and Dr. J. Fry for stilistic revision of the manuscript.

MEMBRANE CONSTITUENTS AND THE MECHANISMS OF MORPHINE ACTIONS

Horace H. Loh and Robert J. Hitzemann
Langley Porter Institute and
Departments of Pharmacology and Psychiatry
University of California
San Francisco, California 94143

Introduction

The past five years have witnessed exciting advances in our understanding of narcotic pharmacology. The characterization of the opiate receptor has been quickly followed by the isolation of several endogenous opioid peptides. However, despite these significant discoveries, we still do not understand the fundamental mechanisms of acute or chronic narcotic action. In the present article, we have attempted to summarize the roles various membrane constituents appear to play in narcotic effects in an attempt to provide a common foundation upon which future studies into the mechanisms of narcotic action can be constructed. Since pharmacological and biochemical evidence suggests that the acute and chronic effects of narcotics can be disassociated (Way, 1973, 1974; Takemori, 1975), we have accordingly divided our presentation into two sections, one dealing with the narcotic receptor and the second dealing with tolerance and dependence development.

The Narcotic Receptor

The narcotic receptor is a membrane-bound entity found only in the vertebrate nervous system. Given that invertebrates contain all putative and established neurotransmitters, Pert et al. (1974a) have suggested "that during the course of evolution the vertebrate nervous system acquires a qualitatively new type of synaptic function responsible for opiate receptor interactions." Within the vertebrate central nervous system, narcotic receptors show a unique regional distribution. The highest regional density of receptors is found in the limbic system, in particular the amygdala, the hypothalamus, the periaqueductal grey and the head of the caudate (Kuhar et al., 1973; Hiller et al., 1973). In discrete subcellular fractions both the microsomal and synaptic membrane fractions are particularly enriched in receptors (Pert et al., 1974b; Smith and Loh, 1976). On the basis of this bifractional distribution, Smith and Loh (1976) have suggested that stereospecific opiate receptors are distributed diffusely on the entire surface of the appropriate nerve cells and are not concentrated at the synaptic region as previously reported (Pert et al., 1974b).

The question of whether or not the receptor is located on the cytoplasmic or external surface of the membrane has not been definitively answered, but the available evidence supports an external localization. For example, Kosterlitz et al. (1975) have found that the quarternary antagonist, N-methyl-nalorphinium, which will not readily cross membranes, rapidly blocks the inhibitory effects of narcotic antagonists on evoked contractions of the longitudinal muscle of the guinea pig ileum. Similarly, N-β-(p-azidophenyl)-ethylnorlevorphanol is a pharmacologically active narcotic (Winter and Goldstein, 1972) which should not easily cross nerve membranes. Using a research strategy developed to localize proteins in the erythrocyte ghost membrane, we examined changes in stereospecific binding after mild tryptic digestion of intact and lysed nerve endings (Hitzemann and Loh, 1975). The data revealed that the opiate receptor in both intact and lysed nerve endings is equally accessible to tryptic digestion (figure 1). Interestingly, NaCl protected the receptor from digestion (figure 2), an effect which was not associated with a general decrease in tryptic activity (data not shown). The mechanism of this protective effect is unknown, but it is probably not related to the Na^+-induced receptor conversions described by Pert et al. (1973) and Simon et al. (1975), since significant conversions are observed at very low Na^+ concentrations. For example, Pert et al. (1973) have reported that 0.5 mM NaCl significantly increases opiate antagonist binding and that 5 mM Na^+ significantly inhibits opiate agonist binding.

The values obtained for the density of opiate receptors in the CNS have varied depending on whether or not high or low affinity binding was measured, on which subcellular fraction was used, on whether or not whole brain or discrete brain regions were used, and on the composition of the reaction buffer. Generally, though, it would appear that the maximum density is no greater than 1 pmole/mg of membrane protein (e.g., Hitzemann et al., 1974). Given this density and assuming the receptor has a molecular weight of at least 100,000, there would be no more than 0.1 µg of receptor/mg of membrane protein.

It is possible to obtain some idea of the physical nature of the cationic site within the receptor by studying the pH profile of the binding reaction. Pert and Snyder (1973) reported that [^{3}H]naloxone binding was maximal at a pH of 7.4; a sharp drop in binding activity was found at lower pH values and a more gradual drop in binding was found at higher pH values. We have observed similar results for the

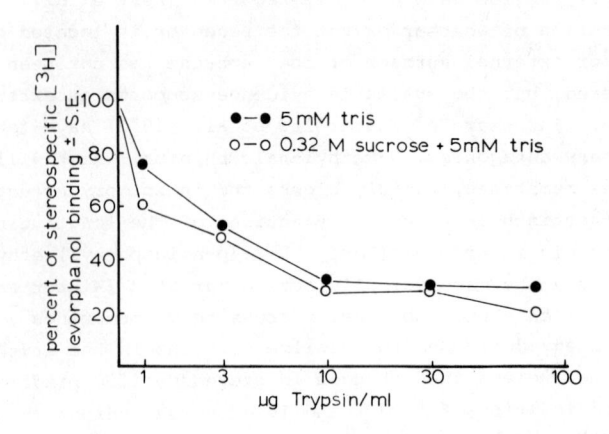

Figure 1. **Effect of various concentrations of trypsin on stereospecific [^{3}H]levorphanol binding to nerve ending membranes.** Nerve ending particles (NEP) were prepared from the whole brain essentially as described by Cotman and Matthews (1971). After harvesting, the NEP were suspended in either 5 mM tris (pH = 7.4 at 33° C), or 0.32 M sucrose plus 5 mM tris (pH = 7.4) at an approximate concentration of 1 mg/ml. The mixture was preincubated for 1 hour at 0-2° C followed by a 15 min incubation at 33° C prior to the addition of various amounts of trypsin (1 to 100 μg/ml final concentration). The incubation was continued for 15 min and then an equal amount of soybean trypsin inhibitor was added. For control NEP, trypsin (100 μg/ml) plus soybean inhibitor were added simultaneously. The mixture was cooled and then centrifuged for 10 min x 30,000 g. The NEP were then lysed for 1 hour in 5 mM tris, centrifuged for 10 min x 30,000 g, and resuspended in 5 mM tris, pH = 7.4 (at 33° C) at a concentration of 0.5 mg NEP protein/ml. Two ml of the mixture were preincubated for 5 min at 33° C before the addition of 2 x 10^{-8} M [^{3}H]levorphanol or [^{3}H]dextrorphan. The incubation was terminated after 15 min by filtering the reaction mixture over Whatman GFC filters under negative pressure, and then washing the filter 3 times with 5 ml of ice-cold tris buffer. Stereospecific [^{3}H]levorphanol binding (SsB) was defined as total [^{3}H]levorphanol bound minus total [^{3}H]dextrorphan bound. Data are presented as percent of control specific binding ± S.E. N = 5 experiments.

stereospecific binding of [^{3}H]levorphanol (data not shown). Cationic sites which can show such a pH profile are likely to contain a phosphate or sulfate group. It is unlikely that a carboxyl group could be involved. In this regard, it is of interest to note that neuraminidase treatment of brain membranes, which removes one source of potential carboxyl binding groups, has no effect on stereospecific binding (Pasternak and Snyder, 1974).

An argument can be developed that the receptor is, in part, proteinaceous. Trypsin, chymotrypsin and a number of protein modifying agents, e.g., N-ethylmaleimide or N-bromosuccinimide, markedly inhibit opiate receptor binding (Pasternak and Snyder, 1974; Pasternak et al., 1975). Interestingly, Pasternak et al. (1975) have observed that by using low concentrations of these agents it is possible to selectively destroy opiate agonist binding and leave antagonist binding largely intact. Since the protein modifying reagents differentiated best between agonist and antagonist binding when the incubations were conducted in the presence of sodium, it was concluded that these reagents interfere with the sodium-induced conversion of the opiate receptor from the agonist to antagonist state.

In addition to protein, the receptor appears to contain some lipid components. For example, phospholipases A and C inhibit receptor binding, suggesting that the receptor contains phospholipids (Pasternak and Snyder, 1974). In line with this hypothesis, it has been found that [^{3}H]morphine binds stereospecifically to phosphatidylserine (PS) (Abood and Hass, 1975) and that the incorporation of PS into either microsomal or synaptic membranes enhances both high and low affinity opiate receptor binding (Abood and Takeda, 1976). Similarly, Wu et al. (1977) demonstrated that opiate agonists and

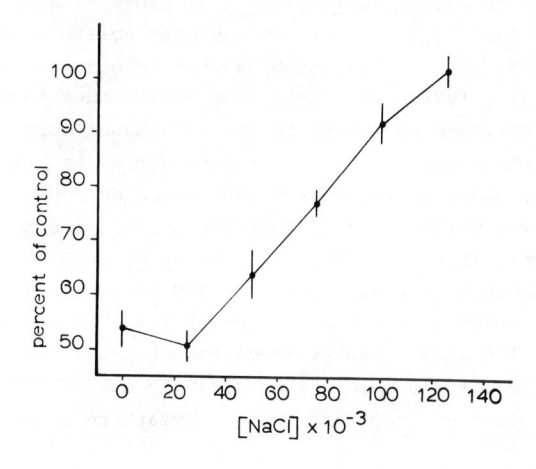

Figure 2. Effect of NaCl on the trypsin-induced inhibition of stereo-specific [^{3}H]levorphanol binding. NEP were incubated in the presence of 10 µg/ml trypsin in 0.32 M sucrose-tris (5 mM) buffer alone or in sucrose-tris buffer where the sucrose-tris was partially replaced by an isotonic amount of NaCl-tris (5 mM) buffer solution. The abscissa indicates the final NaCl concentration. After incubation with trypsin, the NEP were treated as described in the legend to Figure 1. The ordinate gives the percent of control SsB. All data are the mean ± S.E. of five experiments.

antagonists bind stereospecifically to triphosphatidylinositol (TPI). Interestingly, this binding to TPI showed a "Na-effect"; namely, that in the presence of NaCl [^{3}H]levorphanol binding was markedly inhibited while the binding of [^{3}H]naloxone was slightly enhanced.

In addition to phospholipids, research conducted in our laboratory suggests that one glycerolipid, namely, cerebroside sulfate (CS), may be involved in receptor function. Our original interest in CS was based on theoretical considerations, that CS should serve as an excellent model to study opiate-receptor interactions. Briefly, based on molecular models, CS appears to fulfill the structural requirements of the hypothetical opiate receptor as proposed by Beckett and Casey (1954); experimentally, it has been observed that the binding of narcotics to CS is saturable, stereospecific and the binding affinity of a large number of narcotics (> 30) to CS correlates well with their pharmacological potency (Loh et al., 1974, 1975; Cho et al., 1975). More recently we have proven that the partially purified opiate receptor from mouse brain by Lowney et al. (1974) is actually CS (Loh et al. 1975). Interestingly, we have also obtained several pieces of indirect evidence which support the idea that CS is involved in the pharmacologic action of narcotics in vivo: (a) Azure A, a dye which has a high affinity for CS, competitively inhibits opiate receptor binding and increases the AD$_{50}$ for morphine when injected intraventricularly (Law et al., 1978); (b) Jimpy mice, which have low cerebroside sulfate levels, are resistant to the effects of morphine and show a decrease in the number of opiate binding sites (Law et al., 1978). Furthermore, Azure A produced no inhibition of receptor binding in the mutant mice while a significant inhibition of binding was observed in the normal littermate control (Law et al., 1978). (c) The treatment of brain membranes with a purified and specific cerebroside sulfate sulfatase decreases receptor binding at least 50 percent (data in preparation). While these results (a-c) do not prove that CS is a part of the opiate receptor complex, they strongly suggest that CS or a related compound plays an important role in opiate receptor interactions.

There is a precedent for lipids serving important roles in receptor mediated functions. For example, Cutrecasas (1973) has shown that cholera toxin binds specifically to GM$_1$ ganglioside and this binding is responsible for stimulating adenyl cyclase activity. Since the number of GM$_1$ gangliosides is greatly in excess of the number of cyclase molecules, there is no correlation between GM$_1$ levels and the tissue responses. The response depends not only on the presence of receptors but also on the transduction between the receptor recognition site and the effector site. Such a situation may also be involved with regard to CS, since CS is ubiquitously distributed throughout the CNS but "functional" opiate receptors show a specific regional distribution.

Attempts to isolate and purify the opiate receptor have not, to date, been successful. Lowney et al. (1974) attempted to isolate the receptor using a procedure utilized by DeRobertis (1971) to isolate proteolipid binding substances for a number of neurotransmitters. A fraction was isolated using Sephadex LH-20 column chromatography which was able to bind opiates stereospecifically. By analogy with the earlier results of DeRobertis and based on some preliminary analytical determinations, Lowney et al. (1974) suggested the receptor material was a proteolipid. However, subsequent studies (Loh et al., 1975) have shown that the material isolated by Lowney et al. (1974) is identical with CS. Recently, Simon et al. (1975) have solubilized a [^{3}H]etorphine-macromolecular complex using the non-ionic detergent Brig 36T. These authors concluded that the solubilized bound complex has properties consistent with it being an etorphine receptor site complex. The binding exhibits high affinity and stereospecificity and is sensitive to proteolytic enzymes, a sulfhydral reagent and heat suggesting the presence of a protein. However, it has not been possible to demonstrate the binding of opiates to unbound complex isolated in the same way as the bound complex. This and other technical problems have hampered attempts at a more thorough characterization of the receptor.

In conclusion, while many features of the opiate receptor have been described, the question of whether or not the most essential portion of the receptor is protein or lipid has not been answered. Furthermore, we have no information as to how the binding site transmits information to an as yet undescribed effector site. Thus, the role(s) various membrane constituents play in receptor function remains enigmatic.

Narcotic Tolerance and Dependence

Numerous pharmacological studies (see Way, 1973) have amply demonstrated that there is some dissociation between the mechanisms responsible for tolerance and dependence and those responsible for acute narcotic effects. Furthermore, it has been found that there are no specific changes in membrane receptor binding, e.g., an increased affinity towards naloxone, which can be related to tolerance and/or dependence development (Klee and Streaty, 1974; Hitzemann et al., 1974). Therefore, it appears that the study of the membrane constituents involved in chronic narcotic effects will have to proceed with an interrelated but distinct approach from the study of the receptor complex. Given the difficulty that has been encountered in characterizing the receptor, this is perhaps a fortunate circumstance.

During the past 15 years numerous laboratories have shown that RNA or protein synthesis inhibitors block tolerance and dependence development (e.g., Cox and Osman, 1970; Cox, 1973; Loh et al., 1973; Smith et al., 1966; Spoerlein and Scrafini, 1967). These results are not unique for the narcotics. For example, we have found that cycloheximide blocks functional (central) barbiturate tolerance (Hitzemann and Loh, 1976a). However, the narcotic studies do strongly suggest that macromolecule synthesis plays an important, if not unique, role in tolerance and dependence phenomenon. In general, in vitro studies have supported this viewpoint. Several groups of investigators have reported that chronic morphinization stimulates cell-free RNA and protein synthesis (Clouet, 1971; Lee et al., 1973; Datta and Antopol, 1972; Castles et al., 1972). Unfortunately, these in vitro results have not been complemented by similar in vivo results. In fact, in their pioneering studies, Clouet and Ratner (1967) found that acute morphine administration inhibits brain protein synthesis and tolerance does not develop to this acute inhibitory effect.

Several years ago we hypothesized that specific increases in protein synthesis related to tolerance and dependence development may occur only in those brain regions and/or subcellular fractions associated with the antinociceptive actions of narcotics (Loh and Hitzemann, 1974). Our initial studies along these lines are illustrated in table 1. In these experiments we examined the effects of chronic morphinization on the turnover of [^{3}H]protein in six discrete brain regions and three subcellular fractions. The results illustrate at least two features of morphine's effects on protein synthesis. One, the nature of morphine's effects shows marked regional and subcellular differences. Two, in the brainstem and diencephalon, two brain regions associated with antinociception, chronic morphine treatment either inhibits, as in the mitochondrial fraction, or has no effect, as in the microsomal and soluble fraction, on protein turnover. Overall, these data have been interpreted to mean that there is no relevant general increase in protein synthesis associated with tolerance and dependence development. This interpretation was confirmed in two subsequent studies (Hitzemann and Loh, 1976b, 1977c) in which we examined the effects of chronic morphine treatment on the initial incorporation of [^{3}H]lysine into brain protein. Despite the fact that improved subcellular fractionation techniques were employed, we were unable to find a specific chronic morphine induced increase in protein

Table 1. Effect of Chronic Morphine Treatment on the Turnover of [³H]Protein [a]

T½ (Days)

Brain Region	Mitochondrial Control	Mitochondrial Morphine	Microsomes Control	Microsomes Morphine	Soluble Control	Soluble Morphine
Cortex	9	14*	11	30*	11	48*
Cerebellum	11	18*	12	5*	22	17*
Brain Stem	10	15*	18	19	17	19
Diencephalon	10	14*	12	11	11	34
Caudate Nucleus	9	13*	9	7	ND	ND
Hypothalamus	9	12*	12	7*	ND	ND

[a] Animals were given 20 µCi [³H]leucine i.c. Twice daily injections of morphine sulfate were begun 5 days later and continued for 15 days. Injections were begun at 10 mg/kg s.c. and doubled every 5 days. Animals were sacrificed at various times after beginning morphine injection and the specific activity of [³H]protein was determined in the crude mitochondrial, microsomal and soluble fractions prepared from various brain regions. Half-life values (T½) were determined by linear regression.

* Significantly different from control, $p < 0.05$.

ND = Not determined.

synthesis in any subcellular fraction including three different synaptic membrane and synaptic soluble fractions.

Various theories attempting to explain narcotic tolerance and dependence have, in general, suggested that chronic morphinization <u>specifically</u> increases the synthesis of an important regulatory protein, such as a neurotransmitter receptor or biosynthetic enzyme (Shuster, 1961; Goldstein and Goldstein, 1961; Collier, 1965). In an attempt to detect this protein, Hahn and Goldstein (1971) examined the effects of chronic morphine treatment on the incorporation of leucine into various brain protein subunits, as separated by gel electrophoresis. No significant change was observed. Further studies along this line were conducted by Franklin and Cox (1972). These authors examined the effects of chronic morphinization on labeled lysine incorporation into individual synaptic plasma membrane (SPM) proteins. The SPM were prepared using only subcortical brain regions which were known to be involved in morphine's acute effects. However, these authors, like Hahn and Goldstein, (1971), were unable to find any significant specific change in protein synthesis. We hypothesized that perhaps the reason Franklin and Cox (1972) were unsuccessful in their study was because the narcotic effect or effects were only associated with a specific population of SPM. With this idea in mind, we examined the effects of chronic morphine treatment on [^{3}H]lysine incorporation into the proteins of two unique populations of SPM (Hitzemann and Loh, 1977c). It was found that morphine increased the accumulation of labeled protein in the high molecular weight region of SPM-L SDS gels. The SPM-L are SPM prepared from a light (L) population of nerve ending particles (NEP) (see Hitzemann and Loh, 1976a; 1977a). These effects in the SPM-L were dependent on the number of morphine pellets implanted (figure 3) and were not found in any other subcellular fraction. In fact, in the SPM-H gels, the high molecular weight region showed an effect opposite to that seen in the SPM-L gels. Possibly this accounts for the reason why Franklin and Cox (1972), using a more heterogenous population of SPM, were unable to detect a specific change in labeled protein accumulation.

Our attempts to characterize the protein(s) influenced by chronic morphine treatment have been hampered by the fact that the proteins involved are minor SPM constituents (< 1% of the total SPM protein). The proteins affected co-migrate with the high molecular weight doublet found in all brain fractions. The doublet has a molecular weight of approximately 700,000, as determined on Pharmacia 4-30 SDS gradient slab gels and is PAS positive.

Since the NEP-L from which the SPM-L are derived are markedly enriched in GABA containing NEP (data not shown), we decided to

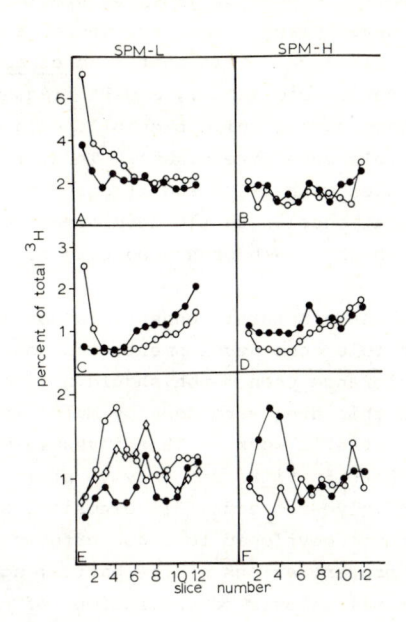

Figure 3. **Effect of chronic morphine treatment on the levels of [³H]protein in the high molecular weight region of SPM-SDS polyacrylamine gels.** Rats were implanted with 1 (◇--◇) or 2 (O--O) morphine pellets or 1 placebo (●--●) pellet 24 hours prior to the intraventricular injection of 20 µCi of [³H]lysine. The animals were sacrificed 24 hours later and synaptic plasma membranes (SPM) were prepared from light (L) and heavy (H) subcortical nerve ending particles. The labeled proteins of the SPM-L and SPM-H were then separated by conventional SDS gel electrophoresis techniques (7% gel, 0.8 x 25 cm). In panels A through D the gels were run at a pH of 7.2 and in panels E and F at a pH of 9.5. After electrophoresis, the gels were sliced (3 mm slices) and counted. No significant changes in label distribution were observed in slices 12 to 80 (see Hitzemann and Loh, 1977). However, in slices 1-8, or the high molecular weight region of the gel (Shapiro et al., 1967), significant changes were found and these data are presented in the figure. The data panels A-B, C-D and E-F were, respectively, obtained from three different experiments. Data are presented as the percent of total [³H] appearing in the gel.

investigate the effects chronic morphinization has on various functional GABA parameters. We have been unable to find any specific change in GABA transport, release or receptor binding that could be related to tolerance development. However, there is evidence that GABA plays a role in tolerance development. Ho et al. (1974) have found that amino-oxyacetic acid (AOAA), a GABA transaminase inhibitor, increases tolerance development while bicuculline, a GABA receptor antagonist, inhibits tolerance development. Thus, it would appear that GABA can be included with c-AMP and tryptophan as substances that accelerate the development of tolerance. The precise mechanism by which GABA neurons regulate tolerance development is, however, unknown.

While there has been a natural tendency to concentrate on the role proteins play in tolerance development, the possible role of membrane lipids in tolerance phenomenon should not be ignored. The pioneering studies in this area were done by Mule´ (see Mule,́ 1971, for a review of this author's work). This investigator found that, in general, morphine both in vitro and in vivo stimulated the incorporation of ^{32}Pi, [^{14}C]glycerol and [^{3}H]myoinositol into cortical phospholipids. Tolerance developed to these effects and the effects were blocked by high concentrations of the partial agonist-antagonist, nalorphine. Interestingly, lower concentrations of nalorphine than those required to block morphine's effects, mimicked some of the effects of morphine on phospholipid synthesis. Mule´ also observed that morphine affected phosphatidylcholine (PC) turnover in a manner somewhat different than the other phospholipids. For example, morphine inhibited [^{14}C]glycerol incorporation into PC while stimulating incorporation into other phospholipids. Morphine also inhibited [^{14}C]choline incorporation into PC.

In an attempt to further characterize the effects of morphine on PC turnover, we examined the drug effects on the accumulation and turnover of [^{14}C]choline-PC in discrete brain region and subcellular fractions (Loh and Hitzemann, 1974). As was observed by Mule,́ we found that morphine inhibited [^{14}C]choline incorporation into PC in the cortex (table 2). However, in the brainstem, diencephalon and hypothalamus morphine stimulated [^{14}C]choline incorporation. Interestingly, these drug effects persisted in chronically morphinized animals. The turnover of mitochondrial [^{14}C]PC was inhibited in the cortex but stimulated in the diencephalon, brainstem and hypothalamus, three brain regions known to play important roles in acute morphine effects (table 3). PC is thought to play an important role in regulating the activity of a number of membrane bound enzymes and through LPC in regulating neurotransmitter release (DaPrada et al., 1972).

Table 2. Effect of Morphine on the Synthesis of [^{14}C]Phosphatidylcholine.[a]

Brain Region	Treatment	[^{14}C]Choline (cpm/g ± S.E. x 10^{-3})	[^{14}C]Phosphatidylcholine (cpm/μmole lipid P ± S.E.)
Cortex	C	12 ± 4	26 ± 4
	M	15 ± 3	11 ± 2*
Cerebellum	C	104 ± 20	187 ± 28
	M	89 ± 13	99 ± 28*
Hypothalamus	C	8 ± 2	39 ± 4
	M	6 ± 2	75 ± 17*
Caudate nucleus	C	7 ± 1	10 ± 3
	M	8 ± 2	11 ± 4
Brain stem	C	114 ± 16	35 ± 4
	M	51 ± 7*	76 ± 10*
Diencephalon	C	36 ± 5	6 ± 2
	M	36 ± 7	24 ± 7*

[a] Animals were given 40 mg/kg of morphine sulfate s.c. 0.5 hr prior to the intracisternal injection of 5 μCi [^{14}C]choline; 0.5 hr later the animals were sacrificed and the levels of [^{14}C]choline and [^{14}C]phosphatidylcholine were determined in discrete brain regions. N equals six animals/determination.

*   Significantly different than control, $p < 0.05$.

   C = control; M = morphine

The possibility that morphine may influence LPC levels in synaptic vesicles is intriguing, since it is well known that morphine inhibits neurotransmitter release and an increase in LPC levels may antagonize this effect.

One mechanism by which choline as well as serine and ethanolamine can be incorporated into PL is through the base-exchange reaction. Base-exchange is a Ca^{++} dependent, ATP independent process which has been found to occur both in vitro and in vivo, to occur at a greater rate in neuronal as compared to glial cells and to be the sole mechanism for the synthesis of PS in the brain (Arienti et al., 1976; Kafner, 1972; Gaiti et al., 1974; Goracci et al., 1973). We decided to investigate the effects of morphine on base-exchange for two reasons (see Natsuki et al., 1978). One, as was found in both our experiments and those of Mule, morphine has some unique effects on the turnover of the choline moiety of PC which could be related to a change in base-exchange. Two, since morphine is known to affect brain Ca^{++} levels (Cardenas and Ross, 1975), it is possible that chronic morphinization could affect the highly Ca^{++} dependent base-exchange reaction and, thus, modify membrane composition. The data in figure 4 illustrate that chronic morphine treatment significantly stimulated the incorporation of [^{14}C]serine into microsomal phospholipids at all Ca^{++} concentrations tested. In contrast to these results, chronic morphine treatment increased [^{14}C]ethanolamine incorporation at only one Ca^{++} concentration (2.5 mM) (figure 5) and markedly inhibited [^{14}C]choline incorporation at all Ca^{++} concentrations (figure 6). Acute morphine treatment or the in vitro addition of morphine was not found to mimic the chronic drug effects.

Table 3. Effect of Chronic Morphine Treatment on the Turnover of [^{14}C]Phosphatidylcholine [a]

Brain Region	$T_{½}$ (Days) Mitochondrial Control	Morphine
Cortex	22	90*
Cerebellum	14	11
Brain stem	13	8*
Diencephalon	17	11*
Caudate nucleus	26	20
Hypothalamus	18	7*

a. Animals were given 5 µCi [^{14}C]choline i.c. Twice daily injections of morphine sulfate were begun 5 days later and continued for 15 days. Injections were began at 10 mg/kg s.c. and doubled every 5 days. Animals were sacrificed at various times after beginning morphine injection and the specific activity of [^{14}C]phosphatidylcholine was determined in the crude mitochondrial fraction prepared from various brain regions. Half-life values ($T_{½}$) were determined by linear regression.

* Significantly different from control, $p < 0.05$.

The results of the base-exchange studies suggest that the mechanism by which chronic morphine treatment increases choline incorporation in vivo does not involve an increase in exchange activity. Secondly, the results suggest that PS levels or turnover may increase during tolerance development. These data are of particular interest since, as previously discussed, PS may play an important role in opiate receptor function (Abood and Hass, 1975; Takeda and Abood, 1976).

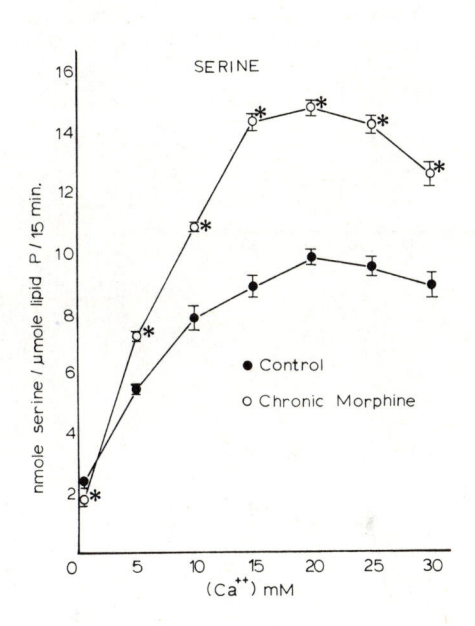

Figure 4. Effects of chronic morphine treatment on the incorporation of [^{14}C]serine into brain microsomal phospholipids via base-exchange. Male Sprague-Dawley rats (180-200 grams) were administered 10 mg/kg of morphine·sulfate or saline between 4 and 5 p.m. The next day the animals were implanted with one 75 mg morphine pellet at 8 a.m., one pellet at 12 p.m. and two pellets at 4 p.m. Control animals received placebo pellets. Forty-eight hours after the first pellet implantation the animals were killed by decapitation and their brains removed and processed immediately. Microsomes were prepared from the brains essentially as described by Gaiti et al. (1974). Base-exchange was measured using saturating concentrations of [^{14}C]serine (1.5 mM) or other substrates, as described by Porcellati and colleagues (Arienti et al., 1976; Gaiti et al., 1974). The concentrations of Ca^{++} appearing in the graph are the final Ca^{++} concentrations in the reaction mixture.

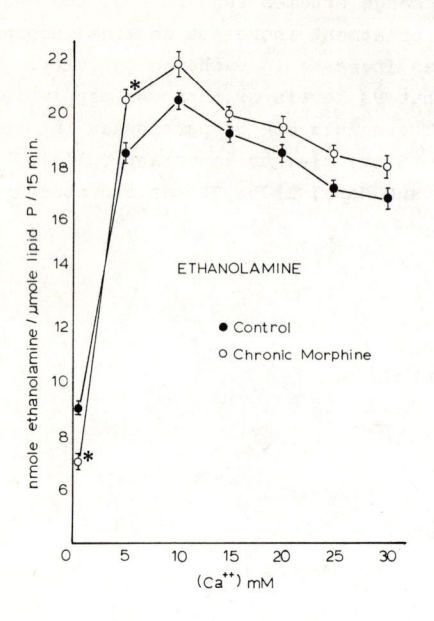

Figure 5. Effects of morphine treatment on the incorporation of [^{14}C]ethanolamine into brain microsomal phospholipids via base-exchange. Details are the same as in the legend to figure 6 except that [^{14}C]ethanolamine (1.7 mM) was used.

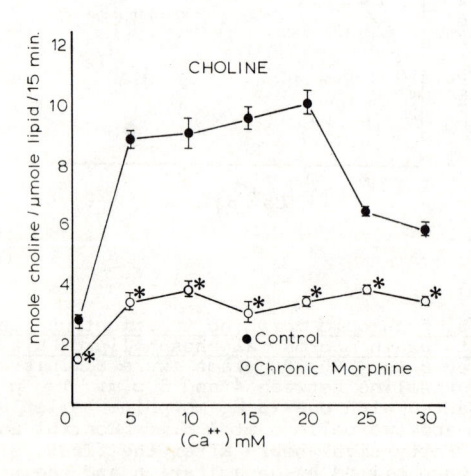

Figure 6. Effects of chronic morphine treatment on the incorporation of [^{14}C]choline into brain microsomal phospholipids via base-exchange. Details are the same as in the legend to figure 5 except that [^{14}C] choline (4.5 mM was used.

Conclusions

In this article we have examined the roles various membrane constituents, notably proteins, phospholipids and glycolipids, may play in the mechanisms of morphine action. Unfortunately, the "state of the art" is not such that firm conclusions can be drawn. Instead, we will only summarize the seemingly most important experimental results and offer their use for futher dialogue and investigation. These results are as follows.

1. The narcotic receptor is an entity unique to the vertebrate nervous system which is diffusely distributed on the external surface of neuronal plasma membranes.
2. The receptor complex appears to have both lipid and protein components. The protein component has only been indirectly characterized in terms of its reactivity to various protein modifying reagents. Three membrane lipids, namely, phosphatidylserine, triphosphatidylinositol and cerebroside sulfate, have been shown in various systems to bind narcotics with properties similar to that of the membrane receptor. To date, a possible role for cerebroside sulfate in the receptor complex has been most clearly demonstrated.
3. Narcotic tolerance and dependence development depends on the brain protein synthesis machinery being intact. However, it has been difficult to demonstrate that chronic morphine treatment induces a qualitative or quantitative change in a protein which could be associated with tolerance and dependence. In one study it was found that chronic morphine treatment increased the apparent synthesis of a high molecular weight synaptic protein (or proteins) whose function is unknown.
4. Chronic morphine treatment has marked effects on brain phospholipid synthesis. Especially interesting effects have been noted in regard to the phospholipid base-exchange reaction. Chronic morphine treatment increases serine exchange into phospholipids but decreases choline exchange.

Overall, it would appear that both the acute and chronic actions of narcotics involve the dynamic interplay between lipids and proteins. For this reason, we conclude that a pluralistic rather than a monistic theory and investigation in terms of membrane components will be required to determine the mechanisms of drug action.

Acknowledgements. The work described in this article was supported in part by grant DA-00564 and DA-01583. H. H. Loh is the recipient of Career Research Scientist Development Award K2-DA-70554. The authors wish to thank Kaye Welch for assistance in preparing the manuscript.

References

Abood, L.G. and Hass, W. (1975) Stereospecific morphine adsorption to phosphatidylserine and other membranous components of brain. Eur. J. Pharmacol. 32:66-75.

Abood, L.G. and Takeda, F. (1976) Enhancement of stereospecific opiate binding to neural membranes by phosphatidylserine. Eur. J. Pharmacol. 39:71-77.

Arienti, G., Brunetti, M., Gaiti, A., Orland, P. and Porcellati, G. (1976) Base-exchange of brain phospholipids. Adv. Exp. Med. Biol. 72:63-78.

Beckett, A. H. and Casey, A. F. (1954) Synthetic analgesics: Stereochemical considerations. J. Pharm. Pharmac. 6:986-1001.

Cardenas, H. L. and Ross, D. H. (1975) Morphine-induced calcium depletion in discrete regions of rat brain. J. Neurochem. 24: 487-493.

Castles, T. R., Campbell, S., Gouge, R. and Lee, C. C. (1975) Nucleic acid synthesis in brains from rats tolerance to morphine analgesia. J. Pharmacol. Exp. Ther. 181:399-406.

Cho, T. M., Cho, J. S. and Loh, H. H. (1976) A model system for opiate-receptor interactions: mechanism of opiate-cerebroside sulfate interaction. Life Sci. 18:231-244.

Clouet, D. H. (1970) The effects of drugs on protein sythesis in the nervous system. In Protein Metabolism of the Nervous System, edited by A. Lajtha. Plenum Press, New York.

Clouet, D. H. (1971) Protein and nucleic acid metabolism. In Narcotic Drugs: Biochemical Pharmacology, edited by D. H. Clouet. Plenum Press, New York, pp. 216-228.

Clouet, D. H. and Ratner, M. (1970) Catecholamine biosynthesis in brains of rats treated with morphine. Science 168:854-856.

Clouet, D. H. and Ratner, M. (1967) The effect of the administration of morphine on the incorporation of (^{14}C)-leucine into proteins of the rat brain in vivo. Brain Res. 4:33-43.

Collier, H. O. J. (1965) A general theory of the genesis of drug dependence by induction of receptors. Nature 205:181-182.

Cotman, C. W. (1974) Isolation of synaptosomal and synaptic plasma membrane fractions. Methods in Enzymology 31:445-452.

Cotman, C. S. and Matthews, D. A. (1971) Synaptic plasma membranes from rat brain synaptosomes: isolation and partial characterization. Biochim. Biophys. Acta 249:398-394.

Cox, B. M. (1973) Effects of inhibitors of protein synthesis in morphine tolerance and dependence. In Agonist and Antagonist Actions of Narcotic Analgesic Drugs, edited by H. W. Kosterlitz, H. O. J. Collier and J. E. Villarreal. University Park Press, Baltimore.

Cox, B. M. and Osman, O. H. (1970) Inhibition of the development of tolerance to morphine in rats by drugs which inhibit ribonucleic acid or protein synthesis. Br. J. Pharmacol. 38:157-170.

Cuatrecasas, P. (1973) Gangliosides and membrane receptors for cholera toxin. Biochemistry 12:3558-3566.

DaPrada, M., Pletscher, A. and Tranzer, J. P. (1972) Lipid composition of membranes of amino-storage organelles. Biochem. J. 127:681-683.

Datta, R. K. and Antapol, W. (1972) Inhibitory effects of chronic administration of morphine on uridine and thymidine incorporating abilities of mouse liver and brain subcellular fractions. Tox. Appl. Pharmacol. 23:75-81.

DeRobertis, E. (1971) Molecular biology of synaptic receptors. Science 171:963-971.

Franklin, G. I. and Cox, B. M. (1972) Incorporation of amino acids into proteins of synaptosomal membrane during morphine treatment. J. Neurochem. 19:1821-1823.

Gaiti, A., DeMedio, G. E., Brunetti, M., Amaducci, L. and Porcellati, G. (1974) Properties and function of the calcium-dependent incorporation of choline, ethanolamine and serine into the phospholipids of isolated rat brain microsomes. J. Neurochem. 23:1153-1159.

Gaiti, A., Goracci, G., DeMedio, G. E. and Porcellati, G. (1972) Enzymic synthesis of plasmalogen and O-alkyl glycerolipid by base-exchange in the rat brain. FEBS Lett. 27:116-120.

Goldstein, D. B. and Goldstein, A. (1961) Possible role of enzyme inhibition and repression in drug tolerance and addiction. Biochem. Pharmacol. 8:48-52.

Goracci, G., Blomstrand, C., Arienti, G., Hamberger, A. and Porcellati, G. (1973) Base-exchange enzymic system for the synthesis of phospholipids in neuronal and glial cells and their subfractions: a possible marker for neuronal membranes. J. Neurochem. 20: 1167-1180.

Hahn, B. and Goldstein, A. (1971) Amounts and turnover rates of brain proteins in morphine tolerant mice. J. Neurochem. 18: 1887-1893.

Hiller, J. M., Peasson, J. and Simon, E. J. (1973) Distribution of stereospecific binding of the potent narcotic analgesic etorphine in the human brain: predominance in the limbic system. Res. Commun. Chem. Pathol. Pharmacol. 6:1052-1063.

Hitzemann, R. J., Hitzemann, B. A. and Loh, H. H. (1974) Binding of [^{3}H]naloxone in the mouse brain: effects of ions and tolerance development. Life Sci. 14:2293-2304.

Hitzemann, R. J. and Loh, H. H. (1975) On the use of tryptic digestion to localize narcotic binding material. In The Opiate Narcotics, edited by A. Goldstein. Pergamon Press, New York, pp. 57-59.

Hitzemann, R. J. and Loh, H. H. (1976a) On the possible role of protein synthesis in functional barbiturate tolerance. Europ. J. Pharmacol. 40(1):163-175.

Hitzemann, R. J. and Loh, H. H. (1976b) Influence of morphine on protein synthesis in discrete subcellular fractions of the rat brain. Res. Commun. Chem. Path. Pharmacol. 14(2):237-248.

Hitzemann, R. J. and Loh, H. H. (1977a) Influence of pentobarbital on synaptic plasma membrane protein synthesis. Life Sci. 20: 35-42.

Hitzemann, R. J. and Loh, H. H. (1977b) Influence of chronic pentobarbital or morphine treatment on the incorporation of ^{32}Pi and [^{3}H]choline into rat brain synaptic plasma membranes. Biochem. Pharmacol. 26:1087-1088.

Hitzemann, R. J. and Loh, H. H. (1977c) Influence of morphine on protein synthesis in synaptic plasma membranes of the rat brain. Res. Comm. Chem. Path. Pharmacol. 17:15.

Ho, I. K., Loh, H. H. and Way, E. L. (1976) Pharmacological manipulation of gamma-aminobutyric acid (GABA) in morphine analgesia, tolerance and physical dependence. Life Sci. 18:1111-1124.

Kafner, J. (1972) Base-exchange reactions of the phospholipids in rat brain particles. J. Lipid Res. 13:468-474.

Kaneto, H. (1971) Inorganic ions: the role of calcium. In Narcotic Drugs: Biochemical Pharmacology, edited by D. H. Clouet. Plenum Press, New York, pp. 300-309.

Klee, W. A. and Streaty, R. A. (1974) Narcotic receptor sites in morphine dependent rats. Nature 248:61-63.

Kosterlitz, H. W., Leslie, F. M. and Waterfield, A. (1975) Rates of onset and offset of action of narcotic analgesics in isolated preparations. Eur. J. Pharmacol. 32:10-16.

Kuhar, M. J., Pert, C. B. and Snyder, S. H. (1973) Regional distribution of opiate receptor binding in monkey and human brain. Nature (London) 245:447-450.

Law, P. Y., Harris, R. A., Loh, H. H. and Way, E. L. Evidence for the involvement of cerebroside sulfate in opiate receptor binding: Studies with Azure A and Jimpy mutant mice. J. Pharmacol. Exp. Therap., in press, 1978.

Lee, N. M., Ho, I. K. and Loh, H. H. (1975) Effect of chronic morphine treatment on brain chromatin template activities in mice. Biochem. Pharmacol. 24:1983-1987.

Loh, H. H., Shen, F. H. and Way, E. L. (1969). Inhibition of morphine tolerance and physical dependence development and brain serotinin synthesis by cycloheximide. Biochem. Pharmacol. 18: 2711-2721.

Loh, H. H. and Hitzemann, R. J. (1974) Effect of morphine on (^{14}C-choline)-phosphatidylcholine and (^{3}H-leucine)-protein synthesis and turnover in discrete regions of the rat brain. Biochem. Pharmacol. 23:1753-1765.

Loh, H. H. and Cho, T. M. (1975) Model system for opiate receptor function. In Tissue Response to Addictive Drugs, edited by D. H. Ford and D. H. Clouet. Spectrum Publishing, New York, pp. 355-371.

Loh, H. H., Cho, T. M., Wu, Y. C., Harris, R. A. and Way, E. L. (1975) Opiate binding to cerebroside sulfate: A model system for opiate-receptor interactions. Life Sci. 16:1811-1818.

Lowney, L. I., Schultz, K., Lowery, P. J. and Goldstein, A. (1974) Partial purification of an opiate receptor from mouse brain. Science 183:749-753.

Mulé, S. J. (1966) Effect of morphine and nalorphine on the metabolism of phospholipids in guinea pig cerebral cortex slices. J. Pharmacol. Exp. Therap. 154:370-383.

Mulé, S. J. (1967) Morphine and the incorporation of ^{32}Pi into brain phospholipids of non-tolerant, tolerant and abstinent guinea pigs. J. Pharmacol. Exp. Ther. 156:92-100.

Mulé, S. J. (1970) Morphine and the incorporation of ^{32}P-orthophosphate in vivo into phospholipids of the guinea pig cerebral cortex, liver and subcellular fractions. Biochem. Pharmacol. 19:581-589.

Mulé, S. J. (1971) Phospholipid metabolism. In Narcotic Drugs: Biochemical Pharmacology, edited by D. H. Clouet. Plenum Press, New York, pp. 190-216.

Natsuki, R., Hitzemann, R. and Loh, H. Effects of morphine on the incorporation of [^{14}C]serine into phospholipid via the base-exchange reaction. Mol. Pharmacol., in press, 1978.

Oguri, K., Lee, N. M. and Loh, H. H. (1976) Apparent protein kinase activity in oligodendroglial chromatin after chronic morphine treatment. Biochem. Pharmacol. 25:2371-2376.

Pert, C. B., Aposhian, D. and Snyder, S. H. (1974) Phylogenetic distribution of opiate receptor binding. Brain Res. 75:356-361.

Pert, C. B., Pasternak, G. and Snyder, S. H. (1973) Opiate agonists and antagonists discriminated by receptor binding in brain. Science 182:1359-1361.

Pert, C. B., Snowman, A. M. and Snyder, S. H. (1974) Localization of opiate receptor binding in synaptic membranes of rat brain. Brain Res. 70:184-188.

Pert, C. B. and Snyder, S. H. Opiate receptor: demonstration in nervous tissue. Science 179;1011-1014, 1973.

Pert, C. B. and Snyder, S. H. (1973) Properties of opiate-receptor binding in rat brain. Proc. Nat. Acad. Sci. 70:2243-2247.

Pert, C. B. and Snyder, S. H. (1974) Opiate receptor binding of agonists and antagonists affected differentially by sodium. Mol. Pharmacol. 10:868-879.

Porcellati, G., Arienti, G., Pirotta, M. and Giorgini, D. (1971) Base-exchange reactions for the synthesis of phospholipids in nervous tissue: the incorporation of serine and ethanolamine into the phospholipids of isolated brain microsomes. J. Neurochem. 18:1395-1417.

Porcellati, G. and deJeso, F. (1971) Membrane-bound activity in the base-exchange reactions of phospholipid metabolism. In Membrane-Bound Enzymes. Plenum Press, New York, pp. 111-135.

Schuster, L. (1961) Repression and de-repression of enzyme synthesis and a possible explanation of some aspects of drug action. Nature (London) 189:314-315.

Shapiro, A. L., Vinuela, E., and Maizel, J. V., Jr. (1967) Molecular weight estimation of polypeptide chains by electrophoresis in SDS-polyacrylamide gels. Biochem. Biophys. Res. Commun. 28: 815-822.

Simon, E. J., Hiller, J. M. and Edelman, I. (1975) Solubilization of a stereospecific opiate-macromolecular complex from rat brain. Science 190:389-390.

Smith, A. A., Karmin, M. and Gabitt, J. (1966) Blocking effects of purmycin, ethanol and chloroform on the development of tolerance to an opiate. Biochem. Pharmacol. 15:1877-1879.

Smith, A. P. and Loh, H. H. (1976) The subcellular localization of stereospecific opiate binding in mouse brain. Res. Comm. Chem. Pathol. Pharmacol. 15:205-219.

Spoerlein, M. T. and Scrafini, J. (1967) Effects of time and 8-azaguanine on the development of morphine tolerance. Life Sci. 6:1549-1564.

Takemori, E. (1975) Neurochemical bases for narcotic tolerance and dependence. Biochem. Pharmac. 24:2121-2126.

Way, E. L. (1973) Brain neurohormones in morphine tolerance and dependence. Proc. Fifth Ing. Cong. Pharmac., Vol. 1, Karger, Basel, pp. 77-83.

Way, E. L. (1974) Some biochemical aspects of morphine tolerance and physical dependence. In Opiate Addiction: Origins and Treatment. Winston, Washington, D. C., pp. 99-110.

Way, E. L., Loh, H. H. and Shen, F. H. (1968) Morphine tolerance, physical dependence and synthesis of brain 5-hydroxytryptamine Science 168:1290-1292.

Way, E. L. and Shen, F. H. (1971) Interaction of morphine with the catecholamines and 5-hydroxytryptamine. In Narcotic Drugs: Biochemical Pharmacology, edited by D. H. Clouet. Plenum Press, New York, pp. 229-253.

Winter, B. A. and Goldstein, A. (1972) A photochemical affinity-labeling reagent for the opiate receptor(s). Mol. Pharmacol. 8:601-611.

Wu, Y. C., Cho, T. M., Loh, H. H. and Way, E. L. (1976) Binding of narcotics and narcotic antagonists to triphosphoinositide. Biochem. Pharmacol. 25:1554-1555.

NEUROTRANSMITTERS AND OPIATE ADDICTION

G. Pepeu, F. Casamenti and F. Pedata
Department of Pharmacology, University of Florence, Florence, Italy

Introduction

Opiate tolerance and dependence can be considered an adaptive process aimed at maintaining homeostasis. Therefore a large number of investigations have been directed at elucidating the adaptive changes in the neurotransmitter systems and the possibility of influencing the development of tolerance and dependence by their manipulation. Seminal to these investigations has been the theory on the mechanism of drug dependence proposed by Collier (1966) which postulates "an interaction of the drug with an endogenous transmitter substance regularly produced in the central nervous system."

The voluminous literature resulting from the investigations aimed at detecting changes in neurotransmitter levels, synthesis and release and in the number and sensitivity of the receptors has been covered by several recent reviews (Clouet,1971; Clouet and Iwatsubo,1975; Lal,1975; Martin and Sloan,1977; Bläsig,1978). This paper will therefore only pinpoint the most relevant findings and present some results from our laboratory on the effect of chronic morphine treatment on brain cholinergic mechanisms.

Comparison of the results obtained by different authors is made difficult by countless variations in the dose, schedule and procedure used in inducing tolerance, dependence and eliciting withdrawal. These factors seem to influence brain amines (Simon et al.,1975; Mehta and Johnson, 1975) and ACh changes (Domino and Wilson,1975) in response to chronic opiate treatment.

Abbreviations:
ACh= acetylcholine; Ch= choline; DA= dopamine; DOPAC= dihydroxyphenyl aceti acid; GABA= gamma aminobutyric acid; HVA= homovanillic acid; 6-OHDA= 6-hydroxydopamine; PCPA= parachlorophenylalanine; NA= noradrenaline; 5HT= 5-hydroxytryptamine.

Catecholamines

A survey of the literature (Pepeu and Nistri,1974) revealed that in most investigations acute opiate administration induced small inconsistent NA changes in rat brain. Similarly from the extensive review by Bläsig (1978) it appears that chronic opiate treatment elicits inconsistent changes in NA and DA levels and turnover in rodent brain. Furthermore the inhibition of catecholamine synthesis by αmethylparatyrosine did not affect the development of dependence as shown by the intensity of the withdrawal syndrome (Bläsig et al.,1975). Degeneration of the catecholamine-containing nerve endings obtained through intraventricular administration of 6-OHDA (Samanin et al.,1975a) or by bilateral destruction of the locus coeruleus (Samanin et al.,1975b) had no effect on the development of dependence in the rat but rather potentiated some of the withdrawal symptoms.

If during the development of tolerance and dependence no relevant changes seem to take place in the presynaptic component of the catecholaminergic systems, definite changes have been observed during the withdrawal syndrome. Although not all results are unanimous, in most experiments a decrease in NA brain level was found in several animal species (see ref. in Martin and Sloan ,1977), including the rat (Gramsch and Bläsig,1976). Furthermore Cicero et al.(1974) showed that phenoxybenzamine and phentolamine but not propranolol suppressed some of the symptoms occurring during withdrawal in the rat.

Following acute opiate administration, a stimulation of DA metabolism has been repeatedly reported (see ref. in Pepeu,1976). The opiate however do not seem to increase DA turnover through a blockade of DA receptors as do neuroleptics (Di Chiara et al.,1977) since they do not inhibit DA sensitive adenylate cyclase which is considered a DA receptor. Direct action of the opiate on dopaminergic neurons could be envisaged since Pollard et al.(1977) have demonstrated the presence of presynaptic enkephalin receptors on dopaminergic nerve terminals in the striatum.

Development of tolerance during chronic opiate treatment not only abolishes the increase in DA turnover observed after a single opiate administration but also reversed the catalepsy and akinesia. Eidelberger and Erspamer (1975) demonstrated that acute administration of haloperidol, a DA receptor blocking agent, potentiated morphine analgesia whereas chronic haloperidol administration enhanced the development of tolerance.

All reports agree that an increase in DA content takes place in the rat striatum during precipitated withdrawal (Mehta and Johnson,1975; Bläsig et al.,1976; Gramsch et al.,1977).Simon et al.(1975) observed a similar increase in the septum, amygdala and hypothalamus. The increase in DA content during withdrawal was associated with a decrease in DA utilization as demonstrated by the reduced αmethylparatyrosine-induced depletion of striatal DA (Gramsch et al.,1977) and by the reduced formation of the DA metabolites HVA, DOPAC and 3-methoxytyramine (Bläsig et al.,1976).

The reduction of DA metabolism observed during withdrawal could indicate that the presence of opiates is necessary in the tolerant rats in order to maintain normal dopaminergic activity.

The changes in DA turnover are not the only sign of the participation of DA-containing neurons in opiate addiction. During chronic morphine administration an increase in DA sensitivity of DA-sensitive adenylate cyclase without changes in basal activity, was observed in the rat striatum (Iwatsubo and Clouet,1975; Meraly et al.,1975). This finding is consistent with the hypothesis that during opiate dependence a supersensitivity of DA receptors in the brain is developing.

Germane to this hypothesis are also the observation that after chronic morphine treatment in the rat the potency of apomorphine, a known DA agonist, in lowering striatal DA turnover was increased and the dose of apomorphine needed to elicit aggression reduced (Gianutsos et al.,1974). Morphine tolerant mice also exhibited enhanced sensitivity to the locomotor actions of L-DOPA and morphine tolerant rats were less sensitive toward the depressant effect of haloperidol(Eidelberg and Erspamer,1975). Shulz and Herz (1977) detected a marked increase in the sensitivity to catecholamine during withdrawal by evaluating the ability of DA, apomorphine and clonidine,injected intra-cerebroventricularly, to reinitiate jumping behavior within 3 hr after naloxone.

The onset of DA receptor supersensitivity during chronic morphine administration was not confirmed by Kuschinsky (1975) measuring the effect of apomorphine on stereotype behavior and of DA on adenylate cyclase 16 - 20 hr after the last injection of morphine.

5-Hydroxytryptamine

According to the large amount of experimental evidence marshaled by Messing and Lytle (1977) an increase in activity of brain and spinal cord 5HT-containing neurons is associated with analgesia and enhanced antinociceptive drug potency, whereas a decrease in the activity of these neurons correlates with hyperalgesia and diminished analgesic drug potency. Single administration of morphine produced little or no change in the level of brain 5HT but increased 5HT turnover rate (see ref. in Martin and Sloan ,1977). Evidence in favor of the implication of 5HT neurons in the development of tolerance and physical dependence is much more controversial. The finding that 5HT levels and turnover were increased in the brain of dependent mice (Way,1972) was not confirmed by several authors (see ref. in Herz and Bläsig,1975). No consistent changes in brain5HT levels and turnover were observed during the withdrawal syndrome (Martin and Sloan,1977).

Also the effect of PCPA, a blocker of 5HT synthesis, on the development of tolerance and dependence is controversial. Tilson and Rech (1974) demonstrated that pretreatment with PCPA reduced the withdrawal symptoms more in a rat strain with high than in a strain with low 5HT turnover. Bläsig et al.(1975) showed that chronic reduction of 5HT levels, achieved by means of PCPA or specific brain lesions during the whole period of morphine exposure, changed withdrawal symptomatology nearly in the same way as did the decrease in 5HT levels during the time of withdrawal only.

5HT does not therefore seem to be involved in the basic process underlying dependence development but only in the pathways mediating some of the withdrawal signs.

Whatever the effect of opiate addiction on 5HT metabolism in the brain, receptor supersensitivity to this neurotransmitter also seems to take place during chronic morphine treatment in the peripheral tissues (Schulz and Herz,1976) and in the central nervous system (Schulz and Herz, 1977).

Acetylcholine

As shown in fig.1A a single administration of morphine to freely moving rats with a subchronically implanted collecting cup, brought about a marked decrease in ACh output from the cerebral cortex. This observation confirms previous findings in anaesthetized cats and rats and unanaesthetized rabbits (see ref. in Pepeu and Nistri,1974) and cats (Labrecque and Domino,1974). The decrease in ACh output can be correlated with an increase in ACh level observed in some rat brain regions (Green et al.,1976) and with the decrease in ACh utilization (Domino and Wilson,1973) and turnover in the rat (Cheney et al,1974; Zsilla et al.,1976). No change, however, was found in the turnover rate of the whole brain in mice after a single injection of morphine (Cheney et al.,1975). As shown in table 1, a slight decrease, not statistically significant, was found in the sodium-dependent choline high affinity uptake in the brain of rats treated with a single dose of morphine.

According to Simon et al.(1976) the rate of Ch high affinity uptake is an indicator of cholinergic neuronal flow and its depression was observed after very large doses of morphine.

Table 1. Effect of acute and chronic morphine administration on Ch high affinity uptake expressed in pmol.4 min.mg protein (Simon et al.,1976)

Treatment	Brain regions	Controls	Treated	% changes	P
Morphine 10 mg/Kg i.p.	cortex	1.38 ± 0.08 (5)	1.21 ± 0.09 (3)	− 12	n.s.
	caudate n.	3.83 ± 0.03 (2)	3.52 ± 0.22 (3)	− 8	n.s.
Morphine 10 days implantation	cortex	1.38 ± 0.08 (5)	1.48 ± 0.13 (2)	+ 7	n.s.
	caudate n.	5.61	5.43 ± 0.07 (2)	− 3	n.s.
Morphine 10 days implantation + Naloxone 1 mg/Kg i.p.	cortex	1.38 ± 0.08 (5)	2.10 ± 0.22 (3)	+ 52	< 0.01
	caudate n.	3.96 ± 0.36 (3)	5.25 ± 0.21 (4)	+ 33	< 0.05

Number of animals in parenthesis
Rats killed 40 min after morphine injection
    "   " 30 "   " naloxone   "
Morphine pellets implantation according to Laschka et al.(1975).

Repeated administrations of morphine induced a tolerance to the decrease in ACh utilization (Domino and Wilson,1973) and to the inhibition of the neocortical ACh release in the cat (Labrecque and Domino, 1974) and,in our experiments, in freely moving tolerant rats. According to Jhamandas and Sutak (1974) the spontaneous ACh output from the cortex in anaesthetized rats, made dependent by daily injection of morphine, was lower than in the control rats. The same observation was made by us (fig. 1C) in freely moving rats made dependent by implantation of morphine pellets according to the method of Laschka et al.(1975). In the same rats we found no change in Ch high affinity uptake (Table 1). An increase in ACh turnover was observed after a short term morphine implantation in mouse brain (Cheney et al.,1974).

The withdrawal syndrome, on the contrary, was accompanied by a marked increase in the activity of the central cholinergic system. As shown in fig. 1C, in unanaesthetized morphine-dependent rats the naloxone-induced withdrawal syndrome was associated with an increase in ACh output from the cerebral cortex and in Ch high affinity uptake (Table 1). The syndrome was characterized by the typical symptoms such as intense exploratory behavior,"wet dog" shakes, jumping, teeth chattering, ptosis, writhing, diarrhea. Naloxone injected in rats implanted with placebo pellets had no effect (Fig.1B). Jhamandas and Sutak (1974) first observed an increase in ACh output from the cerebral cortex during precipitated withdrawal in anaesthetized rats. Domino and Labrecque (1974) found on the contrary that in morphine-dependent cats, transected at midpontine level, the administration of naloxone was followed by a short lasting decrease in ACh output from the cerebral cortex followed by an irregular increase.

The enhancement of central cholinergic activity during withdrawal was confirmed by the changes in brain ACh levels. According to Domino and Wilson (1973, 1975) brain ACh utilization was enhanced during abrupt or precipitated withdrawal when chronic morphine administration lasted for more than 4 days. This could explain why, after 3.5 days of morphine administration, naloxone did not affect ACh turnover in the cerebral cortex, but only abolished the decrease in ACh turnover found in the striatum before naloxone (Cheney et al.,1974). However Bhargava and Way (1975) found a decrease in brain ACh levels during precipitated but not during abrupt withdrawal in rats rendered dependent by morphine pellet implantation for 3 days. In the dependent mice a decrease in brain ACh levels was found 10 min after naloxone-precipitated withdrawal and an increase 6 hr after abrupt withdrawal.

In spite of differences related to species, duration of opiate treatment, the procedure used in eliciting the withdrawal and to the brain regions investigated, some conclusions can be drawn from the experiments reported above. In the rat, acute opiate administration depresses the activity of cholinergic mechanisms at least in the neocortex and striatum. The depression disappears with the onset of tolerance and is substituted by a short-lasting intense stimulation during withdrawal. It should be noted that this pattern seems opposite to that of the dopaminergic system during acute and chronic opiate administration.

Bhargava and Way (1974) reported a relationship between the decrease in brain ACh levels and the jumping observed during the withdrawal syndrome. The existence of a relationship between the increase in striatal DA content and jumping has also been suggested (Iwamoto et al.,1973).

Attempts have also been made to modify the intensity of the withdrawal syndrome by the administration of cholinolytic and cholinomimetic drugs. Jhamandas and Dickinson (1973) found that atropine and mecamylamine reduced the jumping elicited by naloxone in morphine and methadone-dependent mice. However Frederickson (1975) claims that in the rat anticholinergic drugs suppressed the peripheral signs of the naloxone-induced withdrawal syndrome, but eserine reduced some of the symptoms such as "wet dog" shakes, jumping and yawning and suggests that the early phase of the opiate withdrawal response corresponds to a central cholinergic deficit.

In the rat, the electrolytic destruction of the septum is followed by a 40% decrease in cortical ACh level (Pepeu et al.,1973) and by the decrease of the stimulatory effect of DA agonists and of cholinolytic drugs on ACh output from the cerebral cortex (Pepeu et al.,1978). In two rats morphine tolerance and dependence was induced by pellet implantation one month after a large septal lesion. The frequency and intensity of many of the symptoms of the precipitated withdrawal syndrome were reduced and the syndrome was not accompanied by an increase in ACh output from the cerebral cortex (Fig. 1D). In the septal rats sponta-

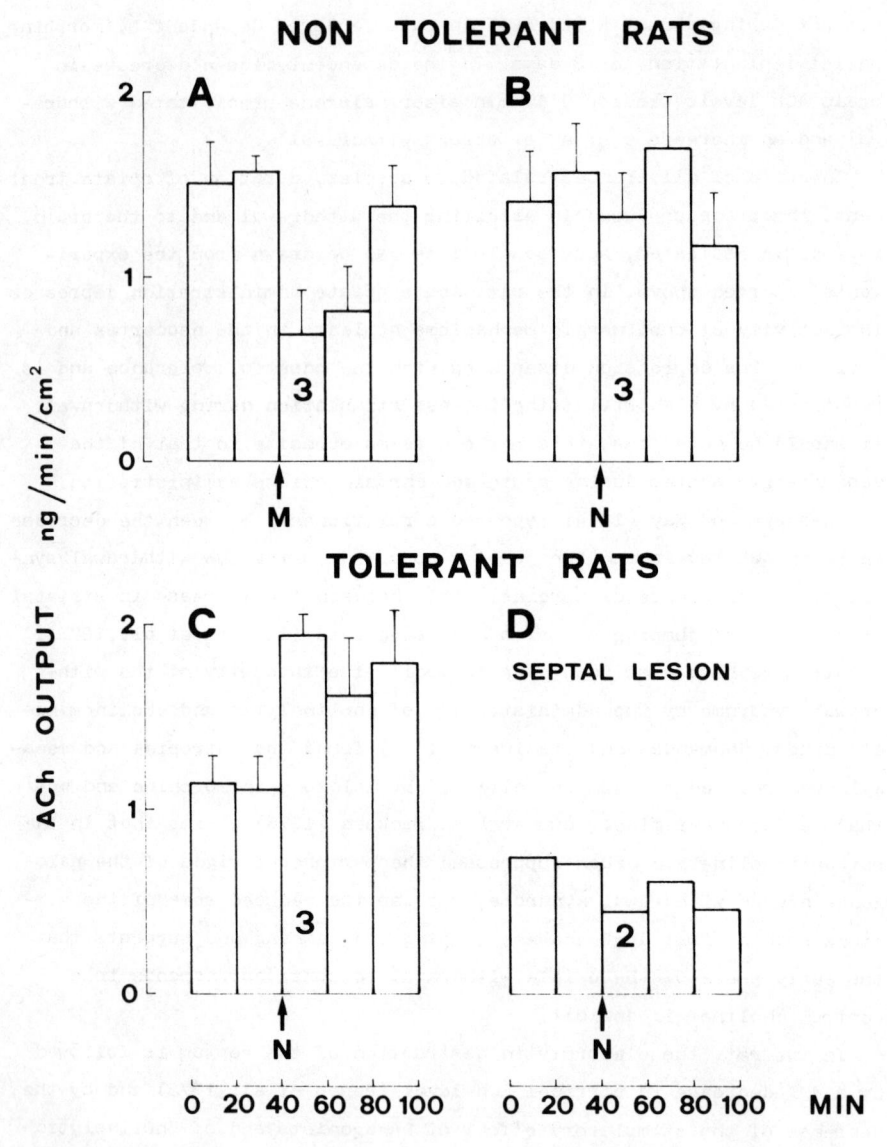

Fig.1 ACh output (ng/min/cm^2) from the cerebral cortex of freely moving rats with a subchronically implanted epidural cup (Beani and Bianchi, 1973).

A: morphine (M) 10 mg/Kg i.p. in non tolerant rats; B: naloxone (N) 1 mg/Kg i.p. in non tolerant rats; C: naloxone (N) in tolerant rats; D: naloxone (N) in tolerant rats with a septal lesion.

Tolerance induced by morphine pellets implantation (Laschka et al.1975) neous ACh output was lower than in the unoperated rats. This observation suggests that the withdrawal syndrome does not depend upon the integrity of the cortical cholinergic system but is modulated by it.

Chronic opiate treatment also seems to elicit a "pharmacological denervation supersensitivity" of the cholinergic receptors. This was detected by Yarbrough (1974) by recording from cortical neurons in the rat and was confirmed by Satoh et al.(1976). According to Vasquez et al.(1974) the directly acting muscarinic agonist pilocarpin depressed the behavior of the morphine-treated rats to a greater degree than that of the controls. However the onset of supersensitivity to ACh in morphine-dependent animals was not always demonstrated either in single brain neurons (Frederickson et al.,1975) or in the intestine (see ref. in Sawynok and Jhamandas, 1977).

Other neurotransmitters

Very little information is available on the effect of chronic opiate administration on other putative neurotransmitters. According to Henwood et al.(1975) histamine concentration in the hypothalamus, brain stem and cortex was not changed after acute treatment with large doses of morphine but chronic treatment for 21 days resulted in a significant decrease in endogenous histamine concentration. Naloxone-induced withdrawal, or abrupt withdrawal, brought about a further decrease. The meaning of this finding is obscure owing to the poor understanding of the role of histamine as neurotransmitter in the brain (Green et al., 1978).

No changes in GABA levels in the brain of dependent dogs were found by Maynert et al.(1962). Ho et al.(1976) found however that the blockade of GABA inactivation, by GABA transaminase or uptake inhibitors, not only antagonized the analgesic action of morphine in mice but also enhanced the development of tolerance and physical dependence. Opposite effects were seen pretreating the mice with bicuculline, a GABA receptor antagonist.

Conclusions

We conclude here, as we did in a recent review on the effect of acute opiate administration (Pepeu,1976), that all neurotransmitters, so far investigated, are involved in opiate addiction. However the changes occurring in the dopaminergic and cholinergic systems seem more consistent than with other neurotransmitters and the effect observed after a single opiate administration opposite to those during withdrawal.

The intensity of the withdrawal symptoms can be reduced by cholinolytic and cholinomimetic drugs (Jhamandas and Dickinson,1973; Fredrickson, 1975), by a decrease in 5HT content (Bläsig et al.,1976) or, in the present study, by a septal lesion which decreases ACh (Pepeu et al.,1973) and 5HT (Harvey et al.,1974) brain levels. "Wet dog" shakes can be reduced by alpha-receptor blocking agents (Cicero et al.,1974). These observations demonstrate that the withdrawal symptoms are a higly integrated response involving several neurotransmitter systems together. It is pertinet to mention that recent evidence suggests that opioid peptide receptors could modulate several neurotransmitter systems since enkephalin receptors have been identified on DA neurons in the striatum (Pollard et al.,1977), beta-endorphine inhibits the release of NA from the cerebral cortex (Arbillo and Langer,1978) and intraseptal beta-endorphine inhibits ACh turnover in the hyppocampus (Moroni et al.,1978).

Acknowledgements

The experimental work was supported by Grant No 77.01670 from Consiglio Nazionale delle Ricerche. F.C. was supported by a fellowship from Fidia Laboratories.

References

Arbilla, S. and Langer,S.Z.(1978) Morphine and beta-endorphine inhibit release of noradrenaline from cerebral cortex but not dopamine from rat striatum, Nature 271: 559-561.

Beani,L. and Bianchi,C. (1973) Effect of amantadine on cerebral acetylcholine release and content in the guinea-pig, Neuropharmacol. 12: 283-289.

Bhargava,H.N. and Way,L. (1975) Brain acetylcholine and choline following acute and chronic morphine treatment and during withdrawal. J. Pharmacol. Exp. Ther. 194: 65-73.

Bläsig,J., Herz,A. and Gramsch,Ch. (1975) Effects of depletion of brain catecholamines during the development of morphine dependence on precipitated withdrawal in rats, Naunyn-Schmiedeberg's Arch. Pharmacol. 286: 325-336.

Bläsig,J., Papeschi,R., Gramsch,Ch. and Herz,A. (1975) Central serotoninergic mechanisms and development of morphine dependence, Drug and Alcohol Dependence 1: 221-239.

Bläsig,J., Gramsch,Ch., Laschka,E. and Herz,A. (1976) The role of dopamine in withdrawal jumping in morphine dependent rats, Arzneim. Forsch. 26: 1004-1006.

Bläsig,J. (1978) On the role of brain catecholamines in acute and chronic opiate action. In: Development in Opiate Research, A. Herz ed. Dekker, New York 279-356.

Cheney,D.L., Trabucchi,M., Racagni,G., Wang,C. and Costa,E. (1974) Effects of acute and chronic morphine on regional rat brain acetylcholine turnover rate, Life Sci. 15: 1977-1990.

Cheney,D.L., Costa,E., Hanin,I., Racagni,G. and Trabucchi,M. (1975) Acetylcholine turnover rate in brain of mice and rats: effects of various dose regimens of morphine. In: Cholinergic mechanisms, P.G. Waser ed. Raven Press, New York 217-227.

Cicero,T.J., Meyer,E.R. and Smithloff,B.R. (1974) Alpha-adrenergic blocking agents: antinociceptive activity and enhancement of morphine-induced analgesia, J. Pharmacol. Exp. Ther. 189: 72-82.

Clouet,D.H. (1971) Narcotic drugs. Biochemical Pharmacology, Plenum Press, New York.

Clouet,D.H. and Iwatsubo,K. (1975) Mechanisms of tolerance to and dependence on narcotic analgesic drugs, Ann. Rev. Pharmacol. 15: 49-71.

Collier,H.O.J. (1966) Tolerance,Physical dependence and Receptors, Advan. Drug Res. 3: 171-188.

Di Chiara,G., Vargiu,L., Porceddu,M.L., Longoni,R., Mulas,A. and Gessa, G.L. (1977) Indirect activation of the DA system as a possible mechanism for the stimulatory effects of narcotic analgesics. In: Adv. in Biochem. Psychopharmacol. 16: 571-575, Raven Press, New York.

Domino,E.F. and Wilson,A. (1973) Effects of narcotic analgesic agonists and antagonists on rat brain acetylcholine, J. Pharmacol. Exp. Ther. 184: 18-32.

Domino,E.F. and Wilson, A. (1975) Brain acetylcholine in morphine pellet implanted rats given naloxone, Psychopharmacol.(Berl) 41: 19-22.

Eidelberger,E. and Erspamer,R. (1975) Dopaminergic mechanisms of opiate actions in the brain, J. Pharmacol. Exper. Ther. 192: 50-57.

Frederickson, R.C.A. (1975) Morphine withdrawal response and central cholinergic activity, Nature 257: 131-132.

Frederickson,R.C.A., Norris,F.H. and Hewes,Ch.R. (1975) Effect of naloxone and acetylcholine on medial thalamic and cortical units in naive and morphine dependent rats, Life Sci. 17: 81-82.

Gianutsos,G., Hynes,M.D., Puri,S.K, Drawbaugh,R.B. and Lal,H. (1974) Effect of apomorphine and nigrostriatal lesions on aggression and striatal dopamine turnover during morphine withdrawal: evidence for dopaminergic supersensitivity in protracted abstinence. Psychopharmacol.(Berl) 34: 37-44.

Gramsch,Ch. and Bläsig,J. (1976) Changes in brain catecholamine turnover during precipitated morphine withdrawal, Naunyn-Schmiedeberg's Arch. Pharmacol. Suppl. 293: R9.

Gramsch,Ch., Bläsig,J. and Herz,A. (1977) Changes in striatal dopamine metabolism during precipitated morphine withdrawal, Europ. J. Pharmacol. 44: 231-240.

Green,J.P., Glick,S.D., Crane,A.M. and Szilagyi,P.I.A. (1976) Acute effects of morphine on regional brain levels of acetylcholine, Europ. J. Pharmacol. 39: 91-99.

Green,J.P., Johnson,C.L. and Weinstein,H. (1978) Histamine as a neurotransmitter. In: Psychopharmacology. A Generation of Progress. M.A. Lipton, A.Di Mascio and K.F. Killam eds. Raven Press, New York 319-332.

Harvey,J.A., Schlosberg,A.J. and Yunger,L.M. (1974) Effect of p.chlorophenylalanine and brain lesions on pain sensitivity and morphine analgesia in the rat, In: Adv.in Biochem. Psychopharmacol. 10: 233-245 Raven Press, New York.

Henwood,R.W. and Mazurkiewicz-Kwilecki,M. (1975) Possible role of brain histamine in morphine addiction, Life Sci. 17: 55-56.

Herz,A. and Bläsig,J. (1975) Serotonin in acute and chronic opiate action. In: Serotonin in Health and Disease. Spectrum Publication, New York.

Ho,I.K., Loh,H.H. and Way,L. (1976) Pharmacological manipulation of gamma-aminobutyric acid (GABA) in morphine analgesia, tolerance and physical dependence, Life Sci. 18: 111-1124.

Iwamoto,E.T., Ho,I.K. and Way,L. (1973) Elevation of brain dopamine during naloxone-precipitated withdrawal in morphine dependent mice and rats, J. Pharmacol. Exper. Ther. 187: 558-567.

Iwatsubo,K. and Clouet,D.H. (1975) Dopamine-sensitive adenylate cyclase of the caudate nucleus of rats treated with morphine or haloperidol, Biochem. Pharmacol. 24: 1499-1503.

Jhamandas,K. and Dickinson,G. (1973) Modification of precipitated morphine and methadone abstinence in mice by acetylcholine antagonists, Nature New Biol. 245: 219-221.

Jhamandas,K. and Sutak,M. (1974) Modification of brain acetylcholine release by morphine and its antagonists in normal and morphine-dependent rats, Br. J. Pharmacol. 50: 57-62.

Kuschinsky, K. (1975) Dopamine receptor sensitivity after repeated morphine administration to rats, Life Sci. 17: 43-48.

Labrecque,G. and Domino,E.F. (1974) Tolerance and physical dependence on morphine: relation to neocortical acetylcholine release in the cat, J. Pharmacol. Exper. Ther. 191: 189-200.

Lal,H. (1975) Narcotic dependence, narcotic action and dopamine receptors, Life Sci. 17: 483-496.

Laschka,E., Herz,A. and Bläsig, J. (1976) Sites of action of morphine involved in the development of physical dependence in rats, Psychopharmacol.(Berl) 46: 133-139.

Martin,W.R. and Sloan,J.W. (1977) Neuropharmacology and neurochemistry of subjective effect, analgesia, tolerance and dependence produced by narcotic analgesics, In: Drug addiction I W.R.Martin ed. Springer Verlag, Berlin 43-127.

Maynert,E.W., Klingman,G.I. and Kaji,H.K. (1962) Tolerance to morphine II. Lack of effects on brain 5hydroxytryptamine and gamma aminobutyric acid, J. Pharmacol. Exp. Ther. 135: 296-299.

Mehta,C.S. and Johnson,W.E. (1975) Possible role of cyclic AMP and dopamine in morphine tolerance and physical dependence, Life Sci. 16: 1883-1888.

Merali,Z., Singhal,R.L., Hrdina,P.D. and Ling,G.M. (1975) Changes in brain cyclic AMP metabolism and acetylcholine and dopamine during narcotic dependence and withdrawal, Life Sci. 16: 1889-1894.

Messing,R.B. and Lytle,L.D. (1977) Serotonin-containing neurons: their possible role in pain and analgesia, Pain 4: 1-21.

Moroni,F., Cheney,D.L and Costa,E. (1978) The turnover rate of acetylcholine in brain nuclei of rats injected intraventricularly with alpha and beta-endorphine, Neuropharmacol. 17: 191-196.

Pepeu,G., Mulas,A. and Mulas,M.L. (1973) Changes in the acetylcholine content in the rat brain after lesions of the septum, fimbria and hippocampus, Brain Res. 57: 153-164.

Pepeu,G. and Nistri,A. (1974) Interaction between morphine and neurotransmitters in the central nervous system. In: Recent advances on Pain, J.J.Bonica, P.Procacci and C.Pagni eds, Charles C. Thomas Pub. Springfield Ill. 64-81.

Pepeu,G. (1976) Involvement of central transmitters in narcotic analgesia, In: Advances in Pain Research and Therapy 1: 595-600, J.J. Bonica and D. Albe-Fessard eds Raven Press New York.

Pepeu,G., Mantovani,P. and Pedata,F. (1978) Drug stimulation of acetylcholine output from the cerebral cortex. In: Cholinergic Mechanisms and Psychopharmacology. D.J.Jenden ed. Plenum Press New York 605-614.

Pollard,H., Llorens-Cortes,C. and Schwartz,J.C. (1977) Enkephalin receptors on dopaminergic neurones in rat striatum, Nature 268: 745-747.

Samanin,R., Bendotti,C., Chezzi,D. and Mauron,C. (1975a) Evidence for an involvement of brain serotonin and catecholamines in morphine analgesia and against a role of these substances in morphine induced physical dependence in rats, In: Neuropsychopharmacology. J.R.Boissier, H.H. Hippius and P.Pichot eds Excerpta Medica, Amsterdam 240-244.

Samanin,R., Bendotti,C., Gradnik,R. and Miranda,F. (1975b) The effect of localized lesions of central monoaminergic neurons on morphine analgesia and physical dependence in rats, In: Problems of drug dependence, 69P-697, Proc.37th Ann. Scientif. Meet.of the Committee on Problems of Drg Dependence, Nat.Acad. of Sciences, New York.

Satoh,M., Zieglgänsberger,W. and Herz,A. (1976) Supersensitivity of cortical neurones of the rat to acetylcholine and l-glutamate following chronic morphine treatment, Naunyn-Schmiedeberg's Arch.Pharmacol. 293: 101-103.

Sawynok,J. and Jhamandas,K. (1977) Muscarinic feedback inhibition of acetylcholine release from myenteric plexus in the guinea-pig ileum and its status after chronic exposure to morphine, Can. J. Physiol. Pharmacol. 55: 909-916.

Schulz,R. and Herz,A. (1976) Aspect of opiate dependence in the myenteric plexus of the guinea-pig, Life Sci. 19: 1117-1128.

Schulz,R. and Herz,A. (1977) Naloxone-precipitated withdrawal reveals sensitization to neurotransmitters in morphine tolerant/dependent rats, Naunyn-Schmiedeberg's Arch. Pharmacol. 299: 95-99.

Simon,J.R., Atweh,S. and Kuhar,M.J. (1976) Sodium-dependent high affinity choline uptake: a regulatory step in the synthesis of acetylcholine, J. Neurochem. 26: 909-917.

Simon,M., George,R. and Garcia,J. (1975) Chronic morphine effects on regional brain amines, growth hormone and corticosterone, Europ. J. Pharmacol. 34: 27-38.

Tilson,H.A. and Rech,R.H. (1974) The effect of p.chlorphenylalanine on morphine analgesia, tolerance and dependence development in two strains of rats, Psychopharmacol.(Berl.) 35: 45-60.

Vasquez,B.J., Overstreet,D.H. and Russell,R.W. (1974) Psychopharmacological evidence for increase in receptor-sensitivity following chronic morphine treatment, Psychopharmacol.(Berl.) 38: 287-302.

Way,E.L. (1972) Reassessment of brain 5-hydroxytryptamine in morphine tolerance and physical dependence. In: Agonist and antagonist actions of narcotic analgesic drugs. H.W. Kosterlitz, H.O.J. Collier and J.E. Villareal eds. Macmillan, London 153-163.

Yarbraugh,G.G. (1974) Actions of acetylcholine and atropine on cerebral cortical neurons in chronically morphine-treated rats, Life Sci. 15: 1523-1529.

Zsilla,G., Cheney,D.L., Racagni,G. and Costa,E. (1976) Correlation between analgesia and the decrease of acetylcholine turnover rate in cortex and hippocampus elicited by morphine, meperidine, viminol R_2 and azidomorphine. J. Pharmacol. Exp. Ther. 199: 662-668.

CALCIUM RELEASE IN PLATELETS: ACTION OF PSYCHOTROPIC DRUGS

H.W. Reading,

MRC Brain Metabolism Unit,

1 George Square, Edinburgh, Scotland.

Response of rabbit platelets to stimulation by ADP and drugs was measured by aggregation (Born 1962) and release of calcium using a fluorescent probe, chlortetracycline (Caswell and Hutchison 1971). In Ca^{2+}-free medium, a graded release of calcium from intracellular sites by differing concentrations of ADP or drug, was obtained. By comparison with internal standards, maximal Ca^{2+} release was $90\text{-}110 \times 10^9$ mols per platelet by $50\text{-}100\,\mu M$ ADP. 2hr incubation of platelet suspensions with drug, in Ca^{2+}-free medium produced $135\text{-}150 \times 10^9$ mols per platelet release of Ca^{2+} by Imipramine $10^{-7}M$ and Amitriptyline $10^{-6}M$ respectively. This corresponds to their efficacies as inhibitors of 5HT uptake (Todrick and Tait 1969). Lithium $10^{-3}M$ released 123×10^{-9} mols Ca^{2+} per platelet.

The significance of the increase in membrane permeability and concentration of intracellular Ca^{2+} may lie in its effect on both uptake and release of neurotransmitter pre-synaptically. Ca^{2+} release is essential for release (Katz and Miledi 1970) and increase in Ca^{2+} ion concentration on the inward side of the membrane will inhibit Na^+K^+ATPase (Skou 1957). Paton et al., (1971) and Nakazato et al. (1978) have provided evidence that inhibition of the sodium pump results in transmitter release both centrally and peripherally and also reduces noradrenaline uptake in synaptosomes (Rodriguez de Lores Arnaiz and de Robertis 1972). It is therefore suggested that, tricyclic antidepressants with structures related to imipramine, increase synaptic cleft transmitter concentration by facilitating release and inhibiting re-uptake pre-synaptically, through increasing membrane permeability to Ca^{2+}. By themselves, they do not inhibit purified membrane preparations of Na^+K^+ATPase. (Scott, personal communication).

Born, G.V.R. (1962) J.Physiol. 162: 67P

Caswell, A.H. and Hutchison, J.D. (1971) Biochem.Biophys.Res.Comm. 42: 43-49.

Katz, B. and Miledi, R. (1970) J. Physiol. 207: 789-801.

Nakazato, Y., Ohga, A. and Onada, Y. (1978) J.Physiol. 278: 45-54.

Paton, W.D.M., Vizi, E.S. and Aboo Zar, M. (1971) J. Physiol. 215: 819-848.

Rodriguez de Lores Arnaiz, G. and de Robertis, E.(1972) 'Current topics in membranes and Transport' Vol.3. Ed Bronner, E. and Kleinzeller, A. Acad.Press, London: 261.

Skou, J.C. (1957) J. Physiol. 267: 261-280.

Todrick, A. and Tait, A.C. (1969) J. Pharm. Pharmacol. 21: 751-762.

DRUG INDUCED DISPLACEMENT OF 5-HT AND LSD FROM SPECIFIC BINDING SITES OF SYNAPTIC MEMBRANES

Nina Weiner and W. Wesemann
Abt. Neurochemie, Physiologisch-chem. Institut II, Philipps-Universität
D 3550 Marburg/Lahn

In order to elucidate possible differences between 5-HT and LSD binding in the central nervous system of the rat, the binding characteristics of synaptic membranes obtained from the mitochondrial fraction (P_2) are compared with the binding sites in the microsomal fraction (P_3) described by SNYDER & BENNETT (1975). The two membrane fractions differ in morphology and protein pattern as shown by electron microscopy and gel electrophoresis (WEINER & WESEMANN, 1978).

Kinetic analysis of LSD binding to mitochondrial membranes reveals two different classes of binding sites with dissociation constants K_{D1}= 7 nM and K_{D2}= 50 nM, contrasting with only one binding site found for 5-HT in the same concentration range, K_D= 6 nM. A biphasic curve was obtained for 5-HT binding to microsomal membranes, K_{D1}= 5 nM and K_{D2}= 20 nM. The high affinity binding of 5-HT to microsomal membranes, K_{D1}, is in the same order of magnitude as the K_D value published by SNYDER & BENNETT (1975) for LSD binding, K_D= 9 - 11 nM. Differences between the two membrane preparations are observed in specific binding: the binding capacity is higher for 5-HT in mitochondrial membranes as compared with LSD, contrary to the binding to microsomal membranes where LSD binding is found to be higher. The differences can be confirmed in displacement studies with LSD analogues, tryptamines, and other psychotropic drugs. Besides quantitative differences in the displacement of LSD and 5-HT from the two membrane preparations, striking qualitative differences are obtained with 5-methoxytryptamine and tryptamine. Both compounds enhance 5-HT binding to mitochondrial membranes but inhibit 5-HT binding to microsomal preparations and LSD binding to both membrane fractions.

In view of the different binding properties of microsomal and mitochondrial membranes the results may be tentatively interpreted that at least not all 5-HT and LSD binding sites are identical.

Snyder, S.H. and Bennett, Jr., J.P. (1975) in "Pre- and Postsynaptic Receptors" (Usdin, E. and Bunney, W.E.Jr., eds.) pp. 191-205, Marcel Dekker, Inc., New York

Weiner, Nina and Wesemann, W. (1978) Hoppe Seyler's Z. Physiol. Chem. 359: 335

ACUTE EFFECTS OF HALLUCINOGENS ON THE DOPAMIN-TURNOVER IN RAT BRAIN REGIONS
L. HETEY, W. OELSSNER
Institute of Pharmacology and Toxicology,
Humboldt University, Berlin, GDR

The characterisation of the effects of neurotropic drugs on distinct transmission systems includes the estimation of changes in synthesis and metabolisation rates of their transmitters. For this purpose a tracerkinetic method, allowing the measurement in unaffected steady-state in vivo, for the quantification of the dopamin (DA)-turnover was developed.

^{3}H-tyrosine is i.v. pulse-injected and the shorttime kinetics of the labelling of tyrosine, DA and their main metabolites dihydroxyphenylacetic acid (DOPAC), homovanillic acid (HVA) and 3-methoxytyramine are determined. From the specific activities of these metabolites the precursor - postcursor flux rates are sequentially calculated.

There is some evidence for the existence of at least two DA-compartiments in normal rat brains, the turnover rate in the metabolic active one is 32,5 nM/g/hr in the caudate nucleus and 36 nM/g/hr in the mesolimbic area; DOPAC and HVA are synthetised to a greater extent from the new formed DA.

Following the i.p. administration of 500 µg/kg LSD and 50 mg/kg Mescalin, respectively, their acute effects on the DA-turnover are determined. Both hallucinogens increase the synthesis rate of DA and decrease the relative part of the apparently metabolic active DA-compartiments in both regions; they show a different influence in respect to the intraneuronally and interneuronally degradation of DA.

These effects are discussed in context with the possible pre- and postsynaptic actions of these hallucinogens on different receptor types.

REGULATION OF CATECHOLAMINE METABOLISM IN RAT BRAIN DURING CHRONIC TREATMENT WITH, AND WITHDRAWAL FROM, METHAMPHETAMINE

M.E. Bardsley and H.S. Bachelard
Department of Biochemistry, University of Bath, Bath, U.K.

Methamphetamine was administered i.p. to rats over a period of 30 days. The drug regime employed was 5mg/kg/24h increasing to 15mg/kg/24h over 15 days. The rats were injected at 12h intervals, and a saline injected control group was used. Eight brain regions were analysed for amino acids, tyrosine hydroxylase, noradrenaline, dopamine and their non-O-methylated metabolites. Tyrosine hydroxylase activity was assayed by the radiometric technique of Hendry and Iversen (1971). Noradrenaline (NA), dopamine (DA) and the non-O-methylated metabolites were assayed by combining the radioenzymatic technique of Cuello et al. (1973) and the thin-layer chromatographic separation of Fleming and Clark (1970).

The results showed a progressive inhibition of tyrosine hydroxylase from 30-40% inhibition (depending on the region) at 15 days, to near-complete inhibition after 30 days. After withdrawal from the drug for 36h, enzyme activity had returned to normal or near-normal levels, again depending on the region. Those regions with the highest original activity showed the highest recovery.

NA and DA levels were depleted in the chronic and withdrawn groups after 30 days of treatment, but were not significantly affected after 15 days. Tyrosine and phenylalanine levels were markedly higher in the withdrawn group.

The differing changes in concentrations of catecholamines and amino acids are discussed in terms of enzyme regulation and of amine pools.

Cuello, A.C., Hiley, R. & Iversen, L.L. (1973) J. Neurochem. 21: 1337-1340
Hendry, I. & Iversen, L.L. (1971) Brain Res. 83: 419-436
Fleming, R.M. & Clark, W.G. (1970) J. Chromatog. 52: 305-312

PARADOXICAL BRAIN MALATE RESPONSE TO MORPHINE

L. J. King, K. H. Minnema, and E. E. Dowdy, Jr.
Departments of Psychiatry and Pharmacology, MCV/VCU
Richmond, Va., USA

The most consistent statistically significant finding in studies of mouse brain energy metabolites after administration of narcotics and their antagonists was decreased malate associated with antinociception (King et al., 1977). Similarly, in mice chronically morphinized with implanted pellets, brain malate was decreased during the first 4-5 hr of implantation when antinociception was present and increased during days 2 and 3 as antinociception diminished (i.e. tolerance developed). However, there was a paradoxical period from 6 to 15 hr during which antinociception was present, but brain malate was elevated rather than decreased (King et al., 1978).

Miller et al. (1972) noted that many acute changes in brain energy metabolites brought about by morphine seemed related to the elevation of arterial CO_2 produced by morphine. Changes in intracardiac blood CO_2 were not correlated with malate changes in our chronic experiments although CO_2 levels did seem paradoxical during the malate paradoxical period. That is, when morphine effect (antinociception) was high, blood CO_2 was lowered, not elevated.

No other citric acid cycle intermediate, related amino acid, or glycolytic pathway substrate has the same pattern or paradoxical period as malate. This paradoxical elevation of malate might result from continuing decreased demand for energy supplies through 15 hr by morphine depressed neurons leading to eventual accumulation of malate in spite of lowered rates of citric acid cycle activity. Consistent with this explanation is a significant ($P < .05$) elevation of the adenylate energy charge at 24 hr, just after the paradoxical malate elevation. Significant ($P < .05$) aspartate and glutamate decrease at 72 hr but not before suggests that transamination reactions did not play a major role in the earlier malate changes.

King, L. J., Minnema, K. H., and Cash, C. (1977) Life Sciences
    21:1465-1474
King, L. J., Minnema, K. H., and Dowdy, E. E., Jr. (1978)
    Submitted for publication
Miller, A. L., Hawkins, R. A., Harris, R. L., and Veech, R. L. (1972)
    Biochem. J. 129:463-469

THE EFFECTS OF ETHANOL ON DEVELOPMENT IN THE RAT

N.K. Detering, P.T. Ozand, and A. Karahasan
Dept. Peds. U. of Md. and W.P. Carter Center
Baltimore, Md. 21201
U. S. A.

The effects of ethanol on body growth and development of the CNS in the offspring of lactating rats were investigated. Dams were fed: 1) regular stock diet (control), 2) liquid diet containing 35% of the calories as ethanol (ETOH), or 3) liquid diet with maltose-dextrins substituted for the calories supplied by ethanol (iso-caloric = IC). Dams fed the ETOH or IC diets consumed an identical amount of calories that was approximately half the amount of calories consumed by the control dams. Diets were administered from the 14th day of gestation until 3 weeks post-partum (pre-plus post-natal exposure) or from birth until 3 weeks post-partum (post-natal exposure). Body weight, crown-rump length, tail length, brain weight, and the specific activities of 5 brain enzymes; tyrosine hydroxylase (TOH), dihydroxyphenylalanine decarboxylase (DOPAdc), dopamine beta-hydroxylase (DBH), monoamine oxidase (MAO), and catecholamine o-methyl transferase (COMT) were followed longitudinally in their pups.

Growth of pups from dams fed ETOH or IC diets was impaired as compared to pups of control dams. The retardation was greatest in the pups from dams fed ETOH diet. When compared to controls, the ETOH pups showed a reduction of 65% in body weight, 30% in crown-rump length, 44% in tail length, and 27% in brain weight ($P \leq 0.001$ in all cases).

The specific activities of MAO and COMT in the brains of ETOH and IC pups were reduced (79% and 88%, respectively, by the third week of life) whereas those of TOH and DOPAdc increased (143% and 135%, respectively, by the third week of life) as compared to controls. These alterations were considered to be nonspecific since no differences were observed between ETOH and IC pups. However, the DBH activity was significantly reduced in ETOH when compared to IC pups (450 nmole/g protein/hr vs 583 nmole/g protein/hr, respectively, at 2 weeks of age, $0.001 \leq P \leq 0.01$). The specific activity of DBH in the brain of IC pups was similar to that in control pups.

It was therefore concluded that administration of ethanol to pregnant or lactating dams impaired growth and altered activities of catecholamine enzymes in their offspring. These differences were more prominent than those caused by nutritional deprivation.

Supported by NIH grant PHS1 RP1 AA03033-01

POTASSIUM ETHYLXANTHOGENATE AS AN INHIBITOR OF ETHANOL METABOLISM

S.G.Yanev
Dept.Pharmacology,Inst. of Physiology,Bulgarian Academy of sciences,
1113,Sofia,Bulgaria

Our earlier work (Mitcheva et al., 1976) has shown that potassium ethylxanthogenate (PEX) inhibits the hexobarbital liver microsomal metabolism. The similar chemical structure of PEX with that of diethyldithiocarbamate (DDC) stimulated additional comparative studies of the effect of these substances on some ethanol actions and metabolism in male albino rats.

The following parameters were studied: ethanol narcosis duration, ethanol toxicity, kinetics of ethanol elimination after intravenous injection, ethanol metabolism by liver microsomes in vitro, liver alcohol dehydrogenase (ADH) and catalase, liver and brain aldehyde dehydrogenase (AlDH), and superoxide dismutase (SOD).

Evidence was obtained to show that after intravenous injection of ethanol in a dose of 1 g/kg the blood ethanol elimination followed a first-order kinetics. PEX and DDC in a dose (0.5 mmoles/kg, 30 min in advance) which inhibited the drug liver metabolism, significantly decreased the ethanol elimination rate as well as the rate of the ethanol metabolism by liver microsomes in vitro. These effects were accompanied by prolongation of ethanol narcosis and potentiation of ethanol toxicity. No changes in liver ADH were observed. The catalase activity was slightly inhibited. The liver and brain AlDH was slightly inhibited too. The liver and brain SOD was strongly suppressed.

The present data suggest the possibility of ethanol peroxidation by liver microsomes. The changes in ethanol narcosis and toxicity after PEX and DDC are probably due to an increased ethanol blood level resulting from the decreased ethanol metabolism as well as to changes in catecholamine level in the brain produced by inhibition of dopamine-beta-hydroxylase activity (Stoytchev et al., 1974).

M.Mitcheva,S.Yanev,T.Stoytchev, 1976, Farmac.,XXVI,4,45-48
T.Stoytchev,S.Yanev,M.Koleva, 1974, Proceedings of Second Congress of the Hungarian Pharmacological Society,Budapest,175-180

Round Table

The Role of Peptides in Brain Function

Chairman:
D. de Wied

THE ROLE OF PEPTIDES IN BRAIN FUNCTION

IMMUNOHISTOCHEMICAL IDENTIFICATION OF EXTRAHYPOTHALAMIC CONNECTIONS OF PEPTIDE
HORMONE PRODUCING NEURONS

A. Weindl and M.V. Sofroniew
Neurologische Klinik der Technischen Universität, und Anatomische Anstalt der
Universität München, 8000 München 80, FRG

The location of peptide neurohormones within perikarya, processes and terminals of secretory neurons was identified with specific antisera in the immunoperoxidase reaction (Sofroniew and Weindl, 1978b; Weindl and Sofroniew, 1976) on serial sections of rat and guinea pig brains.

Vasopressin and oxytocin as well as their related neurophysins are produced in separate magnocellular perikarya of the supraoptic and paraventricular nucleus and the internuclear zone, and are transported to the neural lobe (Weindl and Sofroniew, 1978b). Luteinizing hormone releasing hormone (LRH) or somatostatin is produced in smaller perikarya (Weindl and Sofroniew, 1978b; Weindl et al., 1978). LRH perikarya, widely distributed in the preoptic/anterior hypothalamic area (Weindl and Sofroniew, 1978a; Weindl and Sofroniew, 1978b), send fibers to the lateral parts of the external zone of the median eminence (Weindl and Sofroniew, 1978a) and to permeable vessels of the organum vasculosum of the lamina terminalis (OVLT) (Weindl and Sofroniew, 1978a), an additional outlet for neurohormones into the blood (Weindl and Sofroniew, 1978b). Somatostatin perikarya, located in the periventricular anterior hypothalamic area (Weindl and Sofroniew, 1978b; Weindl et al., 1978), send fibers to the central part of the external zone of the median eminence (Weindl and Sofroniew, 1978a) and to the OVLT (Weindl and Sofroniew, 1978a).

With the same technique, extrahypothalamic fibers of neurons producing vasopressin, oxytocin, neurophysin, LRH or somatostatin were identified. From magnocellular perikarya of the paraventricular nucleus, large caliber fibers can be traced in the stria terminalis to the central amygdala where they form axo-somatic contacts with target neurons (Sofroniew and Weindl, 1978a; Weindl and Sofroniew, 1976). Small caliber fibers ascend from parvocellular vasopressin and neurophysin producing perikarya of the suprachiasmatic nucleus to lateral septum, medial dorsal thalamus, medial part of the lateral habenulae (Weindl and Sofroniew, 1978b), and through the periventricular grey of the mesencephalon and brain stem to the nucleus of the solitary tract (Sofroniew and Weindl, 1978b; Weindl and Sofroniew, 1978a). In these target areas they form axo-somatic contacts. In the rat, fibers descending from the paraventricular nucleus to brain stem and spinal cord react with antisera to oxytocin and neurophysin (Sofroniew and Weindl, 1978a; Weindl and Sofroniew, 1978a); they are also present in the Brattleboro rat (Weindl and Sofroniew, 1978a). LRH perikarya send extrahypothalamic fibers to neurons in the septum, interpeduncular nucleus and brain stem (Weindl and Sofroniew, 1978a). Somatostatin fibers form axo-somatic contacts within the hypothalamus in the infundibular and ventromedial nucleus, outside of the hypothalamus in the dorsal part of the caudal medulla oblongata.

References

Sofroniew, M.V. and Weindl, A. (1978a) Extrahypothalamic neurophysin containing perikarya, fiber pathways and fiber clusters in the rat brain, Endocrinology 102: 334-337.

Sofroniew, M.V. and Weindl, A. (1978b) Projections from the parvocellular vasopressin and neurophysin neurons of the suprachiasmatic nucleus, Am. J. Anat. (in press).

Weindl, A. and Sofroniew, M.V. (1976) Demonstration of extrahypothalamic peptide secreting neurons; a morphologic contribution to the investigation of psychotropic effects of neurohormones, Pharmakopsych. 9: 226-234.

Weindl, A. and Sofroniew, M.V. (1978a) Circumventricular organs and neurohormones: an immunohistochemical investigation. In: Brain-Endocrine Interaction III; Neurohormones and Reproduction; ed. by Scott, D.E., Kozlowski, G.P. and A. Weindl; Karger, Basel, 117-137.

Weindl, A. and Sofroniew, M.V. (1978b) The morphology of the hypothalamo-pituitary unit. In: Advances in the Diagnosis and Treatment of Pituitary Adenomas; ed. by Fahlbusch, R. and Werder, K. von; Thieme, Stuttgart, 10-38.

Weindl, A., Sofroniew, M.V. and Schinko, I. (1978) The distribution of vasopressin, oxytocin, neurophysin, somatostatin and luteinizing hormone releasing hormone producing neurons. In: Neurosecretion and Neuroendocrine Activity: Evolution, Structure and Function; ed. by Bargmann, W., Oksche, A., Polenov, A. and Scharrer, B.; Springer, Heidelberg, 308-315.

NEUROPEPTIDES AS NEUROTRANSMITTERS

L.L. Iversen

MRC Neurochemical Pharmacology Unit, Department of Pharmacology, University of Cambridge, England

The undecapeptide substance P (SP) has been the focus of our own studies in this area, since it appears to fulfil many of the criteria expected of a putative neurotransmitter. The regional distribution of SP in rat brain and spinal cord has been mapped by radioimmunoassay and immunohistochemical techniques (Cuello and Kanazawa, 1978; Cuello et al., 1978). It is present in certain small diameter sensory neurones in dorsal root ganglia, and highly concentrated in the terminals of such neurones in the substantia gelatinosa of spinal cord, and in the sensory nuclei of cranial nerves in brain stem. In addition, SP is present in several neuronal systems within the brain, including striato-pallido-nigral, habenulo-tegmental and amygdalar pathways. SP terminals are concentrated in several brain regions containing the cell bodies of catecholamine neurones, suggesting a possible functional connection with monoaminergic mechanisms. Another intriguing possibility is that SP may co-exist with 5-HT in many if not all of the tryptaminergic neurones in raphe nuclei.

SP release from nerve terminals in rat substantia nigra and trigeminal nerve nucleus has been studied <u>in vitro</u>, using a slice superfusion technique and radio-immunoassay of released peptide. A large increase in SP efflux can be evoked by chemical depolarising stimuli, such as elevated potassium ion concentration or veratridine. This evoked release (but not resting release) is dependent on the presence of calcium ions in the superfusing medium. The potassium-evoked release of SP from nerve terminals in substantia nigra is inhibited by GABA and by muscimol, but is unaffected by opiate analgesics. On the other hand, SP release from sensory nerve terminals in rat trigeminal nucleus can be completely inhibited by opiate analgesics and enkephalins, apparently by an action on opiate receptors located presynaptically on such nerve terminals (Jessell and Iversen, 1977). The latter finding offers a possible model for the analgesic actions of opiates at spinal cord level, and for the possible "gating" function of enkephalins in substantia gelatinosa in controlling the transmission of nociceptive information into supraspinal areas.

A calcium-dependent release of neuropeptides from brain slices <u>in vitro</u> has also been demonstrated for enkephalins (from rat globus pallidus)(Iversen et al., 1978a), somatostatin (from rat hypothalamus and amygdala), neurotensin (rat hypothalamus) (Iversen et al., 1978b) and vasoactive intestinal polypeptide (rat hypothalamus), consistent with the view that these and other neuropeptides are normally released from peptide-containing neurones in CNS. The subcellular distributions of SP and VIP are also reminiscent of the patterns seen with other putative CNS neurotransmitters, with a localization in synaptosome fractions, and an enrichment in synaptic vesicle preparations.

Biochemical studies, after ligation of rat vagus, show that SP is transported rapidly in an orthograde direction in primary sensory neurones in the vagus, and after lesions of the fasciculus retroflexus there is a similar accumulation of SP in fibres of the habenulo-tegmental pathway. Neurophysiological studies have shown that neuropeptides can exert potent effects on CNS neurones, with SP exhibiting consistently excitatory effects. In our own laboratory a potent excitatory effect has also been observed when somatostatin is applied to hippocampal pyramidal cells.

References

Cuello, A.C. and Kanazawa, I (1978) The distribution of substance P immunoreactive fibers in the rat central nervous system, J. Comp. Neurol. 178: 129.

Cuello, A.C., Emson, P., delFiacco, M., Gale, J., Iversen, L.L., Jessell, T.M., Kanazawa, I., Paxinos, G. and Quik, M. (1978) Distribution and release of substance P in CNS. In: Centrally Acting Peptides; ed. by Hughes, J.; Macmillan, London, 135-155.

Iversen, L.L., Iversen, S.D., Bloom, F.E., Vargo, T. and Guillemin, R. (1978a) Release of enkephalin from rat globus pallidus in vitro, Nature 271: 679-681.

Iversen, L.L., Iversen, S.D., Bloom, F.E., Douglas, C., Brown, M. and Vale, W. (1978b) Calcium-dependent release of somatostatin and neurotensin from rat brain <u>in vitro</u>, Nature 273: 161-163.

Jessell, T.M. and Iversen, L.L. (1977) Opiate analgesics inhibit substance P release from rat trigeminal nucleus, Nature 268: 549-551.

OPIOID PEPTIDES

J. Hughes
Department of Biochemistry, Imperial College, London SW7 2AZ, England

Opiate receptors may be subdivided into at least three subgroups denoted by the subscripts µ, κ and δ (Lord et al., 1977). Parallel assay of a series of opiate alkaloids and opioid peptides has revealed variations in the rank order of potency as measured in the mouse vas deferens and guinea-pig ileum bioassays and in the opiate receptor binding assay. Such variations in the order of potency is strong but not conclusive evidence for multiple opiate receptors. However, in the mouse vas deferens more conclusive evidence is provided by the measurement of the affinity constant (K_e) for naloxone as an antagonist of morphine and the opioid peptides. In this tissue naloxone has a K_e of 2 nM against morphine and other opiate alkaloids but this value increases to 20 nM against the enkephalins and endorphins. Thus the mouse vas possesses receptors for morphine (µ-receptor) distinct from those for the opioid peptides (δ-receptor).

Indirect evidence for the neuronal release of enkephalin has been presented by several groups. Waterfield and Kosterlitz (1975) showed that opiate antagonists stereoselectively enhanced the release of acetylcholine from the guinea-pig myenteric plexus during nerve stimulation. Thus the enkephalins may act as inhibitory neuroregulators in the intestine. Naloxone can facilitate the peristaltic reflex of a fatigued segment of intestine (Van Neuten et al., 1976) or enhance the recovery of cholinergic nerve activity after high frequency electrical stimulation of the guinea-pig myenteric plexus (Puig et al., 1977).

We have obtained direct evidence for the release of met-enkephalin and leu-enkephalin from isolated brain slice and synaptosome preparations (Henderson, Hughes and Kosterlitz, 1978). Destruction of the enkephalins was prevented by adding the dipeptides Tyr-Tyr, Leu-Gly and Leu-Leu to the superfusion-medium (1 mM each). The release of enkephalin was increased tenfold on increasing the potassium concentration of the medium from 1.19 mM to 50 mM. This potassium evoked release was enhanced in the presence of raised calcium levels and could be blocked by omitting calcium from the medium. The origin of the enkephalin released into the medium was unlikely to be due to the extracellular breakdown of β-endorphin since no enkephalin activity could be detected when this peptide was added to the superfusing medium.

Veratridine is a more specific stimulus than potassium and this alkaloid (50 µM) also caused the release of enkephalins from superfused guinea-pig striatal slices. This effect could be reversibly blocked with tetrodotoxin (1.6 µM). Veratridine released both met-enkephalin and leu-enkephalin in the ratio of 3.4 to 1. The tissue store met:leu ratio was 3.8 to 1 and thus the enkephalins are released in proportion to their tissue content.

It has been assumed that the enkephalins are derived from proteolytic cleavage of precursor proteins. In the case of met-enkephalin this precursor may well be β-lipotropin. At present there is no information as to the nature of the precursor for leu-enkephalin. In both the isolated guinea-pig myenteric plexus (Sosa et al., 1977) and striatal brain slices there is a lag period of up to one hour in the incorporation of ^{3}H-tyrosine into the pentapeptides. Incorporation then proceeds linearly for several hours. This incorporation was blocked by cycloheximide or puromycin. These results support the hypothesis that the enkephalins are derived from a ribosomally synthesized protein which is then cleaved to form the pentapeptides. In vivo studies also show incorporation of label into the enkephalins; after a single intracisternal injection the endogenous stores are maximally labelled within 2 - 4 hours but the levels of labelled enkephalin do not show a significant decline for up to twelve hours.

References

Henderson, G., Hughes, J. and Kosterlitz, H.W. (1978) In vitro release of Leu- and Met-enkephalin from the corpus striatum, Nature 271: 677-679.

Lord, J.A.H., Waterfield, A.A., Hughes, J. and Kosterlitz, H.W. (1977) Endogenous opioid peptides: multiple agonists and receptors, Nature 267: 495-499.

Neuten, J.M. van, Janssen, P.A.J. and Fontaine, J. (1976) Unexpected reversal effects of naloxone on the guinea-pig ileum, Life Sci. 18: 803-808.

Puig, M.M., Gascon, P., Craviso, G.L. and Musacchio, J.M. (1977) Endogenous opiate receptors ligand: electrically induced release in the guinea-pig ileum, Science 195: 419-420.

Sosa, R.P., McKnight, A.T., Hughes, J. and Kosterlitz, H.W. (1977) Incorporation of labelled amino acids into the enkephalins, FEBS letters 84: 195-198.

Waterfield, A.A. and Kosterlitz, H.W. (1975) Stereospecific increase by narcotic agonists of evoked acetylcholine output in the guinea-pig ileum, Life Sci. 16: 1787-1792.

PEPTIDES IN CONTROL OF BEHAVIOUR AND METABOLISM

K.L. Reichelt, O.E. Trygstad, P.D. Edminson, I. Foss, G. Saelid, J.H. Johansen and J. Bøler
Ped. Res. Inst.Rikshospitalet, Oslo 1, Norway

On the basis of the nature of peptides found and formed in the CNS, a working model, which incorporates the geometric (anatomical) and temporal relationships of the impinging signals on the integrative neurone, can be constructed (Reichelt and Edminson, 1977). It is argued that peptides represent a transcript of the local events also in axo-dendritic sub-systemes, which could be short term memory. Such a model could explain the dependence of the labile phase of short term memory on amino acid uptake (Gibbs and Ng, 1977). The involvement of amines in peptide formation (Reichelt and Edminson, 1977) could link peptide formation to reward and punishment

(Wise et al., 1972) which are involved in all learning, and also to cortical activation or arousal (Kurtz and Palfai, 1977).

Peptide discharge from hypothalamic peptidergic neurones definitely reflects multisignal integration whereby hormonal feed back, neuronal inputs from higher CNS levels and the periphery are mixed at the level of the discharging cell. Likewise the regulation by peptides of bursting pacemaker activity in ganglion cells (Ifshin et al., 1975) and in neurosecretory cells (Barker and Gainer, 1975) following definite afferent inputs, points to a central regulatory role for peptides.

It follows from the above that hyperfunction of master neurones (command neurones) or key ganglion cells should result in either overproduction or frequent release of peptide. If the genetic apparatus for peptide breakdown is insufficient to cope with the increased demand, overflow of peptide material to the body fluids should result. Their activity should also be manifest and "out of control". Thus purely reactive patterns of behaviour could progress into biochemical disease or malfunction.

We have looked at several disease states and found atypical patterns of peptide-carrier-protein complexes in body fluids, and we are able to reproduce the symptoms individually by means of isolated and characterized peptides. For an entirely genetic state, as in congenital generalized lipodystrophy (Foss and Trygstad, 1975), a whole family of peptides is released which individually can reproduce the symptoms of this state in mice, rats and rabbits. In anorexia nervosa we have also found abnormal patterns and the tripeptide Pyroglu-His-GlyOH inducing food refusal and weight loss has been isolated and synthetised. The etiology may, in these syndromes, be genetic insufficiency or the breakdown of a food refusal signal, on functional loading or overloading i.e. slimming. A pathological pattern of peptide-carrier-protein complexes found in severely stressed soldiers clearly shows that hyperfunction, even in healthy persons, can result in hypersecretion of peptides. Data on several psychiatric diseases will also be presented.

References

Barker, J.C. and Gainer, H. (1974) Peptide regulation of bursting pacemaker activity in a molluscan nevrosecretory cell, Science 184: 1371-1373.

Foss, I. and Trygstad, O.E. (1975) Lipoatrophy produced in mice and rabbits by a fraction prepared from the urine from patients with congenital generalized lipodystrophy, Acta Endocrinol. 80: 398-416.

Gibbs, M.E. and Ng, K.T. (1977) Psychobiology of memory towards a model of memory formation, Biobehavioral Reviews 1: 113-136.

Ifshin, M.S., Gainer, H. and Barker, J.C. (1975) Peptide factor extracted from molluscan ganglion that modulates bursting pacemaker activity, Nature 254: 72-73.

Kurtz, P. and Palfai, T. (1977) Mechanisms in retrograde amnesia; a case for biogenic amines, Biobehavioral Reviews 1: 23-33.

Reichelt, K.L. and Edminson, P.D. (1977) Peptides containing probable transmitter candidates. In: Peptides in Neurobiology; ed. by Gainer, H.; Plenum Press, New York, 171-181.

Wise, C.D., Bergner, B.D. and Stein, L. (1973) Evidence for a noradrenergic reward receptors and serotonergic punishment receptors in the rat brain, Biol. Psychol. 6: 3-21.

NEUROPEPTIDES AND BEHAVIOR

D. de Wied

Rudolf Magnus Institute for Pharmacology, Medical Faculty, University of Utrecht, Vondellaan 6, Utrecht, The Netherlands

Pituitary hormones may function as precursor molecules for neuropeptides involved in acquisition and maintenance of adaptive behavior. ACTH is a precursor hormone for neuropeptides which affect motivational, learning and memory processes. The essential elements for these effects are predominantly located in the region $ACTH_{4-7}$ (de Wied et al., 1975). ACTH has a low affinity for opiate receptor sites in the brain and it acts as a partial agonist/antagonist (Terenius et al., 1976). Excessive grooming, stretching and yawning which is elicited following intraventricular administration of relatively high amounts of ACTH is mediated by opiate receptor sites since specific opiate antagonists inhibit these behavioral effects (Gispen et al., 1975).

The opiate-like peptide β-endorphin also elicits excessive grooming following intraventricular administration which can be blocked by opiate antagonists (Gispen et al., 1976). The same compound in much higher amounts via the same route causes naloxone reversible catatonia (Bloom et al., 1976) and other behavioral effects related to its opiate-like activity (Segal et al., 1977). Much lower amounts given either peripherally or centrally delay extinction of active avoidance behavior. Unlike the analgesic effect and the above mentioned catatonia the influence on avoidance behavior is more marked in shorter fragments such as α-endorphin or β-LPH_{61-69} than in β-endorphin (de Wied et al., 1978a). The influence of the endorphins and of ACTH on conditioned behavior is not blocked by opiate antagonists and takes place independently of opiate receptors in the brain (de Wied et al., 1978a). The influence of the endorphins on adaptive behavior is highly specific since γ-endorphin has an effect opposite to that of α-endorphin. Both γ- and α-endorphin might be generated from β-endorphin by brain enzymes (Austen et al., 1977). Removal of the C-terminal amino acid residue tyrosine of γ-endorphin which destroys the opiate-like activity potentiates the influence on avoidance behavior (de Wied et al., 1978b). $[Des\ Tyr^1]$-γ-endorphin (DTγE) appeared to have neuroleptic activity. It was postulated that an inborn error in the generation or metabolism of DTγE is an etiological factor in psychopathology (de Wied et al., 1978b). Support for these assumptions were obtained from studies with DTγE in schizophrenic patients (Verhoeven et al., 1978).

The neurohypophyseal hormones vasopressin and oxytocin modulate memory processes. Vasopressin facilitates while oxytocin attenuates memory consolidation and retrieval (Bohus et al., 1978). These influences are located in different regions of the molecules. Thus, the neurohypophyseal hormones act as precursor molecules for neuropeptides involved in memory processes. The covalent ring structures of both vasopressin and oxytocin mainly affect consolidation, the linear parts, retrieval processes, while nearly the whole oxytocin or vasotocin molecule is needed for attenuation of consolidation and retrieval. The generation of these fragments by the brain is not wholly supported by enzyme studies since no evidence for the generation of the covalent ring structure of either neurohypophyseal hormone has been obtained. More knowledge on the metabolic degradation of the neurohypophyseal hormones in specific areas in the brain is needed to determine the various loci in the molecules which affect consolidation and retrieval.

References

Austen, B.M., Smyth, D.G. and Snell, C.R. (1977) γ-Endorphin, α-endorphin and Met-enkephalin are formed extracellularly from lipotropin C-fragment, Nature 269: 619-621.

Bloom, F., Segal, D., Ling, N. and Guillemin, R. (1976) Endorphins; profound behavioral effects in rats suggest new etiological factors in mental illness, Science 194: 630-632.

Bohus, B., Urban, I., Wimersma Greidanus, Tj.B. van and Wied, D. de (1978) Opposite effects of oxytocin and vasopressin on avoidance behaviour and hippocampal theta rhythm in the rat, Neuropharmacol. 17: 239-247.

Gispen, W.H., Wiegant, V.M., Greven, H.M. and Wied, D. de (1975) The induction of excessive grooming in the rat by intraventricular application of peptides derived from ACTH; structure activity studies, Life Sci. 17: 645-652.

Gispen, W.H., Wiegant, V.M., Bradbury, A.F., Hulme, E.C., Smyth, D.G., Snell, C.R. and Wied, D. de (1976) Induction of excessive grooming in the rats by fragments of lipotropin, Nature 264: 794-795.

Segal, D.S., Bloom, F., Ling, N. and Guillemin, R. (1977) β-Endorphin; endogenous opiate or neuroleptic, Science 198: 411-414.

Terenius, L., Gispen, W.H. and Wied, D. de (1975) ACTH-like peptides and opiate receptors in the rat brain; structure activity studies, Eur. J. Pharmacol. 33: 395-399.

Verhoeven, W.M.A., Praag, H.M. van, Botter, P.A., Sunier, A., Ree, J.M. van and Wied, D. de (1978) [Des-Tyr^1]-γ-endorphin in schizophrenia, Lancet I: 1046-1047.

Wied, D. de, Witter, A. and Greven, H.M. (1975) Behaviorally active ACTH analogues, Bioch. Pharmacol. 24: 1463-1468.

Wied, D. de, Bohus, B., Ree, J.M. van and Urban, I. (1978a) Behavioral and electrophysiological effects of peptides related to lipotropin (β-LPH), J. Pharmacol. exp. Ther. 204: 570-580.

Wied, D. de, Kovacs, G.L., Bohus, B., Ree, J.M. van and Greven, H.M. (1978b) Neuroleptic activity of the neuropeptide β-LPH$_{62-77}$ ([Des-Tyr^1]-γ-endorphin; DTγE), Lancet I: 1046.

PURIFICATION OF BRAIN ENKEPHALINS BY CHROMATOGRAPHY ON LH-20 SEPHADEX

Wideman, J., Medical Research Centre, Polish Academy of Sciences,
00-784 Warsaw, Poland
Stein, S., Roche Institute of Molecular Biology, Nutley, N.J., USA

Increasing interest in enkephalins creates a need for dependent method of their quantification in the tissues. Due to a very low concentration of these endorphins in the brain it is essential for the reliability of biological and radioimmunological assays to use a vigorous preliminary isolation procedure for removing the interfering materials. The presented method of simple, efficient and reproducible isolation of enkephalin fraction of a very high purity can be used in combination with both assays.

Homogenate of guinea-pig striatum in 1 ml of 2.5 M acetic acid with 0.01% thiodiglycol was centrifuged, proteins were precipitated with 10% TCA and post-TCA supernatant was applied on Sephadex LH-20 column /0.7x48 cm/ and run with the same acetic acid. Flow rate was about 5.5 ml/hr. Fluorescamine was used for detecting the primary amines in column effluent/1/. Both enkephalins eluted as a sharp peak between 285-310 min, far behind aminoacids. Elution time was very reproducible when tested by running the solutions of standard "cold" and radioactive enkephalins alone or with tissue extracts. Upon running the single striatum extract and monitoring column effluent for morphine-like activity with binding assay/2/ the activity was eluted in the same portion of effluent as standard enkephalins. All other endorphins co-precipitated with proteins or eluted before aminoacids. The recovery of both enkephalins was about 70%. HPLC of the fraction showed its high purity and enabled to separate Met- from Leu-enkephalin for studying them individually with presently used assays. Depending on the assay radioactive enkephalins or enkephalins amides can be used as internal standards.

Using 0.7x28 cm column shortens more than twice the time of enkephalins elution with purity of the fraction being still satisfactory. The method can be scaled up to gram quantities of the brain.

1. Böhlen, P., Stein, S., Stone, J., Udenfriend, S./1975/ Anal. Biochem., 67: 438-445

2. Diekmann Gerber, L., Stein, S., Rubinstein, M., Wideman, J., Udenfriend, S., Brain Res./ in press /

PURITY OF COMMERCIALLY AVAILABLE RADIOACTIVE ENKEPHALINS AS STUDIED BY HPLC

J. Wideman, Medical Research Centre of Polish Acad. of Sciences,
    00-784 Warsaw, Poland
S. Stein, Roche Institute of Molecular Biology, Nutley, N.J., USA

The reliability of biological and radioimmunological assays depends on the purity of commercially available peptides which are used as primary standards. Combination of HPLC methods with fluorescamine detecting system offers unique opportunities of studying the primary amines with high specificity and sensitivity /1/. One of these techniques was used to test the purity of radioactive enkephalins introduced recently on the market by different producers.

Fresh sample with µCi quantities of ^{3}H-Leu- or ^{3}H-Met-enkephalin /L-enk, M-enk/ dissolved in final concentration of 0.05 M pyridine and 0.25 M acetic acid was analysed using HPLC. 25 x 0.32 cm column with cation exchange resin was run at 65°C with a gradient from the buffer as above to 3.0 M pyridine and 1.5 M acetic acid pumped at 7.5 ml/hr. Pressure changed from about 200% to 320 atm during run. An automatic stream-sampling valve which directs a portion of the column effluent to fluorescamine detection system and the remainder to fraction collector was used. Radioactivity and fluorescence elution profiles were compared. Most of the ^{3}H-L-enk samples showed high radioactive and fluorescent homogenity with at least 80% of total radioactivity co-eluting with L-enk. The remaining radioactivity was evenly distributed among all fractions. However, most of ^{3}H-M-enk samples were impure. The products of one company showed acceptable fluorometric purity and quite reproducible radioactivity elution profile with one eluted earlier extra peak having about 10-15% of cpm found in M-enk peak. The samples from another producer were very impure and different from one sample to another. Up to four marked peaks of radioactivity were eluted and occasionally large extra peaks of fluorescence were observed. It was tested whether the impurities of the most heterogenous sample affect the binding property of opiate receptors. It was found that one peak eluted just before M-enk and accounting for about 20% of total cpm bound to the receptor and was specifically displaced by naloxone. Moreover, a non-radioactive material which was eluted after enkephalins displaced ^{3}H-L-enk bound to opiate receptor. In this sample claimed to have very high specific radioactivity less than 10% of total cpm was in labeled M-enk. The results show that ^{3}H-L-enk should be used preferentially to ^{3}H-M-enk in binding assays, kinetics studies of opiate receptor due to its higher purity similar for different companies.

THE EFFECT OF METENKEPHALIN AND (D-Met^2,Pro^5)-ENKEPHALINAMIDE ON THE ADENYLATE CYCLASE ACTIVITY OF RAT BRAIN

M. Wollemann, A. Szebeni and L. Gráf
Institute of Biochemistry, Biological Research
Center, Hungarian Academy of Sciences, Szeged
and Research Institute for Pharmaceutical
Chemistry, Budapest

(D-Met^2,Pro^5)-Enkephalinamide (Bajusz et al., 1977) inhibited the particulate cell fraction of rat cortex and subcortex adenylate cyclase activity in 10^{-7}M concentration. Metenkephalin stimulated in similar concentration the enzyme activity in the subcortex by 50% provided bacitracin was added and inhibited the adenylate cyclase activity in the cortex. In the absence of bacitracin metenkephalin was effectless.

Naloxon (10^{-6}M) inhibited metenkephalin activation by 50% only in the presence of 100 mM NaCl. These results provide evidences of differences in the action of morphinomimetic peptides in agreement with other data (Algeri et al., 1977; Calderini et al., 1978).

Bajusz, S., Rónai, A. Z., Székely, J. I., Gráf, L., Dunai-Kovács Zsuzsa and Berzétei Ilona (1977) Febs Letters
76: 91-92

Algeri, S., Calderini, G., Consolazione, A. and Garattini, S. (1977) Europ. J. Pharmacol.
45: 207-209

Calderini, G., Consolazione, A., Garattini, S. and Algeri, S. (1978) Brain Research 146: 392-399

DISTRIBUTION AND PROPERTIES OF LIPOTROPIN ACTIVATING PROTEASES IN RAT BRAIN

B.M. Austen and D.G. Smyth
National Institute for Medical Research, Mill Hill, London NW7 IAA

The C-Fragment of lipotropin, a naturally occurring peptide with potent analgesic properties, is released from inactive precursors by proteolytic cleavage at a sequence which contains adjacent basic residues (Bradbury et al., 1975). A second cleavage between methionine and threonine in lipotropin is required for formation of the pentapeptide methionine enkephalin (Hughes et al,,1975).

Peptides Lys-Asp-Lys-Arg-Tyr-Gly (residues 57-62 of lipotropin) and Tyr-Gly-Gly-Phe-Met-Thr-Ser (residues 61-67) were incubated with a series of fractions obtained from rat brain. Activating enzymes were monitored by measuring rates of release of Tyr-Gly or Thr-Ser. Homogenisation in 0.32 M sucrose solubilised most of the Tyr-Gly releasing enzyme; the Thr-Ser releasing protease remained in the crude synaposomal pellet obtained by differential centrifugation. The two cleavages therefore appear to occur in different subcellular locations. Regional dissections show that the trypsin-like enzyme is concentrated in pituitary which is rich in C-Fragment, and in striatum which is rich in enkephalin.

The partially purified soluble protease released Arg and Lys in addition to Tyr-Gly from the synthetic substrate and C-Fragment was formed from ^{125}I-lipotropin in 17% yield. The pH optimum was 8.5 whereas that of the Met-Thr cleaving enzyme was 7.4. The Tyr-Gly releasing protease was not inhibited by soya bean trypsin inhibitor, but was inactivated by phenylmethylsulphonyl fluoride.

The identification of these proteases in brain shows that the opiate peptides can be biosynthesised independently of their formation in the pituitary. Their regulation may be an important controlling factor in the endorphin system.

Bradbury, A.F., Smyth, D.G. and Snell, C.R. (1975) Proc. 4th American Peptide Symp.(Walter and Meienhofer, eds) Ann Arbor Sci. Inc; pp. 609-615

Hughes, J., Smith, T.W., Kosterlitz, H.W., Fothergill, L.A., Morgan, B.A. and Morris, H.R. (1975) Nature 258, 577-579

THE LOCALIZATION AND CHARACTERIZATION OF AMINOPEPTIDASES IN THE CNS AND THE HYDROLYSIS OF NEUROPEPTIDES.

S. G. Shaw and W. F. Cook,
Oxford University,
Department of Pharmacology,
South Parks Road, Oxford.

The possible neurotransmitter or neuromodulatory roles of several peptides found in the mammalian central nervous system has prompted interest in peptidase enzymes which might be involved in their formation and degradation.

Although a number of studies have shown that enzymes do exist in the CNS with the capability to hydrolyse peptides, no assessment has been made of the physiological significance of individual enzymes in the inactivation or formation of specific peptides at discreet sites in the CNS. Using a histochemical method and a polyacrylamide gel electrophoresis technique (Shaw 1978) we have demonstrated selective localization of some aminopeptidases to particular cells and regions of the CNS. The difference in substrate specificities suggests that particular enzymes could play a select role in the metabolism of particular peptides whereas others are probably involved more generally with protein metabolism.

Further application of these techniques should provide important information regarding the characteristics and localization of other enzymes involved in regulating the activity of neuropeptides.

Shaw, S. G. (1978) D.Phil. Thesis, Oxford University.

PEPTIDES AND MEMORY: A SPECULATIVE WORKING MODEL

K.L. Reichelt, P.D. Edminson, G. Sælid
Pediatric Res. Inst., Rikshospitalet, Oslo 1, Norway

Peptides are related to memory acquisition and extinction (DeWied, 1977). Specific peptides have been reported isolated in learning experiments within narrow time limits (Ungar, 1974). Acquisition of new behaviour must depend on arousal, possibly peptide modified (DeWied, 1972) and aminergic systems of reward and punishment (Wise et al, 1973). Any memory model must also account for the relationship and integration of multiple inputs impinging on the postsynaptic neurone, their geometric and temporal relationships and also the postsynaptic responses.

Short term memory depends in its labile phase on amino acid uptake (Gibbs and Ng, 1978) and it has been pointed out that the endogenous peptides and peptides formed in nervous tissues contain transmitter candidates (Reichelt and Edminson, 1977). Also amino acids not transmitters leak out during nervous transmission (Florey and Woodcock, 1968, Mitchell et al, 1969).

We propose that the amine dependent (reward dependent?) peptide formation starting with acetyl-Asp-N-terminally and incorporating transmitter candidates could represent the transmitters released on the neurone. Their sequence of the peptide directly represents the temporal sequence and also the geometry on the dendrites and perikaryon. As IPSP and EPSP are stereotyped unspecific responses, and the same transmitter may, depending on ion equilibria, induce either IPSP or an EPSP (Kuffler and Nicholls, 1977): we propose that the IPSP and EPSP are represented in the peptide formed on reuptake by amino acids leaked from interneuronal pool. The postsynaptic event is thus represented as peptide, which again could control protein synthesis in long term memory (Gibbs and Ng, 1978).

DeWied, D. (1977) Life Sciences 20: 195-204
Florey, E. and Woodcock, B. (1968) Canad. J. Biochem. Physiol. 26: 65-69
Gibbs, M.E. and Ng, K.T. (1977) Biobehavioral Reviews 1: 113-136
Kuffler, S.W. and Nicholls, J.G. (1977) From neurone to Brain Sinauer Assoc. 177-191
Mitchell, F., Neal, M.J. and Srinnas, V. (1969) Brit. J. Pharmacol. 36: 201-206
Reichelt, K.L. and Edminson, P.D. (1977) in Peptides in neurobiology (Gainer, H., ed.) Plenum Press, New York, 171-181
Ungar, G. (1974) Liefe Sciences 14: 595-604
Wise, C.D., Berger, B.D. and Stein, L. (1973) Biol. Psych. 6: 3-21

PEPTIDES AND RAT BRAIN MEMBRANE PHOSPHOPROTEINS

H. Zwiers, V.M. Wiegant, A.B. Oestreicher, P. Schotman and W.H. Gispen

Div.of Mol. Neurobiology, Rudolf Magnus Institute, Lab.of Physiol.Chemistry, Inst.of Molecular Biology, University of Utrecht, The Netherlands

Behaviorally active peptides derived from ACTH inhibit *in vitro* the phosphorylation of at least five protein bands from rat brain synaptic plasma membranes (SPM), as separated on SDS-polyacrylamide slab gel electrophoresis (Zwiers et al., 1976). A direct interaction of ACTH with SPM-protein kinase(s) is likely to be responsible for the effect on phosphorylation (Zwiers et al., 1978). The structure-activity relation found, closely resembles that observed for the effect of these peptides on grooming behavior in the rat (Gispen et al., 1975).

Under similar *in vitro* conditions cAMP stimulates the phosphorylation of three other SPM protein bands with higher molecular weight whereas cGMP is without effect. Attention is focussed on one ACTH-sensitive SPM protein band (B-50, M.W. 48,000), in particular, and on a procedure to isolate the native protein kinase responsible for the phosphorylation of that protein. Treatment of SPM with 0.5% Triton X-100 in 75 mM KCl solibilized 80% of the protein. In this solubilized fraction ACTH again inhibits the endogenous phosphorylation of band B-50. The inhibition is Mg-dependent but not dependent on the presence of Ca. Exogeneously added B-50 obtained after preparative SDS-polyacrylamide gel electrophoresis and extraction of B-50 out of the gel, is also phosphorylated by the SPM extract. Column chromatography using DEAE cellulose and a K-phosphate gradient yields endogenous B-50 phosphorylating activity at about 150 mM K-phosphate. Presumably the B-50 phosphorylating activity resides in a complex of B-50 and protein kinase which after application on Biogel P200 runs in the void volume.

Preliminary data on the effects of various fragments of β-LPH (endorphins) are shown.

If SPM protein phosphorylation is involved in certain types of synaptic transmission (Greengard, 1976) then the present data support the notion that behaviorally active peptides may act as modulators of neurotransmission.

Gispen, W.H., Wiegant, V.M., Greven, H.M. and De Wied, D. (1975) Life Sci. 17: 645-652

Greengard, P. (1976) Nature 260: 101-108

Zwiers, H., Velthuis, D., Schotman, P. and Gispen, W.H. (1976) Neurochem. Res. 1: 56-58

Zwiers, H., Wiegant, V.M., Schotman, P. and Gispen, W.H. (1978) Neurochem. Res. in press

SUBSTANCE P: TWO DIFFERENT BINDING SITES IN RAT BRAIN

A. Saria, N. Mayer, R. Gamse, F. Lembeck
Institut für experimentelle und klinische Pharmakologie
Universitätsplatz 4, A-8010 Graz, Austria

Binding of ^{125}J-Tyr^8-substance P was studied in synaptic vesicle(SV) and in synaptic plasma membrane (PM) preparations. Specific binding with high affinity was found in both preparations. In the SV-fraction only one population of binding sites was found (K_D=0.3 nM, maximum binding = 800 fmol per mg protein), whereas in the PM-fraction two populations of binding sites could be calculated from Scatchard plots, one of high (K_D=0.26 nM) and one of low affinity (K_D=1.9 nM).

The number of low affinity sites was 90 fmol per mg protein. The low number of high affinity sites (9 fmol per mg protein) could be explained by contamination of PM with SV as the K_D-values of the high affinity site of PM and the SV-binding site are identical.

The regional distribution of binding sites per mg protein of the SV-fraction and of the PM-fraction is different: medulla oblongata > midbrain > hypothalamus > striatum > cortex > cerebellum for SV, striatum > medulla oblongata, midbrain, cortex, cerebellum > hypothalamus for PM. The regional distribution of binding sites of SV fairly correlates with the substance P-content (radioimmunoassay) of the regions. The only exception is the hypothalamus, where the highest substance P-content contrasts with relatively few binding sites. No correlation, however, was found between the regional distribution of substance P and the binding sites of PM.

The substance P-related peptide physalaemin in concentrations of 10^{-6} - 10^{-5} M displaces labeled substance P from PM, increases, however, the number of binding sites for substance P up to 300 % in SV, presumably by positive cooperativity.

Conclusion: The results indicate the existence of two different binding sites for substance P localized in SV and in PM. The regional distribution of SV-binding sites correlates with the substance P-content of the brain regions.

ENDORPHINS RELEASE FROM PITUITARY CELLS

Rabi Simantov
Department of Genetics, Weizmann Institute of Science.
Rehovot, ISRAEL

Endorphins in the pituitary gland are concentrated in the median eminance and the anterior lobe, whereas the posterior lobe show residual endorphin activity (Bloom et al., 1977; Simantov and Snyder, 1977). The AtT-20 pituitary cell line originated from an adenocarcinoma of the anterior pituitary of LAF_1 mouse and whole rat pituitary glands were used to study the mechanism of endorphins release. When incubated in low potassium (5.6 mM), AtT cells release endorphins to the medium. This release is calcium dependent and therefor it is unlikely that this basal release evolve from damaged or broken cells. Furthermore, EGTA decreases this basal endorphin release in low potassium.

Depolarization evoked by increased potassium concentrations (50 mM) increased endorphins release more than two folds. The potassium stimulated release was calcium dependent in dose dependent fashion and was also inhibited by addition of EGTA in the presence of both high potassium and calcium. Basal and potassium stimulated calcium dependent endorphins release was also observed in-vitro in isolated whole rat pituitaries. The results may suggest that the basal endorphin release in-vitro from isolated whole pituitary glands or from cultured pituitary cells reflect some spontaneous activity which could be suppressed in-vivo by inhibitory hypothalamic system in similar way to the function of prolactin release-inhibiting and MSH release inhibiting factors. The potassium evoked endorphins release both from normal rat pituitary glands and from cultured pituitary cell line suggest for the existance of another regulatory mechanism, possibly a stimulatiry neurohumoral system that may control endorphins secretion from the pituitary as a response to environmental stimuli in analogy to the secretion of ACTH which recently have been showen to share common regulatory systems of synthesis and secretion with endorphins.

Bloom, F., Battenberg, E., Rossier, J., Ling, N., Leppaluoto, J., Vargo, T.M. and Guillemin, R. (1977) Life Sci. 20 : 43-48

Simantov, R. and Snyder, S.H. (1977) Brain Research 124 : 178-184

HUMORAL CONTROL OF APETITE: ISOLATION AND CHARACTERISATION OF A PEPTIDE INDUCING OBESITY IN MICE

P.D. Edminson, I. Foss, O. Trygstad, K.L. Reichelt
Institute for Pediatric Research, Rikshospitalet, Oslo 1, Norway

We have previously proposed a working hypothesis that peptide formation in the CNS can represent the temporal and geometrical sequence of signals impinging on master (command) neurones (Reichelt & Edminson, 1977). One consequence of this hypothesis is that in hyperfunctional disease states a surfeit of specific peptides should be produced, saturating breakdown systems, and thus emerging into the body fluids. We should thus be able to isolate the peptide(s) and induce the corresponding syndrome in experimental animals. This has been accomplished for a tripeptide isolated from patients with anorexia nervosa (Trygstad et al, 1978, Reichelt et al, 1978) which induced body weight reduction and reduced feed intake in mice.

We have also isolated a peptide from patients with metabolic obesity which induces a long-lasting obese state in mice with increased food intake. Following isolation of a protein carrier-peptide complex by gel filtration, the peptide was further purified by a number of gel filtration; ion exchange and partition chromatography steps. The purified peptide led to an increase in feed intake from the normal level of 5,5 g/mouse/day to 10 g after 20 weeks. Body wights increased from 35 g to 70 g.

Further presentation of other disease states will be given by Reichelt et al (this meeting).

Reichelt, K.L. and Edminson, P.D. (1977)
  Peptides in Neurobiology (Ed: H. Gainer) Plenum Press 171-181

Reichelt, K.L., Edminson, P.D., Foss, I., Trygstad, O., Johansen, J. and Bøler, J. (1978)
  Neuroscience, submitted

Trygstad, O., Foss, I., Edminson, P.D., Johansen, J. and Reichelt, K.L. (1978)
  Acta. Endocrinol. in press

SEA ANEMONE TOXINS : TOOLS IN NEUROBIOLOGY AND PHARMACOLOGY

L. Béress
Institut für Meereskunde an der Universität Kiel
D-2300 Kiel, West Germany
Prof. Dr. Carl Schlieper in honour of his 75th birthday

The sea anemones Anemonia sulcata and Condylactis aurantiaca contain highly toxic polypeptides which are utilized for capture of prey or for defence. Seven toxins were recently purified and characterized (Béress, L. et al., 1975; Béress R. et al., 1976). The molecular weights of the toxins range from 2678-5630 Dalton.

The effect of the toxins of A. sulcata (ATX I, II and III) have been investigated on the giant axons of the isolated nerve cord of the crayfish Astacus leptodactylus. ATX I, II and III prolong the duration of the action potential (Rathmayer and Béress, 1976). For ATX II it has been shown at Ranvier nodes of isolated frog nerves and at crayfish giant axons, that the prolongation of the action potential is due to a selective slowing of the Na-channel inactivation (Bergmann et al. 1976; Romey et al., 1976). ATX II and scorpion toxin, which has a similar effect on sodium channels, share a common receptor associated with the action potential sodium channel (Catterall and Béress, 1978).

ATX I and ATX II act also on mammalian cardiac muscle. ATX II evokes a dose dependent positive inotropic effect in the nanomolar (2-100nM) range (Alsen et al., 1976). In contrast to cardiac glycosides ATX II does not inhibit the action of the enzyme Na/K-ATP-ase (Alsen and Reinberg, 1976). The radioactive labeling of ATX I and ATX II has been recently accomplished (Hucho et al., 1978). The small size (Mw. 4770) of ATX II (Wunderer et al., 1976), its potent action on the mammalian heart muscle, its specific affinity to the sodium channel renders it one of the most interesting substances for the study of the molecular mechanisms accompanying exitation of cell membranes and for the characterization of the sodium channel.

References : Alsen, C., Béress, L., Fischer, K., Proppe, D., Reinberg, T. and Sattler, R. W. (1976) Naunyn-Schmiedeberg's Arch. Pharmacol. _295_, 55-62; Alsen, C. and Reinberg, T. (1976) Bull. Inst. Pasteur _74_, 117-119; Béress, L., Béress, R. and Wunderer, G. (1975) FEBS Lett. _50_, 311-314; Béress, R., Béress, L. and Wunderer, G. (1976) Hoppe-Seyler's Z. physiol. Chem. _357_, 409-414; Bergmann, C., Dubois, J. M., Rojas, E. and Rathmayer, W. (1976) Biochim. Biophys. Acta. _455_, 173-184; Catterall, W. A. and Béress, L. (1978) J. Biol. Chem., in press; Hucho, F., Stengelin, S., Rathmayer, W. and Béress,L. (1978) Hoppe-Seyler's Z. physiol. Chem. _359_, 278; Rathmayer, W. and Béress, L. (1976) J. comp. Physiol. _109_, 373-382; Romey, G., Abita, J. P., Schweitz, H., Wunderer, G. and Lazdunski, M. (1976) Proc. Natl. Acad. Sci., USA _73_, 4055-4059; Wunderer, G., Fritz, H., Wachter, E. and Machleidt, W. (1976) Eur. J. Biochem. _68_, 193-198.

Round Table

Dendritic, Axonal and Transneuronal Transport

Chairman:
M. Cuénod

DENDRITIC, AXONAL AND TRANSNEURONAL TRANSPORT: AN INTRODUCTION

M. Cuénod
Brain Research Institute, University of Zurich
CH-8029 Zurich

Axonal transport is a special case of intracellular transport which appears in the neuronal processes, axon as well as dendrites. Communication is required between the perikaryon and the periphery of the neuron: the materials, necessary for the growth and the maintenance of neuronal processes and essential for the performance of synaptic transmission, are synthetized in the perikaryon and transported to the periphery. Reciprocally, the periphery seems to provide the cell body with transported material, involved at least partly in control functions. Some of the transported molecules have been shown to be transferred to neighboring neurons or glial cells. Transports in two directions, at different speeds and of various materials, have been reported.

Best investigated is the <u>somatofugal</u> transport, often called anterograde, in which the materials transported migrate from the perikaryon toward the endings of the axon or the dendritic extremities. Two main speeds of migration have been described: the "slow" transport moves at a rate of one or a few mm per day (micron/minute) while the "fast" transport progresses at a rate of 100 to 500 mm per day (microns/second). The <u>fast</u> somatofugal axonal transport conveys macromolecules (proteins, glycoproteins, phospholipids), possibly also small molecules (amino acids, nucleotides), which appear to be associated with the membraneous compartments (SER) of the axon and after cell fractionation concentrate in the particulate fractions. This material plays an important role in the maintenance of cell membranes and the machinery necessary for synaptic transmission. The <u>slow</u> transport conveys the bulk of the soluble proteins responsible for the renewal of the axonal matrix. Mitochondria seem to migrate at a slow speed, although they have been observed to move rapidly over short distances and might be found in the terminals labelled by fast transport. The <u>somatopetal</u> transport, often called retrograde, carries materials from the nerve terminals to the perikaryon. It has been demonstrated for both foreign proteins, like horseradish peroxidase, certain lectins or tetanus toxin and for physiological proteins as nerve growth factor (NFG). Toxin and NGF are specifically picked up by the neuronal membrane.

THREE ASPECTS OF THE AXONAL TRANSPORT: MOLECULAR SELECTIVITY,
INTRAAXONAL COMPARTMENTATION AND INTERCELLULAR EXCHANGE

B. Droz and J.-Y. Couraud
Département de Biologie, Commissariat à l'Energie Atomique, C.E.N. de
Saclay, F-91190 Gif-sur-Yvette et U.E.R. de Physiologie, Université
Pierre et Marie Curie, Paris

Most of the neuronal macromolecules required for the maintenance of
the neural circuitry are synthesized in the nerve cell body. Newly synthesized proteins, glycoproteins and phospholipids are then translocated
to the most remote areas of the neuronal expansions in which they become
physiologically operative. The dispatching of specific molecules aimed
at definite sites raises basic questions.

1. <u>Molecular selectivity</u>. When a neuronal enzyme exists under
different molecular forms, does the axonal transport reflect or not a
different behavior of these molecules? In extracts of chicken sciatic
nerve, acetylcholinesterase (AChE) is represented by 4 molecular forms
which are separated on sucrose gradient according to their sedimentation
coefficients: two light forms 5s and 7s and two heavy forms 11s and 20s.
After a nerve transsection, 20% of the 11s and practically all the 20s
forms pile up rapidly in the interrupted stumps whereas the 5s and 7s
forms accumulate much more slowly. Thus the various molecular forms of
AChE are selectively transported along the axons: the 11s and 20s forms
are conveyed with the fast axonal transport (about 400 mm per day) and
the 5s and 7s forms sweep down the axons with the slow axoplasmic flow
(5 - 10 mm per day).

2. <u>Intraaxonal compartmentation</u>. Does the selective convection of
molecules imply the existence of specialized channels for the axonal
traffic? Ultrastructural studies reveal that three main compartments are
present in the axon: a/ the endoplasmic reticulum (SER), which forms a
complex system of pipes and tubes, ensures the fast axonal transport of
most membrane-bound components in ortho- and retrograde direction;
b/ the axoplasm, which includes the axonal cytoskeleton, constitutes a
huge pathway for the slow axonal transport of most proteins unbound to
membranes; c/ the mitochondrial compartment, which is the site of an
autonomous synthesis and of exchange of material with the SER, displays
a net displacement of these organelles at a slow rate.

3. <u>Intercellular exchange</u>. Is there or not a supply of glial cells with neuronal molecules in the course of their axonal transport? Radioautographic results indicate that a neuron-glia exchange of intact macromolecules such as proteins or glycoproteins does not seem to occur except for phospholipids. After administration of ^{3}H-choline or ^{3}H-glycerol, there is a transfer of label from the axon to the encompassing myelin sheath. The accumulation of label in myelin results from two mechanisms: a transfer of phospholipid and a glial incorporation of labeled choline released from the axon. Such an intercellular exchange of molecules could play an important role in the maintenance of the myelin sheath and reflect a chemical interaction between axon and glial cell.

CORRELATIVE ASPECTS OF INTRA- AND TRANSNEURONAL TRANSPORT

G.W. Kreutzberg and P. Schubert
Max Planck Institute for Psychiatry, Kraepelinstrasse 2
8000 Munich 40, F.R. Germany

Intracellular injection of radioactive amino acids reveals cellulifugal transport of proteins in dendrites and in the axon (reviewed in Kreutzberg and Schubert, 1975). In short-term experiments proteins are seen exclusively within the processes of the injected cell. This confirms the view that most of the protein of the neuron is destined to maintain the expanded cellular structures. Further, if adenosine or other nucleosides are injected into nerve cells labelled material can be demonstrated to be not only transported but also released from dendrites and axons. The most active release seems to occur at synaptic terminals where also a transneuronal transfer to the postsynaptic cells occurs (P. Schubert et al., 1977).

It is assumed that adenine nucleotides e.g. ATP are preferentially released. This release is dependent on synaptic activity and increases with stimulation (P. Schubert et al., 1976). Nucleotides appearing in the extracellular cleft cannot pass through membranes. Therefore, to explain the impressive transneuronal transfer of adenosine derivatives, enzymatic mechanisms must be provided. Such mechanisms have been discovered in the form of ATPases and, in particular, 5'-nucleotidase, the enzyme producing membrane-permeable adenosine by hydrolysing AMP. 5'-Nucleotidase is an ectoenzyme located on glial plasma membranes. It is especially active on juxtasynaptic astroglial lamellae and on astrocytic footplates of the vasculature (Kreutzberg et al., 1978).

In both locations formation of adenosine could have important functions such as, supression of neuronal electrical activity, stimulation of cAMP formation or dilatation of local capillaries.

Kreutzberg, G.W. and Schubert, P. (1975) In: The Use of Axonal Transport for Studies of Neuronal Connectivity, Cowan W.M. and M. Cuénod (Eds.) pp. 83-112, Elsevier, Amsterdam
Kreutzberg, G.W., Barron, K.D. and Schubert, P. (1978) Brain Research, in press
Schubert, P., Lee, K., West, M., Deadwyler, S. and Lynch, G. (1976) Nature 260: 541-542
Schubert, P., Rose, G., Lee, K., Lynch, G. and Kreutzberg, G.W. (1977) Brain Res. 134: 347-352

RAPID AXOPLASMIC TRANSPORT OF LOW MOLECULAR WEIGHT SUBSTANCES

D.G. Weiss
Institute for Zoology, University Munich, Luisenstr. 14 and
Max Planck Institute for Psychiatry, 8000 Munich 40, F.R. Germany

Low molecular weight compounds such as amino acids or sugars have widely been used to study axoplasmic transport. They are mainly incorporated into macromolecules whose transport is well documented. Further investigation has shown that these macromolecules are themselves integrated into particulate material (MW above 1 million), which seems to be preferentially transported at the most rapid rates (Gross and Weiss, 1977).

Rapid transport of the precursors themselves is highly controversial (Karlsson, 1977). We have studied this problem in the olfactory nerve of the pike (Esox lucius) which is well suited for studies on axoplasmic transport and which is morphologically and biochemically well characterized (Kreutzberg and Gross, 1977; Weiss et al., 1977).

It was possible to show that rapid transport is not restricted to high molecular weight (HMW) compounds and their complexes, but that a considerable amount of free amino acid, i.e. leucine, is immediately passed into the axon where it is rapidly transported, attaining essentially the same maximal velocity as the HMW material. The amino acid peak is followed two hours later (at 19°C) by the proteins (Gross and Kreutzberg, 1977). After several hours we found that the amount of free amino acid which still travels at maximum velocity has decreased and the bulk of the amino acid has moved only very short distances. The bulk of the HMW material has continued its rapid movement, whereas only a small proportion has been retarded.

It is concluded that the low molecular weight (LMW) compounds more readily equilibrate between the moving compartment and the stationary compartments. Therefore, and in accordance with the chromatographic concept of axoplasmic transport, LMW substances are in bulk retarded. Rapid transport of LMW substances is, however, observed over short distances (10 or 20 mm). Rapid transport of selected LMW compounds over longer distances may require binding to macromolecules or organelles which seem to be less prone to leave the transport compartment.

Gross, G.W. and Weiss, D.G. (1977) Neurosci. Lett. 5: 15-20
Karlsson, J.O. (1977) J. Neurochem. 29: 615-617
Kreutzberg, G.W. and Gross, G.W. (1977) Cell Tiss.Res.181: 443-457
Weiss, D.G., Krygier-Brévart, V., Gross, G.W. and Kreutzberg, G.W. (1978) 139: 77-87
Gross, G.W. and Kreutzberg, G.W. (1978) Brain Res. 139: 65-76

RETROGRADE AXONAL AND TRANS-SYNAPTIC TRANSPORT OF MACROMOLECULES

M.E. Schwab, M. Dumas and H. Thoenen
Dept. of Pharmacology, Biocenter of the University
CH-4056 Basel

Many macromolecules, such as horseradish peroxidase (HRP) or Evans-blue-albumin, are taken up by nerve terminals and transported retrogradely if they are injected in very large quantities. In fact, retrograde axonal transport of these macromolecules has become a very popular and useful tool in neuroanatomy during the past few years. In contrast to the rather unspecific uptake of these tracer molecules, which may largely occur as a consequence of transmitter release by exocytosis and "accidental" uptake of extracellular material, there exists a group of macromolecules which are taken up and transported by a much more selective and efficient mechanism. This group of macromolecules includes nerve growth factor (NGF), tetanus and cholera toxins, antibodies to dopamine B-hydroxylase (DBH; only in adrenergic neurons) and various lectins, i.e. molecules with a wide range of molecular weights and physico-chemical properties. Their common characteristic seems to be their ability to bind with high affinity to specific components of nerve terminal membranes. NGF receptors, DBH, monosialogangliosides (binding sites for cholera toxin), di- and trisialogangliosides (binding sites for tetanus toxin), and various sugar moieties of cell surface glycoproteins and glycolipids (binding sites for lectins) are known to be located in the surface membrane of nerve terminals. EM studies using tracer-coupled tetanus toxin have shown that the toxin - in contrast to albumin - becomes immediately associated with nerve terminal membranes in the rat iris and is subsequently internalized. Biochemical studies have shown saturability of the retrograde transport system which varies from macromolecule to macromolecule and probably reflects the limited number of binding sites available for uptake and retrograde transport. Due to this highly efficient uptake and retrograde transport, labelled tetanus toxin and wheat germ agglutinin can be used as very sensitive and reliable retrograde tracers for neuroanatomy in the central and peripheral nervous system.

EM autoradiographic and ultrahistochemical studies using ^{125}I-labelled and HRP-coupled NGF, tetanus toxin or wheat germ agglutinin have shown that within the cell the material is always present in a membrane-delineated compartment. Transport in the axon occurs in smooth vesicles, tubules or cisternae (transport rate 3-10 mm/h). After arrival at the cell body, the majority of the label is finally incorporated into secondary lysosomes. In peripheral adrenergic neurons, retrogradely transported NGF triggers its specific effects, i.e., the increased synthesis of tyrosine hydroxylase and DBH.

On the other hand, a part of the retrogradely transported tetanus toxin is released from the dendrites of peripheral sympathetic ganglion cells or spinal cord motoneurons and is taken up by closely apposed pre-synaptic terminals. This retrograde trans-synaptic transfer of tetanus toxin is highly selective and does not occur to a significant extent in case of NGF, cholera toxin or the lectins. Studies with HRP-tetanus toxin suggest that this selective trans-synaptic transfer is due to selective release by the dendrites.

These observations explain how tetanus toxin reaches its probable site of action, the inhibitory nerve terminals ending on spinal cord motoneurons, where it blocks the release of glycine and GABA and thereby provokes hyperactivity of the motoneuron. They also demonstrate the existence of pathways for macromolecules (MW of tetanus toxin 150'000) from a target organ by retrograde transport through a first order neuron and subsequent trans-synaptic migration to a second order neuron. Transfer of information coded by macromolecules via such pathways may be of importance during embryonic development as well as for regulation and maintenance of the adult pattern and functional state of the nervous system.

FAST AXONAL TRANSPORT IN RAT SCIATIC NERVES IN VITRO.

M. Hanson; A. Edström and S.Gershagen
Dept. of Zoophysiology, Univ. of Lund, Sweden.

The dorsal ganglia-sciatic nerve of the rat was used to study the migration of a pulse of (^{3}H)leucine-labelled protein in vitro. The preparations were incubated in an apparatus where the ganglia could be separated from the nerve by a glass partition and silicone grease. The transport of newly synthesized protein was reversibly blocked by low temperature in the nerve compartment while synthesis in the ganglia was allowed to continue for 2-3 h at 37°C. Subsequently the temperature in the nerve compartment was raised and a pulse of labelled protein migrated distally. Since the starting-point and -time are known, the rate can be calculated with high precision even in short nerves. The rate of transport in the peripheral and central branches of the ganglion cells were found to be closely similar; 18 mm/h at 37°C. An identical rate has also been found for AChE in the same nerve in vivo with a modification of the present technique (Hanson, 1978). The shape and behavior of the pulse corresponds exactly with that obtained in frog sciatic nerves under the same conditions (Edström and Hanson, 1973).

The extremely low background radioactivity due to local incorporation into Schwann cells, obtained with this technique, makes the preparations ideally suited for high-resolution, 2-dimensional electrophoresis of transported material. Preliminary electrophoretic studies of (^{14}C)labelled material, transported in the sciatic nerve and dorsal root will be presented.

Hanson, M. (1978) Brain Res. in press.
Edström, A. and Hanson, M. (1973) Brain Res. 58: 345-354.

Supported by Statens Naturvetenskapliga Forskningsråd grant no. 2535-15.

AXONAL TRANSPORT AND INCORPORATION OF RADIOACTIVITY AFTER INJECTION OF A SIALIC ACID PRECURSOR INTO THE RED NUCLEUS, AN AUTORADIOGRAPHIC STUDY.

L.D. Loopuijt
Medical Chemistry, Faculty of Medecine, Vrije Universiteit
van der Boechorststraat 7, Amsterdam, THE NETHERLANDS.

Being interested in the incorporation of sialic acid in neuronal cells of the mammalian central nervous system, rats were injected with N-^{3}Hacetyl-mannosamine, a sialic acid precursor, into the red nucleus. The red nucleus is a group of neuronal cell bodies, that is localized in the brainstem. Nerve endings, localized on cell bodies or dendrites of the red nucleus, come from different groups of cell bodies. Of these groups, we mention the cerebellar nuclei, the nucleus dentatus and interpositus (Massion,Physiol. Rev. 47 (1967) 383-436).

After the red nucleus injection, performed with a stereotactic apparatus, the rats survived 24 hours after injection of the radioactive N-acetyl-mannosamine, were perfused with phosphate buffered formalin 4% and autoradiograms of the brain were prepared for light microscopy.

After treatment of the tissue with phosphate buffered formalin, 95% of the low molecular radioactive molecules are not fixated. This can be washed away, so that in our autoradiograms silver grains represent almost exclusively sialoglycomacromolecules.
The following observations were made:
1) In the injection site not only the red nucleus was labeled, but also the central gray matter and the reticular formation.
2) Over the neuronal cell bodies of the red nucleus, a lower concentration of silver grains can be found, than over the neuropil inbetween those cell bodies.
3) Accumulation of silver grains over cell bodies of the cerebellar nuclei, the nucleus dentatus and interpositus can be observed: the concentrations silver grains over some cell bodies being much higher than the surrounding neuropil.
Interpretations of the results:
a) The relative low concentration of silver grains over the neuronal cell bodies of the red nucleus in relation to the surrounding neuropil after a survival time of 24 hrs. can mean that non-neuronal cells have incorporated radioactivity, and/or that the radioactivity after being taken up by the neurons is transported into axons and/or dendrites within 24 hrs.
b) The accumulation of silver grains over the neurons of the cerebellar nuclei can be explained by retrograde axonal transport of radioactivity.

AXONAL TRANSPORT. INTERACTION OF RAPIDLY TRANSPORTED GLYCOPROTEINS WITH LECTINS AND GLYCOPROTEINS.

J.-O. KARLSSON
Intitute of Neurobiology, University of Göteborg, Göteborg, Sweden

Albino rabbits were injected intraocularily with ^{3}H-fucose 20 hours prior to sacrifice. The lateral geniculate bodies and superior colliculi containing rapidly transported fucose-containing glycoproteins were dissected out and homogenized in 50 mM tris-HCl (pH 7.2) containing 2% of the non-ionic detergent Berol 172. Following centrifugation at 10^5 g for 1 hour at 15°C, the supernatant was subjected to gel filtration on Sepharose CL-6B equilibrated in the same buffer containing 0.2% detergent. The eluate containing the void volume and the low mol.wt. region was discarded. The retarded fraction containing the major labelled components was collected and subjected to affinity chromatography in the same buffer. Proteins to be immobilized were coupled to CNBr-activated Sepharose 4B.

The labelled material was allowed to adsorb to the affinity column. Following washing with buffer the adsorbed glycoproteins were eluted with the appropriate haptene sugar. The result is indicated in the table (mean of 2-5 expts).

Immobilized protein	Not bound to column	Bound and eluted with the appropriate sugar	Irreversibly bound
Asialo-fetuin	98	0	2
Ovomucoid	92	5	3
Peanut agglutinin	92	1	7
Soybean lectin	96	2	2
Concanavalin A	50±10	22±8	28±10
Wheat germ lectin	50±8	45±5	5 ±1

PERCENT OF APPLIED RADIOACTIVITY

The results indicate that rapidly transported fucose-labelled glycoproteins may contain terminal N-acetyl-D-glucosamine, glucose or mannose residues but not exposed D-galactose or N-acetyl-D-galactosamine residues.

AXONAL TRANSPORT OF TAURINE IN THE VISUAL SYSTEM OF THE DEVELOPING RABBIT

J. A. Sturman

Developmental Neurochemistry Laboratory, Institute for Basic Research in Mental Retardation, Staten Island, N. Y. 10314, U. S. A.

Taurine is the major constituent of the free amino acid pool in developing mammalian neural tissue but its biological functions are still unclear. It has been proposed as an inhibitory neurotransmitter, or neuromodulator, stabilizing axonal membrane potential. Taurine is clearly involved in the structural integrity of the retina, of the cat at least, and has been implicated in the development of the brain. More recently, taurine has been demonstrated to be transported axonally in the goldfish visual system (Ingoglia et al., 1977; Ingoglia et al., 1978).

The present experiments compared the axonal transport of taurine in the visual system of 10-day-old, 20-day-old and adult rabbits. The arrival of the peak of axonally transported [^{35}S]taurine in the optic tract, lateral geniculate body and superior colliculus (measured by radioactivity in the region contralateral to the injected eye less radioactivity in the region ipsilateral to the injected eye) was earliest in the youngest rabbits and latest in the adult rabbits. Only one component of transported radioactivity could be distinguished in each case. The amount of [^{35}S]taurine arriving at the peak was greatest in the 10-day-old rabbits (4.26, 7.26, 13.73 times greater than adult in optic tract, lateral geniculate body and superior colliculus, respectively) (cf. 1.97, 4.13, 10.45 times greater than adults for the same regions of 20-day-old rabbits). These results suggest that axonally transported taurine may have a special importance in the development of the rabbit visual system. Synaptic connections begin in the rabbit about day 7 with the greatest formation occurring after day 10 until day 20, i.e. after natural eye opening. Taurine may be acting as a modulator of electrical activity in the developing axon, a function which might facilitate development of axons and the formation of synaptic connections. These results, obtained exclusively for the regions of the brain in which the optic axons terminate, could be representative of the brain as a whole, and perhaps provide the explanation for the high concentrations of taurine in developing mammalian brain.

(Supported by New York State Department of Mental Hygiene)

Ingoglia, N.A., Sturman, J.A., Lindquist, T.D. and Gaull, G. E. (1977)
 Brain Res. 115: 535-539

Ingoglia, N.A., Sturman, J.A., Rassin, D.K. and Lindquist, T.D. (1978)
 J. Neurochem. in press

EFFECT OF SULFHYDRYL REAGENTS ON MICROTUBULE-ASSOCIATED ATP-ASE ACTIVITY AND AXONAL TRANSPORT

Margareta Wallin[*], Håkan Larsson[*] and Anders Edström[**]
Departments of Zoophysiology, University of Göteborg[*] and Lund[**], Sweden

It has often been suggested that rapid axonal transport depends on the function of microtubules (MT), although the mechanism is unknown. Several enzyme activities have been found in MT preparations, some of which could be involved in the transport process.

Dynein, a protein with ATP-ase activity, is seen as arms on the MT wall of cilia and flagella. A similar arm-like structure has been found on MT from brain and a brain MT-associated ATP-ase activity has also been found. The function of these enzymes has not been elucidated. In cilia and flagella the ATP-ase activity has been proposed to be the driving force for cilia and flagella movement.

We have earlier reported that small amounts of zinc and sulfhydryl reagents stimulate rapid axonal transport in frog sciatic nerve in vitro. (Edström and Matsson, 1976). At this concentration of reagents colchicine binding was unaffected. At higher concentrations the transport as well as the colchicine binding was inhibited. It was therefore of interest to study if the sulfhydryl reagents could affect the MT-associated ATP-ase activity at low concentrations.

The Mg^{2+}- ATP-ase activity was studied in a MT preparation prepared from bovine brain in a glycerol-free PIPES-buffer (pH 6.8) by two cycles of an assembly-diassembly procedure (Larsson et al., 1976). The samples were incubated with Mg^{2+} (2mM) and ATP (1mM) at 37°C during 15 min. Low concentrations of NEM, PCMBS, Cd^{2+} and Zn^{2+} (10^{-6} and 10^{-7}M) were found to stimulate the ATP-ase activity. Higher concentrations (10^{-3} and 10^{-4}M) inhibited the activity. In contrast, sulfhydryl reagents at low concentrations (10^{-6} and 10^{-7}M) do not affect the polymerization of MT in vitro in a similar preparation (Wallin et al., 1977). It is tempting to speculate that transport stimulation by critical concentrations of sulfhydryl reagents is due to their interaction with a MT associated ATP-ase.

Supported by a grant from the Swedish Natural Research Council (2535:15)

Edström, A. and Mattson, H. (1976) Brain Res. 108: 381-395
Larsson, H., Wallin, M. and Edström, A. (1976) Exp.Cell.Res.107: 219-225
Wallin, M., Larsson, H. and Edström, A. (1977) Exp.Cell.Res.100: 104-110

MIGRATION OF RNA AND RNA PRECURSORS ALONG NEONATAL AND YOUNG ADULT RAT OPTIC AXONS

M.J. Politis and N.A. Ingoglia
Depts. Physiol. and Neurosci., New Jersey Medical School
Newark, New Jersey, USA 07103

In an attempt to determine metabolic requirements of neurons during development, the axonal migration of RNA and RNA precursors(RP) were compared in neonatal and young adult rat optic axons. In this system a large extent of development occurs postnatally, the majority of synapses being formed between 5 and 12 days after birth. ^{3}H uridine was injected into right eyes of developing (1 or 4 day old) and young adult (40 day old) rats. Rats were sacrificed at various times after injection ranging from 6 hours to 20 days. Right and left geniculates(RLG and LLG) were removed and assayed for trichloroacetic acid(TCA) soluble (containing RP) and RNA radioactivity. Left minus right geniculate(L-RLG) activity was used as an index of axonally migrating radioactivity. Metabolism of ^{3}H-RP was determined by lyophilizing the TCA soluble fractions, while the nucleotide composition of the nonlyophilizable radioactivity was determined by thin layer chromatography. Greater than 90% of RLG TCA soluble radioactivity was metabolized to volatile substances (probably ^{3}H H_2O) by three days after injection, leaving ∼3% of the neonatal and ∼10% of the adult activity as ^{3}H-RP. However, in LLG fractions (containing transported material) ∼15% and 40% of total TCA soluble activity was present as ^{3}H-RP in the neonates and adults, respectively, indicating that axonally migration ^{3}H-RP may have been stored in a metabolically protected pool not occupied by blood born ^{3}H-RP. The levels of L-RLG ^{3}H RNA in the neonates were 10 times higher than in adults. In neonates peaks of ^{3}H RNA occurred at ∼5 and 9 days after birth in rats injected at 1 and 4 days of age, indicating that this RNA may have been linked to developmental events. Polyacrylamide gel electrophoresis of neonatal LLG RNA showed a small but significantly higher ratio of 4S/total RNA radioactivity following intraocular (IO) ^{3}H uridine than intraventricular (IV) ^{14}C uridine injection. The labelling pattern of RLG RNA following IO and IV injection was similar. Based on these data we suggest that some of the ^{3}H RNA in the neonates may be axonally transported ^{3}H 4S RNA. Since the ratio of L-RLG ^{3}H-RP/^{3}H RNA in the neonates was ∼ 0.2 at peaks of ^{3}H RNA activity, it is likely that the remainder of L-RLG neonatal ^{3}H RNA was due to a rapid and efficient incorporation of axonally transported ^{3}H-RP into extraaxonal LLG RNA.

A STUDY ON THE AXONAL FLOW OF PHOSPHOLIPIDS IN THE CILIARY GANGLION OF THE CHICKEN

M.Brunetti, B.Droz*, L.Di Giamberardino* and G.Porcellati
Department of Biochemistry, University of Perugia, Italy and *Département de Biologie, Commissariat à l'Energie Atomique, C.E.N. Saclay, Gif-sur-Yvette, France.

2-^{3}H glycerol and Me-^{14}C-choline were injected simultaneously into the cerebral aqueduct of two weeks-old chickens near the Edinger-Westphal nucleus which contains the perikaria of preganglionic neurons. These neurons terminate into the ciliary ganglion forming the giant presynaptic calices. At different time intervals, 1-240 hours, the ciliary ganglion (containing both pre- and post-ganglionic axons) was extracted and the distribution of radioactivity in the phospholipid and hydrosoluble compounds measured. Specific activities were also calculated, when possible.

For all the phospholipids examinated (Ptd-Choline, Ptd-Ethanolamine, Ptd-Inositol, Ptd-Serine), a labelling maximum is observed at 20 hours after injection; only sphingomyelin shows a gradual increase with time. The most heavily labelled phospholipids, for both isotopes, are, at any time, choline phosphoglycerides; the corresponding values for ^{14}C and ^{3}H are about 70% and 55%, respectively, of the total radioactivity present in the ganglion, at least for the time intervals between 6 and 72 hours.

The amount of the radioactivity found into the hydrosoluble compounds represents only 6-8% of total ^{3}H at any time of observation, while that of ^{14}C shows a gradual decrease during time (about 30% at 6 h, 18% at 20 h, 6% at 240 h).

The isotopic ratio ^{14}C/^{3}H shows during time an increase of the ^{14}C which may indicate a reutilization of ^{14}C -choline for the synthesis of new PtdCho either by base-exchange or by net synthesis in the post-ganglionic neurons (B.Droz et al., 1978).

Time from injection (hr)		1	6	10	20	40	72	240
PtdCho*	^{14}C	0.03	0.37	0.48	1.00	0.57	0.59	0.29
	^{3}H	0.14	0.83	0.62	0.91	0.46	0.37	0.08
	^{14}C/^{3}H	0.21	0.44	0.77	1.09	1.23	1.59	3.62

The results are expressed as nCi/ganglion, and are the mean obtained from analyses carried out on 5 ganglions.

The small labelling of hydrosoluble compounds suggests the possibility that the phospholipids may be transported from the nerve cell bodies to the nerve endings by axonal flow.

Data are also presented which indicate that an active catabolism of the transported phospholipids occurs.

B.Droz et al., This Congress.

SUBSTRATE SPECIFICITY OF γ-AMINOBUTYRIC ACID TRANSPORT SYSTEMS IN THE RAT THYROID

H. Gebauer
Department of Zoology, University of Graz,
Graz, Austria

Functional relations between the sympathetic nervous system and the thyroid are well known (Melander et al., 1975). Furthermore, investigations have also shown in the thyroid the occurence of GABA (Haber et al., 1970) as well as the presence of two active carrier systems transporting GABA with K_T-values of 1,8 µM and 800 µM resp. (Gebauer, 1977). The present study is aimed to characterize the substrate specificity of these high- and low-affinity uptake systems.

High- and low-affinity GABA transport is inhibited by structural analogues of GABA while α-amino acids do not interfere. Strongest antagonists of high-affinity GABA uptake are trans-4-aminocrotonic acid and ß-alanine; taurine is only poorly effective. Contrary, taurine (K_i = 84 µM) and ß-alanine (K_i = 130 µM) are much stronger competitive inhibitors of low-affinity GABA uptake than compounds of closer structural relationship to GABA. Moreover, taurine and ß-alanine are not only competitors for GABA carrier sites but they are also transported with K_T-values similar to their inhibition constants of GABA uptake. Considering physiological conditions, these data lean support to the hypothesis that the high-affinity GABA transport system described above may indeed function as GABA carrier whereas the low-affinity system might serve as taurine carrier.

Gebauer, H. (1977) Annual Meeting of the Austrian Biochemical Society, Vienna: 14

Haber, B., Kuriyama, K. and Roberts, E. (1970) Biochem. Pharmacol. 19: 1119-1135

Melander, A., Ericson, L.E., Sundler, F. and Westgren, U. (1975) Rev. Physiol. Biochem. Pharmacol. 73: 39-71

Round Table

The Value of Nerve Cell Cultures for Neurochemical Research

Chairman:
P. Mandel

THE VALUE OF NERVE CELL CULTURES FOR NEUROCHEMICAL RESEARCH

P. Mandel, M. Sensenbrenner and J. Ciesielski-Treska
Centre de Neurochimie du CNRS
11, Rue Humann, 67085 Strasbourg Cedex, France

Cell cultures afford model systems for studying molecular aspects of nervous cell differentiation and function, intracellular regulatory mechanisms, cellular interactions and drug effects. The difficulty of inducing isolated normal neurons to divide in culture limits the use of these cells. Moreover, normal neuroblasts or neurons can be obtained only in mixed populations of primary cultures. Recently, new methods for isolating and culturing neurons were developed, but the survival time of these cells is rather short. Astrocytes can be maintained in primary and secondary cultures for long periods of time, although alterations due to the absence of neurons should be considered. Morphological and biochemical investigations have shown that extracts of embryonic, and even adult, brain promote neuronal differentiation as well as glial proliferation and differentiation. Increases in neuronal GMP cyclase and in glial carbonic anhydrase and S100 protein have been induced by these extracts. Moreover, a number of continuous cell lines derived from spontaneous or induced neoplasms, and from treatment of primary cultures with chemical or viral agents, have become available in the past few years. The characterization of their neuronal or glial specificity by morphological, biochemical and immunological methods is not always easy. Neuronal markers, such as enzymes involved in neurotransmitter metabolism, neurotransmitters, specific proteins (14.3.2) or surface antigens (Schachner, 1974) are often missing. Similarly, glial markers like S100 protein, carbonic anhydrase, 2'3'-nucleoside phosphate 3'-phosphohydrolase are not always present in glial type cells.

When cultured under suitable conditions, the dissociated cells of primary cultures of embryonic or newborn animal brain, dorsal root ganglia, and retina show histological organization which may resemble the tissue of origin. Purified neurons from sympathetic ganglia, and neuroblastoma clonal cell lines, have provided adequate models for the study of catecholamine metabolism. Tyrosine hydroxylase, dopamine-β-hydroxylase, and an inhibitor of this latter enzyme, as well as high amounts of catecholamines and their metabolites, have been detected in some clones. An increase of the activities in parallel to that of catecholamine biosynthesis was observed under conditions of cell maturation induced by bromodeoxyuridine, cAMP, or agents which increase adenylate cyclase or inhibit phosphodiesterase. A pronounced release of newly synthesized catecholamines in culture medium was observed. Similarly, neurosecretory events can be investigated either in primary cultures or in transformed cells of the hypothalamus (De Vitry et al., 1974). Cultured cells and tissues also release some differentiation and growth factors; conditioned medium allows isolation of these particular molecules. Some enzymes, including nicotinamide deamidase (M. Wintzerith, personal communication), are also released to the culture medium.

β-Adrenergic agonists raise the intracellular concentration of cAMP in cultures of mouse brain glial cells. This rise is prevented by a β-adrenergic antagonist and lowered by a coculture of these glial cells with neurons. Cell culture offers unique possibilities for studying ecto-enzymes: calcium and magnesium dependent ATPases, inorganic pyrophosphatases, a part of cellular acetylcholinesterase and 5'-nucleotidase. Striking differences in ecto-ATPase activities have been observed in cocultures of neuroblasts and glioblasts and in reisolated cells of each type during several generations, suggesting a cell selection during the coculture. Similarly, desialylation produces changes of some ecto-enzyme activities. Regulation of S100 protein synthesis in C6 cell line, as well as in rat astroblasts, was also investigated. Proteins with known functional characteristics, such as tubulin, filamentous protein, and actin have been found in tissue culture, and may be studied in relation to regulatory mechanisms.

Neurons may become quite mature cytologically and form abundant synapses. Thus, synapse formation has been shown to occur between clonal neuroblastoma and glioma hybrid cells (NG 108-15)(Nelson et al., 1976) or retinal (Puro et al., 1977) cells and dissociated striated muscle cells in culture. Following the discovery of opiate receptors in neuroblastoma-glioma hybrid (NG 108-15) (Klee and Nirenberg, 1974; Traber et al., 1975) alterations suggesting the establishment of a drug tolerance were reported. Similarly, barbiturate dependence and tolerance accompanied by an induction of mitochondrial biogenesis, were described in glial cells (Roth-Schechter and Mandel, 1976). However, tissue culture has its advantages and disadvantages, and we have to learn to choose the particular model system that will best answer our questions. Thus, energy metabolism of cell cultures will suffer from a relative hypoxia, and only one lactate dehydrogenase isozyme exists in most neuroblastoma cells. High affinity uptake systems for neurotransmitters and their precursors, though present in some clones, are missing in many others. Induction of lysozymal enzymes by hydrocortisone and thyroxine obtained in tumoral or transformed cells could not be produced in primary cultures. Striking differences in the response of adenyl cyclase to usual inducer molecules were found between tumoral or transformed cells and primary cultures. The fatty acid compositions of cultured cells also differ greatly from normal. An apparent lack of specificity of synapse formation has been noted in many types of cultures. A serious pitfall encountered in the study of clonal cell lines is the frequency of chromosomal abnormalities in number and structure. In this respect, diploid primary cultures undoubtedly offer a great advantage. Although over 100 mammalian glial lines have been studied, relatively few have been well characterized with regard to their capacity of expressing differentiated biochemical parameters. Qualitative and quantitative differences in the genetic programs for neuroblast and glioblast differentiation in a cell culture system have been reported. In this field, like in any other, the main problem is to ask the right questions with the right system to know what the system can give, not to ask of it more that it can give, and to be critical. Obviously, whatever is discovered in tissue cultures must also be demonstrated *in vivo*.

   Finally, it should be emphasized that such studies should not be aimed at repeating, in tissue culture, what we already have learned from direct investigations on the nervous system or what can be assessed much better with other methods. The simplified system offered by cells in culture should be focused on obtaining a clean answer using a clean system.

De Vitry, F., Camier, M., Czernichow, P., Benda, P., Cohen, P. and Tixier-Vidal, A. (1974) Proc. Nat. Acad. Sci. USA 71: 3575-3579.

Klee, W.A. and Nirenberg, M. (1974) Proc. Nat. Acad. Sci. USA 71: 3473-3477.

Nelson, P., Christian, C. and Nirenberg, M. (1976) Proc. Nat. Acad. Sci. USA 73: 123-127.

Puro, D.G., De Mello, F.G. and Nirenberg, M. (1977) Proc. Nat. Acad. Sci. USA 74: 4977-4981.

Roth-Schechter, B. and Mandel, P. (1976) Biochem. Pharmacol. 25: 563-571.

Schachner, M. (1974) Proc. Nat. Acad. Sci. USA 71: 1795-1799.

Traber, J., Reiser, G., Fischer, K. and Hamprecht, B. (1975) FEBS Lett. 51: 327-332.

See for complementary bibliography :

Gispen, W.H., Editor (1976) Molecular and Functional Neurobiology, Elsevier, Amsterdam, 499 pp.

Fedoroff, S. and Hertz, L., Editors (1977) Cell, Tissue, and Organ Cultures in Neurobiology, Academic Press, New York, 696 pp.

Sato, G., Editor (1973) Tissue Culture of the Nervous System (in Current Topics in Neurobiology, Vol. 1), Plenum Press, New York, 288 pp.

PRIMARY CULTURES OF ASTROCYTES FROM MAMMALIAN BRAIN HEMISPHERES AS A TOOL IN NEUROCHEMICAL RESEARCH.

Arne Schousboe

Department of Biochemistry A, Panum Institute,
University of Copenhagen, Denmark.

During the last few years the astrocytic culture originally described by Booher and Sensenbrenner (1972) has been further developed and characterized and it seems to constitute a valid model by which information about the metabolism and function of astrocytes in vivo may be obtained (Hertz, 1977; Schousboe, 1977). A variety of established glioma cell lines have also been used to study the function of glial cells (Pfeiffer et al., 1977) but generally care should be taken in drawing quantitative conclusions from such studies with regard to the in vivo situation (Pfeiffer et al. 1977; Schousboe, 1977). Thus the results obtained using primary cultures of astrocytes from mouse brain on uptake of putative amino acid transmitters show great quantitative differences from similar studies using the C-6 astrocytoma cell line (Schousboe, 1977). Moreover, the primary cultures have been shown to express a high carbonic anhydrase activity (Hertz and Sapirstein, 1978) a function which the C-6 has failed to exhibit (De Vellis and Brooker, 1973). Primary cultures of astrocytes have also recently been used to study enzymes involved in glutamate and glutamine metabolism (Svenneby et al. this meeting; Fosmark et al. this meeting) and have been shown to have a very high activity of glutamine synthetase and a relatively high activity of phosphate activated glutaminase indicating that astrocytes are deeply involved in glutamate metabolism in vivo. Furthermore, such cultures may prove useful in studies of possible receptor sites on glial cells for biogenic amines or other neurotransmitters. Such studies have already been initiated (van Calker et al. 1977; Bræstrup et al. 1978).

Booher, J. and Sensenbrenner, M. (1972) Neurobiology 2: 97-105.
Bræstrup, C., Nissen, C., Squires, R.F. and Schousboe A. (1978) Neurosci. Lett. In press.
De Vellis, J. and Brooker, G. (1973) in Tissue Culture of the Nervous System (Sato, G., ed.) pp. 231-245. Plenum Press. New York.
Hertz, L. and Sapirstein, V. (1978). In Vitro. In press.
Hertz, L. (1977) in Cell, Tissue and Organ Cultures in Neurobiology (Fedoroff, S. and Hertz, L., eds.) pp. 39-71. Academic Press. New York.
Pfeiffer, S.E., Betschart, B., Cook, J., Mancini, P and Morris, R. (1977) *ibid*. pp. 287-346.
Schousboe, A. (1977) *ibid*. pp. 441-446.
Van Calker, D., Müller, M. and Hamprecht, B. (1978) J. Neurochem. 30: 713-718.

TRANSMEMBRANE POTENTIALS OF CULTURED GLIA AND GLIOMA CELLS

M. Kanje, P. Arlock, B. Westermark and J. Pontén
Department of Zoophysiology, University of Lund
S-223 62 Lund, Sweden

We have previously reported that a cultured human glioma cell-line 138 MG undergo morphological alterations after treatment with derivatives of c-AMP (1). In addition the cells increase their transmembrane potential from -30 mV to -50 mV after such treatment (1). Primary cultures of rat brain allows the investigation of normal glia cells in culture. In the present report the response of cultured glioma cells and primary glia cells to dbc-AMP with respect to morphology and membrane potentials has been investigated.

In cultures of rat brain several celltypes could be distinguished. Layers of astrocytes, identified by their content of glial fibrillary protein, size and morphology, were surrounded by whirls of elongated cells. Large cells with a flattened irregular morphology could also be observed together with small rounded cells. The transmembrane potentials of the GFA-positive cells were around -45 mV. After treatment with 1 mM dbc-AMP in serum containing medium the GFA-positive cells obtained the stellate morphology characteristic of astrocytes. The dbc-AMP treated astrocytes exhibited membrane potentials of -55 to -60 mV.

A variety of human glioma cell-lines were investigated in the same way. Only 138 MG responded to dbc-AMP with an astrocyte-like morphology whereas 178 MG, 251 MG and 105 MG remained flattened and irregular as the control cells. All cell-lines exhibited a membrane potential around 35 mV. After dbc-AMP exposure the potentials were around 40-50 mV.

Cultured glia cells exhibit lower potentials than glia _in vivo_ and glioma cell potentials are still lower (2, 3). Thus it appears that differentiated cells exhibit higher membrane potentials than less differentiated. The increase in potentials observed after dbc-AMP treatment is in accord with this notion since dbc-AMP induce differentiated characteristics to cultured neoplastic and embryonic cells.

The glia-glioma membrane potential is regulated by potassium. It is feasible to assume that alterations in potassium concentrations and/or potassium permeability are responsible for the effect exterted by dbc-AMP.

The present work was supported by grant no. 78:143 from Riksföreningen mot cancer to Dr. A. Edström.

1. Arlock, P. and Kanje, M. (1977) Exp cell res 109:105
2. Hild, W. and Tasaki, I. (1962) J Neurophysiol 25:277
3. Hamprecht, B., Kemper, W. and Amano, T. (1976) Brain res 101:129

MOLECULAR PROPERTIES OF THE ACTION POTENTIAL Na$^+$ IONOPHORE. INTERACTIONS WITH NEUROTOXINS

Y. Jacques, M. Fosset and M. Lazdunski
Centre de Biochimie, Faculté des Sciences, Parc Valrose,
06034 NICE, France

   Three clones of nervous cells were used in this study; two of them originated in the peripheral murine tumor C 1300 (1) (neuroblastoma clones NIE 115 and N 18); the third is derived from a peripheral metastasis from a rat brain tumor (clone C9) (2). The polypeptide neurotoxin ATX_{II} isolated from the sea anemone Anemonia Sulcata and the alkaloïd neurotoxin veratridine stimulate the initial rate of ^{22}Na uptake by these cells by factors ranging from 2 to 6. $K_{0.5}(ATX_{II})$ is observed at 0.2 µM (neuroblastoma cells) or 37 nM (C9 cells), while $K_{0.5}$(veratridine) is always close to 50 µM. In the three systems analysed, the effects of ATX_{II} are synergically amplified by veratridine and vice versa, the apparent affinity of one toxin being shifted towards higher values in the presence of the other. Moreover, two families of membrane sites for ATX_{II} are observed in the presence of veratridine. The apparent affinity of ATX_{II} for the tight site is 100 times higher than that for the loose site. These results have been fitted with a model suggesting that there are two populations of toxin sensitive Na$^+$ ionophore in the excitable membrane of these nervous systems. The use of the mixture of ATX_{II} and veratridine to stabilize a permeable conformation of the Na$^+$ channel has also permitted a study of the competition between Na$^+$ and TTX for the selectivity filter of the channel. The analysis of the specificity of this channel for other monovalent cations has enabled to propose an allosteric model in which the channel has two tightly coupled binding sites for Na$^+$ or its agonists.

(1) Augusti-Tocco, G., and Sato, G., Proc.Nat.Acad.Sci. 64, 311 (1969).
(2) Gregory West (personnal communication).

THE EFFECTS OF MUSCARINIC AGONISTS AND ANTAGONISTS ON CYCLIC GMP LEVELS IN MOUSE NEUROBLASTOMA CELLS

Philip G. Strange

Department of Biochemistry, University Hospital and Medical School, Clifton Boulevard, Nottingham NG7 2UH, U.K.

The effect of muscarinic ligands on cyclic GMP levels in mouse neuroblastoma cells (clone N1E 115, Amano et al., 1972) has been investigated. All potent agonists tested e.g. carbamoylcholine, elevate cyclic GMP levels 5-10 fold after a one minute stimulation. Some correlation is observed between the agonist concentration that produces a 50% response and the dissociation constant for binding to the low affinity site of the muscarinic receptor (Strange et al., 1977). Weak agonists e.g. pilocarpine, arecoline, produce no measurable elevation of cyclic GMP levels so that agonist potency seems to be reflected in the cyclic GMP response. Weak agonists do however interact functionally with the muscarinic receptor since they block the cyclic GMP response produced by potent agonists and bind to the receptor in ligand binding studies.

The cyclic GMP elevation produced by potent agonists is inhibited by muscarinic antagonists e.g. atropine, but not nicotinic antagonists e.g. tubocurarine. There is a good correlation between the antagonist concentration giving 50% inhibition and the antagonist dissociation constant for receptor binding derived by ligand binding studies.

The cyclic GMP elevation produced by potent agonists is dependent on the presence of Ca^{2+} ions in the external medium and the calcium ionophore A23187 (kindly provided by Eli Lilley & Co.) also elevates cyclic GMP levels in these cells. These results suggest that potent muscarinic agonists elevate cyclic GMP levels in neuroblastoma cells by binding to the muscarinic acetylcholine receptor; Ca^{2+} ions probably act as a second messenger activating guanylate cyclase.

Amano, T., Richelson, E. and Nirenberg, M. (1972) Proc. Natl. Acad. Sci. USA, 69: 258-263

Strange, P.G., Birdsall, N.J.M. and Burgen, A.S.V. (1977) Biochem. Soc. Trans. 5: 189-191

CYCLIC NUCLEOTIDES AND ATP CONTENT IN A CULTURED HUMAN GLIOMA CELL LINE 138 MG

Y. Sommarin, A. Wieslander and Bo Cederholm, Institute of Zoophysiology, Lund, Sweden

Serum withdrawal, dbc-AMP and PGE_1 induce morphological changes and growth inhibition of cultures human glioma cells (138 MG) (1). Serum withdrawal causes a bipolar morphology and further treatment with dbc-AMP or PGE_1 an astrocytic appearance. Tumour cells are known to exhibit changes in both energy metabolism and cAMP-mediated processes. We have studied cyclic nucleotides and ATP in relation to morphological changes. Serum withdrawal did not affect cAMP but increased ATP content by 62% during a 2 hr incubation in Ham's F10 at 37°C with 5% CO_2. DbcAMP and PGE_1 both increased cAMP and cGMP contents.

The results suggest that, (i) an elevates ATP per se does not affect cAMP, (ii) morphological alterations after serum withdrawal are related to energy metabolism rather than to the cyclic nucleotide system, (iii) both cAMP and cGMP could be involved in the morphological changes induced with dbcAMP or PGE_1.

	pmol/mg prot		nmol/mg prot
	c-AMP	c-GMP	ATP
Serum	17.4 ± 2.5[a]	-	17.2 ± 1.1
Serum withdrawal	18.3 ± 0.6	0.29 ± 0.07	27.8 ± 2.4
" " + 1mM dbc AMP	-	25.63 ± 3.6	25.3 ± 1.5
" " + 10μg/ml PGE_1	558 ± 49	1.22 ± 0.2	26.2 ± 1.6

a mean value ± S.E.M. n=3

This work was supported by a grant from the Swedish Cancer Research 78:143 (to Anders Edström)

Edström, A., Kanje, M. and Walum E. (1974) Exp. cell res. 85: 217-223

TRANSPORT AND METABOLISM OF GLUCOSE IN C-1300 NEUROBLASTOMA (N2A) AND GLIOMA (C-6) CELLS

K. Keller, K. Lange, M. Zeitz
Pharmakologisches Institut der Freien Universität Berlin
D-1000 Berlin 33

Determination of the intracellular glucose content and of the uptake rates indicate different properties of the glucose transport system in C-6 glial cells and N2A neuroblastoma. In contrast to C-6 glioma the glucose utilization of the neuroblastoma cells does not seem to be limited by membrane transport under various experimental conditions.

The glycogen content of C-6 glioma changes in dependence on cultivation time, cell density, medium pH, and Glc 6-Pase activity. The influence of these conditions on the glycogen metabolism is not observed in C-1300 neuroblastoma.

Microsomal specific activity of Glc-6-Pase in C-6 glioma increases by B_2cAMP, with time after subcultivation and with cell density. No differences of specific Glc 6-Pase activity could be observed in C-1300 neuroblastoma during differentiation.

TRANSPORT OF [³H]L-GLUTAMATE AND [³H]L-GLUTAMINE BY DISSOCIATED GLIAL AND NEURONAL CELLS IN PRIMARY CULTURE

V. J. Balcar and K. L. Hauser
Research Department, Pharmaceuticals Division, CIBA-GEIGY Ltd.,
CH-4002 Basel, Switzerland

Transport of L-glutamate and L-glutamine was studied in primary cultures containing exclusively glioblasts (G) and primary cultures containing morphologically differentiated neurones attached to the glioblast monolayer (GN) (Hauser and Heid, 1978).

Both types of cultures accumulated [³H]L-glutamate by a saturable, structurally specific transport system with relatively high V_{max} (4-6 nmol/mg prot/min) and low K_m (10-20 µM). The transport was strongly dependent on the external [Na^+], the kinetic studies indicating that two Na^+ were required for the transport of one molecule of L-glutamate. Since both K_m and V_{max} values varied with external [Na^+] it would seem that sodium ions were involved in binding of L-glutamate to the transport site (carrier) as well as in the process of translocation across the membrane.

There were no significant differences in the values of kinetic parameters with respect to either L-glutamate or Na^+ between G and GN, suggesting that similar transport systems were operating in both types of culture. In addition, autoradiographic analysis showed that [³H]L-glutamate was accumulated by both glioblasts and neurons.

The transport of L-glutamine was less dependent on the external [Na^+]. Although a saturable component with high V_{max} (2 nmol/mg prot/min) was clearly present, the values of K_m were much higher (120-150 µM) than those for the transport of L-glutamate. Comparison of kinetic parameters obtained from G and GN cultures respectively, as well as autoradiographic studies showed that [³H]L-glutamine was transported by both glioblasts and neurons.

Hauser, K.L. and Heid, J. (1978) Vol. 1 of Proceedings of the European Society for Neurochemistry (Abstracts of the Second Meeting of the ESN, Göttingen, 1978).

KINETICS OF GLUTAMATE, GLUTAMINE AND LEUCINE TRANSPORT IN CULTURED
NEUROBLASTOMA AND GLIOMA CELLS

E. Walum and C. Weiler

National Defence Research Inst., Dept 4, S-172 04 Sundbyberg, Sweden and
Inst. of Neurobiology, University of Göteborg, S-400 33 Göteborg, Sweden

High affinity uptake systems for the transmitter amino acid L-glutamine have been demonstrated in various preparations of nervous tissue
(Bennet et al., 1972; Blacar and Johnston, 1973; Henn et al., 1974;
Balcar et al., 1977). These systems may regulate in part the extracellular level of the compound and play a major role in terminating its
synaptic activity.

In the present study the transport kinetics of L-[^{3}H]glutamate have
been compaired to those of the non-transmitter amino acids L-[^{3}H]gluta-
mine and L-[^{3}H]leucine in cultured neuroblastoma (line C-1300, clone
41A$_3$ and glioma (line 138 MG) cells. It was found that L-glutamate was
taken up into both cellines by a dual-affinity (high and low) system,
whereas L-glutamine and L-leucine were taken up by single-affinity
(low) systems (table 1).

Table 1: Kinetic parameters of transport of [^{3}H]amino acids into C-1300
and 138 MG cells.

Celline	Amino acid	Km$_H$ µM^H	Vmax$_H$ pmoles/min/10^{6}cells	Km$_L$ µM^L	Vmax$_L$ pmoles/min/10^{6}cells
C-1300	L-glutamate	33	530	520	1,620
	L-glutamine	-	-	700	41,100
	L-leucine	-	-	140	12,000
138 MG	L-glutamate	65	620	750	2,530
	L-glutamine	-	-	490	13,100
	L-leucine	-	-	275	10,200

The results are in agreement with the view that high affinity transport systems exist only for those amino acids, which have a postulated
neurotransmitter function. The results also support the notion that
glia cells, as well as nerve cells, are of importance for the removal
of transmitter amino acids from the synaptic clefts.

Balcar, V.J. and Johnston, G.A.R. (1973) J. Neurochem. 20: 529-539
Balcar, V.J., Borg, J. and Mandel, P. (1977) J. Neurochem. 28: 87-93
Bennet, J.P., Logan, W.J. and Snyder, S.H. (1972) Science 178: 997-999
Henn, F.A., Goldstein, M.N. and Hamberger, A. (1974) Nature, Lond.
    249: 663-664

LITHIUM TRANSPORT AND TOXICITY IN BRAIN CELL CULTURES

I. Szentistványi, Z. Janka, F. Joó*, A. Juhász and Á. Rimanóczy
Clinic of Neurology and Psychiatry, Medical University, Szeged and
*Laboratory of Molecular Neurobiology, Institute of Biophysics,
Biological Research Center, Szeged, Hungary

It is generally accepted that the lithium salts are effective in the treatment of the manic depressive psychoses. Toxic side effects may occur during the maintenance therapy, even independently of the optimal therapeutic range. Several clinical and experimental pharmacological data suggest that the toxic symptoms are determinated by the intracellular concentration of lithium. In our study the steady state distribution of lithium and the morphological characteristics of its toxicity were investigated in primary cultures prepared from 7-day-old chicken embryonic cerebral hemispheres.

In the transport experiments, the 7-day-old cultures were incubated in Tyrode media of different Na^+-concentrations. The media were supplemented with 20 mM LiCl and occasionally with 10^{-5} M ouabain or 10^{-4} M floretin. We have found that the steady state distribution of lithium between the intra- and extracellular spaces ensues by 30 minutes of incubation time. Decreasing the extracellular Na^+ concentration to 20 mM by substituting for choline increased lithium uptake from 19.5 nmol Li^+/(mg protein x min.) to 40 nmol Li^+/(mg protein x min). It has been measured a slight ouabain-sensitive lithium influx in Na^+-media but a marked one in choline-media. The same difference was observed in the floretin-sensitive component of lithium uptake. The short-time exposition of 20 mM Li^+ used in transport experiments did not alter the morphological appearance of the cultures.

The effect of 5 and 10 mM Li^+ on the structural properties of the nerve cell cultures was studied after exposition time of 2 and 5 days. Marked, dose-dependent reduction in total protein has been observed in the 6-day-old cultures when using lithium from day 1 *in vitro* compared with the sodium treated controls. The neural elements showed a reduction in number and a profound decrease in process length measured with morphometric means. When exposing of lithium at day 4 for 48 hours, considerable reduction in the process length of neural cells has been revealed while the morphological characteristics of the glial elements have not changed. Ultrastructurally, we have found swollen, degenerating neural processes in the Li^+-treated brain cell cultures.

ON THE HISTOCHEMISTRY OF CULTURED ENDOTHELIAL CELLS FROM DISSOCIATED RAT BRAIN

P. Panula, F. Joó and L. Rechardt
Department of Anatomy, University of Helsinki, Finland

The cerebral hemispheres from 3-day-old rats were dissociated in sterile conditions by pushing the minced brain tissue through nylon sieves of 250 and 125 mesh pore sizes. Capillary fractions which attached to the sieve were suspended in the tissue culture medium and plated out immediately on the coverslip according to the culture method described previously (Hervonen and Rechardt, 1974). The homogenate obtained was centrifuged and processed further as described by Joó and Karnushina (1973). The pellet was resuspended and in part seeded. The ultrastructure of the pellet was examined. The capillary fractions were characterized by the presence of typical endothelial cells and basal laminae.

From the fragments of dissociated brain tissue several different types of cells started to grow. Among the neuronal and glial cells large (about 25 µ in diameter) round or elongated flat cells were growing. The same type of cell was also present in those cultures which derived from the pellet of the centrifugations. The large and flat cells originated from the smaller elongated cells of the capillary tubes themselves. Many of these cells exhibited a strong alkaline phosphatase activity which covered the cells and their processes. However, many similar cells were without activity. L-DOPA (0.1 mM) was taken up to a varying extent by almost every cell regardless to its nature and origin.

Alkaline phosphatase and DOPA-decarboxylase enzyme activities confined to the capillary wall have been regarded as characteristic markers for the endothelial cells of the brain capillaries. It is possible that the flat cells in the cultures were composed of two different types of cells. The endothelial cells are characterized by the strong alkaline phosphatase activity, whereas the other cell type of unknown origin is devoid of this activity.

Hervonen, H. and Rechardt, L. (1974) Acta physiol. Scand.
    90:267-277
Joó, F. and Karnushina. I. (1973) Cytobios
    8: 41-48

MORPHOLOGICAL AND BIOCHEMICAL DIFFERENTIATION OF NEURONAL CELLS FROM CHICK EMBRYO BRAINS CULTIVATED ON POLYLYSINE-COATED SURFACES

B. Pettmann, A. Porte and M. Sensenbrenner
Centre de Neurochimie du CNRS, 11, Rue Humann,
67085 Strasbourg Cedex, France

Dissociated cells from the cerebral hemispheres of 7 day-old chick embryo were cultivated in polylysine-coated plastic Petri dishes in minimal Eagle medium supplemented with 20 % fetal calf serum.

The polylysine substrate was found to favor growth of the neuronal cells, whereas glioblast proliferation was inhibited. Indeed, the majority of the cells grew fibers and differentiated into bipolar and multipolar neurons. These neuronal cells were histochemically identified by the presence of Nissl bodies, of neurofibrils and of acetylcholinesterase activity. Under our culture conditions the neurons survived for 10-12 days.

An electron microscopic study has demonstrated the ultrastructural maturation of the neuronal bodies and the formation of synapses in absence of any glial cell contact.

The enzymatic activities of acetylcholinesterase and of choline acetyltransferase as well as the amount of tubulin in the isolated neurons increased during the first 7 days in culture and paralleled the growth of fibers and the development of synapses.

In conclusion, polylysine-treated surfaces can be used to obtain near homogeneous neuronal cultures from the chick embryo central nervous system. These neurons, without contact with glial cells can undergo morphological and biochemical differentiation. This system should allow studies on the effects of growth factors on the maturation of isolated neurons as well as investigation on neuron-glial interrelationship.

MORPHOLOGY AND BIOCHEMISTRY OF RAT CORTICAL NEURONS
IN DISSOCIATED CELL CULTURE

K.L. Hauser and J. Heid
Research Department, Pharmaceuticals Division, CIBA-GEIGY Ltd.,
CH-4002 Basel, Switzerland

Neurons isolated from newborn rat cortex and cultured in the presence
of FUdR and uridine (Godfrey, E.W., et al., 1975) on a pre-formed
monolayer of undifferentiated glioblasts of the same origin, develop
both morphologically and biochemically with increasing time in
culture. Phase contrast microscopy revealed early attachment of small
(7-30 u) neurons to the glioblast monolayer, and development of short
dendrites within one day after plating. Transmission electron microscopy of ultra-thin sections cut in the plane of the culture dish
showed the presence of well-developed synapses after only five days
in culture. Characteristically, clear vesicles of different sizes, and
both pre- and post-synaptic densities were visible at the synapse.
The neuronal perikarya contained large numbers of Golgi bodies; long,
well-ordered microtubules, as well as occasional dense core granules
could be seen in the dendrites. In contrast, the glioblasts of the
supporting monolayer were filled with microfilaments. After ten days
in culture, scanning electron microscopy revealed a complex dendritic
network, and both the neuronal cell bodies and the glioblast background appeared free of surface structure. Analysis of pure glioblast
cultures and mixed glioblast and neuronal cultures maintained in
parallel revealed that only neurons synthesize significant amounts
of gamma-aminobutyric acid and cyclic GMP.

Godfrey, E.W., Nelson, P.G., Schrier, B.K., Breuer, A.C. and
Ransom, B.R. (1975) Brain Research 90: 1-21.

RAT GLIAL CELLS IN PRIMARY CULTURE. EFFECTS OF BRAIN EXTRACTS ON ASTROGLIA DIFFERENTIATION AND ON OLICODENDROGLIA PROLIFERATION

J. P. Delaunoy, B. Pettmann, G. Roussel, A. Porte and M. Sensenbrenner
Centre de Neurochimie du CNRS, 11, Rue Humann
67085 Strasbourg Cedex, France

When cells dissociated from brains of newborn rats were grown in Falcon plastic dishes in minimal nutrient medium (MEM + 20 % fetal calf serum) the neuroblasts degenerated rapidly and only the glial cells developed. After 10 days they formed a monolayer mainly composed of flat polygonal cells and a few small fusiform cells. The flat polygonal cells have been identified as astroglial cells evidenced by the presence of gliofilaments and of glial fibrillary acidic protein (Bock et al., 1977).

In minimal nutrient medium the flat cells retain a polygonal shape and resemble immature astroglial cells. The few small fusiform cells present in the culture disappear after 2 weeks. When extracts from brains of newborn or adult rats as well as from the brains of 12 day-old chick embryo were added to the medium after 4 days of culture, a morphological alteration of the astroblasts was observed and most cells resembled mature astrocytes. In the presence of the extracts of either the adult rat or the chick embryo brain there was simultaneously an increase in the number of the small fusiform cells.

We measured by radioimmunoassay the quantities of carbonic anhydrase (CAII) in the cultures and found that the level remained low in the control cultures, but was significantly increased in presence of all three brain extracts. Furthermore, by an immunohistochemical procedure for localization of CAII at light microscopic level the small fusiform cells were more heavily stained than the astroglial cells. By electron microscopy these cells were identified as oligodendroglial cells.

Immunohistochemical methods for detecting more specific markers (Wolfgram protein and myelin basic proteins) of oligodendroglial cells are under investigation.

The results demonstrate that in culture proliferation and differentiation of glial cells can occur under the appropriate conditions.

Bock, E., Møller, M., Nissen, C. and Sensenbrenner, M. (1977) FEBS Lett. 83: 207-211.

CHARACTERIZATION OF SYNAPTOSOMES ISOLATED FROM BRAIN EXPLANT CULTURES

M. Giesing, K. Kriesten*, R. Müller and F. Zilliken
Institut für Physiologische Chemie der Universität
53 Bonn, Nussallee 11, BRD
*Institut für Anatomie und Physiologie der Haustiere der Universität Bonn

The biological capacity of in vitro systems of nervous tissue is related to the maintenance or development of synaptic connectivity and to the presence of neurotransmitters essential for the excitability of nerve cells. Passonneau et al. (1977) reported that substrates and enzyme activities in the GABA shunt are not comparable in C-6 glioma and C-1300 Neuroblastoma lines with in vivo conditions. We have found considerable differences in taurine content between glia explants and dissociated cultures of astrocytes (Schousboe et al., 1976). In this context neocortex explants were used to study the subcellular distribution of some free amino acids. Nerve ending particles were prepared from the crude mitochondrial fraction derived from 10 days old cultures in a discontinuous gradient from ficoll-sucrose yielding two synaptosomal subfractions MA and MB the latter representing a very pure population partially including subsynaptic structures. Particulate amino acids were: Tau, Thr, Ser, Gln, Glu, Gly, Ala, GABA, Orn, Lys, His, Arg. MA plus MB contained 23% of particulate taurine, corresponding to 75% of the mitochondrial subfractions, 46% glutamate (87%), 32% glycine (72%) and 50% GABA (96%). The distribution between MA/MB was for taurine 0.4, for glutamate 1.6, for glycine 1.3 and for GABA 14.6. The subcellular distribution of exogenous taurine (incubation conditions at 37°C: 1h; 100 µM), glycine (10 min; 11.5 µM) and of GABA (10 min; 0.5 µM) was also measured. MA exhibited the highest specific activities in glycine and taurine, MB in GABA.

The results indicate that possibly two functionally different nerve endings can be isolated from grey matter explants of brain cortex at the 10 days stage of cell differentiation in vitro. It will be shown that the amino acid distribution in explant cultures is well comparable to in vivo conditions.

Passonneau, J.V., Lust, W.D. and Crites, S.K. (1977) Neurochemical Research 2: 605-617
Schousboe, A., Fosmark, H. and Svenneby, G. (1976) Brain Research 116: 158-164

DISTRIBUTION OF ACRIDINE ORANGE ACCUMULATING PARTICLES IN NEUROBLASTOMA CELLS DURING DIFFERENTIATION AND THEIR CHARACTERIZATION BY SUBCELLULAR FRACTIONATION

M. Zeitz, K. Lange, K. Keller, H. Herken
Pharmakologisches Institut der Freien Universität Berlin
D-1000 Berlin 33

Vital staining of neuroblastoma monolayer cultures with the fluorescent dye acridine orange leads to a characteristic pattern of red fluorescent particles in the cytoplasm depending on the stage of differentiation. In undifferentiated cells these particles appear as a perinuclear spot. In more differentiated cells the red fluorescent structures are also seen in the growth cones and in the endings of the cell processes. During mitosis and under the influence of colcemid the localized red fluorescence is disintegrated. The acridine derivate mepacrin displaces acridine orange leading to green fluorescent structures.

The ultrastructure of neuroblastoma monolayer cultures is demonstrated in connection with the above-mentioned results: a well developed Golgi area with numerous vesicles is found near the nucleus. Vesicles with an electron dense matrix are dispersed in the cytoplasm. They appear concentrated at the basis of the processes and in the endings of the processes.

Neuroblastoma cells are stained vitally with acridine orange, homogenized, and fractionated into primary mitochondrial (M) and microsomal (P) fractions. The M-fraction is further subfractionated by centrifugation on a discontinuous sucrose density gradient. Acridine orange, measured spectrophotofluorometrically, is enriched at the 0.8/1.0 mole/l sucrose interphase, a fraction also exhibiting the highest specific activity of acid phosphatase. Ultrastructural examination of the acridine orange enriched fraction exhibits vesicular structures mostly with an electron dense matrix.

Thus acridine orange seems to be concentrated in a fraction which is mainly composed of structures derived from the Golgi apparatus, especially lysosomes. The microscopic investigations underline the importance of these structures for the process of differentiation of neuroblastoma cells in vitro.

All electronmicroscopic investigations were carried out with support of Prof. Dr. H.-J. Merker, Sonderforschungsbereich 29 at the Freie Universität Berlin.

MUTUAL INFLUENCE ON GLYCEROLIPIDS INDUCED BY NEURONAL CONTACTS IN ORGANOTYPIC CULTURES OF RAT CEREBRUM

M. Giesing, K. Tischner* and F. Zilliken
Institut für Physiologische Chemie der Universität
53 Bonn, Nussallee 11, BRD
*Max-Planck-Institut für Hirnforschung, Neuropathologische Abtlg.
6ooo Frankfurt, Deutschordenstraße, BRD

The performance of explant cultures is based on the development of a functional network between component cells in vitro integrating both neuron-neuron and neuron-glia interrelationships. Neuritic fibres extended from the tissue specimen towards the periphery will establish contacts with target cells under appropiate conditions, i.e. receptiveness. A coculture assembly of cerebral explants providing innervating (IE) and target tissues (TE) was used to investigate the interaction of different nerve cell complexes via neuritic tracts formed in vitro thus mimicking interhemispheric activities (Giesing and Zilliken, 1977). Lipid biosynthesis in IE and TE was studied from various exogenous precursors being applied to living cells. The following findings were made
1) IE exhibited an increase in the de novo formation practically in all glycerolipids predominantly in diacylglycerol via the glycerol-3-phoshate pathway. Choline-, ethanolamine- and inositol incorporation was not changed. The fatty acid pattern did not differ from the controls.
2) TE showed a decrease in phosphatidylcholine and phsophatidylinositol radioactivity derived from 2-^{3}H glycerol. Choline- and ethanolamine incorporation was lowered, whereas inositol uptake into phosphatidylinositol was doubled. Moreover the formation of arachidonate and docosahexaenoate was enhanced.
The results show that nerve cell explants cooperate via neuritic tracts established in vitro. The morphology of innervating nerve fibres will be discussed on EM level. The increase in diacylglycerol formation seems to occur in nerve endings. Functional innervation might result in regulating the microsomal capacity of target cells to synthesize hydrophobic membrane constituents.

Giesing, M. and Zilliken, F. (1977) in Cell Culture and its Application (R.T.Acton and J.D.Lynn,Eds.), Academic Press, pp. 417-432

ACQUISITION OF CELL SURFACE COMPONENTS IN DEVELOPING RAT CEREBRAL CELLS IN TISSUE CULTURE

E. Yavin[1], Z. Yavin[2] and Y. Dudai[1]

Departments of Neurobiology[1] and Isotope Research[2]
The Weizmann Institute of Science, Rehovot, Israel

The formation and localization of several neuronal membrane associated components in cerebral cell cultures was studied. Antisera, raised in rabbits against mature mouse brain, reacted with 16-day-old, freshly dissociated, fetal cerebral cells and enabled the identification of two distinct immunofluorescent cell populations (Yavin and Yavin, 1978). A second antiserum raised against embryonic brain cells displayed immunofluorescent labeling patterns similar to those observed with the adult brain antiserum when reacted with the same cells. With progressive differentiation, large neuronal cells were labeled on the neuritic cell membranes by the antiserum to adult brain, but not by the antiserum to embryonic cells. This suggests that the former antiserum possesses a class of antibodies reacting with determinants localized specifically on the neuritic plasma membrane.

Distinct changes in the profile and content of gangliosides during growth in culture were observed (Yavin and Yavin, 1978). Thus prior to in vitro synapse formation, GD_3, a ganglioside characteristic to immature nervous tissue, was reduced from a value of about 50% to a value of about 12% by 9 days. At that time, a several-fold increase in disialoganglioside (GD_{1a}) relative proportion was encountered. Tri- and tetrasialo-gangliosides reached a level of about 40% by 3 weeks. The localization of gangliosides on the cell surface was detected by an antiserum raised against GD_{1a}. With advancing maturation, only neuronal cells and their neuritic protrusions were extensively labeled.

The pharmacological profile and the ontogenesis of muscarinic binding sites in cell homogenates has been studied by the use of the muscarinic antagonist [^{3}H]-quinuclidinyl benzylate [QNB] (Dudai and Yavin, 1978). The apparent dissociation constant (K_D) of [^{3}H] QNB was found to be 0.5×10^{-9}. Several muscarinic antagonists including dexetimide, scopolamine and atropine, inhibited [^{3}H] QNB-binding with apparent inhibition constants (K_i) of 1×10^{-10} M, 2×10^{-10} M and 7×10^{-9} M, respectively. After 3 weeks in culture, specific binding of [^{3}H]QNB reached a value of 0.24 pmol/mg protein which represent a 5-fold increase over the 5th day value.

The above findings suggest that brain cells in culture are a valuable model system for studying neurochemical aspects of development.

Yavin, Z. and Yavin, E. (1978) Develop.Neurosci. (in press).
Yavin, E. and Yavin, Z. (1978) submitted for publication.
Dudai, Y. and Yavin, E. (1978) submitted for publication.

PRIMARY CULTURES FROM NEWBORN MOUSE BRAIN. AN ASTROGLIAL CELL MODEL?

Elisabeth Hansson and Åke Sellström
Institute of Neurobiology, Fack
S-400 33 GÖTEBORG 33, Sweden

Cerebral hemispheres of newborn mouse were passed through nylon meshes. The suspended cells were grown for 2 weeks in Eagle (MEM) supplied with double concentrations of amino acids, quadruple concentrations of vitamins, 250,000 IU/litre penicilline and 20% foetal calf sera (Booher and Sensenbrenner, 1972).

During these 2 weeks period the increase of protein and DNA content was monitored as a measure of the growth rate. Following an initial lag phase of 6 days, the cultures reached confluency at approximately 14 days. The confluent culture has been characterised with respect to certain astrocytic and non-astrocytic markers. On using autoradiography to localise cellular ^{3}H-γ-aminobutyric acid accumulation, two populations of cells were seen. Low levels of alkaline phosphatase and S-100 were seen at this stage of development.

As the cultures reached confluency, attempts were made to differentiate them. Dibuturylcyclic-AMP and brain extracts were added to the cultures after 14 days, and parameters such as the S-100 content and the ^{3}H-GABA accumulation were measured to optimize conditions for expressing astrocytic behaviour.

This investigation was supported by a grant No. B78-12X-164-148 from the Swedish Medical Research Council.

Booher, J. and Sensenbrenner, M. (1972) Neurobiol. 2: 97-105.
Schousboe, A. Forsmark, H. and Svenneby, G. (1976) Brain Research 116: 158-164.

EFFECTS OF GABA ANALOGUES OF RESTRICTED CONFORMATION ON GABA RECOGNITION SITES INVOLVED IN THE SYNAPTIC TRANSMISSION PROCESS.

A. Schousboe[1], P. Thorbeck[2], L. Hertz[3], G. Svenneby[4] and P. Krogsgaard-Larsen[2].

1. Department of Biochemistry A, Panum Institute, Univ. of Copenhagen, Denmark.
2. Department of Chemistry BC, Royal Danish School of Pharmacy, Copenhagen, Denmark.
3. Department of Anatomy, Univ. of Saskatchewan, Saskatoon, Canada.
4. Neurochemical Laboratory, The University Psychiatric Clinic, Vinderen, Oslo, Norway.

The major recognition sites involved in the GABA mediated transmission process in the central nervous system are the postsynaptic receptor and the transport carriers in the presynaptic and astrocytic membranes. Since a considerable degree of conformational flexibility is inherited in the GABA molecule we have studied the effects of several conformationally restricted GABA analogues on those recognition sites. In this way information about differences in structural requirements between the receptor and the transport sites might be obtained. The interaction of a variety of aliphatic and 5- and 6-membered heterocyclic compounds with sodium-independent GABA binding and sodium-dependent GABA transport in cultured astrocytes and brain cortex mini-slices was tested. Among the aliphatic compounds tested R- and S-trans-4-methyl-4-amino-crotonic acid proved to be the most interesting. The isomers of this compound were highly selective in the way that the S-form was a relatively potent GABA agonist on the GABA receptor (IC_{50} 4.1 µM) whereas the R-form was much less potent. On the contrary the S-form had no effect on the transport systems whereas the R-form inhibited both the glial (IC_{50} 500 µM) and the neuronal GABA uptake (IC_{50} 160 µM), when tested at an external GABA concentration of 1 µM. Of the five-membered rings tested muscimol was an extremely powerful GABA agonist (IC_{50} 0.02 µM) but had only very little effect on any of the transport systems. β-Proline was found to have no effect on the neuronal GABA transport but was a relatively potent inhibitor of the uptake into astrocytes (IC_{50} 320 µM) although it was less potent than the six-membered rings nipecotic acid and its derivatives. Contrary to the latter compounds, β-proline was found to interact with GABA receptor sites. The GABA agonists isonipecotic acid and isoguvacine, on the other hand, had no effect on the transport systems. It was also shown that S-nipecotic acid was a selective inhibitor of neuronal GABA transport whereas the R-isomer was an equally potent inhibitor of the glial and neuronal GABA transport.

NEURON SPECIFIC UPTAKE OF ^{3}H-GABA IN CELL CULTURES OF CEREBELLUM AND OLFACTORY BULB

D. Neil Currie and Gary R. Dutton
Brain Research Group, Open University, Walton Hall,
Milton Keynes, MK7 6AA, U.K.

As part of a larger project on brain development we are studying primary monolayer cultures of cells isolated from specific brain areas of the 1-8 day old rat using a modified version of an earlier cell preparation (Wilkin et al, Brain Research 115, (1976), 181). We are interested in obtaining cultures containing well defined and restricted populations of cell types and have found that commonly-accepted criteria for morphological identification using phase-contrast microscopy can be quite misleading. Here we report some findings which bear firstly on our search for more reliable means of identifying cell types in culture and, secondly on the ability of these cells to retain their normal in vivo characteristics.

Cultured perikarya isolated from postnatally developing rat cerebellum and olfactory bulb have been studied using autoradiography of the high-affinity uptake of ^{3}H-GABA (2×10^{-7}M). Specific inhibitors of glial uptake (β-alanine) and neuronal uptake (aminocyclohexane carboxylic acid, ACHC) were used to distinguish cell types. (ACHC was a gift from Dr. N.G. Bowery.)

The predominant cell population in the cerebellar cultures was of excitatory granule neurons which showed no ^{3}H-GABA uptake. A minority population of larger, process-bearing cells showed neuron-specific ^{3}H-GABA uptake, just as did the small proportion of inhibitory interneurons as seen in other in vivo studies. The majority of the flattened background cells showed glial-specific ^{3}H-GABA uptake.

By contrast, olfactory bulb cultures showed heavy neuron-specific labelling of the main population of small neurons present. These correspond in size and number to the expected yield of olfactory bulb granule neurons, and may reflect the inhibitory in vivo GABA-ergic character of these cells.

Thus the granule neurons of the cerebellum and of the olfactory bulb, which have been isolated as immature neuroblasts, each express their characteristic in vivo GABA uptake properties as they differentiate in vitro. Neuronal-specific high-affinity GABA uptake is restricted in vivo to GABA-ergic inhibitory neurons. Our results indicate that this relationship remains true in vitro and this method may therefore be a useful in marking cell types in culture. (This work was supported by the M.R.C.).

GABA METABOLISM IN CULTURED GLIA CELLS

M TARDY, J. BARDAKJIAN and P. GONNARD
Departement de Biochimie, CHU Henri MONDOR
94010 Créteil FRANCE

It is now well estabished that the principal mode of inactivation of neurotransmitters is their reuptake into presynaptic terminals (SELLSTROM 1975). However, the close spatial relationship of neurons and glia and the glia ensheatment of axons and synapses suggest that glia as well as neurons may participate in the uptake of synaptically released neurotransmitters. GABA has been shown by autors to be transported by a high affinity transport mechanism to glia cells. Accordingly, the astrocytes not only have the ability to transport GABA but also to metabolize the GABA taken up. A high GABA-T activity has been found in glia cells obtained by dissociation of newborn mice hemispheres. This transminase activity increases with cell growth during two weeks and could be induced by GABA added to the culture medium. This is the first time that GABA-T has been shown to be an inducible enzyme. Induced GABA-T has the same Km, pH_o, S_{20} w, pHi, as the normal glia enzyme. No GAD activity could be found in these cells.

Althought the present results point towards a participation of glia cells in the inactivation of GABA by removal from the synaptic cleft, they do not exclude possibility that these processes in glia cells may be related to GABA metabolism, independant of the transmitter function of this amino acid.

A. SELLSTROM and A. HAMBERGER (1975) J. of Neurochem.
24 : 847-852

CHARACTERISATION OF TAURINE UPTAKE BY NEURONAL AND GLIAL CELLS

J. BORG, V.J. BALCAR, J. MARK and P. MANDEL
Centre de Neurochimie du CNRS, Institut de Chimie Biologique
11, rue Humann, 67085 Strasbourg Cedex FRANCE

Taurine is a putative synaptic transmitter in the mammalian CNS. Some evidence of this comes from the presence of a high affinity system in various preparations of CNS (see for ref. Oja, 1976). In view of the heterogeneity of this material, the increasing use of cultured glial and neuronal cells to study high affinity uptake is becoming more recognized ; previous studies have reported the existence of such a system in glial and neuronal cells of different origins (Schousboe, 1976 and Borg, 1976).

In the present report, we were interested in a further characterisation of that system, concerning its ionic dependency and its structural specificity. The whole series of experiments were performed in parallel on glial and neuronal cells maintained in continuous culture, including neuronal transformed cells. Both glial and neuronal taurine uptake systems were concentrative, highly sodium dependent and inhibited by close related structural analogues such as hypotaurine, β-alanine and GABA. Strychnine was found to be a potent inhibitor of taurine uptake especially in the glial cells, while parachloromercuriphenylsulphonate was more efficient on the neuronal clones. The glial transport was dependent on the presence of calcium in the incubation medium, in contrast with uptake by neuroblastoma cells. It is concluded that these data are consistent with the possibility for taurine to be a neurotransmitter or a neuromodulator in the CNS.

Borg, J., Balcar, V.J. and Mandel, P. (1976) Brain Res.
    118: 514-516.
Oja, S.S., Kontro, P. and Lähdesmäki, P. (1976) in Transport phenomena in the
    nervous system (Levi, Battistin and Lajtha eds), p. 237-252.
Schousboe, A., Fosmark, H. and Svenneby, G. (1976) Brain Res.
    116: 158-164.

EFFECT OF VARIOUS NEUROTRANSMITTERS ON THE DE NOVO BIOSYNTHESIS OF BRAIN GLYCEROLIPIDS DURING DEVELOPMENT IN CULTURE

M. Giesing, U. Gerken and F. Zilliken
Institut für Physiologische Chemie der Universität
53 Bonn, Nussallee 11, BRD

It is well established that incorporation of phosphorus into phosphoglycerolipids is affected by extracellular stimuli in a variety of organs such as by acetylcholine and epinephrine in brain (Lapetina and Michell, 1973). This study was undertaken to examine the question of whether the formation of glycerolipids via the glycerol-3-phosphate pathway is governed by exogenous neurotransmitters applied to viable explants of brain grey matter. The experiments were performed in 8 days old cultures prior to glia maturation and in 16 days tissues after termination of synapse formation. The following findings were made:

1) 8 days cultures: Acetylcholine, epinephrine, GABA and taurine stimulated the biosynthesis of phosphatidylcholine, -ethanolamine and -inositol. Glycine did not exert any effect. In some cases diacylglycerol radioactivity was raised.

2) 16 days cultures: Acetylcholine and carbamylcholine, epinephrine and propranolol and GABA and glycine reduced the de novo formation of the major phosphoglycerolipids. Only minor changes were induced by glutamate and taurine. Cysteinsulfinic acid -immediate precursor of taurine- lowered lipid formation.

The effects observed were concentration dependent between 10^{-5} and 10^{-4} M. Differences among the stimulatory or inhibitory activity characteristic for neurotransmitters and among individual cell lipid classes were also found.

The results show that the glycerol-3-phosphate pathway is influenced by extracellular neurotransmitters. At least some of them may play a role as trophic substances in regulating the metabolism of membrane constituents prior to synapse maturation. The differential development of neurotransmitter receptors in vitro may explain the specific responses detected in the glycerolipid moiety.

Lapetina, E.G. and Michell, R.H. (1973) FEBS Letters
31: 1-10

THE PHOSPHATE ACTIVATED GLUTAMINASE ACTIVITY AND GLUTAMINE UPTAKE INTO ASTROCYTES IN PRIMARY CULTURES.

G. Svenneby[1], A. Schousboe[2], L. Hertz[3] and E. Kvamme[4]

1. Neurochemical Laboratory, University of Oslo, Preclinical Medicine, Oslo, Norway,
2. Department of Biochemistry A, Panum Institute, University of Copenhagen, Copenhagen, Denmark,
3. Department of Anatomy, University of Saskatchewan, Saskatoon, Canada,
4. Neurochemical Laboratory, University of Oslo, Preclinical Medicine, Oslo, Norway.

Uptake and release of glutamine were measured in primary cultures of astrocytes together with the activity of phosphate activated glutaminase (EC 3.5.1.2.). In contrast to previous findings of an effective, high affinity uptake of other amino acids (e.g. glutamate, GABA) no such uptake of L-glutamine was observed, though a nonsaturable, concentrative uptake mechanism did exist. The phosphate activated glutaminase activity was comparable to the activity of this enzyme in whole brain, which was unexpected in view of previous findings of a higher activity of the glutamine synthetase (EC 6.3.1.2) in astrocytes than in whole brain. The observations are compatible with the hypothesis of an in vivo flow of glutamate (and GABA) from neurons to astrocytes where it is metabolized, and a compensatory flow of glutamine from astrocytes to neurons although the former cell type may be more deeply involved in glutamine metabolism than previously thought.

DEVELOPMENT OF ENZYMES INVOLVED IN GLUTAMATE METABOLISM IN PRIMARY CULTURES OF MOUSE ASTROCYTES.

H. Fosmark[1], L. Hertz[2], I. Damgaard[1] and A. Schousboe[1].

1. Department of Biochemistry A, Panum Institute, University of Copenhagen, Denmark.
2. Department of Anatomy, University of Saskatchewan, Saskatoon, Canada.

Glutamate which is considered to be a major excitatory transmitter in the central nervous system (Krnjević, 1974) plays a key role in the metabolic relation between the tricarboxylic acid cycle and amino acid metabolism (van den Berg, 1973). High affinity uptake of this amino acid has been observed in brain slices (Balcar & Johnston, 1972) and evidence has recently been presented that a considerable fraction of this uptake occurs into astrocytes (Schousboe et al. 1977; Hertz et al. 1978). It seemed therefore of interest to study the development of the glutamate metabolizing enzymes (L-glutamate decarboxylase (GAD), glutamate-oxaloacetate transaminase (GOT), glutamate dehydrogenase (GLDH) and glutamine synthetase (Glu-S)), which all show a considerable increase in activity in the brain in vivo during postnatal ontogenesis, in astrocytes cultured from brains of newborn mice. The activities of all these enzymes except GAD, whose activity was too low to be measured, were studied during three weeks of culturing. GLDH showed a rapid increase in activity during the first week of culture, whereas the observed increase in the GOT activity occurred more slowly. Both of these developmental patterns mimick those in the brain in vivo. In contrast, the activity of the Glu-S increased considerably slower than in vivo although the activity observed in 3 weeks old cultures was at least as high as the activity in the adult mouse brain. Experiments are now in progress to find out whether alterations of the culture medium such as the addition of dibutyryl-cyclic AMP or hormones might affect the development of this enzyme in the cultures.

Balcar, V.J. and Johnston, G.A.R. (1972). J. Neurochem. 19: 2657-2666.

Hertz, L., Schousboe, A., Boechler, N., Mukerji, S. and Fedoroff, S. (1978). Neurochem. Res. 3: 1-14.

Krnjević, K. (1974) Physiol. Rev. 54: 418-540.

Schousboe, A., Svenneby, G. and Hertz, L. (1977). J. Neurochem. 29: 999-1005.

van den Berg, C.J. (1963) in Metabolic Compartmentation in the Brain (Baláзs, R. and Cremer, J.E., eds.) pp. 137-166. Macmillan, London.

ALKALINE PHOSPHATASE ACTIVITY IN CULTURED GLIA AND GLIOMA CELLS

M. Kanje, F. Joo[+] and A. Edström
Department of Zoophysiology, University of Lund   [+]Max Planck Institut fur biophys.
Lund 223 62 Lund Sweden   Chemie, Göttingen, W. Germany

Primary cultures of rat and mouse brain yields layers of astrocytes. This cell-type may be identified by immuno-histochemical staining for glial fibrillary acidic protein (GFA). However, also other cell-types can be identified in such cultures, in some instances to such an extent that they are likely to interfere with expected astrocytic responses. Cells derived from brain capillaries are likely contaminants of these cultures (1). These cells reacts strongly for histochemical staining of alkaline phosphatase (1). In contrast, glia cells as opposed to neurons seem to exhibit low alkaline phosphatase activity (2). Thus alkaline phosphatase activity may be a useful marker for contaminating cell-types in layers of astrocytes prepared from rat or mouse brain. In the present report alkaline phosphatase activity has been studied in primary cultures from mouse and rat brain and in glioma and neuroblastoma cells.

Alkaline phosphatase activity, measured at pH 9 in 0.1 M Tris-Hcl with paranitrophenylphosphate as substrate was around or lower than 1 U (umoles phosphate released per hour/mg protein) in 3 human glioma cell-lines and in 3 weeks rat brain cultures. Mouse brain cultures and neuroblastoma cells exhibited higher activities.

The Gomoritechnique (3) was used to demonstrate alkaline phosphatase histochemically. Positive staining was obtained in whirls of elongated cells surrounding GFA-positive cells in mouse and rat brain cultures. The glioma cells, as opposed to the neuroblastoma cells, failed to stain. The rat brain cultures exhibited fewer alkaline phosphatase positive areas than mouse brain cultures. In some mouse brain cultures more than 50 % of the surface area of the culture dish was covered by stained cells.

Alkaline phosphatase activity was confined to cells of non-glial origin it thus appear to be a useful marker for detection of contaminating cell-types in astrocyte cultures prepared from rat and mouse brain.

The present study was supported by grant no. 78:143 from Riksföreningen mot cancer.

1, Panula, P., Joo´and Rechardt, L Experentia (1978) 34:95
2, Kanje, M. Exp cell res (1977) 109:407
3, Pearse, A G E.,Histochemistry theoretical and applied, (1961) Churchill, London

ACTIVITY AND ISOENZYME PATTERN OF LACTATE DEHYDROGENASE IN ASTROCYTES
CULTURED FROM BRAINS OF NEWBORN MICE.

Claus Nissen and Arne Schousboe

Department of Biochemistry A, Panum Institute, University of Copenhagen
DK-2200 Copenhagen N, Denmark.

The newborn mammalian brain tolerates anoxic conditions for a longer period of time than does the adult brain (Fazekas et al., 1941; Hansen, 1977). This difference could partly be due to differences in isoenzyme pattern and activity of lactate dehydrogenase (LDH) in newborn and adult brain. The isoenzyme pattern of whole brain has been shown to change towards a higher proportion of H (oxygen dependent) subunit types as the rat matures (Bonavita et al., 1964); microdissected neurons (Hazama and Uchimura, 1970) and bulk-prepared neurons and glial cells from mammals (Nagata et al., 1974) have shown a similar trend. In the present investigation lactate dehydrogenase activity and isoenzyme distribution was determined in primary cultures of astrocytes as a function of the culture period. The total activity increased during this period with a peak value (1.91 ± 0.18 µmoles × min^{-1} × mg^{-1} cell protein) after two weeks in culture. The isoenzyme pattern changes during three weeks in culture towards a higher proportion of the H_4 isoenzyme which is analogous to the in vivo pattern. Omission of serum and addition of dBcAMP to the culture medium during the third week of culture further enhanced this prominence of the H_4 isoenzyme and the total activity in such cultures (1.58 ± 0.06 µmoles × min^{-1} × mg^{-1} cell protein) was close to the activity in the adult brain.

Bonavita, V., Ponte, F. and Amore, G. (1964). J. Neurochem. 11, 39-47.
Fazekas, J.F., Alexander, F.A.D. and Himwich, H.E. (1941). Am. J. Physiol. 134, 281-287.
Hansen, A.J. (1977). Acta physiol. scand. 99, 412-420.
Hazama, H. and Uchimura, H. (1970). Brain Res. 23, 288-292.
Nagata, Y., Mikoshiba, K. and Tsukuda, Y. (1974). J. Neurochem. 22, 493-503.

INFLUENCE OF REDUCED CHOLESTEROL SYNTHESIS ON THE ACTIVITY OF CEREBRO-
SIDE-SULFOTRANSFERASE IN CULTURED GLIOBLASTOMA CELLS (C_6)

H.P. Siegrist, T. Burkart, U. Wiesmann and N. Herschkowitz
Dept. of Pediatrics, University of Berne
CH-3010 Bern, Switzerland

The regulation of enzyme activities due to alteration of the membrane lipids could be a system involved in the synthesis of structure components in the living cell. Former work in our laboratory gave evidence, that modifications of the lipids surrounding the microsomal enzyme cerebroside-sulfotransferase (CST, EC. 2.8.2.11) lead to a modulation pattern, that mimikries the developmental enzyme activity pattern (1). This modulation effect is due to an age dependent change of the cholesterol/phospholipid ratio in the microsomes during the myelination period in mouse brain. In order to investigate the mechanisms we established a system, in which the cholesterol/phospholipid ratio can be manipulated <u>in vivo</u>. Since steroids like estradiol are known to inhibit cholesterol synthesis in cultured cells (2), we treated glioblastoma (C_6) cells in culture with different amounts (0.05 to 10 µg/ml medium) of 17β-estradiol for 6 to 72 hours. Thereby the cholesterol content is lowered down to 60% of normal. Concomitantely CST activity could be raised up to 200 % of normal after an inoculation period of 24 hours. This effect could be found either in cell homogenates or microsomal subfractions. The effect is further fully reversible: After 48 hours, cholesterol as well as CST activity are back to normal values. We therefore suggest, that CST activity is modulated in cultured cells by the changing cholesterol/phospholipid ratio in the cells.

1. H.P. Siegrist, H. Jutzi, A.J. Steck, T. Burkart, U. Wiesmann and
   N. Herschkowitz (1977) Biochim. Biophys.Acta
   <u>489</u>, 58-63
2. A.A. Kandutsch and R.M. Packie (1970) Arch. Biochem. Biophys.
   <u>140</u> 122- 130

CERAMIDE GALACTOSYLTRANSFERASE ACTIVITY OF RAT C6 GLIAL CELLS

N. M. Neskovic, J. P. Delaunoy, G. Rebel and P. Mandel
Unité 44 de l'INSERM (Section Neurochimie), and Centre de Neurochimie du CNRS
11, Rue Humann, 67085 Strasbourg Cedex, France

Rat C6 glioma cells have generally been described as astrocyte-like cells. However, biochemical investigations have shown that this poorly differentiated cell type has characteristics of both astrocytes and oligodendroglial cells. In addition to protein S100, characteristic of glial cells, the presence of 2',3'-cyclic nucleotide 3'-phosphohydrolase, the enzyme marker of myelin, and of basic protein and proteolipid protein, characteristic of myelin sheath, has been reported in C6 cells.

Previously we have reported (Sarliève et al., 1976) that the homogenates of C6 glial cells grown in culture possess the activity of UDP-galactose:ceramide galactosyltransferase (EC 2.4.1.45) (CGalT), the enzyme which catalyses the synthesis of cerebrosides in central nervous system and which activity is attributed to the myelin-forming oligodendroglia.

In the present work the evolution of the CGalT activity during the proliferation of C6 cells and the influence of the conditions of growing were studied. The specific activity of CGalT increases steadily for the first few days of proliferation and attains a stable level when the cells reach confluency. The properties of the enzyme isolated from C6 cultured cells were compared to those of the purified CGalT of rat brain.

Sarliève, L.L., Neskovic, N.M., Freysz, L., Mandel, P. and Rebel G. (1976)
   Life Sci. 18: 251-260.

COULD GANGLIOSIDE PATTERNS DURING NEUROBLASTOMA DIFFERENTIATION BE AN ARGUMENT FOR A NON-NERVOUS ASPECT OF THIS TYPE OF CELL ?

H. Dreyfus, L. Freysz, J. Robert, S. Harth, P. Mandel and G. Rebel
Unité 44 de l'INSERM (section Neurochimie), and Centre de Neurochimie du CNRS,
11, Rue Humann, 67085 Strasbourg Cedex, France

Neuroblastoma cell is often utilized as a model for studying neuronal differentiation. In undifferentiated cells, the ganglioside pattern shows a lack in tri- and tetra-sialogangliosides (Dawson *et al.*, 1971). Recent studies showed the effect of differentiation on the ganglioside composition of neuroblastoma cells. The results are compared to those obtained in the whole brain as well as in cellular fractions isolated from brain.

Ganglioside analysis of four different proliferating clones showed (a) the lipid-bound sialic acid (expressed per g dry weight) of all clones was very low. It represented about 1/10th of the levels of whole brain; (b) neither tri- nor tetra-sialogangliosides were detected. Only the M1 clone had few amounts of G_{D1b} (1-2 % of total ganglioside-sialic acid); (c) the four clones were characterized by high amounts of G_{M3} and G_{M2} (40 %). After differentiation by dibutyryl cAMP, bromodeoxyuridine or serum deprivation neuroblastoma, cells showed a lot of processes. The total amount of lipid-bound sialic acid increased but not much as compared to brain values. The ganglioside pattern did not show any of tri- and tetra-sialogangliosides. In M1 clone, the amount of G_{D1b} was not increased. However the percentage variations were seen in other gangliosides. Nevertheless G_{M2} and G_{M3} remained at high levels.

The whole brain ganglioside analysis showed that G_{M3} and G_{M2} were the species present at much lower concentrations (about 5 %) whereas tri- and tetra-sialogangliosides represented 15 % and G_{M3}-G_{M2} were present at very low amounts. The comparison of ganglioside distributions between neuroblastoma and neuronal cells showed that even after differentiation the ganglioside patterns of the culture cells were strikingly different. The lack of the higher ganglioside species here may be due to a viral transformation as it has been shown in non-nervous cells (Brady and Fishman, 1974). This study showed that differentiated neuroblastoma cells could not be used as a true normal neuronal cell.

This investigation was supported by grants from the Institut National de la Santé et de la Recherche Médicale (Contracts n° 78.4.017.6 and n° 78.5.260.6).

Brady, R.O. and Fishman, P.H. (1974) Biochim. Biophys. Acta 355: 121-148.
Dawson, G., Kemp, S.F., Stoolmiller, A.C. and Dorfman, A. (1971) Biochem. Biophys. Res. Commun. 44: 687-694.

THE INVOLVEMENT OF GANGLIOSIDES IN GROWTH AND DEVELOPMENT OF CULTURED NEURONAL AND GLIAL CELLS

J.Morgan and W.Seifert
Max-Planck-Institute, Friedrich-Miescher-Laboratorium
Tübingen, W-Germany

In order to study the central question of the control of growth and differentiation in neural tissue we have used permanent cell lines of either neuronal (B104, B103) or glial origin (B12, C6). Under conditions of serum starvation these cells cease division and undergo biochemical and morphological changes indicative of an increase in their differentiated state. For our neuronal cell lines this resting state is relatively unstable and the cells will die within two weeks.
This serum-arrested quiescent state is not irreversible, since the transition from resting to growing state can be effected by addition of either serum or a glial growth factor to the medium (Seifert, 1977; Morgan and Seifert, 1977).

Gangliosides apparently play at least two roles in the various transitions of the cell developmental programme. Firstly under conditions of serum arrest exogenously added gangliosides are stably incorporated into neuronal cells whereby they profoundly enhance cell survival time in culture. They also produce a new morphological appearance under these conditions in at least one cell type (B104).

A second property of gangliosides is that they synergize with submaximal levels of serum to promote the initiation of DNA-synthesis and growth in serum-arrested cultures. Thus one or more of the gangliosides may in fact be a receptor for a serum growth factor. Evidence will also be presented concerning the metabolism of gangliosides under various growth conditions and on the influence of gangliosides upon a number of biochemical parameters.

Morgan, J. and Seifert, W. (1977) Hoppe-Seylers Z. Physiol. Chemie
  358: 1248 - 1248
Seifert, W. (1977) Hoppe-Seylers Z. Physiol. Chemie
  358: 307 - 307

INCORPORATION OF UNSATURATED FATTY ACIDS IN NERVE CELLS MEMBRANES : EFFECTS ON CHOLINE TRANSPORT AND METABOLISM

T. Y. Wong, C. Froissart, J. Robert, P. Mandel and R. Massarelli
Centre de Neurochimie du CNRS, Institut de Chimie Biologique
11, rue Humann, 67085 Strasbourg Cedex, FRANCE

Nerve cells (clone M1 of mouse Neuroblastoma C 1300) were grown in the presence of linoleic (18:2) and linolenic (18:3) acids. After nine days of incubation the incorporation and transformation of the fatty acids in cell phospholipids checked by gas-chromatography as a control. Cells were then incubated in Krebs Ringer phosphate pH 7.2 and with various concentrations of Me [^{14}C] Choline. At different time intervals of incubations the cultures were washed with 0.147 M NaCl and cold 1 N/formic acid acetone (15/85, v/v) was immediately added to the Petri dishes. Cells were afterwards scraped, homogenized, centrifuged and the supernatant was rapidly evaporated. The dry material was dissolved in 0.04 N HCl and aliquots spotted on TLC cellulose plates for the separation of the radioactive compounds.

Incorporation of radioactivity was decreased in total acid soluble extract of 18:2 and 18:3 treated cells. No change was however observed in the choline free compartment, suggesting no effect of the increased membrane fluidity on the transport of choline. Radioactivity in the phosphorylcholine compartment instead was considerably decreased explaining thus the decrease observed in the total acid extract. A surprising result was that the lipid compartment and the phosphatidylcholine compartment contained more radioactivity in 18:2 and 18:3 treated cells when compared to controls. Two possible explanations may be advanced.

1) There is an increase of base exchange phenomena in the lipid compartment and a decrease in choline kinase activity (EC. 2.7.1.32).
2) The turnover of phosphorylcholine is increased while that of phosphatidylcholine is decreased.

These interpretations are however still speculative and further work is in progress in order to analyze the role of choline kinase and of phosphotransferases and/or the differences in specific activities of the compartments mainly affected by the treatment.

Round Table

Techniques for the Assay of Amines and their Metabolism

Chairman:
E. Änggård

TECHNIQUES FOR THE ASSAY OF CATECHOLAMINES (CA), INDOLAMINES (IA) AND THEIR METABOLITES

E. Änggård
Department of Alcohol and Drug Addiction Research, Karolinska Institutet, 104 01 Stockholm, Sweden

Invited discussants:

M. da Prada, Pharmaceutical Research
　　　　　　Hoffmann-La Roche, Basel

J. Korf,　Inst. of Biological Psychiatry
　　　　　University of Groningen, Groningen

B. Sjöquist, Dept. of Alcohol and Drug Addiction Research,
　　　　　　Karolinska Institutet, Stockholm, Sweden

The catecholamines are neurotransmitters both in the central and periphereal nervous system. Most of the progress in our knowledge both in health and disease about these substances has depended on the development of appropriate methods for their measurement in tissues and body fluids. Highlights in this evolution have been the use of column chromatographic techniques for the isolation of CAs and their determination by fluorometry, the mapping of CAs and IAs by fluorescence histochemistry, and more recently radioenzymatic and mass spectrometric methods of analysis. It is the object of this workshop to identify the preferred methods for the analysis of the above mentioned biogenic amines and their metabolites in brain, cerebrospinal fluid (CSF), plasma and urine. The following main questions will be covered during the workshop:

1. What should we analyze for?

The monoamine neurotransmitters are part of neurochemical systems involving precursors, transmitters and transmitter metabolites and the enzymes responsible for their formation. The synthesis and metabolism of the catecholamines and of 5-hydroxytryptamine (5-HT) are shown in fig. 1 and 2.

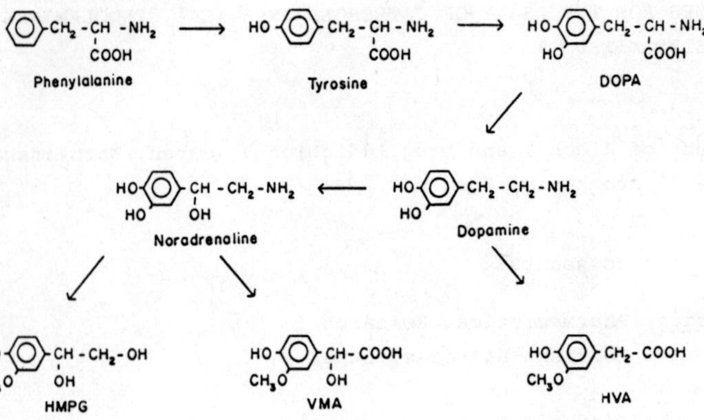

Fig. 1. Synthesis and metabolism of dopamine and noradrenaline

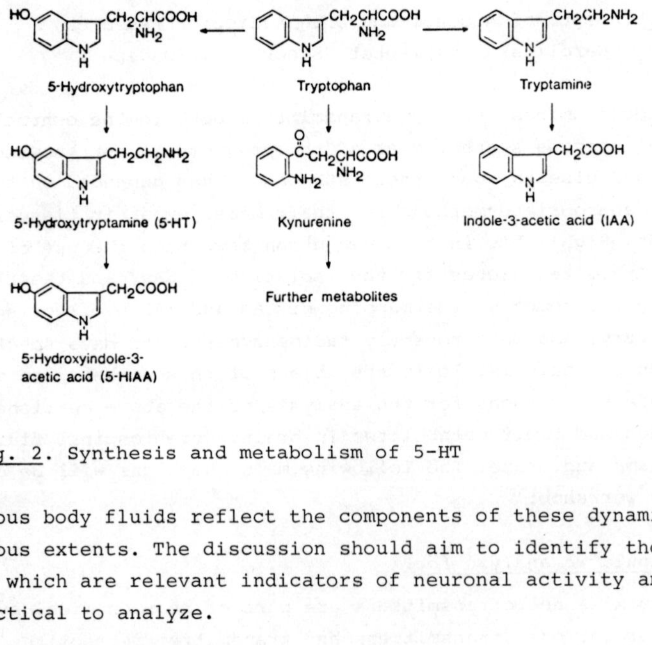

Fig. 2. Synthesis and metabolism of 5-HT

Various body fluids reflect the components of these dynamic systems to various extents. The discussion should aim to identify those metabolites which are relevant indicators of neuronal activity and which are practical to analyze.

2. <u>What methods should we use</u>?

The available methods should be discussed in terms of specificity, sensitivity and precision. To what tissues and body fluids have they been successfully applied? What are the factors (age, physical activity, diet etc) which influence the levels of metabolites? Guide-lines should be given for sample collection (volume, time, special precautions etc).

The following methods will be discussed:

a) Extraction_and_isolation. Techniques such as chromatography on XAD, ion exchange resins, Sephadex, lipophilic gels, alumina, ion pair extraction, will be briefly and critically discussed. Particular emphasis will be placed on evaluation of recent results with high performance liquid chromatography (HPLC) systems.

b) Radioenzymatic_methods. These techniques can detect catecholic amines and their deaminated metabolites (e.g. dihydroxyphenylacetic acid) by the use of $S-^3H$-methyl adenosyl methionine and the enzyme catechol-O-methyltransferase (COMT). Noradrenaline can also be determined by N-methylation to adrenaline using phenylethanolamin-N-methyl transferase (PNMT). These techniques have been successfully applied to assay of catecholamines in plasma and tissue in the femtomole range.

c) Fluorimetric_methods. Fluorimetric methods have more than any other technique contributed to our knowledge about catechol- and indolamines. In the discussion particular emphasis will be placed on current problems and new possibilities concerning the analysis of CAs and 5-HT, and their acidic metabolites in brain and cerebrospinal fluid using fluorimetry.

d) Gas_chromatographic_methods. Gas chromatography (GC), particularly using capillary columnes, is one of the most powerful separation tools available today. It can be combined with a variety of detectors, some nonspecific e.g. the flame ionization detector, others sensitive to particular properties of the molecular species, e.g. the electron capture detector. The most specific and sensitive method available today is the combined use of GC and mass spectrometry (MS). In the GC-MS technique the outflow of the GC column is continously monitored for selected fragments emerging from the molecules of interest. Thus even low concentrations of amines and their metabolites can be analyzed in a complex mixture of biological material. The disadvantage of this technique is that it demands an expensive equipment.

The discussion should aim at describing recent advances in GC-MS of monoamines and their metabolites, point out advantages as well as disadvantages of this technology and finally to make comparisons with other methods.

3. <u>Attempts to evaluate dynamic aspects of monoamine metabolism in man</u>

Time permitting there will be a brief discussion on the state of the art in evaluating turnover of catecholamines in man. Of interest is transport blockade using probenecid and following the increase of acidic metabolites in CSF. Other techniques are the use of synthesis blockade with α-methyl-tyrosine and pulse labelling of amines and their metabolites with stable isotopes.

A selected bibliography covering analytical techniques relevant to the monoamines and their metabolites will be handed out during the workshop.

A METHOD FOR MEASURING CATECHOLAMINES AND THEIR NON-O-METHYLATED METABOLITES IN ISOLATED BRAIN REGIONS

M.E. **Bardsley** and H.S. Bachelard
Department of Biochemistry, University of Bath, Bath, U.K.

This method was developed by combining the radioenzymatic method of Cuello et al (1973) with the thin-layer chromatographic procedure of Fleming & Clark (1970). Noradrenaline (NA), dopamine (DA), 3,4-dihydroxyphenylacetic acid (DOPAC), dihydroxyphenethanol (DHPET), 3,4-dihydroxyphenethylene glycol (DOPEG) and 3,4-dihydroxymandelic acid (DOMA) are converted to their respective O-methylated derivatives in the presence of rat liver catechol-O-methyl transferase and $[^3H]$ S-adenosyl methionine. The O-methylated products are then separated by thin-layer chromatography.

The method was found to be sufficiently sensitive to measure the endogenous levels of noradrenaline, dopamine and their non-O-methylated metabolites in various isolated brain regions.

The standard curve for each substrate showed that the amount of label incorporated into the products from the alcohol substrates was several orders of magnitude greater than for the products from noradrenaline and dopamine. The incorporation of label per nanogram, into the products of the following substrates, in decreasing order, was: DHPET, DOPEG, DA, NA, DOPAC, DOMA.

Rates were linear for all substrates; the linearity of rates with substrate concentration suggests that initial rates are being measured (V/Km). The ratios of V/Km for the various substrates gives an index of the relative affinities of the substrates for the enzyme. Similar studies could be carried out for cerebral catechol-O-methyl transferase.

Cuello, A.C., Hiley, R. & Iversen, L.L. (1973) J. Neurochem. 21: 1337-1340
Fleming, R.M. & Clark, W.G. (1970) J. Chromatog. 52: 305-312

ON THE METABOLISM OF TRYPTAMINE AND SEROTONIN:CONCURRENT EXTRACTION AND ASSAY OF TRYPTOPHAN, TRYPTAMINE, INDOLE-3-ACETIC ACID, SEROTONIN AND 5-HYDROXYINDOLEACETIC ACID BY A NEW GC-MS (MID) METHOD.

F. Artigas and E. Gelpí
Instituto de Biofísica y Neurobiología.
Avda. S. Antonio Mª Claret 171. Barcelona-25. Spain.

Serotonin (5-HT) is one of the few well established neurotransmitters, while tryptamine (T), whose endogenous levels in rat brain have been subjected to controversy during the past years, (Philips et al. 1974) (Snodgrass and Horn,1973), seems to be likewise implicated in neurotransmission, either modulating the action of other compounds(i.e.5-HT) or acting as a classic neurotransmitter in specific neurons. Also, the deactivation pathway of T in brain was virtually unexplored until a very recent report on the normal occurrence of indole-3-acetic acid (IAA) in rat brain (Warsh et al. 1977).

Of the analytical methods reported for TP,T,IAA,5-HT and 5-HIAA in mammalian brain, most are relatively complex and none is suitable for the simultaneous assay of all of these compounds. Thus,in line with our interest in the relationships between neuronal T and 5-HT metabolic pathways we have set up a simple and sensitive method to analyze TP,T,IAA, 5-HT and 5-HIAA in the same sample of brain tissue. The method is based on the adsorption of the indolyl compounds from a single rat brain homogenate(plus deuterated standards) on a short column of XAD-2(Segura et al. 1976) followed by direct GC-MS(MID) analysis of the derivatized (Methyl-PFP) methanol eluate.The endogenous values found in whole rat brain are:TP=4.16 $\pm$ 0,23 µg/g,T=less than 0.38ng/g,IAA=13.1 $\pm$ 2.0 ng/g, 5-HT=526 $\pm$ 81ng/g and 5-HIAA=442 $\pm$ 24ng/g (values for 6 animals,5 for 5-HT). The sensitivity is such that the IAA in 100-150 mg of tissue can be easily analyzed. Preliminary results on its applications to the study of T effects on 5-HT metabolism are also presented.

Philips S.R., Durden D.A. and Boulton A.A.(1974) Can.J.Biochem.52,447.
Segura J.,Artigas F.,Martínez E. and Gelpí E. (1976) Biomed. Mass Spectrometry. 3, 91.
Snodgrass S.R. and Horn A.S. (1973) J. Neurochem. 21, 687.
Warsh J.J., Chan P.W., Godse D.D., Csocina D.V. and Stancer H.C.(1977) J. Neurochem. 29, 955.

THE BIOSYNTHESIS OF DOPAMINE AND OCTOPAMINE BY SCHISTOCERCA GREGARIA NERVOUS TISSUE.

A.K. Mir and P.F.T. Vaughan
Biochemistry Department, Glasgow University, Glasgow G12 8QQ, Scotland.

Previous studies (Vaughan and Neuhoff, 1976) have shown that N-acetyltyramine, N-acetyloctopamine and N-acetyldopamine are formed when tyrosine or tyramine are incubated with cerebral or thoracic ganglia of S. gregaria. Evidence was also presented to suggest that tyramine, rather than L-3, 4-dihydroxyphenylalanine (dopa), was an intermediate in the biosynthesis of dopamine.

This study shows that the aromatic amino acid decarboxylase inhibitor NSD 1015 (3-hydroxybenzyl hydrazine) prevents the conversion of tyrosine to tyramine, octopamine and N-acetylated monoamines, but did not result in accumulation of radioactivity in L-dopa. These results support the hypothesis that dopamine biosynthesis in S. gregaria nervous tissue involves a decarboxylation step to tyramine, followed by hydroxylation to dopamine. This represents a different pathway from that present in mammals (Blaschko, 1973).

Incubation of meso- and metathoracic ganglia with (^{3}H) tyrosine results in recovery of 3, 17, and 9% of the label in N-acetyltyramine, N-acetyldopamine and N-acetyloctopamine, respectively, whereas the corresponding figures for incubation with tyramine are 25, 2 and 1%. These results are consistent with the hypothesis that tyramine is extensively N-acetylated during uptake by the ganglia, whereas tyrosine is metabolised by octopaminergic (Hoyle and Baker, 1975; Evans and O'Shea, 1977) or dopaminergic neurons to the respective amine, which is N-acetylated subsequent to release from the neurons.

Blaschko, H. (1973). Brit. Med. Bull. 29, 105-109.
Evans, P.D. and O'Shea, M. (1977). Nature (Lond.), 270, 257-259.
Hoyle, G. and Barker, D.L. (1975). J. Ex. Zool. 193, 433-439.
Vaughan, P.F.T. and Neuhoff, V. (1976). Brain Res. 117, 175-180.

6-HYDROXY-DOPAMINE AND THE OPIATE RECEPTORS

F. Andrasi and A. Ujvari
Institute for Drug Research
Budapest, Hungary

6-hydroxy-dopamine /6OH-DA/ deteriorates dopaminergic nerve terminals. On the other hand, the morphine molecule contains the dopamine moiety in a rigid structure and the N-terminal tyrosine plays a key-role in the activity of endorphins.

We have investigated the effect of 6-OH-DA on the morphine-induced analgesia and dependence. The 2oo µg/rat dose was injected intracerebroventricularly.

According to our observations 6OH-DA by itself did not influence on the normal pain-perception in the conventional tail-flick test. But the analgetic effect of morphine / 4 mg/kg sc./ was reduced. At the same time the 6OH-DA pretreatment mitigated the withdrawal symptoms of addicted CFLP mice.

MAO AND COMT ACTIVITIES IN FIBROBLASTS FROM NORMAL PERSONS AND PATIENTS WITH GENETICALLY CONTROLLED METABOLIC DEFECTS WITH NEUROLOGICAL INVOLVEMENT

S. Singh, I. Willers, E.-M. Kluß, H. W. Goedde
Institute of Human Genetics, University of Hamburg, FRG

Monoamine oxidase activities have been compared by measuring in fibroblasts grown from 10 normal subjects, from 5 patients with Lesch-Nyhan syndrome (X-linked recessive disease with choreoathetosis, self aggression, and deficiency of HGPRT enzyme), and from 3 patients with classical type of maple syrup urine disease due to defective decarboxylation of branched chain amino acids. In both of these diseases the underlying enzyme defects are also expressed in cultured fibroblasts and the patients are, apart from other specific clinical features, mentally retarded. The exact mechanism and the cause of the mental illness has not been elucidated so far. Because of the difficulty of obtaining brain samples, we have looked for the activity level of some enzymes implicated in the neuronal function also in our cultured fibroblasts, as HGPRT and branched chain ketoacid decarboxylase activities are routinely measured in our laboratory in this tissue. Both A and B type of MAO activities are present in the cultured fibroblasts. Our comparative studies in normal and mutant cells have shown that in fibroblasts of patients suffering from Lesch-Nyhan syndrome significantly reduced MAO activities are present, while in those of patients with maple syrup urine disease a different situation is noted: one patient has a significantly reduced activity, one a moderately reduced and another a significantly elevated MAO activity. These results seem to be of great interest in connection with the neurological involvement in these diseases.

Regarding the activities of COMT, a large amount of variation within the various cell lines was observed and no exact differences could be stated among normal and mutant cell lines.

Acknowledgements: We acknowledge the expert technical assistance of Miss G. Knaack and Mrs. B. Ressler and the Deutsche Forschungsgemeinschaft for financial support.

EFFECTS OF THE MONOAMINE OXIDASE INHIBITORS TRANYLCYPROMINE, PHENELZINE AND PHENIPRAZINE ON THE UPTAKE OF CATECHOLAMINES IN SLICES FROM RAT BRAIN REGIONS.

G. B. Baker, H. R. McKim, D. G. Calverley and W. G. Dewhurst
Department of Psychiatry, University of Alberta, Edmonton, Alberta, Canada

It has been suggested that ability to inhibit reuptake of catecholamines in brain may be an important factor in determining the clinical efficacy of monoamine oxidase inhibitors (MAOIs) as antidepressants (Hendley and Snyder, 1968; Horn and Snyder, 1972). In the present study we have compared three MAOIs with regard to their effects on the uptake of radiolabelled noradrenaline (NA) or dopamine (DA) in brain slices. Phenylethylamine, a cerebral amine structurally related to the three drugs, was also included in the study.

Slices (0.1 mm x 0.1 mm x 2.0mm) were prepared from hypothalamus (NA experiments) or striatum (DA experiments) using a McIlwain tissue chopper. The tissue was suspended in incubation medium, preincubated at 37° for 15 min. and incubated subsequently for 10 min with ^{3}H-labelled NA or DA (concentration: 0.02 μM). Varying concentrations of the drugs were added simultaneously with the radiolabelled catecholamine. The incubation medium contained nialamide at a concentration (12.5 μM) that inhibited MAO but did not interfere with catecholamine uptake (Hendley and Snyder, 1968). After the 10 min. incubation period the tissue was isolated and washed using a Beckman microfuge. The pellet was then solubilized and radioactivity determined by liquid scintillation counting.

The drugs, in decreasing order of strength as inhibitors of NA uptake, were tranylcypromine (TCP) >phenylethylamine (PE)> pheniprazine (PPZ)> phenelzine (PLZ). The order for strength of inhibition of DA uptake was PE >PPZ >TCP >PLZ. IC_{50} values (μM) for inhibition of NA and DA uptake respectively were as follows: PE (0.48; 1.45), TCP (0.43; 4.78), PPZ (1.24; 2.57) and PLZ (2.42; 9.20).

Financial support of the MRC and the University of Alberta Hospital and the expert technical assistance of Mrs. L. Hiob are gratefully acknowledged.

Hendley, E. D. and Snyder, S. H. (1968) Nature (Lond.) 220: 1330-1331
Horn, A. S. and Snyder, S. H. (1972) J. Pharmacol. Exp. Ther. 180: 523-530

STUDIES ON THE BINDING OF THE DOPAMINE PRECURSOR L-DOPA TO tRNA.

H. Bernheimer, G. Högenauer and G. Kreil
Neurochem. Abt., Neurol. Inst. d. Univ. Wien; Sandoz Forschungsinstitut, Wien;
Inst. f. Molekularbiologie der Österr. Akad. d. Wissenschaften, Salzburg, Austria.

The amino acid L-dihydroxyphenylalanine (L-DOPA) is administered in doses of up to several grams per day to patients suffering from Parkinson's disease (1, 4, 5) in order to compensate for the dopamine deficiency in the brain of such patients (2, 3, 6, 7). However, L-DOPA might also be incorporated into proteins under conditions of vast excess of DOPA. We tested this possibility by assaying whether L-$[^3H]$ DOPA can be bound enzymatically to tRNA. Transfer RNA derived from either E. coli, mouse liver or ascites cells, and enzyme preparations derived from E. coli and mouse liver, respectively, were used in our experiments.

In the E. coli system a significant amount of $[^3H]$ DOPA was bound to tRNA. The binding of $[^3H]$ DOPA to tRNA was strongly inhibited in the presence of tyrosine. The affinity of tyrosine was much higher than that of DOPA. Only tyrosine, but none of the other 19 amino acids could suppress the activation of DOPA. $[^3H]$ DOPA was bound also to a tRNA fraction enriched for tRNA^{tyr}; this binding again was inhibited by L-tyrosine. In the mouse system, on the other hand, enzymatic binding of $[^3H]$ DOPA to tRNA was not detected. Obviously, this animal system has a more efficient mechanism for the discrimination between L-tyrosine and L-DOPA. Since a similar situation can be expected to exist in human cell systems, the formation of DOPA-containing proteins even in the presence of large concentrations of DOPA as used in the therapy of Parkinson's disease is most unlikely. (Supported by Fonds zur Förderung der wissenschaftlichen Forschung, Austria, Project No. 1082).

1) Barbeau, A., Sourkes, T. L. and Murphy, G. F. (1962) in: Monoamines et Systeme Nerveux Central (de Ajuriaguerra, J. ed) p. 247, Georg, Geneva
2) Bernheimer, H., Birkmayer, W. and Hornykiewicz, O. (1963) Klin. Wschr. 41, 456-469
3) Bernheimer, H., Birkmayer, W., Hornykiewicz, O., Jellinger, K. and Seitelberger, F. (1973) J. Neurol. Sci. 20, 415-455
4) Birkmayer, W. and Hornykiewicz, O. (1961) Wien. Klin. Wschr. 73, 787-788
5) Cotzias, G. C., Van Woert, M. H. and Schiffer, L. M. (1967) New Engl. J. Med. 276, 374-379
6) Ehringer, H. and Hornykiewicz, O. (1960) Klin. Wschr. 38, 1236-1239
7) Hornykiewicz, O. (1963) Wien. klin. Wschr. 75, 309-312

ASSOCIATION OF SOME BIOGENIC AMINES AND RELATED PSYCHOACTIVE DRUGS WITH ADENOSINE-5'-TRIPHOSPHATE IN AQUEOUS SOLUTION

H.Sapper, W.Gohl, M.Matthies, I.Haas-Ackermann, W.Lohmann
Institut für Biophysik der Universität,
D-6300 Giessen, FRG

The physico-chemical properties of association between biogenic amines and adenosine-5'-triphosphate (ATP) may be of great importance for storage and release of the amines in the neuronal synapses and, furthermore, for a better understanding of the activity of some psychoactive drugs. Therefore, type and stability of binding of several biogenic amines (e.g. dopamine, adrenaline, and nor-adrenaline) and related drugs (e.g. amphetamine, methamphetamine, and mescaline) with ATP have been investigated in aqueous model systems by means of proton magnetic resonance and infrared spectroscopy.

It has been found that the associations studied are determined mainly by contributions of a ring-ring-attraction as well as of an interaction between the protonated side chain of the amines and the negatively charged phosphate group of the nucleotide. The attraction of the aromatic ring systems has been evaluated from the upfield shift of the ring proton resonances. The interaction between the amino groups and the terminal phosphate groups has been characterized by the pH-dependence of the symmetric PO_3^{2-} and the P-O-P vibrations of ATP and of the chemical shifts of the amine side chain protons. The extent of this interaction and the influence of substituents and/or metal ions (Mg^{2+}) as well are demonstrated by the comparison of the association constants evaluated from the spectroscopic quantities.

THE SENSITIVITY OF EMBRYONIC CENTRAL MOTOR OUTPUT TO MONOAMINERGIC
TRANSMITTERS AND RELATED DRUGS

J. Sedláček
Research Laboratory of Psychiatry, Charles University,
Prague, Czechoslovakia

1. The central motor output in chick embryos, tested by the spontaneous motility, is sensitive to monoaminergic transmitters- noradrenaline (NA), dopamine (DA), serotonin (5-HT) - at first on day 15 of incubation. The effects of all transmitters are generally depressive.
2. The depressive effect of NA is exclusively connected with the maturation of supraspinal parts of the embryonic CNS (ineffectivity of NA in spinal embryos), when the effects of DA and 5-HT contain spinal and supraspinal components (effectivity both in normal and spinal embryos with prevalence of the supraspinal component).
3. The embryonic onset of sensitivity to monoamines is in a good agreement with the development of sensitivity to drugs related to metabolism, storage and release of these transmitters. Chlorpromazine, reserpine and haloperidol are till day 15 of incubation without any effect on the spontaneous motility of chick embryos.
4. In contrary, the drugs, acting on synaptic monoaminergic receptors as agonists - clonidine (NA), apomorphine (DA), LSD (5-HT) - affect the activity of central motor output already before day 15 of incubation (in 11- and 13-day-old embryos). This finding admits a direct, non-synaptic effect of these drugs on embryonic motoneurons at early stages of maturation.
5. As revealed by these experimental results, the monoaminergic systems are involved into the embryonic motility of chick embryos around day 15 of incubation as an integral component of the developing supraspinal control of the spontaneous activity of central motor output.
6. The developmental involvement of central monoaminergic systems corresponds with the maturation of some other components of the descendent supraspinal influences: with the onset of spinal shock and with the maturation of glycine- and GABA-sensitive inhibition in embryonic spinal cord (Sedláček, 1977 , 1978; Sedláček and Doskočil, 1978).

Sedláček, J.(1977) Physiol.bohemoslov. 26: 9-12
Sedláček, J.(1978) Physiol.bohemoslov. 27: 105-115
Sedláček, J. and Doskočil M.(1978) Physiol. bohemoslov.,27: 7-14

STEREOCHEMICAL ASPECTS IN THE METABOLISM OF 4-AMINOBUTYRATE IN MOUSE BRAIN.

M. Galli Kienle[+], E. Bosisio[++], A. Manzocchi[+] and E. Santaniello[+]
[+]Institute of Chemistry, School of Medicine and
[++]Institute of Pharmacology and Pharmacognosy, School of Pharmacy, University of Milan, Milano, Italy

The metabolism of glutamic acid to ɣ-amino butyrate (GABA) and to ɣ-hydroxybutyrate has received much recent interest due to the peculiar pharmacological properties of these compounds. In particular, some work has been devoted to the inhibition of the enzymes involved in the process in order to either enhance or decrease the tissue concentrations of these two neuroregulators. In this respect, the knowledge of the stereoselectivity of glutamate decarboxylase and GABA transaminase would provide helpful informations in the design of new stereospecific inhibitors. The decarboxylation of glutamate to GABA implies the introduction of a hydrogen atom in position 2, and a hydrogen from the same carbon atom is lost when GABA is metabolized to succinic semialdehyde (SSA). The stereochemistry of these reactions has been investigated by following the fate of the tritium label of $[2(S)2-^{3}H,U-^{14}C]$ glutamate and of $[4(S)4-^{3}H,U-^{14}C]$ GABA obtained from the doubly labelled glutamic acid with E.coli glutamate decarboxylase, in incubations with mouse brain homogenates. When $[4(S)4-^{3}H,U-^{14}C]$ GABA was incubated with mouse brain homogenates to yield 4-hydroxybutyrate, a complete loss of tritium was observed. This result indicate that the transamination of GABA to SSA occurs in brain with the loss of 4 proS hydrogen of GABA. Retention of tritium was instead observed in 4-hydroxybutyrate formed from $[2(S)2-^{3}H,U-^{14}C]$ glutamate with brain enzymes. This may indicate that the decarborylation of glutamic acid occurs with inversion of configuration of C_2 atom of glutamic acid.

STIMULATION OF RAT STRIATAL Na,K-ATPase BY CATECHOLAMINES: NATURE AND LOCALIZATION

J. A. Van der Krogt, R. D. M. Belfroid and W. F. Maas
Department of Pharmacology, Leiden University Medical Centre,
Leiden, The Netherlands

Na,K-ATPase activity of rat striatal homogenates is stimulated by dopamine and noradrenalin. This stimulation is concentration-dependent, with a maximum of more than 100 percent at 0.1 - 1 mM. Also serotonin stimulates the enzyme, but to a smaller extent (about 60 percent at 1 mM).

In order to further characterize the stimulation of striatal Na,K-ATPase by catecholamines we have studied the effect of some drugs, known to act on dopaminergic and noradrenergic receptors. Like dopamine, the dopamine receptor agonist apomorphine brings about a large stimulation, even with a greater potency viz. a maximal stimulation of about 100 percent at 0.01 mM (whereas 0.1 mM apomorphine inhibits the enzyme nearly totally). However, since stimulation by dopamine and apomorphine is not counteracted by the dopamine receptor antagonist haloperidol, the obvious conclusion that these effects take place via the dopaminergic receptor had to be rejected. Involvement of a (nor)adrenergic receptor has to be excluded too, since neither the alpha-receptor blocker phentolamine, nor the beta-receptor blocker propranolol diminish the stimulation by noradrenalin. So an explanation for the effect of catecholamines and apomorphine has to be sought in another direction, possibly in a complexation of inhibitory metal ions e.g. Ca^{2+}, as suggested by Godfraind et al. (1974), or Fe^{2+}, as suggested by Hexum (1977). Such an explanation is in accordance with the fact that EDTA too can activate rat striatal Na,K-ATPase activity.

To see whether striatal Na,K-ATPase that can be stimulated by dopamine is specifically localized in dopaminergic neurons we have studied the effect of implantation of 6-hydroxydopamine in the nigrostriatal tract. Degeneration of the dopaminergic nerve terminals following this lesion did not affect striatal Na,K-ATPase, neither the basic activity, nor the dopamine stimulation. So we must conclude that Na,K-ATPase that can be stimulated by dopamine is not exclusively or for the greater part localized in dopaminergic neurons.

Godfraind, T., Koch, M.C. and Verbeke, N. (1974) Biochem. Pharmacol.
    23: 3505-3511
Hexum, T.D. (1977) Biochem. Pharmacol.
    26: 1221-1227

INHIBITION OF ETHANOLAMINE AND CHOLINE PHOSPHOGLYCERIDE SYNTHESIS BY CMP.

A. Radomińska-Pyrek, Z. Dąbrowiecki and L.A. Horrocks
Polish Academy of Sciences, Warsaw, Poland and
Dept. of Physiol. Chem., Ohio State Univ., U.S.A.

Activities of ethanolamine phosphotransferase /EC 2.7.8.1/ and choline phosphotransferase /EC 2.7.8.2/ were assayed in microsomal fractions from rat brain with CDP/^{14}C/ethanolamine and CDP/^{14}C/choline in the presence of alkylacylglycerols or diacylglycerols /Radomińska-Pyrek et al., J. Lipid Res. 18:53-58, 1977/.
In the present study, the protein content was reduced to 80 μg and effects of Mn^{2+} substitution for Mg^{2+} were determined. CMP, a product of the reactions, inhibited both phosphotransferases in brain tissue. The inhibition was greater for ethanolamine phosphotransferase than for choline phosphotransferase. The degree of inhibition was highest with alkylacylglycerols in the presence of Mn^{2+} and lowest with diacylglycerols in the presence of Mg^{2+}. The inhibition by CMP is presumably due to reversibility of the phosphotransferases as suggested by Goracci et al. /FEBS Letts. 80:41-44, 1977/ from an experiment with labelled alkylacylglycerols and brain microsomes. The present results provide the first evidence for reversibility of the alkylacylglycerol phosphotransferases in brain assayed with CDP/^{14}C/ethanolamine and CDP/^{14}C/choline.

Radomińska-Pyrek, A., Strosznajder, J., Dąbrowiecki, Z., Goracci, G., Chojnacki, T. and Horrocks, L.A., /1977/ J. Lipid Res. 18:53-58

Goracci, G., Horrocks, L.A. and Porcellati, G. /1977/ FEBS Letts. 80:41-44

HIGHER 5-HT LEVELS IN DIFFERENT BRAIN AREAS OF MUTANT HAN-WISTER RATS

N.N. Osborne, V. Neuhoff, H. Cremer and K.-H. Sontag
Forschungsstelle Neurochemie und Abteilung Biochemische Pharmakologie
Max-Planck-Institut für experimentelle Medizin
3400 Göttingen, FRG

A mutant strain of Han-Wistar rats characterised by spastic paresis and partial paralysis has recently been described (Pittermann et al. 1976). Because of the positive reaction of the mutants to L-Dopa, we thought at first that the animals might serve as a model for studying Parkinson's disease. But, surprisingly, a subsequent study revealed that the striata of the mutants contain more and not less dopamine (Osborne et al. 1977), which indicated that these animals can be eliminated as possible models for examining Parkinson's disease. Recent results have shown that the potassium-induced release of dopamine from the mutant's striata is less than that associated with control animal striata (Osborne et al. 1978). These later studies thus suggest that the rigidity of the mutants is probably due to dopamine deficiency at the post-synaptic dopamine receptors, caused by an insufficient release. A similar occurrence is typical of sufferers from Parkinson's disease.

An analysis of the 5-HT (5-hydroxytryptamine) levels in different areas of the brain (striata, nerve cord, mid-brain, whole brain) revealed a statistically higher amount of this amine in all brain areas of mutants compared with control littermates. Furthermore, injection of 5-HTP into the mutants temporarily reversed the rigidity associated with the animals as did L-Dopa and apomorphine. These results demonstrate clearly that transmitter substances other than dopamine, viz. 5-HT, have to be considered as possible causes for the spastic paresis and paralysis associated with the mutant rats.

Pittermann, W., Sontag, K.-H., Wand, S. Rapp, K. and Deerberg, F. (1976) Neuroscience Letters 2: 45-49

Osborne, N.N., Coelle, E.-F., Neuhoff, V. and Sontag, K.-H. (1977) Neurosciense Letters 6: 251-254

Osborne, N.N., Neuhoff, V., Cremer. H. and Sontag K.-H. (1978) (in prep.)

IMMUNOCHEMICAL RELATIONSHIP OF MONOAMINE OXIDASE FROM HUMAN LIVER, PLACENTA, PLATELETS AND BRAIN CORTEX

S. M. Russell, J. Davey and R. J. Mayer, Department of Biochemistry, University of Nottingham Medical School, Queen's Medical Centre, Nottingham NG7 2UH.

Antiserum was raised in sheep to human liver monoamine oxidase (MAO), purified by the method of Dennick and Mayer (1977). The purified enzyme gave a single band on polyacrylamide gel electrophoresis in the presence of SDS. Homogenised gel containing the subunit of MAO was used for the immunization procedures. The specificity of the IgG prepared from the antiserum (Walker and Mayer, 1977) was assessed by Ouchterlony diffusion analyses against purified human liver MAO, a Triton X-100 extract (1.5% w/v) of human liver mitochondrial membranes and a particle-free supernatant obtained by centrifugation (6 x $10^6 g_{av}$.min.) of a human liver homogenate. Multiple immunoprecipitin lines were observed in all cases. The immunoreactive component of the particle-free supernatant was MAO as evidenced by the loss of immunoinhibitory capacity of MAO activity after adsorption of the IgG with particle-free supernatant and detection of MAO activity in the particle-free supernatant (assayed with [^{14}C]-tyramine; Otsuka and Kobayashi, 1964). Immuno-titration of the MAO activity in Triton depleted Triton X-100 extracts of human liver mitochondrial membranes gave a maximum inhibition of 40% of the MAO activity. Adsorption of the antiserum with liver mitochondria in iso-osmotic conditions (0.32 M sucrose) and subsequent elution (with 0.1 M glycine-HCl buffer, pH 2.8) resulted in a loss of 90% of the total IgG but a retention of 75% of the immuno-inhibitory capacity.

Reactions of identity have been carried out by Ouchterlony diffusion analysis of unadsorbed and mitochondrially-adsorbed antiserum with Triton X-100 extracts of mitochondria from liver, platelets, placenta, brain cortex non synaptosomal and synaptosomal mitochondria and purified liver MAO. Multiple immunoprecipitin lines were obtained. MAO from all samples reacted either partially or completely with the purified human liver MAO. A reaction of identity of purified MAO with the antigen in Triton X-100 extracts of liver, placenta, platelets and non-synaptosomal mitochondria could also be identified by histochemical staining of MAO activity (Glenner et al., 1957).

Dennick, R.G. and Mayer, R.J. (1977) Biochem. J. 161: 167-174
Glenner, G.C., Burton, H.T. and Brown, E.W. (1957) J. Histochem. Cytochem. 5: 591-600
Otsuka, S. and Kobayashi, Y. (1964) Biochem. Pharmacol. 13: 995-1006
Walker, J.H. and Mayer, R.J. (1977) Biochem. Soc. Trans. 5: 1101-1103

VECTORIAL ORIENTATION OF MONOAMINE OXIDASE IN THE MITOCHONDRIAL OUTER MEMBRANE.
IMMUNOCHEMICAL STUDIES ON MITOCHONDRIAL PREPARATIONS FROM HUMAN LIVER AND BRAIN
CORTEX

S. M. Russell, J. Davey and R. J. Mayer, Department of Biochemistry, University of
Nottingham Medical School, Queen's Medical Centre, Nottingham NG7 2UH.

Monoamine oxidase (MAO) is an integral protein of the mitochondrial outer membrane (Ernster and Kuylenstierna, 1970). The precise localisation of the enzyme on one or both sides of the membrane is unknown but may be fundamental to the different drug and substrate specificities of monoamine oxidase (Tipton et al., 1976). We have used the lytic principle of the method of Kuylenstierna et al. (1970) to study the localisation of monoamine oxidase but have chosen immunochemical accessibility of the enzyme as a measure of its distribution on the inside and outside of the outer mitochondrial membrane. The lytic principle was used so that the complication of adventitious lysis of mitochondrial preparations in iso-osmotic conditions could be allowed for. Mitochondrial lysis was estimated by the accessibility of the inter-membrane space marker, adenylate kinase to trypsin.

Monoamine oxidase was assayed by a radiochemical method (Otsuka and Kobayashi, 1964). Experiments with tyramine as substrate show that monoamine oxidase activity is found on both the inner and outer surface of the outer mitochondrial membrane. The distribution of activity was different with preparations of liver mitochondria, brain cortex non-synaptosomal and synaptosomal mitochondria. A larger proportion of tyramine-oxidizing activity was associated with the inner surface of liver mitochondrial outer membrane (50-80%) than with either the non-synaptosomal or synaptosomal mitochondrial outer membrane (30-50%). The ratio of β-phenylethylamine-oxidizing activity to 5-hydroxy tryptamine-oxidizing activity was higher in the brain cortex mitochondria than in the liver mitochondria. Both β-phenylethylamine-oxidizing activity and 5-hydroxy tryptamine-oxidizing activity were inhibited by the antiserum (although the degree of inhibition was greatest with β-phenylethylamine). From immunoinhibition studies we tentatively conclude that the β-phenylethylamine-oxidizing activity is located on the outer surface of the mitochondrial outer membrane and the 5-hydroxy tryptamine-oxidizing activity is distributed on both the inner and outer surfaces of the outer membrane in liver and brain cortex mitochondrial preparations.

Ernster, L. and Kuylenstierna, B. (1970) in Membranes of Mitochondria and Chloroplasts (eds. Racker, E.) pp.172-212, Van Nostrand Reinhold Comp., New York

Kuylenstierna, B., Nicholls, D.G., Hovmöller, S. and Ernster, L. (1970) Eur. J. Biochem. 12: 419-426

Otsuka, S. and Kobayashi, Y. (1964) Biochem. Pharmacol. 13: 995-1006

Tipton, K.F., Houslay, M.D. and Mantle, T.J. (1976) in Monoamine Oxidase and its Inhibition. Ciba Foundation Symp. 36 (New Series) Elsevier, Amsterdam

COMPARISON BETWEEN THE EFFECT OF THE HYPOTHALAMIC RELEASING HORMONES, TRH and MIH, AND AMPHETAMINE ON PRESYNAPTIC STRIATAL DOPAMINERGIC MECHANISMS

L.M. Shapiro and P.F.T. Vaughan
Biochemistry Department, Glasgow University, Glasgow G12 8QQ, Scotland.

A number of studies have shown that thyroloberin (TRH) and melanostatin (MIH) have a direct action on adrenergic neurons, distinct from their role in regulating the release of pituitary hormones. For example in vivo administration of TRH has been reported to enhance cerebral noradrenaline turnover (Keller et al, 1974) and MIH to stimulate striatal dopamine biosynthesis (Friedman et al, 1973). TRH has also been reported to stimulate noradrenaline and dopamine release (Horst and Spirt, 1974) which has led to suggestions that TRH has a similar mechanism of action to amphetamine (e.g. Green et al, 1976).

The present study compares the effect of MIH, TRH and amphetamine on dopamine metabolism and release from rat brain striatal preparations. 10^{-4}M TRH or MIH did not affect monoamine oxidase activity, in either tissue chops or sucrose homogenates. Up to 10^{-3}M TRH and MIH did not alter tyrosine hydroxylase activity in P_2 synaptosomal preparations, under conditions in which 10^{-4}M-amphetamine causes a 30% stimulation of dopamine formation.

10^{-3}M-MIH and TRH cause at 10% stimulation of dopamine release from P_2 synaptosomal preparations, which is dependent upon the presence of calcium ions in the incubation medium. In comparison 10^{-4}M amphetamine results in a 40% stimulation of dopamine release, 25% of which is inhibited by 2mM EGTA.

It is suggested that amphetamine stimulates dopamine release mainly by a calcium insensitive process (possibly from an extra-vesicular pool), whereas TRH and MIH stimulate a calcium dependent release, possibly via exocytosis from a vesicular store.

Friedman, E., Friedman, J., and Gersham, S. (1973).
  Science, 182, 831-832.
Green, A.R., Heal, D.J., Grahame-Smith, D.G. and Kelly, P.H. (1976).
  Neuropharmacol. 15, 591-599.
Horst, W.D. and Spirt, N. (1974). Life Sci. 15, 1073-1082.
Keller, H.H., Martholini, G. and Pletscher, A. (1974).
  Nature, 248, 528-529.

FORMATION OF TETRAHYDRONORHARMANE AND 6-OH-TETRAHYDRONORHARMANE: A NEW PATHWAY FOR INDOLEALKYLAMINES IN MAMMALS

H. Rommelspacher, H. Honecker, and B. Greiner

Institut Neuropsychopharmacology, Free University of Berlin,
D-1000 Berlin 19, Ulmenallee 30, FRG

Tetrahydronorharmane (THN, tetrahydro-ß-carboline, THBC) is formed in vitro by a Pictet Spengler type of reaction from tryptamine with formaldehyde. The existence of THN and its derivatives is conceivable in vivo, too, since both the substrates and the other conditions of the reaction are present in the mammalian organism.

After having developed a sensitive and specific method to extract THN from tissue preparations and to measure its concentration radiometrically, we were able to demonstrate the presence of the compound in the urine of man. During a collection period of 12 h male volunteers excreted 280-800 ng (n=4). The variations may be explained by the fact that the diet was not standardized. After treatment with 4 g Tryptophan at the onset of the collection period, the amount of THN excreted with the urine increased 40-fold. The concentration of THN in human platelets was found in the range of 1-10 ng per 10^8 platelets.

Most of 6-OH-THN - the concentration of which is about one tenth of that of THN - was excreted with the urine as glucuronide. After loading with 300 mg L-5-HTP 100-400 ng (n=4) of the ß-carbolines could be detected at the end of a 12 h period, when the specimen was treated with ß-glucuronidase. In pharmacokinetic experiments with (^{14}C)-THN is was demonstrated that at least in rats the source for 6-OH-THN is not only 5-HTP but also THN itself. THN is hydroxylated by the arylhydroxylase in the liver. Furthermore, THN as well as 6-OH-THN were found in the urine, the platelets and the brain of rats.

The identity of the extracted substances was ascertained by isotopic dilution technique, thin layer chromatography in 7 solvent systems and mass spectrometry. With the latter method THN and 6-OH-THN were detected in the urine of man and rats. The present findings suggest a new metabolic pathway for indolealkylamines besides oxidation by the monoamine oxidase. Since THN as well as 6-OH-THN exert 5-HT-like effects in pharmacological experiments and inhibit the action of the putative dopamine-receptor agonist apomorphine, the ß-carbolines may act as modulating agents of neuronal mechanisms in vivo.

TETRAHYDRONORHARMANE (TETRAHYDRO-ß-CARBOLINE) MODULATES THE K^+ - EVOKED RELEASE OF 5-HYDROXYTRYPTAMINE AND DOPAMINE

H. Rommelspacher

Institute of Neuropsychopharmacology, Free University of Berlin,

D-1000 Berlin 19, Ulmenallee 30, FRG, and

N. Subramanian

Institute of Pharmacology and Toxicology, University of Erlangen

Tetrahydronorharmane (THN), the condensation-product of tryptamine and formaldehyde acts as 5-HT-like drugs and inhibits the stimulation induced by apomorphine in various pharmacological tests (Rommelspacher et al., 1977). The hypothesis was put forward that THN modulates neuronal mechanisms. As one possibility, inhibition of high affinity uptake of putative neurotransmitters was investigated. THN caused reduction of the transmembranal transport of (^{3}H)-5-HT and (^{3}H)-noradrenaline but not of (^{14}C)-dopamine, (^{3}H)-choline and (^{3}H)-GABA (Rommelspacher et al., 1978).

Modulation of the release-mechanism is conceivable as another mode of action to explain the pharmacodynamic effects of THN. In the present communication it is reported about release-experiments with slices from rat brain using superfusion technique as described elsewhere (Subramanian et al., 1977). The ß-carboline stimulated the efflux of (^{3}H)-5-HT and reduced that of (^{3}H)-dopamine (THN concentration 1×10^{-4}M, p 0.01; with 1×10^{-5}M, p 0.05). This finding supports the notion that THN modulates neuronal mechanisms. The site of action remains to be elucidated, since (^{14}C)-THN enters the cell and can be released from brain slices by high concentrations of K^+ (56 mM). Replacement of the transmitters can be ruled out as mode of action because injection of THN increases the concentration of 5-HT in the forebrain of rats and leaves the level of dopamine unchanged.

The relevance of the findings may be assessed by the fact that the existence of THN could be demonstrated in several organs of man and rats (Rommelspacher et al., this meeting).

Rommelspacher, H., Kauffmann, H., Heyck Cohnitz, C., and Coper, H. (1977),
 Arch. Pharmacol. 298:83-91

Rommelspacher, H., Strauss, S., and Rehse, K. (1978),
 J. Neurochem. 30: 1573-1578

Subramanian, N., Mitznegg, P., Sprügel, W., Domschke, W., Domschke, S., Wünsch, E., and Demling, L. (1977)
 Arch. Pharmacol. 299: 163-165

METABOLISM OF γ-AMINOBUTYRIC ACID (GABA) IN RAT-BRAIN MITOCHONDRIA

M.Lopes-Cardozo and R.W.Albers
Laboratory of Neurochemistry, NINCDS, NIH
Bethesda, Maryland 20014, USA

The GABA shunt bypasses GTP formation in the Krebs cycle. Therefore the interrelation of GABA- and nucleotide-metabolism was studied. Isolated brain mitochondria were incubated with [^{32}P]phosphate, α-ketoglutarate (αKG) and GABA. Nucleotides were separated by high-performance liquid chromatography. GABA lowers $^{32}P_i$ incorporation into GTP and ATP in the presence of valinomycin and oligomycin; this effect is reversed by aminooxyacetic acid (Rodichok, 1977). [2,3-^{3}H]GABA and [5-^{14}C]αKG were used as tracers to determine their relative oxidation rates. Succinate, the main reaction product in the presence of malonate, was isolated by anion exchange chromatography. The rates of the reaction αKG → succinate and of the GABA shunt were estimated from the incorporation of ^{14}C and ^{3}H in the isolated succinate. The inhibitory effect of GABA on $^{32}P_i$ incorporation could not be explained by a decreased flow through the succinate thiokinase reaction, since GABA did not affect αKG oxidation. However, αKG had a slight inhibitory effect on GABA oxidation.

In the absence of phosphate acceptor, the GABA shunt accounted for 30 % of the succinate produced. Addition of ADP or valinomycin stimulated αKG oxidation much more than GABA metabolism and the contribution of the shunt was below 10 % under these conditions. The apparent K_m for GABA oxidation is 5 mM.

It has been suggested (Balasz et al., 1972) that GABA, produced in nerve endings, is mainly metabolized in the surrounding glia. It would be interesting to know whether glial mitochondria have a higher capacity for GABA oxidation than synaptosomal mitochondria. A synaptosomal fraction was separated from the mitochondrial fraction on a discontinuous iso-osmolar gradient of Ficoll-400. Comparison of 'free' and synaptosomal mitochondria did not reveal a significant difference between these two populations of mitochondria with regard to the activity of the GABA shunt.

GABA stimulates the energy dependent uptake of divalent cations by brain mitochondria. Elemental analysis of individual mitochondria can be performed on air-dried specimens using electron microprobe - X ray fluorescence (Silbergeld et al., 1977). This method was used to study a possible heterogeneity of brain mitochondria towards GABA metabolism. A marked variability in the content of calcium phosphate granules was observed.

Balasz, R., Patel, A.J. and Richter, D. (1972) in Metabolic Compartmentation in the Brain (Balasz,R. and Cremer,J.E., eds) MacMillan Press, London, pp: 167 - 184
Rodichok, R. (1977) unpublished observations
Silbergeld, E.K., Adler, H.S. and Costa, J.L. (1977) Res. Commun. Chem. Path. Pharm. 17: 715 - 725

THE ACTIVITY OF SUBSTITUTED 4-AMINOCROTONIC ACIDS AS ANALOGUES OF
THE NEUROTRANSMITTER GABA

R. D. Allan, G. A. R. Johnston and B. Twitchin
Department of Pharmacology, John Curtin School of Medical Research,
P.O. Box 334, Canberra City, A.C.T., 2601. Australia.

γ-Aminobutyric acid (GABA) acts as an inhibitory neurotransmitter in the mammalian central nervous system and over the last few years there has been a substantial increase in neurochemistry devoted to its study. However, the number of GABA analogues currently available for structure-activity studies on various GABA processes is small compared to the numbers used for studies of other neurotransmitters such as acetylcholine or adrenaline. We have embarked on a program to extend the range of useful GABA analogues by preparing conformationally restricted analogues of the flexible GABA molecule. It has previously been found that restricting the conformation by the introduction of a double bond produces an analogue; trans-4-aminocrotonic acid, with activity comparable to that of GABA on various GABA processes such as cellular uptake and enzymic degradation (Johnston et al., 1975).

This communication reports the synthesis of further double bond analogues of GABA with methyl, bromo, chloro and fluoro substituents. Their activity on the sodium-independent binding of GABA to rat brain membranes, on the cellular uptake of GABA and on GABA-transaminase is compared to the corresponding saturated analogues. In general substitution was found to lower the activity of trans-4-aminocrotonic acid. 2-Fluoro-4-aminocrotonic acid was one of the most potent analogues on all three assays while 3-bromo-4-aminocrotonic acid was a potent time-dependent inhibitor of GABA-transaminase.

Johnston, G. A. R., D. R. Curtis, P. M. Beart, C. J. A. Game,
R. M. McCulloch and B. Twitchin. (1975) J. Neurochem., 24: 157-160.

HIGH-AFFINITY, BICUCULLINE-SENSITIVE GABA BINDING PROCESSES IN A SYNAPTOSOME-ENRICHED FRACTION OF RAT CEREBRAL CORTEX.

F.V. DeFeudis, M. Maitre, L. Ossola, A. Elkouby and P. Mandel
Centre de Neurochimie du C.N.R.S.
67085 Strasbourg, Cedex, France.

Bicuculline sensitive, GABA binding processes have been demonstrated using subcellular particles of the vertebrate CNS (DeFeudis, 1977). However, the extents to which such processes are related to synaptic GABA-receptors and GABA transport (uptake) systems or to the <u>in vivo</u> convulsant or GABA potentiating actions of bicuculline remain unknown. Therefore, the effects of bicuculline methiodide (BMI) on GABA (^{3}H-GABA ; 1.5×10^{-9} to 5.3×10^{-3} M) binding to a synaptosome-enriched fraction of rat cerebral cortex were determined in both the presence and absence of a physiological concentration of Na^+. BMI (10^{-3} M) and unlabelled GABA (10^{-3} M) were used to determine BMI-sensitive and GABA-sensitive components. In Na^+-free medium, GABA binding was best resolved into two components. For the highest affinity component, likely related mainly to GABA-receptors, BMI-sensitive sites had $K_B \cong 2 \times 10^{-5}$ M, $B_{max} \cong 26$ p-moles/mg protein, and GABA-sensitive sites had $K_B \cong 10^{-5}$ M, $B_{max} \cong 40$ p-moles/mg protein. Thus, about 65 % of the GABA-displaceable ("specific") sites were sensitive to BMI. In Na^+-containing medium, GABA binding was best resolved as a triple-affinity system which possessed two high-affinity components. For the lower affinity component, likely related mainly to GABA transport, BMI-sensitive sites had $K_B \cong 10^{-5}$ M, $B_{max} \cong 0.5$ n-mole/mg protein, whereas GABA-sensitive sites had $K_B \cong 10^{-5}$ M, $B_{max} \cong 2.6$ n-moles/mg protein. For the higher affinity component, likely related mainly to GABA-receptors, BMI-sensitive sites had $K_B \cong 4 \times 10^{-8}$ M, $B_{max} \cong 2$ p-moles/mg protein, and GABA-sensitive sites had $K_B \cong 10^{-7}$ M, $B_{max} \cong 30$ p-moles/mg protein. Thus, about 7 % and 20 %, respectively, of the "specific" binding sites were sensitive to BMI.

These data have revealed three different high-affinity, BMI-sensitive binding processes for GABA that might be related to synaptic GABA-receptors and GABA transport sites. An effect of BMI on GABA transport sites could explain its potentiation of the depressant action of GABA that has been shown in recent microiontophoretic studies (Hill et al, 1973 ; Krnjević et al, 1977), whereas its effect on synaptic GABA-receptors could explain its convulsant action. Although bicuculline-like compounds exert many non-specific effects on CNS preparations, such effects do not appear to be involved to any great extent in explaining our findings.

DeFeudis, F.V., (1977) Progr. Neurobiol. 9:123-145.
Hill, R.G., Simmonds, M.A. and Straughan, D.W. (1973) Brit. J. Pharmac. 49:37-51
Krnjević, K., Puil, E., and Werman, R. (1977) Can. J. Physiol. Pharmacol. 55:670-680.

Free Communications

A SIMPLE, SENSITIVE AND VOLUME-INDEPENT METHOD FOR QUANTITATIVE PROTEIN-DETERMINATION WHICH IS INDEPENDENT OF OTHER EXTERNAL INFLUENCES

V. Neuhoff, K. Philipp and H.-G. Zimmer
Max-Planck-Institut für experimentelle Medizin, 3400 Göttingen (FRG)

Quantitative determination of protein is a common prerequisite for many biochemical statements. The different methods currently available often have a drawback, in that solvents used for preparation of biological tissues, e.g. SDS, Triton X 100, mercaptoethanol, urea, etc., interfer with the analyses. The method described here avoids these disadvantages and furthermore, allows quantitative determinations of small amounts of protein, even if the volume used (in the range between 0.5 and 5 µl) is unknown. The method is performed using cellulose acetate electrophoresis strips (Sartorius Membranfilter) of defined humidity on which the protein solution is routinely applied with a capillary of 0.5, 1, 2 or 5 µl volume. The round spot is stained thereafter with Amidoblack 10 B in methanol/acetic acid (9/1, v/v). The acetate strip is then placed on a glass slide and made transparent. The spectrophotometric evaluation is performed at 510 nm wavelength with a Zeiss ZK 4 gel scanning device using a microgel adaptor developed by Zimmer and Neuhoff. On the corresponding densitogram one observes the diameter of the spot and the optical density of the stained protein. Since there is a linear correlation between the diameter of a spot and the applied volume which is only influenced by the protein concentration (which in turn is determined by the optical density) there is no problem to determine a protein concentration of an unknown volume using a set of calibration curves. The lower limit of detection is in the range of 5×10^{-8} g for Amidoblack staining of 0.5 µl of a solution containing 0.1 mg protein/ml. The presence of SDS in a protein solution leads to a stoichiometric decrease in staining and thus requires calibration curves to correct for the SDS concentration in the solution. Triton X 100, mercaptoethanol or urea and many other commonly used reagents have no influence. If any influence on the characteristics of the staining is detected (which may be observed for reagents firmly bound to the protein) the effect can be compensated for by preparing suitable calibration curves. Since the excess of most reagents is washed out during the staining and destaining procedure, most problems are avoided. For the described optical evaluation, the spots have to be ideally round. This can be a problem if high concentrations of salts, etc. are present. For evaluation of these spots a method is being developed which uses a suitable fluorophore which after staining and destaining can be evaluated by measuring the integrated fluorescence with a specially devised instrument.

Zimmer, H.-G. and Neuhoff, V. (1975) G-I-T Fachz. Labor 19: 481-484

DETERMINATION OF AMINO ACIDS USING DANSYL CHLORIDE

C. Neubach, E. Schulze and V. Neuhoff
Max-Planck-Institut für experimentelle Medizin, 3400 Göttingen (FRG)

Dansyl chloride (5-dimethylaminonaphthalene-1-sulfonyl-chloride) reacts with various amines to form the corresponding sulfonic acid amides, which fluoresce under UV light. These characteristics can be utilized to determine the quantity of many substances at very low concentrations.

Earlier in this laboratory radioactive dansyl chloride was used in an attempt to quantitatively analyse a whole spectrum of amino acids (Briel et al., 1972). We report here a method which does not require radioactive dansyl chloride, yet achieves an at least equal sensibility as before. Substances were reacted with excess dansyl chloride at pH 9.0 for precisely 30 minutes at 37°C in a water bath. The reaction was terminated by addition of a drop of diethylamine, which reacts rapidly with excess dansyl chloride and has no effect upon the chromatography. In each analysis defined amounts of sarcosine and α-phenylglycine as internal standards (Schulze et al., 1976) and dansyl-diethanolamine as external standard were added to the sample before dansylation. The ratio between external and internal standards allows conclusions on the efficiency of the reaction procedure. Dansylated products were separated by ascending chromatography on 3x4 cm micropolyamide layers (Schleicher & Schüll, A 1700, Dassel, Germany). In the first dimension (along 4 cm) 2% formic acid in water was used while in the second dimension 20% acetic acid in toluene was employed.

The chromatograms are evaluated by scanning microscope fluorometry. The area of 3x4 cm is scanned in a meander-like pattern yielding 300x400 measuring values, which are nonlinearely transformed to obtain a dynamic range of 1:1000 and stored as 12 bit words in a laboratory computer (PDP 12). These represent a picture of 120000 points, from which the intensities and positions of the fluorescent spots can be determined by digital image processing. Automatic pattern recognition relates these figures to the corresponding amino acids. Data acquisition, calculation and documentation need 18 minutes for one chromatogram. In the range from 2×10^{-13} to 10^{-10} moles of dansylated amino acid the mass and fluorescence are linearly correlated (Kronberg et al., 1978).

Briel, G. and Neuhoff, V. (1972) Hoppe-Seyler's Z. Physiol. Chem. 353: 540-553

Kronberg, H., Zimmer, H.-G. and Neuhoff, V. (1978) Fresenius Z. Anal. Chem. 290: 133-134

Schulze, E. and Neuhoff, V. (1976) Hoppe-Seyler's Z. Physiol. Chem. 357: 593-600

SOME BIOCHEMICAL AND MORPHOLOGICAL CHARACTERISTICS OF BULK ISOLATED NEURONES FROM ADULT RAT BRAIN

H.H. Althaus, P.J. Gebicke, R. Meyermann[*], W.B. Huttner[**], V. Neuhoff
Max-Planck-Institut für experimentelle Medizin, Forschungsstelle Neurochemie, [*]Neuropathologische Abteilung der Neurologischen Universitätsklinik, 3400 Göttingen (FRG) and [**]Department of Pharmacology, Yale University School of Medicine, New Haven, Conn. (USA)

Neurones from adult rat brain can be obtained by a bulk isolation method which is based on a perfusion of the rat brain with a hypertonic medium (containing 0.1% Collagenase/Hyaluronidase) (Althaus et al. 1977). The neurones retain the proximal part of their processes, reveal a preserved plasma membran, to which synaptic boutons are attached, and contain intact intracellular structures (as observed by SEM and TEM).

Furthermore, these neurones are capable of regenerating their fibers in vitro. A cinematographic time sequence of regenerating fibers will be shown.

In addition, some biochemical data will be presented: protein pattern, lipid composition, some enzymatic activities, the amino acid composition and adenyl cyclase activity of neurones from different areas of the rat brain.

Althaus, H.H., Huttner, W.B. and Neuhoff, V. (1977) Hoppe Seyler's Z. Physiol. Chem. 358: 1155-1159

ISOELECTROFOCUSING OF CRUDE AND PURIFIED CHOLINE ACETYL TRANSFERASE FROM HUMAN PLACENTA

C. Froissart, P. Basset, T. Y. Wong, P. Mandel and R. Massarelli
Centre de Neurochimie du CNRS, Institut de Chimie Biologique
11, rue Humann, 67085 Strasbourg Cedex, FRANCE

Choline acetyltransferase (EC. 2.3.1.6., ChAcT) from human placenta was purified as follows : the homogenate (pH 7.4) was acidified (to pH 5.1) with citric acid and the enzyme (bound to membranes at this pH) resuspended in a Tris buffer, pH 7.4 containing 0.3 M NaCl. An ammonium sulfate precipitation will recuperate 70 % of the enzymatic activity between 30 and 60 % of saturation. The precipitates is then dissolved in Tris buffer pH 8.2, dialyzed and chromatographed on a DEAE cellulose column. Dithio-bis-nitro-benzoic acid-sepharose, blue-sepharose and phenyl-sepharose were further used (in the order) to purifying the enzyme to a final 1,600 folds and with a recovery of about 2 % of the enzymatic activity.

Polyacrylamide gel electrophoresis showed one major band at pH 4.5 and 8.2. Several bands, however, were detected in denaturating conditions (gels containing 0.08 % SDS). Preparative isoelectric focusing (pH 6-10) of this purified enzymatic preparation showed only one main, broad (one pH unit) peak of enzymatic activity around pH 8.0. When the isoelectric focusing was performed on the crude enzyme, only one enzymatic peak was detectable at pH 8.0. It however the homogenate was concentrated (under mild conditions : in a dialysis tube under vacuum) a second peak appeared at an acidic pH (between pH 6.0 and 7.0). An acid (pH 6.0 - 7.0) and a basic (pH 8.0) peaks have also been observed after ammonium sulfate precipitation and on the effluate concentrated after DEAE chromatography.

The results show that concentration of ChAcT may give rise to aggregates and/or to interactions with other proteins. Further work is in progress in order to elucidate the mechanism(s) underlying ChAcT aggregation.

POSSIBLE ROLE OF MEMBRANE-BOUND GLUCOSIDASE IN THE PROCESSING OF CALF BRAIN GLYCOPROTEINS

M.G. Scher and C.J. Waechter
Department of Biological Chemistry
Univ. of Maryland School of Medicine
Baltimore, Maryland 21201
U. S. A.

Earlier studies (Scher et al., 1977; Scher and Waechter, 1978; Waechter and Scher, 1978) have established that calf brain membranes catalyze the transfer of a [Glc-^{14}C] oligosaccharide unit, containing glucose and N-acetylglucosamine, from dolichyl diphosphate to endogenous glycoproteins. We now show that the [Glc-^{14}C] oligosaccharide unit, released from the carrier lipid by mild acid, is partially degraded by an α-mannosidase. Thus, the dolichol-linked oligosaccharide chain formed by calf brain membranes contains mannose, as well as glucose and N-acetylglucosamine.

Since glucose is rarely found in mammalian glycoproteins containing mannose and N-acetylglucosamine, we have investigated the possibility that glucosyl residues serve as a regulatory signal, and are excised during further processing of the oligosaccharide chain. In a pulse-chase experiment, the addition of a large excess of unlabeled UDP-glucose, after [Glc-^{14}C] glycoprotein was labeled by incubating calf brain membranes with UDP-[^{14}C] glucose, resulted in a time-dependent loss of label from the glycoprotein fraction. Direct proof for the presence of glucosidase activity associated with calf brain membranes was obtained by showing that [^{14}C] glucose was released from the free [Glc-^{14}C] oligosaccharide during incubation with membranes. Similarly, [^{14}C] glucose was enzymatically liberated from the [Glc-^{14}C] glycopeptide, produced by Pronase digestion of the [Glc-^{14}C] glycoprotein fraction labeled by incubation with UDP-[^{14}C] glucose or exogenous [Glc-^{14}C] oligosaccharide lipid.

These results provide evidence for a membrane-associated glucosidase activity capable of excising glucosyl residues from glycoprotein glycosylated via the lipid intermediate pathway. An "excision - revision" process for the assembly of complex asparagine-linked oligosaccharide chains in central nervous tissue is proposed.

Scher, M.G., Jochen, A. and Waechter, C.J. (1977) Biochemistry
    16: 5037-5044
Scher, M.G. and Waechter, C.J. (1978) Trans. Am. Soc. Neurochem.
    9: 170
Waechter, C.J. and Scher, M.G. (1978) Arch. Biochem. Biophys.
    188: 385-393

PROPERTIES OF CHOLINE ACETYLTRANSFERASE AND ITS PRESENCE IN INDIVIDUAL IDENTIFIED NEURONES IN THE CNS OF THE SNAIL HELIX POMATIA

N.N. Osborne
Forschungsstelle Neurochemie
Max-Planck-Institut für experimentelle Medizin
3400 Göttingen, Germany

Choline acetyltransferase (Ch.Ac.) is associated with the nervous system and specifically localised in certain neurones in the snail CNS. Neurones which contain Ch.Ac. do not possess other putative transmitter substances such as 5-hydroxytryptamine, histamine, octopamine, dopamine or noradrenaline. Moreover, only the neurones containing Ch.Ac. can form radioactive acetylcholine from the precursor ^{3}H-choline and lack enzymes to produce other transmitter substances from their radioactive precursors. The data thus support Dale's principle that neurones contain, produce and utilise single transmitter substances (see Osborne, 1977).

The enzyme Ch.Ac. from the snail CNS was also partially purified by $(NH_4)_2SO_4$ fractionation and its properties were analysed. NaCl, KCl, histamine, imidazole, 5-hydroxytryptamine and dopamine all activated the enzyme while the amino acids were without effect. $CaCl_2$ and $MgCl_2$ inhibited the enzyme. The enzyme is fairly specific for choline as substrate when compared with choline analogues; Kmax for choline, monoethylcholine, diethylcholine, triethylcholine and pyrrolcholine were respectively 0.37 mM, 1.5 mM, 1.66 mM, 3.12 mM and 1.7 mM. The Vmax values for choline and mono-, di- and triethylcholine with the snail enzyme indicated a direct relationship between enzyme activity and N-alkyl substitution.

The transport of Ch.Ac. along nerves is in a proximal-distal direction (orthograde), although the existence of a retrograde transport mechanism cannot be excluded. This conclusion is drawn from experiments carried out on the visceral nerve of the snail which received either single or double ligatures for periods of up to 48 hrs. The main finding was that after 48 hrs. there was an increase of about 40% Ch.Ac. in the first 4 mm proximal to the ligature. The increase became apparent only after a ligature period of 16 hrs. In parts of the nerve distal to the ligature, no change in enzyme activity was observed. Similar results were achieved with double ligature experiments, although the Ch.Ac. activity proximal to the distal ligature was reduced in comparison to that associated with the proximal ligature.

Osborne, N.N. (1977) Nature 270: 622-623

EFFECT OF BENZENE INHALATION ON THE LEVELS OF GABA AND ITS METABOLIZING ENZYMES IN VARIOUS REGIONS OF RAT CNS

G.K. Kadyrov
Laboratory of Neurochemistry, Institute of Physiology,
Bacu, USSR

The influence of a high concentration of benzene vapor (35 mg/l) on the levels of GABA and its metabolizing enzymes (GAD, EC 4.1.1.15; GABA-T, EC 2.6.1.19) in various regions of the CNS (sensorimotor cortex, cerebellum, pons Varolii, spinal cord) were studied using Wistar rats aged 1, 10 and 21 days, and 3, 12 and 24 months. Animals were exposed to benzene (C_6H_6, specific gravity - 0.879) vapor in a 100 liter chamber for 3,5 hours.

This treatment decreased the level of GA (Glutamic acid) in homogenates and mitochondrial fractions of cortex, cerebellum, pons Varolii and spinal cord at all ages studied, while the levels of GABA increased. The activity of GAD increased in homogenates and mitochondrial fractions of all areas studied at the stages of postnatal development, and similar increases were observed after benzene inhalation.

Benzene treatment did not alter the activity of GABA-T in sensorimotor cortex of newborn and three month old rats, whereas a decrease was observed in 10 and 21 day old rats, and an increase in 12 and 24 month old animals. In cerebellum, pons Varolii and spinal cord of newborn, 3, 12 and 24 month old rats, a considerable increase in GABA-T activity was observed, and a decrease in 10 and 21 day old rats.

The increased levels of GABA and the changes in activity of GAD and GABA-T may represent an adaptive-compensatory reaction of the neural cells to extreme external influences.

TISSUE DIFFERENCES IN THE HUMAN N-ACETYL-β-D-HEXOSAMINIDASE ISOENZIMATIC FORMS.

Pàmpols, T., Codina, J., Girós, M., Sabater, J. and González-Sastre, F.
Dep. Neuroquimica. Inst. Prov. de Bioquímica Clínica. Univ. Autónoma de Barcelona - Cerdanyola. Barcelona. Spain.

By cellulose acetate electrophoresis at pH 6.5 (1) the heterogenity of the Hexosaminidase activity can be demostrated. The differences in electrophoretic mobility and the proportion of the several forms are indicated in the table. The values were obtained by densitometry.

	Serum	Leucocit	Brain	Liver	Spleen	Kidney
activity nM/mg.p/h.	9.0	414.9	230.3	1161.1	721.2	4737.0
% C	-	4.1	26.1	2.2	3.8	2.6
% A	69.9	59.8	48.3	45.2	50.8	45.4
% B_2	23.0	-	-	-	-	-
% B_1	5.7	-	-	-	-	-
% B	-	36.1	25.5	52.6	45.4	52.1

elec.mobility: C Tissues, A Serum-Tissues, B_2 Serum, B_1 Serum, B Tissues

The A and C forms are heat labels while the B forms in serum and tissues are heat stables. The B forms from serum and tissues are clearly distinct: a mixture of B forms obtained from both origins by DEAE column chromatography shows two clearly separated bands beeing the more anodic the seric one.

Hex A isolated from tissues by DEAE chromatography, experiment a slow espontaneous transformation (kept at -20°C) in a B form with the electrophoretic mobility of the tissue B form and a new band more anodic than Hex A. This phenomenon can be observed after 6 hours at 37° C, pH 6.0. The A form from serum does not show this conversion.

By isoelectric focusing in PAG plate further differences can be assesed. The Hex A from tissues has an isoelectric pH lower than Hex A from tissues. Hex B from tissues has an isoelectric pH markedly higher than Hex B from serum. The heterogenity of the form A from tissues is clearly shown in kidney and leucocits. The heterogenity of the forms B is also seen in kidney, spleen and leucocits. The Hexosaminidase pattern of brain is caracterized by the predominance of the A form and the presence of a B band spread in a pH zone less basic than the region of the B forms of the other tissues. See the figure.

(1) POENARU, J. and DREYFUS, J.C. Clin.Chim Acta 43, 439 (1973)

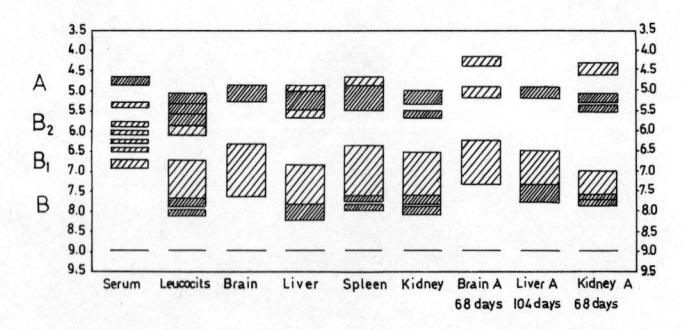

POSSIBLE SYNAPTIC FUNCTION OF GLUTAMATE DECARBOXYLASE

Alfred Fleissner

Neurochem. Abteilung der Psychiatrischen Universitätsklinik
Martinistrasse 52, 2000 Hamburg 20, GFR

Glutamate decarboxylase (GAD) is the key enzyme involved in the synthesis of the inhibitory transmitter γ-aminobutyric acid (GABA) from glutamic acid, which is supposed to be an excitatory transmitter in the CNS. Experiments designed to purify GAD by affinity chromatography resulted in a very strong fixation of the enzyme to azobenzene groups without loss of enzymatic activity (Fleissner, 1978). The K_s value for pyridoxal 5'-phosphate (PLP) of bound GAD compared with the soluble enzyme remained unchanged. When bound to the adsorbent in the presence of 1 μM PLP the activity of the enzyme was slightly higher than the activity of the dissolved enzyme determined at the same PLP concentration. In the presence of sufficient PLP (0.5 mM) attachment of GAD to the adsorbent resulted in so strong a fixation of the coenzyme that after washing the material several times with PLP-free buffer full enzymatic activity was recovered in a PLP-free incubation medium. After addition of PLP the activity was even slightly lower.

GAD is usually thought to be a soluble enzyme. On the other hand it binds to a high degree to certain chemical groups or membrane constituents resulting in a stabilization of its activity. A role of GAD activity as a regulatory mechanism of cerebral excitability has been postulated (Tapia et al., 1975). Conceptually, GAD might be released together with GABA during synaptic transmission by an exocytotic mechanism. Once released into the synaptic cleft, GAD might continue to produce GABA while concomitantly reducing the concentration of the excitatory transmitter glutamate thereby possibly facilitating the consolidation of short term memory contents.

Fleissner, A. (1978) in: O. Hoffmann-Ostenhof et al. eds., Affinity Chromatography, Proceedings of a Symposium held in Vienna, September 20-24, 1977, p 81-84, Pergamon Press, Oxford

Tapia, R., Sandoval, M.E. and Contreras, P. (1975) J. Neurochem. 24: 1283-1285

THE SEPARATION OF EXTRA-AND INTRACELLULAR ACETYLCHOLINESTERASE
ACTIVITY OF RAT SYMPATHETIC GANGLION BY MILD PROTEOLYTIC TREATMENT

Biba Klinar and M. Brzin
Institute of Pathophysiology, Medical Faculty,
61000 Ljubljana, Yugoslavia

The acetylcholinesterase activity in an innervated sympathetic ganglion is localized on presynaptic axons and in postsynaptic nerve cells, but the individual contribution of both structures to the total ganglion acetylcholinesterase is not known. Because of the possible influence that the nerve may exert on the intracellular enzyme activity, the activity of acetylcholinesterase remaining after preganglionic denervation may not reflect the amount of acetylcholinesterase present in an innervated ganglion cell (Somogyi and Chubb, 1976; Chang, 1977; Gisiger et al., 1978).

The cytochemical and biochemical findings of this study show that the mild proteolytic treatment of rat sympathetic ganglion withpapain (0,7 mAnson unit per ml of Ringer solution) removes selectively the acetylcholinesterase activity localized extracellularly and leaves intracellular acetylcholinesterase apparently unchanged. Thus, the separation between the extra- and intracellular acetylcholinesterase activity gives the proportion of enzyme activity at the different ganglion structures and makes possible the investigation of the localization of different molecular forms of ganglion cell acet ylcholinesterase.

Chang, Ping-Lung, (1977) Cell Tiss. Res.
    179: 111-120
Gisinger, V., Vigny, M., Gautron, J. and Rieger, F. (1978) J. Neurochem.
    30: 501-516
Somogyi, P. and Chubb, I.W. (1976) Neuroscience
    1: 413-421

EXPOSURE OF 7 WEEK OLD VISUALLY NAIVE RATS TO DIFFUSE AND STRUCTURED LIGHT:
EFFECTS ON VISUAL CORTEX ACETYLCHOLINESTERASE.

N. Wood and S.P.R. Rose
Brain Research Group, Open University, Milton Keynes, U.K.

An organized pattern of neurochemical changes has been observed at the retinal, geniculate and cortical stages of the visual sensory system, resulting from an initial exposure to light in the 50 day old rat which has been reared from birth in darkness (Rose, 1977). The potential cellular function of these changes has been studied by observing the alteration of enzyme activity in the visual cortex, including acetylcholinesterase, which are known to be involved in cellular processes of connectivity (Sinha and Rose, 1976). The question then posed was: are the observed changes contingent primarily upon the sensory stimulation involved, or the onset of perceptual function, or the associated events which involve the acquisition of information in a novel sensory modality?

Experiments involving the application of transparent, light diffusing and opaque contact occluders during three hour periods of exposure of visually naive rats to controlled lighting conditions indicate that exposure to diffuse light, involving sensory stimulation of the visual system, results in increases of 15-20% in visual cortex acetylcholinesterase, compared to dark controls. No changes were found in a control cortical area (motor cortex). This increase was not significantly different from the elevation associated with exposure to environmental lighting conditions permitting perceptual processing.

The question arising is: are there inter or intra-cellular levels of organization, and neuro-anatomical areas in which the effects of visual stimulation are neurochemically dissociable from the associated events involving visual information processing?

Sinha, A.K. and Rose, S.P.R. (1976) J. Neurochem. 27, 4: 921-926
Rose, S.P.R. (1977) Phil. Trans. R. Soc. Lond. B 278: 307-318

LIGHT EXPOSURE AND NON-SPONTANEOUS LOCOMOTOR ACTIVITY IN VISUALLY DEPRIVED AND NORMALLY REARED RATS: CORTICAL EFFECTS IN THREE COMPONENTS OF THE CHOLINERGIC SYSTEM.

N. Wood and S.P.R. Rose
Brain Research Group, Open University, Milton Keynes, U.K.

Three hour periods of visual stimulation in the 50 day old dark reared rat have been shown to be associated with increases of up to 30% in visual cortex acetylcholinesterase and cholinacetyltransferase activity (Sinha and Rose, 1976) and in the muscarinic cholinergic receptor (Rose and Stewart, 1978). The effects of 3h light exposure on acetylcholinesterase levels have been shown to be partly dependent on the time of day of exposure, and further that motor cortex acetylcholinesterase levels in dark reared rats fluctuate diurnally with a low amplitude rhythm which correlates with the behavioural activity cycle (Wood and Rose, 1978).

Subsequent experiments involved exposing rats reared for 30-60 days from birth in darkness, to three hour periods of light alone, or with simultaneous activity in a motor driven running wheel (rate of motion = 4cm/s at the circumference). Three components of the cholinergic system- the synthetic and degradative enzymes, and ^{3}H - quinuclidinyl benzilate (QNB) binding to the muscarinic cholinergic receptor, by a method developed from that of Yamamura and Snyder (1974)- have been studied in unfractionated homogenates of visual and motor cortices. The results show that in the dark reared animal, the ratio of visual to motor cortex acetylcholinesterase and QNB binding increases with light exposure, due to an elevation in the visual cortex, and increases further with simultaneous exposure and movement in the activity wheel, due to a decrease in the motor cortex. There were no significant effects in cholinacetyltransferase and no significant effects in normally reared animals. Additionally, dark reared animals in the age range 30-45 days show significantly lower levels of AChE (25%) CAT (12%) and QNB binding (41%) than animals in the range 45-60 days. An interpretation is formulated in terms of ontogenetic and neurochemically plastic developmental processes.

Rose, S.P.R. and Stewart, M.G.S. (1978) Nature, 271:169-170.
Sinha, A.K. and Rose, S.P.R. (1976) J. Neurochem. 27, 4: 921-926.
Wood, N. and Rose, S.P.R. (1978) Submitted for publication.
Yamamura, H.I. and Snyder, S.H.(1974) Proc. nat. Acad. Sci. (Wash.) 71: 1725-1729.

THE ACTION OF ANESTHETICS ON THE KINETICS AND ACTIVITY OF ACETYL-CHOLINESTERASE FROM SYNAPTOSOMAL MEMBRANES

A. Pastuszko
Department of Neurochemistry, Medical Research Centre, Pol.Acad.Sci,
00-784 Warsaw, Poland.

The anesthetics depress synaptic transmission of the synapses in the central nervous system. It is possible anesthetics accumulate in the nerve membrane and produce changes in the organization of the membrane, usually referred to as swelling, packing or stabilization. The perturbation in cell membranes caused by anesthetics may have influence on the change in kinetics and activity of membrane bound enzyme. In this study, the effects of anesthetics - thiopental and brietal on the membrane-bound acetylcholinesterase were examined. This enzyme / EC 3.1.1.7 / has been the object of intense interest, because of its paramount importance in the control of electrical activity of excitable membranes.

Male Wistar rats /180-200 g/ were injected intraperitoneally with thiopental /100mg/kg body/ or brietal /60mg/kg body/. After 30 min of sleep rats were decapitated and synaptosomal membranes from brain were prepared. The preparation was similar to the reported by Whittaker et al, 1964, and Gurd et al, 1974. Enzyme activity was determined by the metod of Ellman et al, 1961.

These anesthetic agents: thiopental and brietal caused reversible inhibition of AChE from synaptosomal membranes. The inhibition by anesthetics is not the same. Brietal decreases the Km for acetyltiocholine, without changes in Vmax. An uncompetitive type of inhibition is produced by thiopental. These results suggest a conformational change of the enzymic protein, but do not prove that the effects are mediated through the lipids. Brietal caused no changes in Arrhenius plots and only insignificant changes of the activation energies. Thiopental, in vivo, decreases the temperature break above $3^{\circ}C$. and the activation energy is increased at all temperatures to values in the range of those below the break.
All of these results suggested that this anesthetic compounds produce disturbances in lipid-protein interaction in biological membranes.

Ellman, G.L., Courtney, D.K., Andersen, Jr.V., Featherstone, R.M. /1961/ Biochem.Pharmacology. 7: 88-95.
Gurd, J.W., Jones, L.R., Mahler, H.R., Moore, W.J. /1974/ J.Neurochem.22: 281-290
Whittaker, V.P., Michaelson, A., Kirkland, R.I.A. /1964/ Biochem.J.90:293-298.

IS ACETYLCHOLINE LOCALIZED IN THE CORTICO-STRIATAL PATH?

P. Haug, N. Schröder, J. Kim, K. Paik, and R. Hassler
Neurobiologische Abteilung, Max-Planck-Institut für Hirnforschung
6 Frankfurt/M.-Niederrad (G.F.R.)

The present investigation was undertaken to determine if frontal cortex lesion has any effect on the level of acetylcholine in the striatum. The left frontal cortex of Male Wistar albino rats was removed by suction. After 1 week the animals were decapitated and the striata removed by dissection. The striatum of the unopperated right side served as a control. Acetylcholine was derivatized to B-dimethylaminoethylacetate (and choline was derivatized to B-dimethylaminoethylproprionate) according to the modified Jenden Procedure.[1] The concentration of derivatized acetylcholine and choline was determined by comparison with deuterium labeled internal standards by multiple-ion-monitoring gas chromatography-mass spectrometry.

The acetylcholine level was shown to undergo no significant change. This correlates with the findings that 1) frontal cortex lesions caused no significant effect on choline acetyltransferase activities,[2] 2) glutamic acid is significantly decreased in the striatum one month after ablation of the frontal cortex,[3] and 3) glutamic acid diethyl ester antagonized cortical excitation of the striatum.[4]

1 Karlen, B., Lundgren, I., Nordgren, I. and Holmstedt, B. (1974) Choline and Acetylcholine: Handbook of Chemical Assay Methods Edited by I. Hanin. Raven Press, New York: 163-179
2 McGeer, P.L., McGeer, E. G., Fibinger, H. C. and Wickson, V., (1971) Brain Research 42: 308-314
3 Kim, J., Hassler, R., Haug, P., and Paik, K. (1977) Brain Research 132: 370-374
4 Spencer, H. J. (1976) Brain Research 102: 91-101

PURIFICATION OF CYSTEINE SULFINATE TRANSAMINASE

M. RECASENS   P. MANDEL
Centre de Neurochimie
11, rue Humann
67085 STRASBOURG CEDEX
FRANCE

The major pathway of taurine biosynthesis in mammalian brain is the one which leads to taurine from cysteine via cysteine sulfinic acid and hypotaurine. Injection of ^{14}C-cysteine produces mainly ^{14}C pyruvate (Yamaguchi et al, 1973). This suggests the possibility of a regulation of the cysteine sulfinate level through the cysteine sulfinate transaminase (CSA-T), and consequently a regulatory mechanism for the taurine biosynthesis. This hypothesis was also supported by the high activity of this transaminase, found in different brain areas (Recasens et al, 1978) (about 1000 times higher than cysteine sulfinate decarboxylase activity). These results prompted us to study this enzyme.

CSA-T was purified 500 times from rat brain after five steps. (Ammonium sulfate precipitation and DEAE-sephadex; hydroxyapatite; octylsepharose and sephacryl chromatography). SDS slabgel electrophoresis of 20 µg of protein shows only one band. The molecular weight of the subunit is about 41 000, almost the same as the glutamate oxaloacetate transaminase subunit, sometimes considered identical as CSA-T. Comparative studies on homogenate (Recasens et al, 1978, Gabellec et al, 1978) are rather in favour of two different enzymes. However, studies on purified enzyme will be necessary to solve the problem of one or two distinct enzymes.

Yamguchi, K., Sakakibara, S., Asamizu, J., and Ueda, I. (1973) Biochim. Biophys. Acta, 287 :48-59
Recasens, M., Gabellec, M.M., Austin, L., Mandel, P. (1978) submitted.
Recasens, M., Gabellec, M.M., Mack, G., and Mandel, P. (1978) Neurochem. Res. 3, 27-35
Gabellec, M.M., Recasens, M., Mandel, P. (1978) Life Science (in press)

DIFFERENCES IN THE PROPERTIES OF HUMAN LEUCOCYTE AND FIBROBLAST CEREBROSIDE SULPHATE SULPHATASE

A. Poulos and K. Beckman,
Department of Chemical Pathology, Adelaide Children's Hospital (Inc.),
North Adelaide,   S.A.,   5006,   AUSTRALIA.

The measurement of cerebroside sulphate sulphatase in leucocytes and fibroblasts is important in the diagnosis of metachromatic leukodystrophy. Few studies have been reported concerning the properties of the leucocyte enzymes although the fibroblast enzyme has been examined in some detail (Porter etal, 1971). In this communication we describe some of the properties of the former and compare these with data obtained with the fibroblast enzyme.

The leucocyte enzyme reaction was linear with time up to a period of at least six hours. In contrast, the plot of product formed versus time for the fibroblast enzyme was non-linear. The amount of substrate hydrolysed was linearly related to the leucocyte protein concentration up to about 1mg/ml. Under the similar incubation conditions the fibroblast enzyme activity was also directly proportional to the protein concentration but only up to 0.1mg/ml; beyond this concentration an apparent activation was observed. The leucocyte enzyme had an apparent Km of 1.0-1.4mmol/l. Substrate saturation was achieved at 6.0mmol/l. Determinations for the fibroblast enzyme were complicated by the non-linear time course. However the double reciprocal plots of velocity, calculated at different incubation periods, and substrate concentration, was linear. No evidence for inhibition, suggested by the non-linear time course of the reaction, was detected. Km values (0.05-0.15mmol/l) were considerably lower than for the leucocyte enzyme.

These data indicate considerable differences in the properties of cerebroside sulphate sulphatase in the two tissue sources most frequently used for the diagnosis of metachromatic leukodystrophy and emphasises the dangers involved in using the optimum assay conditions established for one particular tissue for the same assay in another tissue. The differences may be due to the presence of different enzymes although the presence of other factors in the tissue extracts, such as activators or inhibitors, which could alter some of the enzyme properties, cannot be discounted.

Porter, M.T., Fluharty, A.L., Trammel, J. and Kilhara, H., (1971), Biochem. Biophys. Res. Comm., 44: 660-666.

DIFFERENCES IN PROTEIN KINASE ACTIVITY BETWEEN NEURONAL AND GLIAL NUCLEI FROM HUMAN CEREBRAL CORTEX

K. Reichlmeier, H. P. Schlecht, K. Citherlet and M. Ermini*
Department of Basic Medical Research, SANDOZ LTD, CH-4002 Basel
*Inst.Pharmacol.Biochem., Vet.Med.Fac., Univ.Zurich, CH-8057 Zurich

Chemical modifications of chromosomal proteins as catalyzed by nuclear protein kinases (PKases) are thought to be related to the process of genome activation. Neuronal RNA-synthesis is also known to exceed considerably that of glial cells. Therefore, we investigated the possibility that similar differences might exist in the activity of nuclear PKases among the two cell types by comparing whole PKase activity of preparations enriched in either neuronal or glial nuclei. The nuclei were isolated according to Thompson (1973) from human autoptic cerebral cortex. The individuals from which the brain specimens were taken were divided into two age-groups: an adult one (19 to 59 years) and an old one (60 to 91 years).

Utilizing either endogenous substrate or calf thymus histone, a significantly higher PKase activity was observed in neuronal nuclei than in glial nuclei. This corresponds to the findings of Lee and Loh (1977) on PKase activity in mouse brain nuclei. Addition of calf thymus histone nearly doubled the phosphorylation rate in both nuclear populations, while cAMP had no stimulatory effect in either case. However, due to the heterogeneity of nuclear PKases (Kish and Kleinsmith, 1974), a differential response to cAMP cannot be excluded. The nuclear PKase activity does not appear to be subject to age-related changes in both cell types.

Cortex nuclei assayed before separation by density gradient centrifugation into neuronal and glial nuclei, however, showed a cAMP-dependent PKase activity which decreased with age. Since electron-microscopic examination of the mixed nuclei fraction revealed some cytoplasmic contamination, the observed cAMP- as well as the age-effect might be due to extranuclear factors.

Kish, V.M. and Kleinsmith, L.J. (1974) J. Biol. Chem.
  249: 750-760

Lee, N.M. and Loh, H.H. (1977) J. Neurochem.
  29: 547-550

Thompson, R.J. (1973) J. Neurochem.
  21: 19-40

ROLE OF NON-COVALENT BONDS FOR THE HOLDING OF MITOCHONDRIAL BRAIN HEXOKINASE

B. Broniszewska-Ardelt
Polish Academy of Sciences Medical Research Centre,
Warsaw, Poland

There have been no investigations of the role of non-covalent bonds for holding of mitochondrial hexokinase in the brain. According to Eggar and Rapoport (1963) such bonds play an important role in the holding, activation and release of mitochondrial enzymes.
Based on the pH-dependent liberation of mitochondrial hexokinase it is possible to postulate a role of imidazole bonds in holding this enzyme. It has also been shown that the mitochondrial hexokinase of brain may be activated and liberated by dilute urea solutions suggesting the participation of peptide-hydrogen bonds in the holding of the enzyme.
The importance of electromagnetic interactions and hydrophobic bonds is suggested by the observation that liberation of proteins, as well as activation and release of hexokinase, is much greater in an ionic, as compared with a non-ionic medium. The greater release in ionic media is probably caused by the breaking of polar and hydrophobic bonds. Some differences in the liberation of proteins between mature and immature mitochondria by ionic and non-ionic media were noticed.

Eggar, E. and Rapoport, S. (1963) Nature 200:240-242

MEMBRANE-BOUND A_4 LACTATE DEHYDROGENASE OF RAT BRAIN AND ITS POSSIBLE RELATIONSHIP TO ANAEROBIC GLYCOLYSIS

H.H.Berlet, T.Lehnert and B.Volk
Institute of Pathochemistry and General Neurochemistry, and
Institute of Neuropathology,
Heidelberg, German Federal Republic

Studies of soluble lactate dehydrogenase (EC 1.1.1.27, LDH) of brain tissue have consistently shown that the homotetramer A_4, if present at all, is only a minor component among the five isoenzymes. Because of its low K_m value for lactate compared to pyruvate and the differential inhibitory effects of pyruvate the A_4 enzyme is thought to be concerned with anaerobic glycolysis mainly (Cahn et a., 1962).

We have investigated the LDH activity and isoenzyme patterns of subcellular particles and particle fragments ("membranes") of rat brain and found substantially higher proportions of A_4 than in cytoplasm. The particle fragments contained about 15% of the total tissue activity of LDH, and on electrophoresis A_4 comprised as much as 30% of the overall activity of isoenzymes. The relative percentages of A_4 of fragments derived from subcellular particle fractions ranged from 12% (mitochondria, myelin) and 25% (synaptosomes) up to as much as 45% (microsomes). Particle fragments were then treated with a detergent to solubilize the structurally bound LDH (Berlet and Lehnert, 1978). An appreciable portion of the enzyme resisted this treatment, and more than 70% of the residual activity was found to be attributable to the A_4 enzyme.

The results suggest that the A_4 enzyme may be typical of some specialized membranes of nervous tissue. As it is more tightly bound than the other forms of LDH it appears to be an essential structural component fulfilling a specific function in handling lactate and pyruvate, respectively, preferably at subcellular sites of lower oxygen supply. Therefore the effect of chronic hypoxia on the subcellular distribution of LDH isoenzymes is being investigated and results of this work will be presented.

Berlet, H.H. and Lehnert, T. (1978) FEBS Lett., In Press
Cahn, R.D., Kaplan, N.O., Levine, L. and Zwilling, E. (1962)
   Science 136: 962-969

SUBFRACTIONATION OF MOUSE BRAIN MICROSOMES ON A CONTINOUS SUCROSE GRADIENT: ISOLATION AND IDENTIFICATION OF THE MEMBRANES CONTAINING CERAMIDE GALACTOSYLTRANSFERASE AND CEREBROSIDE-SULFOTRANSFERASE

H.P. Siegrist, T. Burkart, U. Wiesmann and N. Herschkowitz
Dept. of Pediatrics, University of Berne
3010 Bern, Switzerland

Several attempts have been made in order to identify the membranes containing the microsomal enzymes ceramide-galactosyltransferase (CGalT, EC 2.1.2.48) and cerebroside-sulfotransferase (CST, EC 2.8.2. 11). All the systems described so far were based on a discontinous gradient used to subfractionate microsomes and are established for kidney (1). Since the composition of the membranes surrounding these enzymes is of importance for their activity during development (2) we worked out a system, which allows the submicrosomal localisation of CGalT and CST in mouse brain. Homogenates of 16 day old mouse brains were centrifuged for 20 minutes at 17'000 xg and the resulting supernatant was fractionated further on a continous sucrose gradient ranging from 0.8 to 1.5 M sucrose. The gradient was collected in 15 1 ml-fractions and assayed for marker enzymes of myelin, plasma membranes, cytosol, Golgi membranes, endoplasmic reticulum and lysosomes. In addition protein and lipid distribution was studied. The fractions were further characterised by electronmicroscopy. The results suggest, that as in kidney CGalT and CST are both located in the Golgi membranes of mouse brain.

1. Fleischer B., Zambrano F. (1973), Biochem. Biophys. Res.Com. 52, 951-958
2. Siegrist H.P., Jutzi H., Burkart T. Steck A.J., Wiesmann U. and Herschkowitz N. (1977) Biochem. Biophys.Acta 489, 58 - 63

PURIFICATION AND SOME PROPERTIES OF RAT BRAIN GUANYLATE CYCLASE

J. ZWILLER, P. BASSET and P. MANDEL
Centre de Neurochimie du CNRS, 11 rue Humann, 67085 Strasbourg Cedex FRANCE

There is a considerable amount of evidence that cyclic GMP is involved in several cellular events (cell proliferation, hormone receptors). In order to understand the regulatory mechanisms involved in these events, it was necessary to purify guanylate cyclase, to investigate its properties and its interactions with membrane receptors. In most mammalian cells, the enzyme is usually detectable in both particulate and soluble subcellular fractions following cell disruption. Thus, we started our purification from the soluble fraction.

A new method for purifying rat brain soluble guanylate cyclase is presented. The procedure involves chromatography on Blue Sepharose 6B, isoelectric focusing and gel filtration. Two peaks of activity were detected by isoelectric focusing. The main peak corresponds to a pHi of 5.9 and the minor to a pHi of 6.2. It is not possible to relate these peaks to the different forms of the enzyme, or to an isoelectric focusing artefact. The most active peak of guanylate cyclase was further purified on Ultrogel ACA 34.

The purified enzyme which was found to be a glycoprotein required Mn^{++} in excess of GTP for maximum activity, although the enzyme can also utilize Mg^{++} to a lesser extent. However, no change in the affinity for GTP was observed when optimal concentration of Mn^{++} was used. Heavy metals (Zn^{++}, Hg^{++}, Cd^{++}, Cu^{++}) were found to be strong inhibitors while Ca^{++} was found to have a stimulatory effect at optimal Mn^{++} concentration. Guanylate cyclase activity was also inhibited by several nucleotides ; the triphosphonucleotides ATP, CTP and UTP were found to be the most potent inhibitors.

Purified enzyme was activated by lysolecithin, a naturally occurring detergent. Nitroprusside and, to a lesser extent, hydroxylamine were also found to activate purified guanylate cyclase but to lesser degree than in the whole homogenate.

BIOSYNTHESIS AND SUBCELLULAR DISTRIBUTION OF DISATURATED PHOSPHATIDYLCHOLINE IN RAT BRAIN

L. Freysz[*] and H. Van den Bosch
Laboratory of Biochemistry, Transitorium 3, Padualaan 8, Utrecht, The Netherlands
and [*]Centre de Neurochimie du CNRS, 11 Rue Humann, 67085 Strasbourg Cedex, France

Recent studies have led to the finding that brain contains high amounts of phosphatidylcholine with two saturated fatty acids (DPC), especially 1-2 dipalmitoylphosphatidylcholine. Its function in brain is not understood, but it is known to constitute part of the functional entity of the lung surfactant. In the present study we have investigated the distribution and the biosynthesis of disaturated phosphatidylcholine in different subcellular fractions of rat brain.

The determination of DPC in the subcellular fractions showed that it represents 40 to 45 % of total phosphatidylcholine in microsomes and synaptosomes and only 23 to 25 % in myelin and mitochondria. In synaptosomes the plasma membranes contained the highest amount of DPC.

The biosynthetic pathway of DPC was investigated and the activities of choline phosphotransferase, lysophosphatidylcholine transacylase, and lysophosphatidylcholine acylase were determined in microsomes and synaptosomes. Both choline phosphotransferase and lysophosphatidylcholine acylase were able to catalyse the synthesis of DPC, but the brain did not exhibit any lysophosphatidylcholine transacylase activity. Therefore the biosynthesis of DPC in brain differs from that in lung where these molecular species are not synthesized by the Kennedy's pathway but by the lysophosphatidylcholine transacylase.

Effect of L-leucine on glutamate synthesis in mitochondrial and synaptosomal fractions from rat brain.

W. Łysiak, A. Szutowicz and S. Angielski
Department of Clinical Biochemistry, Institute of Pathology,
Medical Academy, 80-211 Gdańsk, Poland

It is known that L-leucine has a stimulatory effect on glutamate dehydrogenase /GDH/ activity from rat liver mitochondria. The glutamate synthesis in GDH reaction is the first step for NH_4^+ detoxification in nervous tissue. It is also possible that this reaction in brain may be regulated in the similar manner as in liver.

The incorporation of NH_4^+ into glutamate, in rat brain mitochondria metabolizing pyruvate + malate, was negligible. Under these conditions leucine did not change neither glutamate nor oxoglutarate and citrate production.

Addition of ATP /5 mM/ and Mg^{+2} /2 mM/ resulted in twofold increase in pyruvate uptake, glutamate production rised from 0 to 4.9 nmoles/mg protein/min. In the presence of leucine /2.5 mM/ glutamate synthesis increased to 8.2 nmoles/mg protein/min and that of aspartate from 1.8 to 2.6 nmoles/mg protein/min. This increase was accompanied by about 20% decrease in oxoglutarate and citrate accumulation. L-Leucine affected the GDH reaction; in the presence of aminooxoacetate its stimulatory effect on glutamate synthesis was still observed. The increase in mitochondrial NAD and NADP reduction, after addition of oligomycin resulted in relatively higher /in relation to pyruvate uptake/ glutamate synthesis.

Stimulatory effect of L-leucine on glutamate synthesis was also observed in rat brain synaptosomes. Moreover, L-leucine at concentrations from 0.05 to 5.0 mM increased NADH and NADPH-dependent GDH activities from mitochondrial and synaptosomal fractions from rat brain.

It is concluded that:
a/ L-leucine stimulates glutamate synthesis in nervous tissue
b/ the regulation of GDH activity by L-leucine may take place under physiological conditions.

SOME PROPERTIES OF SYNAPTIC MEMBRANES IN BINDING AMINO ACIDS

P. Lähdesmäki and E. Kumpulainen
Department of Biochemistry, University of Oulu, SF-90100 Oulu 10, Finland

A possible specific interaction of amino acids with pre- or post-synaptic receptors is an important criterion for a transmitter role of the amino acids. The isolated synaptosomes take up glutamate, GABA and taurine through a high-affinity mechanism (Lähdesmäki et al., 1975). The same amino acids interact specifically also with synaptic membranes (Lähdesmäki et al., 1977) having binding constants close to those obtained for the active transport into the intact synaptosomes. The interaction in the latter case may represent in part post-synaptic binding and in part pre-synaptic transport into empty membrane pouches derived from hypo-osmotically ruptured synaptosomes. Sac-like sealed structures are indeed abundant in electron microscopic figures (Lähdesmäki et al., 1977).

We have further studied certain physicochemical characteristics of taurine binding using chicken brain synaptic membranes, and also binding of taurine, glutamate, lysine and norleucine with synaptic membranes in vivo after intracerebral injection in mice. The binding of taurine was also in the present case complicated partially by a dissociation into empty membrane-pouches showing a binding constant of 67 µM. The binding had a pH optimum of pH 7 and a temperature optimum of 310 $^\circ$K. The activation energy of binding was 12.15 kJ/mol, calculated from the temperature course, and its free energy 21 kJ/mol, calculated from the equilibrium constant between the bound and free taurine. The number of specific binding sites for taurine was very low, 37 pmol/g protein, showing thus only 1/20 as many taurine binding sites as GABA binding sites.

Intracerebrally injected $[^3H]$lysine, $[^{14}C]$glutamate and $[^{35}S]$taurine penetrated nerve terminals in vivo lysine and glutamate being bound also at synaptic membranes and synaptic vesicles while taurine remained mainly in the soluble synaptoplasm. $[^{14}C]$Norleucine penetrated the brain cell membranes slowly and was not attached at synaptic membranes.

This study was supported by the Research Foundation of the Orion Corporation.

Lähdesmäki, P., Pasula, M. and Oja, S.S. (1975) J. Neurochem. 25: 675-680.
Lähdesmäki, P., Kumpulainen, E., Raasakka, O. and Kyrki, P. (1977) J. Neurochem. 29: 819-826.

INTERACTION OF ISOLATED SYNAPTIC VESICLES AND PRESYNAPTIC CELL MEMBRANES IN VITRO AND MODULATION OF VESICULAR RELEASE BY CALCIUM IONS

R. Schmidt and H. Zimmermann
Max-Planck-Institut für biophysikalische Chemie, Abteilung für Neurochemie, Postfach 968, D-3400 Göttingen, F.R.G.

Similar to storage granules of other secretory systems, cholinergic synaptic vesicles isolated from the electric organ of Torpedo marmorata contain 0.2 molecules of ATP per molecule of acetylcholine (ACh) (Dowdall et al., 1974). It has been postulated that the release of vesicular constituents occurs by interaction of synaptic vesicles with the cell membrane and that this process is triggered by Ca^{2+} ions. Therefore we investigated whether interaction of synaptic vesicles and nerve terminal membranes isolated from tissue previously frozen in liquid nitrogen would induce release of vesicular components in vitro and whether this process could be influenced by Ca^{2+} ions (Schmidt, 1978).

Synaptic vesicles and nerve terminal membranes were incubated at $4°C$ for 1 min and then acidified using 25 mM $HClO_4$. On immediate analysis neither ACh nor ATP were found to be hydrolysed. Ca^{2+} (up to 10 mM) had no further effect. When acidified samples were assayed after storage for 60 min ($4°C$) the majority of vesicular ATP was hydrolysed after either previous elevation of the Ca^{2+} concentration in the vesicle fraction or previous addition of synaptic membranes. Thus ATPase activities were not completely inhibited by 25 mM $HClO_4$. When the acidity of samples was increased to 70 mM $HClO_4$ no hydrolysis could be observed.

The experiments suggest that synaptic vesicles and nerve terminal membranes isolated from frozen electric tissue do not contain all components necessary to reproduce exocytotic transmitter release in an in vitro system.

Dowdall, M.J., Boyne, A.F. and Whittaker, V.P. (1974) Biochem. J. 140: 1-12

Schmidt, R. (1978) Hoppe Seyler's Z. Physiol. Chem. 359: 316

COMPARISON OF PRESYNAPTIC PLASMA MEMBRANE AND SYNAPTIC VESICLE PROTEINS

H. Stadler, T. Tashiro
Department of Neurochemistry
Max-Planck-Institute for Biophysical Chemistry
Göttingen, G F R

There is increasing evidence that neurotransmitter release involves exocytosis of synaptic vesicles (Zimmermann and Whittaker, 1977). Details at the molecular level of this process are largely unknown. In an attempt to study possible mechanisms of vesicle - presynaptic plasma membrane interaction or fusion a comparison of their respective protein compositions could elucidate underlying events.

The cholinergically innervated electric organ of Torpedo marmorata offers the opportunity to isolate synaptosomes and synaptic vesicles of a single transmitter type in high purity. Vesicles were isolated from it as described in detail elsewhere (Tashiro and Stadler, 1978); the presynaptic plasma membrane was purified by density gradient centrifugation of lysed synaptosomes prepared according to Israel (1976). The purity was established by marker enzyme distribution and electron micrographs. The protein composition of the two types of membranes was compared by SDS gel electrophoresis.

The patterns obtained are substantially different. Beside a low molecular weight band, comigration occurs only in a band that was recently identified in vesicle preparations as an actin of the non muscular type (Tashiro and Stadler, 1978).

We conclude that the synaptic plasma membrane is not composed primarily of vesicle membranes ready for retrieval and reformation. The results suggest instead that vesicular release of acetylcholine might be followed by immediate recovery of the vesicle membrane into the synaptic plasma. The results are parallel to a recent comparison of proteins of proteins of the chromaffin granule membrane and plasma membrane of the adrenal gland (Zinder et al., 1978). In this system transmitter release via exocytosis of secretory granules is widely accepted.

Israel, M., Manaranche, R., Mastour-Frachon, P. and Morel, N. (1976) Biochem. J. 160: 113-115
Tashiro, T. and Stadler, H. (1978), manuscript submitted for publ.
Zimmermann, H. and Whittaker, V.P. (1977) Nature 267: 633-635
Zinder, O., Hoffmann, P.G., Bonner, W.M. and Pollard, H.B. (1978) Cell Tiss. Res. 188: 153-170

PENETRATION OF GLUTAMINE IN RAT BRAIN NON-SYNAPTOSOMAL MITOCHONDRIA

A. Minn and J. Gayet
Laboratoire de Physiologie Générale
Université de Nancy 1 - C.O. 140
54037 Nancy Cedex, France

The compartimentation of glutamine and glutamate metabolisms in the nervous tissue is the consequence of both the plurality of their cerebral functions and the biochemical heterogeneity of the divers cell populations. Glutamine is synthesized exclusively into the glial cytosol (Martinez-Hernandez et al., 1977), whereas its breakdown occurs only in synaptosomal (Bradford and Ward, 1975) or non-synaptosomal mitochondria (Salganicoff and De Robertis, 1965). Although the rate of degradation of glutamine was estimated in the brain (Van den Berg and Garfinkel, 1971), no experimental data on the characteristics of its transport into mitochondrial matrix are available.

This communication is concerned with the study of glutamine uptake by brain non-synaptosomal mitochondria, which was estimated by both the osmotic swelling technique and the centrifugation-stop procedure, as previously described for the study of glutamate penetration (Minn and Gayet, 1977). L-glutamine, but not the D-isomer, promoted an osmotic swelling of brain mitochondria ; this swelling was reversibly inhibited by mersalyl, an ionic non-penetrating thiol reagent. The selective oxidation of 1 mM succinate in the presence of rotenone increased both osmotic swelling and ^{14}C uptake during incubation of mitochondria with 1 mM (^{14}C)L-glutamine. These results show that the penetration of glutamine into brain mitochondria and its access to the site of glutaminase are probably controlled by the activity of a specific, energy-requiring translocator.

(Supported by grants of the Centre National de la Recherche Scientifique, E.R.A. 331).

Bradford, H.F. and Ward, H.K. (1975) Biochem. Soc. Trans.
   3: 1223-1226
Martinez-Hernandez, A., Bell, K.P. and Norenberg, M.D. (1977) Science
   195: 1356-1358
Minn, A. and Gayet, J. (1977) J. Neurochem.
   29: 873-881
Salganicoff, L. and De Robertis, E. (1965) J. Neurochem.
   12: 287-309
Van den Berg, C.J. and Garfinkel, D. (1971) Biochem. J.
   123: 211-218

EFFECTS OF BARBITURATES ON CALCIUM METABOLISM IN SYNAPTOSOMES VISUALIZED BY CHLOROTETRACYCLINE AS A FLUORESCENT CHELATE PROBE

J.W. Łazarewicz, A. Pastuszko, E. Bertoli and K. Noremberg
Department of Neurochemistry, Medical Research Centre, Polish Academy of Sciences, Warsaw, Poland and Institute of Biochemistry, University of Ancona, Ancona, Italy

The inhibitory effect of barbiturates in vitro on depolarization-induced calcium influx into synaptosomes has been suggested as the mechanism of barbiturate general anesthesia /Blaustein and Ector,1975/. In this paper will be described a use of fluorescent Ca-chlorotetracycline /CTC/ chelate probe in studies of effects of anesthetics administered in vitro and in vivo on calcium metabolism in rat brain synaptosomal fraction and isolated synaptosomal membranes.

A large fluorescence was observed when CTC was added to a synaptosomal suspension containing no exogenous calcium. It was neither influenced by changes in K^+/Na^+ ratio in the medium nor by EGTA or EDTA. This fluorescence was attributed to binding of CTC to endogenous intrasynaptosomal divalent cations. Phenobarbital and brietal added in vivo and in vitro had no significant effect on the basic fluorescence, but in vivo administration of thiopental markedly lowered endogenous divalent cations-CTC chelate fluorescence. Addition of exogenous calcium enhanced very much CTC fluorescence. This effect was more pronounced when synaptosomes were incubated in low Na^+ high K^+ or in Na^+ free medium, and was reversed by EGTA. This suggests that chelate probe visualises mainly calcium binding to synaptosomal plasma membranes. Phenobarbital, thiopental and brietal given in vitro significantly increased Ca-CTC chelate fluorescence. This effect was even more pronounced in isolated synaptosomal membranes. It seems that anionic drugs bound to membrane phospholipids increase the net negative charge in the region of phospholipid polar groups and this importantly increases binding of divalent cations.

In these studies, in agreement with Schaffer and Olson/1976/, fluorescent calcium chelate probe appeared to be particularly useful in investigation of Ca binding to plasma membranes. Presented results suggest also important changes of intrasynaptosomal calcium metabolism of barbiturate-treated animals, however until it is not clear what is the location of CTC and calcium in intrasynaptosomal membranes particular caution must be taken in interpreting these fluorescent data.

Blaustein M.P. and Ector A.C. /1975/ Mol. Pharmacol. 11 : 369 - 378
Schaffer W.T. and Olson M.S. /1976/ J. Neurochem. 27 : 1319 - 1325

INFLUX OF CITRATE TO SYNAPTOSOMES AND SYNAPTOSOMAL MITOCHONDRIA FROM RAT BRAIN

H.Księżak, U.Rafałowska, J.W.Łazarewicz
Medical Research Centre,Pol.Acad.Sci.,00784 Warsaw, Poland.

Citrate produced in mitochondria may serve as a donor of acetyl group for AcCoA which is used for acetylcholine synthesis in nerve endings. Van den Berg et al./1975/ suggested that the activity of tri carboxylic acids pathway in synaptosomal and pericarional mitochondria may be different. In our recent paper the mechanism of citrate uptake to pericarional mitochondria was characterized/Rafałowska,1978/. The present study deal with citrate influx through synaptosomal and synaptic mitochondrial membranes. Synaptosomes and synaptosomal mitochondria were obtained using Ficol gradient as described by Lai and Clark,1976./1,5-^{14}C/citrate was used. It was found that the decarboxylation of ^{14}C-citrate in synaptosomes was linear during 60 min at 37°C. In the absence of malate the value of 2.4 nmole of citrate/mg prot./hr was obtained. Adding 1mM malate increased the value to 3.6 nmole of citrate/mg prot./hr. These results indicated the transport of citrate through the synaptosomal membrane with its influx and metabolism in mitochondria being dependent on the presence of malate. In synaptosomes the accumulation of ^{14}C derived from radioactive citrate was maximal after 15 min in the absence of malate. The maximum accumulation in the presence of malate was 4 times higher and occured after 10 min. It may be assumed that higher accumulation of ^{14}C in additionally shorter time in the presence of malate is probably due to th the citrate uptake by synaptosomal mitochondria with the known mechanism of malate/citrate exchange being involved. The obtained fraction of synaptosomal mitochondria showed the intact outer membrane and small impurification with synaptosomes. It oxidized the 12 nmole of citrate/mg prot./10 min to $^{14}CO_2$. When formation of $^{14}CO_2$ was inhibited by adding the rotenone and oligomycin the maximum accumulation of citrate in the presence of malate occured after 1 min and was 11.0 nmole/mg prot. Synaptosomal mitochondria as compared to pericarional one showed slower decarboxylation of citrate and its quicker accumulation on the way of exchange with malate.

Van den Berg et al. in Metabolic Compartmentation and Neurotransmission, 1975, pp. 515-544.
Rafałowska/1978/Bull.Pol.Acad.Sci., in press
Lai and Clark /1976/Biochem.J. 154: 423-432.

UPTAKE AND INCORPORATION OF ADENOSINE INTO ADENINE NUCLEOTIDES BY GUINEA PIG NEOCORTEX SYNAPTOSOMES

C. Barberis, A. Minn, and J. Gayet
Laboratoire de Physiologie Générale - Université de Nancy 1,
54 037 NANCY Cedex - France

It has been previously demonstrated that synaptosomes isolated from guinea pig neocortex were able to take up {^{14}C} adenosine which was incorporated into the 5'-adenine nucleotides (Kuroda and Mc Ilwain, 1974). In the present communication, the kinetic properties of the intrasynaptosomal transport system was investigated, and the incorporation of adenosine into the nucleotides was examined.

Penetration of the synaptosomal cytoplasm by adenosine was measured by an adaptation of the centrifugation-stop procedure (Minn and Gayet, 1977). Synaptosomes were incubated with {^{14}C} adenosine (0.5-10 µM) for up to 30 s in glucose bicarbonate saline containing $^{3}H_2O$. Assays were run in parallel in which {^{14}C} inulin was added instead of {^{14}C} adenosine. After addition of unlabelled adenosine (final concentration 1.0 mM) and rapid centrifugation, the radioactivity of the pellet and the supernatant was estimated. It was assumed that the synaptosomal space represents the difference between the total water space (permeable to $^{3}H_2O$) and the extra synaptosomal space (permeable to {^{14}C}inulin). At 30°C, the penetration of {^{14}C}-adenosine was linear for about the first 2 s of incubation and then decreased. A mean apparent km of 5 µM was estimated.

The incorporation of {^{14}C} adenosine into the synaptosomal nucleotides was examined by incubating the synaptosomes as "beds" in glucose bicarbonate medium containing {^{14}C} adenosine (16.7 µM) for up to 60 min. After 1 min and 5 min of incubation, as much as 31 and 48 per cent of the ^{14}C incorporated by the beds was respectively found as {^{14}C}-ATP. Preincubation of the synaptosome beds with oligomycin or atractyloside did not affect the subsequent 1 min incorporation while the 5 min incorporation was diminished to 39 per cent with oligomycin but not with atractyloside.

The above data suggest that the uptake and incorporation of adenosine into adenine nucleotides is a very rapid process occurring through pathways which remain to be precised.

Kuroda, Y. and McIlwain, H. (1974) J. Neurochem.
  22 : 691 - 699
Minn, A. and Gayet J. (1977) J. Neurochem.
  29 : 873 - 881
(Supported by grants from the Centre National de la Recherche
  Scientifique, ERA 331)

DO MEMBRANE GLYCOPROTEINS PLAY A ROLE IN SYNAPTOGENESIS ?

A. Reeber, J-P. Zanetta, G. Vincendon and G. Gombos
Institut de Chimie Biologique, Faculté de Médecine de l'Université Louis Pasteur et
Centre de Neurochimie du CNRS, 11 rue Humann, Strasbourg, France.

Con-A binding sites, detected in cerebellum sections by the FITC-Con A or the HRP-Con-A methods, are very abundant on the "growing" axons of granule cells (parallel fibers) and undetectable on parallel fibers of adult animals (1), (2). This massive accumulation of Con-A binding sites on parallel fibers is a transient phenomenon, detectable only between the 7th and the 16th postnatal day (1), (2). These sites correspond to 4 transient glycoprotein bands as shown by polyacrylamide gel electrophoresis of Con-A binding glycoproteins present in cerebellar particulate fractions. These transient glycoproteins which accumulate until the 12th postnatal day and disappear after the 16th day, are insoluble in 1% Triton X-100 whereas all other Con-A binding glycoprotein bands which appear throughout development are present in the adult and are Triton X-100 soluble. The accumulation of transient glycoproteins occuring some days before parallel fibers synaptogenesis and their disappearance after synapses are formed, suggests that these molecules could be involved in formation and specification of the interneuronal junction.

In order to test this hypothesis, the transient Con-A binding glycoproteins were solubilized, covalently bound to HRP and incubated with cerebellar slices. They apparently bind to the "partner" cells of parallel fibers. This binding, present on the cellular body, the axon and on the dendrites, is inhibited by 0.3 M mannose or by pretreatment of the complex (HRP-transient glycoproteins) with α-mannosidase. We have also shown that this binding is age dependent, since it is only present between the 10th and 25th postnatal day. The same results have been obtained when cerebellar slices were incubated with glycopeptides derived from Con-A binding glycoproteins, showing that the binding specificity is due to the glycan part of the glycoproteins.

Our results suggest that "partner" cells of parallel fibers contain "receptors" for the transient Con-A binding glycoproteins of parallel fibers. These "receptors" appear to recognize the glycan of these glycoproteins. The timing of the appearance and disappearance of transient Con-A binding glycoproteins and of their "receptors", as well as the cellular and subcellular localization of these molecules, suggest that they may play a role in phenomena which precede synapse formation, but apparently they are not involved in synapse maintenance and stabilization, since they disappear simultaneously with synapse formation. They seem rather to be involved in an early phase of synaptogenesis, forming a specific "early connection" between complementary surfaces of neuronal processes, when these processes contact each other.

(1) Gombos, G. et al. (1978) In "Maturation of Neurotransmission", (E. Giacobini and A. Vernadakis, Eds.) 3, S. Karger, A.G., Basel, in press.
(2) Zanetta, J-P. et al. (1978) Brain Res., 142: 301-319.

PROTEIN COMPONENTS OF SYNAPTIC MEMBRANES BINDING TAURINE

E. Kumpulainen, M. Olkinuora and P. Lähdesmäki
Department of Biochemistry, University of Oulu, SF-90100 Oulu 10, Finland

Specific proteolipid fractions have been prepared from mammalian cerebral cortex with chloroform-methanol mixtures, binding GABA, glutamate and aspartate by a high-affinity mechanism (De Robertis and Fiszer de Plazas, 1976; Fiszer de Plazas and De Robertis, 1975, 1976), and specific glycoproteins with sodium dodecyl sulphate (Zanetta et al., 1975) or Triton X-100 (Michaelis, 1975), the latter retaining their glutamate binding capacity, but nothing is known about the membraneous taurine receptor sites in the synaptic region. We therefore extracted calf brain synaptic membranes with 0.5 % Triton X-100 solution or with chloroform-methanol mixtures (2:1) and studied whether the macromolecules in the extracts are able to bind specifically $[^{35}S]$taurine.

The above extracts as such were incubated with $[^{35}S]$taurine (0.2 μM) at room temperature for 30 min. After the incubation the Triton-extract was applied to a Sephadex G-100 column and eluted with 10 mM potassium phosphate buffer (pH 7.5) containing 0.25 % Triton X-100. The chloroform-methanol extracts were fractionated with chloroform and then with various chloroform-methanol mixtures in a Sephadex LH-20 column. Protein content of the fractions was measured with the Folin phenol reagent, and their radioactivity in a scintillation spectrometer. One large protein peak binding specifically $[^{35}S]$taurine was obtained in the Triton X-100 extract, and one additional small peak, which was completely shadowed by free radioactivity. Hypotaurine inhibited the binding about 10 % and β-alanine about 15 %, while incubation at 4 °C reduced the binding about 40 %. In the Sephadex LH-20 chromatography we obtained near the void volume of the column a protein component binding relatively high amounts of $[^{35}S]$taurine, as compared with the other component being eluted in the volume twice to the first one. The further chemical characterization of the components in the extracts is in progress.

This study was supported by the Research Foundation of the Orion Corporation.

De Robertis, E. and Fiszer de Plazas, S. (1976) J. Neurochem. 26: 1237-1243.
Fiszer de Plazas, S. and De Robertis, E. (1975) J. Neurochem. 25: 547-552.
Fiszer de Plazas, S. and De Robertis, E. (1976) J. Neurochem. 27: 889-894.
Michaelis, E.K. (1975) Biochem.Biophys.Res.Commun. 65: 1004-1012.
Zanetta, J.P., Morgan, I.G. and Gombos, G. (1975) Brain Res. 83: 337-348.

ARTIFICIAL TAURINE RELEASE AFTER ELECTRICAL STIMULATION OF RETINA IN VITRO

E. Schulze and V. Neuhoff

Max-Planck-Institut für experimentelle Medizin, 3400 Göttingen (FRG)

Several findings led to the discussion of taurine as a possible inhibitory neurotransmitter, e.g. marked release of radioactively labelled taurine after electrical stimulation of nervous tissue in vitro.

We studied taurine release from retina using different stimulus currents, pulse widths, polarities, electrode materials and temperatures, and the following results were obtained:
1. No effect of stimulation was observed, when weak biphasic stimuli (2 mA) were used instead of monophasic stimuli with identical pulse shape, whereas we observed substantial release of ^{3}H-taurine using stronger biphasic stimuli. 2. Release of ^{3}H-taurine could not be reproduced on repeated stimulation. 3. ^{3}H-taurine release could be reproduced when the second stimulation was done via another electrode in another part of the retina preparation. 4. ^{3}H-taurine release was also observed after stimulation at 0°C. 5. Significant decrease of ^{3}H-taurine release was observed when the tissue was cooled to 0°C immediately after stimulation at 37°C. 6. Release of lactate-dehydrogenase (LDH) in a peak simultaneously with ^{3}H-taurine release was found after stimulation with monophasic or biphasic stimuli. 7. When platinum or carbon electrodes are used, highly toxic chlorine can be generated at the anode. The activity of native LDH is abolished by stimulation of the LDH-solution via platinum electrode. Addition of a diluted chlorine solution to the incubation medium for a two minute period instead of electrical stimulation resulted in a strong release of ^{3}H-taurine from rat retina. 8. Heavy metal ions are known to be very toxic for many enzymes. Activity of native LDH is abolished by addition of very small amounts of silver ions to the solution, and as well by stimulating the LDH-solution via Ag/AgCl electrodes (operated under nonreversible conditions as usually done in in vitro stimulations). Addition of very small amounts of silver ions to the incubation medium for two minutes instead of electrical stimulation resulted in ^{3}H-taurine release from rat retina.

These results suggest that ^{3}H-taurine release from retinal tissue after electrical stimulation in vitro is at least partially due to toxic effects or tissue damage. Exogenous release evoked by electrical stimulation is therefore not a suitable criterium for taurine to be considered a neurotransmitter.

BRAIN CAPILLARY PERMEABILITY AND AMINO ACID LEVELS IN PLASMA AND CEREBRAL REGIONS OF THE RAT AFTER PORTO-CAVAL SHUNT.

G. Zanchin, P. Rigotti, P. Vassanelli and L. Battistin,
University of Padua Medical School, Padua, Italy.

Amino acid automated analysis was performed in plasma and in four cerebral regions of rats submitted to chronic porto-caval shunt (Mondino, 1969).
Most plasmatic amino acids were significantly lowered (Asp, Thr, Ser, Glu, Ala, Val, Met, Ile, Leu, Lys) or unchanged (Tau, Gly, Pro, Trp, Orn, His, Arg). Asn and Gln levels were significantly higher and a net increase of Tyr and Phe was observed. Among cerebral amino acids, some were not modified (Asp, Thr, Ser, Glu, Gly, Ala, GABA, Lys); Tau was slightly but significantly lowered. Branched chain amino acids were either increased or unchanged, whereas an enormous increase of Tyr, Phe, Trp, His, uniform in the different cerebral regions, was evident.

Cerebral capillary permeability to L-amino acids was studied in vivo, using a short-term intracarotidal injection technique (Oldendorf, 1970).
Neutral amino acid permeability appeared to be greatly increased whereas basic amino acids showed a net decrease in their rate of passage from blood to the brain. No changes were observed for GABA and glutamic acid. These data suggest an altered permeability of the cerebral capillary membranes in the rat with chronic porto-caval anastomosis, that seems to be selective for the different amino acid transport classes. Competitive inhibition decreased the permeability of Trp and Met at the same level both in control and in experimental animals: therefore, the increased brain permeability to neutral amino acid after porto-caval shunt is due to an enhancement of the saturable transport.

The sharp increase of some essential neutral amino acids (Phe, Tyr, Trp, His) into the brain largely exceeds their changes in plasma and is consistent with our observation of a primitive enhancement of the saturable transport for the neutral class in this experimental model.

In hepatic encephalopathy, correction of the altered plasmatic amino acid levels can improve the clinical status (Fischer, 1976). If this result is connected to the concomitant correction of the brain amino acid levels, carefully selected competitive inhibition among various plasmatic amino acids could be a useful therapeutic tool in this pathological condition.

Fischer J.E. et al. (1976) Surgery 80: 77
Mondino A. (1969) J. Chromatogr. 39: 262
Oldendorf W.H. (1970) Brain Res. 24: 372

CHARACTERIZATION OF EXPERIMENTAL PHENYLKETONURIA

J.D. Lane, B. Schöne and V. Neuhoff
Max-Planck-Institut für experimentelle Medizin
Forschungsstelle Neurochemie
3400 Göttingen, FRG

Rodents treated with combinations of phenylalanine and phenylalanine hydroxylase inhibitors exhibit characteristics similar to untreated patients suffering from phenylketonuria (PKU), and provide a model with which to study mechanisms of neurological damage in the developing nervous system. Utilizing the model(s) of Greengard et al., (1976), suckling rats were given daily injections of phenylalanine plus the hydroxylase inhibitors α-methylphenylalanine or p-chlorophenylalanine. These treatments produced hyperphenylalaninemia during a major portion of each day, and resulted in the urinary excretion of large quantities of phenylalanine and its phenylketo-acid metabolites. Compared to saline controls, animals treated with α-methylphenylalanine appeared to develop relatively normally, while those treated with p-chlorophenylalanine exhibited greatly reduced body weights, high mortality, cataracts and loss of body hair. Both PKU models resulted in a decrease in brain (wet) weight. During development, protein yield in total brain homogenates was similar to saline controls, while myelin yield was substantially reduced. The composition of the five major myelin proteins (as determined by SDS microgel electrophoresis) did not appear to be altered. Experimental PKU also induced several changes in the content of amino acids in the cerebrospinal fluid, and in the cerebral cortex (changes in brain amino acids were more profound at early ages, probably a consequence of the later developing "blood-brain barrier" and uptake systems). Treatment with phenylalanine and α-methylphenylalanine during early development brought about long term behavioral changes in adult animals (hyperactivity and hampered performance in maze discrimination for food reinforcement). Several other parameters such as brain lipids and biogenic amines were also evaluated. A detailed analysis of such animal models may give insight into the mechanisms of mental retardation brought about by neonatal insult.

Greengard, O., Yoss, M.S. and Del Valle, J.A. (1976) Science (Wash.) 192:1007-1008.

BRAIN PROTEIN METABOLISM AND AMINO ACID TRANSPORT ACROSS THE BLOOD-BRAIN BARRIER IN HYPERPHENYLALANINEMIC RATS

R. Berger, Th. Dias and F.A. Hommes,
Laboratory of Developmental Biochemistry, Department of Pediatrics,
University of Groningen, Groningen, The Netherlands.

In order to gain more insight into the mechanisms which are responsible for brain dysfunction in phenylketonuria we studied the effect of phenylalanine on brain protein metabolism and amino acid uptake in rats during development. Protein metabolism was measured using the method developed by Dunlop et al. (1975). Rats of various ages were injected intraperitoneally with 1 M (4,5 - ^{3}H) Lysine (dose 10 µmoles/gr body weight; sp.act. about 100 dpm/ n mole). Incorporation into cerebral proteins was found to be lineair for up to 2 hours after injection. Fractional rates of total protein synthesis expressed as % replacement of protein-bound amino acid per hour were calculated from the amount of radioactivity incorporated, the specific activity of the lysine solution and the lysine content of crerebral proteins. The highest rate of protein synthesis was measured in the newborn rat (1.8 %/hr) declining to 0.8%/hr in the 25-days old rat. The fractional rate of net protein deposition was 1.1 %/hr in the newborn approaching very low values in the 25-days old rat (0.06 %/hr). Throughout development there is a high rate of degradation, confirming the results of Dunlop et.al. (1978).

Calculations from the protein accretion rate and the amino acid composition of cerebral proteins show that the newborn rats require the net uptake of **methionine** at a rate of about 3 nmoles/min x gr.w.w. whereas in the 25-days old rats this net uptake is about 0.2 nmoles/min x gr.w.w. The rate of transport of methionine accross the blood brain barrier was measured according to the method of Toth and Lajtha (2) 1977. It was found that the uptake of methionine by the brain is severly inhibited in the presence of elevated plasma phenylalanine levels (0.5-0.6 mM).

These results indicate that phenylalanine can profoundly influence protein metabolism in the newborn and infant rat by inhibiting the net uptake of essential amino acid needed for <u>net</u> protein synthesis during growth. Inhibition of the rate of exchange of essential amino acids across the blood brain barrier (<u>net</u> influx = 0) in mature rats by phenylalanine probably affects protein metabolism to a much lesser extent. However, whether phenylalanine exerts a direct effect on brain protein synthesis remains to be established.

Dunlop, D.S., Van Elden, W. and Lajtha, A. (1975) J. Neurochem. 24: 337-344
Dunlop, D.S., Van Elden, W. and Lajtha, A. (1978) Biochem. J. 170: 637-642
Toth, J. and Lajtha, A. (1977) Neurochem. Res. 2: 149-160.

EVIDENCE FOR SYNTHESIS OF ACIDIC PROTEINS FOLLOWING LEARNING IN DAY OLD CHICKS

A. Longstaff and S.P.R. Rose
Brain Research Group, The Open University, Walton Hall,
Milton Keynes, MK7 6AA, U.K.

In day old chicks exposed for an hour to a flashing orange light as an imprinting stimulus there is an increased incorporation of lysine into TCA precipitable protein in the anterior forebrain roof (Haywood, Hambley, Rose & Bateson, 1977). This area includes the hyperstriatum (Wulst), which various studies have implicated in learning processes. This increased protein synthesis is not dependent on precursor pool size and occurs in a soluble, cytoplasmic fraction.

In the experiments reported here the nature of the proteins synthesized following exposure to the imprinting stimulus was investigated and the relationship of the rate of incorporation of amino acid to the behavioural measure of learning was studied.

A chromatographic fitter assay was used to study the incorporation of ^{14}C leucine into acidic proteins in birds trained for 1 hour on either a red or a yellow imprinting stimulus on two successive days, compared with birds kept in the dark over this period. Following training on the second day, birds were given a two choice discrimination test (Bateson, 1974) which gave a measure of their preference for the familiar. Dark birds were not tested. The birds were injected with 5 µCi ^{14}C leucine and killed 20 minutes later. A second group of birds were similarly trained, but not tested, injected and killed on the first day. Brains were dissected into four regions, anterior and posterior forebrain roof, forebrain base and midbrain.

No significant differences in incorporation were found in the day 1 imprinted birds, indeed animals tested at this time show no evidence of having become imprinted with our procedure. Day 2 yellow trained chicks, which behaviourally were better learners than red trained birds, showed an increased incorporation in the anterior roof compared to both the dark and red trained birds. Since both trained groups received similar visual experience the altered protein synthesis in the yellow birds may be a concomitant of learning. Consistent with this, in the anterior roof of the yellow birds a correlation exists between incorporation of amino acid and the learning measure.

Hambley, J., Haywood, J., Rose, S.P.R. and Bateson, P.P.G. J. Neurobiol.
   8 (1977) 109-118.
Bateson, P.P.G. Behaviour
   42 (1974) 279.

PURIFICATION OF HUMAN S-100 PROTEIN

D.A. Hullin,
Dept. of Medical Biochemistry,
Welsh National School of Medicine,
Heath Park,
Cardiff, Wales.

R.J. Thompson,
Dept. of Clinical Biochemistry,
School of Clinical Medicine,
Addenbrooke's Hospital,
Cambridge, England.

Since its original discovery several years ago (Moore, 1965) the structure and possible heterogeneity of the nervous-system specific S-100 protein have generated conflicting reports (Dannies and Levine 1969, Stewart 1972). We have purified a protein from human brain which appears to be analogous to bovine S-100 protein and have developed an immunoradiometric assay for its detection. Human S-100 protein has a molecular weight of approximately 24,000 daltons, migrates close to the dye front on electrophoresis in 7% polyacrylamide gels, shows calcium-binding activity, and appears to be specific to nervous tissue when measured by immunoradiometric assay. On sodium dodecyl sulphate polyacrylamide gel electrophoresis human S-100 protein shows no evidence of a subunit structure. The purification scheme for human S-100 protein used includes phenylmethylsulphonyl fluoride (PMSF) as a protease inhibitor during the isolation procedure. Omission of PMSF results in an increased proportion of protein of similar molecular weight which also migrates close to the dye front on electrophoresis in 7% polyacrylamide gels, shows an approximately 9,000 subunit structure on sodium dodecyl sulphate polyacrylamide gel electrophoresis, and has much less calcium binding activity. Both proteins show altered electrophoretic mobility under non-denaturing conditions in the presence of calcium, and both proteins cross-react in the immunoradiometric assay. Amino acid analysis is consistent with the subunit form being derived from the single chain form. Possibly previous reports of a subunit structure for S-100 protein and some of the reports of its heterogeneity may have been due to limited proteolysis during purification.

Moore, W. (1965) Biochem.Biophys.Res.Comm. 19, 739-744.
Dannies, P.S. and Levine, L. (1969) Biochem.Biophys.Res.Comm. 37, 587-592.
Stewart, J.A. (1972) Biochem.Biophys.Res.Comm. 46, 1405-1410.

THE INFLUENCE OF EARLY PROTEIN - CALORIE MALNUTRITION ON LEVELS OF A GLIAL BRAIN SPECIFIC PROTEIN (S-100) IN DISCRETE BRAIN AREAS.

K.G. Haglid, L. Rosengren, L. Rönnbäck, P. Sourander and A Wronski
Department of Neurobiology and Department of Neuropathology, University of Göteborg, Sweden

The most active period of cell mitosis of the fore-brain i.e. the brain growth spurt occurs in the rat during the first two postnatal weeks. During this time undernutrition leads to an irreversible reduction of weight, DNA and protein content of the brain (Winick and Noble, 1965; Winick, 1969). However, protein synthesis studied with radiolabelled precursors is either unaffected or slightly depressed after early undernutrition (Ogato et al., 1967; Patel et al., 1975).

Although various classes of cells and different brain areas may vary in vulnerability (cf. Shoemaker et al., 1977) with regard to early undernutrition little attention has been paid to the involvement of cell specific molecules of defined brain areas.

Previously we have shown that shorttime protein restriction in young adult rats results in a significant decrease of amino acid incorporation into the brain specific protein S-100. However, the amount of this protein in certain brain regions did not decrease until four weeks after restricting the alimentary protein supply (Wronski et al., 1977; Rosengren et al., 1977). Thus, the brain reveals a higher degree of resistance against a reduced protein supply than what has been demonstrated for other organs e.g. liver and muscles (Von der Decken and Omstedt, 1970).

In this study undernutrition (PMC) was produced in rats according to the method of Chow and Lee (1964). This implies a pre- and postnatal restriction of nutriments amounting to 50% of the normal consumption. Although brain weights of undernourished rats at 4 weeks of age were 20% less than those of control animals, the brains of PCM animals started to catch up with the controls towards 90 days.

The levels of S-100 protein in different cerebral and cerebellar regions of PCM animals were significantly lower than those of controls 4 weeks postnatally. These differences seemed to decrease in frontal cerebral cortex at the age of 12 weeks. At this age however, significantly lower levels of S-100 protein in the PCM animals as compared to controls were found in central and parieto-occipital cortex, hippocampus, brain stem and the anterior part of cerebellar vermis. Since the S-100 protein is synthesized and located mainly in glial cells, the results indicate that the glial cell populations in certain brain regions are affected by undernutrition.

Chow, B.F. and Lee, C.J. (1964) J. Nutr. 82: 10-18
Decken, A. von der and Omstedt, P.T. (1970) J. Nutr. 100: 623-630
Dobbing, J. (1968) In: Applied Neurochemistry Eds. A.N. Davison and J. Dobbing, pp 287-316
Ogato, K., Kido, H., Abe, S., Furusawa, Y. and Satake, M. (1967) In: Malnutrition, (MASS., USA)
Patel, A.J., Atkinson, D.J. and Balazs, R. (1975) Develop. Psychobiol. 8: (5) 453-464
Rosengren, L., Wronski, A., Haglid, K.G., Jarlstedt, J. and Rönnbäck, L. (1977) Neurosci. Res. 3: 153-162
Shoemaker, W.J., Bloom, F.E.: In Nutrition and the Brain Vol. 2. Eds. Wurtman, R.J. and Wurtman, J.J., Raven Press, New York, 1977
Winick, M. and Noble, A. (1965) Develop. Biol. 12: 451-466
Winick, M. and Noble, A. (1966) J. Nutr. 89: 300-306
Winick, M. (1969) J. Pediatrics 74: 667-679
Wronski, A., Decken, A. von der and Haglid, K.G. (1977) Brain Res. 129: 187-191

IDENTIFICATION OF A GROUP OF OCTOPUS BRAIN PROTEINS AS HISTONES

P. Cimarra and A. Giuditta
Zoological Station and International Institute of Genetics and Biophysics,
Naples, Italy

A group of proteins insoluble in Tris-glycine buffer and solubilized in SDS-buffer has been found to be several-fold more concentrated in the vertical and optic lobes of octopus brain than in the suboesophageal lobe. The proteins have been separated by electrophoresis on SDS-gels and identified as the most rapidly moving species (Giuditta et al., 1975).

We have recently investigated the subcellular localization of these proteins using hypertonic sucrose homogenates of the optic lobe. High speed centrifugation of the homogenate separates a nuclear pellet from an intermediate zone containing soluble material and from a floating layer enriched in cytoplasmic particulates. Upon electrophoretic analysis of the resulting fractions the proteins have been found associated with the nuclear pellet. Some protein bands with similar electrophoretic mobilities present in the floating layer may be attributed to the presence of contaminating nuclear material.

Differential extraction of purified nuclear preparations of the optic lobe controlled by electron microscopy has shown that the proteins are specifically extracted in acid media. Similar results have been obtained from other regions of octopus brain and from several non-nervous organs. The proteins extracted with this procedure from the optic lobe behave as calf thymus or rat brain histones when subjected to electrophoresis in acid conditions and in the presence of urea.

The identification of these proteins as histones is in agreement with their preferential distribution in the vertical and optic lobes whose cell density is far greater than in the suboesophageal lobe. In addition, the decrease in concentration of the proteins found in the optic lobe with increasing body weight may also be explained in terms of a progressively declining cell density (Giuditta et al., 1971).

Giuditta A., Libonati M., Packard A. and Prozzo N. (1971) Brain Res.
    25: 55-62
Giuditta A., Cimarra P. and Prozzo N. (1975) J. Neurochem.
    24: 1131-1133

RELEASE OF CALCIUM FROM MITOCHONDRIA IN THE ELECTRIC ORGAN OF TORPEDO MARMORATA DURING NERVE ACTIVITY

R. Schmidt, H. Zimmermann and F. Joó
Max-Planck-Institut für biophysikalische Chemie, Abteilung für Neurochemie, Postfach 968, D-3400 Göttingen, F.R.G.

An increase in the cytoplasmic concentration of Ca^{2+} is essential for the coupling of stimulation and transmitter release. Based on studies of isotope fluxes both extracellular as well as intracellular Ca sources have been considered to be responsible for the elevation of cytoplasmic Ca levels. At the present little is known about the Ca content of subcellular organelles inside the nerve terminal.

The electric organ of Torpedo has a purely cholinergic innervation. We have used this tissue for the quantitative analysis of Ca and other metal ions in subcellular fractions by atomic absorption spectrophotometry. Among others, a mitochondrial fraction was isolated by a combination of several homogenization and centrifugation steps and purified on density gradients. The Ca concentration of this fraction was much higher than that of any other subcellular fraction (834 nmol Ca / mg protein).

When tissue was stimulated at a high frequency of 5 Hz via the electromotor nerves, the Ca concentration of the mitochondrial fraction was reduced proportionally to the decrease in transmitter content of a fraction of isolated nerve endings (correlation coefficient r = 0.97). After 5000 pulses 80 % of the mitochondrial Ca content was released. When tissue was perfused with d-tubocurarine, until the response of the electrocytes to single test pulses was abolished, 20-40 % of the mitochondrial Ca content was released on repetitive stimulation, whereas transmitter release was unaffected. It is concluded that on stimulation Ca is released from both nerve terminal mitochondria and mitochondria of electrocytes. The participation of intraterminal mitochondria in the control of cytoplasmic Ca^{2+} concentration is further supported by an ultrastructural Ca staining technique (Parducz and Joó, 1976). After perfusion with d-tubocurarine and stimulation only the presynaptic mitochondria were stained. Preferential Ca staining of intraterminal mitochondria is interpreted in terms of uptake of Ca^{2+} from the fixation solution by mitochondria which have lost Ca during nerve activity.

Parducz, A. and Joó, F. (1976) J. Cell Biol.
   69: 513-517

SPECIES AND REGIONAL VARIATIONS IN ENDOGENOUS PROSTAGLANDIN FORMATION
BY BRAIN HOMOGENATES

E. Änggård and M.S. Abdel-Halim
Department of Alcohol and Drug Addiction Research and department of
Pharmacology, Karolinska Institutet, 104 01 Stockholm, Sweden

Recent work has indicated the possibility of considerable species differences in the formation of prostaglandins (PGs) in the brain. Early work had shown $PGF_{2\alpha}$ to be the dominant PG in the bovine brain (Samuelsson 1964). Later thromboxane B_2 was shown to be formed in higher amounts than $PGF_{2\alpha}$ in guinea pig brain (Wolfe et al. 1976). In the rat brain PGD_2 was found to be more abundant than all other PGs (Abdel-Halim et al. 1977). In the present paper a systematic comparison is made between the PGs formed in different regions of the brain of rat, guinea pig, rabbit and cat.

The major prostanoids, were in the rat PGD_2, in the guinea pig PGD_2 and $PGF_{2\alpha}$, in the rabbit $PGF_{2\alpha}$ and in the cat PGE_2. A consistent pattern of content of total PGs was found when different regions (cortex, cerebellum, hippocampus, striatum, pons + medulla obl., nucl. accumbens) were compared. In all species the limbic system (hippocampus and nucleus accumbens) were the highest, and cerebellum the lowest with the other regions intermediate. These results should serve as a biochemical basis for further studies on the functional aspects of the brain prostaglandins.

Abdel-Halim, M.S., Hamberg, M., Sjöquist, B. and Änggård, E. (1977)
   Prostaglandins 14: 633-643
Samuelsson, B. (1964) Biochim. Biophys. Acta 84: 218-219
Wolfe, L.S., Rostworowski, K. and Marion, J. (1976) Biochem. Biophys.
   Res. Comm. 70: 907-913

THE LOCALIZATION AND OF ATP-CITRATE LYASE /CCE/ IN CHOLINERGIC NEURONS OF RAT BRAIN

A. Szutowicz, W. Łysiak
Department of Clinical Biochemistry, Institute of Pathology,
Medical Academy, Gdańsk, Poland

CCE is one of the enzymes involved in acetyl-CoA synthesis in brain cytoplasm. The activity of this enzyme in frog, chick embryo, guinea pig, mouse and rat brain varied from 18 to 30 μmoles/h/g of tissue, being much higher than that of choline acetyltransferase /ChAT/. This indicates that CCE activity may not be a rate limiting step in ACh synthesis in brain provided that the enzyme is located in cytoplasm of cholinergic neurons.

The activities of ChAT and CCE in rat cerebrum which contains about 10% of cholinergic neurons, were 10 and 4 times higher, respectively than in cerebellum where the density of cholinergic innervation is negligible. On the other hand the activities of pyruvate dehydrogenase, citrate synthase, acetyl-CoA hydrolase, carnitine acetyltransferase and acetyl-CoA synthetase were similar in both parts of brain. Subcellular fractionation showed that over 70% of activities of both CCE and ChAT in secondary fractions is found in synaptosomes, where they are preferentially located in synaptoplasm as indicated by RSA values above two. The CCE activity in cerebellum, like that of fatty acid synthetase, decreased about 70% during development. On the contrary in developing cerebrum CCE activity did not change significantly, what agrees with the reciprocal developmental patterns of fatty acid synthetase and ChAT activities. Developmental profiles of other enzymes of acetyl-CoA metabolism were similar in both parts of brain. Data presented indicate particular link of CCE with cholinergic neurons in brain. Thus it was possible to calculate that 100% of ChAT, 80% of CCE and only 10% of activities of other enzymes of acetyl-CoA metabolism in whole synaptosomal fraction may be found inside cholinergic nerve endings.

Citrate was a main product of pyruvate + malate metabolism in synaptosomes /2.5 nmoles/min/mg of protein/. This synthesis was markedly reduced /80%/ by CCE activator Mg-ATP and then increased threefold by its specific inhibitor /-/hydroxycitrate /1 mM/. Bromopyruvate /0.5 mM/ inhibiting 75% pyruvate utilization brought about 98% decrease in citrate synthesis. Both inhibitors caused decrease in ACh synthesis by about 45 and 70%, respectively. Data presented show that CCE pathway is operative in cholinergic nerve endings and provides up to 45% of acetyl units for ACh synthesis.

A DYNAMIC APPROACH TO STUDY THE INTERACTION OF GABA WITH PYRIDOXAL KINASE

J. E. Churchich and Francis Kwok
Department of Biochemistry
Knoxville, Tenn., USA

The formation of P-pyridoxal from ATP, pyridoxal and a divalent metal ion is catalyzed by pyridoxal kinase, an enzyme which has been detected in various tissues. This enzyme may play an important role in the regulation of vitamine B_6 metabolism in brain.

Although it has been reported that biogenic amines, i.e., serotonin and dopamine, inhibit pyridoxal kinase activity, the mechanisms by which these inhibitors influence the catalytic activity of the kinase are unknown. It thus appears of interest to investigate the binding properties of purified pyridoxal kinase from pig brain and to determine the influence of several ligands on the affinity of pyridoxal and ATP for the catalytic site. The binding of pyridoxal to the kinase is paralleled by a large increase in the emission anisotropy (r = 0.16). This change in emission anisotropy was used to determine the dissociation constant of the enzyme pyridoxal complex (K_D = 20μM). The neurotransmitter GABA and its dansylated derivative (DNS-GABA) inhibit pyridoxal kinase activity with K_i values of 3mM and 2μM respectively, but they are unable to displace pyridoxal from the catalytic site. DNS-GABA binds strongly to the enzyme and the attachment of the inhibitor is accompanied by changes in fluorescence lifetime (19 ns) and emission anisotropy (r = 0.18).

The inhibitor DNS-GABA is displaced from the enzyme by ATP, indicating that both ligands compete for the same hydrophobic site. This interpretation is supported by steady-state kinetic measurements which reveal that DNS-GABA and GABA are competitive inhibitors with respect to ATP. It is postulated that inhibitors of kinase activity, GABA and DNS-GABA, exert their effects by binding to the ATP site.

A STUDY OF THE MECHANISM OF ACTION OF 2-OXO-1-PYRROLIDINE ACETAMIDE ON RNA AND PROTEIN METABOLISM IN RAT BRAIN

Z. S. Tencheva, S. V. Tuneva and N. Tijtijlkova
Brain Research Laboratory, Bulgarian Academy of Sciences and Drug Research Institute,
Sofia, BULGARIA

It has been shown that 2-oxo-1-pyrrolidine acetamide (2-OPA) enhances learning aquisition and resistance to impairing agents (Sara, S.J. et al. 1976; Wolthuis, O.L. 1971; Mindus, P. et al. 1976). The present study reports results of <u>in vivo</u> uptake of ^{32}P-phosphate by the nuclear and cytoplasmic brain RNA after treatment (100 mg/kg orally 3 times during an 18h period) with 2-OPA.

A significant increase of the relative specific radioactivity (R.S.A.) of the nuclear RNA from the treated animals was observed. After chronical administration (14 days) of the drug, no significant differences of the R.S.A. of RNA from the treated animals were found.

<u>In vitro</u> studies with brain tissue suspensions showed that 2-OPA in concentrations 35 to 50 mM stimulates the ^{3}H-uridine incorporation in the total RNA.

The action of 2-OPA on nucleoplasmic RNA polymerases (RNA polymerases B) and nucleolar RNA polymerases (RNA polymerases A) in <u>in vitro</u> experiments was examained.

The uptake of ^{3}H-leucine into total protein of brain suspensions shows rates 40% higher in the presence of 30 to 50 mM 2-OPA. No significant effect of amphetamine and centrophenoxine in the same <u>in vitro</u> system was observed.

The significance of the in vivo and in vitro results will be discussed.

Mindus, P., Cronholm, B., Levander, S.E. and Schalling, P. (1976) Acta psychiat. scand. 54: 150-160
Sara, S. J. and Lefevre, D. (1972) Psychopharmacologia (Berl.) 25: 32-40
Wolthuis, O.L. (1971) Europ.J. Pharmacol. 16: 283-297

DOES THE FIRST VISUAL STIMULATION EVOKE AN INCREASE OF INCORPORATION OF LABELLED LEUCINE INTO CEREBRAL CORTEX OF KITTENS?

Jolanta Skangiel-Kramska, Małgorzata Kossut, Katarzyna Mitros
Nencki Institute of Experimental Biology, Warsaw, Poland

Visual deprivation and first visual experience offer favourable conditions for studying the effect of a new sensory experience upon the biochemical functioning of the brain.

The antecedents of the present experimental model were the studies of Blakemore (1977) and Garey and Pettigrew (1974). One-months old kittens, binocularly deprived from birth of visual input with hoods were used. The experiments on the effect of visual stimulation were performed on kittens with transected brain stem at the pretrigeminal level. One of the hemispheres was stimulated by visual patterns, while the other one was used as a control. During the stimulation period $(1-^{14}C)$leucine or $(4,5-^{3}H)$leucine was injected intravenously (90-210 µCi per 100 g of body weight). Various time of stimulation and of incorporation (30-180 min) were applied. Independently of the experimental design visual stimulation evoked in some cases an increase (up to 148% of control) of incorporation of labelled leucine into the proteins of the striate cortex, but not of the somatosensory cortex. In spite of the fact that in another series of experiments uniform experimental conditions were used (90 min of stimulation and 30 min of incorporation) only in a part of kittens and increase of incorporation of labelled leucine to protein of stimulated visual cortex was observed.

The lack of the investigated effect appeared not to be connected either with cortical EEG activity or with ocular behavior. It seems therefore that some not yet determinated physiological factors (attention?) may affect the sensitivity of the visually deprived kittens to visual stimulation.

Blakemore, C. (1977). Phil. Trans. R. Soc. Lond. 278: 425-434.
Garey, L.J. and Pettigrew, D.J. (1974). Brain Res. 66: 165-172.

N,N-DIMETHYLAMINOETHANOL ACETATE (P_2-AGENT) — THE ACTIVE AGENT OF THE VEGETATIVE NERVOUS SYSTEM

E. F. P. Szabó
Mediator Forschungslaboratorium,
D-6601 Riegelsberg/Saar, Hochstraße 30, W. Germany

Dimethylaminoethanol acetate (P_2-agent): $(CH_3)_2N-CH_2-CH_2-O-CO-CH_3$ was isolated out of the adrenal gland. Procedure: extraction of the gland (removed out of the rabbit in urethane narcosis) with a solution of acetone and conc.HCl 100:1, shaking the extract with ether (after evaporation to a small volume), drying the extract in a desiccator in vacuo, resolving the rest in water, ascending paper chromatography (control: the synthetic P_2-agent; paper: Schleicher & Schül 2043 b Mgl) with the solvent n-butanol/ethanol abs./water (4:1:1.5) or n-butanol/water, development of the paper in an atmosphere of iodine.

The effect of the synthetic P_2-agent on the blood pressure of the rabbit (1-2 ml/kg of a 1:1000 solution i.v.) and on the isolated frog's heart (2-8 drops of a 1:500 solution to 1.5 ml Ringer's solution) is identical with that of the acetylcholine.

The result is further evidence for the author's opinion: the vegetative nervous system is unitary, that is it does not possess any separated sympathetic or parasympathetic parts.

Szabó, E.F.P. (1972) Fifth International Congress on Pharmacology, San Francisco, abstract: 1359

Szabó, E.F.P. (1972) Sechsundzwanzigster Österreichischer Ärztekongreß, Wien, Tagungsbericht: 84-92

Szabó, E.F.P. (1973) Ninth International Congress of Biochemistry, Stockholm, abstract: Cc 20

Szabó, E.F.P. (1973) Le Progrès Médical, 101: 273

Szabó, E.F.P. (1974) Zeitschrift für Immunitätsforschung, Supplemente, 1: 111-116

Szabó, E.F.P. (1975) Sixth International Congress of Pharmacology, Helsinki, abstract: 1350

Szabó, E.F.P. (1975) Fifth International Meeting of the ISN, Barcelona, abstract: 466

Szabó, E.F.P. (1976) Tenth International Congress of Biochemistry, Hamburg, abstract: 12-3-004

Szabó, E.F.P. (1976) First Meeting of the ESN, Bath, abstract: 81P

STRUCTURAL FEATURES OF PORCINE BRAIN α-TUBULIN

H. Ponstingl, E. Krauhs, M. Little, R. Hofer-Warbinek and T. Kempf
Institut für Zellforschung, Deutsches Krebsforschungszentrum,
D-6900 Heidelberg, FRG

The α-chain of tubulin from porcine brain was cleaved with thrombin and the digest fractionated on DEAE-cellulose with an ammonium bicarbonate gradient in 8 M urea. The electrophoretically homogeneous fragments were characterized by molecular weight, amino acid composition and partial sequences. Their tentative ordering by overlapping chymotryptic peptides revealed some structural features of the chain: (1) The aminoterminal part is about neutral, whereas the carboxyterminal half is rich in glutamic acid and aliphatic residues with small hydrophobic side chains. This region shows a strong aggregation tendency and may play a particular role in tubulin polymerization. (2) In a few positions amino acid exchanges are found. They can be alleles or reflect gene duplications carrying subsequent mutations. (3) Preliminary work on ß-tubulin indicates that both polypeptides are similar throughout their sequence, some regions even being virtually identical. (Supported by Deutsche Forschungsgemeinschaft).

POLYPEPTIDE PATTERN OF MAMMALIAN SKELETAL MUSCLE MEMBRANE BEFORE AND
AFTER DENERVATION

Giovanni Savettieri and Oscar Lo Verde
Department of Neurology, University of Palermo
Via G. La Loggia, 1 - Palermo (Italy)

Emploing the tecnique of Barchi et al. (1977) the isolation of sarcolemmal membrane from normal and 7-days denervated muscle of rat was carried uot. Such a method of muscle membrane isolation avoid both prolonged salt extraction and incubation at elevated temperatures. Our aim was to verify wether denervation can modify the polypeptide composition of mammalian sarcolemma.

The isolated muscle membranes were investigated by SDS-polyacrylamide disc gel electrophoresis and by linear gradient polyacrylamide slab gel electrophoresis. A very high resolution occurs when the latter method was used: this technique resolves more than 40 bands.

Although the high resolution, no clear differences were found in polypeptide composition of sarcolemma before and after denervation.

The results will be discussed.

Barchi R.L., Bonilla E., and Wong M. (1977) Proc. Natl. Acad. Sci. USA 74: 34 - 38

EFFECTS OF POLYAMINES ON RNA SYNTHESIS IN CELL NUCLEI ISOLATED FROM RAT BRAIN

S. Štípek, J. Crkovská, S. Trojan, J. Prokeš
1st Department of Medical Chemistry, Department of Physiology,
Department of Toxikology, Faculty of General Medicine, Charles University,
Prague, Czechoslovakia

In authors oppinion in brain polyamines putrescine, spermidine and spermine play a similar role in regulation of cell differentiation and in nucleic acid synthesis as in other tissues /for rev. Russel, 1973; Seiler, Lamberty, 1975/. Mechanism of their action is not known. In vitro polyamines stimulate the activity of each purified DNA-dependent RNA polymerases /Jacob and Rose, 1975/ until now described. If added to cell nuclei, which were isolated from brain hemispheres of the rat by the method of Thompson /1973/, spermine /1-20 mM/ stimulated only Mn^{2+} dependent RNA polymerase activity estimated in presence of 0,15 M ammonium sulphate, whereas Mg^{2+} dependent RNA polymerase was significantly inhibited by spermidine /20 mM/ and spermine /5-20 mM/, respectively. Both the inhibitory and activating effects of spermine were pronounced than those of spermidine. The data suggest that polyamines may control the synthesis of different RNA species in nervous tissue in different ways.

Jacob, S.T. and Rose, K. /1975/ Biochem. Biophys. Acta 425: 125-128
Russell, D. H. /Ed./ Polyamines in normal and neoplastic growth, Raven Press /1973/
Seiler, N. and Lamberty, U. /1975/ J. Neurochem. 24: 5-13
Thompson, R.J. /1973/ J. Neurochem. 21: 19-40

BRAIN NOREPINEPHRINE INVOLVEMENT IN THE ANTIHYPERTENSIVE EFFECT OF PARGYLINE

J. A. Fuentes and A. Ordaz

Institute of Medicinal Chemistry, C.S.I.C., c/Juan de la Cierva, 3
Madrid-6. Spain.

Evidence has accumulated indicating that blood pressure is regulated by an α-receptor located in the brainstem. In 1974, De Jong demonstrated that norepinephrine (NE) inyected into the brainstem lowered blood pressure, its effect being antagonized by previous treatment with the α-blocking agent phentolamine at the same site.

Several hypothesis have been suggested to explain the antihypertensive effect of the monoamine oxidase (MAO) inhibitor pargyline (Fuentes and Neff, 1977). However, all of them implied that it acted peripherally rather than centrally. Interestingly, after treatment with pargyline we found a positive correlation between the level of brain MAO activity against NE and the blood lowering effect of the inhibitor drug in spontaneously hypertensive rats (SHR). Pretreatment with intraventricular 6-hydroxy-dopamine (2 × 1.63 μmol/rat), that diminished brainstem NE by about 50 %, prevented the decrease in blood pressure induced by a dose of 10 mg/kg i.v. of pargyline. Moreover, phentolamine injected intraventricularly (100 μg/rat) either prevented or antagonized the fall in blood pressure produced by the above indicated dose of pargyline.

These results indicate that the integrity of the central noradrenergic systems is important for the antihypertensive effect of pargyline in this model of hypertension.

De Jong, W. (1974) European J. Pharmacol.
    29: 179-181
Fuentes, J.A. and Neff, N.H. (1977) Biochem. Pharmacol.
    26: 2107-2112

ANTIBODIES TO MESODERMAL CELLS - A TOOL IN CELL DIFFERENTIATION STUDIES?

L. Rönnbäck and L. Persson
Institute of Neurobiology, University of Göteborg, and Department of Neurology, Sahlgren's Hospital, Göteborg, Sweden.

An antiserum against rat peritoneal macrophages was prepared in rabbits [2]. Cross reaction against soluble components of lung, kidney, spleen, liver, thymus and brain was seen. By immunofluorescence techniques, the antibodies bound to blood monocytes, to mononuclear reticular cells at the margin of germinal centres in the thymus and to ovoid or fusiform mononuclear cells similar to reticular cells in the spleen red and white pulp. Cells in the adventitial sheet of larger vessels in liver, spleen, kidney, thymus and leptomeninges of brain were positive with the antiserum. Siderophages of the spleen and Kupffer cells of the liver showed no cross reaction.

The results obtained indicate a common mesodermal origin of phagocytic mononuclear cells in various organs, irrespective of their differentiation into blood monocytes, histiocytes, reticular cells or brain microglial cells observed in cerebral stab wounds [2], and irrespective of their morphological differentiation into a variable ultrastructural appearance.

The use of the anti-macrophage antiserum in cerebral stab wounds, during maturation of nervous tissue or in the studies of different neurological diseases allows a positive identification of reactive microglial cells in various types of brain lesions. Together with antisera to the brain specific proteins S-100 (mainly localized to glial cells [1,4]) and 14-3-2 (which is localized to neurons[3]), all reactive cells in a brain wound can be identified, except neutrophilic leucocytes which can easily be identified by their morphology.

1. Moore, B.W. (1965) Biochem. biophys. Res.Commun
   19: 739-744.
2. Persson, L. and Rönnbäck, L. (1978) in press.
3. Persson, L. Rönnbäck, L., Grasso, A., Haglid, K.G., Hansson, H.-A., Dolonius, L. Molin, S.-O. and Nygren, H. (1978) J. Neurol. Sci.
   35: 381-390.
4. Rönnbäck, L.(1975)   Thesis, Göteborg.

IMMUNOCHEMICAL RELATIONSHIP BETWEEN CYTOSOL AND SYNAPTOSOMAL RAT BRAIN GLYCOPROTEINS[**]

G. Gennarini[*], D. Iannelli[o], P. Corsi, C. Di Benedetta
Istituto di Fisiologia Umana Università di Bari
Istituto di Produzione Animale Università di Napoli[o]

Antibodies were raised in rabbits against soluble con A-binding glycoproteins of rat brain. At least ten antigens can be detected in this group of glycoproteins by crossed immunoelectrophoresis, only two of which (n. 2 and 6) are not shared by rat serum. The organ specificity of these two antigens and their subcellular localization was investigated. Antigen n. 6 can be identified, although in different amount, in several rat tissues. Its concentration is highest in brain and spleen and is lowest in muscle and heart. This antigen is also present in the soluble as well as in the membrane bound synaptosomal fractions.

Antigen n. 2 is only present in brain and liver cytosol, but not in the synaptosomes.

In conclusion rat brain con A-binding glycoproteins have a low degree of organ specificity. Since one of these antigens (n. 6) shows the same concentration in the cytosol and synaptosomal soluble fractions, its axonal flow is hypothesized. Moreover its presence also in the synaptosomal plasma membranes suggests a role in the morphological organization of the nervous tissue.

[**] Partially supported by CNR (n. 76.01388.04); Nato (n. 706) and Ministero P.I. grants.
[*] Borsista CNR.

ISOLATION OF BRAIN PROTEINS REACTING IN VITRO WITH ANTI-NEURONAL ANTIBODIES IN PATIENTS WITH HUNTINGTON´S DISEASE.

E. Wedege and G. Husby
Neurochemical Laboratory, University of Oslo, P.O. Box 1115, Blindern, Oslo 3, Norway and Department of Rheumatology, Institute of Clinical Medicine, University of Tromsø, Tromsø, Norway.

We have recently described IgG antibodies to neuronal antigens occurring in approximately 50% of patients with Huntington´s Disease (HD) (Husby et al., 1977). Several protein extracts from human and monkey brain were tested for their ability to absorb out anti-neuronal antibody activity in sera from HD patients as detected by the indirect immunofluorescence technique. No absorption was seen with water, Triton X-100 or SDS-soluble protein extracts. Complete absorption was found with a neutral protein fraction extracted from brain with perchloric acid. This fraction consisted of two proteins of mol. wt. 13,000 and 11,000 as revealed by SDS-gel electrophoresis. Preliminary experiments showed that both proteins were necessary for antibody absorption. No absorption was obtained with corresponding extracts from human and monkey liver indicating that the proteins involved were specific to the central nervous system.

Husby, G., Li, L., Davis, L.E., Wedege, E., Kokmen, E. and Williams, Jr., R.C. (1977) J. Clin. Invest. 59: 922-932.

EXOGENOUS (1-^{14}C) STEARIC ACID UPTAKE BY NEURONS AND ASTROCYTES

O. Morand, N. Baumann and J.M. Bourre
Laboratoire de Neurochimie INSERM U.134, Hôpital de la Salpêtrière,
75634 Paris Cédex 13, France

Saturated fatty acids necessary for brain membrane formation are obtained endogenously or exogenously. It has been shown that radioactive saturated long chain fatty acids are taken up by the brain, either when fed (Dhopeshwarkar and Mead, 1969) or injected (Bourre et al., 1974). Moreover we have shown that the label of stearic acid, after subcutaneous injection, is incorporated into lipids of subcellular particles including myelin (Gozlan-Devillierre et al., 1978) or after being metabolized in brain. Thus this work was undertaken to study the abilities of neurons and astrocytes to incorporate an exogenous saturated fatty acid.

15 days old animals under standard diet were injected with 26 µCi albumin bound stearic acid ; neurons and astrocytes were isolated after trypsination and screening according to Norton and Poduslo. Lipids were analyzed with two dimensional thin-layer chromatography.

The specific radioactivity (cpm/mg protein) found in neurons increases up to 20 hours and then levels off. In astrocytes this increase is important for 2 hours and less prominent afterwards up to 40 hours (at 20 hours 2.200 cpm/mg protein in astrocytes and 800 in neurons). The table shows that 20 hours after injection phospholipids contain high amount of radioactivity (80 % in astrocytes and 68 % in neurons). It must be pointed out that both cells have low amount of radioactivity in glycosphingolipids. Thus exogenous stearic acid is needed to synthesize the membranes of neurons and astrocytes.

	pc	pe	ps	pi	sph	pa	cer	sulf	front
Neurons %	39	17	4	6	7	1,8	4	3	17
Astrocytes %	38	23	9	9	8	1,1	1,3	0,8	9

pc : phosphatidyl-choline, pe : -ethanolamine, ps : -serine, pi : -inositol, sph : sphingomyelin, front : cholesterol, cholesterol esters and free fatty acids.

Dhopeshwarkar, G.A. and Mead, J.F. (1969) Biochim. Biophys. Acta 187 :461-467

Bourre, J.M., Bouchaud, C. and Baumann, N. (1974) Arterial Wall 1 : 58

Gozlan-Devillierre, N., Baumann, N. and Bourre, J.M. (1978) Biochim. Biophys. Acta 528 : 490-496

CONTENT OF BRAIN SIALOGLYCOCONJUGATES IN MUCOLIPIDOSIS I AND II

B. Berra, S. di Palma and C. Lindi
Chair of Biochemistry, College of Pharmacy,
University of Milano, ITALY

Mucolipidosis I is a rare congenital disorder characterized clinically by coarse facial features, skeletal dysplasia, neurodegeneration and mental retardation. An abnormal accumulation of sialic acid-containing compounds and a diminished activity of sialidase in cultured fibroblasts suggested a defect in the catabolism of sialoglycopeptides and/or gangliosides as the metabolic basis of this disease (Michalski et al. 1977).

The most significant elements of diagnosis in Mucolipidosis II are the cytoplasmatic inclusions in cultured fibroblasts, the increase of activity of several lysosomal hydrolases in extracellular fluids, an elevated excretion of urinary sialyl-oligosaccharides, with a total lack of α-neuraminidase activity in leukocytes (Strecker et al. 1978).

Total gangliosides were extracted in a 23 yrs old patient with Muco I and in a 4 yrs old patient with Muco II with the method of Tettamanti et al. (1973) and fractionated by TLC. Protein bound sialic acid was determined after hydrolisys on defatted residue and purification of the free NeuNac by Dowex 2x8 column chromatography.

The results in the Mucolipidosis I brain as compared with age matched control demonstated a 34% increase of total sialic acid. The relative increase of lipid bound NeuNac was about 27% and of protein bound NeuNac of 56%. The ganglioside distribution was practically normal.

In Muco II the amount of NeuNac, both lipid- and protein-bound, was normal, with some slight changes in ganglioside pattern.

Michalski, J.C., Strecker, G., Fournet, B., Cantz, M. and Spranger, J. (1977)
 FEBS Letters 79: 101-104
Strecker, G. and Michalski, J.C. (1978)
 FEBS Letters 85: 20-24
Tettamanti, G., Bonali, F., Marchesini, S. and Zambotti, V. (1973)
 Biochim. Biophys. Acta 296: 160-170

OCCURRENCE IN MOUSE BRAIN OF GANGLIOSIDES CARRYING ALKALI LABILE LINKAGE

S. Sonnino, R. Ghidoni, N. Baumann[+], M. L. Harpin[+] and G. Tettamanti
Department of Biochemistry, The Medical School, University of Milan, Milan, Italy, and [+] Laboratoire de Neurochimie INSERM U. 134, CNRS ERA 421, Hôpital de la Salpêtrière, Paris, France.

In the course of a systematic work on fractionation of brain gangliosides by thin-layer chromatography, using the solvent system chloroform/methanol/aqueous 0.25 % $CaCl_2$, 60/42/11, by vol, we observed that minor components changed their chromatographic behaviour after exposure of the ganglioside mixture to alkaline conditions. Two novel gangliosides were isolated, from the total ganglioside extract of mouse brain by column chromatography on Kieselgel 100 using chloroform/methanol/aqueous $CaCl_2$ 0.3 %, 60/35/8, by vol, as the eluting solvent. They are coded provisionally ALG_I and ALG_{II}.

After alkaline treatment ganglioside ALG_I behaved, chromatographically, exactly as ganglioside GT1b. The alkali treated ganglioside ALG_I was proved to contain glucose, galactose, N-acetylgalactosamine and sialic acid in the molar ratios : 1.00/2.10/0.93/3.08 , the same composition and molar ratios being recorded on the untreated material. The treatment of ganglioside ALG_I with α-neuraminidase caused, besides the release of NeuAc, the formation of an alkali labile ganglioside having faster TLC migration than ALG_I ; after alkali treatment this ganglioside behaved as disialoganglioside GD1b. On prolonged treatment with α-neuraminidase another ganglioside was formed which showed the same chromatographic behaviour as ganglioside GM1 and a second species of sialic acid released ; this residue of sialic acid gave origin to N-acetylneuraminic acid , after alkali treatment. These evidences indicate that the alkali labile linkage present in ganglioside ALG_I is most likely carried by a single sialic acid residue.

Ganglioside ALG_{II}, not yet totally studied, behaved chromatographically exactly as ganglioside GQ1 after alkali treatment. The alkali treated ganglioside ALG_{II} was proved to contain glucose, galactose, N-acetylgalactosamine and sialic acid in the molar ratios : 1.00/2.15/0.90/3.92, the same composition and molar ratios being recorded on the untreated material.

The aim of our further research is to establish the total structure of gangliosides ALG_I and ALG_{II}.

GANGLIOSIDES IN DIFFERENT BRAIN AREAS OF INBRED MICE STRAINS

H. Dreyfus[1], A. Giuliani-Debernardi, G. Mack, S. Harth, P.F. Urban and P. Mandel
Unité 44 de l'INSERM (Section Neurochimie), and Centre de Neurochimie du CNRS,
11, Rue Humann, 67085 Strasbourg Cedex, France

Studies of gangliosides in different brain areas of inbred mouse strains (Balb, C57BL/6J and DBA/2J) have been performed using various analytical methods ; direct thin layer chromatography of the total lipid extract, analyses of ganglioside content and patterns after purification through sephadex G25 superfine columns and also after alkaline hydrolysis. All the methods showed the different brain areas to contain the mono- (G_{M3}, G_{M2}, G_{M1}), di- (G_{D3}, G_{D1a}, G_{D1b}), tri- (G_{T1}) and tetra-sialogangliosides (G_{Q1}). However, after the alkaline hydrolysis one compound, migrating between G_{D1b} and G_{T1}, was lost. This ganglioside could be a lactonic form of G_{T1} (G_{T1L} ; Svennerholm, personal communication). The ganglioside concentrations in various brain areas (cerebellum, pons and medulla, cortex, hypothalamus, caudate nucleus and hippocampus) of the Balb and C57 strains are similar, whereas those of the DBA strain are higher. The distribution of gangliosides in the areas did not show significant differences between the three strains. When gangliosides were not subjected to an alkaline hydrolysis G_{T1} + G_{T1L} amounted for 40-45 % of total ganglioside-sialic acid in cerebellum and pons-medulla, whereas they represented no more than 30-38 % in other areas which are largely of grey matter. G_{D1a} is the major ganglioside in cortex and caudate nucleus. The lowest amounts of G_{M1} are seen in cerebellum. After the alkaline hydrolysis G_{Q1} varied between 10 to 19 % in the different areas whereas it represented no more than 5 % without hydrolysis. This observation led us to think that the compound migrating between G_{D1b} and G_{T1} coudl also contain a lactonic form of G_{Q1}. Analysis after hydrolysis showed the cerebellum to contain the largest quantities of G_{T1} whereas G_{D1a} was the major ganglioside species in cortex and caudate nucleus. Pons-medulla contained the highest levels of G_{M1}.

[1]This investigation was supported by a grant from the Institut National de la Santé et de la Recherche Médicale (Contract n° 78.4.017.6).

UPTAKE AND INTRACELLULAR DISTRIBUTION IN DIFFERENT TISSUES OF GANGLIOSIDE GM1, INTRAVENOUSLY INJECTED IN THE MOUSE

P. Orlando[o], A. Leon[oo], R. Ghidoni[ooo], P. Massari[o], and G. Tettamanti[ooo]

[o] Department of Pharmacology, The Medical School, Catholic University, Rome, Italy; [oo] Fidia Research Laboratories, Abano, Italy; [ooo] Department of Biological Chemistry, The Medical School, University of Milan, Milan, Italy.

Pure ganglioside GM1, tritium labeled in the external galactose, was intravenously injected (10 ug, carrying 10 uCi) in adult male mice of the Swiss strain. After different times (from 1/2 to 24 hours) from injection, the animals were sacrificed, perfused with isotonic washing solution, and the distribution of radioactivity recorded in different tissues (blood, liver, kidney, muscles, spinal cord, full brain) also at subcellular level. The main results were the following:

a) The incorporated radioactivity was, in all cases, linked partly to ganglioside GM1, partly to water.

b) The incorporation of ganglioside GM1 reached a maximum level : in blood after 15 min from injection; in liver and kidney after 30-40 min; in muscles after 60 min; in spinal cord and brain after 60-90 min. Then it decreased, reaching, after 6-8 hours, a plateau, which was maintained (with the exception of blood) till at least 24 hours.

c) The radioactivity carried by water reached the maximum value after 4 hours in blood, liver, kidney and muscles; after only 2 hours in spinal cord and brain.

d) After tissue homogenization a portion of radioactivity, linked to both ganglioside GM1 and water, distributed in the soluble fraction, containing the cytosol; a portion, linked solely to ganglioside GM1, in the total particulate fraction. The radioactivity linked to particulate material, as % of total incorporated radioactivity, was 3-6 % in muscles, 10-15 % in brain, 20-25 % in kidney and 30-35 % in liver, regardless of the time after injection.

e) The subcellular fractions providing the highest degree of radioactivity incorporation were, in brain and liver, those containing plasma membranes. Purified preparations of plasma membranes (obtained from nerve endings, after hypoosmotic shock, in the case of brain; from the crude nuclear fraction in the case of liver) had by far the highest specific radioactivity. Conversely, purified preparations of nuclei, and of mitochondria, from either brain and liver, had the lowest specific radioactivity.

In conclusion, ganglioside GM1, injected in vivo, is uptaken by the various tissues, with different rate and kinetics. It appears to be incorporated into plasma membranes, in a rather tight and stable way, and, in a relevant portion to be internalized into the cell, where it undergoes degradation following the trend predicted by in vitro experiments (release, as the first step, of galactose which is rapidly metabolized producing water).

EFFECT OF EXOGENOUSLY ADDED GANGLIOSIDES ON NEURONAL MEMBRANE PROPERTIES.

D.BENVEGNU', A.C. BONETTI, A. LEON and G. TOFFANO
Fidia Research Laboratories, Abano Terme - Italy

Gangliosides, components of the cell membranes, are particularly abundant in neuronal synaptic preparations (Hansson et al., 1977). Gangliosides, moreover, have been shown to be taken up by cells from the culture medium (Callies et al., 1977) and to become associated as functional components of the cellular plasma membranes (e.g. increased cholera toxin binding).

Attempts were therefore done to supplement surface glycolipids of neuronal membrane preparations by incorporation of exogenously added pure gangliosides (GM_1, GD_{1a}, GD_{1b}, GT). The kinetics of each single ganglioside interaction with the membranes was studied in detail. The ganglioside interaction was seen to be characterized as a rapid, high-affinity, saturable process resulting in functional modification of membrane components such as ATPase and adenylate cyclase activity. These results are in accord with data demonstrating modification of liposome permeability properties after addition of gangliosides to liposome structure.(Ohsawa et al., 1977)

Gangliosides therefore bind to neuronal membranes in a finite manner affecting neuronal membrane properties which may perhaps be related to the role played by gangliosides in neuronal development and synaptogenesis (Obata, 1977 and Purpura et al., 1977).

Ohsawa T., et al. (1977) Japan J. Exp. Med. 47: 221
Hansson, H.A., et al. (1977) Proc.Nat.Acad.Sci.USA 74: 3782
Callies, R., et al. (1977) Eur.J.Biochem. 80: 425
Obata, K. (1977) Nature 266: 369
Purpura, D.P., et al. (1977) Nature 266: 553

CEREBROSPINAL FLUID TOTAL FATTY ACID PATTERN

J.Tichý and I.Skorkovská
Department of Neurology, Charles University, Prague, Czechoslovakia

The spectrum of total fatty acids in the CSF was determined by means of GLC on Perkin Elmer F-30 apparatus with FID on glass column packed with 13 per cent DEGS on Chromosorb W isothermicaly at 175°C, carrier gas N_2 75 ml/min. Evaluation by means of SIP 1 integrator. The preparation of methylesters has been done by means of 10 per cent BF_3 in methanol 20 min.at 100° C.

28 different FAMEs has been identified according to their retention time logarithms in comparison to commercial standards. The majority of FAMEs appeared irregularly and in less than in 10 per cent of cases. As main fatty acids in the CSF of control patients (n=12) there were found following FAMEs (values in per cent): 18:1(27,3), 16:0(23,2), 18:0(12,2), 18:2(7,7), 14:0(5,0) and 16:1(4,5).

In the group of neurologic patients (16 cerebrovascular accidents 11 degenerative, 7 discopathies and 3 hypophyseal adenoms) there were found some changes in the FAMEs pattern. A significant increase of all odd saturated FAMEs and of 18:0 and 20:0 has been found in hypophyseal adenoms. In the group of atrophic processes of the CNS 16:0 and 18:0 predominated to the detriment of 18:1. The composition of FAMEs in the CSF of patients with lumbar discopathy and with acute ischemic cerebrovascular accidents don´t differ from the control group.

COMPARATIVE ASPECTS OF ENERGY METABOLISM IN BRAIN TISSUE FROM INSECTS

G. Wegener
Institut für Zoologie der Johannes-Gutenberg-Universität
D-6500 Mainz, Germany

Significant progress in our understanding of the nervous system has resulted from studies using lower animals as model systems. To utilize these animals in an adequate manner some knowledge of their basic metabolism is a pre-requisite. Moreover knowledge about the insects, this highly adaptable and most successful class of all recent animals is meaningful per se and may be helpful for a general understanding of metabolic processes in nervous tissues.

To elucidate the significance of various fuels and metabolic pathways in insect brains we determined the activities and regulatory features of enzymes representative for different energy yielding pathways in members of four insect orders: Locust (Locusta migratoria, Orthoptera), silkmoth (Bombyx mori, Lepidoptera), honey bee (Apis mellifica, Hymenoptera), and blowfly (Calliphora erythrocephala, Diptera). The activity of PFK from insect brains is regulated by adenylnucleotides. Contrary to vertebrates and other groups of invertebrates, the brains of insects have a very limited capacity to derive energy anaerobically by the formation of lactate. On the other hand some insect brains (locust and moth) have high enzymatic activities of the fatty acid oxidizing enzyme HOADH, which are very low in all other brains tested till now (Wegener and Pfeifer, 1975). O_2-Consumption and respiratory quotient of isolated cerebral ganglia measured by microrespirometry point to a physiological significance of the enzymatic outfit.

The unusual features of metabolic organization are probably adaptations to the special demands of adult insects' living conditions. During metamorphosis, which in the holometabolic forms includes drastic changes in morphology, environment, diet, and behavioural performance, the characteristic enzyme patterns of the adult brains of Calliphora and Bombyx are formed. A rapid switch in metabolism seems to occur in the pupal phase: The potency of aerobic pathways is increased (about tenfold), the ability of lactate-formation is lost (LDH decreases to 2% in Calliphora) and that of FA-oxidation is established (in Bombyx).

To summarize: a) Brain tissues from insects have metabolic capacities for energy production unknown in other animals b) The metabolic organization of their brains undergo rapid and drastic changes during the metamorphis.

Wegener, G. and Pfeifer, J. (1975) Verh. dtsch. Zool. Ges. 1975:266-271 α-ALBUMIN (GFA): DOSAGE AND LOCALISATION IN HUMAN NERVOUS TISSUE AND CEREBROSPINAL FLUID

M. Noppe, D. Karcher, J. Gheuens, A. Lowenthal
Neurochemistry, Born-Bunge Stichting, Universitaire Instelling Antwerpen, Antwerp, Belgium

α-albumin is a brain specific protein which has been described in normal and pathological conditions in man (Chamoles, Karcher, Zeman, Lowenthal, 1970; Zeman, Karcher, Chamoles, Lowenthal, 1970). At a later date GFA (glial fibrillary acidic protein) studied mainly in animal was described and found to be closely related to α-albumin (Gheuens, Lowenthal Karcher, Noppe, 1977). Both proteins are localised in astrocytes. 2-site immunoradiometric assay (IRMA) was applied to the dosage of α-albumin in human tissue extracts. The outcome of this study confirmed the topographical distribution of α-albumin observed in the human control nervous system after agar gel electrophoresis. In Tay Sachs, endematous brains and peritumoral tissue, a marked increase of α-albumin was reported. In cerebrospinal fluid, α-albumin was detected by 2 site IRMA in 67 out of 649 cases (10.3%) and in 13.7% of the CSF with increased α globulin after agar gel electrophoresis. The concentration values ranged from 4.6ng/ml to 182ng/ml, the diagnosis for the positive CSF being intoxication, MS bouts and mainly epilepsy. The presence of α-albumin in CSF may indicate an acute lesion of the nervous tissue. However the quantitative values obtained with the 2 site IRMA and the histoimmunological reaction with peroxidase do not correlate, observation which speaks in favour of a soluble and an insoluble form of the protein.

Chamoles, N., Karcher, D., Zeman, W. and Lowenthal, A. (1970) Brain Research
17: 315-324
Zeman, W., Karcher, D., Chamoles, N. and Lowenthal, A. (1970) Brain Research
17: 325-334
Gheuens, J., Lowenthal, A., Karcher, D. and Noppe, M. (1977) Dynamic Properties of Glial Cells
Pergamon Press, in press. Satellite Symposium to the sixth intenational meeting of the International Society For Neurochemistry, August 29-31, 1977.

PRESENCE, METABOLISM AND UPTAKE OF PIPECOLIC ACID IN THE MOUSE BRAIN

T. Schmidt-Glenewinkel, E. Giocobini, Y. Nomura, Y. Okuma and T. Segawa
Dept. Biobehav. Sci., UConn., Storrs, CT 0628 and Dept. Pharmacol.,Inst.
Pharmac. Sci., Hiroshima Univ., Sch. Med., Kasumi 1-2-3,Hiroshima,Japan

Pipecolic acid, an imino acid related to lysine metabolism, has recently been identified in the mouse brain in our laboratory by means of TLC and mass spectrometry. Endogenous levels of pipecolic acid are 18 ± 4 nmoles/g in whole brain homogenate of adult mice. Several organs of the adult mouse, including brain, showed *in vitro* formation of pipecolic acid from L-lysine. The kidney demonstrated the highest rate followed by the brain (Schmidt-Glenewinkel et al., 1977). The synthesis of pipecolic acid from L-lysine was studied in whole embryo and brain of mouse and chick. In the mouse brain formation of pipecolic acid could be detected as early as at day 17 of gestation and in the chick embryo head at day 5 of incubation. Synaptosomes isolated from mouse brain showed a temperature dependent uptake of (^{3}H)-pipecolic acid at a concentration of 2×10^{-7}M. The uptake was Na$^+$ dependent, ouabain sensitive and showed a $K_m = 4.2 \times 10^{-6}$M. Structural analogues of pipecolic acid showed a significant inhibitory effect on the uptake at a concentration of 10^{-4}M. Release of pipecolic acid could also be demonstrated in rat brain slices. The demonstration of presence, biosynthesis in adult and embryonal brain and high affinity uptake of pipecolic acid, suggests a physiological role of this substance in the central nervous system of the mouse.

Schmidt-Glenewinkel, T. et al. (1977) Neurochem. Res. 2: 619-637

IN VIVO ETHANOLAMINE INCORPORATION INTO BRAIN LIPIDS BY BASE-EXCHANGE AND NET SYNTHESIS.

By: P. Orlando (1), G. Arienti (2), L. Corazzi (2), P. Massari (1), S. Roberti (1) and G. Porcellati (2).

(1): Department of Pharmacology, Università Cattolica del S.C., Via Pineta Sacchetti 644, ROME, Italy.

(2): Department of Biological Chemistry, The Medical School, Perugia, Italy.

Ethanolamine can be incorporated, both in vivo and in vitro (1), by net synthesis which requires the intermediate formation of PE (phosphorylethanolamine) and CDP-ethanolamine (cytidine-5'-diphosphate-ethanolamine) and by base-exchange reactions (2) whose operative existence in vivo is, however, still a matter of discussion.

In order to ascertain whether or not some ethanolamine could enter lipid bypassing the formation of PE, i.e. by base-exchange (or some other unknown pathway), [^{3}H]-ethanolamine and [^{32}P]-orthophosphate were simultaneously injected into brain ventriculi of rats. At selected time intervals (1-10 min), the animals were sacrificed and ethanolamine, PE and EPG (ethanolamine phosphoglycerides) extracted, separated and counted for radioactivity; the velocity constant (K) for PE incorporation into EPG was then calculated according to Zilversmit et al.(3) from both [^{32}P] and [^{3}H] data.

If net synthesis was the only way operative in brain in vivo to introduce ethanolamine into EPG, the Ks should have been identical (obviously the phosphate and ethanolamine moieties of the same PE molecule are incorporated into lipid at the same speed) and their ratio equal to 1. If some ethanolamine entered lipid bypassing the formation of PE, the K calculated on [^{3}H] data would appear to exceed the one calculated on [^{32}P] data. The table shows that this is the case, at least at short times after the injection. The decrease of the specific radioactivity of free ethanolamine could explain why base-exchange is no more detectable after 6 min (and onwards) from the injection.

TABLE. Ratios between the velocity constants for PE incorporation into EPG calculated on [^{3}H] and [^{32}P] data.

time (min)	K[^{3}H] / K[^{32}P]	time (min)	K[^{3}H] / K[^{32}P]	time (min)	K[^{3}H] / K[^{32}P]
1	2.9	4	7.0	8	0.85
2	4.4	5	5.2	9	0.9
3	7.5	6	1.7	10	0.9
		7	0.9		

The Ks have been calculated from the data obtained from, at least, 4 animals.

Base-exchange appears to be several times faster than net synthesis (table), but in this work it is probably underevaluated:

(1) Ansell, G.B. and Spanner S., J. Neurochem.14, 1967,873-885.
(2) Porcellati,G.,Arienti,G.,Pirotta,M. and Giorgini,D.,J.Neurochem.18,1971,1395.
(3) Zilversmit,D.B.,Enteman,G.,Fischler,M.C.,J.Gen.Physiol.26,1943,325-331.

MODEL SYSTEMS FOR THE STUDY OF ANTIPARKINSON DRUGS - SYNAPTOSOMES AND BLOOD PLATELETS

I. von Pusch, G. Muschalek, H. Stöltzing, W. Wesemann
Abt. Neurochemie, Physiol.-chem. Institut II, Philipps-Universität
N. Paul
Institut für Cytobiologie und Cytopathologie, Philipps-Universität
D 3550 Marburg/Lahn

The antiviral substance 1-aminoadamantane, D_1, and its C-alkyl derivative 1-amino-3.5-dimethyladamantane, D 145, exhibit antiparkinson activity. Both compounds inhibit non-competitively 5-HT uptake into synaptic vesicles and synaptosomes isolated from rat brain and 5-HT binding to synaptic membranes. Though high concentrations of these drugs ($>10^{-4}$M) are necessary to release 5-HT or DA from synaptosomes, the electrical stimulated release is increased by subthreshold concentrations ($5 \times 10^{-6} - 5 \times 10^{-5}$M) which are too low to liberate the transmitters.

In an attempt to test the possible central activity of 1-aminoadamantane and its C-alkyl derivatives (D 145: 3.5-dimethyl-; D 176: 3-methyl-; D 177: 3-isopropyl-; D 178: 3-butyl-; D 191: 3.5.7-trimethyl-) by means of a peripheral model system blood platelets are used to study the interaction of these drugs with 5-HT uptake, with 5-HT, ATP, and ADP release and with platelet aggregation. The effect on 5-HT uptake into blood platelets parallels the effect observed with synaptosomes: a non-competitive type of inhibition is found for both uptake mechanisms with almost identical K_I values (about 10^{-4}M). Increasing concentrations of the drugs first liberate 5-HT from the storage organelles and subsequently ATP and ADP in addition. The release can be followed by electron microscopy which shows a decrease in number of storage organelles. Under the experimental conditions used 1-aminoadamantane releases only 5-HT but no adeninnucleotides.

After preincubation with C-alkyl-adamantanes the ADP-induced platelet aggregation is blocked. Thus the release reaction and/or the platelet aggregation, which can be measured even more easily, provide a mean to screen drugs like 1-aminoadamantanes.

DI-, MONO-, AND NONPHYTANYL SERUM TRIGLYCERIDES IN REFSUM'S DISEASE AND THEIR DISTRIBUTION IN THE LIPOPROTEINS

B. Molzer [a], H. Bernheimer [a] and E. Koller [b]

[a] Department of Neurochemistry, Neurological Institute, University of Vienna, Vienna, Austria

[b] Physiological Institute, University of Vienna, Vienna, Austria

The relative proportions of the diphytanyl (TG 2), monophytanyl (TG 1) and nonphytanyl (TG 0) triglycerides ("triglyceride pattern"), the phytanic acid content of the triglycerides and the phytanic acid levels in the serum of 3 patients with Refsum's disease (heredopathia atactica polyneuritiformis, phytanic acid storage disease) were estimated by thin-layer chromatography, densitometry and gas chromatography, respectively. The individual triglyceride patterns were clearly dependent on the phytanic acid content of the triglycerides: the more phytanic acid in the triglycerides, the higher the percentage of the diphytanyl and the lower the percentage of the nonphytanyl triglycerides. The monophytanyl triglycerides were also related to the phytanic acid content of the triglycerides, although in a complex manner.

Since the serum triglycerides are associated to the lipoproteins, we investigated the distribution of TG 0, TG 1 and TG 2 in the different lipoprotein fractions (VLDL, LDL, HDL_2 and HDL_3) which were prepared by standard flotation techniques. At high percentages of TG 0 in the total serum triglycerides, TG 0 accumulated in the VLDL and LDL fractions and thus showed a distribution similar to that of the triglycerides of normal controls. At lower percentages of TG 0 in the total serum triglycerides, TG 0 was found predominantly in the LDL fraction. TG 1 and TG 2 accumulated mainly in the LDL fraction irrespective of their percentage in the total serum triglycerides. This particular accumulation might be the consequence of a poor metabolic degradation of TG 1 and TG 2. (Supported by Fonds zur Förderung der wissenschaftlichen Forschung, Austria, Project No. 1082 and 2779/S).

THE OESTROGEN RECEPTOR IN THE RAT HYPOTHALAMUS AND UTERUS: THE EFFECTS OF OESTRADIOL ADMINISTRATION

J.O. WHITE, A.C. NEETHLING[*] & L. LIM
Department of Neurochemistry
Institute of Neurology (Queen Square)
33 John's Mews, London WC1N 2NS, UK

[*] Department of Chemical Pathology
University of Stellenbosch
Tygerberg Hospital, South Africa

In the rat hypothalamus and uterus, the specific oestrogen receptors translocate from the cytosol into the cell nucleus as a result of increases in circulating oestrogen, particularly at the pro-oestrus phase of the female reproductive cycle. Increases in nuclear receptor content also occur after injection of oestradiol. The increase is maximal within 1 h., the content subsequently declines. We have measured nuclear content after injection of ^{3}H-oestradiol by determining radioactivity in the nuclei as well as by an in vitro 37°C exchange procedure to assay for receptor. In the hypothalamus, at early time periods (1 h.) these different procedures yielded similar values; however at later time periods (6 h.) measurements by exchange gave higher values than those determined on the basis of radioactivity in the nucleus. Similar results were obtained with the uterus although in this case results were more variable. This apparent "deficit" in receptor when assayed with label alone can be explained by the presence in the nucleus of oestrogen receptor free of oestradiol. The in vitro exchange procedure can also be used to determine "empty" receptor by measurements at 4°C instead of 37°C. Using this exchange procedure, we found an increase in the empty receptor; amount of this increase being most pronounced at later times e.g. 6 h. The metabolism and function of the oestrogen receptor in the nucleus are discussed in relation to these two classes of receptor.

EFFECTS OF HISTAMINE-AGONISTS AND -ANTAGONISTS (H_1 AND H_2) ON GANGLIONIC TRANSMISSION AND ON CYCLIC NUCLEOTIDES (cAMP AND cGMP) ACCUMULATION IN RAT SUPERIOR CERVICAL GANGLION IN VITRO.

Toni Lindl, Fachbereich Biologie, Universität Konstanz, P.O.B. 7733, D-7750 Konstanz, GFR

There is substantial evidence that histamine is able to alter ganglionic transmission (Brimble, M.J. & Wallis, D.J. 1973) and to increase cyclic nucleotide levels in peripheral nervous tissue (Lindl, T. & Cramer, H. 1974 and Kebabian, J.W., Steiner, A.L. & Greengard, P. 1975). It has been proposed that alterations in cyclic nucleotide metabolism may reflect changes in the expression of certain transmission processes in the ganglion (Greengard, P. 1976). It is, therefore, of special interest to combine electrophysiological data with measurements of intraganglionic cyclic AMP and cyclic GMP accumulation after incubation of ganglia with histamine and its related agonists and antagonists. The rat superior cervical ganglion was used for in vitro-observations of extracellularly recorded fast and slow potentials and for concurrent measurement of cyclic AMP and cyclic GMP accumulation. In electrophysiological experiments histamine exhibited a potent ganglionic blocking ability. This was shown by the reduction of the compound action potential (C.A.P.) to about 45 % of its initial value after superfusion of ganglia with histamine or with 4-methylhistamine, a H_2-receptor agonist. The H_1-agonist, 2-pyridylethylamine did not alter ganglionic transmission. In addition, pre-incubation with metiamide and subsequent incubation of ganglia with 4-methylhistamine increased, under certain stimulus conditions, the C.A.P. to about 25 % over control values. Cyclic AMP and cyclic GMP levels increased when the ganglia were incubated with histamine or 4-methylhistamine. The H_1-agonist, 2-pyridylethylamine, was able to increase only cyclic AMP concentrations significantly. Histamine-induced accumulation of cyclic AMP and of cyclic GMP were sensitive to H_1- and H_2-blockade. The H_2-agonist, 4-methylhistamine, showed the same potency as histamine in increasing cyclic nucleotides in the ganglion. It is concluded that histamine-induced alterations in ganglionic transmission could be interpreted as H_2-mediated inhibition and facilitation. These effects are accompanied by significant changes in cyclic AMP and cyclic GMP also via H_2-receptors.

Brimble, M.J. & Wallis, D.I. (1973) Nature (London) 246, 156 - 158
Greengard, P. (1976) Nature (London) 260, 101 - 108
Kebabian, J.W., Steiner, A.L. & Greengard, P. (1975) J. Pharmacol. Exp. Therap. 193, 474 - 488
Lindl, T. & Cramer, H. (1974) Biochim. Biophys. Acta 343, 182 - 191

ISOLATION AND LABELING OF CHONDROITIN-4-,-6-SULFATE, HEPARAN SULFATE
AND HYALURONATE IN RAT BRAIN WITH RADIOSULFATE AND ^{3}H-GLUCOSAMINE
IN VITRO
T. O. Kleine and U. Schwadtke
Klinisch-chemisches Labor, Universitäts-Nervenklinik
D-3550 Marburg / Lahn, W.Germany

Isolation of glycosaminoglycans from rat brain was performed after their labeling in vitro: brains prepared free from blood vessels and membranes were cut in small pieces and incubated aerobically with phosphate buffer, pH 7.1, containing glucose, amino acids, electrolytes, vitamins and $^{35}SO_4$ plus D-glucosamine-6-^{3}H. The labeled brain pulp was exhaustively extracted with chloroform-methanol (1/1;v/v), followed by 6 M guanidine-chloride in 0.05 M acetate buffer, pH 6.0. The guanidine extract and the 6000 r.p.m. residue were dialysed salt-free, digested by CM-cellulose-bound papain and their supernatants analysed acc. to Kleine and Stephan, 1976. Labeling of the fractions with both radioactive precursors increased linearly over an incubation period of 90 min exhibiting in the extract 4 to 5 times higher values for ^{35}S-labeling and twice as high values for ^{3}H-glucosamine (glcN) as those of the residue. Both fractions isolated from the guanidine extract and from the residue contained 5 to 10 times more amino sugars than uronic acids (UA) besides amino acids (AS). The guanidine extract which consisted of 3 to 5 times more UA and AS than the residue fraction proved to be heterogeneous. It was separated on Sepharose-6B into 4 fractions of different mol.wt.: A $>7.5 \times 10^4$, B $6-7.5 \times 10^4$, C $1.7-6.0 \times 10^4$, D $0.3-1.7 \times 10^4$. Their constituents were: hexosamine-N / peptide-N: A 0.5, B 1.8, C 0.6, D 0.1; glcN / hexosamine: A 0.4, B 0.3, C 0.4, D 0.8; main 4 AS were for A, C Asp, Glu, Ser, Gly, for B Gly, Ser, Ala, Asp, for D Glu, Asp, Gly, Ala. Labeling with ^{3}H-glcN increased with decreasing mol.wt. of the fraction, whereas fraction C exhibited highest ^{35}S-labeling and A none. Low-labeled hyaluronate was separated from fractions A and B after specific degradation with leech hyaluronidase (cf. Kleine, 1976). From fractions C and D highly labeled glycopeptides with low mol.wt. ($< 3 \times 10^4$) could be separated off on Sephadex G-75. The excluded fractions from A,B,C,D were treated with chondroitinase ABC showing the presence of chondroitin-4-,-6-sulfate and heparan sulfate which were labeled with $^{35}SO_4$ and ^{3}H-glcN in the ratio 2 : 1. The data point to the presence of chemically and metabolically different glycosaminoglycans in rat brain.

Kleine, T.O. (1976) Pharm.Res.Commun. 8: 219-228
Kleine, T.O. and Stephan, R. (1976) Biochim.Biophys. Acta 451: 444-456

EFFECT OF DI-N-PROPYLACETATE ON RAT BRAIN GABA-T AND SSA-DH

J.W. van der LAAN, Th. de BOER, J. BRUINVELS
Department of Pharmacology, Faculty of Medicine, Erasmus University,
Rotterdam, The NETHERLANDS

The kinetic constants for 4-aminobutyric-2-oxoglutaric acid-transaminase (GABA-transaminase) and succinic semialdehyde dehydrogenase (SSA-DH) have been determined using rat brain homogenate.
The kinetics of GABA-T have been shown to be consistent with a Ping Pong Bi Bi mechanism. Using the appropriate HALDANE relationship, a K_{eq} of 0.01 for GABA-T was found, indicating that the reaction was strongly biased towards GABA.
The effect of di-n-propylacetate (DPA) on both GABA-T and SSA-DH was measured. DPA inhibited SSA-DH competitively with respect to SSA, giving a K_i of 0.5 mM. Hardly any effect of DPA was observed on the reaction of GABA-T, using GABA and 2-OG as substrates (forward reaction), whereas the reaction using glutamate and SSA as substrates (reverse reaction) was slightly inhibited. The K_i for this latter reaction (15.2 mM with respect to GLU and 22.9 mM with respect to SSA) was 30-40 times higher than that for SSA-DH. These results suggest that GABA-accumulation in the brain, after administration of DPA *in vivo*, is caused by SSA-DH inhibition. Two mechanisms for this accumulation are indicated by the data. (1) The higher level of SSA, which results from inhibition of SSA-DH, initiates the reverse reaction of GABA-T, thus increasing the level of GABA via conversion of SSA. (2) The degradation of GABA is inhibited by the higher level of SSA, since SSA has a strong inhibitory effect on the forward reaction as calculated from the present data.
Recently, it has been shown that DPA administered to naive rats *in vivo* (300 mg/kg, i.p.) evokes a behavioural syndrome resembling a morphine-abstinence syndrome, which could be suppressed by GABA antagonists (De Boer et al., 1977). The underlying mechanism for this action of DPA might be that described above.
Also the anticonvulsive action of DPA, so far not clearly understood, can be explained in this way.

De Boer, Th., Metselaar, H.J. and Bruinvels, J. (1977).
    Life, Sciences 20: 933-942.

Author Index

A

Abdel-Halim, M.S.	596
Abulaban, F.	124
Alberghina, M.	304
Albers, R.W.	549
Allan, R.D.	550
Allen, J.N.	312
Althaus, H.H.	557
Andrasi, F.	534
Änggård, E.	527,596
Angielski, S.	577
Annunziata, P.	220,226,236
Arienti, G.	619
Arlock, P.	493
Artigas, F.	532
Austen, B.M.	460

B

Bachelard, H.S.	303,442,531
Baker, G.B.	536
Balcar, V.J.	498,513
Barberis, C.	584
Bardakjian, J.	512
Bardsley, M.E.	442,531
Bartolomé, M.	235
Basset, P.	558,575
Battistin, L.	588
Baumann, N.	48,128,129 131,232,609,611
Beckmann, K.	560
Belfroid, R.D.M.	541
Di Benedetta, C.	607
Benvegnu', D.	614
Béress, L.	467
Berlet, H.H.	573
Berger, R.	590
Bernheimer, H.	537,621
Berntman, L.	253
Berra, B.	610
Bertoli, E.	582

Biesold, D.	199
Bigl, V.	199
Bird, T.D.	131
Bläsig, J.	386
Blindermann, J.M.	225
Bock, E.	216,217
Boer, Th. de	625
Boggan, W.O.	228
Du Bois, H.	119
Bøler, J.	453
Bonett, A.C.	614
Borg, J.	513
Bosch, H. van den	576
Bosisio, E.	540
Bourre, J.M.	128,131,137 609
Broniszewska-Ardelt, B.	572
Bruinvels, J.	625
Brunetti, M.	484
Brzin, M.	231,564
Buniatian, H. Ch.	117
Burkart, T.	519,574

C

Calverley, D.G.	536
Cantrill, R.C.	136,233
Carey, E.M.	138,233
Carlsson, A.	266
Casamenti, F.	425
Caspary, E.A.	78
Cassagne, C.	137
Cederholm, B.	496
Chao, S.-W.	126
Churchich, J.E.	598
Christova, M.	125
Chronwall, B.M.	158
Ciesielski-Treska, J.	489
Cimarra, P.	594
Citherlet, K.	571
Clapshaw, P.	121

Clausen, J.	135	Elkouby, A.	551
Clos, J.	218	Ermini, M.	571
Codina, J.	562		
Collier, H.O.J.	374	**F**	
Cook, W.F.	461	Fagg, G.E.	139
Corazzi, L.	619	Federico, A.	220,226,236
Corona, M.	226,236	Filippi, D.	216,217
Corsi, P.	607	Finesso, M.	310,311
Couraud, J.-Y.	472	Fini, C.	285,311
Cremer, H.	543	Fleissner, A.	563
Crkovská, J.	604	Floridi, A.	285,311
Cuénod, M.	471	Fosmark, H.	516
Currie, D.N.	511	Foss, I.	453,466
		Fosset, M.	494
D		Francescangeli, E.	309
Dabrowiecki, Z.	542	Freysz, L.	521,576
Damgaard, I.	516	Froissart, C.	523,558
D'Amore, I.	236	Fuentes, J.A.	605
Darriet, D.	137		
Daudu, O.	128	**G**	
Davey, J.	544,545	Galli, C.	271
Davison, A.N.	91	Galli, G.	271
Debuch, H.	234	Galli Kienle, M.	540
DeFeudis, F.V.	551	Gamse, R.	464
Delaunoy, J.P.	504,520	Gayet, J.	581,584
Detering, N.K.	444	Gebauer, H.	485
Dewhurst, W.G.	536	Gebicke, P.J.	557
Dias, Th.	590	Gelpí, E.	532
Domańska-Janik, K.	305,306,307	Gennarini, G.	607
Dowdy, E.E. Jr.	443	Geremia, E.	304
Dreyfus, H.	521,612	Gerken, U.	514
Droz, B.	472,484	Gershagen, S.	478
Drukker, J.	221	Ghandour, M.S.	174,216,217,218
Dudai, Y.	508	Gheuens, J.	617
Dum, J.	386	Ghidoni, R.	611,613
Dumas, M.	476	Di Giamberardino, L.	484
Dupouey, P.	129,131	Giesing, M.	505,507,514
Dutton, G.R.	511	Giocobini, E.	618
		Giorgi, P.P.	118,119
E		Girós, M.	562
Edgar, A.D.	312	Gispen, W.H.	463
Edminson, P.D.	453,462,466	Giuditta, A.	594
Edström, A.	478,482,517	Giuffrida, A.M.	304

Giuliani-Debernardi, A.	612	Husby, G.	608
Goedde, H.W.	535	Huttner, W.B.	557
Gohl, W.	538		
Gombos, G.	174,216,217 218,235,585	**I**	
Gonnard, P.	512	Iannelli, D.	607
Gonzáles-Sastre, F.	224,562	Ingoglia, N.A.	483
Goracci, G.	285,309,311	Iversen, L.L.	450
Goranov, I.	125		
Gorgani, M.	229	**J**	
Gráf, L.	459	Jacque, C.	131
Greiner, B.	547	Jacques, Y.	494
Guazzi, G.C.	220,236	Janka, Z.	500
Gunawan, J.	234	Jelínek, R.	227
Gustke, H.H.	219	Johansen, J.H.	453
		Johnston, G.A.R.	550
H		Joó, F.	500,501,517,595
Haas-Ackermann, I.	538	Juhász, A.	500
Haglid, K.G.	593		
Hanson, M.	478	**K**	
Hansson, E.	509	Kadyrov, G.K.	561
Harpin, M.L.	611	Kalant, H.	317
Hassler, R.	568	Kanje, M.	493,517
Harth, S.	521,612	Karavasan, A.	444
Haug, P.	568	Karcher, D.	617
Hauser, K.L.	498,503	Karlsson, I.	189
Hauw, J.J.	232	Karlsson, J.-O.	480
Heid, J.	503	Kastrapeli, N.	133
Hetey, L.	441	Keller, K.	497,506
Herken, H.	506	Kempf, T.	602
Herschkowitz, N.	519,574	Kerr, L.	136,233
Hertz, L.	510,515,516	Khatchatrian, L.	307,308
Herz, A.	386	Kiauta, T.	231
Hitzemann, R.J.	404	Kim, J.	568
Högenauer, G.	537	King, L.J.	443
Höllt, V.	386	Kleine, T.O.	624
Hofer-Warbinek, R.	602	Klinar, B.	564
Hommes, F.A.	590	Kluss, E.-M.	535
Honecker, H.	547	Koller, E.	621
Horrocks, L.A.	285,309,312,542	Konat, G.	135
		Kossut, M.	600
Hughes, J.	452	Kowalewski, S.L.	219
Hullin, D.A.	592	Krauhs, E.	602
Hunt, G.G.	303	Kreil, G.	537

Kreutzberg, G.W.	474	Majewska, M.D.	308
Kriesten, K.	505	Mallol, J.	235
Krogsgaard-Larsen, P.	510	Mandel, P.	225,489,513 520,521,523,551 558,569,575,612
Van der Krogt, J.A.	541		
Krzalič, Lj.	133	Manukian, K.H.	117
Ksiezak, M.	583	Manzocchi, A.	540
Kumpulainen, E.	578,586	Mark, J.	513
Kvamme, E.	515	Massarelli, R.	523,558
Kwok, F.	598	Massari, P.	613,619
		Matthies, M.	538

L

Laan, J.W. van der	625	Matthieu, J.-M.	127,130
Lähdesmäki, P.	578,586	Mayer, N.	464
Lane, J.D.	589	Mayer, R.J.	544,545
Lange, K.	497,506	Mazzari, S.	310,311
Larsson, H.	482	McKim, H.R.	536
Laurent, G.	216,217	De Medio, G.E.	285,311
Lazarewicz, J.W.	285,308,582,583	Meisami, E.	229,230
Lazdunski, M.	494	Meyermann, R.	557
Legrand, J.	218	Middaugh, L.D.	228
Lehnert, T.	573	Minn, A.	581,584
Lembeck, F.	464	Minnema, K.H.	443
Leon, A.	613,614	Mir, A.K.	533
Liljequist, S.	359	Mitros, K.	600
Lim, L.	622	Molzer, B.	621
Lindi, C.	610	Monge, M.	129,232
Lindl, T.	623	Morand, O.	609
Linington, C.	115	Morgan, J.	522
Little, M.	602	Mousavi, R.	230
Loh, H.H.	404	Müller, R.	505
Lohmann, W.	538	Mugnaini, E.	3
Longstaff, A.	591	Murphy, S.	222,223
Loopuijt, L.D.	479	Muschalek, G.	620
Lopez-Cardozo, M.	549		

N

Lovelidge, C.A.	124	Neethling, A.C.	622
Lowenthal, A.	617	Nesham, M.E.	312
Lunt, G.G.	303	Neskovic, N.M.	520
Lysiak, W.	577,597	Neubach, C.	556
		Neuhoff, V.	120,139,543,555, 556,557,587,589

M

Maas, W.F.	541	Nissen, C.	518
Mack, G.	612	Nomura, Y.	618
Maitre, M.	225,551	Noppe, M.	617

630

Noremberg, K. 582

O

Oelssner, W. 441
Oestreicher, A.B. 463
Oey, J. 121
Offner, H. 135
Okuma, Y. 618
Olkinuora, M. 586
Ordaz, A. 605
Orlando, P. 613,619
Osborne, N.N. 543,560
Ossola, L. 255,551
Ozand, P.T. 444

P

Paik, K. 568
di Palma, S. 610
Palmerini, C.A. 285,311
Palo, J. 132
Panula, P. 501
Pàmpols, T. 562
Pastuszko, A. 567,582
Pedata, F. 425
Pepeu, G. 425
Persson, L. 605
Pettmann, B. 502,504
Philipp, K. 555
Pieringer, R.A. 140
Pollet, S. 128,232
Politis, M.J. 483
Poulos, A. 570
Pontén, J. 493
Ponstingl, H. 602
Porcellati, G. 64,285,309
311,484,619
Porte, A. 502,504
Privat, A. 131
Prokes, J. 604
von Pusch, I. 620
Przewłocki, R. 386

R

Radomińska-Pyrek, A. 542
Rafalowska, U. 583
Reading, H.W. 439
Rebel, G. 520,521
Recasens, M. 569
Rechardt, L. 501
Reeber, A. 174,585
Reichelt, K.L. 453,462,466
Reichlmeier, K. 571
Reiber, H. 123
Rickmann, M. 158
Riekkinen, P. 132
Rigotti, P. 588
Rimanóczy, A. 500
Rivett, A.J. 332
Robert, J. 521,523
Roberti, S. 619
Rodés, M. 224
Rönnbäck, L. 593,605
Rommelspacher, H. 547,548
Rose, S.P.R. 565,566,591
Rosengren, L. 593
Roussel, G. 504
Rumsby, M.G. 115,116,126
Russel, S.M. 544,545
Rychter, Z. 227

S

Sabater, J. 224,562
Saelid, G. 453,462
Safail, R. 230
Sanderson, C. 222
Santaniello, E. 540
Sapper, H. 538
Saria, A. 464
Sarlieve, L.L. 140
Sarraga, C. 235
Sautebin, L. 271
Savettieri, G. 603
Le Saux, F. 232
Scher, M.G. 559
Schiffer, D. 219

Schipper, H.I.	139	Stipek, S.	604
Schlecht, H.P.	571	Stöltzing, M.	620
Schmidt, R.	579,595	Strange, P.G.	495
Schmidt-Glenewinkel, T.	618	Strosznajder, J.	285,308,313 314
Schöne, B.	589	Sturman, J.A.	481
Schotman, P.	463	Subba Rao, G.	140
Schousboe, A.	492,510 515,516,518	Suda, H.	122
Schröder, N.	568	Svenneby, G.	510,515
Schubert, P.	474	Symon, L.	239
Schulze, E.	556,587	Szabó, E.F.P.	601
Schwab, M.E.	476	Szebeni, A.	459
Schwadtke, U.	624	Székely, G.	143
Sedláček, J.	539	Szentistványi, I.	500
Sedzik, J.	134	Szutowicz, A.	577,597
Segawa, T.	618		
Seifert, W.	121,522	T	
Seil, F.J.	100	Tardy, M.	512
Sellström, A.	509	Tashiro, T.	580
Sensenbrenner, M.	489,502,504	Tauber, H.	120
Serra, I.	304	Tencheva, Z.S.	599
Shapiro, L.M.	546	Tettamanti, G.	611,613
Shaw, S.G.	461	Thoenen, H.	476
Siegrist, H.P.	519,574	Thompson, R.J.	592
Siesjö, B.K.	253	Thorbeck, P.	510
Simantov, R.	465	Tichý, J.	615
Singh, S.	535	Tijtijlkova, N.	599
Sinha, A.	223	Tischner, K.	507
Skangiel-Kramska, J.	600	Tipton, K.F.	332
Skorkovská, I.	615	Toffano, G.	614
Smyth, D.G.	460	Trojan, S.	604
Smith, I.L.	332	Trovarelli, G.	285,311
Snyder, W.R.	312	Trygstad, O.E.	453,466
Sofroniew, M.V.	449	Tsukada, Y.	122
Sommarin, Y.	496	Tuneva, S.V.	599
Sonnino, S.	611	Turpin, J.C.	232
Sontag, K.-H.	543	Twitchin, B.	550
Sourander, P.	593		
Spagnuolo, C.	271	U	
Stadler, H.	580	Urban, P.F.	612
Stastný, F.	227	Ujvari, A.	534
Staykova, M.	125		
Stein, S.	457,458		

V

Vassanelli, P.	588
Vaughan, P.F.T.	533,546
Verde, O., Lo	608
Vierbuchen, M.	234
Virtanen, P.	346
Vincendon, G.	174,216,217 218,235,585
Vitiello, F.	218
Volk, B.	573
Voss, W.	123

W

Waechter, C.J.	559
Waehneldt, T.V.	32,120,127
Walker, A.G.	116
Wallgren, H.	346
Wallin, M.	482
Walum, E.	499
Wedege, E.	608
Wegener, G.	616
Weiler, C.	499
Weindl, A.	449
Weiner, N.	440
Weiss, D.G.	475
Welsum, R.A. van	221
Wesemann, W.	440,620
Westermark, B.	493
van der Westhuyzen, J.	136
White, J.O.	622
Wideman, J.	457,458
Wied, D., de	455
Wiegant, V.M.	463
Wiesmann, U.	519,574
Wieslander, A.	496
Wikström, J.	132
Willers, I.	535
Wise, H.	303
Woelk, H.	64
Wolff, J.R.	158
Wollemann, M.	459
Wong, T.Y.	523,558
Wood, N.	565,566

Wronski, A.	593

Y

Yanev, S.G.	445
Yavin, E.	508
Yavin, Z.	508

Z

Zalc, B.	129
Zalweska, T.	305,306
Zanchin, G.	588
Zanetta, J.-P.	174,216 217,235,585
Zaprianova, E.	125
Zeitz, M.	497,506
Zemp, J.W.	228
Zgorzalewicz, B.	134
Zilliken, F.	505,507,514
Zimmer, H.-G.	555
Zimmermann, H.	579,595
Zwiers, H.	463
Zwiller, J.	575

Subject Index

A

Absorption with cerebroside, antimyelin activity 104
Acetaldehyde
-, and alcohol dehydrogenase 332
-, and alcohol metabolism 336
-, oxidation 334
Acetylcholine
-, during development 230
-, effect of morphine 429
-, influence by naloxone 430
-, in cortico-striatal path 568
-, in olfactory bulb 230
-, under olfactory deprivation 230
-, vesicular release 580
Acetylcholinesterase
-, and lipid-protein interactions 229
-, during development 229, 230
-, effect of anesthetics 567
- -, light 565
-, in neuronal membranes 229
-, in olfactory bulb 230
-, in sympathetic ganglion 564
-, in visual cortex 565
-, synaptic somal membranes 567
-, under olfactory deprivation 230
Acid hydrolase activity in multiple sclerosis 68
Acid phosphatase
-, during development 221
-, in anterior horn cells 221
-, under experimental conditions 221
-, under normal conditions 221
Acid proteinase
-, and β-glucoronidase in MS 68

-, lysosomal 67
Acidic proteins
-, synthesis 591
- -, and learning 591
Active transport, role of ethanol 348
Adenine nucleotides 584
Adenosine uptake 584
Adenyl cyclase
-, effect of enkephalins 459
- -, visual deprivation 208, 209
-, inhibition by opiates 380
-, see also cyclic AMP
Adhesion
-, intralamellar 21
Age groups, lipid microanalysis 232
Aging
-, human brain 226
-, rat brain 226
Alcohol
-, effect on membranes 346
-, metabolism 332
-, oxidation 332
-, relationship to acetaldehyde levels 336
-, see also ethanol
-, tolerance 321
Alcohol dehydrogenase
-, atypical isoenzyme 333
-, formation of acetaldehyde 332
α-Albumin in nervous tissue 617
Alkaline phosphatase in cell cultures 501, 509, 517
Alkaloids
-, formation of tetrahydroisoquinoline 338
-, isoquinoline and alcohol effects 351
Alkanes
-, biosynthesis in PNS 137

-, in quaking PNS myelin	137
Allergic encephalomyelitis, demyelination	79
Amines, uptake of biogenic	206,207
Amino acids	
-, availability	223
-, blood brain barrier	590
-, capillary permeability	583
-, cerebral cortex	223
- -, regions	588
-, determination	556
-, in separated cells	223
-, membrane binding	578
- -, transport	223,485,498,499 510,511,513,515
-, phenylketonuria	589,590
-, pools of brain	224
- -, of plasma	224
- -, in undernutrition	224
-, under development	223
γ-aminobutyric acid see GABA	
Amyotrophic lateral sclerosis, axonal degeneration	109
Anaemic anoxia	240,256
Anaerobic glycolysis, LDH	573
Anesthesia, membrane expansion theory	347
Anesthetics, effect on acetylcholinesterase	567
Anlage	
-, cortical	161,169
-, genetic	199
-, pallial	161
Anterior horn cells	
-, and acid phosphatase	221
-, during development	221
-, under experimental conditions	221
- -, normal conditions	221
Antibodies	
-, anti-neuronal	608
-, glycoproteins	607
-, to mesodermal cells	606

Antimyelin antibodies, serum	100
Antiparkinson drugs	620
Anoxia	
-, clinical aspects	239
- -, classification	240,275
-, see also Hypoxia and Ischemia	
Apposition of cytoplasmic faces in myelin	116
Arachidonic acid metabolism	271 275,286,291,297
Aromatic amino acid hydroxylases, hypoxia	266
Arrhenius plot	
-, of acetylcholinesterase	229
-, of Na^+-K^+ ATPase	229
Arterial oxygen content, signs and symptoms	253
Arylsulphatase A, in EAE	75
Astrocytes, stearic acid uptake	609
ATP, citrate lyase	597
Autoradiography of neuron	160,161
Axonal growth	174
- -, capacity	177
- -, guiding signals	175
- -, in agranular rat cerebella	177
- -, influence of target cells	177
- -, intrinsic and extrinsic factors	174,175
- -, stop after contact	177
Axonal transport	471,472 474,475,479
- -, and microtubules	482
- -, in vitro	478
- -, neuron-glia interaction	473
- -, amino acids	481
- -, glycoproteins	480
- -, neuropeptides	451
- -, phospholipids	484
- -, RNA	483
- -, Wolfgram proteins	118

- -, retrograde 476,479
- -, selectivity 472

B

Barbiturates
-, calcium metabolism 582
-, effect of p-chloro-
  phenylalanine 326
-, tolerance 319,321
-, treatment following
  hypoxia 249
Basal lamina 6
Behaviour
-, modulation by neuro-
  peptides 435,455
-, tolerance 319
Belt, outer and inner 8
Benzene, effect on GABA 561
Benzodiazepine, receptors 317
Bergmann glia 175
Binding studies, of opiate
  receptors 387
Biochemical changes
- -, aging human brain 226
- -, aging rat brain 226
-, studies of glial
  cells 216,217
-, theories of opioid
  dependence 374
Blood
-, brain barrier, amino
  acid transport 590
-, platelets, test system
  for drugs 620
-, pressure, pargyline 605
-, vessel
- -, contact with spongioblasts 164
- -, intracerebral
  radial 164,169
Body growth
- -, and malnutrition 195
- -, protein-calorie deficient diet 192
Brain
-, aging 226

-, amino acid pool 224
-, concentration of opiates 389
-, development 219,220,236
- -, ganglioside concentration 193
- -, sensory deprivation 199-215
-, effect of undernutrition 224
-, impairment, neuro-
  chemical study 236
-, lipids, ethanolamine
  incorporation 619
-, metabolism
- -, ethanol 335
- -, phosphoethanolamine 234
-, prostaglandin 596
-, protein metabolism,
  phenylketonuria 590
- -, Huntington's disease 608
-, regional patterns of
  glycolytic enzymes 219
-, regulation of GABA levels 225
-, specific protein 592,595
- - -, α-Albumin 617
- - -, see also S-100
-, stem, precerebellar nuclei
  and neurogenesis 152
-, studies of glycosidases 220
-, subcellular fractions 234
Bulk isolation, neurons 557

C

C_{18} elongase, in mutant
  mice 55
Calcium
-, binding, alteration by
  ethanol 350
-, metabolism 582
-, nerve activity 595
-, release from platelets 439
-, synaptosomes 582
-, vesicular release 579
Camera lucida drawing
-, glial cells 165
-, spongioblasts 165
Cation transport, relation
  to central depressants 323

Catecholamines
-, assay of 527,531
-, effects on ATPase 541
-, effect of gamma-butyrol-
   acetone 366
- -, methamphetamine 442
-, interaction with ATP 538
-, membrane transport 536
-, metabolism, in genetic
   disorders 535
- -, effect of ischemia/
   hypoxia 262,266,310
-, opiate influence 426
-, see also specific compunds
Cell
-, compatibility 184
-, cultures
- -, as model systems 489,492
- -, development
- - -, biochemical 502,503,504
   508,509,514,518,520,522
- - -, morphological 502,503
   504
- -, glial
   see glial cell cultures
- -, glucose metabolism 497
- -, histochemistry 501,502
   506,517
- -, membrane transport
- - -, of amino acids 498,499
   510,511,513,515
- - -, of glucose 497
- - -, of lithium 500
- - -, of sodium 494
- -, neuronal
   see neuronal cell cultures
-, cycles, in nervous system 158
-, death and nervous system 158
-, differentiation 606
-, division, visual deprivation 211,212
-, migration
- -, neocortex 158,173
- -, nervous system 158
- -, synaptogenesis 158,173

-, number, based on DNA
   estimation 189
-, recognition 144
-, surface specifity, glycoproteins and glycolipids 182
Central depressants
-, biochemical aspect of
   tolerance 317
-, cyclic nucleotides 322
-, definition 317
-, effect on
- -, cation transport 323
- -, membrane structure 323
- -, protein synthesis 324
-, neurochemical mechanisms
   of tolerance 321
-, neurotransmitter turnover 321
-, physical dependence 317
-, tolerance 318
Central nervous system
-, cultures, bioelectric
   activity 111
-, myelin-cerebroside organisation 115
-, myelin isolation 32
-, myelin proteins 32
-, tolerance 320
Centrifugation techniques,
CNS myelin 32
Ceramide galactosyltransferase 579
-, in cultured glia 520
Cerebellum
-, development 216,235
-, glia cells 216,217
-, hypothyroidism 218
-, migration of neuroblasts 163
-, muscarinic receptor 235
-, neurogenesis 152
-, ontogenesis 218
Cerebellar granule cell,
migration of 175
Cerebral circulation
- -, experimental stroke 244,245
- -, pathology 241
-, cortex
- -, glutamate binding 222

- -, kainic acid binding 222
-, energy state, hypoxia 254
  255,257
-, insult, symptoms 242
-, oxygen consumption, catecholamines 260,266
-, swelling, clinical picture 242
Cerebroside
-, immunohistochemical localization 129
-, molecular disposition in CNS myelin 115
-, sulfotransferase 574
- -, in cultured glia 519
-, sulphate sulphatase
- -, in leukocytes 570
- -, in fibroblast 570
-, EAE 104
-, effect of malnutrition 195
-, estimation of myelin 184
Cerebrospinal fluid
-, α-Albumin 617
-, fatty acids 615
-, levels of cholesterol ester hydrolase 123
-, lysosome concentration in MS 95
-, pressure
- -, effect of cortisol 227
- -, under development 227
Chelate probe 582
Chemioaffinity theory 178
p-Chlorophenylalanine 589
-, effect on 5-HT depletion 326
-, influence on ethanol and barbiturates 326
Chlorotetracycline 582
Cholesterol
-, in Jimpy mouse 56
-, ester
- -, in Wallerian degeneration 72
- -, hydrolase in CSF 123
- - -, of MS patients 123
Choline acetyl transferase
-, from placenta 558

-, in snail neurones 560
-, isoelectric focusing 558
-, phosphotransferase, regulation by CMP 309,542
Cholinergic neurons
- -, citrate lyase 597
-, system, effect of light 566
Cholinesterases
-, activity of 231
-, in diapragm 231
-, location of 231
Chondroitin sulfate 624
Choroidal Na$^+$K$^+$-ATPase
-, effect of cortisol 227
-, under development 227
Citrate
-, influx 583
-, lyase 597
-, synaptosomes 583
Cobalt staining of spinal motoneurons 153
Compact myelin,'pure myelin' 3
Compaction of myelin-layers, defect in neurological mutants 50
Compartmentation 581
Compound eyes, retinotectal connections 146
Compressed projection and retino-tectal connections 147,148
Concentric sclerosis, demylination 78
Contacts on undifferentiated neuroblasts 176
Cortex
-, ganglioside concentration 193
-, form-function relationship 153
-, visual of rat 166
Cortical electrical activity, experimental stroke 245
-, period of visual deprivation 202
Corticosterone
-, development 228

-, plasma concentration 228
-, phenobarbital 228
Cortico-striatal path, acetylcholine 568
Cortisol, effect on
-, choroidal Na^+K^+-ATPase 227
-, CSF pressure 227
Cyanide encephalopathy, myelin proteins and lipids 134
Cyclic AMP 623
-, hypoxia/ischemia 277,307
-, prostaglandins 274,275,278
-, GMP, ischemia 278,623
2',3'-cyclic nucleotide 3'-phosphohydrolase (CNP)
-, in cultured cells 121
- -, myelin 66,121
- -, optic nerve 120
- -, PNS myelin subfractions 127
- -, white matter 121
- -, zonal fractions 41
-, Km 121,122
-, molecular weight 121,122
-, purification from white matter 122
Cyclic nucleotides 623
-, in glial cell cultures 496
- -, neuronal cell cultures 495
-, effect of central depressants 322
Cysteine sulfinate transaminase, purification 569
Cytodifferentiation, chemioaffinity theory 178
Cytoplasmic incisures 7
-, protein in myelin 13
-, reticulum 10

D
Dansyl chloride 556
-, GABA 598
Demyelinating
-, activity, comparison between MS and EAE 108
-, capacity, absorption with myelin 110

-, diseases, other than MS 78
-, factor in MS induction 111
Demyelination
-, activity in IgG2 of serum 100,101
-, allergic, CNS antigens 78
-, attack by macrophages 79
-, causes 78
-, in EAE, mononuclear cells 94
-, of myelinated cultures 100
-, possible mechanisms 79
-, spinal cord 80
-, virus-induced 80
-, with cerebrosides, cerebellar cultures 104
Denervation, polypeptide pattern 603
Densities, postsynaptic before fiber arrival 176
Density of myelin, zonal-centrifugation 127,128
Dependence
-, effect of narcotics 409
-, occurence with tolerance in one cell 376
-, opioid sensitive neurones 375
-, relationship to tolerance 326
Depressants, 317
-, see also central depressants
Deprivation
-, monocular and visual cortex 200,201
-, reversibility of changes 201
-, sensory,
- -, brain development 199,215
- -, electrophysiological aspects 199
- -, functional aspects 199
- -, morphological aspects 199
- -, visual system 199
-, visual cell division 211,212
- -, critical period 202
- -, effect on adenylate cyclase 208,209
- - -, transmitterspecific enzymes 204,208

– –, glia proliferation 211
– –, macromolecular synthesis 209,211
– –, protein-glycoprotein synthesis 209,210
– –, RNA synthesis 209
– –, tubulin synthesis 211
– –, uptake of biogenic amines 206,207
Deposition of postmitotic neurons 163
Determination
–, amino acids 556
–, protein 555
Development
–, acetylcholine content 230
–, acetylcholinesterase activity 229,230
–, activity of
– –, acid phosphatase 221
– –, cholinesterase 231
–, anterior horn cells 221
–, availability of amino acids 223
–, cell separation 223
–, cerebellum 216,235
–, cerebrospinal fluid pressure 227
–, diaphragm 231
–, effect of
– –, corticosterone 228
– –, ethanol 444
– –, experimental conditions 221
– –, phenobarbital 228
– –, glycolytic enzymes 219
–, GABA level 225
–, ganglioside concentration 193
–, glia cells 216
–, glutamate binding 222
–, kainic acid binding 222
–, lipid-protein interactions 229
–, localization of cholinesterases 231
–, muscarinic receptor 235
–, myelin 124

–, Na^+K^+-ATPase 227
– –, activity 229
–, neuronal membranes 229
– –, network 158,159
–, neurochemical study 236
–, olfactory deprivation 230
–, optic nerve 120
–, order of neural connections 143,157
–, sensory deprivation 199,215
–, studies of glycosidases 220
–, tolerance 325
Diacylglycerols, effects of ischemia 286,291
Diaphragm
–, activity of cholinesterase 231
–, localization of cholinesterases 231
Diet, protein-calorie deficient 192
Differentiation of
–, glia cells 160
–, neurons 160
Dihydroxyphenylalanine metabolism, hypoxia 266
see also DOPA
N,N-Dimethylaminoethanol acetate 601
Disease and brain development 236
Displacement
–, mechanism of 163
– –, postmitotic neurons 160,163
DOPA, binding to tRNA 537
Dopamine
–, biosynthesis 533
–, effect of halucinogens on turnover 441
–, metabolism 546
–, release 546
–, sensitive mechanism, effect of ethanol 366
–, ß-hydroxylase, hypoxia 267
Dorsal horn, form-function relationship 153

Drugs
-, action of psychotropic drugs on calcium release 439
-, antiparkinson 620
-, dependence, definition 374
- -, producing
- - -, effects on neuronal activity 368
- - - -, neurotransmitters 359
-, effect of ethanol 340
- -, on neuronal activity 359
- - -, transmitter depletion 325
-, influence on monoamine neurons 359
Dysmyelination, neurological mutants 49,130

E
EAE
-, acquired tolerance 85
-, acute form 84
-, and MS, comparison 85,94
-, chronic form 84
-, controll by S-cells 85
-, humoral factor 94
-, immunology 84
-, induction 84,101
-, "model" for multiple sclerosis 84
-, perivascular inflammation 93
-, relapsing form 84
-, T-cell mediated condition 94
-, transfer with T-cells 85
-, treatment with MBP on synthetic polymer 84
EEG
-, experimental hypoxia 253,255
-, experimental stroke 246
Electric organ 595
Electrical stimulation, taurine release 587
Electrolytic destruction of septum 431
Electrophysiological aspects of sensory deprivation 199

β-Endorphin, relationship with β-lipotropin 397
Endorphins
-, and behaviour 455,456
-, in opiate addiction 395
Energy
-, charge, in hypoxia 254
-, metabolism, insect brains 616
-, utilization during hypoxia 253, 306,310
Enkephalins
-, biosynthesis 453,460
-, effects on adenyl cyclase 459
-, morphine effects 395
-, purification 457
-, purity of commercial material 458
-, release 452
Ethanol
-, effect of p-chlorophenyl-alanine 326
- -, ingestion 335
- -, active transport 348
- -, development 444
- -, dopamine-sensitive mechanisms 366
- -, membranes 339
- -, Na^+K^+-ATPase 349
- -, neurotransmitters 336
- -, sodium influx 347
-, induction by enzyme 334
-, influence on phospholipids, Ca^{2+} binding and membrane composition 350
- -, sleep 339
-, interaction with drugs 340
-, metabolic inhibitor 445
-, metabolism in
- -, brain and other organs 335
- -, liver 332
-, pathways of oxidation 333
-, potassium ethylxanthogenate 445
-, relationship to opiate dependence 352

-, role of isoquinoline
   alkaloids   351
-, see also alcohol
-, tolerance   318
-, transmitter metabolism   362,364
Ethanolamine   619
Excitatory messengers, re-
ceptors   379
Experimentel stroke
-, animal responses   244
-, attempts at treatment   250
-, blood flow   244
-, effect of free radicals   249
-, methods   243
-, structural damage   246
Explant cultures, myelin
proteins   139

F
Fatty acids
-, activation by thyroid
   hormone   233
- -, during myelination   233
-, effect of ischemia   286,290
                                       297
-, in spinal fluid   615
Features common to CNS and PNS
myelin   10
Focal ischemia, consequences   243
Forebrain, development and
malnutrition   192,195
Form-function relationship
-, in cortex   153
-, in dorsal horn   153
-, in retina   153
-, neuronal morphology   153
Free postsynaptic
thickenings   167,169
Freeze-cleavage, hydrophobic
domain   5
Freeze-fracture, myelin   5
Functional aspects of sensory
deprivation   199
Fusion, major dense line   21

G
GABA
-, accumulating cells, distribution of   167
-, binding   510,550,551
-, dansyl   598
-, effect of benzene   561
-, level in developing brain   225
-, regulation of level   225
-, metabolism   512,540,549
-, role in tolerance   414
-, pools and synaptogenetic
   activity   167
-, production and non-pyramidal
   cells   166
-, propylacetate effect   625
-, transaminase   625
-, transport   485,510,511
                                       551
-, uptake and glial cells   166
GABA-ergic neurones   166
Galactosyltransferase, and
ceramide   574
Gamma-butyrolactone
-, influence on catecholamine
   synthesis   366
- -, tyrosine hydroxylase   367
Ganglion cells,
-, in cat retina   201
-, superior cervical and free
   synaptic thickenings   167
-, superior cervical and
   GABA application   167
Ganglionic transmission   623
Gangliosides
-, chemical structure of   189
-, concentration   190
- -, at birth   193
- -, cerebral cortex   193
- -, during postnatal
      development   193
-, development in cell cultu-
   res   508,521,522
-, effect of malnutrition   195-197
-, effect on membranes   614
-, GM_1 613

–, in brain areas	612
– –, development, effect of malnutrition	189-198
– –, glial cell fraction	190-191
– –, synaptosomal plasma membrane	190-191
– –, unmyelinated white matter	190
–, marker for neuronal membranes	189
–, microscopic visualization	191
–, topographic localization	190-191
Glial cell cultures	492
–, amino acid metabolism	512, 515, 516
–, cerebroside-sulfotransferase	519
–, cyclic nucleotide levels	496
–, membrane potential	493
Glia cells	
–, Bergmann glia	163
–, biochemical studies	216, 217
–, development	216
–, differentiation	160
–, early damage during MS	93
–, GABA uptake	166
–, ganglioside concentration	190, 191
–, immunohistochemical studies	216, 217
–, in adult animals	217
–, in rat cerebellum	216, 217
–, long radial	163, 165
–, marker	217
–, measles-like inclusions in MS	93
–, short radial	163, 164
Glial inhibition, in cultures	101
Glial processes and migrating neuroblasts	163
Glia proliferation, visual deprivation	211
Glucosamine	624
Glucose	
–, in cell cultures	497
–, metabolism in hypoxia	305, 306
–, uptake into brain cells	126
Glucosidase	559
Glutamate	
–, binding to synaptic membranes	222
– –, and developing cortex	222
– – –, kainic acid	222
–, synthesis	577
–, decarboxylase	
– –, immunohistochemical demonstration	166
– –, synaptic function	563
Glutamine, penetration	581
Glycans and glycoproteins	181, 182
Glycoconjugates in aging brain	226
Glycolipids	182, 183
Glycolytic enzymes	
–, development	219
–, in rat brain	219
–, regional patterns in brain	219
Glycoproteins	
–, antibodies	607
–, coding capacity	181, 182
–, developing nerve cells	183
–, extracellular space	5, 14, 15
–, glycan moiety	181
–, recognition molecules	181
–, role of glucosidase	559
–, transient Con A binding	183, 184
–, synaptogenesis	585
–, synaptosomal	607
–, synthesis and visual deprivation	209, 210
Glycosidases	
–, during development	220
–, in human brain	220
Glycosyltransferases, specific	182
Grafting experiments	143, 145
–, of Miner	143
– –, Sperry	143
– –, Weiss	143
–, neuronal specificity theory	145

Granule cells
-, cerebellar migration of 175
-, neurogenesis in cerebellum 152
Growth, cone 174
-, of body protein deficient diet 192
- -, forebrain, protein deficient diet 192
Guanylate cyclase
-, properties 575
-, purification 575

H

Halucinogens, effect on dopamine turnover 441
Hebbian coincidence, principle and neuronal specificity 149
Heparan sulfate 624
Hexokinase, mitochondrial 572
Histamine
-, agonists 623
-, antagonists 623
Histones 594
Histotoxic anoxia 240
Hope's arrow model 151
Hormone, thyroid
-, effect on fatty acid activation 233
- -, myelination 233
5-HT see serotonin and 5-Hydroxytryptamine
5-HTP see 5-Hydroxytryptophan
Human cerebral cortex and ganglioside concentration 193
Huntington's disease 608
Hyaluronate 624
Hydrophobic protein, DM-20 35
6-Hydroxydopamine and morphine receptors 534
5-Hydroxytryptamine
-, binding to synaptic membranes 441
-, depletion by p-chlorophenylalanine 326
-, effect of morphine 428
-, in spastic rats 543
-, metabolism, hypoxia 267
5-Hydroxytryptophan metabolism, hypoxia 266
Hyperphenylalaninemia 589,590
Hypertension, model 605
Hypoglycemia, anoxia 240
Hypo/hypercapnia, aromatic amino acid hydroxylases 266
Hypomyelination, dysmelination 130
Hypothalamic releasing hormones, effects on dopamine release 546
Hypothalamus, oestrogen receptor 622
Hypothyroidism
-, effect on cerebellum 218
- -, ontogenesis 218
Hypoxia
-, acetate metabolism 305
-, amino acid metabolism 260,268
-, arachidonic acid metabolism 271,277
-, arterial oxygen content 253
-, carbohydrate metabolism 257,260 305
-, catecholamines 266
-, cerebral blood flow 253,261 262
- -, energy state 254,255,257
-, citric acid cycle metabolism 259
-, clinical features 241
- -, management following insult 249
-, effect of catecholamine neuronal activity 268
-, effect on lipid synthesis 286,304
-, enzymes requiring molecular oxygen 266
-, 5-hydroxytryptamine 267
-, induced behavioral changes 268
-, see also anoxia and ischemia

I

IEF see isoelectric focusing
Imino acid 618

Immunity
-, cellular in demyelinating
 diseases 78,80
-, humoral in demyelinating
 diseases 78,83
Immunohistochemical demonstration of glutamate decarboxylase 166
-, localization of cerebroside in normal and Quaking mouse 129
-, studies of glial cells 216,217
Impairement of brain by two diseases 236
Indoleamines
-, assay of 527,532
-, biosynthesis 547
-, see also specific compounds
Ingestion of ethanol 335
Inhibition of sulfatide synthesis in cultures 101
Inner glial loop, growth cone 29
Insect brain, energy metabolism 616
Inside-out layering
-, in cortex and autoradiography 160,161
-, migrating neurons in cortex 160,161
-, mode of deposition 163
-, non-pyramidal neurons 161
-, pyramidal neurons 161
Intraperiod lines, separation 10
Ischemia
-, animal models 285,287
-, calcium accumulation by mitochondria 298
-, changes in free fatty acids 290,297
-, cholinergic neurons 247
-, cyclic nucleotides 277
-, diacylglycerols 286
-, effects on acyl-phopholipids (phosphoglycerides) 292
- -, triglycerides 292
-, GABA uptake 247
-, incorporation of radiolabelled choline 296
- - -, fatty acids 294,304
-, lipid catabolism 286,293,299
-, oedema 248
-, phospholipid metabolism 285 295,297
-, potassium fluxes 247
-, prostaglandins 275
-, release of adenosine 278
-, see anoxia and hypoxia
Ischemic anoxia 240,275
- -, free fatty acids and lipids 275
- -, prostaglandin release 276
-, 'penumbra' 246
Isoelectric focusing, choline acetyl transerase 558
Isoenzymes
-, atypical form of alcohol dehydrogenase 333
-, N-acetyl-ß-0-hexosaminidase 562
Isoquinoline, alkaloid influence by ethanol 351

J
Jimpy mouse
-, low sulfotransferase activity 53
-, neurological mutant 49
-, protein composition of PNS myelin 130

K
Kainic acid binding
-, developing cortex 222
-, glutamate 222
-, to synaptic membranes 222

L
Lactate dehydrogenase
-, anaerobic glycolysis 573
-, membrane bound 573
Labelling
-, of myelin basic protein 116
- -, membrane faces 115

Lactate/pyruvate ratio, hypoxia	254
LDH see lactate dehydrogenase	
Learning, protein synthesis	591
Lectin-like molecules	178
L-leucine, glutamate synthesis	577
Light	
-, cortical effects	566
-, effect on acetylcholinesterase	565
- -, cholinergic system	566
Lipid-protein interactions	
-, estimation of acetylcholinesterase	229
- -, Na^+K^+-ATPase	229
-, in neuronal membranes	229
-, studies using Arrhenius plots	229
Lipid structure, ischemic anoxia	308
Lipids	
-, effect of morphine	416
-, in different age groups	232
- -, nerve biopsies	232
-, methods of analyses	287,288 289
-, micro-analysis	232
-, role in tolerance	414
-, synthesis from various substrates	304
Lipoproteins, triglycerides	621
ß-lipoprotein, relationship with ß-endorphin and methionine-enkephalin	397
Lithium	
-, toxicity	500
-, transport in cell cultures	500
Liver, ethanol metabolism	332
Localization of cholinesterases	231
Long chain fatty acids, Jimpy mouse	55
LSD, binding to synaptic membranes	441
Lysoglycerophospholipids, acylation during hypoxia	314

Lysophosphatides, action on myelin	73
Lysosomal hydrolases in multiple sclerosis	132
Lysosomes, myelin degradative enzymes	67

M

Macrophages, role in demyelination	70
Major dense line	7,10,21
-, myelin basic protein	21
Malate, influence of morphine	443
v.d. Malsburg model of striate cortex	150
Malnutrition	
-, effect on body growth	195
- -, cerebrosides	195
- -, forebrain development	195
- -, ganliosides	195,196,197
- -, ganglioside development	189,198
-, S-100 protein	593
Marginal belt	7
Marker	
-, for glial cells	217
-, of neuronal membranes, gangliosides	189
Measles antibody, visual infection and demyelination	79
Mechanisms of morphine actions	404
Mechanical stress in PNS myelin	12
Medulla of fish embryos, transplanted	175
Membrane	
-, composition, effect of ethanol	350
-, effect of alcohol	339,346
- -, ethanol	339,346
- -, morphine on constituents	404
-, expansion theory in anaesthesia	347
-, glycoproteins	585
- -, in synaptogenesis	174
-, potential of cultured glia	493

647

-, structure, effect of central depressants 323
-, transport system and amino acid uptake 223
- -, in separated cells 223
Memory, modulation of neuropeptides 453,455,462
Metabolism
-, of alcohol 332
-, pipecolic acid 618
Metachromatic leukodystrophy 570
Methamphetamine, catecholamine metabolism 442
Methionine-enkephalin, relationship with β-lipotropin and β-endorphin 397
α-Methylphenylalanine 589
Microanalysis of lipids 232
Microglial cells, reactive 606
Microsomes, subfractionation 574
Migration of
-, cerebellar granule cells 175
-, neuroblasts in cerebellum 163
- -, in neocortex 163
Migrating neuroblasts and glial processes 163
-, neurons, ^{3}H-Thymidine labelling 160,161
Mitochondria
-, calcium accumulation 313
-, calcium release 595
-, hexokinase 572
-, non-synaptosomal 589
-, synaptosomal 583
Mode of deposition of postmitotic neurons 163
Modules, structural and neural organization 150
Molecular disposition of myelin proteins 42
-, mechanism of opioid action 377
Monkey mechanisms, adhesion of myelin lamellae 22
Monoamine neurons, effect of drugs 359

Monoamine oxydase 544,545
-, hypoxia 267
-, inhibitors, effects on catecholamine uptake 536
Monoamines
-, and motor control 539
-, membrane transport 548
-, inhibition by tetrahydronorharmane 548
-, see also specific compounds
Morphine
-, actions 404
-, brain malate response 443
-, effect on acetylcholine 429
- -, on enkephalins 395
- -, 5-hydroxytryptamine 428
- -, lipids 416
- -, membrane constituents 404
- -, protein synthesis 410
- -, specific proteins 412
-, influence on phospholipid metabolism 416
Morphological aspects of sensory deprivation 199
Morphology
-, neuronal and form-function relationship 153
-, neuronal specificity 153,154
Motoneurons
-, spinal, cobalt staining 153
- -, different types 153
Motor control, effect of nonoamines 539
Multiple sclerosis
-, active plaques 93
-, agent aquired during childhood 91
-, antigen on human leucocyte (HLA) 91
-, antimyelin factors 107
-, auto-immun process 78
-, changes in lymphocyte activity 95
-, cholesterol ester hydrolase in CSF 123
-, demyelinating disease 78

-, EAE as a model 78
-, genetic predisposition plus infective agent 92
-, HLA markers 82
-, humoral immunity 83
-, IgG in the CSF 92
-, immune mediated pathogenesis 80
-, immunology 91
-, inherited factor 91
- -, immunity defect 91
-, initiation by viral infection 78
-, leucocyte sensitivity 83
-, localization of proteinase activities 95
-, loss of myelin constituents 93
-, lymphocyte response to mitogen 81
-, lysosomal enzyme changes in glial cells 93
- -, hydrolases (in lymphocytes and granulocytes) 132
-, macrophage electrophoretic mobility test 82
-, oligoclonal bands of antibody 92
-, oligodendrocytes 107
-, paramyxovirus as causative agent 93
-, polyunsaturated fatty acids 93
-, possible therapies 97
-, population differences in B-lymphocytes 91
-, prealbumin in CSF 133
-, proteinase activities in leucocytes 96
-, rabbit eye, demyelinating model 83
-, radioimmunoassay to myelin basic protein 83
-, rates in different populations 91
-, S-cells 82
-, serum demyelinating activity, absorption on myelin, absorption on oligodendrocytes 107

-, steroid treatment 81
-, T and B lymphocytes 80
-, T-G cells and IgG immune complex 81
Mucolipidosis 610
Muscarinic ligands
-, binding in cell cultures 508
-, effect on cyclic nucleotides 495
-, receptor
- -, cerebellum 235
- -, development 235
Myelin
-, amount and cerebrosides 189
-, associated glycolipids, in C_6 and brain cells 140
- -, glycoprotein
- - -, changes during development 39
- - -, glycopeptides from 39
- - -, in neurological mutants 39
- - -, labelling with fucose, glucosamine, N-acetylmannosamine, ^{35}S-sulfate 39
- - -, molecular weight 39
- - -, purification 34,40
-, axo-glial junction 10
-, breakdown, cooperativity of proteinases and phosphohydrolases 76
-, basic protein
- -, autolytic degradation 37
- - -, proteolysis 66
- -, complexes with anionic lipids 65
- -, encephalitogenicity 64
- -, heterogeneity 36,38
- -, labelling 116
- -, location on cytoplasmic side 65
- -, lymphocyte sensitisation, autoimmune response 93
- -, microheterogeneity 37
- -, molecular weight 33,36,38
- -, selective loss in early MS 93
- -, sequence 37

649

-, catabolism	64
-, cholesterol : phospholipid : galactolipid ratio	64
-, composition during ontogeny	56
- -, in vitamin B_{12} deficiency	136
-, continuity with glial plasma membrane	65
-, cytoplasmic incisures, dynamic aspects	25
-, deficient mutants	49
-, degradative enzymes, transport	67
-, development	124
-, ganglioside concentration	190,191
-, hydrophobic proteins, aggregates	35
-, in experimental cyanide encephalopathy	134
-, intramembrane particles, distribution	10,14
-, like	3
-, like, SN 4	41
-, lipid content	48
-, lipid/protein ratio	64
-, lipids	
- -, biosynthesis	48,50
- -, catabolism	70
- -, ceramide	52
- -, cholesterol	54
- -, classes	48
- -, criteria for involvement in myelination	55
- -, deposition	48,56
- -, fatty acids	50
- -, galactocerebrosides	52
- -, gangliosides	53
- -, phospholipids	54
- -, regulation of biosynthesis	54
- -, sequence of addition	57
- -, slow and fast turnover	70
- -, sulfatides	53
- -, synthesis	126
- - -, genetic control	49

-, membrane domains	18
- -, inhomogeneity	18
-, module	25
-, optic nerve	118,119
-, phospholipase A_2 action	72
-, phospholipids, turnover	70
- -, turnover and exchange proteins	138
-, preparations, cytoplasmic components	28
-, presence of cytoplasmic spaces with lysosomal properties	75
- -, hydrolyzing enzymes	56
-, proteins	
- -, amino acid composition	36
- -, catabolism	65
- -, charge ratio	33
- -, classification	33,34
- -, content and density during development	124
- -, differential extraction	32
- -, half-lives	65,66
- -, heterogeneity, lipid/ protein ratio	40
- -, high molecular weight	34
- -, in explant cultures	139
- -, isolation	32
- -, localization on external side	75
- -, low molecular weight	34
- -, molecular weights	33
- -, phenylketonuria	589
- -, proportions	33
- -, SDS-PAGE	34,35
- -, synthesis	126
- - -, neuronal contribution	118
- - -, triethyllead intoxication	135
-, proteolipids, molecular weight	117
-, related membranes	
- -, contamination with non-myelin membranes	41
- -, SN 4	41
-, segments per glial cell	12
-, structure, CNS and PNS	3

-, synthetising enzymes, location in microsomes 55,56
-, tight junctions 11
Myelination
-, fatty acid activation 233
-, inhibition, myelin proteolipid protein 105
- -, of unmyelinated cultures 100
- -, with anti-CNS antisera from subhuman primates 103
- - -, gangliosides 104
- - -, purified oligodendrocytes 104
- - -, rat SBP 105
- - -, synthetic galactocerebrosides 104
-, metabolism of phosphoethanolamines 234
-, subcellular fractions 234
-, thyroid hormone 233
Myelinogenesis, contribution of glia vs. neuron 125

N

N-acetyl-ß-D-hexoaminidase
-, isoenzymes 562
-, tissue differences 562
Na^+K^+-ATPase
-, during development 229
-, effect of cortisol 227
- -, ethanol 349
-, lipid-protein interactions 229
-, neuronal membranes 229
NADH/NAD ratios see Redox changes
Naloxone, effect on acetylcholine 430
Narcotic
-, receptor 404
-, tolerance and dependence 409
Neocortex
-, cell migration 158
-, migration of neuroblasts 163
-, synaptogenesis 158
Neonatal oxygen deprivation, persistent effects 269
-, see also hypoxia

Nerve activity, calcium release 595
-, biopsy and lipid microanalysis 232
Nervous system
-, cell death 158
-, cell migration 158
-, number of cell cycles 158
Neural connections, development of order 143
Neurite growth and synapse formation 174
Neuroblasts
-, contacting 176
-, migration and synapse formation 174
- -, in cerebellum 163
- -, in neocortex 163
Neurochemical study of brain development 236
Neurogenesis
-, basic stem precerebellar nuclei 152
-, in cerebellum 152
-, sequential patterns 152,153
Neuroleptics, receptor 317
Neurological mutants
-, alkanes in Quaking 137
-, effect on myelination in CNS vs. PNS 49
-, protein composition of PNS myelin 130
-, Shiverer mouse dysmyelination 131
Neuromyelitis optica, demyelination 78
Neuronal
-, activity, influence by dependence-producing drugs 359,368
-, cell cultures 489
- -, cyclic nucleotide levels 495
- -, ganglioside content 521
- -, membrane transport of choline 523
- - -, sodium 494
-, cell surface 176
- -, and synapse formation 176

-, differentiation, synapse specification 176
-, interconnections 144
-, membranes, acetylcholinesterase activity 229
- -, development 229
- -, ganliosides as marker 189
- -, lipid-protein interactions 229
- -, and Na$^+$K$^+$-ATPase activity 229
-, morphology, form-function relationship 153
- -, specificity 144,149 153,154
-, network, development 158,159
-, organization, structural modules 150
-, specificity, eye rotation 144
- -, Hebbian coincidence principle 149
- -, skin rotation 144
- -, theory 143,145,149
- - -, grafting experiments 145,149
Neurons
-, bulk isolation 557
-, choline acetyl transferase 560
-, developing, change of glycoproteins 183
-, differentiation 160
-, GABA-ergic 166
-, in cell culture 489
-, inside-out layering in cortex 160,161
-, non-pyramidal, characterization 164
- -, inside-out layering 161
-, migrating, temporospatial distribution 160
-, opioid sensitive 375
-, postmitotic displacement mechanism 160
-, primitive pyramidal 164
-, pyramidal and non-pyramidal 162
-, snail 560

-, stearic acid uptake 609
Neuron-glia cooperation in myelin synthesis 125
Neuropeptides
-, and behaviour 453,455
- -, membrane protein phosphorylation 463
- -, memory 455,462
-, antisera to 449
-, axonal transport 451
-, formation and degradation 453 460,461
-, localization of neurons 449,450
-, receptors 452
-, release 451,452,465
-, see also specific compounds
Neurotransmitter
-, effect of ethanol 336
-, GABA 598
-, influence by dependence-producing drugs 359
-, opiate addiction 425
-, release 580
-, taurine 586,587
-, turnover, effect of central depressants 321
Neurotoxins
-, interaction with sodium ionophore 494
-, sea anemone 467
Neutral proteinase, myelin constituent 66
Nodal microville 6
Non-myelin contaminants 3
Non-pyramidal cells
-, and GABA accumulation 166
- -, production 166
-, neurons
- -, characterization 164
- -, diagram of position 162
- -, inside-out layering 161
Noradrenaline 605
-, development of tolerance 326

O

Octopamine, biosynthesis 533
Octopus, brain proteins 594
Oedema, ischemia 248
Oestradiol 622
Oestrogen receptor 622
Old dog distemper, relation to MS 79
Olfactory bulb
-, acetylcholinesterase activity 230
-, content of acetylcholine 230
-, during development 230
Olfactory deprivation
-, development 230
-, effect on acetylcholine content 230
- -, acetylcholinesterase activity 230
Oligodendrocytes
-, enzymic machinery for myelination 55
-, impairment in neurological mutants 50
-, lipid deficiency 50
Ontogenesis
-, cerebellum 218
-, effect of hypothyroidism 218
Opiate
-, addiction, influence of neurotransmitters 425
- -, involvement of endorphins 395
-, brain concentration 389
-, definition 374
-, dependence, relationship to ethanol dependence 352
-, effect on adenylate cyclase 380
-, influence on catecholamines 426
-, isolation of opiate receptor 409
-, receptors 317
- -, binding studies 387
- -, composition 407
- -, distribution in CNS 405
- -, in addiction 386
Opioid
-, action, molecular mechanism 377
-, definition 374
-, dependence
- -, biochemical theories 374
- -, protein kinase 381
-, receptor 378
-, sensitive neurons 375
-, tolerance 381
Optic fibers, tectal connections 146
-, nerve
- -, biochemical heterogeneity 119
- -, CNP 120
- -, ontogenetic development 120
- -, organisation of axons, myelin, glia 119
Organotypic cultures, MS serum 83
Oxidation
-, of acetaldehyde 334
- -, alcohol 332
-, various pathways for ethanol 333
2-Oxo-1-pyrrolidine acetamide 599

P

P_2-agent 601
P and E faces 5
Pargyline 605
Parkinson, drugs 620
Particulate components, freeze fracture 4
Peptides
-, as neurotransmitters
-, influence on tolerance 327
-, sea anemone toxins 467
-, see also neuropeptides
Permeability, capillary 588
Phenobarbital
-, development 228
-, effect on corticosterone 228
Phentolamine 605

Phenylketonuria 589,590
-, myelin proteins 589
-, brain proteins 590
Phosphatidylcholine
-, biosynthesis 576
-, subcellular distribution 576
Phosphocreatine, hypoxia 254
Phosphodiesterase, quasi-morphine syndrome 380
Phosphoethanolamines
-, in brain 234
-, in myelinating brain 234
-, in subcellular fractions 234
Phosphoglycerides, effect of ischemia 292,295,297,311
Phospholipidase
-, A_2, in MS 75
-, A and A_2, in neuron- and glia-enriched fractions 73
-, cellular and subcellular localization 73
-, cyclic AMP effects 278
-, effect of ischemia 286,296 297,299,312
Phospholipids
-, changes during ischemia 295,297
-, effect of ethanol 350
- -, morphine 416
-, exchange proteins, role in myelin formation 138
-, metabolism, in EAE 75
-, role during hypoxia 275,285 296,313
Physical dependence, relationship to tolerance 319
Pipecolic acid 618
PKU, see Phenylketonuria
Plasma
-, amino acids 224,588
-, concentration of corticosterone 228
-, effect of undernutrition 224
-, membrane, presynaptic 580
Plasmalemmal vesicles (caveolae) in PNS myelin 13

Plasticity and retino-tectal connections 148
Platelets, calcium release 439
PNS myelin
-, protein distribution 127
-, subfractions by zonal centrifugation 127
Polyamines 604
Polylysine, use in cell culture 502
Polypeptide pattern, after denervation 603
Polyphosphoinositides, linkage to myelin basic protein 68
Porto-caval shunt 588
Postsynaptic densities, before fiber arrival 176
Potassium ethylxanthogenate 445
- -, ethanol metabolism inhibitor 445
Prealbumin, in CSF of MS patients 133
Precursor-product
-, density of myelin fractions 56
-, microsomes-myelin 56
Pre-myelin membrane, in Quaking mouse 57
Presynaptic cell membranes 579
Progressive multifocal leukoencephlopathy, demyelination 79
Propylacetate 625
Prostaglandins
-, biological role 273
-, brain levels 273
-, during development 274
-, elevation by trauma and drugs 273
-, endogenous 596
-, metabolism during ischemia 275,277
-, relationship to cyclic AMP 275,278
-, synthesis and metabolism 271-273
-, vasodilation 276
Prostaglycin 279
Protein
-, 14-3-2 606

-, CNS myelin 32
-, determination 555
-, glycoprotein synthesis and visual deprivation 209,210
-, kinase 571
- -, opioid dependence 381
-, metabolism, effect of pyrrolidine acetamide 599
-, of PNS myelin of Quaking and Jimpy 130
-, phosphorylation, effects of neuropeptides 463
-, proportions, in myelin-related membranes 41
-, S-100 592,593,606
- -, see also S-100 protein
-, synthesis, effect of central depressants 324
- -, effect of morphine 410,412

Proteolipid protein 37
-, disposition in membrane 65
-, partical sequence 35
Psychotropic drugs, action on platelets 439
Purification, CNP 122
Putrescine 604
Pyramidal neurons
-, diagram of position 162
-, inside-out layering 161
-, primitive 164
Pyridoxal kinase 598
-, inhibitors 598
Pyrrolidine acetamide 599

Q
Quaking mouse
-, dense fraction 57
-, myelin, alkane synthesis 137
- -, zonal centrifugation 128
-, neurological mutant 49
-, protein composition of PNS myelin 130
Quasi-morphine syndrome, phosphodiesterase induced 380

R
Radial component in myelin 13,16
Ranvier nodes in myelin 13
Receptor
-, benzodiazapine 317
-, for excitatory messengers 379
-, in developing cerebellum 235
-, isolation of opiate receptor 409
-, narcotic 404
-, neuroleptic 317
-, oestrogen 622
-, opiate 317,386
- -, composition 407
- -, density in CNS 405
- -, isolation 409
-, opioid 378
-, relationship to tolerance 378
Recognition molecular 180,181,185
-, complementary 184
Redox changes, hypoxia 258,260
Refsum's disease 621
Remyelination
-, failure of 101
-, prevention of 101
Respiration, synaptosomes 303
Retina
-, classes of ganglion cells 201
-, form-function relationship 153
-, taurine release 587
Retino-tectal
-, connections 146
- -, and plasticity 148
- - -, compressed projection 147,148
- -, diagram of different 147
- -, of compound eyes 146
RNA
-, metabolism, effect of pyrrolidine acetamide 599
-, synthesis, effect of polyamines 604
- -, visual deprivation 209

S

S-100 protein 592,593,606
-, purification 592
-, human 592
Schmidt-Lanterman
-, clefts 12
-, incisure 8
Schilder's disease, demyelination 79
Schwann cell 6
Sclerosis, in cultures 101
SDS-PAGE, and myelin proteolipids 117
Semicompact domain, in PNS myelin
Sensory deprivation
-, brain development 199,215
-, electrophysiological aspects 199
-, functional aspects 199
-, morphological aspects 199
-, visual system 199
Separated cells
-, amino acid uptake 223
-, cerebral cortex 223
-, membrane transport system 223
Septum, electrolyte destruction 431
Serotonin, see 5-hydroxytryptamine
Shiverer mouse
-, dysmyelination 131
-, myelin, lipids and proteins 131
-, major dense line 50,57
-, neurological mutant 49
Sialoglycoconjugates, mucolipidosis 610
Simulation studies 150,152
-, Hope's arrow model 151
-, v.d. Malsburg model 150
Sleep, influence by ethanol 339
Snail, neurons 560
Sodium
-, movements, influenced by ethanol 347

-, transport, effect of neurotoxins 467,494
Specific receptor proteins 144
Spermidine 604
-, in myelination 22
Spermine 604
Spinal motoneurons
-, differnt types 153
-, stained with cobalt 153
Spongioblasts 164
-, arborizing 165
-, contact with blood vessels 164
Stearic acid 609
Subacute sclerosing panencephalitis, demyelination 79
Subcellular fraction
-, brain 234
-, during myelination 234
-, metabolism of phosphoethanolamines 234
Substance P 450,464
Succinic semialdehyde dehydrogenase 625
Sulfatide synthesis, inhibition in cultures 104
Sulfotransferase
-, cerebroside 574
-, in mutant mice 55
-, in oligodendroglia 53
Sympathetic ganglion
-, acetylcholinesterase 564
-, GABA application 167
Synapse
-, differentiation
- -, critical evaluation of theories 180,181
- -, cytodifferentiation 178
- -, selective stabilization of synapses 179,180
- -, timing theory 179
-, formation 174,177
- -, at neuronal cell surface 176
- -, in culture 174
- -, GABA pools 167
- -, neurite growth 174

– –, neuroblast migration 174
–, specification 178,179,180
– –, chemical guidance 178
– –, neuronal differentiation 176
Synaptic
–, function, glutamate decarboxylase 563
–, membranes, amido acid
   binding 578
– –, binding of 5-HT and LSD 440
– –, glutamate binding 222
– –, kainic acid binding 222
– –, taurine binding 586
–, thickenings 167
–, vesicles 579
– –, proteins 580
Synaptogenesis 160,585
–, cell migration 158,173
–, F-post 169
–, membrane glycoproteins 174,188
–, neocortex 158,173
–, visual system 179
Synaptosomal membranes
–, acetylcholinesterase 567
–, ganglioside concentration 190,191
Synaptosomes
–, adenine nucleotides 584
–, adenosine uptake 584
–, calcium metabolism 582
–, citrate influx 583
–, from tissue cultures 505
–, respiration 303
–, test system for drugs 620
Synthesis, of myelin lipids
and proteins 126

T
Taurine 481,513,586,587
–, release 587
Tay-Sachs, α-albumin 617
Tectal connections, with optic
fibers 146
Tectum of Xenopus levis, transient
functional contacts 176

Temporo-spatial distribution of
migrating neurons 160
Tetrahydroisoquinoline, formation 338
Tight junctions
–, in myelin development 29
–, in myelin lamellae 13,16
Tissue cultures
–, development, biochemical 507
– –, morphological 507
–, EAE 100
–, isolation of synaptosomes 505
Thromboxanes 277,279
^{3}H-Thymidine labelling of
migrating neurons 160,161
Thyroid hormone
–, effect on fatty acid activation 233
– –, myelination 233
Tolerance
–, alcohol 318,321
–, barbiturate 321,388
–, CNS adaptions 320
–, development 325
–, effect of narcotics 409
– –, norepinephrine 326
–, ethanol 318,321
–, influence by behaviour
   and environment 319
–, neurochemical mechanisms 321
–, occurence with dependence
   in one cell 376
–, relation to central depressants 308
– –, number of receptors 378
– –, physical dependence 319,326
–, role of GABA 414
– –, lipids 414
–, single or multiple process 320
–, use of endogenous brain and
   hypophysical peptides 327
Torpedo marmorata 580,595
Transaminase
–, cysteine sulfinate 569
–, GABA 625

Transmitter
-, accumulation to opioid
    tolerance and dependence   381
-, effect of drugs   325
-, metabolism, effect of acute
    ethanol administration   362,364
-, specific enzymes   204,208
Transneuronal transport   474,476
Transport
-, ethanol effects   348
-, myelin components   56
Triethyllead, effects on myelin
protein synthesis   135
Triglycerides
-, in Refsum's disease   621
-, ischemia   292
Triiodothyronine, effect on
myelin glycolipids   140
Trypsin, action on myelin   68
α-Tubulin   602
-, synthesis, visual
    deprivation   211
Tyrosine hydroxylase, effect
of gamma-butyrolactone   367

U
Undernutrition
-, effect on amino acid pool
    of brain   224
- - -, of plasma   224
-, protein-calorie deficiency   224
Uterus, oestrogen receptor   622

V
Vegetative NS, active system   601
Vesicular release, modulation   579
Visual
-, cortex
- -, acetylcholinesterase   565
- -, of rat   166
- -, monocular deprivation   200,201
-, deprivation
- -, cell division   211,212
- -, critical period   202
- -, effect on adenylate cyclase
                        208,209
- -, glia proliferation   211
- -, macromolecular synthesis   209
                        210,211
- -, protein-glycoprotein
    synthesis   209,210
- -, RNA synthesis   209
- -, transmitter-specific
    enzymes   204,208
- -, tubulin synthesis   211
- -, uptake of biogenic
    amines   206,207
-, stimulation   600
-, system
- -, receptor field properties
                        202
- -, timing theory of synaptoge-
    nesis   179
Vitamin B_{12} deficiency,
myelin composition   136

W
Wallerian degeneration, in rat
optic nerve, in sciatic nerve   72
Wolfgram proteins
-, axonal transport   118
-, content of acidic amino
    acids   39
-, heterogeneity on SDS-PAGE   38
-, isolation by SDS-PAGE   39
-, molecular weight   39
White matter, unmyelinated
human brain, ganglioside concentration   190

Z
Zonal centrifugation
-, human myelin   128
-, infantile neuronal ceroid
    lipofuscinosis   128
-, mouse myelin   128
-, myelin density   127,128
- -, heterogeneity   40
-, Pelizaeus-Merzbacher-dis-
    ease   128
-, PNS myelin   127
-, Quaking myelin   128
Zonular occludens, in myelin   10